Reinforced Concrete with Worked Examples

AF888252

Franco Angotti · Matteo Guiglia · Piero Marro ·
Maurizio Orlando

Reinforced Concrete with Worked Examples

Springer

Franco Angotti
DICEA
University of Florence
Florence, Italy

Matteo Guiglia
Studio AITEC
Turin, Italy

Piero Marro
DISEG
Politecnico di Torino
Turin, Italy

Maurizio Orlando
DICEA
University of Florence
Florence, Italy

ISBN 978-3-030-92838-4 ISBN 978-3-030-92839-1 (eBook)
https://doi.org/10.1007/978-3-030-92839-1

Translation from the Italian language edition: *Progetto delle Strutture in Calcestruzzo Armato* by Franco Angotti, Matteo Guiglia, Piero Marro and Maurizio Orlando, © Hoepli 2019. Published by Hoepli. All Rights Reserved.
© The Editor(s) (if applicable) and The Author(s), under exclusive license to Springer Nature Switzerland AG 2022
This work is subject to copyright. All rights are solely and exclusively licensed by the Publisher, whether the whole or part of the material is concerned, specifically the rights of reprinting, reuse of illustrations, recitation, broadcasting, reproduction on microfilms or in any other physical way, and transmission or information storage and retrieval, electronic adaptation, computer software, or by similar or dissimilar methodology now known or hereafter developed.
The use of general descriptive names, registered names, trademarks, service marks, etc. in this publication does not imply, even in the absence of a specific statement, that such names are exempt from the relevant protective laws and regulations and therefore free for general use.
The publisher, the authors and the editors are safe to assume that the advice and information in this book are believed to be true and accurate at the date of publication. Neither the publisher nor the authors or the editors give a warranty, expressed or implied, with respect to the material contained herein or for any errors or omissions that may have been made. The publisher remains neutral with regard to jurisdictional claims in published maps and institutional affiliations.

This Springer imprint is published by the registered company Springer Nature Switzerland AG
The registered company address is: Gewerbestrasse 11, 6330 Cham, Switzerland

Preface

This volume deals with the design of normal and prestressed reinforced concrete structures in the light of the most recent developments both in the field of concrete and steel materials and in the field of structural modelling. The consolidated part of this evolution has been implemented by the most modern and advanced codes. Eurocode 2 "Design of concrete structures" (EC2), which is under revision, is certainly among these.

The arguments are presented starting from their theoretical approach and then moving on to the regulatory feedback represented by EC2.

The text is accompanied by numerous numerical examples chosen from professional practice. Through this approach, the text will be of interest to both structural engineers and students of engineering and architecture.

The text is developed in 12 chapters, where topics are dealt with according to the same order of EC2. Only Chap. 6, dedicated to prestressed concrete structures, does not have a corresponding section in EC2, where prestressed concrete is considered as a particular case of reinforced concrete and dealt with in various sections of the code.

It will be noted that a lot of space has been given to the most innovative parts such as second-order effects, punching shear and strut-and-tie models.

Eurocode 2, like all other Eurocodes, is applicable in the European Union (EU) countries only if it is accompanied by the "National Annexes", which must be approved by the competent national authorities. They provide values of national determined parameters (NDP), the choice of which is left to the responsibility of each EU country.

In the text, references to paragraphs and formulas taken from EC2 are indicated in square brackets; for example, [(3.1)] indicates the formula (3.1) of EC2.

Chapters 1, 3, 6, 9, 10 and 12 were authored by Franco Angotti and Maurizio Orlando.

Chapters 2, 4, 5, 7, 8 and 11 were authored by Piero Marro and Matteo Guiglia.

Information for Students and Instructors

The book stands as an ideal learning resource for students of structural design and analysis courses in civil engineering, building construction and architecture, as well as a valuable reference for concrete structural design professionals in practice.

All topics are presented starting from their theoretical bases and passing to corresponding EC2 formulations. The textbook contains twelve chapters, matching the same structure of EC2; only Chap. 6, dealing with prestressed concrete structures, does not match any chapter of EC2, as prestressed concrete is presented in EC2 as a particular case of reinforced concrete, and corresponding formulations are shed over different chapters.

Each chapter presents an organic topic, which is eventually illustrated by worked examples useful for the student who is not familiar with the design of reinforced and prestressed concrete structures by the limit state method. Examples have been chosen among the most frequent cases of the professional practice, so they are also useful for concrete structural design professionals. Moreover, all chapters contain tables, which allow the reader to develop his calculations not only in dimensional form but also in nondimensional form using tabulated values; some tables for punching verification are contained in a specific appendix at the end of Chap. 9.

The appendix at the end of the book provides tables and diagrams for the adimensional calculation at both ULS and SLS of rectangular, T and circular sections subjected to simple bending or to bending combined with axial force.

Florence, Italy — Franco Angotti
Turin, Italy — Matteo Guiglia
Turin, Italy — Piero Marro
Florence, Italy — Maurizio Orlando

Structural Eurocodes and Product Standards

In addition to Structural Eurocodes, the following product standards are mentioned in the text:

EN 206:2021—Concrete. Specification, performance, production and conformity
EN 197-1:2011—Cement—Composition, specifications and conformity criteria for common cements

Units of Measurement

The following units of measurement were used in the examples:

- *Length*: m, cm, mm
- *Area*: m^2, cm^2, mm^2
- *Force*: N, kN
- *Stress*: N/mm^2, MPa
- *Elastic modulus*: N/mm^2, kN/mm^2, GPa

Contents

About the Authors

Franco Angotti is professor emeritus of Strength of Materials at the University of Florence, Italy. He is President of the Italian Mirror Committee to the European Committee CEN/TC250/SC2 "Concrete Structures", member of the Committee for Technical Standards of the Italian National Research Council (CNR) and the Task Group of the Italian High Council of Public Works for the definition of National Determined Parameters of Eurocodes. He is President of the Italian Association for Reinforced and Prestressed Concrete (AICAP).

Matteo Guiglia has a PhD in Structural Engineering from the Turin Polytechnic, Italy. He delivers both consultancy and structural design services at both national and international level. He is member of the Task Group T2.1 "Serviceability models" of the Fédération Internationale du Béton (fib).

Piero Marro is professor emeritus of Strength of Materials at the Turin Polytechnic, Italy. As an affiliated to the Comité Euro-International du Béton (CEB) since 1975, he participated in the drafting of Model Code 1978 and Model Code 1990 and in the conversion of EC2 from provisional to definitive standard in 2004. He is a "fib Life Member" and corresponding member of the Turin Academy of Sciences.

Maurizio Orlando is associate professor of Structural Analysis and Design at the School of Engineering of the University of Florence, Italy. His research focuses on theoretical and experimental analysis of reinforced concrete structures under both static and seismic loading. He is member of the Working Group CEN/TC 250/SC 2/WG 104 "Shear, punching, torsion" dealing with the revision of section 6 of Eurocode 2, Vice President of the Italian Mirror Committee to the European Committee CEN/TC250/SC2 "Concrete structures" and Secretary of the Italian Association for Reinforced and Prestressed Concrete (AICAP).

Chapter 1
General Structural Design Criteria

Abstract The chapter deals with structural safety, that is the definition of the safety margins to assure that failure of a structure will occur or that specific criteria will be exceeded for very low probability levels. These levels depend on the type of construction and consequences on the safety of people and damage to property. Any dangerous situation for a building is denoted as a "limit state": it represents a condition at which the building is no longer able to perform the functions for which it was designed. The limit states are classified into ultimate limit states (ULS) and serviceability limit states (SLS) according to the severity of their consequences. ULS are associated with the collapse of the whole structure or some of its parts. SLS affect the functioning of the structure, comfort of people and appearance of the structure and can be reversible (no residual effect remains once the loads are removed) or irreversible (some effects remain). The structural safety verification method indicated in EN1992-1-1, as well as in all Eurocodes, is the partial factor method or semi-probabilistic method. The method does not require the designer to have any knowledge of the probabilistic methods for the analysis of structural safety, because the probabilistic aspects of the problem are already considered in the calibration process of the method itself, i.e., in the choice of the characteristic values, the partial safety factors, etc.

1.1 Introduction

Given the random nature of the quantities involved in the design of a structure (loads, geometry, constraint conditions, material strengths, etc.), the assessment of structural safety cannot be expressed in deterministic terms, but it requires a probabilistic analysis. The verification of structural safety consists in controlling that the probability of reaching or exceeding a dangerous condition for the structure remains below an assigned value. This value is fixed according to the type of construction and the consequences on the safety of people and damage to property.

This chapter was authored by Franco Angotti and Maurizio Orlando.

© The Author(s), under exclusive license to Springer Nature Switzerland AG 2022
F. Angotti et al., *Reinforced Concrete with Worked Examples*,
https://doi.org/10.1007/978-3-030-92839-1_1

Any dangerous situation for a building is called a "limit state": it represents a condition at which the building is no longer able to perform the functions for which it was designed. The limit states are classified into ultimate limit states (ULS) and serviceability limit states (SLS) according to the severity of their consequences: ULS are associated with the collapse of the whole structure or some of its parts, SLS affect the functioning of the structure, comfort of people and appearance of the structure. The serviceability limit states can be reversible or irreversible: for the former, there is no residual effect once the loads are removed, while for the latter some effects remain. Examples of reversible limit states are the deformations and vibrations of the floors produced by short-term service loads, while on the contrary, the permanent viscous deformations of the floors under long-term loads with damage to the finishing elements represent a typical example of an irreversible limit state.

The structural safety verification method indicated in EN1992-1-1, as well as in all Eurocodes, is the partial factor method or semi-probabilistic method.

This method verifies the reliability of the structure by respecting a set of rules in which the "characteristic values" of the problem variables and a series of "safety coefficients" are used; these are the so-called partial safety factors γ which cover the uncertainties on actions and materials and additive elements Δ for the uncertainties on geometry, for example, to take into account the uncertainty of the value of the concrete cover and therefore of the value of the useful height of the section of a reinforced concrete beam.

The method does not require the designer to have any knowledge of the probabilistic methods for the analysis of structural safety, because the probabilistic aspects of the problem are already considered in the calibration process of the method itself, i.e., in the choice of the characteristic values, the partial safety factors, etc., given in the technical standards.

In summary, the limit state verification method is based on the following assumptions.

- strength and load are independent random variables,
- the characteristic values of strength and load coincide with the fractile of 0.05 or 0.95 of the corresponding probability density function.

- the other uncertainties are considered by transforming the characteristic values into design values, by applying partial safety factors,
- the safety measure is positive if the design stresses do not exceed the corresponding design resistances.

When a few statistical data are available, the nominal values indicated in the standards or indicative values are taken as characteristic values of the actions, as in the case of accidental actions (shocks, explosions, etc.).

Regardless of the structural safety method, a structure is defined as reliable if it has positive safety measures for each of its limit states during the entire design working life (or nominal life) T_u. T_u is defined as the period during which it is assumed that the structure is used for its intended purposes, with scheduled maintenance, but

Table 1.1 Indicative design working life

Type	Description	Indicative design working life V_N (years)
1	Temporary structures	10[a]
2	Replaceable structural parts, e.g. gantry girders, bearings	10 ÷ 30
3	Agricultural and similar structures	15 ÷ 25
4	Building structures and other common structures, not listed elsewhere in this table	50
5	Monumental building structures, highway and railway bridges, and other civil engineering structures	120

[a] Structures or parts of structures that can be dismantled with a view of being re-used should not be considered temporary

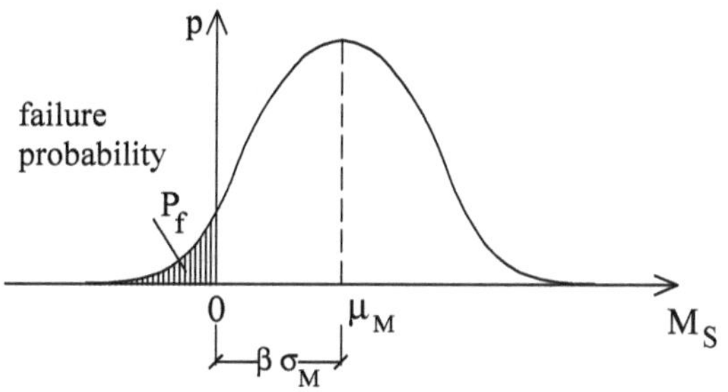

Fig. 1.1 Gaussian probability density function of the safety margin $M_S = R - S$

without substantial repairs being required. Table 1.1 provides the indicative values of the design working life for different types of structures.

EN1990 defines three reliability classes (RC1, RC2, RC3) according to the probability of P_f reaching the crisis during the design working life. S and R indicate the two random variables that represent stress and resistance and with M_S the safety margin ($M_S = R - S$), P_f can be expressed as a function of the reliability index β ($\beta = \mu_M / \sigma_M$ where μ_M and σ_M are respectively the mean and the standard deviation of the safety margin), if S and R have a normal distribution and the safety margin M_S depends on S and R linearly.

If R and S are statistically independent Gaussian random variables,[1] the safety margin $M_S = R - S$ is also a Gaussian variable (Fig. 1.1); if the distribution functions of M_S and m (being m the normalized variable $m = (M_S - \mu_M)/\sigma_M$) are denoted by Φ_M and Φ_m, the probability of failure P_f is given by the following expression:

[1] *Statistical independence*: two events A and B are defined statistically independent if the probability of one occurring is not influenced by the occurrence of the other, in symbols: P (A/B) = P (A) and P (B/A) = P (B), where P (A/B) is the probability that A occurs, provided that B has already occurred. If two random variables X and Y are statistically independent, their joint density function is equal to the product of their density functions.

Table 1.2 Relationship between β and P_f

P_f	10^{-1}	6.7×10^{-2}	10^{-2}	1.9×10^{-3}	10^{-3}	10^{-4}	10^{-5}	10^{-6}	10^{-7}
β	1.28	1.5	2.32	2.9	3.09	3.72	4.27	4.75	5.20

Table 1.3 Minimum recommended values of the reliability index β (ultimate limit states)

Reliability class	Reference period	
	1 year	50 years
RC3	5.2	4.3
RC2	4.7	3.8
RC1	4.2	3.3

$$P_f = P\{M_S \leq 0\} = \Phi_M(0) = \Phi_m[(0 - \mu_M)/\sigma_M] = \Phi_m(-\beta)$$

that is, the probability of collapse coincides with the distribution function of the normalized variable m calculated adopting $m = -\beta$ (Table 1.2).

It also appears that the distance of the mean value μ_M from the origin is equal to $\beta\,\sigma_M$ (Fig. 1.1).[2]

If R and S are not Gaussian this result is not valid, and the probability of failure can be estimated with the following formula

$$P_f = \int_{-\infty}^{+\infty} p_E(x)\,\Phi_R(x)\,dx$$

where $p_E(x)$ is the density function of the effect and $\Phi_R(x)$ is the distribution function of the resistance.

The recommended minimum values of β for the ultimate limit states are shown in Table 1.3, for reference periods of 1 and 50 years.

A design based on Eurocodes generally leads to a structure with a value of β greater than 3.8 for a reference period of 50 years, that is, with a probability of reaching the ULS in 50 years of 7.2×10^{-5} (less than $10^{-4} = 1/10{,}000$, P_f value corresponding to $\beta = 3.72$ in Table 1.2).

For irreversible serviceability limit states, which are less dangerous and do not affect the safety of people, the reliability index β is 2.9 for a reference period of 1 year and 1.5 for 50 years (Table 1.4). For these values of β, from Table 1.2 it results that the probability of reaching an irreversible serviceability limit state for structural elements of class RC2 is equal to 1.9×10^{-3} ($\cong$1/500) in 1 year and 6.7×10^{-2} ($\cong$ 1/15) in 50 years.

[2] If the distance of the mean value μ_M of M from the origin is expressed as a multiple λ of the standard deviation σ_M ($\mu_M - 0 = \lambda\,\sigma_M$), then $= \mu_M/\sigma_M$ or equivalently $\lambda = \beta$.

Table 1.4 Reference values of the reliability index β for structural elements belonging to class RC2

Limit state	Reference period	
	1 year	50 years
Ultimate	4.7	3.8
Fatigue	–	da 1.5 a 3.8
Serviceability (irreversible)	2.9	1.5

A design that uses the partial factors provided by EN1992-1-1 and the partial factors provided in the appendices of EN1990 leads to a reliability class RC2.

1.2 Design Principles for Limit States

The Limit State Method (LSM) is based on the use of structural and loading models for each limit state and has the objective of verifying that no limit state is exceeded when the design values of the actions, the properties of materials and products, and the geometric characteristics of the structural elements are used in the calculation model.

The verifications should be carried out for all design situations and all relevant load cases. As already mentioned above, according to the severity of the consequences associated with the achievement of a limit state, two categories of limit states are defined.

- ultimate limit states (ULS),
- serviceability limit states (SLS).

The verification must be performed for both categories, being able to accept the verification for one of the two, only if it can be shown that it is implicitly satisfied by the verification for the other.

The ultimate limit states are associated with the collapse of the whole structure or some of its parts, so they are directly related to the safety of people.

Table 1.5 shows the classification of the ULS according to EN1990 [(6.4.1)-EN1990].

The serviceability limit states correspond to situations in which the structure or one of its parts can no longer be used, even if strength and stability are still preserved. They cause limited damage but make the structure no longer usable for the needs defined in the design: functional requirements (not only of the structure but also of machines and equipment supported by the structure), user comfort, aesthetic requirements (high deformations, extensive cracking, etc.), damage to non-structural components. Normally functional requirements are set in the contract and/or in design documents.

EN1990 indicates three different types of combinations of actions for serviceability limit state verifications: characteristic combination (already called rare combination in previous versions of the technical standards), frequent combination, and quasi-permanent combination.

Table 1.5 Classification of ultimate limit states (ULS)

EQU	Loss of static balance of the structure or any part of it considered as a rigid body, when • small variations in the intensity or spatial distribution of the actions caused by a single source are significant (e.g., variations in self-weight), • the strengths of building materials or the soil are generally not determinant Note: [point 2.4.4] specifies that the EQU limit state also applies to the verification of anti-lifting devices or lifting devices of support devices in the case of continuous beams
STR	Internal collapse or excessive deformation of the structure or structural elements, including foundations, poles, retaining walls, etc., when the collapse is governed by the resistance of the construction materials of the structure
GEO	Excessive collapse or deformation of the soil when the resistance of the soil or rock is decisive in guaranteeing resistance
FAT	Fatigue collapse of the structure or structural elements

The combinations of actions to be considered are linked to the distinction between reversible and irreversible limit states: for the former, the frequent or quasi-permanent combination is used and for the latter the characteristic combination.

The definition of the limit states of interest for a reinforced concrete structure first requires an analysis of the different situations in which the structure could be. The situations chosen for the design must cover both those that may occur at the time of construction and those associated with the use of the structure. In current cases, design situations are classified into.

- persistent design situations, connected to conditions of normal use,
- transient design situations, relating to the temporary conditions applicable to the structure, for example during the construction or repair phase,
- accidental design situations, concerning accidental conditions that may affect the structure such as exposure to a fire, an explosion, a collision, etc.
- seismic design situations, related to the conditions of the structure during a seismic event.

1.3 Actions

Each design situation is characterized by the presence of different types of actions, which can be classified according to

1. variation over time,
2. origin,
3. variation of spatial position,
4. nature and/or type of response induced in the structure.

Depending on the variation over time, the actions can be divided into

- permanent actions (G), whose duration of application is continuous and equal to the duration of the design life of the structure, or whose variations are minimal and therefore negligible over time (for example the self-weight). Permanent actions are also those which, such as prestressing or shrinkage of concrete, show a monotonous variation over time which tends to a limit value,

- variable actions (Q), among which variable actions with discrete occurrences more or less punctual in time (e.g., the weight of people and in general short-term loads on the floors of a residential building) and variable actions with intensity and/or direction variable over time and not monotonous (for instance: snow, wind, temperature, sea waves);
- accidental actions (A), which are difficult to predict and are of short duration (for instance: explosions, knocks, fire).

Depending on the origin, the actions can be divided into

- direct actions: actions directly applied to the structure (own weight, carried permanent loads, service loads, snow load, wind load, etc.) and their schematization is, generally, independent of the characteristics of the structure or the structural response,
- indirect actions: the withdrawal of concrete is an example of indirect action, it induces effects on the structure where it cannot manifest itself freely; differential settlements and compulsions, in general, are further examples of indirect actions because they give rise to effects on the structure where the structure is hyperstatic.

Depending on the variation of spatial position, the actions can be divided into

- mobile actions: an action is mobile when its position can vary within an assigned spatial interval (for example the length of a bridge beam). Examples of mobile actions are traffic loads: they are applied on the deck of a bridge in the most unfavourable positions to obtain the most severe effects on the structure,
- fixed actions: actions applied in fixed positions in space.

Depending on the nature and/or the type of response induced in the structure, they are classified into

- static actions: if they do not produce significant accelerations of the structure,
- dynamic actions: if they produce significant accelerations of the structure.

1.3.1 Permanent Actions (G)

A unique characteristic value G_k is attributed to each permanent action with reduced variability. This is the case of actions due to self-weight: they are generally represented by a nominal value calculated from the design data (dimensions of the structural and non-structural elements) and the mean weight per unit volume of the materials ($G_k = G_m$).

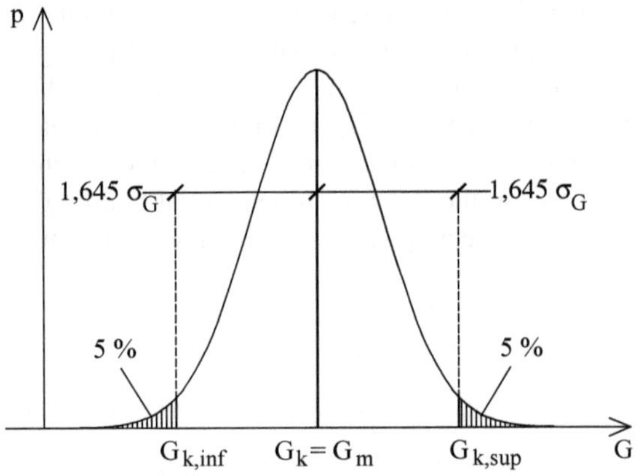

Fig. 1.2 Characteristic value of a permanent action; if the coefficient of variation is negligible $G_k = G_m$, if the coefficient of variation is high, a lower characteristic value $G_{k,inf}$ and an upper one $G_{k,sup}$ are defined

When a permanent action has significant variations, i.e. a coefficient of variation δ_X[3] greater than 10%, and sufficient statistical information is available, two characteristic values are defined (upper $G_{k,sup}$ and lower $G_{k,inf}$), equal respectively to fractile of 0.95 and 0.05 of the probability density function (Fig. 1.2).

The fractile of 0.95 has a 5% probability of being "overcome", while the fractile of 0.05 has a 5% probability of not being "overcome", in other words, the probability that the actual value of the action exceeds the characteristic value on the most unfavourable part is less than 5%.

1.3.2 Variable Actions (Q)

Each variable action is characterized by four representative values. The main one is its characteristic value Qk; the other representative values, in descending order, are as follows

- the combination value, indicated with the symbol $\psi_0 Q_k$;
- the frequent value, indicated with $\psi_1 Q_k$;
- the quasi-permanent value indicated with $\psi_2 Q_k$.

For the sake of simplicity, each of these last three values is defined as a fraction of the characteristic value obtained by applying a reduction coefficient to Q_k. In reality, the frequent and quasi-permanent values are intrinsic properties of the variable action, and the coefficients ψ_1 and ψ_2 are nothing more than the ratios between these values and the characteristic value. On the other hand, the coefficient ψ_0, called the combination coefficient, fixes the intensity level of a variable action when it is taken into account, in the calculation, at the same time as another variable action,

[3] The coefficient of variation δ_X of a random variable X with non-zero mean is defined as the ratio between the standard deviation σ_X and the mean value μ_X: $\delta_X = \sigma_X/\mu_X$.

Table 1.6 Partial safety factors ψ for loads on buildings

Loads on buildings by category	ψ_0	ψ_1	ψ_2
Category A: housing, residential areas	0.7	0.5	0.3
Category B: offices	0.7	0.5	0.3
Category C: conference areas	0.7	0.7	0.6
Category D: commercial areas	0.7	0.7	0.6
Category E: stores	1.0	0.9	0.8
Category F: area open to traffic, vehicle weight ≤ 30 kN	0.7	0.7	0.6
Category G: area open to traffic, 30 kN < vehicle weight ≤ 160 kN	0.7	0.5	0.3
Category H: roofs	0.0	0.0	0.0
Snow on buildings (sites above 1000 m a.s.l.)	0.7	0.5	0.2
Snow on buildings (sites below 1000 m a.s.l.)	0.5	0.2	0.0
Wind loads on buildings	0.6	0.2	0.0
Temperature (not fire) in buildings	0.6	0.5	0.0

called dominant, considered instead with its characteristic value. The coefficient ψ_0, therefore, considers the reduced probability of simultaneous occurrence of the most unfavourable values of several independent variable actions. It is used for both ULS and irreversible SLS verifications.

The frequent value ($\psi_1\, Q_k$) and the quasi-permanent value ($\psi_2\, Q_k$) appear in the combinations for ULS verifications and those for reversible SLS combinations. The quasi-permanent value ($\psi_1\, Q_k$) is also used in seismic verifications. Table 1.6 shows the values of the coefficients ψ recommended by EN1990.

The characteristic value of a variable action is defined as the value that presents a probability, accepted a priori, of being exceeded by the most unfavourable values during a period taken as the reference period and normally coincident with the useful life of the structure. For most of the climatic variable actions as well as for service loads on the floors of the buildings, the characteristic value has a probability of being exceeded of 0.02 (2%) in a year, or, in other words, which has a return period T = 1/0.02 = 50 years.

The frequent and quasi-permanent value of the loads on the floors of the buildings are fixed in such a way that the fraction of time during which they are exceeded is on average equal to 10% respectively (ratio between the sum of the sections GH, IL and MN, and the reference life T_{ref} in Fig. 1.3) and 50% of the reference duration of 50 years.

For accidental actions, a single nominal value is directly determined because, due to their character, insufficient data are available to properly apply the statistical methods.

To take into account the uncertainties on the choice of the characteristic values of the actions as well as a part of the uncertainties related to the modelling of the action, the characteristic values are not used in the calculations, but amplified values

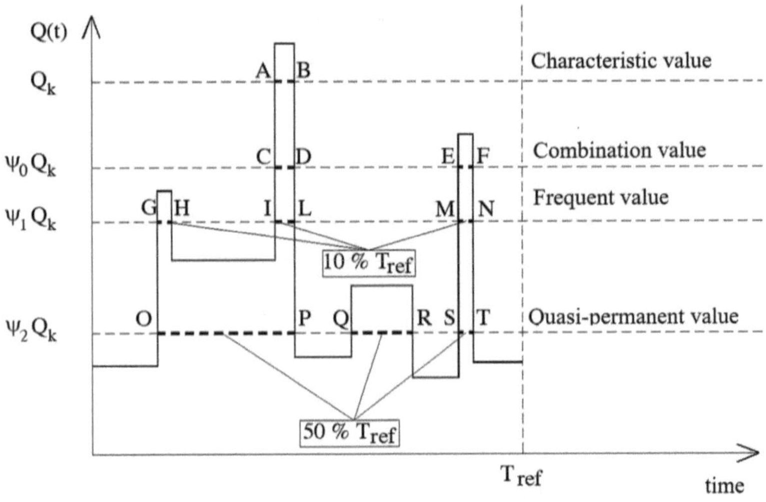

Fig. 1.3 Schematic illustration of the representative values of the variable actions

indicated as design values, obtained by multiplying the first for a partial safety factor greater than or equal to 1.

The symbols that represent the design values have the subscript *d* (design).

Table 1.7 lists the steps to be followed to move from the representative values of the actions to the design values of the effects they produce on the structure, as part of common linear analysis.

Concerning the partial factors of the actions, it should be noted that

- the partial factors of the $\gamma_{F,i}$ actions, the ones intended to cover only the uncertainties on the choice of the characteristic values of the actions and those related to the modelling of the action (e.g. schematization of the working loads on the floors as uniformly distributed loads) are marked from a tiny subscript (*g* for permanent loads, *q* for variable loads);
- the partial factors of the actions $\gamma_{F,i}$, which include both the uncertainties on the characteristic values and the schematization of the actions and the model ones, are distinguished by an upper case subscript (G for the permanent loads, Q for the variable ones).

The relationship between the two types of coefficients is expressed by the following relationship [(6.2b) EN1990]: $\gamma F = \gamma_{Sd} \cdot \gamma_f$, where γ_{Sd} is the partial factor of the model.[4]

The procedure indicated in Table 1.7 can be used in linear analyses, while in non-linear analyses the following steps must be followed.

[4] γ_{Sd} assumes values between 1.05 and 1.15, as indicated in Note 4 of Table A1.2 (B) of EN1990.

Table 1.7 Procedure for determining the design values of the effects on the structures starting from the representative values of the actions (model uncertainties are included in the partial safety factors of the actions)

Expression	Comment
Self-weight (G_1) Permanent load (G_2) Service load (Q) Wind (W), snow (S), earthquake (E)	The actions that act on the construction are identified
$G_{k1}, G_{k2}, Q_k, V_k, N_k, S_k$ (characteristic values) Or[a] G_{k1}, G_{k2}, 0.7 Q_k, 0.6 V_k, 0.5 N_k (combination values) G_{k1}, G_{k2}, 0.5 Q_k, 0.2 V_k, 0.2 N_k (frequent values) G_{k1}, G_{k2}, 0.3 Q_k, 0 V_k, 0 N_k (quasi-permanent values)	Their representative values are assigned to the actions: characteristic values or other values (of combination, frequent, quasi-permanent)
$F_{d,i} = \gamma_{F,i}\, F_{k,i}$ (o $\gamma_{F,i}\, \psi\, F_{k,i}$) where $\psi = \psi_0, \psi_1, \psi_2$ $G_{d1} = \gamma_{G1}\, G_{k1}, G_{d2} = \gamma_{G2}\, G_{k,2}$, etc.	The design values of the actions are determined by multiplying the representative values $F_{k,i}$ or $\psi\, F_{k,i}$ (where $\psi = \psi_0, \psi_1, \psi_2$) by a partial factor $\gamma_{F,i}$ $\gamma_{F,i} = f\,(\gamma_{Sd}, \gamma_{f,i})$ is a partial factor that takes into account both a coefficient $\gamma_{F,i}$ intended to cover, in a general way, the uncertainties on the choice of the characteristic values of the actions and, at times, a part of the uncertainties related to the modelling of the action (e.g. schematization of the working loads on the floors as uniformly distributed loads) and of a partial factor (γ_{Sd} which mainly covers the uncertainties of the structural model In the case of permanent actions, when it is necessary to divide the action into a favourable part and an unfavourable part, two different partial factors are used, indicated as $\gamma_{G,\mathrm{sup}}$ and $\gamma_{G,\mathrm{inf}}$
$M_{\mathrm{Ed}}\,(G_{d1}; G_{d2}; Q_d; V_d; N_d; S_d; a_d)$	The design value of the effects is obtained directly from the calculation of the structure: the actions that can occur simultaneously are considered, the combinations of the actions are constructed and the effects of these combinations on the construction are calculated (for example the bending moment in the section of a beam) a_d represents the design value of the set of geometric data (in general the values indicated in the design drawings); a_d can also represent data that considers the possibility of geometric imperfections, which can cause 2nd order effects

[a] For snow load, the coefficients ψ_i valid for sites at an altitude below 1000 m have been considered: $\psi_0 = 0.5$; $\psi_1 = 0.2$; $\psi_2 = 0.0$ (for sites at an altitude above 1000 m, coefficients take on the following values: $\psi_0 = 0.7$; $\psi_1 = 0.5$; $\psi_2 = 0.2$)

- the design actions are defined through the partial factors γ_f with a small subscript, which do not include the model uncertainties,
- these actions apply to the facility,
- the effects obtained are multiplied by the partial factor γ_{Sd} of the model.

1.4 Properties of Materials and Products

Various properties of the materials intervene in the design of the structures, of which the main one is resistance, which is the ability to bear loads without collapsing.

The resistance of the materials is a random quantity; therefore, it is represented through a characteristic value, indicated with the symbol f_k.

This value is the one that has an assigned probability of not being reached or exceeded during a hypothetical unlimited series of tests. It is defined as the 5% fractile of the relative probability distribution when a "low" resistance value is unfavourable (general case), and with the 95% fractile, when a" high "value is unfavourable.

For an assigned property of material *X*, the following statistical parameters are usually considered: the mean value μ_X, the standard deviation σ_X and the asymmetry coefficient α_X.[5] For a symmetric distribution—(e.g., Gaussian) $\alpha_X = 0$ and only the mean and the standard deviation are considered.

If variable *X* has an asymmetric distribution, its probability density function has a qualitative trend of the type shown in Fig. 1.4 (dashed line). In the same figure, the solid line represents the probability density function of a normal distribution (Gulvanessian et al. 2006).

The fractile x_P of X corresponding to a given probability *P* can be calculated with the following relationship

$$X_P = \mu_X - k_{P,\alpha}\sigma_X$$

[5] Following central moments of a random variable can be defined: moments related to the position (*mean, mode, median*), moments measuring the degree of dispersion (*standard deviation, coefficient of variation*), moments measuring the asymmetry (*skewness*) or the heaviness of the tails of the distribution (*kurtosis*). The asymmetry coefficient or *skewness* is defined as $\alpha_X = \left[\int_{-\infty}^{+\infty}(x-\mu_X)^3 p_X(x)dx\right]/\sigma_X^3$, where $p_X(x)$ is the probability density function of *X*, μ_X its mean value and σ_X its standard deviation; α_X is a measure of the asymmetry of the probability density function. If $p_X(x)$ is symmetric (like the normal distribution) the mode of *X* that is the most probable value, which gives the maximum value of $p_X(x)$, is coincident with the mean value μ_X and the "skewness" α_X is equal to zero. If $p_X(x)$ is asymmetric and the mode of *X* is higher than its mean value μ_X then $\alpha_X < 0$, while if the mode is lower than the mean value $\alpha_X > 0$.

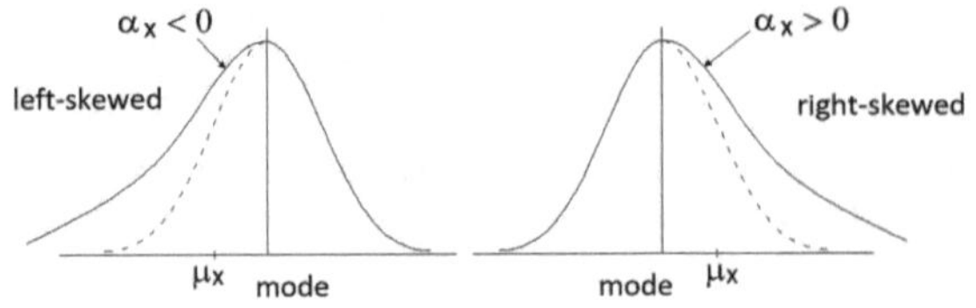

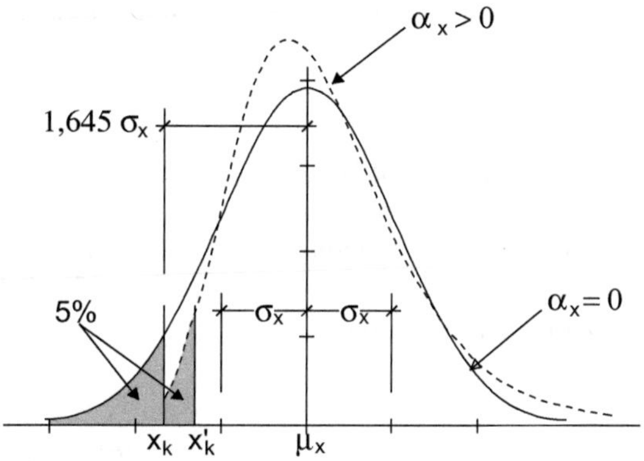

Fig. 1.4 Normal (solid line) and log-normal (dashed line) distributions

Table 1.8 Values of $k_{P,\alpha}$ for $P = 5\%$ at varying the coefficient of skewness α_X

Coefficient of skewness α_X	– 2.0	– 1.0	– 0.5	0.0	0.45	0.5	1.0	2.0
Coefficient $k_{P,\alpha}$ for $P = 5\%$	1.89	1.85	1.77	1.645	1.465	1.49	1.34	1.10

where $k_{P,\alpha}$ depends on the probability P and, for a log-normal distribution, also on the asymmetry coefficient α_X. Table 1.8 shows the values of $k_{P,\alpha}$ for P = 5% at varying α_X.

The lower fractile calculated for a log-normal distribution with $\alpha_X < 0$ is less than that obtained considering the normal distribution having the same mean and standard deviation. Therefore, if the variable X follows a log-normal distribution with a negative coefficient of skewness, the use of the Gaussian distribution overestimates the characteristic value, for safety reasons. On the contrary, the Gaussian distribution underestimates the characteristic value, and therefore it is on the safe side if the coefficient of skewness is positive.

A widely used distribution for material properties is the log-normal one with positive values, whose asymmetry coefficient is positive and is given by the following relationship

$$\alpha_X = \delta_X^3 + 3\delta_X$$

where δ_X is the coefficient of variation of X, already mentioned in note 3 ($\delta_X = \sigma_X/\mu_X$).

For a log-normal distribution, the lower fractile can be calculated in one of the following two ways:

1. α_X is calculated starting from the value of the coefficient of variation δ_X: $\alpha_X = \delta_X{}^3 + 3\delta_X$, then the corresponding value of $k_{P,\alpha}$ is taken from Table 1.8 and finally x_P is calculated

$$x_P = \mu_X - k_{P,\alpha}\sigma_X$$

2. x_P is calculated using the following approximate formula, valid for $\delta_X < 0.2$[6]

$$x_P = \mu_X \; e^{-k_{P,0}\,\delta_X}$$

where $k_{P,0}$ is the coefficient $k_{P,\alpha}$ valid for $\alpha_X = 0$; for $P = 5\%$ it results that $k_{P,0} =$ 1645 (Table 1.8).

For cast in situ concretes, $\delta_X \cong 0{,}15$ can be assumed and therefore the asymmetry coefficient is $\alpha_X = 0.15^3 + 3 \cdot 0.15 \cong 0.45 > 0$; Table 1.8 gives $k_{P,\alpha} = 1465$ for $\alpha_X = 0.45$ and $k_{P,0} = 1645$ for $\alpha_X = 0$, therefore the first expression of x_P becomes

$$\begin{aligned} x_P &= \mu_X - k_{P,\alpha}\sigma_X = \mu_X(1 - k_{P,\alpha}\delta_X) \\ &= \mu_X(1 - 1.465 - 0.15) \cong 0.78\mu_X \end{aligned}$$

and the same result is reached with the second expression:

$$x_P = \mu_X \; e^{-k_{P,0}\,\delta_X} = \mu_X \; e^{-1{,}64 \cdot 0{,}15} \cong 0.78\,\mu_X$$

On the other hand, if the lower fractile is calculated assuming a Gaussian distribution for X, a lower value is obtained

$$\begin{aligned} x_P &= \mu_X - k_{P,0}\sigma_X = \mu_X(1 - k_{P,0}\delta_X) \\ &= \mu_X(1 - 1.645 - 0.15) \cong 0.75\mu_X < 0.78\mu_X \end{aligned}$$

as $k_{P,0} > k_{P,\alpha}$ for log-normal distributions with positive asymmetry coefficient (see Table 1.8). Table 1.9 lists the steps for calculating the design resistances using the partial factors of the materials which only consider the uncertainties on the resistances.

Concerning the modulus of elasticity, viscosity coefficient, thermal expansion coefficient, etc., the characteristic value is taken equal to the mean value because, according to the cases, these parameters can be favourable or unfavourable.

The partial safety factor of the resistance γ_{mat} of the *i-th* material considers both unfavourable deviations of the resistance from its characteristic value and the random part of the conversion coefficient η, where η is a coefficient that considers volume and scale effects, relative humidity, and temperature effects, etc. (see note 7).

As for the actions, also for the resistances, the design values are obtained by applying a partial factor γ_{Rd} (which considers the model uncertainties of the resistance[7] and the geometric deviations if not explicitly modelled) to the values calculated using the individual properties of the materials (design strengths of concrete

[6] The lower fractile of a log-normal distribution is given by the expression $x_P = \mu_X \dfrac{e^{-k_{P,0}\sqrt{\ln(1+\delta_X^2)}}}{\sqrt{1+\delta_X^2}}$, which could be approximated as $x_P = \mu_X \; e^{-k_{P,0}\,\delta_X}$ for $\delta_X < 0.2$.

[7] An example of model uncertainty of resistance is represented by the resistant scheme for punching shear verification, for which a comprehensive analytical formulation is still missing.

Table 1.9 Procedure for the calculation of design values of the resistance using partial factors γ_c e γ_s (with tiny subscript) which consider only the uncertainties on the resistances

Expression	Comment
f_c, f_y	The resistance of the materials and products involved in the verifications are identified
f_{ck}, f_{yk}	The characteristic values of the resistances of materials and products are introduced
$f'_{cd} = \alpha_{cc} f_{ck}/\gamma_c$ $f'_{yd} = f_{yk}/\gamma_s$	The design value of the resistance of a material is determined from its characteristic value with the following two operations • first, the characteristic value is divided by a partial factor γ_{mat} (γ_c for concrete, γ_s for steel, etc.) which takes into account the uncertainties, in the unfavourable sense, on the characteristic value of this property and any localized defects, • if necessary, the result of the first operation is multiplied by a conversion coefficient η which considers volume and scale effects, relative humidity and temperature effects, and any other significant parameter[8] (for concrete and steel η it is already incorporated in γ_{mat} so that this coefficient does not appear explicitly in the calculation of f'_{cd} and f'_{yd}; in f'_{cd} a coefficient α_{cc} appears which considers the duration of the loads)
$N_R(a_d)$, $M_R(a_d)$, $V_R(a_d)$, ecc	The structural strength is determined from the design values of the individual properties of the materials and the geometric data
$N_{Rd}(a_d) = N_R(a_d)/\gamma_{Rd}$ $M_{Rd}(a_d) = M_R(a_d)/\gamma_{Rd}$ $V_{Rd}(a_d) = V_R(a_d)/\gamma_{Rd}$	According to the same procedure used to derive the design value of effects of actions, the design value of the structural strength is determined according to the individual properties of the materials and the geometric data; γ_{Rd} is a coefficient that covers the model uncertainties of the resistance and the variations of the geometric data if these are not explicitly considered in the model

and steel) and using the geometrical data (cross-section dimensions, position, and area of reinforcing bars, etc.).

As an alternative to the procedure described in Table 1.9, the design values of the structural resistances (N_{Rd}, M_{Rd}, V_{Rd}) coincide with the values calculated if the model uncertainties of the resistances are included in the partial safety factors of

[8] η is the mean value of the conversion coefficient between the potential laboratory resistance and the actual resistance of concrete in the structure; η takes on the value 0,85. As allowed by EN1990 at point 6.3.3 (2), in EC2 the coefficient η of concrete, like that of steel, is already incorporated in the partial safety factor of the material (see § 1.4.1). Moreover, from a conceptual point of view, the coefficient η should also consider effects of the duration of the load, but in practice these effects are considered through other coefficients (α_{cc} and α_{ct}), which appear in the definition of the design resistance to compressive and the tensile loads (see Chap. 3).

Table 1.10 Procedure for calculating the design values of the resistances using the partial factors γ_C and γ_S which incorporate the model uncertainties

Expression	Comment
f_c, f_y	The resistance of the materials and products involved in the verifications are identified
f_{ck}, f_{yk}	The characteristic values of the resistances are introduced
$f_{cd} = \alpha_{cc} f_{ck}/\gamma_C = 0.85 f_{ck}/1.5$ $f_{yd} = f_{yk}/\gamma_S = f_{yk}/1.15$	The design value of the resistances is determined from the characteristic value with the following two operations • the characteristic value is divided by partial safety factor γ_{Mat} (γ_C for concrete, γ_S for steel, etc.) which accounts for uncertainties, in the unfavourable sense, on the characteristic value of this property and any localized defects; as in the case of the effects of the actions, the coefficient for model uncertainties γ_{Rd} is integrated into the overall safety coefficient which divides the characteristic resistance of the material: $\gamma_{Mat} = f(\gamma_{Rd}, \gamma_{mat})$ (note the capital subscript to distinguish this safety coefficient from the one listed in Table 1.9) • if necessary, the result of the first operation is multiplied by a conversion coefficient η mainly intended to consider the scale effects (see note 7)
$N_{Rd}(a_d)$, $M_{Rd}(a_d)$, $V_{Rd}(a_d)$, ecc	The design value of the structural strength (internal forces) is determined directly from the design values of the strengths of the materials and the geometric data

the resistances of the materials. In this case, the partial factors of the resistances are indicated with a capital subscript (C for concrete, S for steel) and the procedure for calculating the design resistances is modified as indicated in Table 1.10.

1.4.1 *Partial Factors for Concrete and Steel*

Partial factors for concrete and reinforcing steel bars consider uncertainties in the properties of materials and the modelling of the structural strength. They are indicated with symbols γ_C and γ_S, respectively, where capital letters are used as subscripts.

The characteristic value of the material strength f_k is defined as the 5% fractile of the probability density function of the strength

$$f_k = \mu_R - k_{P,\alpha}\sigma_R$$

where

μ_R mean value of the resistance,
σ_R standard deviation of the resistance,

$k_{P,\alpha}$ multiplicative coefficient of the standard deviation able to define the characteristic value starting from the mean value; for $\delta_R = 0.15$ and $\alpha_R = 0.45$, from Table 1.8 $k_{P,\alpha} = 1465$ is obtained.

The design value ("design") f_d of the resistance of the material is obtained instead with the following expression

$$f_d = \mu_R - \beta\sigma_R$$

where β is the reliability index ($\beta = \mu_M/\sigma_M$ being μ_M and σ_M mean and standard deviation of the safety margin $M = R - S$, respectively, see § 1.1).

Finally, the partial safety factor γ_{Mat} of the material is given by the ratio between the characteristic resistance and the design resistance

$$\gamma_{\text{Mat}} = f_k/f_d$$

Using the expressions of f_k and f_d, γ_{Mat} is expressed by the following formula

$$\gamma_{\text{Mat}} = (\mu_R - k_{P,\alpha}\sigma_R)/(\mu_R - \beta\sigma_R) = (1 - k_{P,\alpha}\delta_R)/(1 - \beta\delta_R)$$

where $\delta_R = \sigma_R/\mu_R$ is the coefficient of variation of the resistance of the material.

Alternatively, assuming a log-normal distribution at positive values for the resistance, f_k can be expressed in the form (see note 6)

$$f_k = \mu_R e^{-k_{P,0}\,\delta_R}$$

valid for $\delta_R < 0.2$ and where the coefficient appears $k_{P,0}$ (=1645 per $P = 5\%$) instead of $k_{P,\alpha}$.

The design resistance can instead be expressed as (Table C3 of EN1990)

$$f_d = \mu_R e^{-\alpha\beta\,\delta_R}\ (\text{valid for}\ \delta_R < 0.2)$$

where α is the sensitivity coefficient of the resistance parameters on the considered limit state so that the coefficient γ_{Mat} takes the following form

$$\gamma_{\text{Mat}} = \frac{f_k}{f_d} = \frac{\mu_R e^{-k_{P,0}\,\delta_R}}{\mu_R e^{-\alpha\beta\delta_R}} = e^{(\alpha\beta - k_{P,0})\,\delta_R}$$

Considering also the randomness of geometry (δ_G) and model of resistance (δ_O) and the difference between the potential laboratory resistance and the actual resistance of the structure, the expression of γ_{Mat} changes in the following

Table 1.11 Parameter values for the determination of the partial safety factors for the ULS of concrete and steel for persistent and transient design situations

Material	δ_O	δ_G	δ_R	δ	β	α	η
Concrete	0.05	0.05	0.15	0.16583	3.8	0.80[a]	0.85
Steel	0.05	0.05	0.05	0.08660	3.2	0.80[a]	1.0

[a] The value is given at point C7(3) of EN1990

$$\gamma_{\text{Mat}} = \frac{f_k}{f_d} = \frac{1}{\eta} e^{(\alpha\,\beta\,\delta - k_{P,0}\,\delta_R)}$$

where $\delta = \sqrt{\delta_R^2 + \delta_G^2 + \delta_O^2}$.

with

δ_G coefficient of variation related to geometric tolerances,
δ_O coefficient of variation related to the section calculation model,
η conversion coefficient between the potential laboratory resistance and the effective resistance of the structure.

For concrete and steel reinforcement, the values shown in Table 1.11 can be assumed.

Substituting these values in the expression of γ_{Mat}, the following values are obtained:

$$\gamma_C = 1.5218 \text{ for concrete}, \ \gamma_S = 1.1496 \text{ for steel}.$$

Table 1.12 shows the values of the partial factors of concrete and steel to the ULS according to EN1992-1-1. The value of the partial factors γ_C and γ_S at the serviceability limit states is 1.0. Next to the partial factors for materials, EC2 also provides partial factors for the action of shrinkage, prestressing, and fatigue-inducing loads (Table 1.13).

Table 1.12 Partial factors γ for concrete and steel for ULS verifications

Design situation	γ_C for concrete	γ_S for ordinary reinforcement	γ_S for prestressing steel
Persistent and transient	1.5	1.15	1.15
Accidental	1.0	1.0	1.0

Table 1.13 Partial factors γ for shrinkage, prestressing and fatigue

Shrinkage	(For instance, in the verification for the ULS of stability, where the effects of the second order are important)	γ_{SH}	1.0
Prestressing	• favourable in persistent and transient design situations	$\gamma_{P,\text{fav}}$	1.0[a]
	• ULS of stability with external prestressing is an increase of the prestressing could be unfavourable	$\gamma_{P,\text{unfav}}$	1.3
	• Local effects	$\gamma_{P,\text{unfav}}$	1.2
Fatigue		$\gamma_{F,\text{fat}}$	1.0

[a] $\gamma_{P,\text{fav}} = 1.0$ could be utilized in combinations for the fatigue limit state

1.5 Combinations of Actions

Below are the general formats for the combinations of actions for the ultimate and serviceability limit states, as defined in EN1990—Section 6.

It should be stressed that according to EN1990 the partial factor γ_Q of variable actions could also be zero, although $\gamma_Q = 0$ has no probabilistic meaning. Therefore, $\gamma_Q = 0$ only represents a way to remove from the load combination those variable actions which give favourable effects (e.g., to maximize the positive moment in the central span of a three-span continuous beam, the variable load is only applied on the central span, which is equivalent to placing $\gamma_Q = 0$ on the two lateral spans).

1.5.1 Combinations of Actions for ULS Verification

For the ultimate limit states, three different types of action combinations are considered: fundamental, accidental and seismic.

Combinations of actions for persistent or transient design situations (fundamental combinations)[9]

$$\sum_{j\geq 1} \gamma_{G,j} G_{k,j} + \gamma_P P + \gamma_{Q,1} Q_{k,1} + \sum_{i>1} \gamma_{Q,i} \psi_{0,i} Q_{k,i}$$

where

$\gamma_{G,j}$ partial factor for permanent action j.
$G_{k,j}$ characteristic value of permanent action *j*.
γ_P partial factor for prestressing actions.
P relevant representative value of a prestressing action
$\gamma_{Q,1}$ partial factor for variable action *1*.

[9] In this combination formula, as in the other combination formulas, the symbol Σ means "combined effect of", while the symbol + means "to be combined with".

$Q_{k,1}$ characteristic value of the leading variable action *1*
$\gamma_{Q,i}$ partial factor for variable action *i*.
$\psi_{0,i}$ factor for combination value of variable action *i*
$Q_{k,i}$ characteristic value of the accompanying variable action *i*.

Combinations of actions for accidental design situations

$$\sum_{j\geq 1} G_{k,j} + P + A_d + (\psi_{1,1} \text{ or } \psi_{2,1}) Q_{k,1} + \sum_{i>1} \psi_{2,i} Q_{k,i}$$

where A_d represents the accidental design action (in the case of a fire it represents the design value of the indirect thermal action due to the fire).

If a variable action can be present on the structure at the time when an accidental action occurs, its frequent value ($\psi_{1,1}\ Q_{k,1}$) is used in the combination, otherwise, its quasi-permanent value is used ($\psi_{2,1}\ Q_{k,1}$); the other variable actions are introduced in the combination with their quasi-permanent values ($\psi_{2,\mathrm{i}}\ Q_{k,\mathrm{i}}$).

Combinations of actions for seismic design situations

$$\sum_{j\geq 1} G_{k,j} + P + A_{Ed} + \sum_{i\geq 1} \psi_{2,i} Q_{k,i}$$

where A_{Ed} represents the design value of the seismic action.

Note that the seismic action is combined with the quasi-permanent value of the variable actions, while the permanent actions $G_{k,j}$ appear with their characteristic value and prestressing with its representative value.

1.5.2 Combinations of Actions for SLS Verifications

The combinations of actions for the serviceability limit states are of three types: characteristic, frequent and quasi-permanent.

Characteristic combination

$$\sum_{j\geq 1} G_{k,j} + P + Q_{k,1} + \sum_{i>1} \psi_{0,i} Q_{k,i}$$

Frequent combination

$$\sum_{j\geq 1} G_{k,j} + P + \psi_{1,1} Q_{k,1} + \sum_{i>1} \psi_{2,i} Q_{k,i}$$

Quasi-permanent combination

$$\sum_{j\geq 1} G_{k,j} + P + \sum_{i\geq 1} \psi_{2,i} Q_{k,i}$$

Detailed expressions of load combinations are given in normative annexes of EN1990 (Annex A1 for buildings, A2 for bridges, etc.), together with recommended values of the partial safety factors γ_F and combination factors ψ.

The characteristic combination should normally be considered for short-term limit states, linked to the achievement of an assigned value of the studied effect only once. It corresponds to those effects whose probability of exceeding is close to the probability of exceeding the characteristic value of the dominant variable action $Q_{k,1}$. In other words, it can be said that the characteristic combination should be considered for irreversible SLS verifications, associated with unacceptable permanent damage or deformation.

The frequent combination must be considered for medium-term limit states, linked to the achievement of a certain value of the effect studied for a small fraction of the reference duration or for an assigned number of times. It corresponds to the effects whose durations or exceeding frequencies are close to those of the frequent value ψ_1 $Q_{k,1}$ of the dominant variable action.

The quasi-permanent combination must be considered for the study of long-term effects, linked to the achievement of an assigned value of the analysed effect for a long fraction of the reference duration.

Frequent and quasi-permanent combinations must be considered for reversible SLS verifications, for which effects cease as soon as loads are removed.

Combination of actions for fatigue verifications

The combination of actions for fatigue verifications is given in [6.8.3(3)P]

$$\left(\sum_{j\geq 1} G_{k,j} + P + \psi_{1,1} Q_{k,i} + \sum_{i\geq 1} \psi_{2,i} Q_{k,i}\right) + Q_{\text{fat}}$$

where Q_{fat} is the load that produces fatigue, such as the load produced by vehicular traffic on a bridge, or the cyclic load induced by a vibrating machine on the supporting structure.

Reference

Gulvanessian, H., Calgaro, J.-A., & Holický, M. (2006). *Designer's guide to EN1990 Eurocode: Basis of structural design.* Thomas Telford.

Chapter 2
Materials

Abstract The chapter examines the strength and deformation characteristics of the two basic materials: concrete and steel. The values are contained in EC2, here they are commented and discussed relating to their correct use according to the Standard. As regards the concrete, we comment on the diagrams parabola rectangle and parabola generalized rectangle, bilinear and rectangular stress block. For ordinary steels that EC2 prescribes with 400 < f_{yk} < 600 N/mm^2, two types are considered: B450C, with f_{yk} > 450 N/mm^2 and high ductility and B450A, with f_{yk} > 450 N/mm^2 and medium ductility; both types are covered by EN10080. For prestressed steels: wires, strands, and bars in accordance with EN 10138.

2.1 Concrete

Paragraph [3.1] of EC2 gives principles and rules for normal weight concrete, i.e., according to EN 206–1, with weight per unit volume between 2000 and 2600 kg/m^3. Rules for lightweight aggregate concrete are given in [Section 11]. Self-compacting concretes (SCC) are not discussed.

In [3.1.2(1)P] compressive strength f_{ck}, according to EN 206–1, is defined by the characteristic value (5% fractile of the statistical distribution) obtained by processing the results of the compressive tests at 28 days hardened for cylindrical specimens with diameter 150 mm and height 300 mm. As many countries refer the experimentation to cubes with side 150 mm, EC2 also provides the strength $f_{ck,\mathrm{cube}}$.

Strength classes are defined using the letter C followed by two numbers representing cylindrical characteristic strength and cubic strength expressed in N/mm^2, for example C30/37. EC2 defines 14 classes: from C12/15 to C90/105, recalled in Table 2.1.

This chapter was authored by Piero Marro and Matteo Guiglia.

© The Author(s), under exclusive license to Springer Nature Switzerland AG 2022

F. Angotti et al., *Reinforced Concrete with Worked Examples*,
https://doi.org/10.1007/978-3-030-92839-1_2

Table 2.1 Strength classes

C12/15	C45/55
C16/20	C50/60
C20/25	C55/67
C25/30	C60/75
C30/37	C70/85
C35/45	C80/95
C40/50	C90/105

Table [3.1] of EC2 collects the numerical values of mechanical and strain characteristics associated with strength classes and analytical relationships expressing (these) values as a function of f_{ck}.

To make the reading easier, an excerpt from Table [3.1] is recalled in Table 2.2, collecting strength and strain values and n exponents of σ–ε laws for the ULS checks which will be used in the application examples of the following Chapters.

Figure 2.1 shows the values of medium compressive strength f_{cm}, tensile strength $f_{ctm,}$ and elasticity modulus E_{cm} as a function of f_{ck} as shown in [Figure 3.2]; E_{cm} is defined by the slope of the secant line of the σ–ε diagram between points $\sigma = 0$ and $\sigma = 0.4.\ f_{cm}$.

Point [3.1.2(5)] indicates the following conventional relationship between f_{cm} and f_{ck} at time t between 3 and 28 days.

$$f_{ck}(t) = f_{cm}(t) - 8\ (\mathrm{N/mm^2})\ \text{for}\ 3 < t < 28\ \text{days}$$
$$f_{ck}(t) = f_{ck}\ \text{for}\ t \geq 28\ \text{days}$$

Point [3.1.2(6)] discusses the evolution of compressive strength with time. Formulas [(3.1)] and [(3.2)] allow to calculate the mean compressive strength f_{cm} at time t (days) as a function of the value at 28 days (which can be obtained by Table [3.1]) and employed cement class with the medium temperature of 20 °C and hardening conditions (*curing*) according to EN 12390. Cement classes, according to EN 197, and relevant coefficients s of the following formula [(3.2)] are:

R (fast hardening) for cements CEM 42.5R, CEM 52.5N and CEM 52.5R; $s = 0.20$
N (normal hardening) for cements CEM 42.5N and CEM 32.5R; $s = 0.25$
S (slow hardening) for cements CEM 32.5N; $s = 0.38$.

$$f_{cm}(t) = \beta_{cc}(t) \cdot f_{cm} \quad [(3.1)]$$

$$\beta_{cc}(t) = \exp\left\{ s \cdot \left[1 - \left(\frac{28}{t} \right)^{0,5} \right] \right\} \quad [(3.2)]$$

Table 2.2 Mechanical and strain characteristics associated with strength classes. Excerpt from Table [3.1] of EC2

f_{ck} (N/mm^2)	12	16	20	25	30	35	40	45	50	55	60	70	80	90
$f_{ck,\text{ cubo}}$ (N/mm^2)	15	20	25	30	37	45	50	55	60	67	75	85	95	105
f_{cm} (N/mm^2)	20	24	28	33	38	43	48	53	58	63	68	78	88	98
f_{ctm} (N/mm^2)	1.6	1.9	2.2	2.6	2.9	3.2	3.5	3.8	4.1	4.2	4.4	4.6	4.8	5.0
$f_{ctk,0.05}$ (N/mm^2)	1.1	1.3	1.5	1.8	2.0	2.2	2.5	2.7	2.9	3.0	3.1	3.2	3.4	3.5
$f_{ctk,0.95}$ (N/mm^2)	2.0	2.5	2.9	3.3	3.8	4.2	4.6	4.9	5.3	5.5	5.7	6.0	6.3	6.6
E_{cm} (kN/mm^2)	27	29	30	31	33	34	35	36	37	38	39	41	42	44
ε_{c1} (‰)	1.8	1.9	2.0	2.1	2.2	2.25	2.3	2.4	2.45	2.5	2.6	2.7	2.8	2.8
ε_{cu1} (‰)	3.5	3.5	3.5	3.5	3.5	3.5	3.5	3.5	3.5	3.2	3.0	2.8	2.8	2.8
ε_{c2} (‰)	2.0	2.0	2.0	2.0	2.0	2.0	2.0	2.0	2.0	2.2	2.3	2.4	2.5	2.6
ε_{cu2} (‰)	3.5	3.5	3.5	3.5	3.5	3.5	3.5	3.5	3.5	3.1	2.9	2.7	2.6	2.6
n	2.0	2.0	2.0	2.0	2.0	2.0	2.0	2.0	2.0	1.75	1.6	1.45	1.4	1.4
ε_{c3} (‰)	1.75	1.75	1.75	1.75	1.75	1.75	1.75	1.75	1.75	1.8	1.9	2.0	2.2	2.3
ε_{cu3} (‰)	3.5	3.5	3.5	3.5	3.5	3.5	3.5	3.5	3.5	3.1	2.9	2.7	2.6	2.6

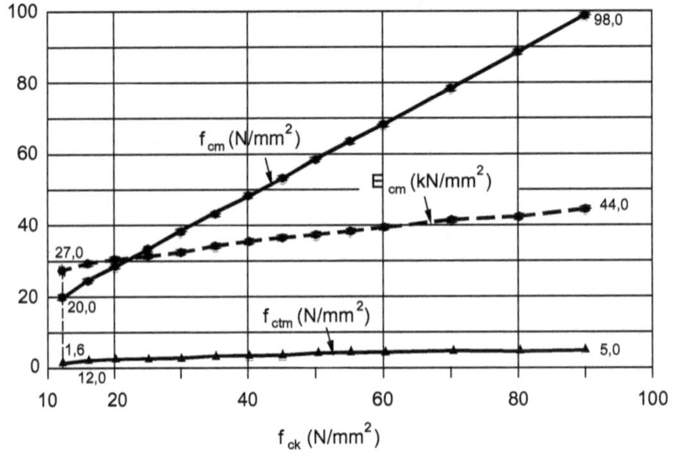

Fig. 2.1 f_{cm}, f_{ctm} and E_{cm} as a function of f_{ck}

Figure 2.2 shows the time evolution of coefficients $\beta_{cc}(t)$ for concretes with the three cement classes.

Paragraph [3.1.3] discusses the elastic deformation represented by the values of the elastic modulus E_{cm}. It is worth noting that the values of modulus E_{cm} collected in Fig. 2.1 are referred to concretes made of silica aggregates, hardened for 28 days. For concretes made of calcareous and sandstone aggregates the predicted values shall be reduced by 10% and 30% respectively. For concretes made of basaltic aggregates, the values shall be increased by 10%.

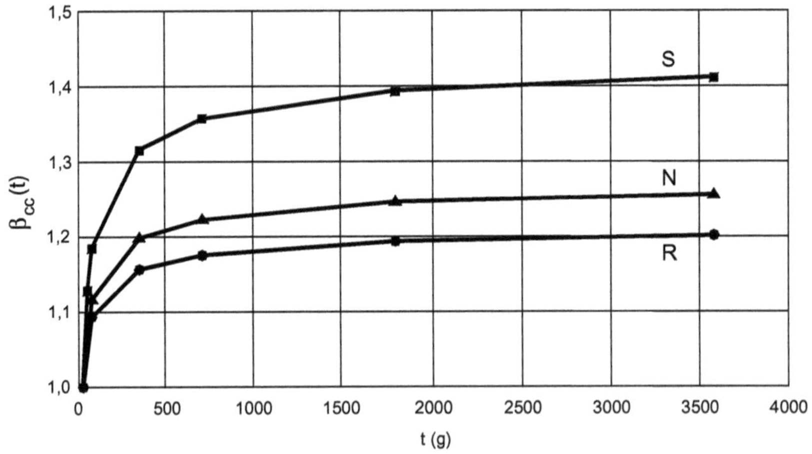

Fig. 2.2 β_{cc} as a t function (days)

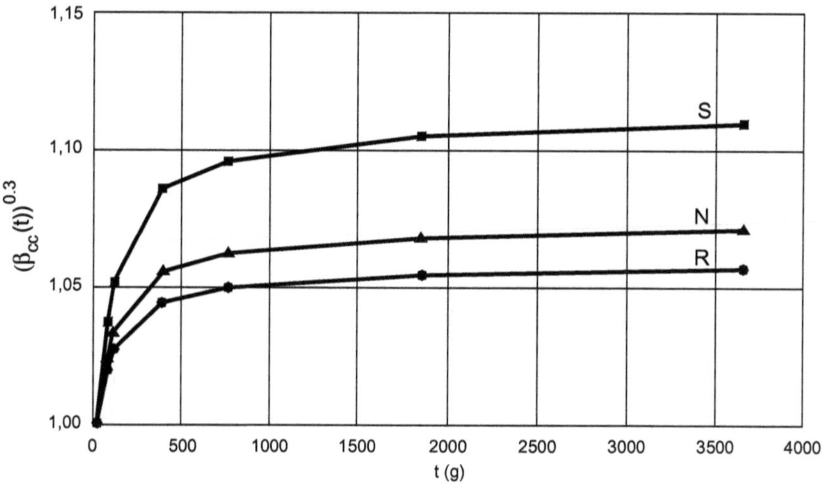

Fig. 2.3 $(\beta_{cc})^{0.3}$ as a *t* function (days)

The evolution of modulus by time is obtained from the modulus corresponding to f_{cm} using the following formula

$$E_{cm}(t) = \left(\frac{f_{cm}(t)}{f_{cm}}\right)^{0.3} \cdot E_{cm} = (\beta_{cc}(t))^{0.3} \cdot E_{cm} \quad [(3.5)]$$

Figure 2.3 shows the evolution of $(\beta_{cc}(t))^{0.3}$ as a function of time for the three clement classes (R, N, S).

Paragraph [3.1.4] treats creep and shrinkage.

These two phenomena are peculiar for concrete. They are consisting, in the first case of the deformation developing in time after load application and, in the second case, of a spontaneous shortening deformation.

The evolution of both phenomena is depending on environment humidity, structural element dimensions and concrete composition. Creep is also dependent on the hardening level of the concrete, the duration and intensity of the load.

Both phenomena—creep and shrinkage—influence the evolution of prestressing as indicated in paragraph [5.10.6] and deflection at SLS [7.4.3(5)] and [7.4.3(6)]. Moreover, creep influences the second-order effects [5.8].

2.1.1 Creep

Creep deformation $\varepsilon_{cc}(\infty, t_o)$ (the first index means *concrete*, the second one *creep*) of concrete at time t_∞, induced by a compressive stress σ_c, not variable during time, applied at the time t_o, is given by the expression

$$\varepsilon_{cc}(\infty, t_0) = \phi(\infty, t_0) \cdot \left(\frac{\sigma_c}{E_c}\right) \quad [(3.6)] \tag{2.1}$$

where φ (∞, t_0) is the creep coefficient referred to E_c, tangent modulus at the origin of the diagram σ_c–ε_c which can be taken equal to 1.05 E_{cm} (this last one collected in Table [3.1]). Annex B reports detailed information about the evolution of creep in time. When the stress σ_c at the time t_0 is not greater than $0.45 f_{ck}(t_0)$, the evaluation of the creep coefficient can be performed utilizing the graphs of Figure [3.1].

The reported values are valuable for environmental conditions characterized by relative humidity between 40 and 100% and temperature between −40 and +40 °C. The graphs are moreover function of concrete age t_o expressed in days at the time of load application, of conventional dimension $h_o = 2A_c/u$, where A_c is the area of the cross-section and u, is the perimeter of the portion exposed to drying, of the concrete class (e.g. C30/37) and of type and class (R, N, S) of employed cement as specified at [3.1.2(6)].

If the stress at the time t_0 is greater than $0.45 f_{ck}(t_o)$, the proportionality expressed by the formula (2.1) is not existing and it is necessary to use the exponential expression [(3.7)].

2.1.2 *Shrinkage*

Negative deformation due to shrinkage ε_{cs} (the second index means *shrinkage*) is the sum of two terms.

- ε_{cd} strain due to drying shrinkage (d = *drying*), developing slowly in time and being a function of the water migration from the hardening concrete to the environment;
- ε_{ca} autogenous shrinkage strain developing in the first days after the cast, in the hardening phase.

Shrinkage is essentially a function of the environmental humidity and $h_o = 2A_c/u$. Paragraph [3.1.4(6)] reports formulas and current tabular values as a function of the relative humidity and strength class. Annex B (part B2) gives further information.

2.1.3 *Stress-Deformation Diagram for Structural Analysis*

Paragraph [3.1.5] gives the stress-deformation relationship for the nonlinear structural analysis described by [Figure 3.2] and expression [(3.14)]. The stress–strain diagram is known as "Sargin parabola".

2.1.4 *Compressive and Tensile Design Strength*

Compressive design strength f_{cd} is given by

$$f_{cd} = \alpha_{cc} f_{ck} / \gamma_C \quad [(3.15)] \tag{2.2}$$

where

α_{cc} (the second index means compression) is a coefficient taking into account the strength decrease occurring for a long duration load application or due to unfavourable effects due to the way of application of load. The recommended value by EC2 is equal to 1.0, however, it should lie between 0.8 and 1.0 depending on the National Annex of the Country of reference.

γ_C a safety coefficient equal to 1.50 [Table 2.1N].

Tensile design strength, f_{ctd}, is given by

$$f_{ctd} = \alpha_{ct} f_{ctk,0,05} / \gamma_C \quad [(3.16)] \tag{2.3}$$

where

α_{ct} a coefficient analogous to α_{cc}, which takes the value 1 (as recommended by EC2);

$f_{ctk\,0.05}$ the tensile characteristic strength, fractile lower than 5%, which can be deduced by Table [3.1].

2.1.4.1 Stress–Strain Diagrams for the Cross-Sectional Design at ULS

For strength f_{ck} until 50 N/mm^2, paragraph [3.1.7(1)] identifies the parabola-rectangle diagram and, for greater strengths, parabola-rectangle "generalized" diagrams characterized by exponents n lower than 2 as a function of f_{ck}. In both cases, the type diagram is made by a curvilinear part, defined by the exponent n from the origin ($\varepsilon_c = 0$) to the abscissa ε_{c2} with a horizontal tangent, followed by a segment with constant ordinate until the abscissa ε_{cu2}. For concrete C90/105, there is an exception because the diagram is made only of the curvilinear portion: in this case, $\varepsilon_{c2} = \varepsilon_{cu2} = 2.6/1000$. The curvilinear part is defined by

$$\sigma_c = f_{cd}\left[1 - \left(1 - \frac{\varepsilon_c}{\varepsilon_{c2}}\right)^n\right] \quad [(3.17)] \tag{2.4}$$

and the horizontal portion by

$$\sigma_c = f_{cd} \, [(3.18)] \tag{2.5}$$

Table 2.2 collects the values of n, ε_{c2} and ε_{cu2}, as a function of f_{ck}; the same values are shown in Fig. 2.4.

Figure 2.5 shows the diagrams of formulas (2.4) and (2.5) for four concrete classes.

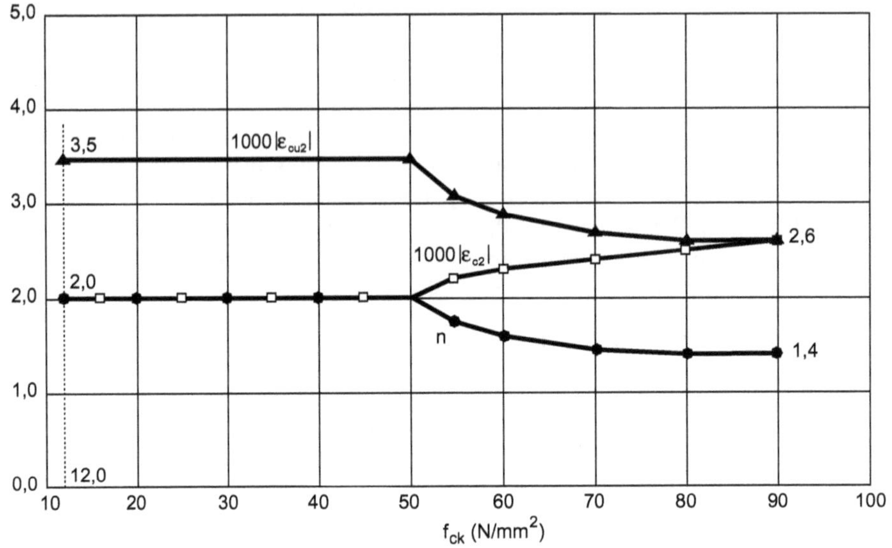

Fig. 2.4 n, ε_{c2} and ε_{cu2} as a function of f_{ck}

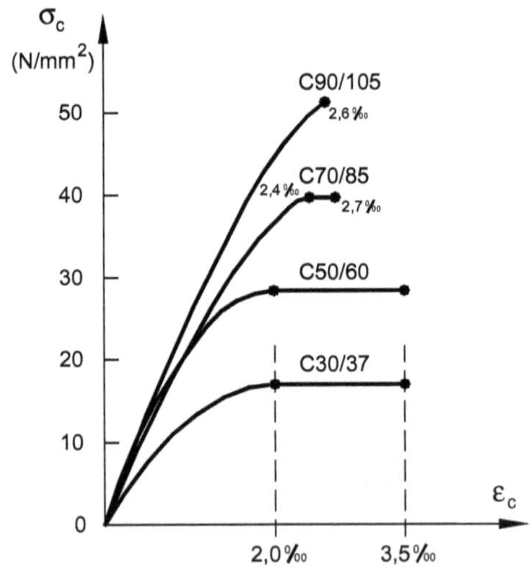

Fig. 2.5 Parabola-rectangle diagrams for C ≤ 50/60 and generalized parabola-rectangle diagrams for C > 50/60

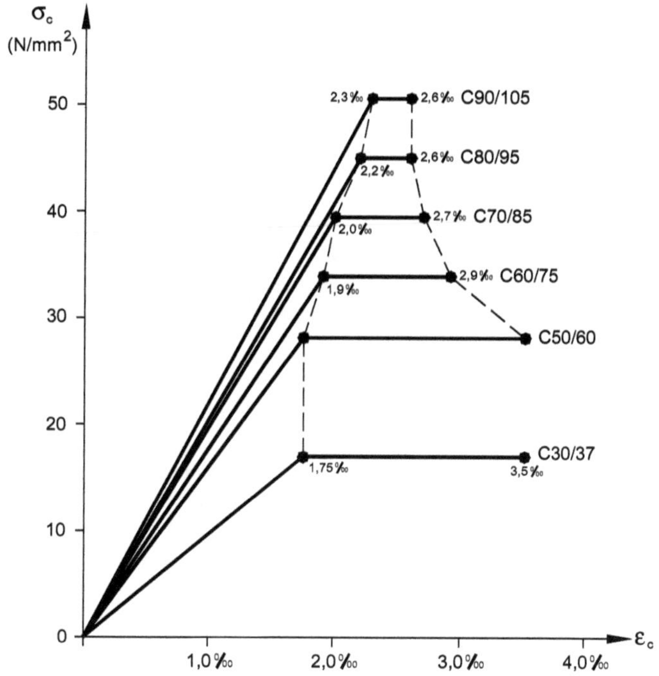

Fig. 2.6 Bilinear diagrams

Alternatively, paragraph [3.1.7(2)] provided the employ of simpler diagrams σ–ε under the assumption that they are equivalent or more conservative of those of paragraph [3.1.7(1)]; it recalls, for example, a bilinear diagram with a first inclined segment until the abscissa ε_{c3} followed by a second horizontal segment with ordinate f_{cd} until the abscissa ε_{cu3} (Fig. 2.6). Values of ε_{c3} and ε_{cu3} are collected in Table 2.2.

Paragraph [3.1.7(3)] accounts for the possibility of employing a uniform stress block. In the case of linearly variable deformation (plane section conservation) with the neutral axis inside the cross-section (Fig. 2.7), the uniform stress block is defined by two coefficients: λ limiting the extension of the block to the depth x of the compressed area and η reducing the strength f_{cd}, both function of f_{ck}, as it results from Table 2.3.

2.1.5 Flexural Tensile Strength

Paragraph [3.1.8] defines flexural tensile strength, which depends on the mean axial tensile strength of the material and the depth h of the deflected element. For values of h not greater than 600 mm, strength is given by:

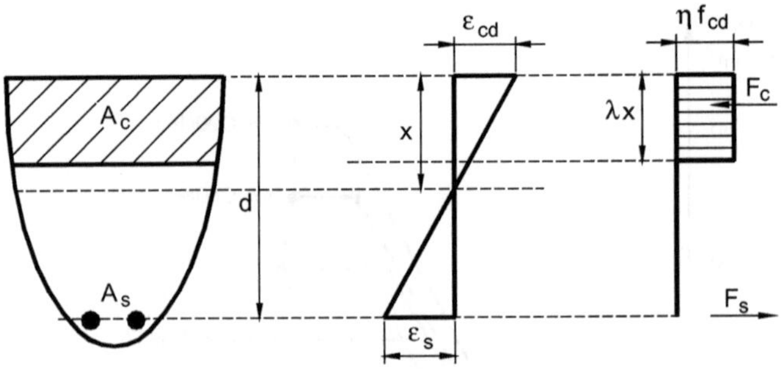

Fig. 2.7 Stress rectangular distribution [Figure 3.5]

Table 2.3 λ and η

f_{ck} (N/mm^2)	λ	η
≤50	0.8000	1.0000
55	0.7875	0.9750
60	0.7750	0.9500
70	0.7500	0.9000
80	0.7250	0.8500
90	0.7000	0.8000

$$f_{ctm,\text{fl}} = f_{ctm}(1.6 - h/1000) \text{ with h in mm} \quad [(3.23)]$$

while, for h greater than 600 mm, $f_{ctm,\text{fl}}$ can be identified with f_{ctm}.

2.1.6 *Triaxial Compressive Strength*

Under the assumption of two identical transverse compressive stresses ($\sigma_2 = \sigma_3$), strength is given by:

$$f_{ck,c} = f_{ck}[1.000 + 5.0(\sigma_2/f_{ck})] \text{ if } \sigma_2 \le 0.05 f_{ck} \quad [(3.24)] \qquad (2.6)$$

$$f_{ck,c} = f_{ck}[1.125 + 2.5(\sigma_2/f_{ck})] \text{ if } \sigma_2 > 0.05 f_{ck} \quad [(3.25)] \qquad (2.7)$$

Axial strength increases up to 50% if transverse compressive stresses are equal to 15% of f_{ck}, and doubles if compressions are equal to 35% of f_{ck}. Also, the ultimate deformation increases substantially.

Transverse stresses can be realized through confinement or compression on the element. The index c, added to f_{ck}, signifies "confined".

2.2 Ordinary Reinforcement

Paragraph [3.2.2(3P)] affirms that design rules provided by EC2 are valid for values of the characteristic yield strength f_{yk} in the range $400 \div 600$ N/mm^2. In the present discussion two types of steel in ribbed bars, weldable, are considered: B450C and B450A, with the following characteristics:

- *B450C (included in C class)*
  - characteristic (tensile strength) $f_{tk} \geq 540$ N/mm^2
  - characteristic (yield strength) $f_{yk} \geq 450$ N/mm^2

$$1.15 \leq \left(\frac{f_t}{f_y}\right)_k < 1.35$$

  - Characteristic elongation per unit length under the maximum trial load

$$\varepsilon_{uk} = \left(\mathrm{A_{gt}}\right)_{\mathrm{k}} \geq 0.075$$

  - Diameters from 6 to 40 mm
- *B450A (included in class A)*
  - characteristic collapse and yielding stress: as B450C

$$1.05 \leq \left(\frac{f_t}{f_y}\right)_k$$

  - Characteristic elongation per unit length under the maximum trial load

$$\varepsilon_{uk} \geq 0.025$$

  - Diameters from 5 to 10 mm.

Safety coefficient γ_S for ULS takes the value 1.15 according to Table [2.1N].

The design stress–strain diagram is shown in Fig. 2.8.

For the cross-section verification, it is possible to assume one of the following models:

(a) an inclined branch with strain limit given by ε_{ud} and maximum stress $k\frac{f_{yk}}{\gamma_S}$ in ε_{uk}, where $k = \left(\frac{f_t}{f_y}\right)_k$. The value of ε_{ud} recommended by EC2 is 0.9 ε_{uk}.

(b) a horizontal top branch without the need to check strain limit.

The value of elasticity modulus E_s can be assumed equal to 200,000 Nmm^{-2}.

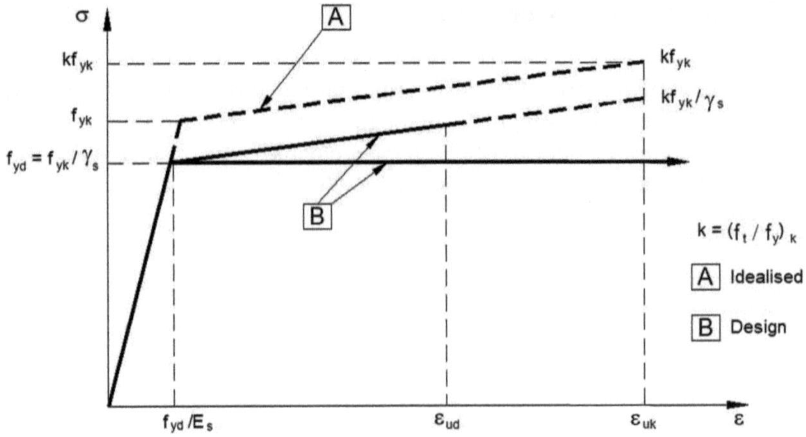

Fig. 2.8 Idealised and design diagrams for ordinary reinforcement [Figure 3.8]

Figure 2.9 shows, for example, the diagram stress—percentage elongation for a steel bar C with diameter 18 mm (cross-section area 256 mm^2) controlled in a factory production. Tests provide the following results:

$$\text{(yield strength)}\, f_y = 472\,\text{N/mm}^2$$
$$\text{(tensile strength)}\, f_t = 589\,\text{N/mm}^2$$

$$\left(\frac{f_t}{f_y}\right) = 1.24$$

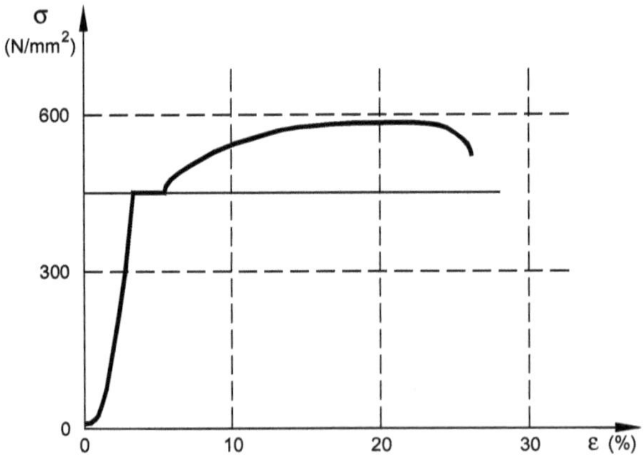

Fig. 2.9 Experimental stress–strain diagram for steel rebar

Elongation at maximum trial load $\varepsilon_u = 220/1000$

Effectively, the diagram of Fig. 2.9 shows in ordinates the values F/A (the ratio between the trial load and the area of the starting cross-section of the sample), and in abscissas, the elongation of the sample referred to its starting length. The descending branch beyond the maximum value is a consequence of the striction interesting a short portion of the specimen: in this portion, the product of the contracting cross-sectional area due to the increasing stress remains unchanged or decreases. The diagram of Fig. 2.8 does not take into account this phenomenon, not interesting for design purposes.

2.3 Prestressing Reinforcement

Here we collect the data contained in EC2 integrated with those of prEN 10138 recalled by EC2. Prestressing tendons, considering the geometric point of view, are classified as follows:

- wires with smooth or indented surface with diameter between 3.0 and 11.0 mm
- strands formed by two wires helically wound concentrically in a helix; nominal diameter between 4.5 and 5.6 mm
- strands formed by three wires helically wound concentrically in a helix; nominal diameter between 5.2 and 7.7 mm
- 7-wire strands made of six wires helically wound about a core wire; nominal diameter between 6.4 and 18.0 mm
- ribbed bars; nominal diameter between 15.0 and 50.0 mm.

For each type, reinforcements are classified according to the following properties:

- strength, identified by tensile strength f_p and stress $f_{p0,1}$ denoting the value of the 0.1% proof stress;
  - ductility, defined by the ratio $(f_{pk}/f_{p0,1\,k}) \geq 1.1$. Another ductility parameter is the elongation under maximum load (ε_{uk}). This value, not numerically defined in EC2, according to EN10138 shall be equal at least to 0.035.
- class, defining the relaxation behaviour. EC2 defines three classes:

  Class 1: ordinary relaxation wires, strands.
  Class 2: low relaxation wires, strands.
  Class 3: hot-rolled and processed bars.

ρ_{1000} is defined as the loss due to relaxation at 1000 h after tensioning and at a mean temperature of 20 °C, expressed as a percentage of the initial stress, which is set up at $0.7f_p$, where f_p is the real tensile strength. Values of ρ_{1000} indicated in EC2 for structural design are: 8% for Class 1, 2, 5% for Class 2, 4% for Class 3.

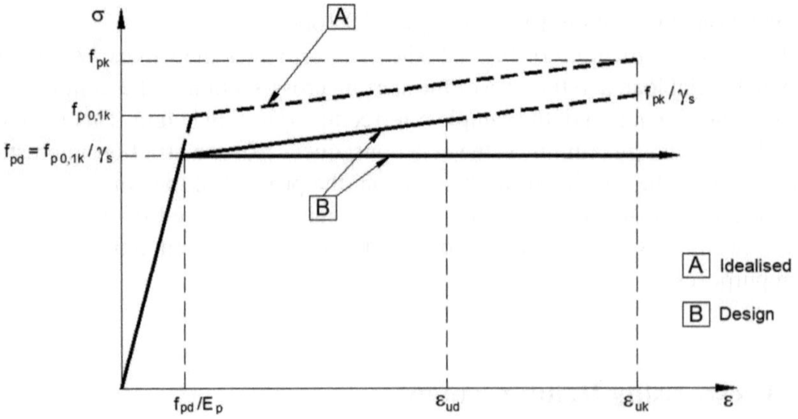

Fig. 2.10 Idealised and design diagrams for prestressing tendons [Figure 3.10]

Paragraph [3.3.2(7)] reports the formulas for the calculation of relaxation at various time t for the previous three Classes. For accurate calculations, [Annex D] provides the necessary elements.

Fatigue. Prestressing tendons can suffer fatigue phenomena. Relevant criteria and methods are discussed in [6.8].

Design data. In addition to strength and ductility, EC2 provides the following design data: modulus of elasticity suggested for wires and strands: 195,000 N/mm^2; for wires and bars: 205,000 N/mm^2.

Design stress-elongation diagrams. As represented in Fig. 2.10, under the assumption of safety factor $\gamma_S = 1.15$, two design diagrams are considered, differing from the plastic branch after a common elastic portion made of a linear segment until the ordinate f_p:

- a hardening straight branch until the deformation $\varepsilon_{ud} = 0.9\ \varepsilon_{uk}$, or 0.02;
- a horizontal branch without deformation limit.

Figure 2.11 shows, e.g., the diagram stress-elongation in % for a 7-wires strand with nominal diameter 15.2 mm, with cross-sectional area 140 mm^2, controlled in a production factory. Tests provide the following data:

- tensile strength $f_{pt} = 1960$ N/mm^2
- $f_{p0.1} = 1780$ N/mm^2.

elongation at failure $\varepsilon_u = 6.2\%$

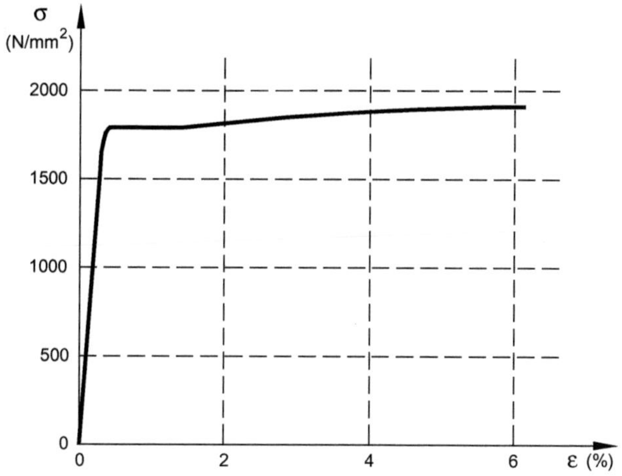

Fig. 2.11 Experimental stress–strain diagram for a strand of a prestressing tendon

Chapter 3
Durability and Cover to Reinforcement

Abstract The chapter presents the requirements of EC2 to assure the durability of reinforced concrete structures during their design working life, without significant loss of utility or major repair. The protection of the structure is established by considering its intended use, design working life, maintenance program and actions. In reinforced concrete structures, durability is mostly dependent on corrosion protection of steel reinforcements, which is related to density, quality and thickness of concrete cover and concrete cracking. The design working life of a structure shall be guaranteed through suitable requirements to protect each structural element from the relevant environmental actions. The traditional method of calculating the concrete cover according to environmental conditions does not account for all the important factors for the evaluation of durability. Besides environmental exposure classes, cement type, concrete quality (controlled by a/c ratio and drying time) and climatic conditions (temperature, humidity) must be considered, too. At the end of the chapter, some case studies are dedicated to the choice of the minimum strength class of concrete and the evaluation of concrete cover for common structural elements.

3.1 Introduction

A durable structure shall meet the requirements of serviceability, strength, and stability during its design working life, without significant loss of utility or major repair. The protection of the structure shall be established by considering its intended use, design working life, maintenance program, and actions. Direct and indirect actions, environmental conditions, and their effects shall be considered.

In reinforced concrete structures, durability is mostly dependent on corrosion protection of steel reinforcements, which is related to the density, quality, and thickness of concrete cover and cracking.

Density and quality of concrete cover are achieved by controlling the maximum water/cement ratio and minimum cement content and may be related to a minimum

This chapter was authored by Franco Angotti and Maurizio Orlando.

© The Author(s), under exclusive license to Springer Nature Switzerland AG 2022
F. Angotti et al., *Reinforced Concrete with Worked Examples*,
https://doi.org/10.1007/978-3-030-92839-1_3

strength class of concrete (indicated as *indicative strength class for durability in Annex E of EC2*).

The design working life of a structure shall be guaranteed through suitable requirements to protect each structural element from the relevant environmental actions.

The traditional method of calculating the concrete cover according to environmental conditions does not account for all the important factors for the evaluation of durability. The most important factors, besides environmental exposure classes, are cement type, concrete quality (controlled by a/c ratio and drying time), and climatic conditions (temperature, humidity).

3.2 Concrete Cover

The concrete cover is the distance between the surface of the reinforcement closest to the nearest concrete surface (including links and stirrups and surface reinforcement where relevant) and the nearest concrete surface (Fig. 3.1). The nominal cover shall be specified on the drawings and is denoted by c_{nom} in EC2. It is defined as a minimum cover c_{min} plus an allowance in design for deviation Δc_{dev}:

$$c_{nom} = c_{min} + \Delta c_{dev}$$

where

c_{min} is the minimum concrete cover, which has to be chosen to fulfil the following three requirements: (1) reinforcement bond, for the safe transmission of bond forces; (2) environmental conditions, related to the protection of the steel against corrosion (durability); (3) adequate fire resistance,

Δc_{dev} is the allowance in design for deviation.

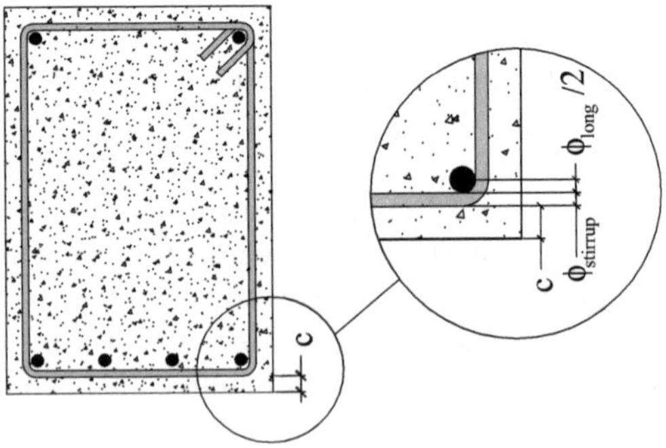

Fig. 3.1 Concrete cover for a beam with rectangular cross-section

EC2 only deals with the first two requirements for the choice of the concrete cover, while the third requirement (fire resistance) is fulfilled by applying the rules of EN1992-1-2.

In this chapter, only requirements on the concrete cover for bond and durability are discussed.

According to EC2, c_{min} takes the maximum value among $c_{min,b}$ fulfilling the bond requirement, $\hat{c}_{min,dur}$ for environmental conditions and 10 mm:

$$c_{min} = \max\{c_{min,b};\ \overset{\frown}{c}_{min,dur};\ 10\ \text{mm}\}$$

The concrete cover $\hat{c}_{min,dur}$ is expressed as the sum of four terms:

$$\hat{c}_{min,dur} = c_{min,dur} + \Delta c_{dur,\gamma} - \Delta c_{dur,st} - \Delta c_{dur,add}$$

where

$c_{min,dur}$	minimum concrete cover due to environmental conditions,
$\Delta c_{dur,\gamma}$	additional safety coefficient,
$\Delta c_{dur,st}$	reduction of the minimum concrete cover when stainless steel or special features are employed,
$\Delta c_{dur,add}$	reduction of the minimum concrete cover when additional protection is used (e.g. the coating of the external surface, cathodic protection, etc.).

EC2 recommends to assume $\Delta c_{dur,\gamma} = \Delta c_{dur,st} = \Delta c_{dur,add} = 0$ mm, so the following equality holds:

$$\hat{c}_{min,dur} = c_{min,dur}$$

and the expression of c_{min} can be rewritten as:

$$c_{min} = \max\{c_{min,b};\ c_{min,dur};\ 10\ \text{mm}\}.$$

3.3 Minimum Concrete Cover $c_{min,b}$ for Bond

To transmit bond forces safely and to ensure adequate compaction of the concrete, EC2 recommends adopting a minimum concrete cover not less than $c_{min,b}$. The minimum value depends on the reinforcement type and other factors (Table 3.1):

Table 3.1 Minimum cover $c_{min,b}$

Ordinary reinforcement

Nominal maximum aggregate size	Arrangement of reinforcing bars	$c_{min,b}$
$d_{max} \leq 32$ mm	Isolated bars	Bar diameter ϕ
	Bundled bars (see Chap. 12)	Equivalent diameter ϕ_n (measured from the external contour line of the group)
$d_{max} > 32$ mm	Isolated bars	Bar diameter $\phi + 5$ mm
	Bundled bars (see Chap. 12)	Equivalent diameter ϕ_n + 5 mm (measured from the external contour line of the group)

Remark. The equivalent diameter for a bundle of bars is given by the following expression [(8.14)]:
$\varphi_n = \varphi\sqrt{n_B} \leq 55$ mm
where n_B is the number of bundled bars with the following limitations:
$n_B \leq 4$ for vertical bars in compression and bars in a lapped joint,
$n_B \leq 3$ for all the other cases

Post-tensioned bonded tendons

Duct section	$c_{min,b}$
Circular ($\phi < 80$ mm)	Duct diameter
Rectangular $b \times h$ (with $h < b \leq 80$ mm)	max (h; $b/2$), where b is the greatest dimension of the cross-section

Remark. EC2 does not give any rule for the cover of circular or rectangular ducts with dimensions greater than 80 mm, so it is required to refer to the ETA (European Technical Approval) of the prestressing system

Prestressing tendons

Type of prestressing tendon	$c_{min,b}$
Strand	1.5 times the strand diameter
Plain wire	1.5 times the wire diameter
Indented wire	2.5 times the wire diameter

- maximum nominal dimension of aggregates,[1] arrangement, and bar diameter for *ordinary reinforcement*,
- duct shape and dimension for *post-tensioned tendons*,
- reinforcement diameter and type for *pre-tensioned tendons*.

For lightweight concrete, $c_{min,b}$ of ordinary reinforcement shall be increased by 5 mm. Finally, for pre-tensioned tendons, the minimum cover of anchorage shall be calculated according to the relevant European Technical Approval (ETA).

[1] According to EN 206-1:2001 "Concrete—Part 1: Specification, performance, production and conformity", nominal maximum aggregate size (d_{max}) shall be chosen accounting for concrete cover and the width of the minimum cross-section.

3.4 Minimum Concrete Cover $c_{min,dur}$ Due to Environmental Conditions

The minimum cover to satisfy the requirements for environmental conditions (durability) is given by EC2 as a function of environmental conditions and the structural class. The procedure to evaluate the cover for durability consists of the following steps:

- evaluation of the environmental conditions of concrete (*exposure class*)
- choice of the concrete strength (*indicative strength classes for durability*)
- *choice of the structural class as a function of the exposure class, concrete strength* and other parameters as the design working life of the building (which depends on the strategic relevance of the structure), quality control for concrete production (it is worth noting that structural class is never defined explicitly)
- evaluation of $c_{min,dur}$ as a function of the structural class and exposure class.

3.4.1 Environmental Conditions (Exposure Classes Related to Environmental Conditions)

The environmental actions, chemical and physical, acting on a structure in addition to the mechanical actions, are indicated in EC2 as environmental *Exposure Classes* and are collected in Table 3.2, based on EN206-1. Environmental exposure classes are defined according to their attack power against concrete and/or reinforcing bars.

The following six main exposure classes can be defined as a function of environmental actions:

- Class 1: (X0) no risk of corrosion or attack
- Class 2: (XC) corrosion induced by carbonation ($C =$ *Carbonation*).
- Class 3: (XD) corrosion induced by chlorides (except seawater) ($D =$ *De-icing*)
- Class 4: (XS) corrosion induced by chlorides from seawater ($S =$ *Seawater*)
- Class 5: (XF) Freeze/Thaw Attack ($F =$ *Freezing*)
- Class 6: (XA) chemical attack (chemically aggressive soils and/or groundwater) ($A =$ *Attack*).

It is worth observing that exposure classes 2, 3 and 4 refer to the reinforcement corrosion risk, while exposure classes 5 and 6 to the concrete attack from external agents.

Except for class 1, all the other five classes are subdivided into subclasses: four subclasses each for class 2 (XC) and 5 (XF) and three subclasses each for class 3 (XD), 4 (XS), and 6 (XA).

Concrete can be subjected to more than one environmental action and in some cases, environmental conditions should be expressed as a combination of more than one exposure class. For example, a wall in contact with chemically aggressive soil is both in class XC4, cyclically wet and dry, as the external face is exposed to rain,

Table 3.2 Exposure classes related to environmental conditions [Table 4.1]

Class designation	Description of the environment	Informative examples where exposure classes may occur
Class 1—No risk of corrosion or attack		
X0	For concrete without reinforcement or embedded metal: all exposures except where there is freeze/thaw, abrasion or chemical attack For concrete with reinforcement or embedded metal: very dry	Concrete inside buildings with very low air humidity
Class 2—Corrosion induced by carbonation		
XC1	Dry or permanently wet	Concrete inside buildings with low air humidity Concrete permanently submerged in water
XC2	Wet, rarely dry	Concrete surfaces subject to long-term water contact Many foundations
XC3	Moderate humidity	Concrete inside buildings with moderate or high air humidity External concrete sheltered from rain
XC4	Cyclic wet and dry	Concrete surfaces subject to water contact, not within exposure class XC2
Class 3—Corrosion induced by chlorides		
XD1	Moderate humidity	Concrete surfaces exposed to airborne chlorides
XD2	Wet, rarely dry	Swimming pools Concrete components exposed to industrial waters containing chlorides
XD3	Cyclic wet and dry	Parts of bridges exposed to spray containing chlorides Pavements Car park slabs
Class 4—Corrosion induced by chlorides from seawater		
XS1	Exposed to airborne salt but not in direct contact with seawater	Structures near to or on the coast
XS2	Permanently submerged	Parts of marine structures
XS3	Tidal, splash and spray zones	Parts of marine structures
Class 5—Freeze/thaw attack		
XF1	Moderate water saturation, without a de-icing agent	Vertical concrete surfaces exposed to rain and freezing
XF2	Moderate water saturation, with a de-icing agent	Vertical concrete surfaces of road structures exposed to freezing and airborne de-icing agents

(continued)

Table 3.2 (continued)

Class designation	Description of the environment	Informative examples where exposure classes may occur
XF3	High water saturation, without de-icing agents	Horizontal concrete surfaces exposed to rain and freezing
XF4	High water saturation with de-icing agents or seawater	Road and bridge decks exposed to de-icing agents Concrete surfaces exposed to direct spray containing de-icing agents and freezing Splash zone of marine structures exposed to freezing
Class 6—Chemical attack		
XA1	Slightly aggressive chemical environment according to EN 206-1, Table 2	Natural soils and groundwater
XA2	Moderately aggressive chemical environment according to EN 206-1, Table 2	Natural soils and groundwater
XA3	Highly aggressive chemical environment according to EN 206-1, Table 2	Natural soils and groundwater

Note The composition of concrete affects both the protection of the reinforcement and the resistance of concrete to external attack. Annex E gives indicative strength classes for environmental exposure classes, which may lead to the adoption of a higher strength class than required for the structural design. In such cases, the calculation of minimum reinforcement and crack width should be performed using the value of f_{ctm} associated with the strength class chosen for the resistance to environmental conditions

both in one of the three classes XA corresponding to the chemical attack, because the base and the internal face are in contact with the chemically aggressive soil.

It is worth to precise that, rigorously, exposure class X0 can be used only if the structure or the structural members are not exposed, before the end of the construction, to more severe environmental conditions for a significant time.

It should be observed that each of the first four classes—class 1 (X0), 2 (XC), 3 (XD) or 4 (XS)—is sufficient to define the environment of a structure, while class 5 (XF) and class 6 (XA) should be specified in combination with one or more of the first four classes because they only refer to the risk of concrete attack.

For the same structure, the environmental exposure class can be different depending on the considered structural part. For example, for the deck of a viaduct, the effects of de-icing agents are stronger where the water tends to stagnate, that is in correspondence with joints and curbs, and slighter in the other parts. The designer should consider the most aggressive exposure class.

Concerning class XA (chemical attack), for which Table 3.2 refers to Table 2 of EN206-1, it is worth to precise that chemical aggressive environments classified in Table 2 of EN206-1 refer to natural soils, groundwater with temperatures

Table 3.3 Increase of the minimum cover $c_{\min}$ to consider abrasion

Abrasion class	Description	c_{min} increase (mm)
XM1	Moderate abrasion like for members of industrial sites frequented by vehicles with air tires	5
XM2	Heavy abrasion like for members of industrial sites frequented by forklifts with air or solid rubber tires	10
XM3	Extreme abrasion like for members industrial sites frequented by forklifts with elastomer or steel tires or track vehicles	15

of the water/ground between 5 and 25 °C and sufficiently low water velocity to be approximated to static conditions.

The heaviest chemical condition defines the exposure class. To define the three classes (XA1, XA2, XA3), the following chemical characteristics are accounted for: SO_4^{2-}, pH,CO_2, NH_4^+, Mg_2^+ for the groundwater; SO_4^{2-} and acidity for the ground soil.

In addition to the environmental conditions associated with the exposure classes, some special aggressive or indirect actions should also be considered, like:

- chemical attack, due for example to the use of the building or of the structure (liquid deposit, etc.), acid or sulfate solutions, chlorides within the concrete, alkali-aggregate reactions,
- physical attack, due for example to temperature variations, abrasion, water penetration.

The concrete abrasion could be considered by increasing the concrete cover (sacrificial layer) according to Table 3.3.

3.4.2 *Indicative Strength Classes for Durability*

To achieve the required corrosion protection for the reinforcement and to protect each structural element against the relevant environmental actions, the concrete choice requires an accurate study about its composition (*mix-design*) and water-cement ratio. It is required to adopt a water-cement ratio as smaller as greater is the level of environmental attack, so the concrete can be compact and waterproof, also guaranteeing suitable reinforcement protection.

Since the concrete strength increases at decreasing the water-cement ratio,[2] the adoption of a maximum value of the water-cement ratio to guarantee the durability can provide a compressive strength also greater than required by the design. In this

[2] Concrete strength increases at decreasing the water-cement ratio, provided that the w/c ratio is not smaller than a limit value which depends on the compaction method. In presence of fluidifying additives, strength increases also for very small water-cement ratios, thanks to the greater compaction guaranteed by additives.

Table 3.4 Indicative strength class for durability versus exposure class [Table E.1 N]

Corrosion										
Type of attack	Carbonation-induced corrosion				Chloride-induced corrosion			Chloride-induced corrosion from seawater		
Exposure class	XC1	XC2	XC3	XC4	XD1	XD2	XD3	XS1	XS2	XS3
Indicative strength class	C20/25	C25/30	C30/37		C30/37		C35/45	C30/37	C35/45	

Damage to concrete							
Type of attack	No risk	Freeze/Thaw attack			Chemical attack		
Exposure class	X0	XF1	XF2	XF3	XA1	XA2	XA3
Indicative strength class	C12/15	C30/37	C25/30	C30/37	C30/37		C34/45

case, the concrete tensile strength to be adopted in the calculation of the minimum reinforcement and crack verification shall be associated with the highest strength class.

Therefore, as environmental conditions (exposure classes) change, an *indicative strength class* for durability can be defined as the minimum concrete strength class to fulfil requirements for durability. Recommended values are collected in [Table E.1 N], which is shown in Table 3.4.

It is, however, worth noting that the approach which relates durability requirements to only the concrete strength, is not completely satisfactory, because concrete durability is also related to the water-cement ratio in a very tight way and only in an indirect way to the concrete strength. A small value of the water-cement ratio reduces the diameter of the concrete pores to $(10 \div 50)\ 10^{-9}$ m and increases the durability because the penetration of the aggressive agents is more difficult. At the same time, moderate values of the water-cement ratio provide greater strength, if all the other parameters do not change. Conversely, a target minimum value of the concrete strength could be attained for different values of the water-cement ratio, which leads to concretes with different porosities. For example, as the water-cement ratio increases, the same concrete strength could be reached at increasing the cement class (32.5–42.5–52.5 N/mm^2).[3]

It is therefore clear that a minimum strength class is not sufficient by itself to guarantee a maximum value of the water-cement ratio and limit the porosity, and therefore the durability requirements. The same discussion can be repeated if different values of the cement amount are accounted for: with the same parameters, a concrete made with better quality cement, and therefore with a smaller amount of cement,

[3] As the cement quality increases, the same strength value can be obtained with small amounts of cement and therefore through an increase of the water-cement ratio.

is less durable, but also this aspect is usually neglected. In the same way, the use of additives, like silica smokes, which provide greater compactness and durability to the hardened concrete, is neglected. For this purpose, it is useful to recall Table F.1 of EN 2061, which has been partially included in [Table E.1 N]. EN 206-1, in addition to the minimum strength class, gives the maximum water-cement ratio, the minimum amount of cement (kg/m^3), and the minimum amount of englobed air (%).

Moreover, EN 206-1 indicates that table values are referred to the use of cement CEM I of class 32.5 according to EN 197-1 and maximum aggregate dimension between 20 and 32 mm.

3.4.3 Strength Class for Durability

Regarding durability requirements, reinforced concrete structures are classified into six *structural classes*, from S1 to S6. The higher structural classes correspond to a greater vulnerability to external attacks, due to one or more of the following causes:

- high environmental attacks to the structure due to its exposure,
- low concrete strength (related, in general, to greater water-cement ratio and greater permeability),
- poor concrete quality (absent or low-quality control in the concrete production)
- significant influence of the constructive process on the reinforcement arrangement (greater uncertainties on the effective values of the concrete cover)
- long design working life.

For the same exposure class and the same type of reinforcement (ordinary or pre-stressing steel), the required cover to fulfil durability requirements increases at increasing the structural class. The minimum structural class is S1.

If the minimum concrete strength of Table 3.4 is adopted and a design working life of 50 years is considered, the conventional structural class is S4.

The structural class shall be increased by 2 (which implies the increase in the concrete cover if all the other parameters remain the same) if the structure is designed for a working life of 100 years, whatever the environmental exposure class. Conversely, the structural class can be reduced, with corresponding cover reduction, for all the cases listed in Table 3.5.

Examples collected at the end of this chapter show how to choose the structural class and the cover for durability $c_{\mathrm{min,dur}}$ for different types of structural elements and environmental conditions.

3.4.4 Values of $c_{min,dur}$

The minimum concrete cover due to environmental conditions is given by $c_{\mathrm{min,dur}}$, whose values are collected in Tables 3.6 and 3.7 for ordinary reinforcement and

Table 3.5 Recommended structural classification [Table 4.3N]

Criterion	Exposure class			
	X0/XC1	XC2/XC3	XC4—XD1—XD2/XS1	XD3/XS2/XS3
Design working life of 100 years	Increase class by 2			
Strength class (Note [a] and [b])	≥C30/37 reduce class by 1	≥C35/45 reduce class by 1	≥C40/50 reduce class by 1	≥C45/55 reduce class by 1
Member with slab geometry (position of reinforcement not affected by construction process)[c]	Reduce class by 1			
Special quality control of the concrete production ensured (as common during the production of precast elements)[d]	Reduce class by 1			

[a] Values of the strength class and w/c ratio are considered to be correlated to each other. A special composition (type of cement, w/c value, fine fillers) with the intent to produce low permeability may be considered
[b] The limit may be reduced by one strength class if air entrainment of more than 4% is applied
[c] This is mainly the case of bidimensional elements (plates and shells) where the design concrete cover can be guaranteed very accurately because the reinforcement arrangement is not affected by the constructive process either by the position of other reinforcing bars, as it occurs for the top longitudinal reinforcement of beams, whose position is associated by stirrups to the position of the bottom longitudinal reinforcement
[d] EC2 does not provide requirements to guarantee special concrete quality control

Table 3.6 Values of minimum cover, $c_{min,dur}$ (mm), requirements concerning durability for reinforcement steel

Structural class	Exposure class						
	X0	XC1	XC2/XC3	XC4	XD1/XS1	XD2/XS2	XD3/XS3
S1	10	10	10	15	20	25	30
S2	10	10	15	20	25	30	35
S3	10	10	20	25	30	35	40
S4	10	15	25	30	35	40	45
S5	15	20	30	35	40	45	50
S6	20	25	35	40	45	50	55

Table 3.7 Values of minimum cover, $c_{min,dur}$ (mm), requirements concerning durability for prestressing steel

Structural class	Exposure class						
	X0	XC1	XC2/XC3	XC4	XD1/XS1	XD2/XS2	XD3/XS3
S1	10	15	20	25	30	35	40
S2	10	15	25	30	35	40	45
S3	10	20	30	35	40	45	50
S4	**10**	**25**	**35**	**40**	**45**	**50**	**55**
S5	15	30	40	45	50	55	60
S6	20	35	45	50	55	60	65

prestressing steel. Tables are referred to exposure classes 1 (X0), 2 (XC), 3 (XD) and 4 (XS), that is only to classes concerning the risk of corrosion, while classes 5 (XF) and 6 (XA) are not considered. The risk of concrete attack has been already accounted for through the choice of the minimum strength class of the concrete (Table 3.4). Values of $c_{min,dur}$ can be applied both to normal-weight concretes and lightweight concretes.

3.5 Special Cases for the Choice of c_{min}

In special situations (for example for provisional or monumental structures, or structures under extreme or unusual actions, etc.) the designer shall consider additional requirements, as indicated in the following paragraphs.

3.5.1 In-Situ Concrete Placed Against Other Concrete Elements

In case of in-situ concrete placed against other concrete elements (precast or in-situ) the minimum cover c_{min} of the reinforcement to the interface should not consider the requirements for durability and could be chosen equal to $c_{min,b}$, that is the minimum value corresponding to the requirement for the bond if the following conditions are fulfilled:

- the strength class of concrete is at least C25/30,
- the exposure time of the concrete surface to the outdoor environment is short (<28 days),
- the interface has been roughened.

3.5.2 Unbonded Tendons

The cover for unbonded tendons should be provided following the relevant European Technical Approval (ETA).

3.5.3 Uneven Surfaces

In case of uneven surfaces, like ribbed finishes or exposed aggregates, the minimum cover c_{min} should be increased by at least 5 mm.

3.5.4 Exposure Classes XF and XA

Where the concrete is subjected to thaw/freeze cycles (class XF) or chemical attack (class XA), cover values given in Tables 3.6 and 3.7 are normally sufficient, while special attention should be given to the concrete composition, according to the prescriptions of EN 206-1 Section 6 (for example, the amount of englobed air to guarantee the required thaw/freeze resistance).

3.6 Allowance in Design for Deviation Δc_{dev}

The effects of the constructive deviations on the cover are accounted for through a design deviation Δc_{dev}, which is added to the minimum cover to obtain the nominal cover, c_{nom}.

The cover deviation Δc_{dev} can be negative (cover reduction) or positive (cover increase): negative values ($\Delta c_{dev,minus}$) shall be used for the evaluation of the nominal cover to be represented in design drawings, while the positive values ($\Delta c_{dev,plus}$) shall be used for the evaluation of the effective depth of the structural member in the strength verifications.

EC2 provides the absolute value of the negative deviation to be considered: Δc_{dev} = 10 mm. Moreover, for the deviation values to adopt for buildings, EC2 refers to EN 13670 "Execution of concrete structures", which gives the allowable values of the cover tolerances to employ for cast-in-situ concrete; for precast elements, reference should be made to product standards.

EN 13670 sets up the negative deviation (cover reduction) to 10 mm, whatever is the effective depth h of the cross-section of the structural element, while the positive deviation (cover increase) varies depending on the depth h (Table 3.8).

Paragraph 6.6 (3) of EN 13670 also guides how to guarantee the nominal cover during all the construction phases: for instance, metallic spacers in contact with

Table 3.8 Permitted deviations of the cover in structural members of buildings

Type of deviation	Description	Permitted deviation Δ class 1[a]	
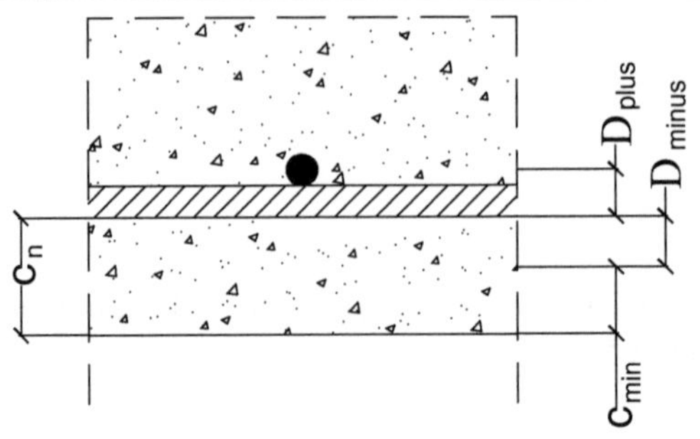	$\Delta_{(\text{minus})}$	For all values of h	− 10 mm
	$\Delta_{(\text{plus})}$[b]	For $h \leq$ 150 mm	+10 mm
		For $h =$ 400 mm	+15 mm
		For $h \geq$ 2500 mm	+20 mm
		For intermediate values of h, the value of $\Delta_{(\text{plus})}$ can be obtained by linear interpolation	
$c_{\min}$ = required minimum cover		Δ = permitted deviation from c_n	
c_n = nominal cover = $c_{min} + \lvert\Delta(\text{minus})\rvert$		h = cross-section depth	
c = actual cover		$c_n + \Delta_{(\text{plus})} > c > c_n - \lvert\Delta_{(\text{minus})}\rvert$	

[a] EN 13670 provides two different classes for structural deviations, considering as structural deviations those having effect on structural safety. Class 1, indicated also as the class of normal deviations guarantees design assumptions and the level of structural safety of EN 1992

[b] For the foundation reinforcement, the positive deviation $\Delta_{(\text{plus})}$ can be increased by 15 mm, while the negative deviation $\Delta_{(\text{minus})}$ remains equal to −10 mm

the concrete surface are allowed only for dry environments, i.e., for exposure class X0. Values of Δc_{dev} are considered at the time of choosing the design nominal cover; deviation values adopted for buildings are usually sufficient for other types of structures, too.

3.6.1 Reduced Values Δc_{dev} for Some Situations

In some situations, according to EC2 reduced values of Δc_{dev} could be adopted, as indicated in Table 3.9.

3.6.2 Uneven Surfaces for Concrete Cast Against Ground

For in-situ concrete cast against uneven surfaces, higher values of Δc_{dev} should be adopted in the choice of the design cover. In every case, the value of the deviation

Table 3.9 Reduced values Δc_{dev} for some situations (recommended values)

Situations where a reduced value could be adopted	Reduced value of Δc_{dev}
Where fabrication is subjected to a quality assurance system, in which the monitoring includes measurements of the concrete cover	5 mm $\leq \Delta c_{dev} \leq$ 10 mm
Where it can be assured that a very accurate measurement device is used for monitoring and non-conforming members are rejected (e.g., precast elements)	0 mm $\leq \Delta c_{dev} \leq$ 10 mm

Table 3.10 Minimum cover c_{nom} for concrete cast against uneven surfaces

c_{min}	Cause of the unevenness
$\geq$40 mm	Concrete cast against prepared grounds or blinding concrete (e.g., foundations on blinding concrete)
$\geq$75 mm	Concrete cast against soil (retaining walls, diaphragms, foundation piles)

Δc_{dev} shall lead to values of the minimum cover c_{min} not smaller than those listed in Table 3.10.

3.7 Examples

In the following paragraphs, some examples for the calculation of the concrete covers for structural elements according to EN1992-1-1 are discussed.

All the examples are developed only considering the rules of EN1992-1-1 for bond and durability, but not the rules of EN1992-1-2 for fire resistance.

Table 3.11 collects the structural elements accounted for in the examples with the design data (environmental exposure class, concrete strength class, bar diameter) and the calculated value of the cover.

It is worth observing that for concrete covers greater than 40 mm it is appropriate to adopt a suitable surface reinforcement to ensure an adequate resistance against the detachment of the cover itself ([see Sect. 9.2.4]). For details on surface reinforcement, see Chap. 12.

Examples are developed under the assumption that bars are isolated and, therefore, under the assumption of a minimum cover for bond (in the calculation according to EN1992-1-1) the isolated bar diameter. Finally, deviation Δc_{dev} has been assumed equal to 10 mm in all the Examples, according to the value recommended by EC2.

However, it is possible to decrease the value of Δc_{dev} until 5 mm for cast elements, when the execution is performed under quality control comprehensive of the cover measurement, and until 0 mm for the pre-cast elements if it is possible to ensure the use of a very accurate measuring device for control and that all nonconforming elements are rejected (Table 3.9). Moreover, for the plastered surfaces it is not possible to decrease the cover dimension by the amount of the plaster thickness. At least it is possible to assimilate the plaster to a protective covering, always if this

Table 3.11 Nominal cover according to EN1992-1-1 for the examples 1 ÷ 12; Δc_{dev} has been assumed equal to 10 mm in all the examples

Example ID	Structural element	Exposure class	Concrete strength class	Bar diameter (mm)	Concrete cover (mm)
					EC2
1	Floor beam	XC1	C30/37	16	26
2	Bridge slab	XC3	C40/50	20 (long.) 16 (transv.)	35
3	Platform roof	XC3, XC4	C30/37	20 (long.) 8 (stirrups)	35 (columns) 40 (beams)
4	Internal beam	XC1	C25/30	18 (long.) 8 (stirrups)	25
5	External beam with exposed face	XC4	C30/37	20 (long.) 8 (stirrups)	40
6	External beam with exposed face near to or on the coast	XC4, XS1	C30/37	20 (long.) 8 (stirrups)	45
7	T-beam foundation	XC2	C25/30	20 (long.) 8 (stirrups)	40
8	Retaining wall	XC4	C30/37	16 (vert.) 12 (horiz.)	40
9	Retaining wall in contact with chemical highly aggressive ground soil	XC4, XA3	C35/45	16 (vert.) 12 (horiz.)	40
10	New Jersey barrier	XC4, XD3, XF2	C45/55	12	50
11	Drilled pile	XC2, XA1	C25/30	20 (long.) 10 (stirrups)	75
12	Prestressed roofing beam near to or on the coast	XC4, XS1	C40/50	12 (long.) 8 (stirrups)	30

fact can be justified through valid documents also indicating the corresponding cover reduction. In every case, the cover cannot be smaller than the cover required for the bond.

3.7.1 Example 1. Floor Beam in a Building with Low Relative Humidity

Calculate the cover for a floor beam assuming the following design data: 16 mm bars, concrete strength class C30/37, nominal maximum aggregate size 32 mm.

Minimum concrete cover for bond

From Table 3.1 for $d_{max} \leq 32\,\text{mm}$, it results that $c_{min,b} = \phi = 16\,\text{mm}$.

Exposure class

XC1: dry or permanently wet environment (concrete inside buildings with low relative humidity).

Minimum strength class

For exposure class XC1, Table 3.4 indicates class C20/25 as the minimum strength class.

Structural class

For a working life of 50 years, the structural class is S4 if the minimum strength class is adopted (C20/25); the design strength class (C30/37) is greater than the required minimum strength class; for exposure class XC1 and strength class ≥ C30/37, Table 3.5 provides a reduction of the structural class by one: S4 → S3.

Minimum concrete cover due to environmental conditions

From Table 3.6, for XC1 and S3, $c_{min,dur} = 10\,\text{mm}$.

Minimum concrete cover

$$c_{min} = \max\left(c_{min,b};\, c_{min,dur};\, 10\right) = \max\,(16;\, 10;\, 10) = 16\,\text{mm}$$

Nominal concrete cover

$$\Delta c_{dev} = 10\,\text{mm},\ c_{nom} = c_{min} + \Delta c_{dev} = 16 + 10 = 26\,\text{mm}.$$

The nominal cover to be required in the design is, therefore, $c_{nom} = 26$ mm. Table 3.12 collects the nominal covers versus slab bar diameter.

Table 3.12 Nominal cover for a building internal slab with low relative humidity (XC1), $d_{max} =$ 32 mm and working life 50 years, versus longitudinal bars' diameter

Bar diameter (mm)	8	10	12	14	16	18	20
Concrete cover (mm)	20	20	22	24	26	28	30

3.7.2 Example 2. Bridge Slab

Calculate the cover for a bridge slab with the following design data: 20 mm internal longitudinal bars, 16 mm external transverse bars, concrete strength class C40/50, nominal maximum aggregate size 32 mm, design working life 100 years. The slab is protected on the upper side by a waterproof sheet.

Minimum concrete cover for bond

From Table 3.1 for $d_{\max} \leq 32$ mm

$$c_{\min,b} = \phi = 20 \text{ mm for longitudinal bars,}$$
$$c_{\min,b} = \phi = 16 \text{ mm for transverse bars.}$$

Exposure class

XC3: environment with moderate humidity, External concrete sheltered from rain (surface protected by a waterproof sheet)

Minimum strength class

For exposure class XC3, the minimum strength class is C30/37 (Table 3.4); for concrete class C30/37, the structural class for a working life of 50 years is S4.

From Table 3.5 it results that:

- for working life of 100 years, the structural class increases by two: S4 → S6;
- for exposure class XC3 and concrete C40/50 > C35/45, the structural class decreases by one (Table 3.5): S6 → S5;
- for slabs, the structural class decreases by one: S5 → S4;
- therefore, the structural class is S4.

Minimum concrete cover due to environmental conditions

From Table 3.6, for XC3 and S4, $c_{\min,dur} = 25$ mm.

Minimum concrete cover

$$c_{\min} = \max\left(c_{\min,b};\, c_{\min,dur};\, 10\right) = \max\,(20;\, 25;\, 10) = 25 \text{ mm for longitudinal bars,}$$
$$c_{\min} = \max\left(c_{\min,b};\, c_{\min,dur};\, 10\right) = \max\,(16;\, 25;\, 10) = 25 \text{ mm for transverse bars.}$$

Nominal concrete cover

$$\Delta c_{dev} = 10 \text{ mm,}$$
$$c_{nom,long} = c_{\min} + \Delta c_{dev} = 25 + 10 = 35 \text{ mm for longitudinal bars,}$$
$$c_{nom,trasv} = c_{\min} + \Delta c_{dev} = 25 + 10 = 35 \text{ mm for transverse bars.}$$

To guarantee a cover 35 mm to longitudinal bars, it is necessary to adopt a cover for transverse bars equal to $35 - 16 = 19$ mm < $c_{\text{nom,trasv}}$. The nominal cover to be assigned in the design is therefore $c_{\text{nom}} = 35$ mm.

3.7.3 Example 3. Platform Roof

Calculate the cover for the platform roof shown in Fig. 3.2 assuming the following design data for beams and columns: 20 mm bars, 8 mm stirrups, concrete strength class C30/37, nominal maximum aggregate size 32 mm.

Minimum concrete cover for bond

From Table 3.1 for $d_{\max} \leq 32$ mm

$$c_{\text{min,b}} = \phi = 20 \text{ mm for bars,}$$
$$c_{\text{min,b}} = \phi = 8 \text{ mm for stirrups.}$$

Exposure class

Columns: XC3—moderate humidity (External concrete sheltered from rain).

Beams: XC4 (Cyclic wet and dry) + XF3 (Horizontal concrete surfaces exposed to rain and freezing).

Minimum strength class

Both for exposure class XC3, XC4 and XF3, the minimum strength class is C30/37 (Table 3.4); adopted strength design is coincident with the minimum strength class and therefore structural class is S4 for a working life of 50 years.

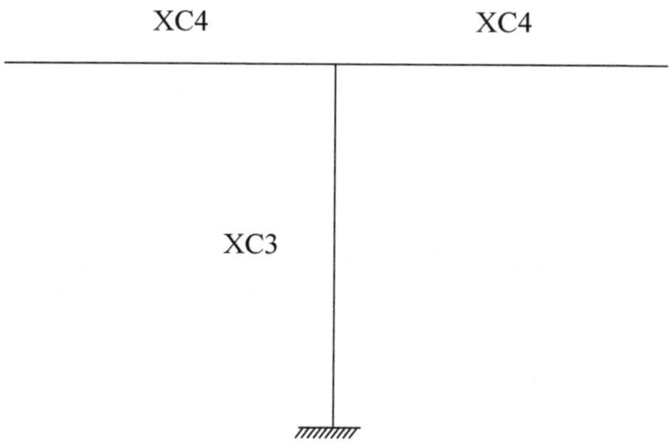

Fig. 3.2 Layout of the platform roof

Minimum concrete cover due to environmental conditions

Columns: from Table 3.6, for XC3 and S4, $c_{min,dur} = 25$ mm.
Beams: from Table 3.6, for XC4 and S4, $c_{min,dur} = 30$ mm.

Minimum concrete cover

Columns

$$c_{min} = \max(c_{min,b};\ c_{min,dur};\ 10) = \max\ (20;\ 25;\ 10) = 25 \text{ mm for bars,}$$
$$c_{min} = \max(c_{min,b};\ c_{min,dur};\ 10) = \max\ (8;\ 25;\ 10) = 25 \text{ mm for stirrups.}$$

Beams

$$c_{min} = \max(c_{min,b};\ c_{min,dur};\ 10) = \max\ (20;\ 30;\ 10) = 30 \text{ mm for bars,}$$
$$c_{min} = \max(c_{min,b};\ c_{min,dur};\ 10) = \max\ (8;\ 30;\ 10) = 30 \text{ mm for stirrups.}$$

Nominal concrete cover

$$\Delta c_{dev} = 10 \text{ mm}$$

Columns

$$c_{nom,long} = c_{min} + \Delta c_{dev} = 25 + 10 = 35 \text{ mm for bars,}$$
$$c_{nom,stirrups} = c_{min} + \Delta c_{dev} = 25 + 10 = 35 \text{ mm for stirrups.}$$

- to guarantee a 35 mm cover of longitudinal bars, it is necessary to adopt a cover for stirrups equal to $35 - 8 = 27$ mm $< c_{nom,stirrups}$;
- nominal cover to be assigned in the design is the cover of stirrups: $c_{nom} = 35$ mm.

Beams

$$c_{nom,long} = c_{min} + \Delta c_{dev} = 30 + 10 = 40 \text{ mm for bars,}$$
$$c_{nom,stirrups} = c_{min} + \Delta c_{dev} = 30 + 10 = 40 \text{ mm for stirrups.}$$

- to guarantee a 40 mm cover of longitudinal bars, it is necessary to adopt a cover for stirrups of $40 - 8 = 32$ mm $< c_{nom,stirrups}$;
- nominal cover to be assigned in the design is the cover of stirrups: $c_{nom} = 40$ mm.

3.7.4 Example 4. Beam Inside a Building with Low Relative Humidity

Calculate the cover of an internal beam inside a building with low relative humidity adopting the following design data: 18 mm longitudinal bars, 8 mm stirrups, concrete strength class C25/30, nominal maximum aggregate size 32 mm.

Minimum concrete cover for bond

From Table 3.1 for $d_{\max} \leq 32$ mm

$$c_{\min,b} = \phi = 18\,\text{mm for longitudinal bars,}$$
$$c_{\min,b} = \phi = 8\,\text{mm for stirrups.}$$

Exposure class

XC1.

Minimum strength class

For exposure class XC1, the minimum strength class is C20/25; therefore, the structural class for a working life of 50 years is S4; for concrete C25/30, the structural class does not change, because Table 3.5 shows that only for classes $\geq$ C30/37 it is possible to decrease by one the structural class.

Minimum concrete cover due to environmental conditions

From Table 3.6, for XC1 and S4, $c_{\min,dur} = 15$ mm.

Minimum concrete cover

$$c_{\min} = \max(c_{\min,b};\, c_{\min,dur};\, 10) = \max\,(18;\, 15;\, 10) = 18\,\text{mm for longitudinal bars,}$$
$$c_{\min} = \max(c_{\min,b};\, c_{\min,dur};\, 10) = \max\,(8;\, 15;\, 10) = 15\,\text{mm for stirrups.}$$

Nominal concrete cover

$$\Delta c_{dev} = 10\,\text{mm,}$$
$$c_{nom,long} = c_{\min} + \Delta c_{dev} = 18 + 10 = 28\,\text{mm for longitudinal bars,}$$
$$c_{nom,stirrups} = c_{\min} + \Delta c_{dev} = 15 + 10 = 25\,\text{mm for stirrups.}$$

To guarantee a 28 mm cover of longitudinal bars, it is necessary to adopt a stirrup cover of $28 - 8 = 20$ mm $< c_{nom,stirrups}$.

The nominal cover to indicate in design drawings is, therefore, $c_{nom} = 25$ mm. Table 3.13 provides values of the minimum cover to adopt for reinforced concrete internal beams in exposure class XC1 versus diameter of longitudinal bars and stirrups (6, 8, 10 mm).

Table 3.13 Nominal cover for an internal beam in environment XC1, $d_{max} \leq 32$ mm and working life 50 years, versus longitudinal bar and stirrup diameter

Stirrup diameter (mm)	Longitudinal bar diameter (mm)			
	$\phi \leq 20$	22	24	26
6	25	26	28	30
8	25	25	26	28
10	25	25	25	26

3.7.5 Example 5. Exposed External Beam

Calculate the cover of an exposed external beam with the following design data: 20 mm longitudinal bars, 8 mm stirrups, concrete strength class C30/37, nominal maximum aggregate size 32 mm.

Minimum concrete cover for bond

From Table 3.1 for $d_{max} \leq 32$ mm

$$c_{min,b} = \phi = 20\,\text{mm for longitudinal bars,}$$
$$c_{min,b} = \phi = 8\,\text{mm for stirrups.}$$

Exposure class

XC4: cyclic wet and dry.

Minimum strength class

From Table 3.4 for exposure class XC4, the minimum strength class is 30/37, coincident with the design one; therefore, the structural class for a working life of 50 years is S4.

Minimum concrete cover due to environmental conditions

From Table 3.6, for XC4 and S4, $c_{min,dur} = 30$ mm.

Minimum concrete cover

$$c_{min} = \max(c_{min,b};\, c_{min,dur};\, 10) = \max(20;\, 30;\, 10) = 30\text{ mm for longitudinal bars,}$$
$$c_{min} = \max(c_{min,b};\, c_{min,dur};\, 10) = \max(8;\, 30;\, 10) = 30\text{ mm for stirrups.}$$

Nominal concrete cover

$$\Delta c_{dev} = 10\text{ mm,}$$
$$c_{nom,long} = c_{min} + \Delta c_{dev} = 30 + 10 = 40\text{ mm for longitudinal bars,}$$
$$c_{nom,stirrups} = c_{min} + \Delta c_{dev} = 30 + 10 = 40\text{ mm for stirrups.}$$

The nominal cover to indicate in design drawings is $c_{nom} = 40$ mm.

Remark: if a different diameter is considered for longitudinal bars and the same diameter for stirrups is accounted for (8 mm), the nominal cover does not change and is equal to 40 mm for $\phi_{long} \leq 38$ mm, while $c_{nom} = 42$ mm for $\phi_{long} = 40$ mm.

3.7.6 Example 6. Exposed External Beam Close to or on the Coast

Calculate the cover of an exposed external beam close to or on the coast with the following design data: 20 mm longitudinal bars, 8 mm stirrups, concrete strength class C30/37, nominal maximum aggregate size 32 mm.

Minimum concrete cover for bond

From Table 3.1 for $d_{max} \leq 32$ mm

$$c_{min,b} = \phi = 20 \text{ mm for longitudinal bars,}$$
$$c_{min,b} = \phi = \ 8 \text{ mm for stirrups.}$$

Exposure class

XC4: cyclic wet and dry,

XS1: Exposed to airborne salt but not in direct contact with seawater.

Minimum strength class

From Table 3.4, both for class XC4 and XS1, the minimum strength class is C30/37, coincident with the design one; therefore, the structural class for a working life of 50 years is S4.

Minimum concrete cover due to environmental conditions

From Table 3.6, for XC4 and S4, $c_{min,dur} = 30$ mm.

From Table 3.6, for XS1 and S4, $c_{min,dur} = 35$ mm, therefore $c_{min,dur} = 35$ mm.

Minimum concrete cover

$$c_{min} = \max(c_{min,b};\ c_{min,dur};\ 10) = \max\ (20;\ 35;\ 10) = 35 \text{ mm for longitudinal bars,}$$
$$c_{min} = \max(c_{min,b};\ c_{min,dur};\ 10) = \max\ (8;\ 35;\ 10) = 35 \text{ mm for stirrups.}$$

Nominal concrete cover

$$\Delta c_{dev} = 10 \text{ mm,}$$
$$c_{nom,long} = c_{min} + \Delta c_{dev} = 35 + 10 = 45 \text{ mm for longitudinal bars,}$$

$$c_{nom,stirrups} = c_{min} + \Delta c_{dev} = 35 + 10 = 45 \text{ mm for stirrups.}$$

The nominal cover to indicate in design drawings is $c_{nom} = 45$ mm.

Remark If a different diameter is considered for longitudinal bars ($\phi_{long} \leq 40$ mm), while the diameter of stirrups is again of 8 mm, the nominal cover does not change and is equal to 45 mm.

3.7.7 Example 7. Foundation T-Beam

Calculate the cover of a foundation T-beam with the following design data: 20 mm longitudinal bars, 8 mm stirrups, concrete strength class C25/30, nominal maximum aggregate size 32 mm.

Minimum concrete cover for bond

From Table 3.1 for $d_{max} \leq 32$ mm

$$c_{min,b} = \phi = 20 \text{ mm for longitudinal bars,}$$
$$c_{min,b} = \phi = 8 \text{ mm for stirrups.}$$

Exposure class

XC2: wet, rarely dry.

Minimum strength class

From Table 3.4 for class XC2, the minimum strength is C25/30, coincident with the design one; therefore, the structural class for a working life of 50 years is S4.

Minimum concrete cover due to environmental conditions

From Table 3.6, for XC2 and S4, $c_{min,dur} = 25$ mm.

Minimum concrete cover

$$c_{min} = \max\left(c_{min,b};\ c_{min,dur};\ 10\right) = \max\ (20; 25; 10) = 25 \text{ mm for longitudinal bars;}$$
$$c_{min} = \max\left(c_{min,b};\ c_{min,dur};\ 10\right) = \max\ (8; 25; 10) = 25 \text{ mm for le stirrups.}$$

Nominal concrete cover

$$\Delta c_{dev} = 10 \text{ mm,}$$
$$c_{nom,long} = c_{min} + \Delta c_{dev} = 25 + 10 = 35 \text{ mm for longitudinal bars,}$$
$$c_{nom,stirrups} = c_{min} + \Delta c_{dev} = 25 + 10 = 35 \text{ mm for stirrups.}$$

The nominal cover to indicate in design drawings should be $c_{\text{nom}} = 35$ mm, nevertheless, as for concrete cast against prepared grounds or blinding concrete a minimum cover of 40 mm is required (see Table 3.10), the nominal cover to indicate in the design drawings is 40 mm.

Remark For ground soils slightly or moderately aggressive, the minimum strength class to be assumed would be C30/37, while the cover value would not change.

3.7.8 Example 8. Retaining Wall

Calculate the cover of a retaining wall with the following design data: 16 mm internal vertical bars, 12 mm horizontal external bars, concrete strength class C30/37, nominal maximum aggregate size 32 mm.

Minimum concrete cover for bond

From Table 3.1 for $d_{\max} \leq 32$ mm

$$c_{\text{min,b}} = \phi = 16 \text{ mm for vertical bars,}$$
$$c_{\text{min,b}} = \phi = 12 \text{ mm for horizontal bars.}$$

Exposure class

XC4: cyclic wet and dry.

Minimum strength class

For exposure class XC4, the minimum strength class is C30/37; adopted design strength is coincident with the minimum strength class and therefore structural class for a working life of 50 years is S4.

Minimum concrete cover due to environmental conditions

From Table 3.6, for XC4 and S4, $c_{\text{min,dur}} = 30$ mm.

Minimum concrete cover

$$c_{\min} = \max(c_{\text{min,b}}; c_{\text{min,dur}}; 10) = \max(16; 30; 10) = 30 \text{ mm for vertical bars,}$$
$$c_{\min} = \max(c_{\text{min,b}}; c_{\text{min,dur}}; 10) = \max(12; 30; 10) = 30 \text{ mm for horizontal bars.}$$

Nominal concrete cover

$$\Delta c_{\text{dev}} = 10 \text{ mm,}$$
$$c_{\text{nom,vertic.}} = c_{\min} + \Delta c_{\text{dev}} = 30 + 10 = 40 \text{ mm for vertical bars,}$$

$$c_{\text{nom,orizz.}} = c_{\min} + \Delta c_{\text{dev}} = 30 + 10 = 40 \text{ mm for horizontal bars.}$$

To guarantee a 40 mm cover of vertical bars, it is necessary to adopt a horizontal bar cover of $40 - 12 = 28$ mm > $c_{\text{nom,orizz}}$.

The nominal cover to indicate in design drawings is, therefore, $c_{\text{nom}} = 40$ mm. This value also fulfills the requirement of 40 mm, the lowest limit of Table 3.10 for concrete cast against prepared grounds or blinding concrete, like the slab of the wall.[4]

3.7.9 Example 9. Retaining Wall in Contact with Chemical Highly Aggressive Ground Soil

Calculate the cover of a retaining wall assuming the following design data: 16 mm internal vertical bars, 12 mm external horizontal bars, concrete strength class C35/45, nominal maximum aggregate size 32 mm.

Minimum concrete cover for bond

From Table 3.1 for $d_{\max} \leq 32$ mm

$$c_{\text{min,b}} = \phi = 16 \text{ mm for vertical bars,}$$
$$c_{\text{min,b}} = \phi = 12 \text{ mm for horizontal bars.}$$

Exposure class

XC4: cyclic wet and dry,

XA3: Highly aggressive chemical environment.

Minimum strength class

For exposure class XC4 minimum strength class is C30/37, while for class XA3 is C35/45, therefore C35/45 is adopted; the adopted strength design is equal to the minimum strength class, and therefore the structural class for a working life 50 years is S4.

Minimum concrete cover due to environmental conditions

The minimum cover for durability is the same as the previous Example, in facts from Table 3.6, for XC4 and S4, $c_{\text{min,dur}} = 30$ mm, while there are no prescriptions for class XA3. The minimum cover is therefore the same as the previous Example.

[4] The calculation is made under the assumption that the cast of the vertical wall is not against ground, but both sides of the wall are provided with formworks; on the contrary, the cover of the surface against ground shall be increased to 75 mm (see Table 3.10).

3.7.10 Example 10. New Jersey Barrier

Calculate the cover for a New Jersey barrier provided with 12 mm bars. The concrete strength class is C45/55 and the nominal maximum aggregate size is 32 mm.

Minimum concrete cover for bond

From Table 3.1 for $d_{\max} \leq 32$ mm

$$c_{\min,b} = \phi = 12 \text{ mm for longitudinal bars,}$$
$$c_{\min,b} = \phi = 12 \text{ mm for transverse bars.}$$

Exposure class

XC4: wet, rarely dry,

XD3: cyclic wet and dry (parts of bridges exposed to spray containing chlorides),

XF2: moderate water saturation, with a de-icing agent.

Minimum strength class

For exposure class XD3, the minimum strength class is C35/45 (for XC4 is C30/37 and for XF2 is C25/30); for concrete C35/45 and design working life 50 years, structural class is S4.

By adopting a strength $\geq$ C45/55, from Table 3.5 we obtain that structural class decreases by one: S4 $\rightarrow$ S3.

Minimum concrete cover due to environmental conditions

From Table 3.6, for XD3 and S3, $c_{\min,dur} = 40$ mm.

Minimum concrete cover

$$c_{\min} = \max(c_{\min,b};\, c_{\min,dur};\, 10) = \max\,(12;\, 40;\, 10) = 40 \text{ mm for vertical bars,}$$
$$c_{\min} = \max(c_{\min,b};\, c_{\min,dur};\, 10) = \max\,(12;\, 40;\, 10) = 40 \text{ mm for horizontal bars.}$$

Nominal concrete cover

$$\Delta c_{dev} = 10 \text{ mm,}$$
$$c_{nom,vertic.} = c_{\min} + \Delta c_{dev} = 40 + 10 = 50 \text{ mm for vertical bars,}$$
$$c_{nom,orizz.} = c_{\min} + \Delta c_{dev} = 40 + 10 = 50 \text{ mm for horizontal bars.}$$

The nominal cover to indicate in design drawings is, therefore, $c_{nom} = 50$ mm.

3.7.11 Example 11. Drilled Pile in a Slightly Aggressive Ground Soil

Calculate the cover of a drilled pile assuming the following design data: 20 mm longitudinal bars, 10 mm stirrups, concrete strength class C25/30, nominal maximum aggregate size 32 mm.

Minimum concrete cover for bond

From Table 3.1 for $d_{max} \leq 32$ mm

$$c_{min,b} = \phi = 20 \text{ mm for longitudinal bars,}$$
$$c_{min,b} = \phi = 10 \text{ mm for stirrups.}$$

Exposure class

XC2: wet, rarely dry.

Minimum strength class

For exposure class XC2, the minimum concrete strength is C25/30; the design strength is coincident with the required minimum strength and therefore the structural class for a working life of 50 years is S4.

Minimum concrete cover due to environmental conditions

From Table 3.6, for XC2 and S4, $c_{min,dur} = 25$ mm.

Minimum concrete cover

$$c_{min} = \max\left(c_{min,b};\ c_{min,dur};\ 10\right) = \max\ (20;\ 25;\ 10) = 25 \text{ mm for longitudinal bars,}$$
$$c_{min} = \max\left(c_{min,b};\ c_{min,dur};\ 10\right) = \max\ (10;\ 25;\ 10) = 25 \text{ mm for stirrups.}$$

Nominal concrete cover

$$\Delta c_{dev} = 10 \text{ mm,}$$
$$c_{nom,long} = c_{min} + \Delta c_{dev} = 25 + 10 = 35 \text{ mm for longitudinal bars,}$$
$$c_{nom,trasv} = c_{min} + \Delta c_{dev} = 25 + 10 = 35 \text{ mm for stirrups.}$$

To guarantee a cover 35 mm for longitudinal bars, it is necessary to adopt a cover of $35 - 10 = 25$ mm $< c_{nom,trasv}$ for stirrups.

Therefore, the nominal cover to indicate in the design drawings should be $c_{nom} =$ 35 mm. Nevertheless, being the concrete of the drilled pile cast against the uneven ground, the cover to adopt is 75 mm (Table 3.10).

It is worth observing that the maximum value of the nominal cover given by Table 3.6 is 55 mm (class XD3/XS3 and structural class S6), which, added to Δc_{dev}

= 10 mm, gives $c_{\text{nom}} = 65$ mm < 75 mm. Finally, for foundation piles, diaphragms and retaining walls, also in presence of very aggressive environments, the calculation of c_{nom} is unnecessary, being forced to assume the minimum cover provided by Table 3.10, equal to 75 mm.

3.7.12 Example 12. Prestressed Roofing Beam for a Precast Building Near to or on the Coast

Calculate the cover accounting for the following design data: 12 mm longitudinal bars, 8 mm stirrups, prestressing strands 0.6″ (15.3 mm), concrete strength class C40/50, nominal maximum aggregate size 32 mm.

Minimum concrete cover for bond for ordinary reinforcement

From Table 3.1 for $d_{\text{max}} \leq 32$ mm

$$c_{\text{min,b}} = \phi = \text{ 12 mm for longitudinal bars,}$$
$$c_{\text{min,b}} = \phi = \text{ 8 mm for stirrups.}$$

Minimum concrete cover for bond for pretensioned tendons

From Table 3.1 for pretensioned tendons

$$c_{\text{min,b}} = 1.5\phi = 1.5 \times 15.3 \text{ mm} = 22.95 \text{ mm}.$$

Exposure class

XC4: cyclic wet and dry,

XS1: exposed to airborne salt but not in direct contact with seawater.

Minimum strength class

From Table 3.4 both for class XC4 and for XS1, the minimum strength class is C30/37; from Table 3.5, for classes ≥ C40/50 it is possible to reduce by one the structural class, from S4 to S3; moreover, as for the precast roofing beam a special quality control of the concrete production is assured, it is possible to reduce again the structural class by one, from S3 to S2.

Minimum concrete cover due to environmental conditions

From Table 3.6, for XS1 and S2, $c_{\text{min,dur}} = 25$ mm for ordinary reinforcement, from Table 3.7, for XS1 and S2, $c_{\text{min,dur}} = 35$ mm for pretensioned strands.

Minimum concrete cover

$$c_{\min} = \max(c_{\min,b}; c_{\min,dur}; 10) = \max(12; 25; 10) = 25\text{ mm for longitudinal bars,}$$
$$c_{\min} = \max(c_{\min,b}; c_{\min,dur}; 10) = \max(8; 25; 10) = 25\text{ mm for stirrups,}$$
$$c_{\min} = \max(c_{\min,b}; c_{\min,dur}; 10) = \max(22,95; 35; 10) = 35\text{ mm for strands.}$$

Nominal concrete cover

Under the assumption that the roofing beam is produced under quality control which includes the cover measurement, Δc_{dev} can be assumed equal to 5 mm

$$\Delta c_{dev} = 5\text{ mm,}$$
$$c_{nom,long} = c_{\min} + \Delta c_{dev} = 25 + 5 = 30\text{ mm for longitudinal bars,}$$
$$c_{nom,stirrups} = c_{\min} + \Delta c_{dev} = 25 + 5 = 30\text{ mm for stirrups,}$$
$$c_{nom,trefoli} = c_{\min} + \Delta c_{dev} = 35 + 5 = 40\text{ mm for strands.}$$

The nominal cover to be indicated in design drawings is $c_{nom} = 30$ mm.

Chapter 4
Structural Analysis

Abstract The chapter examines the calculation methods to identify the resistant schemes of the structures at ULS and SLS. In synthesis: • Linear elastic analysis, valid up to ULS with appropriate limitations. • Linear elastic analysis with limited redistributions: based on the linear elastic solution, limited redistributions are applied, defined as a function to the type of concrete, steel, x/d ratio. • Plastic analysis, that admits unlimited redistributions. It is a limit theory that does not adapt well to the characteristics of reinforced concrete. Therefore, many limitations and precautions are required for use. Applies only to ULS. • Non-linear analysis: this is the general method which considers resistance and ductility in all phases as the load increases. It is a verification method and not a design method, it is complex and requires ad hoc calculation software.

4.1 The Structural Behaviour at Failure of a Beam in Bending

A simply supported reinforced concrete beam under the action of two symmetrical P loads is considered. The central zone has a constant bending moment and null shear (Fig. 4.1). The beam has a rectangular cross-section b width and h depth respectively and it is provided with longitudinal reinforcements A_s. The effective depth is $d()$.

With the load (and therefore bending moment) increase, it is possible to identify three subsequent phases in the behaviour of a cross-section of the central zone (Fig. 4.2).

- Phase I—Entirely reacting cross-section, with both materials in the linear elastic range.
- Phase II—Partialized cross-section in the area in tension and behaviour still elastic (in fact, the transition from Phase I to Phase II involves an apparent plasticization of the concrete in tension corresponding to a unique crack gradually increasing).
- Phase III—Depending on the reinforcement ratio $\rho = A_s/bd$ it is possible to have, for low reinforcement ratio (III a), overcoming the yield strength followed by a

This chapter was authored by Piero Marro and Matteo Guiglia.

© The Author(s), under exclusive license to Springer Nature Switzerland AG 2022
F. Angotti et al., *Reinforced Concrete with Worked Examples*,
https://doi.org/10.1007/978-3-030-92839-1_4

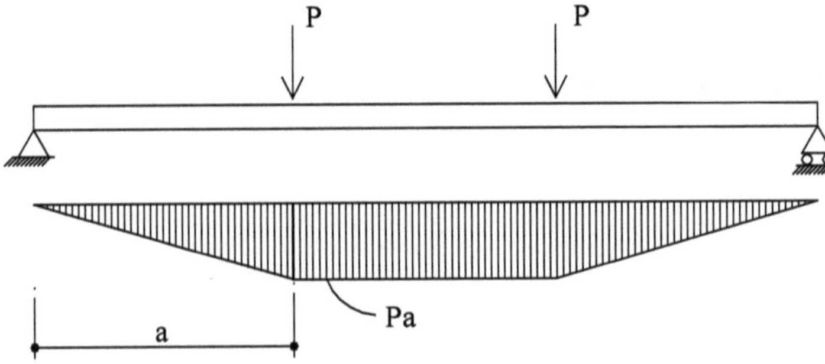

Fig. 4.1 Structural scheme of the beam and bending moment diagram

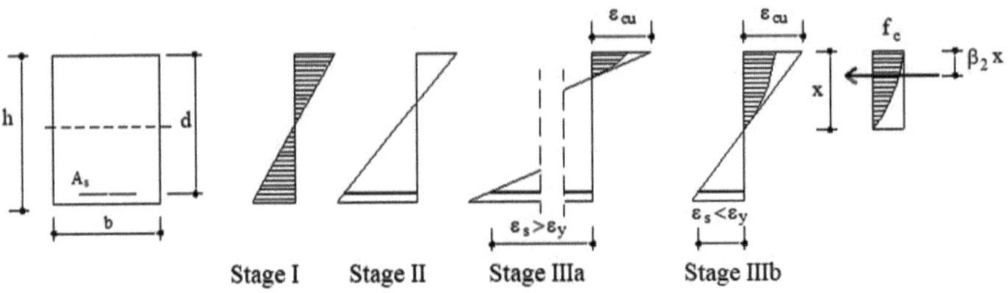

Fig. 4.2 Evolution of the failure behaviour of the cross-section

high elongation and therefore crushing of the concrete on the side in compression (only for very low values of the ρ ratio it is possible to have the failure of bars before of the concrete failure due to compression); otherwise, for high values of ρ (III b), plasticization of the concrete until crushing with the steel not reaching the yield.

The discussion proposed above allows representing the failure configuration of the cross-section as in Fig. 4.2 where the strain diagram (considered plane) is always passing through the limit of the concrete shortening ε_{cu}, while on the side in tension steel is reaching a different final deformation depending on the ratio ρ.

With reference to Fig. 4.2 the following notations are introduced:

β_1 filling coefficient, the ratio between the area of the stress diagram of the concrete in compression and the area of the rectangle which circumscribes it (see Chap. 7).

f_c cylindrical concrete compressive strength. It is known that concrete strength varies with the shape of the considered prism. The strength of a cylinder 15 × 30 cm is smaller than the one of a cube with a side of 15 cm because of the smaller influence of the friction on the sides in contact with the plates of the machine. As a first approximation, $f_{c,\text{cil}} = 0.83 f_{c,\text{cube}}$.

β_2 coefficient which defines the position of the compressive resultant force.

f_y steel yielding stress.

ε_s steel strain at failure, corresponding to σ_s stress on the steel diagram.

Following equations rule the problem:

- planarity of the cross-section (i.e. compatibility strain equations)

$$\frac{x}{d} = \frac{\varepsilon_{cu}}{\varepsilon_{cu} + \varepsilon_s} \tag{4.1}$$

- translational equilibrium (compression in concrete, tension in reinforcements)

$$\beta_1 x b f_c = A_s \sigma_s \tag{4.2}$$

- rotational equilibrium between the external moment M_{Ed} and the internal resistant moment M_{Rd}

$$M_{Rd} = \beta_1 b x f_c (d - \beta_2 x) \tag{4.3}$$

From (4.2) we get

$$\frac{x}{d} = \frac{A_s \sigma_s}{\beta_1 b d f_c} = \rho \frac{\sigma_s}{\beta_1 f_c} \tag{4.4}$$

Associating this equation to (4.1), the following relationship is written

$$\frac{\varepsilon_{cu}}{\varepsilon_{cu} + \varepsilon_s} = \rho \frac{\sigma_s}{\beta_1 f_c} \tag{4.5}$$

Equation (4.5) is a hyperbolic relationship between σ_s and ε_s. If this equation is represented on a graph with the steel strain diagram, Fig. 4.3 is obtained where the most significant hyperboles representative of (4.5) take different positions as a function of the ratio ρ.

It is easy to understand that curve (4.5) moves toward the left with the increase of ρ. We get therefore several typical positions: position (a) is corresponding to a strongly

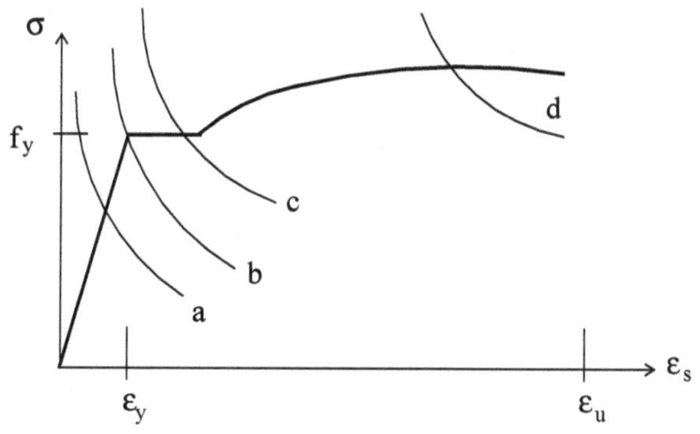

Fig. 4.3 Stress–strain diagram of the reinforcement and failure hyperboles

reinforced cross-section, with failure due to concrete compression and steel still in the elastic range and therefore not completely exploited; position (b) is corresponding to the superior ρ critical value, identifying the transition from the failure of concrete to the failure due to steel yielding, followed by concrete crushing which assumes the character of a reflected phenomenon. Position (c) is corresponding to the "ductile" failure with yielded steel and crushing of concrete: practically, we have the complete exploitation of both materials together with big deformations of reinforcement. For smaller values of ρ (position d), curve (4.5) crosses the steel diagram in the hardening range: in such a case there is a hyper-strength to the yielding. Failure due to pulling out of reinforcement is corresponding to the case when the two curves do not cross.

4.2 General Discussion on the Nonlinear Behaviour of the Structures

Concerning the previous paragraph, it is possible to assess that reinforced concrete structures mainly loaded by bending and axial loads, in the evolution of the load increase, exhibit inelastic rotational deformations due to the following causes:

- cracking due to the exceeding of concrete tensile strength;
- plastic behaviour of reinforcement due to exceeding the elastic limit;
- plastic behaviour of the concrete in compression.

For statically determined structures, such deformations can develop freely because of compatibility with supports, with a simple increase of deflections; for statically undetermined structures, instead, such deformations are not compatible with restrains: therefore reactions and internal forces arise depending on the structural stiffness and resulting elastic deformations. Relevant stresses added to those due to external loads, make the behaviour of the structure nonlinear. Elastic deformations provide to restore compatibility with the total deformations.

Let us analyse what is expressed above considering the case of a continuous beam with two spans of equal length ℓ, rectangular cross-section and subjected to a uniform increasing load distribution q. The elastic response is described by the bending moment diagram shown in Fig. 4.4. For a certain value of the load q_f, the cross-section at the intermediate support, where the moment is maximum in absolute value, cracks. The resulting rotation $\varphi = w/(h–x)$, (w = crack amplitude, h = depth of the cross-section, x = depth of the neutral axis) is compensated by elastic rotations with subsequent static effects. These are calculated by the principle of virtual works, as follows.

X is the moment at the support due to the cracking rotation φ. Compatibility equation can be written as:

$$1 \cdot \varphi = X \int \frac{M_1^2}{EI} ds$$

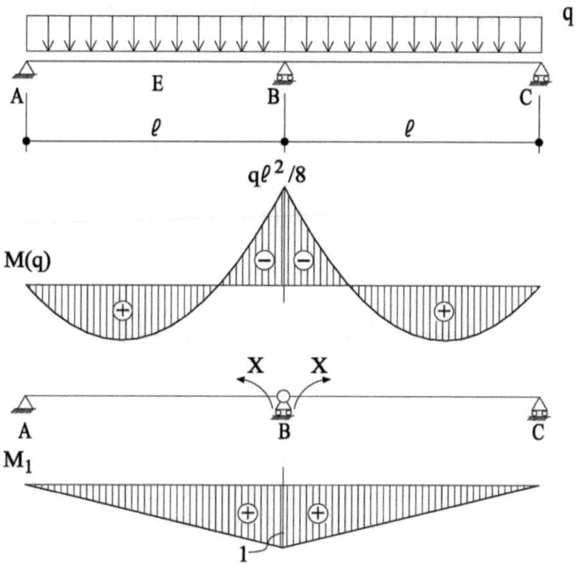

Fig. 4.4 Continuous beam. Cracking of the central support

being M_1 the moment caused by $X = 1$.

By solving, we get

$$X = \frac{3EI\varphi}{2\ell}$$

The effect due to X is a distribution of positive moments which, superimposing itself to the elastic diagram, decreases the negative moments in the area close to the central support and increases the positive moments along the span. On the support, bending moment M is given by

$$M = -\frac{1}{8}q\ell^2 + \frac{3EI\varphi}{2\ell}$$

not proportional to the load q.

With the load increase, cross-sections along the span under a positive moment are subjected to cracking and the new cracks lead to a distribution of negative moments which worsen the situation at the intermediate support and ameliorate the positive moments along the span. Finally, it is possible to say that cracks have an opposite effect compared to the loads which have caused them.

The cracks of both signs and the relevant moments originate a nonlinear structural response.

4.3 Structural Analysis: General

The purpose of structural analysis is to determine the effects of the actions (either forces and moments, or impressed deformations) in terms of internal forces or stresses or strains, for a geometrically and mechanically defined structure.

EC2 provides four analysis methods of internal forces accounting in a different way the phases of structural behaviour described in the previous paragraphs.

4.3.1 Linear Elastic Analysis (L)

Load effect

The method is based on the hypothesis of completely reacting cross-section and linear elasticity until the ultimate limit state. Consequently, internal forces are proportional to loads, and the principle of superposition of effects is valid.

For the design of resistant cross-sections of continuous beams, it is suitable to assume values of x/d not greater than those corresponding to $\delta = 1$ in formulas [(5.10a)] and [(5.10b)]. By exceeding those values, the behaviour of the beam tends to a brittle failure.

Imposed deformations

For imposed deformations due to thermal variations, support displacements with small amplitude, etc., the effects at ULS can be calculated by assuming cracked cross-sections without the contribution of the stiffening concrete between cracks (*tension stiffening*). The effects at SLS are calculated with partially cracked cross-sections [5.4.3].

4.3.2 Linear Elastic Analysis with Limited Redistribution (LR)

At the ultimate limit state, plastic rotations originate in the most stressed cross-sections and locally limit the increase of the bending moment, redistributing in other areas the effect of the subsequent load increments. For the reinforcement design, in those sections, it is possible to assume a moment δM reduced compared to the moment M obtained from the elastic linear calculation, under the assumption that in all the other parts of the structure the corresponding variations to guarantee equilibrium are considered.

This method is a design method.

The redistribution factor δ can be defined as the ratio of the redistributed bending moment and the bending moment evaluated through linear analysis. Paragraph [5.5(4)] relative to continuous beams and slabs with ratios between adjacent spans

Table 4.1 Limits $(x_{u/d})$ for formulas [5.10a] and [5.10b] as a function of the concrete class

Class	$\delta = 1$	$\delta = 0.70$
≤C 50/60	0.450	0.208
C 55/67	0.350	0.127
C 60/75	0.340	0.118
C 70/85	0.329	0.114
C 80/95	0.323	0.112
C 90/105	0.323	0.112

between 0.5 and 2 collects the formulas for the acceptable redistribution factor δ as a function of concrete Class, steel type, and ratio x_u/d once the redistribution has occurred.

For concrete classes until C50/60 and steels Class B and Class C, with medium and high ductility respectively, the following formula is given

$$\delta \geq 0.44 + 1.25\left(\frac{x_u}{d}\right);\ \delta \geq 0.70 \quad [5.10a]$$

For concretes from C55/67 to C90/105, associated to steels Class B and C, the following formula is valid

$$\delta \geq 0.54 + 1.25\left(0.6 + \frac{0.0014}{\varepsilon_{cu2}}\right)(x_u/d);\ \delta \geq 0.70 \quad [5.10b]$$

being ε_{cu2} a function of f_{ck}. Values are collected in Table 2.2.

Values (x_u/d) corresponding to $\delta = 1$ and $\delta = 0.70$ for various classes of concrete are shown in Table 4.1.

It is necessary to consider that a redistribution based on ductility rules guarantees only the equilibrium at ULS. For SLS specific verifications are required. Very strong redistributions, which can be advantageous at the ULS, often have to be reduced to fulfil requirements at SLS.

4.3.3 *Plastic Analysis (P)*

Plastic analysis is valid only at ULS. It can be developed essentially using two methods: static and kinematic. There is also a third method, called "mixed", which is a synthesis of the other two methods.

4.3.3.1 Static (Lower Bound) Method

This method is based on the plasticity static theorem that says: "every load Q corresponding to a statically acceptable distribution of internal forces is lower or equal to the ultimate collapse load Q_u".

The term "statically acceptable" is referred to a forces field fulfilling equilibrium conditions and boundary conditions without exceeding plastic strength. The calculated value of the load gives a lower estimate (*lower bound*) of the ultimate load.

An important application of the method is the analysis through the strut-and-tie method [5.6.4]. Other applications are the analysis of the slabs with the strip method [Annex I] and the analysis of the shear resistant truss for beams with transverse reinforcements [6.2.3].

4.3.3.2 Kinematic (Upper Bound) Method

This method considers that at the ULS the structure becomes a mechanism made of rigid members connected through plastic hinges.

The method is based on the kinematic theorem that says: "every load Q corresponding to a collapse mechanism kinematically acceptable is greater than or equal to the ultimate collapse load Q_u". For this reason, the obtained value gives an overestimate of the ultimate load (*upper bound*).

The method can be applied to continuous beams, frames and slabs, in this last case through the theory of the failure lines.

For beams, in [5.6.2(2)] it is specified that the development of plastic hinges is guaranteed only if the following conditions are fulfilled:

(a) the area of reinforcements in tension is such that in every cross-section

- (x_u/d) is not greater than 0.25 for concretes until Class C50/60
- (x_u/d) is not greater than 0.15 for concretes of Class C55/67 and greater;

(b) reinforcements are Class B or C (medium or high ductility)

(c) ratio between moments at the intermediate supports and in the adjacent spans is between 0.5 and 2.

If the previous conditions are not all fulfilled, it is necessary to control rotation capacity, comparing the required rotations with the acceptable ones according to [(Fig. 5.6N)].

4.3.3.3 Mixed Method

The method comes from the corollary of the two theorems of limit analysis and says: if the collapse mechanism is kinematically acceptable and, at the same time, originates a statically acceptable stress field, it is the true one and the calculations *upper bound* and *lower bound* take to the same result.

A necessary condition for this circumstance is a sufficient ductility of the structure to allow the development of the plastic hinges.

Methods of plastic analysis are valid only for ULS. The needs of SLS shall be fulfilled with specific verifications.

4.3.3.4 Determination of the Formulas Corresponding to Plastic Methods

A beam is considered with span ℓ under a uniform load q and provided of resistant moments M_B and M_C at the end-sections B and C respectively (negative moments) and M_E along the span. We want to evaluate the maximum value q_u and the position of the cross-section E where the maximum positive bending moment arises (Fig. 4.5).

Application of the Kinematic Method

With reference to Fig. 4.6, rotation θ_B is assumed as the parameter of the mechanism.

Rotation θ_C takes the value: $\theta_C = \theta_B \cdot \frac{k}{1-k}$ and rotation in E (at the cross-section at the distance $k\,\ell$ from B) takes the value: $\theta_E = \frac{1}{1-k}\theta_B$.

Considering the energy balance and assuming that the work made by q is equal to the work dissipated in the 3 plastic hinges, we get

$$+M_B\theta_B + M_C\theta_B\frac{k}{1-k} + M_E\theta_B\frac{1}{1-k} = q\ell \cdot \theta_B \cdot k\ell/2$$

It is worth noting that the work dissipated by plastic hinges is always positive, i.e. bending moments and rotations have always the same sign. Therefore in the applications, moments will be introduced in absolute value.

By solving the previous equation we get

$$q = \frac{2}{\ell^2} \cdot \left[\frac{+M_B(1-k) + M_E + M_C k}{k(1-k)}\right]$$

To search the maximum value of q, it is set

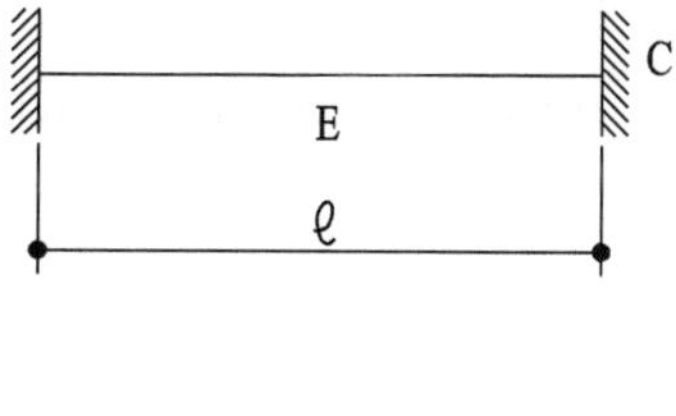

Fig. 4.5 Beam geometry

$$\frac{\partial q}{\partial k} = 0$$

We get

$$k^2 + 2k\frac{(M_E + M_B)}{(M_C - M_B)} - \frac{(M_E + M_B)}{(M_C - M_B)} = 0$$

Given that

$$A = \frac{M_E + M_B}{M_C - M_B}$$

we can write

$$k^2 + 2kA - A = 0$$

the solution is

$$k = -A + \sqrt{A^2 + A}$$

and the distance x_E of the cross-section with the maximum positive bending moment from cross-section B is given by

$$x_E = k\ell$$

If the moments at the end-sections are equals, the equation in k becomes

$$2k - 1 = 0, \text{ from which } k = 0.5$$

If the beam has support in B and a fixed joint in C ($M_B = 0$) the solution is the following

$$A = \frac{M_E}{M_C}$$

$$k^2 + 2k \cdot \frac{M_E}{M_C} - \frac{M_E}{M_C}$$

$$k = -A + \sqrt{A^2 - A}$$

If the beam has support in C and a fixed joint in B ($M_C = 0$) the solution is the following

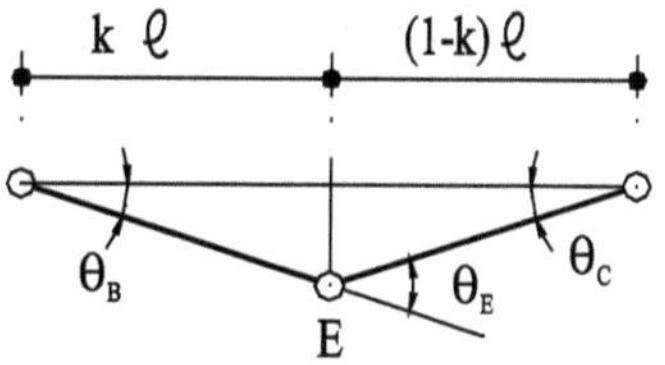

Fig. 4.6 Kinematic model

$$A = \frac{M_E + M_B}{M_B}$$

$$k^2 - 2k \cdot \frac{M_E + M_B}{M_B} + \frac{M_E + M_B}{M_B}$$

$$k = +A - \sqrt{A^2 - A}$$

Application of the Static Method

A field of statically acceptable stresses can be obtained (in a non-exclusive manner) by eliminating the redundant restraints. It is obvious that the ultimate load is smaller than the one of the real structure.

Application of the Mixed Method

Regarding the beam of Fig. 4.5, the problem is the design of the parabolic diagram with moments at the end-sections M_B and M_C and maximum positive moment M_E. To set the last condition, we search the abscissa x_E where the shear is zero.

Formulas are

$$M_B + V_B\ell - q\ell^2/2 = M_C$$

in the cross-section E $V_B - qx_E = 0$.

From which $x_E = \frac{V_B}{q}$

$$M_E = M_B + V_B x_E - q\frac{x_E^2}{2}$$

By substituting, we get $V_B = \sqrt{2q(M_E - M_B)}$

and then $M_B + \ell\sqrt{2q(M_E - M_B)} - q\frac{\ell^2}{2} = M_C$.

By developing, we get

$$q^2 - 2\frac{q}{\ell^2} \cdot 2 \cdot (2M_E - M_B - M_C) + \frac{4}{\ell^4} \cdot (M_C - M_B)^2 = 0$$

which, once solved, gives

$$q = \frac{2}{\ell^2}(2M_E - M_B - M_C) + \frac{2}{\ell^2}\sqrt{(2M_E - M_B - M_C)^2 - (M_C - M_B)^2}$$

Once q is known, x_E can be obtained using the previous formulas.

In case of equal moments at the end-sections, we get $q = \frac{8}{\ell^2} \cdot (M_E - M_C)$.

It is worth to precise that, by this method, resistant elastic moments have to be taken into account with the relevant signs: M_E is positive, M_B and M_C are negative.

If $M_B = 0$ (beam with simple support and a fixed joint), formulas can be simplified as follows

$$V_B = \sqrt{2qM_E}$$

$$q = \frac{2}{\ell^2} \cdot (2M_E - M_C) + \frac{2}{\ell^2}\sqrt{4M_E(M_E - M_C)}$$

If $M_C = 0$ (beam with a fixed joint and simple support)

$$V_B = \sqrt{2qM_E}$$

$$q = \frac{2}{\ell^2} \cdot (2M_E - M_B) + \frac{2}{\ell^2}\sqrt{4M_E(M_E - M_B)}$$

4.3.4 Nonlinear Analysis (NL)

The procedure to calculate internal forces is based on the assumption of an adequate nonlinear behaviour for materials (for concrete see EC2, [3.1.5(1) expression (3.14) and Fig. 3.2]; for steel [3.2.7(1) Fig. 3.8]), for members and structure (cracking, second-order effects). It is required that reinforcements are defined because it is an analysis procedure and it takes to internal forces which are not proportional to the applied loads. Nonlinear methods of analysis may be used for both ULS and SLS.

The procedure shall be developed by a computer providing that equilibrium and compatibility are satisfied at each load step. Compatibility is usually guaranteed by assigning to each cross-section a law moment–curvature and by integrating curvatures along the axis of the members. Inelastic rotations are usually concentrated in critical cross-sections. Deformations due to shear are usually neglected, those related to axial force are accounted for only if they strongly influence the solution. The principle of superposition of effects does not apply because of nonlinearity and calculations have to be carried out for each load condition: for each load condition it can be conventionally imagined to reach the ultimate limit state through subsequent increments of the applied loads.

EC2 does not provide any details about the procedures to be applied, but only general suggestions.

4.3.5 *Examples*

The development of examples requires the methods shown in Chaps. 7, 8 and 11 for ULS and SLS.

4.3.5.1 Example 1: Ductile Beam Which Allows the Application of Plastic Methods

(a) **Ductile beam with two fixed joints (B and C), Fig. 4.5.**

Data

$$\ell = 5.0$$

$$M_B = -100 \text{ kNm}; \; M_C = -70 \text{ kNm}; \; M_E = +120 \text{ kNm}$$

Kinematic method

$$A = \frac{120 + 100}{-100 + 70} = -7.33$$

$$k = 7.33 - \sqrt{7.33^2 - 7.33} = 0.5183$$

$$q_u = \frac{2}{25} \cdot \frac{100 \cdot (1 - 0.5183) + 120 + 0.5183 \cdot 70}{0.5183 \cdot (1 - 0.5183)} = 65.51 \text{ kN/m}$$

$$x_E = 0.5183 \cdot 5.0 = 2.59 \text{ m}$$

Static method

By eliminating the fixed joints, the simply supported beam has the moment M_E at the midspan. The ultimate load is given by:

$$q_u = \frac{8M_E}{\ell^2} = 8 \cdot \frac{120}{25} = 38.4 \text{ kN/m}$$

Mixed method

$$q_u = \frac{2}{25} \cdot (2 \cdot 120 + 100 + 70) + \frac{2}{25} \cdot \sqrt{(2 \cdot 120 + 100 + 70)^2 - (-70 + 100)^2}$$
$$= 32.8 + 32.71 = 65.51 \text{ kN/m}$$

$$V_B = \sqrt{2 \cdot 65.51 \cdot (120 + 100)} = 169.77 \text{ kN}$$

$$x_E = \frac{169.77}{65.51} = 2.59 \text{ m}$$

(b) **Ductile beam with support in B and fixed joint in C ($M_B = 0$)**

Kinematic method

$$A = \frac{120}{70} = +1.71$$

$$k = -1.71 + \sqrt{1.71^2 + 1.71} = 0.4428$$

$$q_u = \frac{2}{25} \cdot \frac{120 + 0.4428 \cdot 70}{0.4428 \cdot (1 - 0.4428)} = 48.95 \text{ kN/m}$$

$$x_E = 0.4428 \cdot 5.0 = 2.21 \text{ m}$$

Static method

$$q_u = \frac{8M_E}{\ell^2} = 8 \cdot \frac{120}{25} = 38.4 \text{ kN/m}$$

Mixed method

$$q_u = \frac{2}{25} \cdot (2 \cdot 120 + 70) + \frac{2}{25} \cdot \sqrt{4 \cdot 120 \cdot (120 + 70)}$$
$$= 24.8 + 24.15 = 48.95 \text{ kN/m}$$

$$V_B = \sqrt{2 \cdot 48.95 \cdot 120} = 108.38 \text{ kN}$$

$$x_E = \frac{108.38}{48.95} = 2.21 \text{ m}$$

(c) **Ductile beam fixed joint in B and support in C ($M_C = 0$)**

Kinematic method:

$$A = \frac{120 + 100}{100} = 2.20$$

$$k = 2.20 - \sqrt{2.20^2 - 2.20} = 0.5752$$

$$q_u = \frac{2}{25} \cdot \frac{100 \cdot (1 - 0.5752) + 120}{0.5752 \cdot (1 - 0.5752)} = 53.19 \text{ kN/m}$$

$$x_E = 0.5752 \cdot 5.0 = 2.87 \text{ m}$$

Static method: see case (b)

Mixed method:

$$q_u = \frac{2}{25} \cdot (2 \cdot 120 + 100) + \frac{2}{25} \cdot \sqrt{4 \cdot 120 \cdot (120 + 100)}$$
$$= 27.20 + 25.99 = 53.19 \text{ kN/m}$$

$$V_B = \sqrt{2 \cdot 53.19 \cdot (120 + 100)} = 152.98 \text{ kN}$$

$$x_E = \frac{152.98}{53.19} = 2.87 \text{ m}$$

4.3.5.2 Example 2: Application of the Linear Method (L), Linear Method with Redistribution (LR), Plastic Method (P) to a Beam Provided of Big Ductility

Cross-section $b = 250$ mm; $d = 500$ mm; $h = 550$ mm. Span 7,50 m. Double fixed joint.

Materials: concrete C30/37 ($f_{cd} = 17$ N/mm^2); steel $f_{yk} = 450$ N/mm^2.
Uniformly distributed load q.
Cross-section of the fixed point and at the mid-span identically reinforced:
A_s (in tension) $= 1200$ mm^2; $A'_s = 0.2\, A_s$.

(a) In dimensionless terms (with reference to Chap. 7) cross-sections do have:

$$\omega = \frac{A_s f_{yd}}{b d f_{cd}} = \frac{1200 \cdot 391}{250 \cdot 500 \cdot 17} = 0.22$$

and being

$$\omega'/\omega = 0.2,$$

Table U1 gives

$$\xi = 0.219 \text{ and } \mu = 0.20$$

Resistant moments are:

$$M_{Rd} = \mu b d^2 f_{cd} \pm 0.20 \cdot 250 \cdot 500^2 \cdot 17 = \pm 212.5 \,\text{kNm}$$

The value of the ultimate load according to the linear elastic method is given by:

$$q_u = 12 \cdot \frac{M_{Rd}}{\ell^2} = 12 \cdot \frac{212.5}{7.5^2} = 45.33 \text{ kN/m}$$

By applying kinematic plastic moment with equal values of the plastic moment at the mid-span and the fixed points, being verified all the rules of [5.6.2(2)], we get:

$$2M_{Rd}\theta + M_{Rd}2\theta = q\ell \cdot \frac{1}{2} \cdot \frac{\ell}{2} \cdot \theta, \text{ from which}$$

$$q = 16 \cdot \frac{M_{Rd}}{\ell^2} = 60.44 \,\text{kN/m}$$

Using the formulas of the plastic mixed method, being $M_E = -M_B = -M_C = 212.5$ kNm, we get

$$q_u = 60.44 \text{ kN/m}$$

By applying the linear method with redistribution, the suitable value of δ is

$$\delta = 0.44 + 1.25 \cdot 0.219 = 0.714$$

To obtain the same moments at the fixed joints and the mid-span it is necessary to limit redistribution to $\delta = 0.75$. In such a case moments take the value $\frac{q\ell^2}{16}$. Therefore, everything happens as initially at the fixed joint the moment would be applied

$$M = -\frac{212.5}{0.75} = -283.33 \text{ kNm}$$

The load q is therefore

$$q = 12 \cdot \frac{283.3}{7.5^2} = 60.44 \,\text{kN/m}$$

Table 4.2 Synthesis of results. Values of q_u (kN/m)

Beam	L	P	LR
a	45.33	60.44	60.44
b	23.50	45.80	33.50

(b) Cross-section over the span as in (a); cross-sections at the fixed joint with half of the reinforcement.

For these:
$A_s = 600\ \text{mm}^2, A'_s = 120\ \text{mm}^2, \omega = 0.11, \omega'/\omega = 0.2, \mu = 0.104$ and therefore

$$M_{Rd} = 110\ \text{kNm}$$

Also, in this case all the rules of [5.6.2(2)] are fulfilled.
The value of the ultimate critical load according to the elastic linear theory is

$$q = 12 \cdot \frac{110.0}{7.5^2} = 23.5\ \text{kN/m}$$

Instead, the kinematic method gives:

$2 \cdot 110 \cdot \theta + 212.5 \cdot 2\theta = q\frac{\ell^2}{4}\theta$ from which

$q = 45.8$ kN/m.

By applying the linear method with the redistribution, the acceptable value of δ is è $\delta = 0.70$. Therefore, everything happens as if it was at the fixed joints

$$M = -\frac{110.0}{0.70} = -157.14\ \text{kNm}$$

Load q is therefore

$$q = 12 \cdot \frac{157.14}{7.5^2} = 33.5\ \text{kN/m}$$

Table 4.2 collects the synthesis of the ultimate lads obtained by the 3 different analysis methods.

4.3.5.3 Example 3: Three-Spans Beam—Design by (LR) and Verification by (P)

Three-span beam designed at ULS with elastic analysis with redistribution and verified through plastic analysis.

The three-span beam shown in Fig. 4.7 is considered. Spans have a length of 11.50 m, the beam is made of concrete C30/37 and steel B450C, and it is subjected to the following loads:

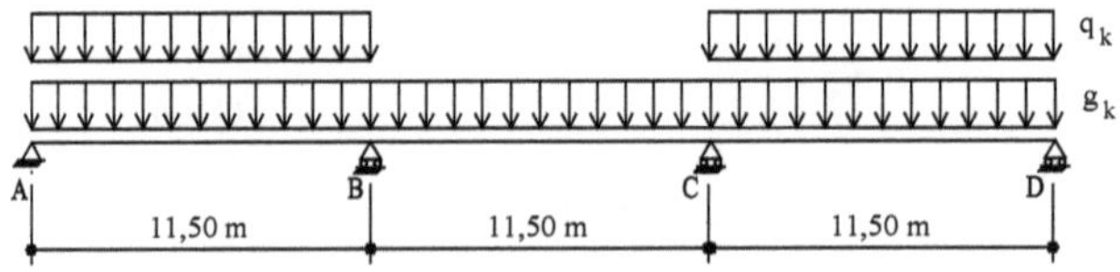

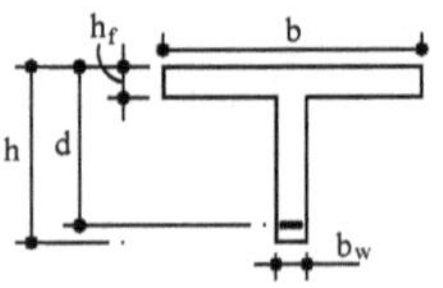

Fig. 4.7 Beam and cross-section geometries

- structural self-weight: $g_{k1} = 6.0$ kN/m
- permanent load: $g_{k2} = 4.0$ kN/m
- variable load: $q_k = 10.0$ kN/m arranged on the first and third span.

Design load at ULS is:

$q_{Ed} = 6.0 \cdot 1.30 + 4.0 \cdot 1.50 + 10.0 \cdot 1.50 = 7.8 + 6.0 + 15.0 = 28.8$ kN/m

Beam has a T-shaped cross-section with $b = 1200$ mm, $h = 550$ mm, $d = 500$ mm, $h_f = 100$ mm, $b_w = 300$ mm.

First, the reinforcement is calculated using linear analysis; later, redistributions are performed on the intermediate supports until $\delta = 0.70$ with the corresponding increases of reinforcements along the spans.

Basic Elements for the Elastic Solution at the ULS

Permanent load originates the following moment on the intermediate supports:

$M = -0.10 \cdot (6.3 \cdot 1.3 + 4 \cdot 1.5) \cdot 11.50^2 = -182.5$ kNm.

Variable load applied on the lateral spans originates on the same supports the following moment:

$$M = -0.05 \cdot (10.0 \cdot 1.50) \cdot 11.50^2 = -100.0 \text{ kNm}$$

Total moment

$$M_{Ed} = -(182.5 + 100.0) = -282.5\text{kNm}$$

Moments in the spans

- *Lateral spans*

The maximum moment is evaluated by calculating the reaction A at the first support through rotational equilibrium equation around B:

$A\ell - q\frac{\ell^2}{2} = -282.5$ kNm from which

$A = \frac{-282.5 + \frac{28.8 \cdot 11.5^2}{2}}{11.5} = 141$ kN.

In the span, shear takes the value zero when

Table 4.3 Design redistributions on the intermediate simple supports, moments on the simple supports and moments redistributed along the span

Δ	M sup (kNm)	ΔM (kNm)	M lat. span (kNm)	M central span (kNm)
1.00	– 282.5	0.0	345.2	– 54.3
0.90	– 254.2	28.3	357.0	– 26.0
0.80	– 226.0	56.5	370.0	+2.2
0.70	– 197.8	84.7	381.3	+30.4

Table 4.4 Reduced moments, reinforcement ratios for bars in tension, relative positions of the neutral axis and amount of tensile and compressive reinforcement

δ	μ	ω	ξ	A_s (mm^2)	A'_s (mm^2)
1.0	0.2216	0.246	0.246	1607	321
0.9	0.1994	0.220	0.219	1438	288
0.8	0.1773	0.194	0.197	1264	253
0.7	0.1559	0.168	0.175	1095	220

$x = 141/28.8 = 4.90$ m, where the moment is maximum.
The moment in that cross-section is given by

$$M_{Ed} = 141 \times 4.90 - 28.8 \cdot \frac{4.90^2}{2} = 345.2 \text{ kNm}$$

- *Mid-span*

Being the load: $q = 6.0 \cdot 1.30 + 4.0 \cdot 1.50 = 13.8$ kN/m the moment at the mid-span is given by:

$$13.8 \cdot \frac{11.5^2}{8} - 282.5 = -54.3 \text{ kNm}$$

Redistributions are applied with $\delta = 0.9$, $\delta = 0.8$, $\delta = 0.7$ and along-the-span moments are calculated. The increase of the positive moment on the lateral spans is given by 4.90/11.50 = 0.42 times the reduction at the simple supports. Values are collected in Table 4.3.

Reinforcements for continuity moments are designed with the assumption that compressive reinforcement is 0.2 times tensile reinforcements. The following dimensionless parameters are adopted

$$\mu = \frac{M_{Rd}}{bd^2 f_{cd}}; \omega = \frac{A_s f_{yd}}{bd f_{cd}}$$

$$\xi = \frac{x}{d}$$

Table 4.5 Reinforcements in tension and compression for lateral spans

Δ	μ	ω	A_s	A'_s
1.00	0.2707	0.3071	2002	400
0.90	0.2800	0.3200	2087	417
0.80	0.2902	0.3331	2172	434
0.70	0.300	0.3463	2258	451

and relevant Tables for ULS contained in the Appendix can be used.

Starting point: moments μ corresponds to moments M_{supp} of Table 4.3. For Example, from

$$M_{supp} = 282.5\text{kNm}$$

$$\mu = \frac{282.5 \cdot 10^6}{300 \cdot 500^2 \cdot 17} = 0.2216$$

Proposed redistributions δ are suitable with the corresponding ξ values, according to the formula

$$\delta = 0.44 + 1.25 \cdot \xi, \text{ valid for concrete Class} \leq \text{C50/60.}$$

Reinforcements in compression at the supports have nominal values for the calculation, but they should be at least equal to the 25% of those placed at the same level in the lateral spans (see [9.2.1.4]).

Reinforcements in lateral spans under positive moment are dimensioned through the formula:

$$A_s = \frac{M}{z \cdot f_{yd}}$$

where

$$z = (d - \frac{h_f}{2}) = (500 - \frac{100}{2}) = 450 \text{ mm}$$

In the central span, which is subjected to low stresses, we arrange a reinforcement at least equal to the minimum suggested by [9.1N].

$A_s = 0{,}26 \cdot \frac{f_{ctm}}{f_{yk}} \cdot b_t \cdot d = 0,26 \cdot \frac{2{,}9}{450} \cdot 300 \times 500 = 252 \text{ mm}^2$.

Moments μ and design reinforcements for lateral spans are collected in Table 4.5.

In the mid-span following reinforcements are arranged: $A_s = 400 \text{ mm}^2$; $A'_s = 400 \text{ mm}^2$.

Table 4.6 Moment at the support and in the span; ultimate load

Δ	M_B (kNm)	M_E (kNm)	q_u (kN/m)
1.00	– 282.5	345.2	28.79
0.90	– 254.2	357.0	28.77
0.80	– 226.0	370.0	28.81
0.70	– 197.8	381.3	28.73

Plastic Analysis

The method is applied to the solutions already designed through redistributions.

The mixed method is applied through the formula evaluating the ultimate load as a function of the resistant moment on the continuity support (M_B) and along the span (M_E) (see Sect. 4.3.4) designed using the LR method:

$$q_u = \frac{2}{\ell^2}(2M_E - M_B) + \frac{2}{\ell^2}\sqrt{4M_E(M_E - M_B)}$$

As example, for $\delta = 1.0$:

$$q_u = \frac{2}{11.50^2}(2 \cdot 345.2 + 282.5) + \frac{2}{11.50^2}\sqrt{4 \cdot 345.2 \cdot (345.2 + 282.5)}$$
$$= 14.71 + 14.08 = 28.79 \text{ kN/m}$$

The values δ corresponding to the four hypotheses of redistribution are collected in Table 4.6.

Results are almost coincident with the design data because of the high ductility of sections.

4.3.5.4 Example 4: Two-Spans Continuous Beam—Calculation of the Ultimate Load with (L), (LR), (P), (NL)

Two-spans continuous beam 8 m length, rectangular cross-section $b = 250$ mm; $h = 550$ mm; $d = 500$ mm. Uniform load q (Fig. 4.8). Conventionally B is the cross-section at the support and E the cross-section (or the cross-sections) along the span.

Given reinforcements and materials, we calculate the value of the ultimate load q_u through the following methods:

- linear elastic (L),
- linear with limited redistribution (LR),
- plastic (P),
- nonlinear (NL).

Two cases (a) and (b) are analysed with different concrete and reinforcement.

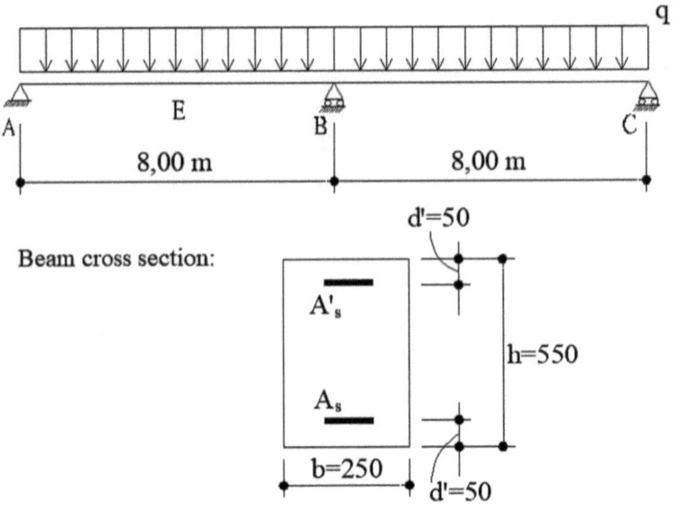

Fig. 4.8 Beam and cross-section geometries

Case (a)

$f_{ck} = 30$ N/mm^2; $f_{cd} = 17$ N/mm^2; $f_{yd} = 391$ N/mm^2.

Reinforcements in B: $A_s = 6\varphi 20 = 1884$ mm^2; $A'_s = 2\varphi 16 = 402$ mm^2.

Reinforcements in E: $A_s = 7\varphi 16 = 1407$ mm^2; $A'_s = 2\varphi 14 = 308$ mm^2.

Reinforcement mechanical ratios:

- in B: $\omega = \frac{1884 \cdot 391}{250 \cdot 500 \cdot 17} = 0.3467$; $\omega' = 0.0736$; $\frac{\omega'}{\omega} = 0.21$
- in E: $\omega = \frac{1407 \cdot 391}{250 \cdot 500 \cdot 17} = 0.2589$; $\omega' = 0.0567$; $\frac{\omega'}{\omega} = 0.22$

Values of ξ and resistant moments μ are calculated through Table U1 as a function of ω. We get: in B $\xi = 0.342$, $\mu = 0.301$; in E $\xi = 0.250$, $\mu = 0.232$. Therefore, we get:

- in B: $M_{Rd} = \mu b d^2 f_{cd} = -318$ kNm;
- in E: 244 kNm.
- *Linear elastic analysis*

  Load q_u is given by the smallest of the two values corresponding to the moments along the span and at the intermediate support: $14.3 \cdot \frac{M_{Rd,E}}{\ell^2}$ and $8 \cdot \frac{M_{Rd,B}}{\ell^2}$.

  From the moment in B, we get $q_u = 39.7$ kN/m.
- *Plastic analysis*

  By applying the mixed method, subject to ductility check, we get:

$$q = \frac{2}{64}(318 + 2 \cdot 244) + \frac{2}{64}\sqrt{4 \cdot 244(244 + 318)} = 48.3 \text{ kN/m}$$

Ductility check. Preliminarily, for $f_{ck} = 30$, it must be

$$\xi = \frac{x}{d} =< 0.45.$$

In this case, the value $\xi = 0.342$ fulfils the condition.

Furthermore, being ξ greater than 0.25, it is necessary to verify the rotational capacity around the support cross-section, according to [5.6.3]. Rotation $\theta_s = 1.2 \cdot \frac{\varepsilon_{cu2}}{x} \cdot h = 1.2 \cdot \frac{0.0035}{171} \cdot 550 = 13.5$ mrad is smaller than the one of 6N], which for $x/d = 0.342$ indicates as acceptable value 15.5 mrad. Therefore, the verification is satisfied.

- *Linear analysis with limited redistribution*
  x/d is known in B, then we can calculate the acceptable value of δ.
  We get

$$\delta = 0.44 + 1.25 \cdot 0.342 = 0.87$$

  The elastic moment in B before redistribution is given by
  $M_{Rd} = \frac{318}{0,87}$ =365.5 kNm and the corresponding load is
  $q = \frac{8 \cdot 365.5}{64} = 45.6$ kN/m.
  The moment in E because of the redistribution is increased by the amount
  $\Delta M = 0.4(365.5 - 318.0) = 19.0$ kNm and therefore it takes the value
  $M_E = \frac{1}{14} \cdot 45.6 \cdot 8^2 + 19.0 = 223$ kNm < 244 kNm, which is the value of the resistant moment.
- *Nonlinear analysis*
  The analysis developed under the assumption of the law [(3.14)] (Sargin parable) gives the following results:

  - design yielding moment at the support: $M_{yd,\mathrm{B}} = 294.6$ kNm
  - design yielding moment at the midspan: $M_{yd,\mathrm{E}} = 224.6$ kNm

Acceptable plastic rotations according to Figure [5.6N] as a function of x/d at ULS are:

- at the midspan, for $x/d = 0.250$: $\theta_{pl} = 21,7$ mrad
- at the support, for $x/d = 0.342$: $\theta_{pl} = 15,5$ mrad

With the increase of the load, the following circumstances occur, as described in Table 4.7.

From tabular results (here not recalled) it is possible to deduce that for a load $q = 0.7 \cdot 47.2 = 33.0$ kN/m, corresponding to the characteristic service combination (see Chap. 11), the bending moment in B is equal to 260 kNm, corresponding to $\sigma_s = 324$ N/mm^2 smaller than 360 ^{N}/mm^2 $= 0.80 \cdot f_{yk}$.

The diagram representing the evolution of load-bending moment is represented in Fig. 4.9.

Table 4.7 Case (a): occurrences and plastic rotation as a function of the applied load

Load q (kN/m)	Occurrence	Plastic rotation at the support (mrad)	Plastic rotation along the span (mrad)
10.0	Cracking in B	–	–
12.0	Cracking in E	–	–
36.8	Yielding at the support	0.0	0.0
45.2	Yielding along the span	7.3	0.0
472	ULS at the support	15.5	9.8

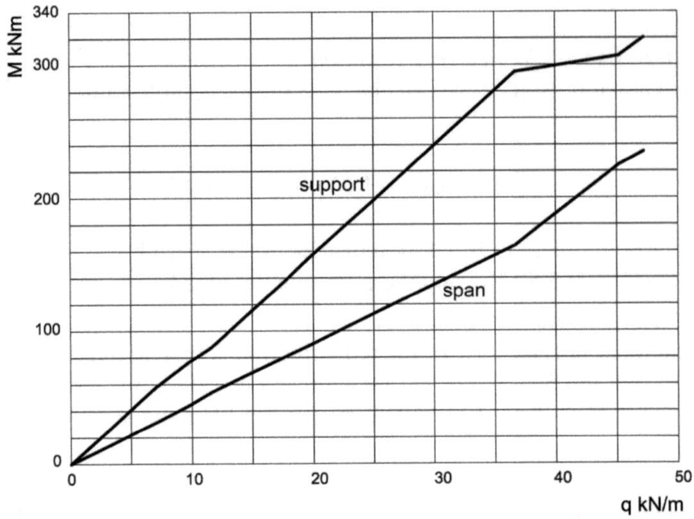

Fig. 4.9 Case (a): load-moment diagrams obtained by NL

Case (b)

$f_{ck} = 60$ N/mm^2 ; $f_{cd} = 34$ N/mm^2; $f_{yd} = 391$ N/mm^2
Reinforcements in B: $A_s = 5\varphi 26 = 2650$ mm^2; $A'_s = 2\varphi 20 = 628$ mm^2.
Reinforcements in E: $A_s = 2\varphi 20 + 3\varphi 26 = 2218$ mm^2; $A'_s = 3\varphi 14 = 462$ mm^2.
Reinforcement ratios:

- in B we have $\rho = 2.12/100$; $\rho' = 0.50/100$
- in E we have: $\rho = 1.77/100$; $\rho' = 0.37/100$

Mechanical reinforcement ratios:

- in B: $\omega = \frac{2650 \cdot 391}{250 \cdot 500 \cdot 34} = 0.2438$; $\omega' = 0.0578$; $\frac{\omega'}{\omega} = 0.237$
- in E: $\omega = \frac{2218 \cdot 391}{250 \cdot 500 \cdot 34} = 0.2041$; $\omega' = 0.0425$; $\frac{\omega'}{\omega} = 0.208$

Values of ξ and resistant moment μ can be calculated as a function of ω using Table U2. We get:

- in B: $\xi = 0{,}280$, $\mu = 0.219$
- in E: $\xi = 0{,}240$, $\mu = 0.185$

from which we get the resistant moment M_{Rd} respectively.

- in B: $M_{Rd} = \mu b d^2 f_{cd} = 465.3$ kNm
- in E: 393.1 kNm
- *Linear elastic analysis*

  Load q_u is the smallest of the following values: $14, 3 \cdot \frac{M_{Rd,E}}{\ell^2}$ and $8 \cdot \frac{M_{Rd,B}}{\ell^2}$. From the moment in B, we get $q_u = 58.2$ kN/m.

- *Plastic analysis*

By applying the mixed method, with the reserve on the verification of ductility, we get

$$q = \frac{2}{64}(465.3 + 2 \cdot 393.1) + \frac{2}{64}\sqrt{4 \cdot 393.1\,(393.1 + 465.3)} = 75.4 \text{ kN/m}$$

Ductility verification:

Preliminarily, it is necessary that for $f_{ck} = 60$, $\xi = \frac{x}{d} \leq 0.35$ 0.35. In this case, $\xi = 0.280$ fulfils this condition.

Moreover, because ξ is greater than 0.15, it is necessary to verify the rotation capacity around the support cross-section according to [5.6.3]. Rotation $\theta_s = 1.2 \cdot \frac{\varepsilon_{cu2}}{x} \cdot h = 1.2 \cdot \frac{0{,}0029}{140} \cdot 550 = 13.7$ mrad is smaller than the one of [Fig. 5.6N], which, for $x/d = 0.28$ indicates as acceptable value 16.0 mrad. Therefore, the verification is fulfilled.

- *Linear analysis with limited redistribution*

Being known x/d in B, the acceptable value of δ can be calculated.

We get

$$\delta = 0.54 + 1.25 \cdot (0.6 + 0.0014/0.0029) \cdot 0.28 = 0.92$$

The elastic moment in B before redistribution is therefore given by

$$\mathrm{M_{Rd}} = \frac{465.3}{0.92} = 505.7\,\text{kNm and the corresponding load is } q = \frac{8 \cdot 505.7}{64} = 63.22\,\text{kN/m}$$

Table 4.8 Case (b): occurrences and plastic rotation as a function of the applied load

Load q (kN/m)	Occurrence	Plastic rotation at the support (mrad)	Plastic rotation along the span (mrad)
18.0	Cracking in B	–	–
21.0	Cracking in E	–	–
52.6	Yielding at the support	0.0	0.0
69.3	Yielding at the midspan	9.5	0.0
73.2	ULS at the support	20.0	12.2

The moment in E due to redistribution is increased by $\Delta M = 0.4(465.3 - 505.7) = 16$ kNmand it is given therefore by $M_E = \frac{1}{14} \cdot 63.22 \cdot 8^2 + 16 = 299$ kNm < 393 kNm, resistant value.

- *Nonlinear analysis*

The analysis developed by adopting for the concrete the law [3.14] gives the following results:

- design yielding moment at the support: $M_{yd,\mathrm{B}} = 421.4$ kNm
- design yielding moment at the midspan: $M_{yd,\mathrm{E}} = 355.4$ kNm

Available plastic rotations given by Figure [5.6N] as a function of x/d at ULS are given by:

- at the midspan, for $x/d = 2.241$: $\theta_{pl} = 22.4$ mrad
- at the support, for $x/d = 0.273$: $\theta_{pl} = 20.0$ mrad

The evolution of occurrences with the load increase is collected in Table 4.8.

From the tabular values we can observe that for the characteristic combination at SLS ($q = 0.7q_u = 0.7 \cdot 73.2 = 51.2$ kN/m), moment in B is 410.0 kNm corresponding to, with $\alpha_e = 9$ (see Chap. 11):

$\sigma_c = 30.3$ N/mm^2 < 36 N/mm^2

$\sigma_s = 359$ N/mm^2 < 360 N/mm^2

and at the midspan $M = 229.7$ kNm.

$\sigma_c = 18.4$ N/mm^2 < 36 N/mm^2.

$\sigma_s = 239$ N/mm^2 < 360 N/mm^2.

The diagram of the evolution load-bending moment is shown in Fig. 4.10.

Synthesis of the Results at ULS in Case a and Case B

Table 4.9 collects the results obtained through the four analysis methods; q_u values are expressed in kN/m.

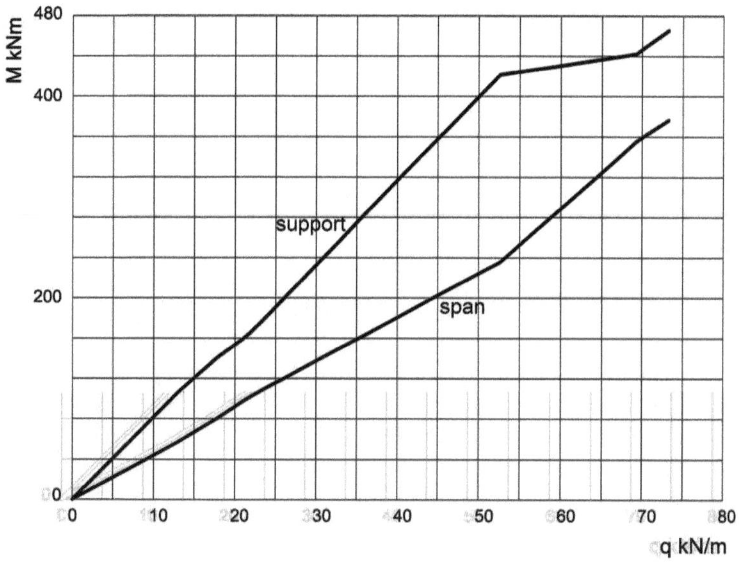

Fig. 4.10 Case (b): load-moment diagrams obtained by NL

Table 4.9 Ultimate load obtained by the 4 analysis methods

f_{ck}	*L*	*LR*	*P*	*NL*
30	39.7	45.6	48.3	47.2
60	58.2	63.2	75.4	73.2

- *Remark*

From the examination of Table 4.9, it can be deduced that the minimum value of q_u can be obtained using linear analysis. This is determined by the reaching of the ULS in the cross-section at the support when the span still has strength supplies that the method does not allow to use. As q_u increases, the linear analysis with limited redistribution is following employing, at least partially, the capacity of plastic rotation of the cross-section at the support.

The nonlinear analysis goes further, taking into account the deformability of all the elements with relevant strain laws, and it describes the evolution of the behaviour with the increase of the load pointing out all the significant phases: cracking, yielding critical cross-section rotations, reaching of the ULS. In the two examples proposed, the end of the evolution arrives when the limit plastic rotation at the supports occurs, while along the spans plastic rotation is only partially developed. This observation justifies the reached value of q_u which is smaller than the one obtained through the plastic method assuming that plastic rotations can develop themselves completely.

Generally, it can be said that in both examined cases the three methods employing plastic resources give results only slightly different from the other one and that differences are clearly identifiable and related to the base hypotheses of the three procedures.

Verifications at SLS for Case (a)

Verification procedures at SLS are described in Chap. 11.

Concrete

$$f_{ck} = 30\ \text{N/mm}^2$$

Ultimate load identified through plastic analysis is given by

$$q_u = 48.3\ \text{kN/m}.$$

For the characteristic service combination, it can be assumed

$$q = 0.7q_u == 33.8\ \text{kN/m}$$

For the quasi-permanent combination, it can be assumed

$$q = 0.5q_u == 24.2\ \text{kN/m}$$

The assumed value of 0.7 for the characteristic combination is generally quite accurate. For the quasi-permanent combination, the ratio can be between 0.45 and 0.55 (CEB 127-1995, 33–34).

Being the maximum internal forces at the intermediate support B, bending moment at that cross-section can be calculated (dimensionless values are also indicated because they will be employed in the verifications):

- characteristic combination:
  $M_B = -\frac{1}{8} \cdot q\ell^2 = -\frac{1}{8} \cdot 33.8 \cdot 64 = -270.4$ kNm; $\mu_B = \frac{M_B}{bd^2 f_{cd}} = 0.2545$.
- quasi-permanent combination:
  $M_B = -193.6$ kNm; $\mu_B = 0.1822$.
- *Stress verifications*

For stress verifications, Table E3 with $\alpha_e = 15$ can be used, with dimensionless values $\frac{\xi}{i}$, and $\frac{1-\xi}{i}$ for the cross-section with $\frac{\rho'}{\rho} = 0.2$ being $\rho = \omega \cdot \frac{f_{cd}}{f_{yd}} = 0.0151$.

They are: $\frac{\xi}{i} = 4.42$; $\frac{i-\xi}{i} = 5.20$.

By applying dimensionless formulas developed in 11.2.1.2 we get:

- Characteristic combination ($f_{cd} = f_{ck} \cdot \frac{0.85}{1.5}$ 0.85/1.5)

$$\frac{\sigma_c}{f_{ck}} = 0.2545 \cdot 4.42 \cdot \frac{0.85}{1.5} = 0.64 > 0.60$$

$$\frac{\sigma_s}{f_{yk}} = 0.2545 \cdot 15 \cdot 5.20 \cdot \frac{17}{450} = 0.75 < 0.80$$

- Quasi-permanent combination

$$\frac{\sigma_c}{f_{ck}} = 0.1822 \cdot 4.42 \cdot \frac{0.85}{1.5} = 0.45$$

$$\frac{\sigma_s}{f_{yk}} = 0.1822 \cdot 15 \cdot 5.20 \cdot \frac{17}{450} = 0.537; \sigma_s = 241 \text{ N/mm}^2$$

Concrete stress for characteristic combination is not acceptable in a corrosive environment (exposure Classes XD, XF, XS of [Table 4.1]), but it could be tolerated in a normal environment. If we want to respect the limitation $\sigma_c = 0.60 f_{ck}$ it is necessary to reduce service load as follows:

- in characteristic combination $q_{es} = 33.8 \cdot \frac{0.60}{0.64} = 31.6$ kN/m;
- in quasi-permanent combination: $31.6 \cdot \frac{0.5}{0.7} = 22.5$ kN/m
  Moments assume the values:
- in characteristic combination:
  $M_B = 252.8$ kNm; $\mu_B = 0.238$;
- in quasi-permanent combination:
  $M_B = 180.0$ kNm; $\mu_B = 0.170$;
  Stresses are, therefore:
- characteristic combination

$$\frac{\sigma_c}{f_{ck}} = 0.238 \cdot 4.42 \cdot \frac{0.85}{1.5} = 0.596 < 0.60$$

$$\frac{\sigma_s}{f_{yk}} = 0.238 \cdot 15 \cdot 5.20 \cdot \frac{17}{450} = 0.70 < 0.80$$

- quasi-permanent combination

$$\frac{\sigma_c}{f_{ck}} = 0.170 \cdot 4.42 \cdot \frac{0.85}{1.5} = 0.43$$

$$\frac{\sigma_s}{f_{yk}} = 0.170 \cdot 15 \cdot 5.20 \cdot \frac{17}{450} = 0.50; \sigma_s = 225 \text{ N/mm}^2$$

- *Verification of cracking in service quasi-permanent load combination*

Preliminary elements (see paragraph 11.3 of Chap. 11). Reinforcement in cross-section B is given by 6 bars in tension with a diameter of 20 mm. By equation [7.10] it can be calculated: $\rho_{\text{eff}} = \frac{6 \cdot 314}{250 \cdot 125} = 0.06$.

The other terms of calculation are given by

$$f_{ct,\mathrm{eff}} = 2.9\ \mathrm{N/mm^2}$$

$$\alpha_e = \frac{E_s}{E_{cm}} = \frac{200{,}000}{33{,}000} = 6$$

$$k_t = 0.4$$

$$c = 30\mathrm{mm}$$

By applying [7.9] we get:

$$\varepsilon_{sm} - \varepsilon_{cm} = \frac{225 - 0.4 \cdot \frac{2.9}{0.06}(1 + 6 \cdot 0.06)}{200,000} = \frac{225 - 26}{200,000}$$
$$= 1.0 \cdot 10^{-3} > 0.6 \cdot \frac{225}{200{,}000} = 0.67 \cdot 10^{-3}$$

and with [7.11] we get

$$s_{r,\max} = 3.4 \cdot 30 + 0.8 \cdot 0.5 \cdot 0.425 \cdot 20/0.0754 = 147\ \mathrm{mm}$$

Finally, by [7.8] we obtain:

$$w_k = 147 \cdot 1.0 \cdot 10^{-3} = 0.147\ \mathrm{mm}$$

which is acceptable in every environmental condition according to Table [7.1N].

- *Verification of deflection in quasi-permanent combination*

In the beginning, the procedure follows paragraph [7.4.2(2)]. The beam can be considered as the final part of a continuous beam for which coefficient k of Table [7.4N] takes the value 1.3. The cross-sections along the span have the following reinforcements ratios: $\rho = 0.0112$ and $\rho' = 0.0025$.

Reference reinforcements ratio is given by: $\rho_0 = 10^{-3}\sqrt{30} = 0.00547$. Because $\rho > \rho_0$, we apply [7.16.b] obtaining the acceptable ratio $\frac{\ell}{d}$:

$\frac{\ell}{d} = 1.3 \cdot \left[11 + 1.5\sqrt{30} \cdot \frac{0.00547}{0.0112-0.0025} + \frac{1}{12}\sqrt{30} \cdot \sqrt{\frac{0.0025}{0.00547}}\right] = 21.40(500/450) = 23.75$

In the case here examined, the ratio takes the value: $\frac{\ell}{d} = \frac{8000}{500} = 16 < 23.75$ 23.75.

This result, according to [7.4.2(2)] allows us to affirm that beam deflection is below the limits indicated in [7.4.1(4) and (5)]. However, we can also calculate deflection according to the method of [7.4.3].

The maximum vertical displacement of the beam simply supported on a side and fully restrained on the other one with constant EI is given by the expression $f = 0.0054q\ell^4/EI$.

In this case the term $0.0054q\ell^4$ takes the value:

$$0.0054 \cdot 22.5 \cdot 8.0^4 = 497 \text{ kNm}^3$$

Equation [7.18] expresses deflection as $\alpha = \zeta\alpha_{II} + (1-\zeta)\alpha_I$ where α_I and α_{II} are deflections calculated with completely reacting cross-section and partialized cross-section respectively. ζ is the coefficient accounting for the stiffening effect due to concrete between cracks, taking the value $\zeta = 1 - \beta\left(\frac{M_{cr}}{M}\right)^2$ being M_{cr} the cracking moment, M the service moment and β a coefficient taking the value 1 for short-duration loads and 0,5 for long-duration or repeated loads. Here we adopt the value 0.5.

In this case study $M_{cr} = W \cdot f_{ctm} = 250 \cdot \frac{550^2}{6} \cdot 2.9 = 36.5$ kNm.

Being $q = 24.2$ kN/m, service moment in the quasi-permanent load combination is given by:

$M = 24.2 \cdot \frac{8^2}{14.3} = 108.3$ kNm.

We get therefore

$$\zeta = 1 - 0.5 \cdot \left(\frac{36.5}{108.3}\right)^2 = 0.94$$

For the uncracked section, neglecting the presence of reinforcements, we get:

$$EI = 33{,}000 \cdot \frac{250 \cdot 550^3}{12} = 114 \cdot 10^{12} \text{ Nmm}^2$$

For the partialized cross-section, from Table E3 we get for $\rho = 0.0112$, $\rho'/\rho = 0.2$ and $\alpha_e = 15$: $i = 0.0836$.

We get therefore

$I = 0.0836 \cdot 250 \cdot 500^3 = 2612 \cdot 10^6 \text{mm}^4$

$$\text{EI} = 86.2 \cdot 10^{12} \text{Nmm}^2$$

and then it results

$$\alpha_I = \frac{497 \cdot 10^{12}}{114 \cdot 10^{12}} = 4.36 \text{ mm}$$

$$\alpha_{II} = \frac{497 \cdot 10^{12}}{86.2 \cdot 10^{12}} = 5.76 \text{ mm}$$

Deflection takes therefore the value

$$\alpha = 0.94 \cdot 5.76 + (1 - 0.94) \cdot 4.36 = 5.41 + 0.26 = 5.67 \text{ mm}$$

If we want to account for concrete creep, it is necessary to apply a correction to the obtained result according to [7.4.3(5)]. More precisely, it is sufficient to substitute the elastic modulus E_{cm} with the elastic modulus given by [7.20]:

$$E_{c,\text{eff}} = \frac{E_{cm}}{1 + \varphi(\infty, t_0)}$$

Under the assumption of an environment with relative humidity 80%, t_0, time from the load expressed in days = 30; being the value of $h_0 = 2A/u = 2 \cdot \frac{550 \cdot 250}{250 + 2 \cdot 550} = 275$ mm(considering that the vertical sides and the superior side of the bema are exposed to the air), according to graphs contained in Figure [3.1-b] it can be deduced that $\varphi(\infty, 30) = 2.0$.

Therefore, according to [7.20] we get $E_{c,\text{eff}} = \frac{33{,}000}{3} = 11{,}000$ N/mm^2.

Correcting the result with the one obtained by the previous calculation, we get $\alpha \cdot \frac{E_{cm}}{E_{c,eff}} = 5,67 \cdot \frac{33{,}000}{11{,}000} = 17.0$ mm.

This value is $\frac{\ell}{471} < \frac{\ell}{250}$.

4.3.6 *Effective Width of Flanges of T-section Beams*

This topic is discussed in [5.3.2], represented in Figure [6.7] of [6.2.4] and recalled in [9.2.1.2(2)].

[5.3.2.1.1(P)] says that in T beams the effective flange width, over which uniform conditions of stress can be assumed (*remark: it should be considered concrete in compression in case of positive bending moment, same stress of reinforcement in tension in case of negative bending moment*), depends on the web and flange dimensions, type of loading, the span, support conditions, and transverse reinforcement.

The following paragraph (2) recommends that the effective width of the flange should be based on the distance ℓ_0 between points of zero moment, as shown in Fig. 4.11. In synthesis, along the span, i.e. where $M(+)$, $\ell_0 = 0{,}7\ \ell$; on the supports: $\ell_0 = 0{,}15(\ell_1 + \ell_2)$ being ℓ_1 and ℓ_2 the two adjacent spans. For the two cases,

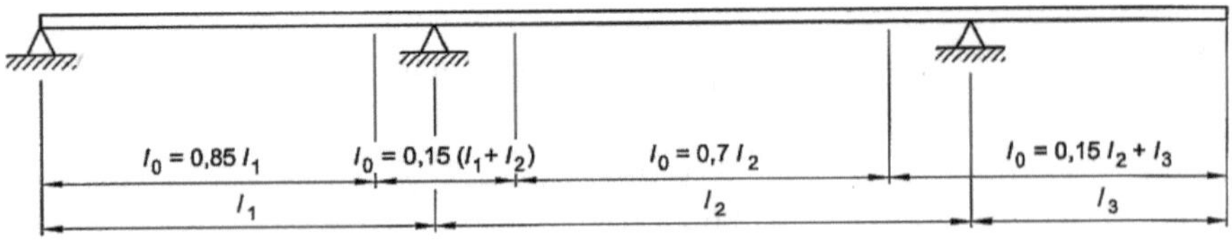

Fig. 4.11 Definition of l_0 for the calculation of the effective width of the flange (Figure [5.2])

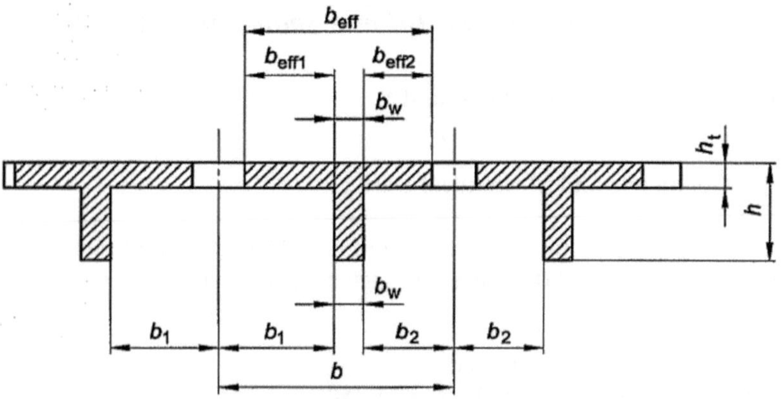

Fig. 4.12 Parameters for the effective width of flange (Figure [5.3])

paragraph (3) shows the formulas for the calculation of b_{eff}. Such width is indicated in Figure [6.7] of [6.2.4] for the calculation of the connection web-flange.

Finally, for cases of negative moment, [9.2.1.2(2)] suggests distributing reinforcement in tension over a width b_{eff} as defined in [5.3.2.1].

Formulas for the effective width b_{eff} (Fig. 4.12)

$$b_{\text{eff}} = \Sigma b_{\text{eff},i} + b_w \leq b \quad [(5.7)]$$

where

$$b_{\text{eff},i} = 0.2b_i + 0.1\ell_0 \leq 0.2\ell_0 \quad [(5.7a)]$$

and

$$b_{\text{eff},i} \leq b_i \quad [(5.7b)]$$

b_{eff} is the effective width
$b_{\text{eff},i}$ is the portion of "effective" slab placed on one side or the other side of the web
b_w is the web width
b_i is half of the net distance between two webs
b is the distance between webs.

4.3.7 Continuous Beam with Infinite Spans 10 m at ULS-SLS

Design at ULS and verifications at SLS.

4.3.7.1 Geometric and Mechanical Data

The transverse cross-section represented in Fig. 4.13 is given by beams with depth 0.75 m at a distance of 4.00 m linked through a top slab with depth 0.15 m.

- Beam width: $b_w = 0.25$ m
- Beam effective depth $d = 0.70$ m
  Materials: concrete: C 30/37; steel: B450C.
  Design strength:

$f_{cd} = 17$ N/mm^2.

$f_{yd} = 450/1.15 = 391$ N/mm^2.

$\nu = 0.5$; $\nu f_{cd} = 0.5 \cdot 17 = 8.50$ N/mm^2 (for ν values see Chap. 8)
Load analysis

- Self-weight of the structure for the centre-to-centre distance (4.0 m):
  $(0.15 \cdot 4 + 0.25 \cdot 0.60) \cdot 25 = 18.75$kN/m.
- Superstructure load
  1.25 kN/m $\cdot\, 4.0 = 5.00$ kN/m
- Permanent load
  $g_k = 18.75 + 5.00 = 23.75$ kN/m
- Variable load
  $q_k = 20.00$ kN/m

Safety factors for internal forces at ULS (EN 1990)

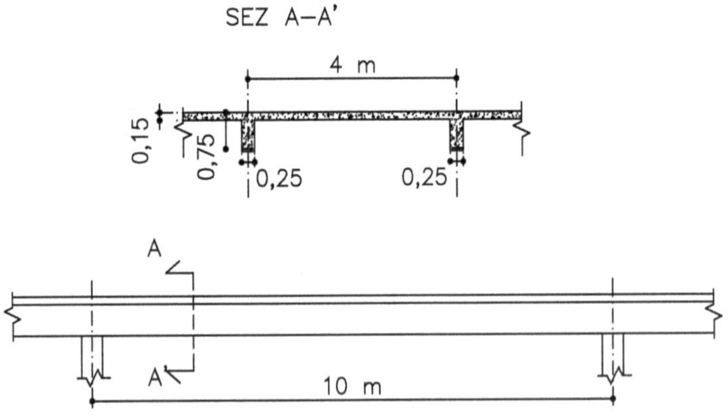

Fig. 4.13 Structural geometry

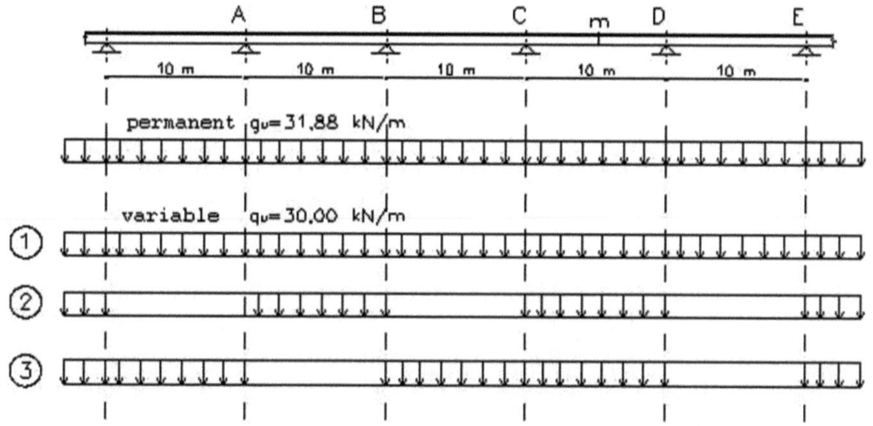

Fig. 4.14 Load conditions for ULS

$$\gamma G1 = 1.30; \gamma G2 = 1.50; \gamma Q = 1.50$$

Permanent load at ULS:

$$g_{k1} \cdot \gamma_{G1} + g_{k2} \cdot \gamma_{G2} = 18.75 \cdot 1.30 + 5.00 \cdot 1.50 = 24.38 + 7.50 = 31.88 \text{ kN/m}$$

Variable load at ULS:

$$qk \cdot \gamma Q = 20.0 \cdot 1.50 = 30 \text{ kN/m}$$

The variable load can be applied on single or multiple consecutive spans. Figure 4.14 shows three examined load conditions.

4.3.7.2 Structural Linear Elastic Analysis

(a) *Coefficients $q\ell^2$ for the calculation of bending moments.* Coefficients are collected in Table 4.10. The values of bending moments are collected in Table 4.11.

Table 4.10 Coefficients for the calculation of bending moments

Load	Condition	M_B	M_C	M_m	M_D	M_E
Permanent	Unique	– 1/12	– 1/12	+1/24	– 1/12	– 1/12
Variable	1	– 1/12	– 1/12	+1/24	– 1/12	– 1/12
Variable	2	– 1/24	– 1/24	+1/12	– 1/24	– 1/24
Variable	3	– 0.022	– 0.114	+0.057	– 0.022	– 0.022

Table 4.11 Bending moments at ULS

Load	Condition	M_B (kNm)	M_C (kNm)	M_m (kNm)	M_D (kNm)	M_E (kNm)
Permanent	Unique	– 267	– 267	+133	– 267	– 267
Variable	1	– 250	– 250	+125	– 250	– 250
Variable	2	– 125	– 125	+250	– 125	– 125
Variable	3	– 66	– 342	+171	– 66	– 66

(b) *Maximum bending internal forces.* Maximum negative moment can be obtained by combining permanent load with variable service load of condition 3: $M_C = -(267 + 342) = -609$ kNm; positive moment for the same combination is given by: $M_m = (133 + 171) = +304$ kNm. Maximum positive moment can be obtained by combining permanent load with variable load of condition 2: $M_m = (133 + 250) = +383$ kNm.

(c) *Reinforcement design using the linear elastic method.* Moment—609 kNm could be supported only by reinforcements in tension; using the verification method of 7.6.1.1.1, the neutral axis is placed at the depth:

$$x = \frac{d}{2\beta_2} - \sqrt{\left(\frac{d}{2\beta_2}\right)^2 - \frac{M_{Ed}}{\beta_1 \beta_2 f_{cd} b}}$$

$$= \frac{700}{2.0,\ 416} - \sqrt{\left(\frac{700}{2.0,\ 416}\right)^2 - \frac{609000000}{0,\ 8095.0,\ 416.17.250}} = 305 \text{ mm}$$

corresponding to $x/d = 0.436$. However, because [9.2.1.5 (1)] requires that, on continuity supports, reinforcements on the compressed side are arranged of an amount at least equal to 25% of the amount placed along the span, it is convenient to arrange a certain amount of reinforcement in compression, employing it as follows.

The moment equal to—609 kNm is supported by two internal couples:

- the first one is made of reinforcement in tension and in compression with the same area $A'_s = 500$ mm^2, lever arm (700 – 50) = 650 mm, which gives the moment $M' = 500\ 391\ 650 = 127$ kNm;
- the second one, equal to $M = 609 - 127 = 482$ kNm, is made of concrete in compression and reinforcement in tension.

Adopting the above-mentioned procedure, with $M = 482$ kNm, we get $x =$ 232 mm.

$$F_c = x \cdot b \cdot \beta_1 \cdot f_{cd} = 232 \cdot 250 \cdot 0.8095 \cdot 17 = 798 \text{ kN}$$

Compressive resultant force is given by:

$$F_c = x \cdot b \cdot \beta_1 \cdot f_{cd} = 232 \cdot 250 \cdot 0.8095 \cdot 17 = 798\ \text{kN}$$

and, for bending $F_c = F_t$, required reinforcement A_s is given by $F_t/f_{yd} = 798{,}000/391 = 2041\ \text{mm}^2$.

Therefore, at ULS it is required that the cross-section is reinforced above by 500 + 2041 = 2541 mm^2 and below by 500 mm^2.

Mid-span cross-section under positive moment $M = 383$ kNm.

Being T cross-section with compressed flange, lever internal arm is given by $d - h_f/2 = 700 - 150/2 = 625$ mm. Tensile force is therefore given by: 383/0.625 = 613 kN, which requires reinforcements $A_s = 613{,}000/391 = 1568\ \text{mm}^2$ provided by 5ϕ 20 mm.

At the supports, it is necessary to arrange at least 25% of this area, i.e. 0.25·1568 = 392 mm^2.

Redistribution of Negative Moments and Reinforcement Design (According to [5.5(4)] and Chap. 7)

(a) We assign a redistribution of 20% ($\delta = 0.80$) to the negative moment in C, under reserve to verify the acceptability of the values of (x/d) and stresses at service conditions.

Redistributed moment: –0.80·609 = –487 kNm. The difference 609 – 487 = 122 kNm increases the positive bending moment in the spans adjacent to the support C. It is worth observing that, in condition 3, bending moments at the adjacent supports B e D take the values—(265 + 66) = –331 kNm, then they are not subjected to redistribution simultaneously to C. Therefore, for the sake of simplicity and on the safety side, the value of half of 122, i.e. 61 kNm, is given to the midspan. Cross-sections are therefore loaded by the moment M_m = 304 + 61 = 365 kNm. This value is lower than the maximum calculated elastic value (+383 kNm). The moment –487 kNm is supported by two internal couples:

- the first one made of reinforcements in tension and compression with equal area $A'_s = 400\ \text{mm}^2$, lever arm (700 – 50) = 650 mm which gives $M' = 400 \cdot 391 \cdot 650 = 101$ kNm;
- the second one, equal to $M = 487 - 101 = 386$ kNm, with concrete in compression and reinforcements in tension. By means of the previous formulas we get: $x = 179$ mm; $F_c = 615.4$ kN and $A_s = 1574\ \text{mm}^2$. Finally, reinforcement required at the top side is (400 + 1574) = 1974 mm^2 and reinforcement required at the bottom side is 400 mm^2. The value $\delta = k_1 + k_2(x/d) = 0.44 + 1.25 \cdot (0.6 + 0.0014/0.0035) \cdot 0.255 = 0.76$ (acceptable redistribution 24%) is corresponding to the neutral axis $(x/d) = 179/700 = 0.255$ because of the [5.10a]. Proposed redistribution (20%) is therefore acceptable.

Table 4.12 Moments and reinforcements in the support cross-section for the ultimate design load

Δ	M_{Ed} (kNm)	A_s (mm²)	A'_s (mm²)
1.00	609	2541	500
0.80	487	1974	400
0.70	426	1693	400

Table 4.13 Bending moments at ULS and reinforcements (tabular solution)

δ	M_{Ed} (kNm)	μ_u	ξ	ω	A_s (mm²)	A'_s (mm²)
1.00	609	0.2915	0.330	0.335	2548	509
0.80	487	0.2334	0.258	0.262	1992	400
0.70	426	0.2041	0.223	0.225	1711	342

(b) If a stronger redistribution is proposed, equal to the maximum allowed value ($\delta = 0.70$), the negative moment is reduced to –0.70·609 = –426 kNm. The difference 609 – 426 = 183 kNm increases the positive moment in the spans adjacent to the support C. As shown in the previous procedure, we get M = 304 + 183/2 = 395 kNm. This value is slightly greater than the maximum calculated value (+383), therefore it will be sufficient for a slight adjustment of reinforcement. For example, 1ϕ22 + 4ϕ20 instead of 5ϕ20. The moment—426 kNm is supported by two internal couples:

- the first one made of reinforcements in tension and in compression with the same area $A'_s = 400$ mm², lever arm (700 – 50) = 650 mm which gives M' = 400 391 650 = 101 kNm;
- the second one, equal to a M = 426 – 101 = 325 kNm, with concrete in compression and reinforcements in tension.

By the previous formulas we get: $x = 147$ mm; $F_c = 506$ kN and $A_s = 1293$ mm². Finally, required reinforcement at the top side is (400 + 1283) = 1693 mm², while required reinforcement at the bottom is 400 mm². The value $\delta = 0.44 + 1.25 \cdot 0.21 = 0.70$ is corresponding to the neutral axis (x/d) = 147/700 = 0.21, equal to the proposed one.

Redistribution of 30% is therefore acceptable. Table 4.12 collects in synthesis reinforcements of the intermediate support cross-section required in the three analysed hypotheses of redistribution.

The mentioned above results can be easily obtained using Table U1 corresponding to the ULS and characterized by solutions with $\omega'/\omega = 0.2$.

Table 4.13 collects the values of: δ, M_{Ed}, $\mu_u = \frac{M_{Ed}}{bd^2 f_{cd}}$, ξ, ω, A_s, A'_s.

We can observe the substantial agreement of results.

Table 4.14 Verifications at SLS: geometric data

δ	$100/\rho$	ρ/ρ'	ξ/i	(1− ξ)/i
1.00	1.45	0.20	4.479	5.382
0.80	1.13	0.20	4.825	6.833
0.70	0.97	0.23	5.211	7.910

4.3.7.3 Shear Verifications and Reinforcements Design (Vertical Stirrups)—Reference to Chap. 8

We consider the span C-D under the variable load condition 3, subject to ultimate moments M_C = –609 kNm and M_D = –333 kNm (we neglect redistribution giving a slight modification of the shear). Acting shear V_{Ed} at ULS on the right of the support C is given by moments equilibrium equation to D

$$M_C + V_{Ed}\ell - (g_k\gamma_G + q_k \cdot \gamma_Q)\ell^2/2 = M_D$$

i.e., by substituting:
$-609 + V_{Ed} \cdot 10 - (31{,}88 + 30)\ 100/2 = -333$, from which $V_{Ed} = 337$ kN.

Diagram V is shown in Figure 4.15.

Maximum resistant shear is given by (formula [6.9]) with $\theta = 45°$, being $z = 0.9 \cdot 700 = 630$ mm and $\nu f_{cd} = 8.50$ N/mm^2:

$V_{Rd,\max} = 0.5 \cdot 250 \cdot 630 \cdot 8.50 = 669$ kN, greater than V_{Ed}.

The angle $\theta = 0.5 \cdot \arcsin\,[2 \cdot 337{,}600/(8.50 \cdot 250 \cdot 630)] = 15.1°$ (see 8.1.3.2) is smaller than 21.80° therefore, under the assumption $\theta = 26.56°$, $\cot\theta = 2.0$, by means of [6.8] we get:

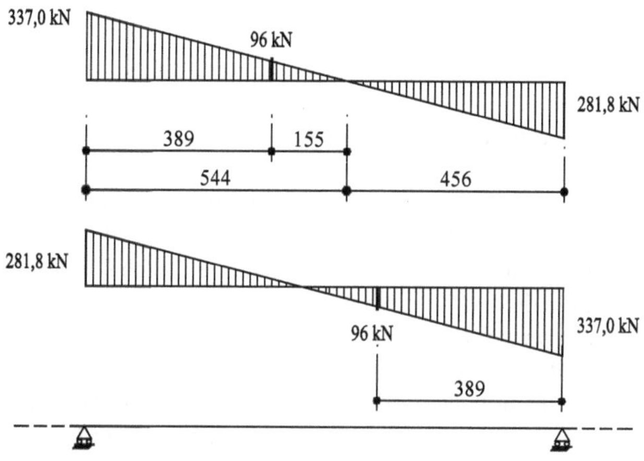

Fig. 4.15 Shear diagrams at ULS

$A_{sw}/s = 337{,}000/(630{\cdot}391{\cdot}2, 0) = 0.684\ \text{mm}^2/\text{mm}$.

Setting $s = 200$ mm we get $A_{sw} = 136\ \text{mm}^2$, largely fulfilled by 2-arm stirrups 10 mm diameter ($A_{sw} = 78 \cdot 2 = 156\ \text{mm}^2$). The number of stirrups can be reduced in the central span, as discussed here.

Value of the shear V_{Rdc} under which we can arrange only minimum reinforcement. Utilizing [6.2a], being $k = 1 + \sqrt{\frac{200}{700}} = 1.54$; $\rho_l = \frac{1568}{250 \cdot 700} = \frac{0.90}{100}$ we get:

$$V_{Rdc} = \left[\frac{0.18}{1.50} \cdot 1.54 \cdot (0.90 \cdot 30)^{1/3}\right] \cdot 250 \cdot 700 = 96\ \text{kN}$$

The segment where $V_{Ed} < V_{Rdc}$ is extended for the central length $10 - 2 \cdot 3.89 = 2.22$ m, as shown by Fig. 4.15 where also the diagram V_{Ed} corresponding to the configuration with switched continuity moments is represented. For such a segment equation [9.5N] requires:

$$\rho_{w,\min} = \left(\frac{A_{sw}}{s \cdot b_w}\right)_{\min} = \left(0.08 \cdot \sqrt{f_{ck}}\right)/f_{yk} = \left(0.08 \cdot \sqrt{30}\right)/450 = \frac{0.0974}{100}$$

Arranging two-arm stirrups 8 mm diameter with $s = 300$ mm we get:

$$\rho_w = \frac{2 \cdot 50}{300 \cdot 250} = \frac{0,133}{100} > \rho_{w,\min}$$

To determine tensile longitudinal force due to shear, we can translate the diagram M of the amount $a_\ell = z\,(\cot\theta - \cot\alpha)/2$, in analogy with the Example of Figure [9.2]. In the present case, being $\cot\theta = 2$ and $\cot\alpha = 0$ (vertical stirrups), we get $a_\ell = z = 630$ mm.

Web-Flange Connection

Application Field

The calculation procedure is discussed in 8.1.7 of Chap. 8 with reference to [6.2.4] of EC2.

For the field $M(-)$ we account for the load combination permanent + variable load scheme 3 with redistribution $\delta = 0.80$ (4.3.7.3 a); for the field $M(+)$ the load combination permanent load + variable load scheme 2 with $\delta = 1.0$ (4.3.7.2).

In the first case, the field $M(-)$ is extended along the span for 2 *m* starting from the support where lever arm z of the internal couple takes the value 0.60 *m*. Therefore, as indicated in [6.2.4(2)] and required in 8.1.7, for the calculation we consider a segment with a length equal to half of the distance between the point with the minimum moment and the point of null moment, i.e. $\Delta x = 1$ m starting from the support. In such a segment the moment varies from the value –487 kNm to the value –209 kNm, with $\Delta M = 278$ kNm. In such a segment the variation of the tensile force is equal to $\frac{\Delta M}{z} = \frac{286}{0.625} = 278/0.60 = 463$ kN.

In the second case, the field $M(+)$ is extended for a length of 3.50 m on both sides of the midspan where lever arm z takes the value 0.625 m. At the distance $\Delta x = 3.50/2 = 1.75$ m from the null point towards midspan, the moment takes the value $M = +286$ kNm and this is the variation that should be taken into account. Over such a segment the variation of the tensile force is given by $\frac{\Delta M}{z} = \frac{286}{0.625} = 286/0.625 = 458$ kN.

Calculation of the widths b_{eff} corresponding to the two examined segments

Field M (–)

$$\ell_o = 0.15(10.0 + 10.0) = 3.0 \text{ m}$$

$b_1 = \frac{1}{2}(4.0 - 0.25) = 1.875$ m
$b_{eff1} = 0.2 \cdot 1.875 + 0.1 \cdot 3.0 = 0.675 \text{ m} > 0.2 \cdot \ell_0 = 0.2 \cdot 3.0 = 0.60$ m [(5.7a)]
Being $b_{eff1} = 0.60$ m, we get
$b_{eff} = 2 \cdot 0.60 + 0.25 = 1.45$ m [5.7]
Field $M(+)$

$$\ell_0 = 0.7 \cdot 10 = 7.0\,\text{m}$$

$$b_1 = \frac{1}{2}(4.0 - 0.25) = 1.875 \text{ m}$$

$$b_{eff1} = 0.2b1 + 0.1\updownarrow_0 =< 0.2 \cdot 1.875 + 0.1 \cdot 7.0 = 1.075\text{m} < 0.2\updownarrow_0 = 1.40\text{m} \quad [(5.7a)]$$

Being

$$1.075 < 1.40 \quad [(5.7b)]$$

we get

$$b_{eff} = 1.075 \cdot 2 + 0.25 = 2.40\text{m} < 4.00 \text{ m} \quad [5.7]$$

Tangential force transferred to a side of the beam in the segment Δx. Stress Verification and Calculation of linking reinforcement

Formulas are developed in 8.1.7.

Field M(–).

For the beam in this field it is assumed, as previously, lever arm $z = 0.60$ m.

Being $\eta = b_w/b_{\text{eff}} = 0.25/1.45 = 0.17$

tangential force transmitted to the web from one of the sides of the flange takes the value:

$\Delta F_d = \frac{\Delta M(1-\eta)}{2z} = [278\ (1 - 0.17)/(2 \cdot 0.60)] = 193$ kN.

and tangential stress is:

$$\nu_E = \frac{\Delta F_d}{h_f \Delta x} = \frac{193000}{150 \cdot 1000} = 1.29\ \text{N/mm}^2$$

If the inclination of the compressed struts is assumed, according to EC2 [6.2.4(4)Remark)], equal to $\theta_f = 38.6°$, corresponding to: sin $\theta_f = 0.6247$; cos $\theta_f = 0.7809$; tan $\theta_f = 0.80$, we get:

$1.29 < \nu \cdot f_{cd} \cdot \sin\theta_f \cdot \cos\theta_f = 0.5 \cdot 17 \cdot 0.6247 \cdot 0.7809 = 4.14\ \text{N/mm}^2$, as requested by the [(6.22)].

Required reinforcement, because of [(6.21)], should be:

$\frac{A_s}{s_f} \geq \frac{\nu_E \cdot h_f \cdot \tan\theta_f}{f_{yd}} = \frac{1.29 \cdot 150 \cdot 0.80}{391}$ $=(1.29 \cdot 150 \cdot 0.80)/391 = \frac{1.29 \cdot 150 \cdot 0.80}{391} = \frac{1.29 \cdot 150 \cdot 0.80}{391}$ $=0.40\ \text{mm}^2/\text{mm}$.

If the reinforcement spacing s_f is assumed equal to 150 mm, we get $A_s \geq 60$ mm^2.

Field M(+).

For the beam in this field, it is assumed (second-to-last paragraph of 4.3.7.2) $z = 0.625$ m.

Being $\eta = b_w/b_{\text{eff}} = 0.25/2.40 = 0.10$.

Tangential force transmitted to the web from one of the sides of the flange takes the value:

$$\Delta F_d = \frac{\Delta M(1-\eta)}{2z} = = [286 \cdot (1 - 0.10)\ /(2 \cdot 0.625)] = 206\ \text{kN}$$

and tangential stress is:

$$\nu_E = \frac{\Delta F_d}{h_f \Delta x} = \frac{206000}{150 \cdot 1750} = 0.79\ \text{N/mm}^2$$

If the inclination of the compressed struts is assumed, according to EC2 [6.2.4(4) Remark)], equal to $\theta_f = 38.6°$, corresponding to: sin $\theta_f = 0.6247$; cos $\theta_f = 0.7809$; tan $\theta_f = 0.80$, we get:

$0.79 < \nu \cdot f_{cd} \cdot \sin\theta_f \cdot \cos\theta_f = 0.5 \cdot 17 \cdot 0.6247 \cdot 0.7809 = 4.14\ \text{N/mm}^2$ as required by [(6.22)].

Required reinforcement shall be not less than:

$$\frac{A_s}{s_f} \geq \frac{\nu_E \cdot h_f \cdot \tan\theta_f}{f_{yd}} = \frac{1.29 \cdot 150 \cdot 0.80}{391} = (0.79\ 150\ 0.80)/391 = 0.25\ \text{mm}^2/\text{mm}$$

If reinforcement spacing is $s_f = 150$ mm, we get $A_s \geq 38$ mm^2.

The arrangement of transversal reinforcements should be associated with the reinforcement for bending of the slab.

4.3.7.4 Slab

Load analysis at ULS

- Self-weight: $0.15 \cdot 25 \cdot 1.3 = 4.875$ kN/m
- Permanent load: $1.25 \cdot 1.5 = 1.875$ kN/m
- Total: 6.75 kN/m
- Variable load: $5.0 \cdot 1.5 = 7.50$ kN/m

Internal forces at ULS, calculated for an internal span of the slab and accounting for the heaviest load combinations, are:

- Negative moment (support): $M = -22.85$ kNm
- Positive moment (span): $M = 14.76$ kNm

Reinforcement design for negative moment

$$\mu = \frac{22850000}{1000 \cdot 125^2 \cdot 17} = 0.0860.$$

From Table U1 of Appendix (Chap. 13), for $\omega'/\omega = 0.2$, we get $\omega = 0.09$.
$A_{s,\text{sup}} = 0.09 \cdot \frac{1000 \cdot 125 \cdot 17}{391} = 489$ mm^2/m, i.e. 73 mm^2/150 mm.
$A_{s,\text{inf}} = 0.2 \cdot 73 = 15$ mm^2/150 mm
Reinforcement design for positive moment
$\mu = \frac{14760000}{1000 \cdot 125^2 \cdot 17} = 0.0556$. From Table U1 for $\omega'/\omega = 0.3$ we get $\omega = 0.0565$
$A_{s,\text{inf}} = 0.0565 \cdot \frac{1000 \cdot 125 \cdot 17}{391} = 307$ mm^2/m, i.e. 46 mm^2/150 mm
$A_{s,\text{sup}} = 0.3 \cdot 46 = 15$ mm^2/150 mm.

Total slab reinforcement at the fully restrained section of the slab is given by the amount related to bending and by the amount corresponding to the amount required for the "web-flange" connection. This last quantity is distributed in equal parts at the top and the bottom of the slab.

Here we collect the results. It is worth noting that in all cases reinforcement spacing is fixed to 150 mm.

(a) Segment where the beam in subject to negative moment:
$A_{s,\text{sup}} = (73 + 60/2) = 105$ mm^2/150 mm; arranged $78 + 50 = 128$ mm^2/150 mm
$A_{s,\text{inf}} = (15 + 60/2) = 45$ mm^2/150 mm; arranged $50 = 128$ mm^2/150 mm

(b) Segment where the beam in subject to positive moment:
$A_{s,\text{sup}} = (73 + 30/2) = 92$ mm^2/150 mm; arranged $78 + 50 = 128$ mm^2/150 mm
$A_{s,\text{inf}} = (15 + 30/2) = 34$ mm^2/150 mm; arranged 50 mm^2/150 mm

(c) Slab fields subject to $M(+)$
$A_{s,\mathrm{inf}} = 46\ \mathrm{mm}^2/150\ \mathrm{mm}$; arranged $50\ \mathrm{mm}^2/150\ \mathrm{mm}$
$A_{s,\mathrm{sup}} = 0.3 \times 46 = 14\ \mathrm{mm}^2/150\ \mathrm{mm}$; arranged $50\ \mathrm{mm}^2/150\ \mathrm{mm}$.

Figure 4.18 shows the layout of the reinforcement as above defined.

4.3.7.5 Reinforcement Arrangement

Figure 4.16 shows an example of reinforcement arrangement according to the envelope of the internal forces (for the midspan: permanent load + variable load scheme 2; for the continuity support: permanent load + variable load scheme 3) with redistribution of the continuity moment $\delta = 0.8$. The figure shows internal forces in upper and lower trusses taking into account the translation of the bending moment diagram and the diagram of the resistant internal forces. It is worth noting that longitudinal reinforcement at the bottom side at supports is made of 2 bars 20 mm diameter, sufficient to fulfil what is required by [9.2.1(5)], i.e. at least the 25% of the reinforcement arranged along the span, which in the present case is made of 5 bars of equal diameter. Reinforcement at the top, over the support, will be distributed partially on the rib, partially on the two adjacent slab zones within the limit of b_{eff}, as predicted in [9.2.1.2(2)] and shown in Figure [9.1] and in 4.20.

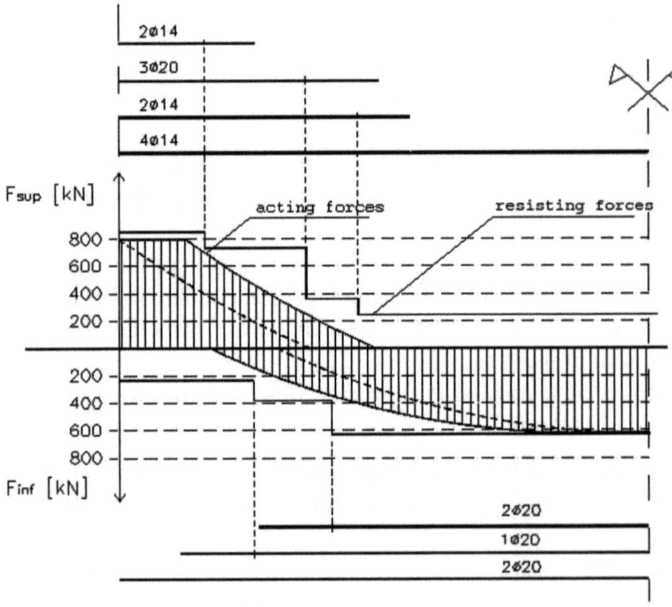

Fig. 4.16 Acting and resistant forces in upper and lower struts of the beam

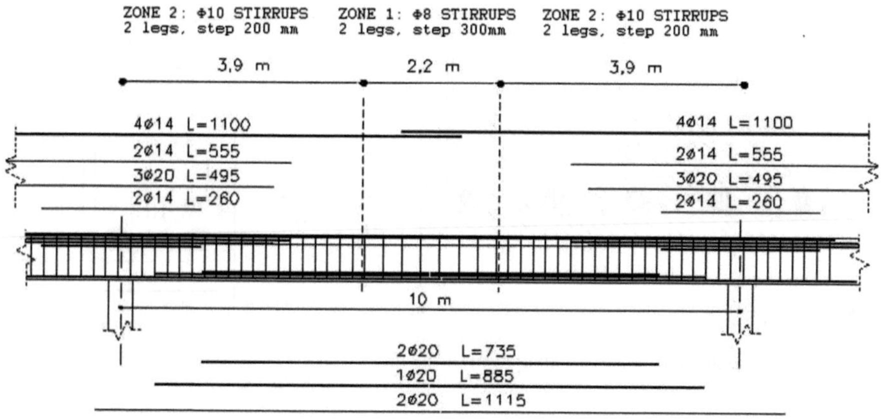

Fig. 4.17 Longitudinal and transverse reinforcement arrangement

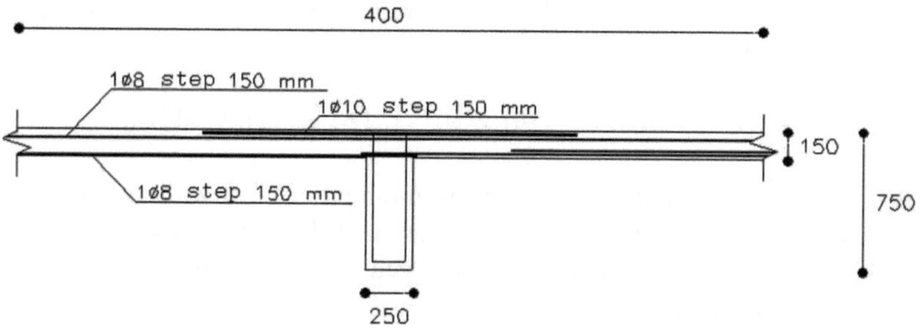

Fig. 4.18 Reinforcement arrangement in the web-flange connection

In the arrangement of the reinforcement, requirements are fulfilled with slight excess. Figure 4.17 shows the arrangement of longitudinal and transverse reinforcement along the span, while Fig. 4.18 shows the reinforcement arrangement for the beam-slab connection.

4.3.7.6 Verifications at SLS

Verification procedures at SLS are shown in Chap. 11.

Stress Verification

According to [7.2(2)] maximum allowable stresses are:

for concrete: $0.6 f_{ck} = 18\ \text{N/mm}^2$
for steel: $0.80 f_{yk} = 360\ \text{N/mm}^2$

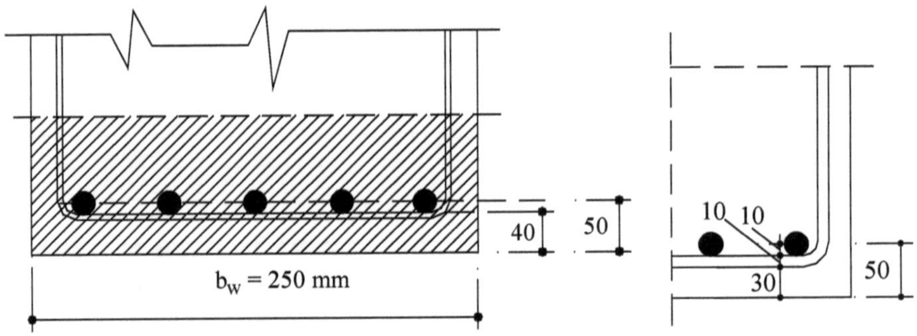

Fig. 4.19 Midspan cross-section: calculation of the area $A_{c,\text{eff}}$

Intermediate support cross-section

We consider the characteristic load combination represented by the permanent and variable load in the most unfavourable configuration (the same assumed for the ULS).

The negative moment at support C is given by:

$$-23.75 \times 10^2/12 - 20 \times 10^2 \cdot 0.114 = -425.9 \text{ kNm}$$

Such a value, if compared to the design moment at ULS without redistribution (−609 kNm), represents the 70%, if compared to the 20% redistributed moment (−487 kNm), the 87% and finally, if compared to the 30% redistributed moment (−426 kNm) is practically 100%.

Verifications are developed through dimensionless notations, using the Tables for the evaluation of $\frac{\xi}{i}$ and $\frac{1-\xi}{i}$, and therefore of stresses.

Characteristic combination service moment: $M = -425.9$ kNm.

Dimensionless moment $\mu = \frac{425.9 \cdot 10^6}{250 \cdot 700^2 \cdot 17}$ =0.2045.

Starting from the ULS design values A_s, we calculate geometric ratios $\rho = \frac{A_s}{bd}$ and $\rho' = 0, 2\rho$.

As a function of those values, using Table E3, we get:

$$\frac{\xi}{i}, \frac{1-\xi}{i}$$

and after $\frac{\sigma_c}{f_{ck}}$ and $\frac{\sigma_s}{f_{yk}}$.

For example, with the required reinforcement for $\delta = 0.80$ (A_s= 1974 mm^2; $\rho =$ 1.13/100; $\rho'/\rho = 0.20$) we get:

$\sigma_c = \mu \frac{\xi}{i} f_{cd} = 0.2045 \cdot 4.825 \cdot 17 = 16.7$ N/mm^2; $\frac{\sigma_c}{f_{ck}} = 0.56$.

$\sigma_s = \mu \frac{1-\xi}{i} \alpha_e f_{cd} = 0.2045 \cdot 6.833 \cdot 15 \cdot 17 = 356$ N/mm^2; $\frac{\sigma_c}{f_{ck}} = 0{,}79$.

If for the same case (redistribution $\delta = 0.80$) the calculation is repeated considering the reinforcements proposed with the scheme of Fig. 4.17 (A_s = 2174 mm^2, A'_s = 628 mm^2), we get slightly lower stresses.

Table 4.15 Verifications at SLS: stresses

δ	σ_c	σ_c/f_{ck}	σ_s	σ_s/f_{yk}
1.00	15.5	0.51	280.0	0.62
0.80	16.7	0.56	356.0	0.79
0.70	18.0	0.59	411.2	0.90

Stresses as a function of δ are collected in Table 4.15.

From Table 4.15 we get that redistribution $\delta = 0.80$ is acceptable, while that one $\delta = 0.70$ would give excessive stresses in tensile reinforcement.

Mid-span cross-section

We assume a width of 1000 mm for the flange, smaller than the one allowed by [5.3.2.1].

Characteristic combination service moment is given by

$$M = 10^2 \cdot \left(\frac{23.75}{24} + \frac{20.00}{12}\right) = 265.6\,\text{kNm}$$

The reduced dimensionless moment is

$$\mu = \frac{265.6 \cdot 10^6}{1000 \cdot 700^2 \cdot 17} = 0.0319$$

T cross-section is reinforced by 1568 mm^2. Geometric ratio is given by

$$\rho = \frac{1568}{1000 \cdot 700} = 0.0022$$

Being

$w = h_f/d = 150/700 = 0.214$ and

$n = b/b_w = 1000/250 = 4$,

we adopt the Table corresponding to $w = 0.20$ and $n = 4$ (Table E6 of Appendix), deducing, for

$\rho = 0.0022 = \frac{\xi}{i} = 9.62$; $\frac{1-\xi}{i} = 32.73$.

Relevant stresses are:

for concrete: $\frac{\sigma_c}{f_{ck}} = 0.0319 \cdot 9.62 \cdot \frac{0.85}{1.50} = 0.17 < 0.60$.

for reinforcements: $\frac{\sigma_s}{f_{yk}} = 0.0319 \cdot 15 \cdot 32.73 \cdot \frac{17}{450} = 0.59 < 0.80$.

Cracking Verifications

Midspan cross-section

For quasi-permanent load combination, with permanent load and variable load according to scheme 2, we get:

$g_k = 23.75$ kN/m
$0.6q_k = 0.6 \cdot 20 = 12$ kN/m

Midspan moment:
$M = 10^2 \cdot \left(\frac{23.75}{24} + \frac{12}{12}\right) = 199$ kNm
The reduced moment is given by:

$$\mu = \frac{199 \cdot 10^6}{1000 \cdot 700^2 \cdot 17} = 0.0239$$

Reinforcement in tension $A_s = 1568$ mm^2 (see the second to last paragraph of 4.3.7.2).

With the same values already assumed for the stress verification in characteristic condition, we get:

$\sigma_s = 0.0239 \cdot 32.73 \cdot 15 \cdot 17 = 200$ N/mm^2.

The calculation of crack opening is performed according to [7.3.4].

Maximum spacing between cracks

$$s_{r,\max} = k_3 c + k_1 \cdot k_2 \cdot k_4 \cdot \frac{\phi}{\rho_{p,eff}} \quad [7.11]$$

in [7.11]:

$k_3 = 3.4$

$c = 40$ mm, concrete cover,

$k_1 = 0.8$ (rebar),

$k_2 = 0.5$ (bending),

$k_4 = 0.425$,

$\phi = 20$ mm,

$\rho_{p,\text{eff}} = \frac{A_s}{A_{c,\text{eff}}}$; $A_{c,\text{eff}} = b_w \cdot h_{c,\text{eff}}$; $h_{c,\text{eff}} = 2.5 \cdot (h - d) = 2.5 \cdot (750 - 700) =$ 125 mm

$$\rho_{p,\text{eff}} = \frac{5 \cdot 314}{250 \cdot 125} = \frac{5}{100}$$

We get:

$$s_{r,\max} = 3.4 \cdot 40 + 0.8 \cdot 0.5 \cdot 0.425 \cdot \frac{20}{0.05} = 136 + 68 = 204 \text{ mm}$$

Steel medium deformation, deduced the deformation of concrete between cracks, can be evaluated through the following expression

$$\varepsilon_{sm} - \varepsilon_{cm} = \frac{\sigma_s - k_t \frac{f_{ct,\mathrm{eff}}}{\rho_{p,\mathrm{eff}}}\left(1 + \alpha_e \cdot \rho_{p,\mathrm{eff}}\right)}{E_s} \geq 0,6\frac{\sigma_s}{E_s} \quad [7.9]$$

In the present case:
$\sigma_s = 200$ N/mm^2
$k_t = 0.4$ (long duration)
$f_{ct,\mathrm{eff}} = f_{ctm}$ which, for $f_{ck} = 30$ N/mm^2, takes the value 2.9 N/mm^2

$$\alpha_e = \frac{E_s}{E_{cm}} = \frac{200000}{33000} = 6.06$$

we get:

$$\varepsilon_{sm} - \varepsilon_{cm} = \frac{200 - 0.4\frac{2.9}{0.05}(1 + 6.06 \cdot 0.05)}{200000} = \frac{200 - 30}{200000} = \frac{0.85}{1000}$$

which is greater than $0.6\frac{\sigma_s}{E_s} = \frac{0,6 \cdot 200}{200000} = \frac{0.60}{1000}$.

The opening w_k, because of [7.8], is equal to:

$$w_k = s_{r\max} \cdot (\varepsilon_{sm} - \varepsilon_{cm}) = 204 \cdot \frac{0.85}{1000} = 0.17 \text{ mm}$$

Cross-section at the continuity support

For quasi-permanent load combination, with permanent load and variable load according to scheme 3, we get:

$$M = -23.75 \cdot \frac{10^2}{12} - 0.6 \cdot 20 \cdot 10^2 \cdot 0.114 = -(198 + 137) = -335.0 \text{ kNm}$$

(for coefficient 0.114 see Table 4.10).

Reduced moment takes the value:

$$\mu = \frac{335.0 \cdot 10^6}{250 \cdot 700^2 \cdot 17} = 0.16$$

With the values already employed for the stress verification of the characteristic combination (Table 4.15), we obtain by proportion:

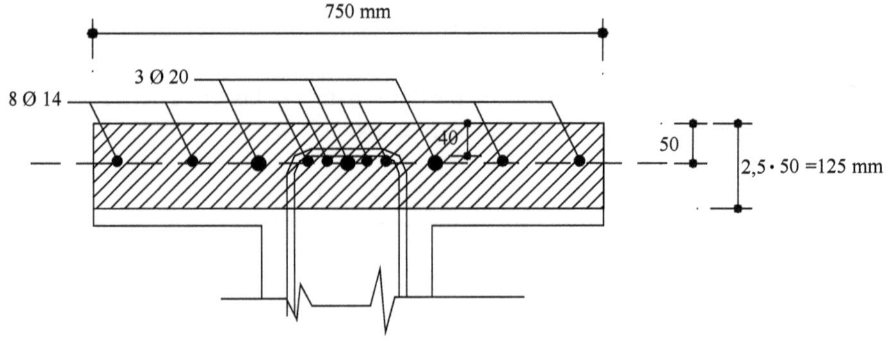

Fig. 4.20 Cross-section at support: calculation of the area $A_{c,\text{eff}}$

$$\sigma_s = \frac{0.16}{0.2045} \cdot 356 = 277\ \text{N/mm}^2$$

Being the area of the reinforcement amount $(3\phi20 + 8\phi14) = 2174\ \text{mm}^2$ as shown in Fig. 4.17 displaying the proposed scheme of reinforcement, we can deduce:

$$\rho_{p,\text{eff}} = \frac{2174}{750 \cdot 125} = \frac{2.32}{100}$$

The width of the slab segment, associated with the rib, to define $A_{c,\text{eff}}$ has been set equal to 750 mm depending on the reinforcement arrangement; such a value is smaller than the one available according to [5.3.2.1(3)] and [9.2.1.2(2)].

Reinforcement distribution of Fig. 4.20 respects the condition that the bar spacing is not greater than $5(c + \phi/2) = 5(40 + 20/2) = 250$ mm because in that context reinforcements rule cracking (Figure [7.2], detail D).

Maximum spacing between cracks and deformation is given by:

$$\begin{aligned} s_{r,\max} &= k_3 c + k_1 \cdot k_2 \cdot k_4 \cdot \frac{\phi}{\rho_{p,\text{eff}}} = 3.4 \cdot 40 + 0.8 \cdot 0.50 \cdot 0.425 \cdot \frac{20}{0.0232} \\ &= 136 + 146 = 282 \end{aligned}$$

$$\begin{aligned} \varepsilon_{sm} - \varepsilon_{cm} &= \frac{277 - 0.4\frac{2.9}{0.0232}(1 + 6.06 \cdot 0.0232)}{200000} \\ &= \frac{277 - 52}{200000} = \frac{1.10}{1000} > \frac{0.6 \cdot 278}{200000} = \frac{0.83}{1000}\text{mm} \end{aligned}$$

The opening w_k, according to [7.8], takes the value:

$$w_k = s_{r\max} \cdot (\varepsilon_{sm} - \varepsilon_{cm}) = 282 \cdot \frac{1,10}{1000} = 0,31\ \text{mm}$$

Deflection Control for Quasi-Permanent Service Load Combination

Midspan cross-section for a generic span of the continuous beam has reinforcement ratio $\rho = 0.0022$ (see 4.3.7.8.1). Considering the discussion of 11.4 (Chap. 11) the comparison of the slenderness with the one proposed in [(7.16 a)] highlights that there are no problems of excessive deflections. However, as an example, we can develop the calculation of deflection through formulas (7.18) and (7.19) of EC2, as shown in 11.4.

More in detail, we can calculate deflection at the midspan through [(7.18)] combining the deflection evaluated for completely reacting cross-sections (f_1) with the one obtained for partialized cross-sections (f_2) at infinite time, accounting for the "tension stiffening" effect.

- *Elasticity modulus*

Concrete C30/37 is corresponding to $E_{cm} = 33{,}000$ N/mm^2. For the evaluation at infinite time, we assume $\varphi = 2$. Therefore, using [(7.20)] we get $E_{c,\infty} = \frac{33000}{(1+2)} = 11,000$ N/mm^2.

For the calculation of the cracking moment, cross-sections should be homogenized with the ratio $\alpha_e = 200{,}000/33{,}000 = 6.06$. Using this moment, we can calculate the coefficient of *tension stiffening* ζ.

For the calculation of deflection at infinite time, all the cross-sections, both uncracked and cracked ones, are homogenized adopting $\alpha_e = 200{,}000/11{,}000 = 18.02$.

- *List of mechanic and geometric characteristics*

Uncracked section ($\alpha_e = 6.06$).

By using formulas of Chap. 11, paragraph 11.2.1.1.2, being

$\rho_1 = \frac{A_S}{b \cdot h} = \frac{1568}{750 \cdot 1000} = 0.0021$

$$n = \frac{b}{b_w} = \frac{1000}{250} = 4$$

$w_1 = \frac{h_f}{h} = \frac{150}{750} = 0.20$

$p = \frac{d}{h} = \frac{700}{750} = 0.93$

$\alpha_e = 6.06$

we get:

$\xi_1 = 0.368\ 0.368$; $i = 0.0405$; from which $x_G = 276$ mm; $(h - x_G) = 474$ mm; $I_1 = 1.70 \cdot 10^{10}$ mm^4.

Uncracked section ($\alpha_e = 18.2$)

We get: $\xi_1 = 0{,}40$; $i = 0.481$; from which $x_G = 300$ mm; $(h - x_G) = 450$ mm; $I_1 = 2028$ mm^4.

Cracked section ($\alpha_e = 18.2$).

Formulas in 11.2.1.2.3
$\rho = \frac{1568}{700 \cdot 1000} = 0.0022$

$$n = \frac{b}{b_w} = \frac{1000}{250} = 4$$

$w = \frac{h_f}{d} = \frac{150}{700} = 0.2143.$
$\alpha_e = 18.2$
By solving we get:

A = 0.79
C = 0.45
$\xi = 0.247$
$i = 0.0277$

from which $x_G = 173$ mm; $I_2 = 0.95 \times 10^{10}$ mm^4.

- *Flexural internal forces*

For the quasi-permanent service load combination, we can assume the following loads:

self-weight: 18.75 kN/m

permanent load: 5.00 kN/m

variable load 0.6 × 20 = 12 kN/m (scheme 2).
Moment at the midspan: $M_m = \frac{10^2}{24}(18.75 + 5.0) + \frac{10^2}{12} \cdot 12 = 99 + 100 = 199$ kNm.
Moment at supports: $M_{\text{app}} = - \frac{10^2}{12}(18,75 + 5,0) - \frac{10^2}{24} \cdot 12 = -198 - 50 = -248$ kNm.
Cracking moment and tension stiffening coefficient.
For a concrete Class C30/37 tensile strength is $f_{ctm} = 2.9$ N/mm^2. Therefore we deduce:
$\frac{M_{cr}(h - x_1)}{I_1} = 2.9$ from which
$M_{cr} = \frac{(1.70 \cdot 10^{10}) \cdot 2.9}{474} = 104$ kNm.
Tension stiffening coefficient.
$\zeta = 1 - 0.5 \cdot \left(\frac{M_{cr}}{M_m}\right)^2 = 1 - 0.5 \cdot \left(\frac{104}{199}\right)^2 = 0.86.$

- *Deflection calculation*

The diagram of the bending moment along the span, being parabolic, can be subdivided in:

parabolic diagram of a simply supported beam with the ordinate at the midspan

$M_{\text{par}} = 248 + 199 = 447$ kNm

rectangular diagram with ordinate $M_{\text{rett}} = —248$ kNm

Deflection at the midspan is given by the sum of the effects of the two solicitations: $f = \frac{5}{48} \cdot \frac{M_{par}}{EI} \ell^2 - \frac{M_{rett}}{8EI} \ell^2 = \frac{\ell^2}{EI} \cdot \left(\frac{5}{48} \cdot 447 - \frac{248}{8}\right) = 15.56 \frac{\ell^2}{EI}$.

and by applying [(7.18)]

$$f = f_1(1 - \zeta) + f_2\zeta$$

Numerical values of stiffness EI are:

$E_\infty I_1 = 11000 \cdot 2.028 \cdot 10^{10} = 22.30 \cdot 10^{13}$ Nmm^2.

$E_\infty I_2 = 11000 \cdot 0.95 \cdot 10^{10} = 10.45 \cdot 10^{13}$ Nmm^2.

By substituting, we get:

$f_1 = \frac{15.56 \cdot 10^6 \cdot 10000^2}{22.30 \cdot 10^{13}} = 6.97$ mm.

$f_2 = \frac{15.56 \cdot 10^6 \cdot 10000^2}{10.45 \cdot 10^{13}} = 14.89$ mm.

Deflection is, therefore:

$f* = 6,97 \cdot (1 - 0,86) + 14,89 \cdot 0,86 = 0,98 + 12,80 = 13,78$ mm.

which is a value equal to $\frac{\ell}{725}$ span, smaller than $\frac{\ell}{250}$.

Reference

(CEB 127-1995) CEB Bulletin No 127 “Analysis of beams and frames”. August 1995.

Chapter 5
Analysis of Second Order Effects with Axial Load

Abstract The aim is to estimate how much the strength of compressed elements is affected by the deformation (deflection) induced by various causes. Since the calculations according to the general refined methods are very complex and time consuming, in the chapter the approximate methods are reported and compared with more general approaches, introducing generally admissible simplifying hypotheses.

5.1 Definitions

"First order effects" are the internal forces N, M, V, T of a structure in the undeformed configuration. "Second order effects" are the further internal forces arising as a consequence of the structural deformation and which are corresponding to the amplification of the first order effects. This phenomenon, mainly significant in slender structures under axial loads, can cause two different situations:

(a) deformed configuration in equilibrium with applied loads and subsequent failure of critical cross-sections at ULS;
(b) indefinite increase of displacements and following buckling.

The subject of the present Chapter is the research of the possible equilibrate solutions at ULS in the deformed configuration.

5.2 Geometric Imperfections

A special case of internal forces due to displacements comes from geometric and constructive imperfections of the structure. To this purpose, we recall the following points of EC2:

- [5.2(1)P] "The unfavourable effects of possible deviations in the geometry of the structure and the position of loads shall be taken into account in the analysis of

This chapter was authored by Piero Marro and Matteo Guiglia.

© The Author(s), under exclusive license to Springer Nature Switzerland AG 2022
F. Angotti et al., *Reinforced Concrete with Worked Examples*,
https://doi.org/10.1007/978-3-030-92839-1_5

members and structures". For these last ones, see the minimum eccentricity given at [6.1(4)].

- [5.2(2)P] "Imperfections shall be taken into account in ultimate limit states in persistent and accidental design situations"
- [5.2(3)] "Imperfections need not be considered for serviceability limit states".

Based on three previous points, accounting for imperfections appear truly necessary for verifying structures in presence of second order effects.

For the evaluation of imperfections, EC2 refers to those related to Execution Class 1 according to EN 13,670. In that case, imperfections may be represented by an inclination θ_i given by

$$\theta_i = \theta_0 \cdot \alpha_h \cdot \alpha_m \quad [(5.1)]$$

where

θ_0 is the basic inclination value, whose recommended value is 1/200 = 0.005;

$\alpha_h = 2/\sqrt{\ell}$ (with ℓ in meters) is the reduction factor for length or height of the column if isolated members are considered or the height of the building if the column elements are part of a building. In both cases, limit takes the value $(2/3) \leq \alpha_h \leq 1$;

α_m when m vertical members are contributing to the total effect, inclinations θ_i of the single members can be different one from the other in value and sign, therefore we introduce this reduction coefficient α_m which derives from a synthesis of statistical data: $\alpha_m = \sqrt{0,5 \cdot (1 + \frac{1}{m})}$.

In [5.2(7)], the effect of imperfection is evaluated for a single column in two different ways:

(a) with the eccentricity e_i expressed as

$$e_i = \theta_i \cdot (\ell_0/2) \quad [(5.2)]$$

being ℓ_0 the effective length.
For columns in braced systems, it can be assumed

$$e_i = \ell_0/400$$

being implicitly $\alpha_h = 1$.

(b) by applying a horizontal force H in the point causing the maximum bending moment.

For an isolated column rigidly fixed at the base and free at the end-section (Figure [5.1-a1]),

$$H_i = \theta_i \cdot N \quad [(5.3a)]$$

being N the axial load acting on the column.

For a vertical column constrained at the two end-sections (Fig. [5.1-a2]),

$$H_i = 2\theta_i \cdot N \quad [(5.3b)]$$

5.3 Isolated Members and Bracing Systems

Second order effects can arise both in isolated members and braced systems; for braced systems, it is worth noting that significant global second order effects can arise also if single members are not characterized, individually, by high slenderness.

Therefore it is not correct to isolate the single elements from a braced system, evaluating individually their sensitivity to second order effects without having previously evaluated the global effects on the overall system.

Moreover in [5.8.2(1)P], it is noticed the possibility that second order effects can arise also in structures provided with bracing elements when they are very flexible and then low-efficient or when, even if bracing elements are not particularly slender, the joined members are loaded by very high axial loads.

Accounting for second order effects is time-consuming, therefore it is suitable to distinguish cases when it is necessary or not. To this purpose, EC2 gives the following general criterion in [5.8.2(6)]: second order effects may be ignored if they are less than 10% of the corresponding first order effects.

Even if applicable, such criterion is effectively inconsistent because it is not possible to define if the condition is fulfilled without having previously performed a complete structural analysis accounting for geometric non-linearity. For this reason, simplified criteria are given for isolated members in 5.8.3.1 and structures in 5.8.3.3.

5.4 Simplified Criteria for Second Order Effects

5.4.1 Slenderness Criterion for Isolated Members

Second order effects for isolated members may be ignored if the slenderness λfulfills the following requirement

$$\lambda < \lambda_{\text{lim}}$$

where λ is the slenderness of the considered element and is evaluated according to the criteria defined in [5.8.3.2], while λ_{lim} is calculated according to the formula

[(5.13N)]

$$\lambda_{\lim} = 20 \cdot A \cdot B \cdot C/\sqrt{n}$$

The value of $\lambda_{\lim}$ to compare with the effective slenderness of the member is calibrated to assure that the total moment is not exceeding 10% of the first order moment. It depends on the maximum acting vertical load (n = non-dimensional normal force) and on three parameters which are functions, respectively, of the creep coefficient (A), of the mechanical ratio of longitudinal reinforcement (B), and the first order bending moment distribution along with the member (C). However, it is possible to define the value of the slenderness limit by taking three constant and prudential values for the three coefficients A, B, C appearing in the previous relationship without a specific definition of them.

For the calculation of the element slenderness λ, it is necessary to evaluate the effective length ℓ_0 taking into account the eventual effective partial restraint due to the members connecting at the end-section of the considered element.

For columns with fixed nodes (braced members), ℓ_0 is between 0.5 and 1 times the geometric length; for columns with mobile nodes (unbraced members), ℓ_0 is, in general, greater than the geometric length. Figure [5.7] defines the value of ℓ_0 for braced members with a combination of simple restraints.

For the calculation of ℓ_0 in presence of more complex restraints, [5.8.3.2(3)] gives two formulas

$$\ell_0 = 0,5\ell \cdot \sqrt{\left(1 + \frac{k_1}{0,45 + k_1}\right) \cdot \left(1 + \frac{k_2}{0,45 + k_2}\right)} \qquad [(5.15)]$$

$$\ell_0 = \ell_{\max}\left\{\sqrt{1 + 10 \cdot \frac{k_1 \cdot k_2}{k_1 + k_2}};\ \left(1 + \frac{k_1}{1 + k_1}\right) \cdot \left(1 + \frac{k_2}{1 + k_2}\right)\right\} \qquad [(5.16)]$$

where [(5.15)] is for columns with fixed nodes and [(5.16)] is for columns with mobile nodes; such formulas have been obtained using refined numerical calculations (Westerberg, 2004).

In the previous expressions, the parameters k_1 and k_2 are the relative flexibilities of rotational restraints at ends 1 and 2 respectively. In general, flexibility k is defined as

$$k = \frac{\theta}{M} \cdot \frac{(EI)_c}{L_c}$$

where θ è is the rotation of restraining members for bending moment M.

Limit values for k are given by: $k = 0$ for the perfect rigid rotational restraint and $k = \infty$ for the support (null rotational restraint).

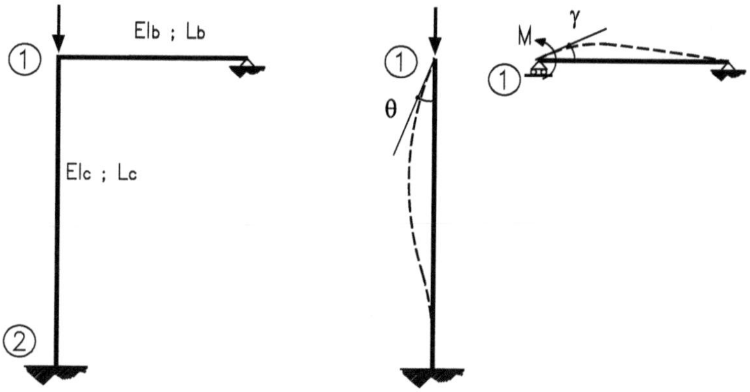

Fig. 5.1 Evaluation of slenderness k

The method to obtain the values of the required flexibility k for the nodes at the end-sections of the member consists in writing down separately the expressions of the node rotation for bending moment M for the column and the restraining beams members and in making them equal.

With reference to Fig. 5.1, the value k_1 for the end-section column node is obtained: the rotation $\theta$of the column end-section for bending moment M is given by

$$\theta = M \cdot k_1 \cdot \frac{L_c}{EI_c}$$

and the rotation γ of the beam supported by the end-section is given by

$$\gamma = \frac{M \cdot L_b}{3EI_b}$$

By making equal the two expressions, the following relationship is obtained

$$k_1 = \frac{EI_c/L_c}{3EI_b/L_b}$$

Figure 5.2 shows the flexibilities for nodes with different types of restraints which allow solving most of the current cases; the node studied in the Figure is the end-section node, but the found flexibility is not depending on the restraint at the base of the column, therefore the result has a generic value.

The schemes of Fig. 5.2 are referred to fixed-node columns. In reality, the flexibility value of the node can be right, in some cases, also for unbraced systems: such situation is for example the one represented in scheme B, if the beam support on the right is substituted by a vertical restraint, obtaining a scheme where the node flexibility is not changing respect to the one indicated. In that case, however, the k_1 value will be used for the evaluation of ℓ_0 with [5.16] and not with [5.15].

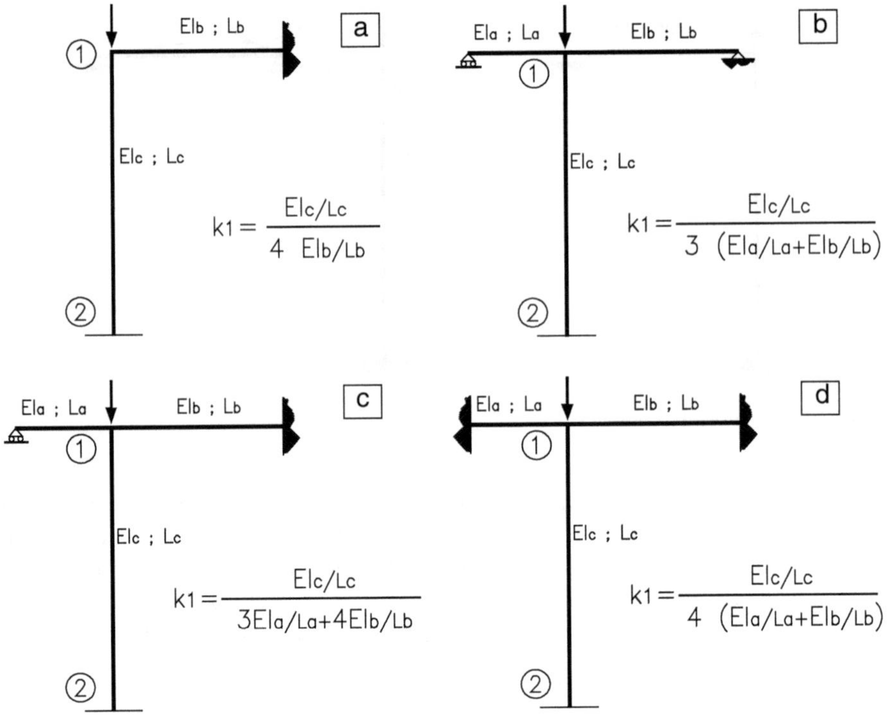

Fig. 5.2 Evaluation of flexibility k for some restraint schemes

If the examined column is connected with a column placed above or below, the evaluation of the node flexibility shall take it into account [5.8.3.2(4) as an application of the rule (3)P expressed in 5.8.2]. It is then necessary to define the behaviour of the structure that can be schematized under two assumptions:

(a) the upper or lower column reaches the instability load together with the considered column (under the same multiplier of vertical loads): in such case, both columns have the same rotational restraint due to the beams convergent in the node. The relevant flexibility of this latter is then obtained according to the criteria previously exposed and following Fig. 5.2, by substituting the quantity at numerator (EI_c/L_c), relative to the column, with the quantity ($EI_{\text{adjacent}}/L_{\text{adjacent}} + EI_c/L_c$), referred to the column and the beam converging in the node. According to this method, the flexibility evaluated is greater, and therefore ℓ_0 increases.

(b) the upper or lower column reaches instability under a vertical load greater than the one which produces the critical load of the considered column: in such case, the upper or lower column can, in general, collaborate with the beans to restrain the node; such contribution can, however, be neglected.

Figure 5.2 shows the schemes that can be adopted to evaluate the flexibility of a node (the node 1 of the Figure, at the head of the column), characterized by convergent beams with different lengths and restraints. It is worth noting that, for the calculation of the flexibility of the upper node, the node at the other section does not give any contribution (node 2 of the Figure, at the column base), and it is not indicated.

For the definition of the effective length of members ℓ_0, it can be suitable to account for the rule of [5.8.2(3)P] which includes in the analysis the effect of adjacent members and foundations. For example, for columns joined at the foundations, the adoption of a perfect rotational restrain at the base ($k = 0$) can be not prudential, therefore it is suggested to assume $k = 0.1$ [see 5.8.3.2(3)].

5.4.1.1 Calculation of Flexibility and Effective Length ℓ_0: Examples

With reference to Fig. 5.2, we calculate the flexibility k_1 of the upper node.

Scheme A

If $EI_c = EI_b$; $L_c = L_b$, we get $k_1 = 0.25$.

If $2EI_c = EI_b$; $L_c = L_b$, we get $k_1 = 0.125$.

Scheme B with null rotational restraint at the end-section of the beam b and B' with simple support (only vertical restraint)

If $EI_a = EI_b = EI_c$; $L_a = 3$ m, $L_b = 6$ m, $L_c = 4$ m

$$\mathrm{k}_1 = \frac{1/4}{3 \cdot \left(\frac{1}{3} + \frac{1}{6}\right)} = 0.1667$$

If $EI_a = EI_b = 2EI_c$; $L_a = 3$ m, $L_b = 6$ m, $L_c = 4$ m

$$k_1 = \frac{1/4}{3 \cdot \left(\frac{2}{3} + \frac{2}{6}\right)} = 0.0833$$

Scheme C

If $EI_a = EI_b = EI_c$; $L_a = 3$ m, $L_b = 6$ m, $L_c = 4$ m

$$k_1 = \frac{1/4}{\frac{3}{3} + \frac{4}{6}} = 0.150$$

If $EI_a = EI_b = 2EI_c$; $L_a = 3$ m, $L_b = 6$ m, $L_c = 4$ m

$$k_1 = \frac{1/4}{3 \cdot \frac{2}{3} + \frac{4 \cdot 2}{6}} = 0.075$$

Scheme D

If $EI_a = EI_b = EI_c$; $L_a = 3$ m, $L_b = 6$ m, $L_c = 4$ m

$$k_1 = \frac{1/4}{4 \cdot \left(\frac{1}{3} + = \frac{1}{6}\right)} = 0.125$$

If $EI_a = EI_b = 2EI_c$; $L_a = 3$ m, $L_b = 6$ m, $L_c = 4$ m

$$k_1 \frac{1/4}{4 \cdot 2 \cdot \left(\frac{1}{3} + \frac{1}{6}\right)} = 0.0625$$

The values of the effective length ℓ_0 for the columns of the schemes A, B, C and D of Fig. 5.2, and B' can be obtained by replacing the null rotational restraint of beam *b* with a simple support (with only vertical restraint).

The effective length ℓ_0 is calculated through [(5.15)] for the schemes A, B, C and D, which are representative of structures with fixed nodes, and using [(5.16)] for the scheme B' because the node can move horizontally. The obtained results, under the two hypotheses that the base restraint is rigid ($k_2 = 0$) or slightly "soft" ($k_2 = 0.10$), are collected in Table 5.1.

Detail of the calculation for two examples

Scheme A

$k_1 = 0.25$; $k_2 = 0.10$.

By means of [(5.15)] we get:

$$\ell_0 = 0.5\ell \sqrt{\left(1 + \frac{0.25}{0.45 + 0.25}\right) \cdot \left(1 + \frac{0.10}{0.45 + 0.10}\right)} = 0.633\ell$$

Scheme B'

$k_1 = 0.1667$; $k_2 = 0.10$.

Using the first of the two formulas [(5.16)], we get:

$$\ell_0 = \ell \sqrt{1 + 10 \cdot \frac{0.1667 \cdot 0.10}{0.1667 + 0.10}} = 1.275\ell$$

(using the second formula, we would get $\ell_0 = 1, 5.247\ell$)

Remarks

The effective length is the quantity employed in the following discussion. Such a quantity is defined for single columns according to Figure [5.7] and according to

Table 5.1 Effective length ℓ_0

Scheme	k_1	k_2	ℓ_0
A	0.25	0.00	0.583ℓ
	0.25	0.10	0.633ℓ
	0.125	0.00	$0,552\ell$
	0,125	0.10	0.600ℓ
B	0.1667	0.00	0.563ℓ
with null rotational restraint at beam b (fixed nodes)	0.1667	0.10	0.613ℓ
	0.0833	0.00	0.538ℓ
	0.0833	0.10	0.584ℓ
B'	0.1667	0.00	1.143ℓ
W with simple vertical restraint at beam b	0.1667	0.10	1.275ℓ
(non-fixed nodes)	0.0833	0.00	1.077ℓ
	0.0833	0.10	1.175ℓ
C	0.15	0.00	0.560ℓ
	0.15	0.10	0.600ℓ
	0.075	0.00	0.534ℓ
	0.075	0.10	0.581ℓ
D	0.125	0.00	0.552ℓ
	0.125	0.10	0.600ℓ
	0.0625	0.00	0.530ℓ
	0.0625	0.10	0.575ℓ
	0.25	0.00	0.583ℓ

the values collected in Table 5.1. Those values have been obtained following the calculations developed using [(5.15)] for the schemes A, B, C and D for structures with fixed nodes (for those it results $\ell_0 \leq \ell$), and using [(5.16)] for the scheme B' corresponding to structure with non-fixed nodes (for those it results $\ell_0 \geq \ell$).

It is worth noting that the first of the [(5.16)] with $k_2 = 0$ and $k_1 > 0$ but not infinite, gives $\ell_0 = \ell$; the second expression gives values between ℓ and 2ℓ, therefore good values. If k_2 is different from zero, also the first expression is valid and its results are comparable with the results of the first one.

The formulas of flexibility given by schemes A, B, C and D of Fig. 5.2 contain the inertial moments of the various members. Point 5.8.3.2(5) of EC2 says that, in the definition of the effective length, the stiffness of members connecting the column shall include the cracking effects, unless it is possible to demonstrate that such members are not cracked at the ULS. Usually, this fact cannot be demonstrated, and therefore the problem of quantifying the cracking effect arises. The solution to such a problem is not easy because of two aspects:

(a) cracking is a common phenomenon but not easily treatable in a reliable way;
(b) cracking is dependent on several quantities, such as load value and load distribution.

Preliminarily, it does not seem consistent to introduce the hypothesis of all cracked cross-sections. Because of the lack of suggestions provided by EC2, we propose to assume the inertial moment of the beam equal to 2/3 of the inertial moment of the complete concrete cross-section, to represent an intermediate situation. This procedure is described in the volume "Applications de l'Eurocode 2", chapter "Poteaux, instabilité", 82–86 (see the Bibliography).

Finally, we consider that [(5.15)] and [(5.16)] have no physical meaning, as stated by Bo Westerberg in ("Second order effects in slender concrete structures", 2004), because they are obtained through mathematical models built to show the results of complicated numerical calculations. Therefore, its application is often demanding and dominated by cracking uncertainties. In the present paragraph, therefore, we introduce the alternative tabular method contained in BS 8110-85, for the definition of the effective length.

Such a method can be described as follows.

The effective length ℓ_0 for a column with length ℓ in an assigned plane is given by $\ell_0 = \beta\ell$. Values β for branched and unbranched columns are collected in the two following Tables, as a function of the 4 restraint conditions at the end-section.

- Condition 1—The end-section of the beam is rigidly connected on both sides of the beam which have height at least equal to the dimension of the column in the considered plane. If the column is connected with a foundation, this should have suitable shape and dimensions.
- Condition 2—The end-section of the column is rigidly connected on both sides to the beams which have a height smaller than the column dimension on the considered plane.
- Condition 3—The end-section of the column is connected to members which are not designed to provide a rotational restraint to the column, but can provide it in some way.
- Condition 4—The end-section of the column is free to rotate and translate (for example, the free end-section of a cantilever column of an unbranched structure).

Top section condition	Base section condition		
	1	2	3
β for branched columns			
1	0.75	0.80	0.90
2	0.80	0.85	0.95
3	0.90	0.95	1.00
β for unbranched columns			
	1	2	3

(continued)

(continued)

Top section condition	Base section condition		
	1	2	3
1	1.2	1.3	1.6
2	1.3	1.5	1.8
3	1.6	1.8	–
4	2.2	–	–

5.4.1.2 Example of Calculation of Slenderness for a Column of a Framework

Here we calculate the slenderness of the column of the framework of Fig. 5.3, in the plane of the framework itself.

The column has a height equal to 3 m and a square cross-section of 35 × 35 cm. All the beams have rectangular cross-sections 60 × 25 cm and span equal to 4.5 m (beams on the left) and 5 m (beams on the right). Lower beans are rigidly connected to the two bracings shown in the Figure.

It is hypothesized that the global structure is not sensible to second order effects and that beams and columns are made of the same concrete.

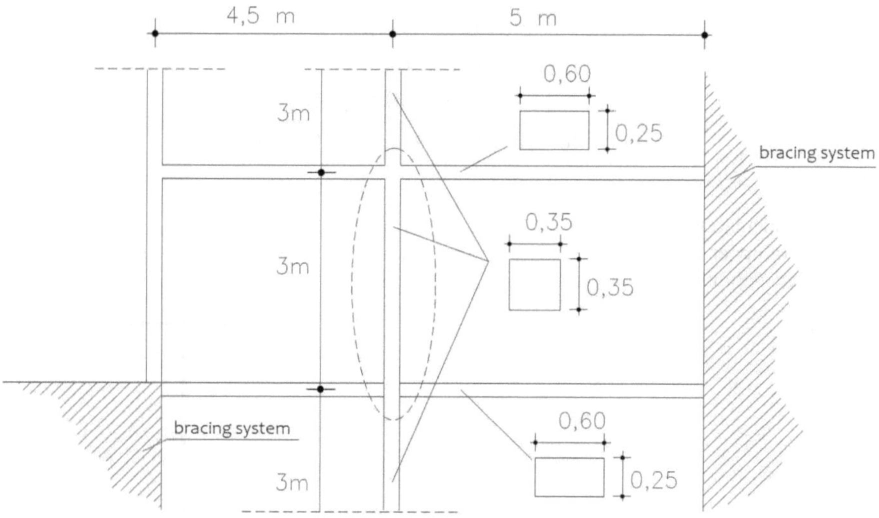

Fig. 5.3 Slenderness calculation: structural geometry

EC2 procedure

We calculate preliminarily the inertial moment of the column cross-section

$$I_c = 350^4/12 = 12.50 \times 10^8 \text{mm}^4$$

For beams, we assume that the inertial moment of the concrete cross-section is reduced to 2/3, as proposed in the already mentioned volume "Applications de l'Eurocode 2"

$$I_{\text{beam}} = \frac{2}{3} \cdot \frac{600 \cdot 250^3}{12} = 5.20 \times 10^8 \text{mm}^4$$

The restraint scheme for the calculation of flexibility of the top node (node 1) and the base node (node 2) of the column is shown in Fig. 5.4. The calculation is performed considering scheme C and scheme D of Fig. 5.2, respectively.

Here we assume that the column above node 1 and below node 2 reaches its instability load together with the considered column: therefore, nodes 1 and 2 shall be a restraint both for the considered column and the adjacent one.

$$k_1 = \frac{(EI_c/L_c) + (EI_{colA}/L_{colA})}{3EI_a/L_a + 4EI_b/L_b} = \frac{\left(12.50 \cdot 10^9/3000\right) \cdot 2}{3 \cdot 5.20 \cdot 10^8/4500 + 4 \cdot 5.20 \cdot 10^8/5000} = 1.09$$

$$k_2 = \frac{(EI_c/L_c) + (EI_{colB}/L_{colB})}{4EI_a/L_a + 4EI_b/L_b} = \frac{\left(12.50 \cdot 10^9/3000\right) \cdot 2}{4 \cdot (5.20 \cdot 10^8/4500 + 5.20 \cdot 10^8/5000)} = 0.95$$

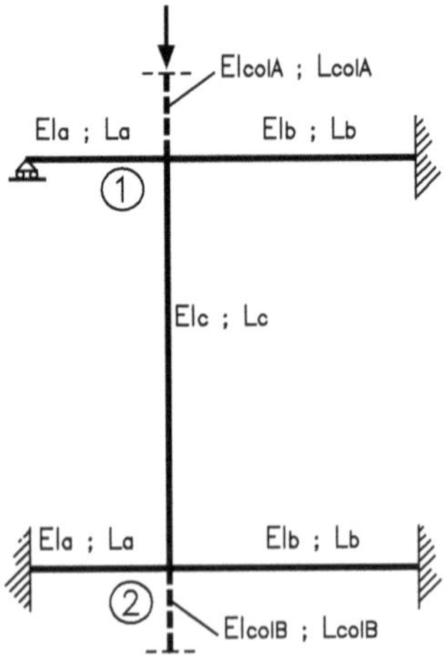

Fig. 5.4 Calculation of flexibility of upper and lower nodes of the column

The calculation of the effective length is performed by [(5.15)]. We get:

$$l_0 = 0.5L \cdot \sqrt{\left(1 + \frac{k_1}{0.45 + k_1}\right) \cdot \left(1 + \frac{k_2}{0.45 + k_2}\right)}$$

$$= 0.5 \cdot 3 \cdot \sqrt{\left(1 + \frac{1,09}{0.45 + 1.09}\right) \cdot \left(1 + \frac{0.95}{0.45 + 0.95}\right)} = 2.54 \text{ m}$$

Being $i = \frac{h}{\sqrt{12}} = \frac{0.35}{3.46} = 0.10$ m, slenderness is given by, $\lambda = \frac{\ell_0}{i} = \frac{2.54}{0.10} = 25.4$.

Calculation Using Tables BS (see 5.4.1.1.1)

The beams have a height of 250 mm, smaller than the side of the column. Therefore, condition 2 is verified both for the top section and the base section. For this one, the first table gives $\beta = 0.85$ and we get $\ell_0 = 0.85 \cdot 3.0 = 2.55$m, as already obtained in the previous calculation.

5.4.2 Global Second-Order Effects in Buildings

It is not possible to analyse second order effects for a complex structure through the analysis of the second order effects of the isolated members of the structure, separated from their context. Therefore, it is always necessary to study preliminarily the global behaviour of the structure, evaluating if it is sensible or not to the global second order effects. If the structure is not sensible, first order internal forces, comprehensive of the geometric imperfection effects, are the starting point for the detailed analysis of each structural member which can be isolated from the rest of the structure and analysed using the criteria typical of single members. If second order effects are significant for the structure, internal forces shall be evaluated by taking into account geometric nonlinearity.

Geometric nonlinearity can be accounted for through the application of a general method, based on the nonlinear geometric and mechanical analysis of the entire structure (on a plane or within the space). Such an approach gives reliable results but unless for very simple structural schemes, it is burdensome from the computational point of view.

A simplified method, developed using linear analysis, is proposed in [5.8.7.3]: with this procedure, the amplification corresponding to the second order effects is searched for through amplification coefficients of first order bending moments.

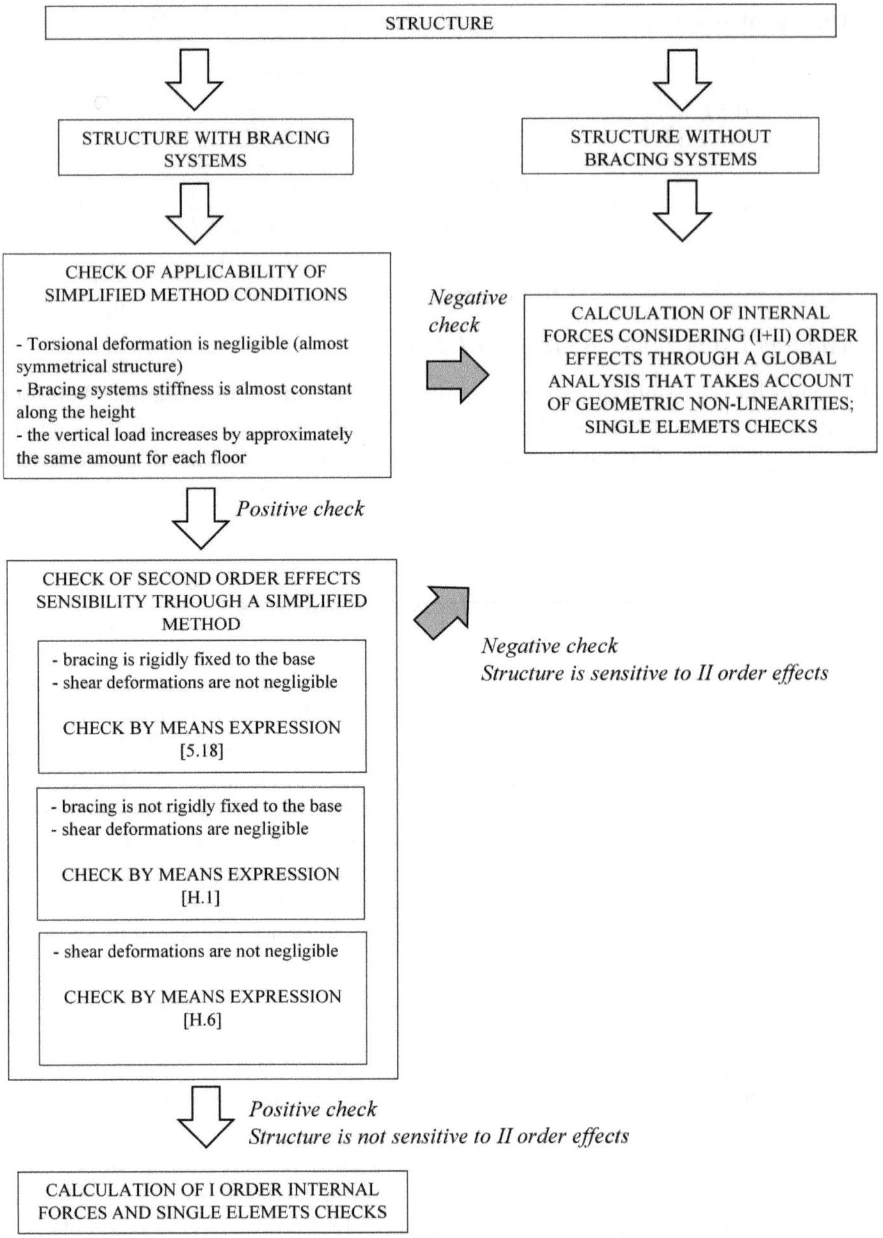

The procedure to define if second order effects are negligible or not requires distinguishing two classes of structures: (a) unbraced structures; (b) braced structures. As for isolated members, also in this case the procedure is calibrated to guarantee that first order internal forces are not exceeded more than 10%.

(a) Unbraced structures: EC2 does not provide any criterion for the simplified analysis to distinguish sensible or not sensible members to second order effects. Therefore, it is necessary to perform a structural analysis taking into account geometric nonlinearity.
(b) Braced structures: it is possible to apply simplified criteria if the following conditions are fulfilled [5.8.8.3(1)]:

- torsional instability is not governing (i.e. structure is reasonably symmetrical or, more generally, the eccentricity between the application axis of the resultant horizontal force is small to the stiffness centroid of the bracing system);
- bracing members stiffness is almost constant along with the height;
- vertical load is increasing approximately the same amount for each store.

If these conditions are fulfilled, for bracing members base-fixed and characterized by negligible shear deformations (without large openings), the relevant expression is [5.18], shown in 5.4.2.1.

If the previous conditions are fulfilled, for bracing members not characterized by significant shear deformations but with base rotation not negligible, the expression to be used is [H.1]. If the bracing members are characterized by significant shear deformations, the right expression is indicated in [H.6].

The described above procedures are synthesized in the scheme of the previous page.

5.4.2.1 Example: Identification of the Importance of the Second Order Effects in a Building

Here we examine the possibility of neglecting the global effects of the second order effects for the office building of Fig. 5.5, provided with bracing walls and core.

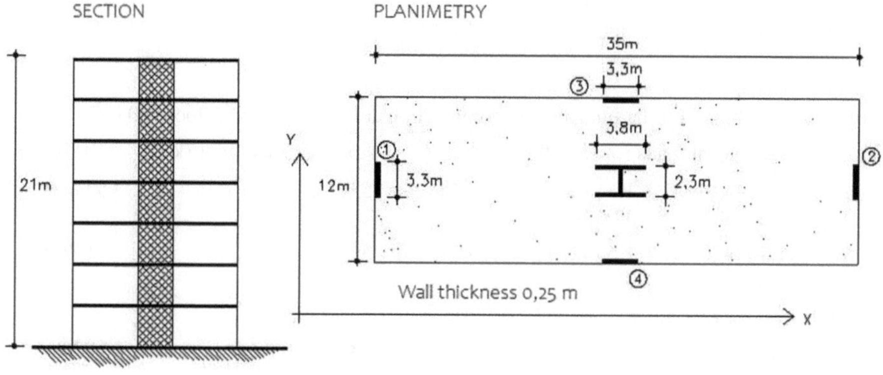

Fig. 5.5 Structural scheme of the building

The building is made of concrete C30/37, with $E_{cm} = 33{,}000\ \text{N/mm}^2$.
Following loads are assumed:

Load	Characteristic value
Permanent loads[a]	
Intermediate slabs and roof	4.0 kN/m^2
Pavements, partition walls, etc	2.0 kN/m^2
External façades (for each storey)	10.0 kN/m
Elevator tower (for each storey)	141.0 kN
Wall 1 (for each storey)	37.4 kN
Wall 2 (for each storey)	37.4 kN
Wall 3 (for each storey)	37.4 kN
Wall 4 (for each storey)	37.4 kN
Elevator	35.0 kN
Variable loads	
Crowd	4.0 kN/m^2
Snow	2.0 kN/m^2

[a]Self-weights and permanent loads

Based on the assigned geometry and the assumed loads, the total vertical load at ULS is calculated considering the fundamental combination with predominant variable load due to the crowd: such a combination, in the analysed case, induces the maximum vertical load (Table 5.2).

The structure has the following properties, described in [5.8.3.3(1)]:

- torsional instability is not governing, because the structure is symmetrical;
- the stiffness of bracing members is reasonably constant along with the height;
- the total vertical load increases by approximately the same amount per storey.

It is therefore possible to apply a simplified criterion to verify the structure sensitivity to the global second order effects.

Table 5.2 Total load for the fundamental combination at ULS (area for 1 floor 420 m^2)

Floor	Permanent loads for each floor	Load for each floor (kN)	Σ(kN)
7	$1.30\,[420 \cdot 4] + 1.5\,[420 \cdot 2 + 10 \cdot 2\,(12 + 35) + 141 + 4 \cdot 37.4 + 35]$	5342.0	5342.0
1–6	$1.30\,[420 \cdot 4] + 1.5\,[420 \cdot 2 + 10 \cdot 2\,(12 + 35) + 141 + 4 \cdot 37.4]$	5290.0	31,740.0
Variable load for each floor			
7	$1.5 \cdot 0.7\,[420 \cdot 2]$	882.0	882.0
1–6	$1.5\,[420 \cdot 4]$	2520.0	15,120.0
		$F_{V,Ed}$	53,084.0

Moreover, the bracing system is made of shear and core walls without openings: therefore, it is possible to neglect the shear deformation.

Under the hypothesis that the bracing system is rigidly fixed at the base, i.e. rotations are negligible, the simplified criterion to adopt is given by [5.18]

$$F_{V,Ed} \leq k_1 \cdot \frac{n_s}{n_s + 1.6} \frac{\Sigma E_{cd} I_c}{L^2} \quad [5.18]$$

where

$F_{V,ED}$	53,084 kN
n_s	7, storey number
E_{cd}	$E_{cm}/\gamma_{CE} = 33{,}000/1.2 = 27{,}500\ \text{N/mm}^2 = 27.5 \times 10^6\ \text{kN/m}^2$
I_c	is the inertial moment
L	21 m (building height).

EN 1992 recommends adopting the value 0.31 for k_1. Moreover, [5.8.3.3(2)] allows substituting k_1 with $k_2 = 0.62$ if it is possible to demonstrate that bracing structures are uncracked at ULS. Because this fact is uncertain, here we adopt the value $k_1 = 0.31$.

If we consider instability in Y-direction (Fig. 5.5), the more strict for the adopted geometry, inertial moments of reacting bracing members (lift tower and shear walls 1 and 2) are:

- Lift tower: $I_c = 2.128\ \text{m}^4$
- Shear walls 1–2: $I_c = 0.749\ \text{m}^4$.

By applying [5.18], we get:

$$0.31 \cdot \left(\frac{7}{7+1.6}\right) \cdot \frac{27,500,000 \cdot (2.128 + 2 \cdot 0.749)}{(21)^2} = 57,054\,\text{kN}$$

This value is greater than the maximum design load previously evaluated, equal to 53,084 kN: therefore, it is not necessary to account for global effects of the second order in the evaluation of internal forces for single structural members.

5.5 Methods of Analysis

EC2 provides three methods of analysis for the second order effects:

(a) a general method;
(b) a method based on nominal stiffness;
(c) a method based on nominal curvature.

The general method is the most complete one, is valid in every case but it is not easy to apply.

The method based on nominal stiffness is a simplified method, which may be used for both isolated members and whole structures.

The method based on nominal curvature is specific for the analysis of isolated members.

Methods (b) and (c) are numerical methods adjusted on the results obtained through nonlinear analyses. Therefore, they may lead also to very different results and a priori it is not possible to identify the most prudential one.

The method based on nominal stiffness is based on the hypothesis of linear elastic behaviour of the structure, through the introduction of a nominal flexural stiffness EI. To be representative of the structural behaviour, flexural stiffness has to be a function of several quantities, i.e. cross-section dimensions, assigned reinforcement, concrete strength and concrete strain (elasticity modulus, creep), axial load, slenderness according to [5.8.7.2(1), (2)] and cracking. The criteria for the evaluation of EI have been selected according to the results of nonlinear analyses.

Under this assumption, the procedures to evaluate total internal forces of first and second order are based on the classical rules of structural mechanics. In [5.8.7.3], EC2 proposes explicitly the calculation of the total moment as the product between the first order bending moment and an amplification factor. This factor is a function of the critical Eulerian load N_B, which is itself a function of EI, and which can be different according to the shape of the first order bending moment M diagram: constant, linear or parabolic, etc.

Other procedures not mentioned by EC2 can be applied: integration of the elastic differential equation, calculation of first and second order displacements of the axial force direction. These methods will be discussed in the following, with corresponding applications.

The method based on nominal stiffness is a verification method, therefore several subsequent attempts are necessary to optimize a design and refine the solution. For example, if the solution is unsatisfactory and reinforcement is increased, EI and N_B increase, while the amplification factor decreases. The opposite occurs if the reinforcement is too much, and its reduction is adopted.

About this method, in [5.8.7.2(1) and (3)] EC2 provides a simplified procedure for the evaluation of the nominal stiffness, not dependent on the reinforcement amount and therefore which can be considered as a design procedure. However, even under the assumption that the reinforcement geometric ratio is not smaller than 0.01, EC2 advises that the method can only work as a first design step, and it has to be later developed using criteria of [5.8.7.2(2)].

The method based on nominal curvature allows calculating the total moment as the sum of first and second order moments $M_{\text{tot}} = N_{Ed}\,(e_1 + e_2)$. If the axial force is not too big ($N_{Ed} < N_{\text{bal}}$), e_2 is depending only on $\varepsilon_{syd} = f_{yd}/E_s$ and the effective depth d. The total eccentricity $(e_1 + e_2)$ is therefore invariant to N_{Ed}. The total moment is proportional to N_{Ed}. Once M and N are known, it is possible to calculate the necessary reinforcement. Then, in this context ($N_{Ed} < N_{\text{bal}}$), the method based on nominal curvature can be considered as a design method. Otherwise, if the reinforcement is assigned, whatever is the value of N_{Ed}, the method allows to calculate the second order moment and then the total internal force. In such a case, it is necessary to verify if the internal force is suitable with the cross-section strength.

5.5.1 *The General Method*

The general method is based on structural nonlinear analysis, including geometric and mechanical nonlinearities.

Such a method is therefore a special case of nonlinear analysis of [5.7].

It is necessary to adopt stress–strain relationships for steel and concrete based on the design values (and not on the mean values) to obtain the design value of the ultimate load at the end of the analysis.

For concrete, EC2 uses the expression given by [(3.14)] modified as follows:

$$\frac{\sigma_c}{f_{cd}} = \frac{k\eta - \eta^2}{1 + (k-2)\eta}$$

where

$\eta = \varepsilon_c / \varepsilon_{c1}$

$$k = \frac{1.05 \cdot E_{cd} \cdot |\varepsilon_{c1}|}{f_{cd}} = \frac{1.05 \cdot \frac{E_{cm}}{1.2} \cdot |\varepsilon_{c1}|}{f_{cd}}$$

The creep effect can be considered by multiplying the values of strain—taken from the mentioned above law σ-ε—by $(1 + \varphi_{ef})$.

The application of the general method can be performed by subdividing the structure into segments and by evaluating preliminarily for each of them the moment—mean curvature law under the effective normal force acting. It is therefore necessary to know a priori not only the geometry but also reinforcement distribution.

For each calculation step, corresponding to a given configuration of first order actions on the structure, an iterative procedure is developed to satisfy equilibrium and compatibility conditions and to obtain the deformation of the structure and the distribution of the total internal forces (first and second order).

5.5.2 *The Method Based on Nominal Stiffness*

5.5.2.1 The Description of the Method

The method based on nominal stiffness is a structural analysis taking into account geometric nonlinearity (i.e. second-order effects), developed under the conventional assumption of mechanical linear behaviour.

Nominal stiffness can be defined through the following expression

$$EI = K_c E_{cd} I_c + K_s E_s I_s \quad [5.21]$$

where

E_{cd} is the reduced value of modulus of elasticity of concrete taken equal to $E_{cd} = \frac{E_{cm}}{\gamma_{CE}}$ with $\gamma_{CE} = 1.2$ (see [5.8.6(3)],

I_c and I_s are respectively the moments of inertia of concrete cross-section and reinforcement,

K_c and K_s are coefficients taking into account, respectively, cracking and creep and reinforcement contribution.

For K_s and K_c, in 5.8.7.2 EC2 provides two calculation procedures.

First procedure

If $\rho = \frac{A_s}{A_c} \geq 0,002$ being A_s the total reinforcement amount of the cross-section, it is assumed.

$K_s = 1$

$$K_c = k_1 \cdot k_2/(1 + \varphi_{ef})$$

being

$$k_1 = \sqrt{\frac{f_{ck}}{20}} \text{ with } f_{ck} \text{ expressed in N/mm}^2 \quad [(5.23)]$$

$$k_2 = n \cdot \frac{\lambda}{170} \leq 0.20 \left(n = \frac{N_{Ed}}{A_c f_{cd}} \right) \quad [(5.24)]$$

$$\varphi_{ef} = \varphi(\infty, t_0) \cdot \frac{M_{0Eqp}}{M_{0Ed}} \quad [(5.19)]$$

where M_{0Eqp} is the first order bending moment in quasi-permanent serviceability conditions and M_{0Ed} is the first order design bending moment at ULS; φ_{ef} can be taken equal to zero if all the following conditions are fulfilled.

$$\varphi(\infty, t_0) < 2$$

$$\lambda \leq 75$$

$$M_{0Ed}/N_{Ed} \geq h$$

This procedure is always applicable (the rule $\rho \geq 0.002$ is not an obstacle, being the geometric ratio, in presence of second order effects, always greater than 0,002 for the prescription of [9.5.2(2)]). It requires the knowledge of reinforcement.

Second procedure

If $\rho = \frac{A_s}{A_c} \geq 0.01$, it is assumed.

$$K_s = 0$$

$$K_c = 0.3/(1 + 0.5 \cdot \varphi_{ef})$$

Theoretically, this second procedure allows the evaluation of the total moment useful to define the amount of the necessary reinforcement. However, being an alternative simplified method, the obtained results are valid only in the first approximation, if the reinforcement, once the calculations are performed, fulfils the condition $\rho \geq 0.01$ and is not too far from this value. If ρ is also slightly smaller than the limit value, the second order effect is underestimated. If ρ is greater than 0.01, the opposite situation occurs. The procedure, as specified in the present discussion, can be employed as a first step, followed by more accurate calculations to be developed using the first procedure. Therefore, in the following examples, it will not be applied.

The structural behaviour is then defined through the nominal stiffness *EI* which implies moment-mean curvature relationships linear and, more in general, relationships between loads and deformations linear themselves.

The second order analysis based on the amplification factor of the moments is a procedure that allows the calculation of the second order effects in closed form, once first order internal forces are given. In this case, total design moment M_{Ed} (I + II order) is given by

$$M_{Ed} = M_{0Ed}\left[1 + \frac{\beta}{(N_B/N_{Ed}) - 1}\right] \quad [(5.28)]$$

where M_{0Ed} is the first order moment, including the effect of geometric imperfections, N_B is the Eulerian critical load based on the nominal stiffness, N_{Ed} is the design normal force; β is a factor taking into account the distribution of the first order bending actions along with the considered member.

5.5.2.2 Deduction of the Formulas

The present paragraph 5.5.2.2 is based on paragraph 7.1 of (Westerberg, 2004).

Considering a column with two supports without rotational rigidity ($\ell = \ell_0$), second order moment can be written as

$$M_2 = N \cdot y$$

being $y = y_0 + y_2$ the total displacement of the generic cross-section due to the first and second order internal forces.

In the middle cross-section (m) it can be written down $y_m = \left(\frac{1}{r}\right)_m \cdot \frac{\ell^2}{c}$ being c a coefficient depending on the moment distribution and being $\left(\frac{1}{r}\right)_m = \frac{M_m}{EI}$ the curvature. In the case of the first order, the displacement in the middle section is $y_{0m} = \frac{M_0\ell^2}{EIc_0}$, and, in the case of the second order, it is $y_{2m} = \frac{M_2\ell^2}{EIc_2}$.
By substituting we get

$$M_2 = N\frac{\ell^2}{EI} \cdot \left(\frac{M_0}{c_0} + \frac{M_2}{c_2}\right)$$

The introduction of c_0 and c_2 allows giving different distributions to moments M_0 and M_2. By solving for M_2 we get

$$M_2 = M_0 \cdot \frac{N\frac{\ell^2}{EIc_0}}{1 - \frac{N\ell^2}{EIc_2}} = M_0 \cdot \frac{c_2/c_0}{[(c_2EI/\ell^2)/N] - 1}$$

A sinusoidal distribution is given to the second order moment, corresponding to the quantity $c_2 = \pi^2$ (see 5.5.2.3). By introducing the expressions

$$\beta = \pi^2/c_0 \text{ and } N_B = \pi^2\frac{EI}{\ell^2}$$

the Eulerian critical load of the column, we get

$$M_2 = M_0\frac{\beta}{(N_B/N) - 1}$$

and the total moment

$$M = M_0 + M_2 = M_0\left[1 + \frac{\beta}{\left(\frac{N_B}{N} - 1\right)}\right]$$

If the first and second order moments have the same distributions (or this can be assumed as a first approximation), then $\beta = 1$ and

$$M = \frac{M_0}{1 - (N/N_B)}$$

5.5.2.3 *Relationship Between the Moment in the Middle Cross-Section and the Displacement in the Same Section. Values of c_0 and c_2*

In the case of columns with two simple supports (with null rotational rigidity), coefficients c_0 and c_2 (here briefly indicated with c), define the relationship between

Table 5.3 Parameters c_0 and β as a function of the distribution of the first order moments

Distribution of bending moment	c_0	β
M_0 uniform	8.00	$\pi^2/8 = 1.23$
M_0 sinusoidal	9.86	1.00
M_0 triangular	12.00	$\pi^2/12 = 0.82$
M_0 parabolic convex	9.6	$\pi^2/9.6 = 1.02$

the bending moment and the relevant displacement in the middle cross-section. The relationship is the following

$$y_m = \frac{M_m}{EI} \cdot \frac{\ell^2}{c}$$

Some examples:
If M = constant (uniform bending moment along the column), we get

$$y_m = \frac{M}{EI} \cdot \frac{\ell^2}{8}, \text{ and then c} = 8$$

If the distribution of M is parabolic with maximum value $M*$ in the middle cross-section

$$y_m = \frac{M*}{EI} \cdot \frac{\ell^2}{(48/5)} \text{and then c} = 48/5 = 9.6$$

If the distribution of M is triangular, symmetrical to the middle cross-section, and with $M*$ in the middle cross-section, we get

$$y_m = \frac{M*}{EI} \cdot \frac{\ell^2}{12} \text{and then c} = 12$$

If the distribution of M is sinusoidal with $M*$ in the middle cross-section, we get

$$y_m = \frac{M*}{EI} \cdot \frac{\ell^2}{\pi^2} \text{and then c} = \pi^2 = 9.86$$

In the Table 5.3, values $c_0 = c$ for fist order moments and corresponding values $\beta = \pi^2/c_0$ are collected.
Values of c_0 above collected are in [5.8.7.3(2)] of EC2.

5.5.2.4 Solving Formulas for the Column Fixed at the Base and Free at the Top

An analogous discussion can be developed by assuming the column as a cantilever fixed at the base and free at the top, with geometric length ℓ and Eulerian critical load $N_B = \pi^2 \frac{EI}{4\ell^2}$.

By linking the transverse displacement of the free end-section with the bending moment at the fixed joint, we get the following relationships:

(a) Column under a constant normal force and bending moment (N_{Ed} with eccentricity e)

$$y = \frac{N_{Ed} \cdot e \cdot \ell^2}{2EI} = \frac{M_0 \cdot \ell^2}{2EI} \text{from which } c_0 = 2$$

(b) Column subjected to a horizontal force H_{Ed} at the top section and normal force N_{Ed}

$$y = \frac{H_{Ed} \cdot \ell^3}{3EI} = \frac{M_0 \cdot \ell^2}{3EI} \text{from which } c_0 = 3$$

(c) Column subjected to an horizontal force q distributed along the height

$$y = \frac{q\ell^4}{8EI} = \frac{M_0 \cdot \ell^2}{4EI} \text{from which } c_0 = 4$$

(d) Sinusoidal distribution M with M^* at the fixed joint

$$y_2 = \frac{4M * \ell^2}{\pi^2 EI} \text{from which } c = \frac{\pi^2}{4}$$

Under the adopted structural scheme, we get: $\beta = \frac{\pi^2}{4c_0}$.
Values of c_0 and β for the examined cases are collected in Table 5.4.

5.5.2.5 Nominal Stiffness and Differential Equation of the Elastic Curve

Once the stiffness EI is established using rules of [5.8.7.2] and the quantity is assumed as constant, internal forces of the column can be obtained through the integration of the differential equation of the elastic curve, by neglecting shear contribution, but taking into account strain with flexural terms.

Under the assumption that EI is constant, it can be considered as valid the principle of superposition effects of transverse actions, under the hypothesis that the same normal force N_{Ed} is superimposed to all of them.

Table 5.4 Parameters c_0 and β as a function of the first order moment distribution

Moment distribution M_0	c_0	β
M_0 uniform	2.0	1.23
M_0 triangular	3.0	0.82
M_0 parabolic concave	4.0	0.62

The analysis is based on the parameter $\alpha = \sqrt{\frac{N_{Ed}}{EI}}$ (which has the dimension of the inverse of a length). The solutions will be presented during the discussion of the following examples.

5.5.2.6 The Nominal Stiffness and the Displacement of the Direction of the Axial Load

This procedure allows calculating the total moment by adding to the initial eccentricity of the axial load (which is a problem datum) elastic transverse displacements due to the different causes, amplified by the factor $\frac{1}{1-\frac{N_{Ed}}{N_B}}$. It is an approximated method, theoretically analogous to the integration of the differential equation of the elastic curve. The approximation is very high (errors are around 1%) and the procedure is not mentioned in EC2.

The calculation of elastic displacements is easy. The only relevant case is the horizontal displacement of the top section of a column with length ℓ fixed at the base and free at the top due to geometric imperfections.

Because of the slope θ, under the action of the vertical force N_{Ed}, the column experiences a bending moment growing up with linear law from the top ($M = 0$) to the base ($M = N_{Ed} \cdot \theta\ell$). If a horizontal virtual force $H = 1$ is associated at the top, a linear diagram M1 analogous to the other one is defined, with a value $M_1 = 1 \cdot \ell$ at the base. By applying the principle of virtual works, horizontal displacement at the top is given by

$$s = \int \frac{M \cdot M_1}{EI} ds = \frac{\ell}{3EI} \cdot [(N_{Ed} \cdot \theta\ell) \cdot \ell] = \frac{N_{Ed} \cdot \theta\ell^3}{3EI}$$

In the following, the procedure is applied to Examples 1, 2, 3, 4 of 5.5.4.

5.5.3 *The Method Based on Nominal Curvature*

The method is applied to isolated members solicited by axial loads.

The first step of the procedure requires the calculation of the total design moment M_{Ed} due to first and second order effects, using the following expression

$$M_{Ed} = M_{0Ed} + M_2 \quad [(5.31)]$$

where

M_{0Ed} is the first order moment, comprehensive of the effect due to imperfections
M_2 nominal second order moment, given by

$$M_2 = N_{Ed} \cdot e_2 \quad [(5.33)]$$

based on the estimation of the member deflection, under the implicit hypothesis on the deformed shape of the column and by estimating the maximum curvature of the critical cross-section.

In this way, through [(5.33)], we get

$$M_2 = N_{Ed} \cdot e_2 = N_{Ed} \cdot \left[(1/r) \cdot \ell_0^2/c\right]$$

where $(1/r)$ is the curvature of the critical cross-section, ℓ_O is the effective length and c is a parameter depending on the curvature distribution along the axis of the member. Under the hypothesis of sinusoidal distribution, c is assumed as equal to 10.

The curvature of the critical cross-section $(1/r)$ is calculated using expressions from [(5.34)] to [(5.37)], collected in [5.8.8.3] as follows:

$$\frac{1}{r} = K_r \cdot K_\varphi \cdot \frac{1}{r_0} \quad [(5.34)]$$

where $K_r = (n_u - n)/(n_u - n_{\mathrm{bal}}) \leq 1$ [(5.36)].

with

$$n = \frac{N_{Ed}}{A_c \cdot f_{cd}}$$

$$n_u = 1 + \omega$$

$$\omega = \frac{A_S \cdot f_{yd}}{A_c \cdot f_{cd}}$$

$$n_{\mathrm{bal}} = 0.4$$

being n_{bal} the reduced axial load associated with the maximum design bending resistance μ of the interaction diagram; it is the axial force associated with the configuration defined by $\varepsilon_c = \varepsilon_{cu2}$ e $\varepsilon_s = \frac{f_{yd}}{E_s}$.

$$K_\varphi = 1 + \beta\varphi_{ef} \geq 1 \quad [(5.37)]$$

with $\beta = 0.35 + f_{ck}/200 - \lambda/150$

$$\varphi_{ef} = \varphi_{(\infty,t_0)} \cdot \frac{M_{0Eqp}}{M_{0Ed}}$$

In the expression [(5.34)], we assume the value of the curvature conventionally defined by imposing the reaching of the deformation of the design elastic limit $\varepsilon_{yd} = \frac{f_{yd}}{E_s}$ for the reinforcement of both sides

$$(1/r)_0 = \frac{\varepsilon_{yd}}{0.45 \cdot d} = \frac{2 \cdot \varepsilon_{yd}}{0.9 \cdot d}$$

If $n < n_{\mathrm{bal}}$, from [(5.36)] we get $K_r > 1$. But being K_r maximum equal to 1, the whole procedure is not depending on ω, including the total moment. It is then possible to define directly, for example through interaction diagrams, the reinforcement necessary to support the total load, given by the sum of the first and second order.

The total acting moment is given by the sum of the first order moments corresponding to various loads (vertical and horizontal forces, geometric imperfections) and of the second order moment corresponding to vertical forces.

5.5.4 Examples

The following examples discuss cases of a column fixed at the base and free at the top section, under the action of eccentric axial loads and/or associated with transverse forces. In Examples 1, 2, 3 and 4, the axial force is contained in a plane of symmetry (combined axial force and uniaxial bending); Examples 5 and 6 are referred to combined axial force and biaxial bending. Example 7 is relevant for a framed column.

All the examples are developed by applying the methods of nominal stiffness and nominal curvature. In both cases, by comparison, the application of the general method is also considered. In every case, the reinforcement is given: it is the result obtained after attempts developed a priori by increasing the reinforcement required by the first order internal forces until a satisfying value is obtained. All the examples collected verify the compatibility of data with the total internal forces of first and second order. In all of the examples, steel is B450C with $f_{yk} = 450$ N/mm^2 and $f_{yd} = 391$ N/mm^2.

The uncertainty about the application point of the axial load of [6.1(4)] is not taken into account because negligible in comparison with all the other eccentricities.

5.5.4.1 Example 1—Column with Length 5 m Under Centred Axial Load and Bending Load (Fig. 5.6)

Problem Data

Square cross-section with side $h = 400$ mm, symmetrical reinforced with 4 ϕ16mm on each side with $d' = 40$ mm

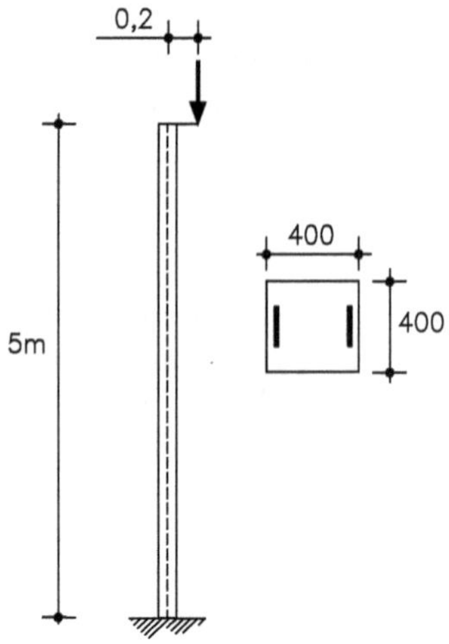

Fig. 5.6 Example 1: structural scheme

$$N_{Ed} = 400 \text{ kN with eccentricity } e = 0.20 \text{ m}; N_{0Eqp} = 200 \text{ kN}$$

Concrete C30/37

$$f_{cd} = 17 \text{ N/mm}^2; E_{cm} = 33{,}000 \text{ N/mm}^2;$$
$$E_{cd} = E_{cm}/1.2 = 27{,}500 \text{ N/mm}^2 \quad [(5.20)].$$

Preliminary Calculations

Effective length

$$\ell_0 = 2 \cdot 5 = 10\text{m}$$

Slenderness

$$\lambda = \ell_0/(h/3.46) = \frac{10}{\left(\frac{0.40}{3.46}\right)} = 86.5$$

Reinforcement geometric ratio

$$\rho = \frac{2 \cdot 800}{400^2} = 0.01$$

Reinforcement mechanical ratio

$$\omega = \rho \frac{f_{yd}}{f_{cd}} = 0.01 \cdot \frac{391}{17} = 0.23$$

First order internal forces at ULS

$$N_{Ed} = 400\text{kN};\ n_{Ed} = \frac{N_{Ed}}{A_c f_{cd}} = \frac{400{,}000}{400^2 \cdot 17} = 0.1471\ \text{N/mm}^2;$$

$$\text{M}_{\text{0Ed}} = 4000 \cdot 0.2 = 80\ \text{kNm}$$

Internal forces due to constructive imperfections.
Vertical imperfection

$$\theta_i = 0.005 \cdot 2/\sqrt{5} = 0.0045 \quad [5.1]$$

$$\theta_i \cdot \ell = 22.5\text{mm}$$

Bending moment due to imperfections

$$N_{Ed} \cdot (\theta_i \cdot \ell) = 9\ \text{kNm}$$

The total first order bending moment at ULS

$$M_{0Ed} = 80 + 9 = 89\ \text{kNm}$$

Internal forces at SLS, quasi-permanent combination

$$N_{0Eqp} = 200\ \text{kN};\ \text{e} = 0.2\ \text{m};$$

$$M_{0Eqp} = 200 \cdot 0.2 = 40\,\text{kNm}$$

Creep

$$\varphi_{(\infty,t_0)} = 2.5$$

$$\varphi_{ef} = \varphi_{(\infty,t_0)} \cdot (M_{0Eqp}/M_{0Ed}) = 2.5(40/89) = 1.12 \quad [5.19]$$

Limit slenderness according to [5.13 N]

$$\lambda_{\text{lim}} = 20 \cdot A \cdot B \cdot C/\sqrt{n}$$

With the above-collected data, we get

$$A = 1/(1 + 0.2 \cdot \varphi_{ef}) = 1/(1 + 0.2 \cdot 1.12) = 0.817$$

$$B = \sqrt{1 + 2\omega} = \sqrt{1 + 2 \cdot 0.23} = 1.20$$

$$C = 0.7$$

and, being the examined column an unbraced member (see remark in [5.8.3.1]),

$$\sqrt{n} = \sqrt{0.1471} = 0.3835$$

and, therefore

$$\lambda_{\text{lim}} = 20 \cdot 0.817 \cdot 1.20 \cdot 0.7/0.3835 = 35.8$$

Being $\lambda = 86.5$ greater than λ_{lim}, it is necessary to take into account second order effects.

Method of Nominal Stiffness

Solution according to the EC2 procedure

The nominal stiffness is calculated according to [(5.21)]:

$$EI = k_c E_{cd} I_c + k_s E_s I_s$$

Steel part

Being $\rho = 0.01 > 0.002$, according to [5.22], $k_s = 1$, and therefore:

$$\begin{aligned} E_s I_s &= E_s \cdot 2A_s \cdot (h/2 - d')^2 = 200{,}000 \cdot 2 \cdot 800 \cdot (200 - 40)^2 \\ &= 8.18 \cdot 10^{12}\ \text{Nmm}^2 \end{aligned}$$

Concrete part:

$$k_c = k_1 k_2/(1 + \varphi_{ef})$$

$$k_1 = \sqrt{\frac{f_{ck}}{20}} = \sqrt{\frac{30}{20}} = 1.22$$

$$k_2 = n_{Ed}\lambda/170 = 0.1471 \cdot 86.5/170 = 0.0748$$

$$k_c = 1.22 \cdot 0.0748/(1 + 1.12) = 0.043$$

$$k_c \cdot E_{cd} \cdot I_c = 0.043 \cdot 27{,}500 \cdot \frac{400^4}{12} = 2.52 \cdot 10^{12}\text{Nmm}^2$$

By adding the terms, we get:

$$EI = 2.52 \cdot 10^{12} + 8.18 \cdot 10^{12} = 10.70 \cdot 10^{12}\text{Nmm}^2$$

Eulerian critical load based on the nominal stiffness is given by

$$N_B = \pi^2 EI/\ell_0^2 = 9.86 \cdot (10.70 \cdot 10^{12})/10^8 = 1055\,\text{kN}$$

Being the values of β in [(5.28)] different with the variation of vertical load (uniform M) and geometric imperfection (triangular M), partial total moments can be calculated by adding corresponding first and second order moments.

Vertical load

$M_{0Ed} = 80$ kNm; $\beta = 1.23$

Imperfections

$M_{0Ed} = 9$ kNm; $\beta = 0.82$

The total moments (first and second order) in the two cases are equal, by applying [(5.28)], to:

$$M_{Ed} = 80 \cdot \left[1 + \frac{1.23}{\frac{1055}{400} - 1}\right] = 140.0\,\text{kNm}$$

$$M_{Ed} = 9 \cdot \left[1 + \frac{0.82}{\frac{1055}{400} - 1}\right] = 13.5\,\text{kNm}$$

and, together, to $M_{\text{tot}} = 140.0 + 13.5 = 153.5$ kNm.

The ratio between the total moment and the first order moment is given by 153.5/89 = 1.72.

Solution of the Problem by Integration of the Differential Equation of the Elastic Curve

If we indicate by y the displacement of the generic cross-section at distance x from the fixed point and by f the displacement of the top end-section ($x = \ell$), the bending moment of a generic cross-section can be written down as

$$M = -N_{Ed}(e + \theta\ell + f - y).$$

and differential equation of elastic curve becomes

$$EIy'' = N_{Ed}(e + \theta\ell + f - y)$$

If $\alpha = \sqrt{\frac{N_{Ed}}{EI}}$.

by integrating twice we get

$$y = C_1 \sin \alpha x + C_2 \cos \alpha x + e + \theta\ell + f$$

Constant parameters C_1 and C_2 can be evaluated by applying at the fixed point ($x = 0$) the conditions $y = 0$ and $y' = \theta$, obtaining:

$$C_1 = (\theta/\alpha);\ C_2 = -(e + \theta\ell + f)$$

The elastic curve is therefore described by the following equation:

$$y = \frac{\theta}{\alpha}\sin\alpha x + (e + \theta\ell + f)\cdot(1 - \cos\alpha x)$$

For $x = \ell$, i.e. at the top end-section, total displacement is equal to $y = (\theta\ell + f)$. By equating it to the value of y calculated for $x = \ell$, we get:

$$e + \theta\ell + f = \frac{e + (\theta/\alpha)\sin\alpha\ell}{\cos\alpha\ell}$$

and the elastic displacement is

$$f = \frac{e(1 - \cos\alpha\ell) + (\theta/\alpha)\sin\alpha\ell - \theta\ell\cos\alpha\ell}{\cos\alpha\ell}$$

Total bending moment is given by

$$M_{Ed} = N_{Ed}(e + \theta\ell + f)$$

being the first two terms in brackets the assigned eccentricities and the third term the elastic displacement. By substituting the f expression, the total bending moment holds

$$M_{Ed} = N_{Ed}\frac{e + (\theta/\alpha)\sin\alpha\ell}{\cos\alpha\ell}$$

By introducing problem given data, results are the following:

$$\alpha = \sqrt{\frac{400{,}000}{10.70\cdot 10^{12}}} = 1.93\cdot 10^{-4}\ \text{mm}^{-1}$$

Therefore, we get

$$\alpha\ell = 1.93\cdot 10^{-4}\cdot 0.5\cdot 10^{4} = 0.97;\ \sin\alpha\ell = 0.82;\ \cos\alpha\ell = 0.565$$

$$(\theta/\alpha) = 45\cdot 10^{-4}/1.93\cdot 10^{-4}$$

$$\theta\ell = 22.5\ \text{mm} = 0.0225\ \text{m}$$

$$f = \frac{200 \cdot (1 - 0.565) + 23.3 \cdot 0.82 - 22.5 \cdot 0.565}{0.565} = 165.3\text{mm}$$

$$M_{Ed} = 400 \cdot (0.200 + 0.0225 + 0.165) = 400 \cdot 0.388 = 155\,\text{kNm}$$

or, by the second expression:

$$M_{Ed} = 400 \cdot \frac{0.20 + 0.0233 \cdot 0.82}{0.565} = 155\text{ kNm}$$

Solution of the Problem by Calculation of Displacements of the Direction of the Axial Load

Calculation of transverse displacements

- Combined axial force and uniaxial bending moment
  $f_1 = \frac{N_{Ed} \cdot e \cdot \ell^2}{2EI} = \frac{400 \cdot 0.20 \cdot 10^6 \cdot 25 \cdot 10^6}{2 \cdot 10.7 \cdot 10^{12}} = 93.45$ mm.
- Geometric imperfection $\theta = 0.0045$; $\theta\ell = 22.5$ mm
  $s = 400{,}000 \cdot \frac{5000^3}{3 \cdot 10.7 \cdot 10^{12}} \cdot 0.0045 = 7.0$ mm.
- Coefficient of displacement amplification
  $N_B = 1055$ kN
  $\frac{1}{1 - N_{Ed}/N_B} = \frac{1}{1 - (400/1055)} = 1.61.$
- Sum of the bending displacement and amplification of the sum:
  $(93.45 + 7.0) \cdot 1.61 = 162$ mm $= 0{,}162$ m.
- The total bending moment at the fixed point:
  $M = 400 \cdot (0.20 + 0.0225 + 0.162) = 153.8$ kNm.

Method of the Nominal Curvature

The second order moment is evaluated as followsThe comparison highlights

$$M_2 = N_{Ed} e_2 \quad [(5.33)]$$

being $e_2 = (1/r) \cdot \ell_0^2/c$, where c is depending on the curvature distribution, but 10 is the value normally employed in case of uniform cross-section.

$$(1/\text{r}) = K_r \cdot K_\varphi \cdot (1/r_0) \quad [(5.34)]$$

$$K_r = (n_u - n)/(n_u - n_{bal}) \leq 1 \quad [(5.36)]$$

with $n = \frac{N_{Ed}}{A_c f_{cd}}$; $n_u = 1 + \omega$; $n_{\text{bal}} = 0.4$

$$K_\varphi = 1 + \beta\varphi_{ef} \geq 1 \text{ being } \beta = 0.35 + f_{ck}/200 - \lambda/150$$

$$1/r_0 = \varepsilon_{yd}/0.45d$$

By substituting in the formulas the corresponding values, we get:
$n = 0.1471$; $n_u = 1.23$; therefore, being $n < n_{\mathrm{bal}}$, $K_r = 1$.
$\beta = 0.35 + 30/200 - 86.5/150 = -0.08$, then $K_\varphi = 1$.
$1/r_0 = 1.96 \cdot 10^{-3}/(0.45 \cdot 360) = 12.10 \cdot 10^{-6}\mathrm{mm}^{-1}$.
$(1/r) = 1 \cdot 1 \cdot 12.10 \cdot 10^{-6} = 12.10 \cdot 10^{-6}\mathrm{mm}^{-1}$
$e_2 = 12.10 \cdot 10^{-6} \cdot 10^8/10 = 121$ mm $= 0.121$ m.
$M_2 = 400 \cdot 0.121 = 48$ kNm.
$M_{Ed} = 80 + 9 + 48 = 137$ kNm.

Verification

Design bending resistance in the cross-section at the fixed point associated with N_{Ed}.
The calculation is developed by two different procedures:

- using the interaction diagram $\mu - \nu$
- through the procedure described in paragraph 7.3 of Chap. 7: "Analysis of the equilibrium configurations at ULS for a rectangular cross-section under axial and bending loads".

Interaction diagram
According to Table U10 for f_{ck} until 50 N/mm^2, with $\omega = 0.23$, in presence of axial load

$$\nu = \frac{400{,}000}{400 \cdot 400 \cdot 17} = 0.1471$$

design bending resistance is equal to $\mu = 0.155$, from which $M_{Rd} = 0.155 \cdot 400^3 \cdot 17 = 168$ kNm $> M_{Ed}$.
Analytical calculation

$$(\beta_1 = 0.8095;\ \beta_2 = 0.416;\ \varepsilon_{cu_2} = 0.0035;\ \varepsilon_{cu_2} E_s = 700\,\mathrm{N/mm^2})$$

The section of the solution is identified with reference to Fig. 7.3 of 7.3.
The 0-deformation line, where stress in the reinforcement A_{s1} is zero, is equal to:

$$N_{Rd} = -0.8095 \cdot 400 \cdot 40 \cdot 17 + 765 \cdot 391 = -220.18 + 299.11 = +78.93\ \mathrm{kN}$$

Line 1: stress of A_{s1} is—391 N/mm^2. We get $x = k'd = 2.27 \cdot 40 = 90.8$ mm

$$N_{Rd} = 0.8095 \cdot 400 \cdot 90.8 \cdot 17 = -499.81\ \mathrm{kN}$$

The solution is therefore placed in sector 1.

The position of the neutral axis is evaluated through the Eq. (7.6), by introducing the compressive force $N_{Ed} = -400{,}000$ N:

$$x^2 - x\left(\frac{765 \cdot 391 + 400000 - 765 \cdot 700}{0.8095 \cdot 400 \cdot 17}\right) - \frac{765 \cdot 700 \cdot 40}{0.8095 \cdot 400 \cdot 17} = 0$$

The solution is equal to $x = 79$ mm, from which we get

$$\sigma_{s1} = 700\left(1 - \frac{40}{79}\right) = 345\ \text{N/mm}^2$$

Finally, using (7.7), we get
$M_{Rd} = 765 \cdot 345\ (200 - 40) + 0{,}8095 \cdot 400 \cdot 79 \cdot 17\ (200 - 0.416 \cdot 79) +$
$+\ 765 \cdot 391\ (200 - 40) = 162.7\ \text{kNm} > M_{Ed}$.

Comparison with the General Method

To compare results obtained by simplified methods, nominal stiffness method, and nominal curvature method, the problem has been solved by applying nonlinear analysis as described in [5.8.6], without considering geometric imperfection (slope θ).

To this purpose, the moment–curvature relationship has been obtained by applying the stress–strain expressions for structural analysis [3.1.5] with the design material laws, as required in [5.8.6]. In Fig. 5.7 we collect the relationships moment–curvature, with and without creep. In the following diagrams, we draw the results of nonlinear analysis under the combination of design loads: mean curvature, horizontal displacement of the axis, and total moment along the column, accounting for first and second order effects and creep. The total bending moment at the base is equal to $M = 124.1$ kNm.

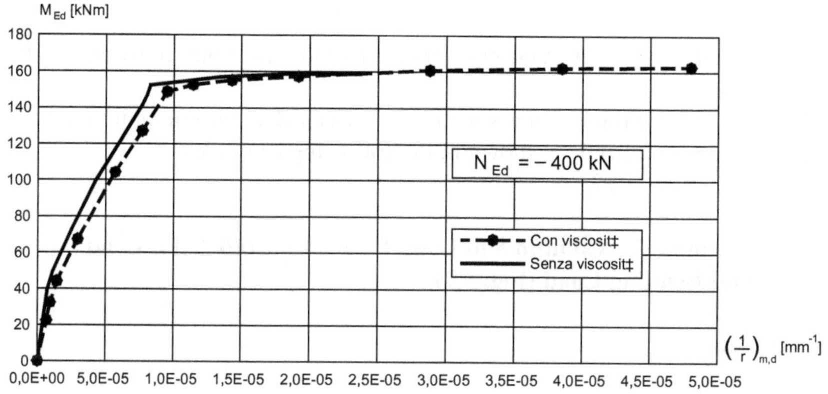

Fig. 5.7 Example 1: moment-mean curvature diagram in presence of design axial load

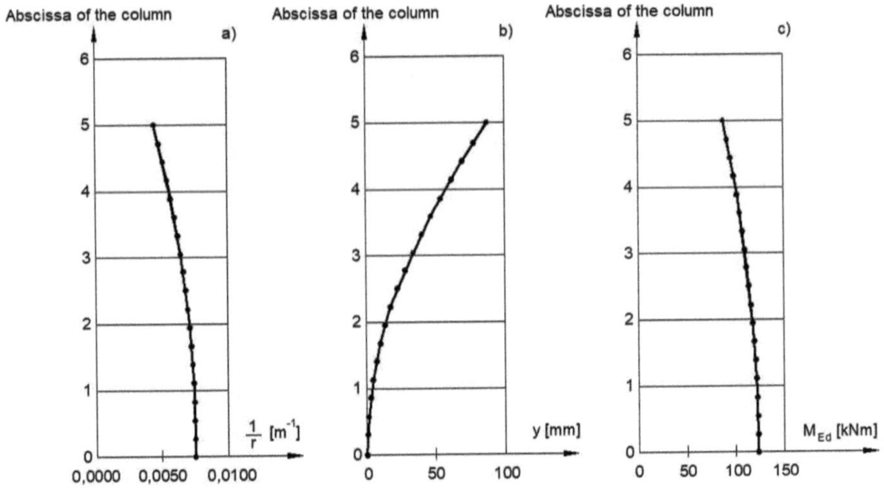

Fig. 5.8 Example 1: **a** mean curvature; **b** horizontal displacement of the axis; **c** total bending moment as the sum of first and second order effects

Table 5.5 Example 1: results of the three analysis methods

	M_{Ed}(I + II order) (kNm)	Variance
Method of nominal stiffness EC2	140.0	+13%
Method of nominal curvature	128.0	+3%
General method	124.1	–

Table 5.5 collects the values of the total bending moment (I + II order) at the bottom of the column obtained with simplified methods and the percent difference to the general method, without accounting for the geometric imperfection θ.

- *Remark*

The comparison highlights that simplified methods are both conservative to the general method.

Moreover, the nominal stiffness method is stricter than the curvature method, but this cannot be generalized, as shown in the following examples.

5.5.4.2 Example 2—Column 8 m Long Under Centred Axial Load and Bending Load (Fig. 5.9)

Problem Data

Circular cross-section 800 mm diameter

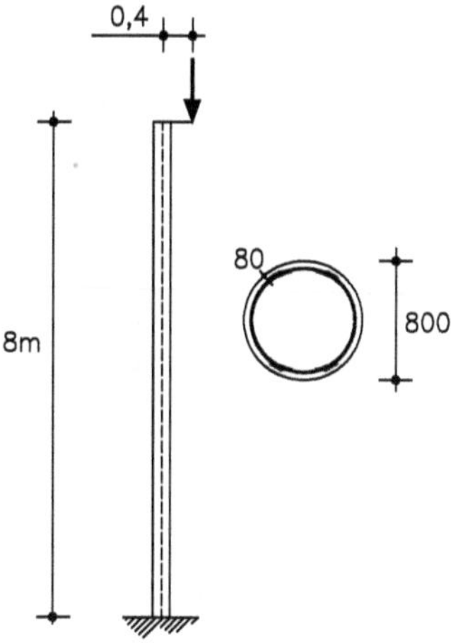

Fig. 5.9 Example 2: structural scheme

Reinforcement

26 ϕ30 arranged along the circumference with diameter 640 mm.

Concrete C 60/75.

$f_{cd} = 34$ N/mm^2.

$E_{cm} = 39{,}000$ N/mm^2; $E_{cd} = E_{cm}/1.2 = 32{,}500$ N/mm^2 [5.20]

Internal force at ULS: axial load.

$N_{Ed} = 3000$ kN with eccentricity 0.40 m.

Internal forces at SLS, quasi-permanent combination.

$N_{0Eqp} = 1500$ kN; $e = 0.40$ m.

Preliminary Calculations

Reinforcement geometric ratio

$$\rho = \frac{18{,}378}{502{,}654} = \frac{3.65}{100} < 0.04, \text{ maximum allowed value.}$$

Reinforcement mechanical ratio

$$\omega = \frac{A_s f_{yd}}{A_c f_{cd}} = \frac{18378 \cdot 391}{502{,}654 \cdot 34} = 0.42.$$

Slenderness

$$\lambda = \ell_0/(r/2) = 16{,}000/200 = 80$$

Internal forces at ULS

$$\mathrm{N_{Ed}} = 3000\,\mathrm{kN};\ \mathrm{n_{Ed}} = \mathrm{N_{Ed}}/(\mathrm{A_c f_{cd}}) = \frac{3.000{,}000}{502{,}654 \cdot 34} = 0.1755$$

$$e = \ 0.40\ \mathrm{m};\ M_{0Ed} = N_{Ed}e = 1200\ \mathrm{kNm}$$

Internal forces due to constructive imperfections.
Verticality imperfection

$$\theta_i = 0.005.2/\sqrt{8} = 0.0035$$

$\theta_i\,\ell = 28.3$ mm. We assume the rounded value of 30 mm.
Moment at the fixed point due to imperfections

$$\mathrm{N_{0Ed}}(\theta_\mathrm{i}) = 3000 \cdot 0,03 = 90\,\mathrm{kNm}$$

First order total moment at ULS

$$M_{Ed} = 1290\ \mathrm{kNm}$$

Internal forces at SLS, quasi-permanent combination

$$N_{0Eqp} = 1500\ \mathrm{kN};\ e = \ 0.40\ \mathrm{m};\ M_{0Eqp} = 600\ \mathrm{kNm}$$

Creep

$$h_0 = \frac{2A}{u} = \frac{2 \cdot 0.50}{\pi \cdot 0.8} = 0.4\mathrm{m} = 400\mathrm{mm}$$

$$\varphi_{(\infty,t_0)} = 2.0$$

$$\varphi_{ef} = \varphi_{(\infty.t0)}(M_{0Eqp}/M_{0Ed}) = 2 \cdot (600/1200) = 1$$

Limit slenderness according to (5.13N)

$$\lambda_{\mathrm{lim}} = 20\ \mathrm{A\ B\ C}/\sqrt{n}$$

With the previous data, we get:

$A = 1/1.2 = 0.833$.
$B = \sqrt{1 + 2\omega} = \sqrt{1 + 2 \cdot 0.42} = 1.35$.
$C = 0.7$ being an unbraced member (see Remark at [5.8.3.1(1)]).
$\sqrt{n} = \sqrt{0.1755} = 0.42$.

and then $\lambda_{\lim} = 20 \cdot 0.833 \cdot 1.35 \cdot 0.7/0.42 = 37.5$
Being $\lambda = 80$ greater than $\lambda_{\lim}$, second order effects have to be taken into account.

Method of the Nominal Stiffness

Solution according to the EC2 procedure

$$I_c = \pi \cdot \frac{r^4}{4} = 3.14 \cdot \frac{400^4}{4} = 2.0 \cdot 10^{10}\text{mm}^4$$

$$E_{cd} I_c = 32,500 \cdot 2.0 \cdot 10^{10} = 6.5 \cdot 10^{14}\text{Nmm}^2$$

$$I_s = \frac{A_s \cdot r_s^2}{2} = \frac{18{,}378 \cdot 320^2}{2} = 941 \cdot 10^6\text{mm}^4$$

$$E_s I_s = 200{,}000 \cdot 941 \cdot 10^6 = 1.88 \cdot 10^{14}\text{Nmm}^2$$

Being then
$\lambda = 80$.

$$k_1 = \sqrt{\frac{f_{ck}}{20}} = 1.73$$

$$k_2 = n_{Ed}\lambda/170 = 0.1755 \cdot 80/170 = 0.0826$$

We get

$$k_c = k_1 \cdot k_2/(1 + \varphi_{ef}) = 1.73 \cdot 0.0826/2 = 0.0714$$

The nominal stiffness is given by

$$EI = k_c \cdot E_{cd} I_c + E_s I_s = (0.0714 \cdot 6.5 + 1.88) \cdot 10^{14} = 2.34 \cdot 10^{14}\text{Nmm}^2$$

Eulerian critical load evaluated by the nominal stiffness is given by

$$N_B = \pi^2 EI/\ell_0^2 = 9.86 \cdot 2.34 \cdot 10^{14}/16{,}000^2 = 9010 \cdot 10^3 N = 9010\,\text{kN}$$

Being the β values different depending on the vertical load (M uniform) and geometrical imperfections (M triangular), we can calculate corresponding moments:

- vertical load $M = 1200$ kNm; $\beta = 1.23$
- imperfections $M = 90$ kNm; $\beta = 0.82$.

Using [5.28], first and second order total moments are given by

$$1200 \cdot \left[1 + \frac{1.23}{\frac{9010}{3000} - 1}\right] = 1936\,\text{kNm};\ 90 \cdot \left[1 + \frac{0,82}{\frac{9010}{3000} - 1}\right] = 127\,\text{kNm}$$

and finally: 1936 + 127 = 2063 kNm.

Solution of the problem by integration of the differential equation of the elastic curve

We introduce the nominal stiffness.

$EI = 2.34 \times 10^{14}\text{Nmm}^2$ and we calculate

$$\alpha^2 = \frac{N_{Ed}}{EI} = \frac{3000 \cdot 10^3}{2.34 \cdot 10^{14}} = 1.382 \cdot 10^{-8}\text{mm}^{-2}$$

therefore $\alpha = 1.13 \cdot 10^{-4}\text{mm}^{-1}$.

The other terms are given by:

$$\alpha\ell = 1.13 \cdot 10^{-4} \cdot 8000 = 0.905;\ \sin\alpha\ell = 0.7869;\ \cos\alpha\ell = 0.6170$$

$$\frac{\theta}{\alpha} = \frac{35 \cdot 10^{-4}}{1.13 \cdot 10^{-4}} = 31;\ \theta\ell = 30.$$

Second order eccentricity f is given by (the expression in symbols is in Example 1):

$$f = \frac{400 \cdot (1 - 0.6170) + 31 \cdot 0.7869 - 30 \cdot 0.6170}{0.6170} = 257\text{mm}$$

Total eccentricity is given by 400 + 30 + 257 = 687 mm.

Acting moment is $M_{Ed} = 3000 \cdot 0.687 = 2061$ kNm.

Solution of the Problem by Calculation of Displacements of the Direction of the Axial Load

Calculation of transverse displacements

Eccentric normal force

$$f_1 = \frac{N_{Ed} \cdot e \cdot \ell^2}{2EI} = \frac{3000 \cdot 0.0 \cdot 10^6 \cdot 64 \cdot 10^6}{2 \cdot 234 \cdot 10^{12}} = 164.0\,\text{mm}$$
$$= \frac{3000 \cdot 0,40 \cdot 10^6 \cdot 64 \cdot 10^6}{2 \cdot 234 \cdot 10^{12}}$$

Geometric imperfection $\theta = 0.0035$; $\theta\ell = 28.3$ mm.

$$s = \frac{3000 \cdot 512 \cdot 10^{12}}{3 \cdot 234 \cdot 10^{12}} \cdot 0.0035 = 7.7\,\text{mm}$$

Coefficient of displacement amplification
$N_B = 9010$ kN

$$\frac{1}{1 - N_{Ed}/N_B} = \frac{1}{1 - (3000/9010)} = 1.50$$

Sum of flexural displacements and amplification of the sum

$$(164.0 + 7.7) \cdot 1.50 = 258 \text{ mm} = 0.258 \text{ m}$$

Total moment at the fixed point:

$$M = 3000 \cdot (0.40 + 0.028 + 0.258) = 2058.0 \text{ kNm}$$

Method of the Nominal Curvature

We introduce the value $\omega = 0.42$ adopted by the stiffness method and we develop the calculation by means of the EC2 procedure.

$n_{bal} = 0.4$
$n_u = 1 + \omega = 1.42$
$K_r = (n_u - n)/(n_u - n_{bal}) = (1.42 - 0.1755)/(1.42 - 0.4) = 1.22.$
By means of [5.36], we assume $K_r = 1$.
$\beta = 0.35 + f_{ck}/200 - \lambda/150 = 0.35 + 0.30 - 0.53 = 0.12$
$K_\varphi = 1 + \beta\, \varphi_{ef} = 1 + 0.12 = 1.12$
$\varepsilon_{yd} = 1.96 \times 10^{-3}$.

Because of reinforcement distribution along a circumference with a diameter of 640 mm, the expression [5.35] defines the effective depth.

$$d = \left(\frac{h}{2}\right) + i_s$$

Being

$$\frac{h}{2} = \frac{800}{2} = 400\text{mm}$$

and

$$i_s = \sqrt{\frac{I_s}{A_s}} = \sqrt{\frac{94{,}100 \cdot 10^4}{18{,}378}} = 226\,\text{mm}$$

we get

$d = (400 + 226) = 626$ mm.

By applying this value, we obtain:

$(1/r)_0 = 1.96 \times 10^{-3}/(0.45d) = 6.96 \cdot 10^{-6}$ mm^{-1}

$1/r = K_r\, K_\varphi\, (1/r)_0 = 6.96 \cdot 10^{-6} \cdot 1.0 \cdot 1.12 = 7.79 \cdot 10^{-6}$mm^{-1}.

and, finally

$e_2 = (1/r)l_0{}^2/10 = 7.79 \cdot 10^{-6} \cdot 16^2 \cdot 10^6/10 = 200$ mm.

$M_2 = N_{Ed}\, e_2 = 3000 \cdot 0.2 = 600$ kNm.

Total acting moment is therefore given by

$$M_{Ed} = M_{0Ed} + M_2 = 1200 + 90 + 600 = 1890 \text{ kNm}$$

Checking of the Cross-Section at the Fixed Point

The calculation is developed considering the interaction diagram $\nu - \mu$ of Table U19, valid for a circular cross-section with $f_{ck} = 60$ N/mm^2.

We assume internal forces evaluated with the stiffness method. Non-dimensional values are given by:

$$\nu = \frac{N_{Ed}}{A_c f_{cd}} = \frac{3{,}000{,}000}{502{,}654 \cdot 34} = 0.1755$$

$$\mu = \frac{M_{Ed}}{A_c \cdot h \cdot f_{cd}} = \frac{2063 \cdot 10^6}{502{,}654 \cdot 800 \cdot 34} = 0.1509$$

For such a couple of values, Table U19 requires $\omega = 0.37$.

In the calculation we employed.

$$\omega = 0.42 > 0.37$$

then the check is fulfilled.

5.5.4.3 Example 3—Column with Length 5 m Under Centred Axial Load and Transverse Load (Fig. 5.10)

Problem Data

Square cross-section with side

$$h = 400\text{mm}$$

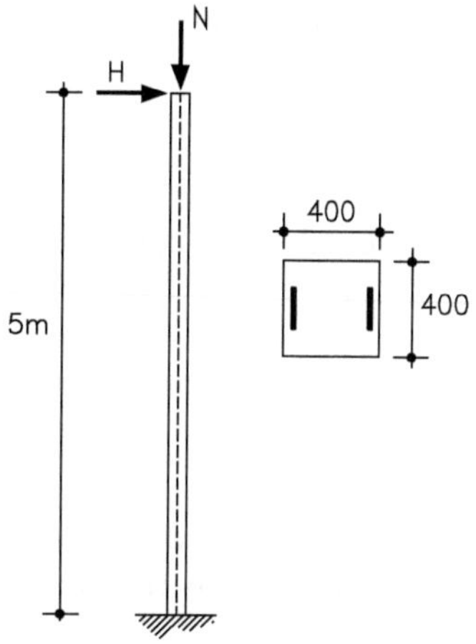

Fig. 5.10 Example 3: structural scheme

Double symmetric reinforcement given by 4 ϕ26 at each side
$d' = 40\,mm$.
Concrete C 70/85
$f_{cd} = 39.6$ N/mm^2.
$E_{cm} = 41{,}000$ N/mm^2; $E_{cd} = E_{cm}/1.2 = 34.166$N/mm^2 [5.20].
Internal force at SLU.

axial force $N_{Ed} = 500$ kN centred;
horizontal force $H_{Ed} = 50$ kN applied at the top end-section.

Internal forces at SLS, quasi-permanent combination

$$NO_{Eqp} = 250/\text{kNcentred}.$$

Preliminary Calculations

Effective length

$$\ell_0 = 2 \cdot 5 = 10\text{m}$$

Slenderness

$$\lambda = \frac{\ell_0}{(h/3.46)} = \frac{10000}{(400/3.46)} = 86.50$$

Reinforcement geometric ratio

$$\rho = \frac{2 \cdot 2120}{400^2} = 0.0265$$

Reinforcement geometric ratio

$$\omega = \rho \frac{f_{yd}}{f_{cd}} = 0.0265 \cdot \frac{391}{39.66} = 0.2617$$

Internal forces at ULS

$$N_{Ed} = 500\text{kN}; n = N_{Ed}/(A_c f_{cd}) = 0.0789$$

$$H_{Ed} = 50\text{kN}; M_{0Ed} = 50 \cdot 5 = 250\text{kNm}$$

Internal forces due to constructive imperfections.
Verticality imperfections

$$\theta_i = 0.005 \cdot 2\sqrt{5} = 0.0045$$

$$\theta_i \ell = 22 \text{ mm}$$

Moment due to imperfections

$$N_{Ed}(\theta i \, \ell) = 11 \text{ kNm}$$

First order total moment at ULS

$$M_{0Ed} = 261 \text{ kNm}$$

Internal forces at SLS, quasi-permanent combination

$$N_{0Eqp} = 250 \text{ kN}$$
$$H_{0Eqp} = 20 \text{ kN}; M_{0Eqp} = 100 \text{ kNm}$$

Creep

$$\varphi_{(\infty,t_0)} = 1.5$$
$$\varphi_{ef} = \varphi_{(\infty,t_0)}(\mathrm{M_{0Eqp}/M_{0Ed}}) = 1.5\ (100/250) = 0.60$$

Limit slenderness according to [5.13 N]

$$\lambda_{\lim} = 20 \text{ A B C} / \sqrt{n}$$

By means of the above collected data, we get:

$A = 1/(1 + 0.2 \cdot 0.6) = 0.89$

$B = \sqrt{1 + 2 \cdot 0.26} = 1.23$

$C = 0.7$ being an unbraced member (see Remark in [5.8.3.1]

$\sqrt{n} = \sqrt{0.0789} = 0.28$

and, therefore

$\lambda_{\lim} = 20 \cdot 0.89 \cdot 1.23 \cdot 0.7/0.28 = 55.$

Being $\lambda = 86.5$ greater than $\lambda_{\lim}$, it is necessary to account for the second order effects.

Method of the Nominal Curvature

Solution according to the EC2 procedure

$E_{cd}\, I_c = 34{,}166 \cdot 400^4/12 = 7.28 \times 10^{13}$ Nmm^2.

$E_s\, I_s = E_s\, A_s\, (h/2 - d\,')^2$.

Being $A_s = 2120$ mm^2 on each side, we get

$$E_s I_s = 2 \cdot 2120 \cdot (200 - 40)^2 \cdot 2000 \cdot 200000 = 21.69 \times 10^{12}\ \mathrm{N/mm^2}$$

$$k_1 = \sqrt{\frac{f_{ck}}{20}} = 1.87$$

$$k_2 = n_{Ed}\lambda/170 = 0.0789 \cdot 86.5/170 = 0.0402$$

$$k_c = k_1 k_2/(1 + \varphi ef) = 1.87 \cdot 0.0402/1.60 = 0.047$$

$$\begin{aligned} EI &= k_c E_{cd} I_c + E_s I_s = 0.047 \cdot 7.28 \cdot 10^{13} + 21.69 \cdot 10^{12} \\ &= 25.11 \cdot 10^{12}\mathrm{N/mm^2} \end{aligned}$$

$$N_B = \pi^2 EI/l_0^2 = 9.86 \cdot 25.11 \cdot 10^{12}/10^8 = 2476\mathrm{kN}$$

Being $\beta = \pi^2/c_0$, with $c_0 = 12$ (M_1 triangular diagram), we get $\beta = 9.86/12 = 0.82$ and therefore:

$$M_{Ed} = M_{0Ed}\left[1 + \frac{\beta}{\frac{N_B}{N_{Ed}} - 1}\right] = 261 \cdot \left[1 + \frac{0.82}{\frac{2476}{500} - 1}\right] = 315.1\,\mathrm{kNm}$$

Solution of the problem by integration of the differential equation of the elastic curve

We write down the solving formulas for the discussed example:

- Slope θ (geometric imperfection)

$$M_{\max}=N_{Ed}\cdot(\theta/\alpha)\cdot\tan(\alpha\ell)$$

- Horizontal force H_{Ed} at the top end-section:
  This force induces the first order moment $M_0 = H_{Ed}\cdot\ell$.
  The total moment in presence of N_{Ed} is given by:

$$M_{\text{tot}}=M_0\cdot\frac{\tan(\alpha\ell)}{\alpha\ell}$$

By introducing the problem data in the previous relationships, we obtain:

$$\cos\alpha\ell = 0.7613$$

$$\tan\alpha\ell = 0.8518$$

$$\alpha=\sqrt{\frac{N_{Ed}}{EI}}=\sqrt{\frac{50\cdot 10^4}{25.11\cdot 10^{12}}}=1.41\cdot 10^{-4}\,\text{mm}^{-1}$$

$$\alpha\ell = 1.41\cdot 10^{-4}\cdot 5000 = 0.705$$

$$\theta/\alpha = 0.0045/(1.41\cdot 10^{-4}) = 3.19\,\text{mm}$$

Geometric imperfection

$$M = 500\cdot 0.032\cdot 0.8518 = 13.6\,\text{kNm}$$

Horizontal force

$$M = 50\cdot 5\cdot\frac{0.8518}{0.705} = 302\,\text{kNm}$$

Total moment: $302.0 + 13.6 = 315.6$ kNm.

Solution of the problem by calculation of displacements of the direction of the axial load

Calculation of transverse displacements

Effect of the horizontal force H_{Ed}

$$f_1 = \frac{H_{Ed}\cdot\ell^3}{3EI} = \frac{50\cdot 10^3\cdot 125\cdot 10^9}{3\cdot 25.11\cdot 10^{12}} = 83.0\,\text{mm}$$

Geometric imperfection $\theta = 0.0045$; $\theta\ell = 22.5$ mm

$$s = \frac{500 \cdot 125 \cdot 10^{12}}{3 \cdot 25.11 \cdot 10^{12}} \cdot 0.0045 = 3.70 \text{ mm}$$

Coefficient of displacement amplification

$$\frac{1}{1 - N_{Ed}/N_B} = \frac{1}{1 - (500/2476)} = 1.25$$

Sum of flexural displacements and amplification of the sum
$(83.0 + 3.7) \cdot 1.25 = 108$ mm $= 0.108$ m
Total moment at the fixed point:
$M = 50 \cdot 5.0 + 500 \cdot (0.022 + 0.108) = 250.0 + 65.0 = 315.0$ kNm

Method of the Nominal Curvature

We introduce the value of $\omega = 0.26$ adopted by the stiffness method and we develop the calculation according to the EC2 procedure.

$n_{bal} = 0.4$
$n_u = 1 + \omega = 1.26$
$k_r = (n_u - n)/(n_u - n_{bal}) = (1.26 - 0.0789)/(1.26 - 0.4)$.

Being n smaller than 0.4, we assume $k_r = 1$.
$\beta = 0.35 + f_{ck}/200 - \lambda/150 = 0.35 + 0.35 - 86.5/150 = 0.123$.
$k_\varphi = 1 + \beta\omega_{ef} = 1 + 0.123 \cdot 0.6 = 1.07$.
$\varepsilon_{yd} = 1.96 \times 10^{-3}$.
$d = h - d' = 360$ mm.
$(1/r)_0 = 1.96 \times 10^{-3}/(0.45\, d) = 12.10 \times 10^{-6}$ mm^{-1}.
$(1/r) = k_r\, k_\varphi\, (1/r)_0 = 1.07 \cdot 12, 10 \cdot 10^{-6} = 12.95 \cdot 10^{-6}$ mm^{-1}.
$e_2 = (1/r)\, \ell_0{}^2/10 = 12.95 \cdot 10^{-6} \cdot 10^8/10 = 130$ mm.
$M_2 = N_{Ed}\, e_2 = 500 \cdot 0.130 = 65$ kNm.
$M_{Ed} = M_{0Ed} + M_2 = 261 + 65 = 326$ kNm.

Verification of the Cross-Section at the Fixed Point

Interaction diagram

The design bending resistance (equal to $\mu = 0.14$) is obtained from Table U12, corresponding to $f_{ck} = 70$ N/mm^2, with $\omega = 0.26$, in presence of the axial load $\nu = \frac{500{,}000}{400 \cdot 400 \cdot 39.6} = 0.0789$; it follows that

$$M_{Rd} = 0.14 \cdot 400^3 \cdot 39.6 = 354 \text{ kNm} > \text{MEd}.$$

Analytical calculation
($\beta_1 = 0.637$; $\beta_2 = 0.362$; $\varepsilon_{cu2} = 0.0027$; $\varepsilon_{cu2}E_s = 540$ N/mm^2)
The sector containing the solution is identified by means of Fig. 7.3 of 7.3. By the line 0 corresponding to a null stress for the reinforcement A_{s1}, we get:

$$N_{Rd} = -0.637 \cdot 400 \cdot 40 \cdot 39.6 + 2120 \cdot 391 = +425\,\text{kN}$$

By the line 1, we get $x = k_1 d' = 3,65 \cdot 40 = 146$ mm and therefore $N_{Rd} = -2120 \cdot 391 - 0.637 \cdot 400 \cdot 146 \cdot 39.6 + 2120 \cdot 391 =$ - 1473 kN

The solution passing through $N = -500$ kN is placed within sector 1. Therefore, we obtain:

$$x^2 - x\left(\frac{2120 \cdot 391 + 500{,}000 - 2120 \cdot 540}{0.637 \cdot 400 \cdot 39.6}\right) - \frac{2120 \cdot 540 \cdot 40}{0.637 \cdot 400 \cdot 39.6} = 0$$

The solution is $x = 77$ mm, and it follows that

$$\sigma_{s1} = 540\left(1 - \frac{40}{77}\right) = 259\,\text{N/mm}^2.$$

Finally, be means of (7.7), we get:

$$M_{Rd} = 2120 \cdot 259(200 - 40) + 0.637 \cdot 400 \cdot 77 \cdot 39.6\,(200 - 0.362 \cdot 77) + 2120 \cdot 391(200 - 40) = 354\,\text{kNm} > M_{Ed}.$$

Comparison with the General Method

To compare the obtained results with the simplified methods, the exercise has been solved through nonlinear analysis, as described in [5.8.6], without accounting for geometric imperfection (slope θ).

The analysis has been driven out by keeping constant the value of vertical load and by increasing the horizontal load, by simulating, in this way, the behaviour of a column under seismic action.

In Fig. 5.11, moment-mean curvature relationships are shown, with and without creep. Diagrams of Fig. 5.12 show the evolution of the mean curvature, of the horizontal displacement of the axis, and the total moment along the column, considering first and second order effects, under the action of design loads.

Table 5.6 collects the values of the total design moment (I + II order) at the base of the column obtained with the three models and the percentage variance of the two simplified methods to the general method, without accounting for geometric imperfection θ.

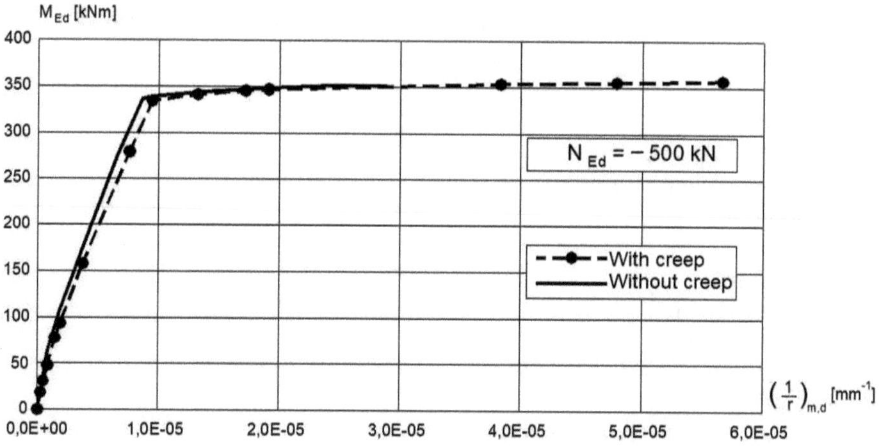

Fig. 5.11 Example 3: moment-mean curvature diagram in presence of design axial load

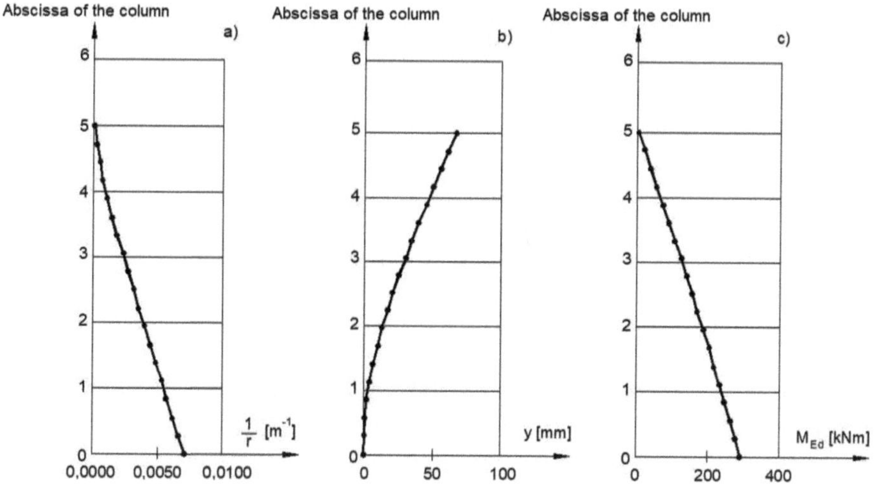

Fig. 5.12 Example 3: **a** mean curvature; **b** horizontal displacement of the axis; **c** total bending moment as the sum of first and second order effects

Table 5.6 Example 3: comparison of the three analysis methods

	M_{Ed}(I + II order) kNm	Variance
Nominal stiffness EC2 method	302.0	+3%
Nominal curvature method	315.0	+7%
General method	292.7	–

5.5.4.4 Example 4—Column 8 m Length Under Eccentric Axial Load and Transverse Load (Fig. 5.13)

Problem Data

Square cross-section with side
$h = 500$ mm.
Double symmetrical reinforcement ($4\phi 28 + 4\phi 26 = 4583$ mm^2 in each side).
$d' = 50$ mm.
Concrete C 90/105.
$f_{cd} = 51$ N/mm^2.
$E_{cm} = 44{,}000$ N/mm^2; $E_{cd} = E_{cm}/1{,}2 = 36{,}666$ N/mm^2 [5.20].
Internal force at SLU.
Axial force $N_{Ed} = 1000$ kN with $e = 0.20$ m.
Horizontal force $H_{Ed} = 50$ kN applied at the top end-section of the column.
Internal forces at SLS, quasi-permanent combination

$$N_{0Eqp} = 500 \text{ kN};\ e = 0.20 \text{ m};\ H_{0Eqp} = 25 \text{ kN}$$

Preliminary Calculations

Effective length

$$\ell_0 = 2 \cdot 8 = 16\text{m}$$

Slenderness

$$\lambda = \frac{\ell_0}{(h/3.46)} = \frac{16,000}{(500/3.46)} = 110.7$$

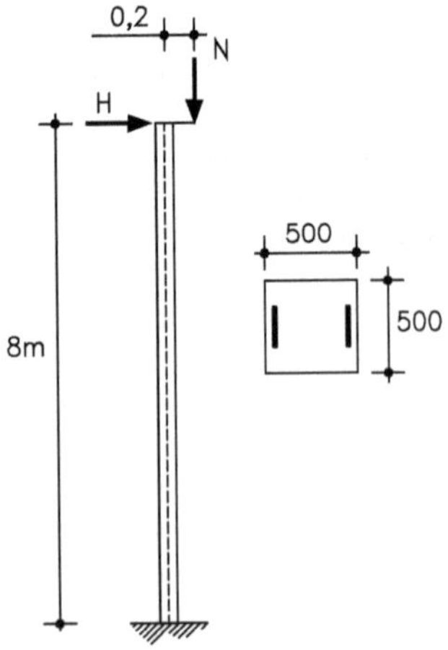

Fig. 5.13 Example 4: structural scheme

Reinforcement

$$\rho = \frac{2 \cdot 4583}{500^2} = 0.0366 < 0.04, \text{ maximum allowed [9.5.2(3)].}$$

Reinforcement mechanical ratio

$$\omega = \rho \frac{f_{yd}}{f_{cd}} = 0.0366 \cdot \frac{391}{51.0} = 0.2809$$

Internal forces at ULS
$N_{Ed} = 1000$ kN; $n = N_{Ed}/(A_c f_{cd}) = 0.0784$.
$e = 0.2$ m; $H_{Ed} = 50$ kN.
$M_{0Ed} = 1000 \cdot 0.2 + 50 \cdot 8 = 200 + 400 = 600$ kNm.
Internal forces due to constructive imperfections.
Verticality imperfection

$$\theta_{\rm i} = 0.005 \cdot 2/\sqrt{8} = 0.0035 \quad [5.1]$$

$$\theta i\, \ell = 0.028 \text{ m}$$

Moment due to imperfections

$$N_{Ed}(\theta_i \ell) = 28 \text{ kNm} \quad [5.2]$$

Moment due to horizontal force and imperfections

$$\mathrm{M_{0Ed}} = 400 + 28 = 428 \text{ kNm}$$

Moment due to vertical force

$$1000 \cdot 0.2 = 200 \text{ kNm}$$

First order total moment at ULS

$$M_{Ed} = 628 \text{ kNm}$$

Internal forces at SLS, quasi-permanent combination
$N_{0Eqp} = 500$ kN; $e = 0.2$ m; $H_{0Eqp} = 25$ kN.
$M_{0Eqp} = 500 \cdot 0.2 + 25 \cdot 8 = 300$ kNm.
Creep

$$\varphi_{(\infty,t_0)} = 1.30$$

$$\varphi_{ef} = \varphi_{(\infty,t_0)}(\mathrm{M_{0Eqp}/M_{0Ed}}) = 1.30 \cdot (300/628) = 0.62 \quad [5.19]$$

Limit slenderness according to [5.13 N]

$$\lambda_{\text{lim}} = 20\ \text{A B C} / \sqrt{n}$$

By means of the above mentioned data, we get:

$A = 1/(1 + 0.2 \cdot 0.62) = 0.89$

$B = \sqrt{1 + 2 \cdot 0.28} = 1.25$

$C = 0.7$, being an unbraced member (see Remark at [5.8.3.1])

$\sqrt{n} = 0.28$.

And, therefore,

$$\lambda_{\text{lim}} = 20 \cdot 0.89 \cdot 1.25 \cdot 0.7/0.28 = 44.5$$

Being $\lambda = 110.7$ greater than λ_{lim}, it is necessary to account for second order effects.

Method of the Nominal Stiffness

Solution according to the EC2 procedure

$E_{cd}\, I_c = 36{,}666 \cdot 500^4/12 = 19.0 \cdot 10^{13}$ Nmm^2.

$E_s\, I_s = E_s\, A_s\, (h/2 - d')^2$.

Being $A_s = 4583$ mm^2 on each side, we get

$E_s\, I_s = 200{,}000 \cdot 4583 \cdot 2 \cdot 200^2 = 7.33 \cdot 10^{13}$ Nmm^2

$$k_1 = \sqrt{\frac{f_{ck}}{20}} = 2.12$$

$k_2 = n_{Ed}\, \lambda/170 = 0.0784 \cdot 110.7/170 = 0.0511$

$k_c = k_1\, k_2/(1 + \varphi_{ef}) = 2.12 \cdot 0.0511/(1 + 0.62) = 0.067$

$EI = k_c\, E_{cd}\, I_c + E_s\, I_s = 0.0675 \cdot 19 \cdot 10^{13} + 7.33 \cdot 10^{13} = 8.60 \cdot 10^{13}$ N/mm^2

$N_B = \pi^2\, EI/\ell_0{}^2 = 9.86 \cdot 86.0 \cdot 10^{12}/256 \cdot 10^6 = 3312$ kN.

First order moment due to vertical load is constant along the height, therefore it is necessary to associate to it the value $\beta = \pi^2/8 = 1.23$. First order moment due to horizontal force and constructive imperfections is instead triangular, therefore we introduce the value $\beta = \pi^2/12 = 0.82$. We get therefore:

$$M_{Ed} = M_{0Ed}\left[1 + \frac{\beta}{\frac{N_B}{N_{Ed}} - 1}\right] = 200 \cdot \left[1 + \frac{1.23}{\frac{3312}{1000} - 1}\right]$$

$$+ 428 \cdot \left[1 + \frac{0.82}{\frac{3312}{1000} - 1}\right] = 886.2\ \text{kNm}$$

Solution of the Problem by Integration of the Differential Equation of the Elastic Curve

We collect below solving formulas of the examined example:

Eccentric axial load:

$$M_{tot} = N_{Ed}\frac{e}{\cos(\alpha\ell)}$$

Slope θ (geometric imperfection)

$$M_{\max} = N_{Ed} \cdot (\theta/\alpha) \cdot \tan(\alpha\ell)$$

Horizontal force H_{Ed} at the top end-section: this force produces the first order moment

$$M_0 = H_{Ed} \cdot \ell$$

The total moment in presence of N_{Ed} is equal to:

$$M_{tot} = M_0 \cdot \frac{\tan(\alpha\ell)}{\alpha\ell}$$

By introducing the problem data in the previous relationships, we get:

$$\alpha = \sqrt{\frac{N_{Ed}}{EI}} = \sqrt{\frac{100 \cdot 10^4}{86.0 \cdot 10^{12}}} = 1.078 \cdot 10^{-4}\,\text{mm}^{-1}$$

$$\alpha\ell = 1.078 \cdot 10^{-4} \cdot 8000 = 0.8627$$

$$\cos\alpha\ell = 0.6504$$

$$\tan\alpha\ell = 1.1678$$

$\theta/\alpha = 0.0035/(1.078 \times 10^{-4}) = 32.4$ mm.

Combined axial force and uniaxial bending: $M = \frac{1000 \cdot 0.20}{0.6504} = 307.7$ kNm.

Geometric imperfection: $M = 1000 \cdot 0.032 \cdot 1.1678 = 37.4$ kNm.

Horizontal force: $M = 50 \cdot 8 \cdot \frac{1.1678}{0.8627} = 541.4$ kNm.

Total moment: $307.7 + 37.4 + 541.4 = 8865$ kNm.

Solution of the Problem by Calculation of Displacements of the Direction of the Axial Load

Calculation of transverse displacements

Combined axial force and uniaxial bending moment

$$f_1 = \frac{N_{Ed} \cdot e \cdot \ell^2}{2EI} = \frac{1000 \cdot 0.20 \cdot 10^6 \cdot 64 \cdot 10^6}{2 \cdot 86 \cdot 10^{12}} = 74.4 \text{ mm}$$

Effect due to the horizontal force H_{Ed}

$$f_1 = \frac{H_{Ed} \cdot \ell^3}{3EI} = \frac{50 \cdot 10^3 \cdot 512 \cdot 10^9}{3 \cdot 86 \cdot 10^{12}} = 99.2\text{mm}$$

Geometric imperfection $\theta = 0.0035$; $\theta\ell = 28.0$ mm

$$s = \frac{1000 \cdot 512 \cdot 10^{12}}{3 \cdot 86 \cdot 10^{12}} \cdot 0.0035 = 6.9 \text{ mm}$$

Coefficient of displacement amplification
$N_B = 3312$ kN

$$\frac{1}{1 - N_{Ed}/N_B} = \frac{1}{1 - (1000/3312)} = 1.43$$

Sum of the flexural displacements and corresponding amplification
$(74.4 + 99.2 + 6.9) \cdot 1.43 = 258 \text{ mm} = 0.258$ m.
Total moment at the fixed point:
$M = 50.0 \cdot 8.0 + 1000 \cdot (0.20 + 0.028 + 0.258) = 886.0$ kNm.

Example 4bis

An analogous exercise may be performed through the application of the solving formulas of 5.5.2.4 (case c).

The force H_{Ed} is substituted by the distribution $q_H = 12.5$ kN/m producing the same moment at the base fixed point $M_0 = 400$ kNm. The column, therefore, experiences three different types of first order M diagrams:

- uniform because of N_{Ed}, with the value of $M = 200$ kNm at the fixed point;
- triangular because of the imperfections with $M_0 = 28$ kNm
- parabolic because of q_H, with $M_0 = 400$ kNm. Because of this distribution, we get $\beta = 0.62$.

Total moment at the fixed point is equal to:

$$M_{Ed} = 200 \cdot \left[1 + \frac{1,23}{\frac{3312}{1000} - 1}\right] + 28 \cdot \left[1 + \frac{0,82}{\frac{3312}{1000} - 1}\right] + 400 \cdot \left[1 + \frac{0,62}{\frac{3312}{1000} - 1}\right] =$$
$$200 \cdot 1.53 + 28 \cdot 1.35 + 400 \cdot 1.27 = 306.0 + 37.8 + 508.0 = 851.8 \text{ kNm}.$$

which is a value smaller than the one obtained with the force H_{Ed} because in this case, the diagram M0 due to q_H has a smaller area.

Solution of the problem by integration of the differential equation of the elastic curve

The moment due to the eccentric force N_{Ed} and to imperfections are those of Example 4.

Uniformly distributed horizontal forces.

$$qH = 12.5\text{kN/m}$$

Fixed point moment[1]:

$$M = \left(q_H \cdot \frac{\ell^2}{2}\right) \cdot \left[2 \cdot \frac{\tan(\alpha\ell)}{\alpha\ell} - 2 \cdot \frac{(1-\cos(\alpha\ell))}{(\alpha\ell)^2 \cdot \cos(\alpha\ell)}\right]$$
$$= 400 \cdot \left[2 \cdot \frac{1.1678}{0.8627} - 2 \cdot \frac{(1-0.6504)}{0.8627^2 \cdot 0.6504}\right] = 505.2\,\text{kNm}$$

Total moment

$$307.7\ + 37.4\ + 505.2\ = 850.3\ \text{kNm}$$

Solution of the problem by calculation of displacements of the direction of the axial load

- Combined axial force and uniaxial bending moment

$$f_1 = \frac{N_{Ed} \cdot e \cdot \ell^2}{2EI} = \frac{1000 \cdot 0.20 \cdot 10^6 \cdot 64 \cdot 10^6}{2 \cdot 86 \cdot 10^{12}} = 74.4\,\text{mm}$$

- Effect of the horizontal distribution q_H

$$f_1 = \frac{q_{hd} \cdot \ell^4}{8EI} = \frac{12.5 \cdot 8^4 \cdot 10^{12}}{8 \cdot 86 \cdot 10^{12}} = 74.4\,\text{mm}$$

- Geometric imperfection
  $\theta = 0.0035$; $\theta\ell = 28.0$ mm.

$$s = \frac{1000 \cdot 512 \cdot 10^{12}}{3 \cdot 86 \cdot 10^{12}} \cdot 0.0035 = 6.9\ \text{mm}$$

- Amplification coefficient of displacements
  Being
  $N_B = 3312$ kN, we get:

---

[1] The formula is deduced by the study of a vertical column with length 2ℓ joined with a vertical restraint and a support (with no rotational rigidity), under the load q_H and with reactions $q_H\ell$ (taken from O. Belluzzi, Scienza delle costruzioni, 1953, vol. 1, pag. 288, formulas (4.31) and (4.34)).

$$\frac{1}{1 - N_{Ed}/N_B} = \frac{1}{1 - (1000/3312)} = 1.43$$

Sum of the flexural displacements and relevant amplification:

$$(74.4 + 74.4 + 6.9) \cdot 1.43 = 222 \text{ mm} = 0.222 \text{ m}$$

Total moment at the fixed point:
$M = 50.0 \cdot 8.0 + 1000 \cdot (0.20 + 0.028 + 0.222 = 850.0 \text{ kNm})$

Method of the Nominal Curvature

We introduce the value $\omega = 0.28$ adopted by means of the stiffness method and we develop the calculation according to EC2 procedure.

$n_{\text{bal}} = 0.4$
$n_u = 1 + \omega = 1.28.$
$k_r = (n_u - n)/(n_u - n_{\text{bal}}) = (1.28 - 0.0784)/(1.28 - 0.4).$
Being n smaller than 0,4, we assume $k_r = 1$.
$\beta = 0.35 + f_{ck}/200 - \lambda/150 = 0.35 + 0.45 - 110.7/150 = 0.062.$
$k_\varphi = 1 + \beta\omega_{ef} = 1 + 0.062 \cdot 0.62 = 1.04.$
$\epsilon_{yd} = 1.96 \times 10^{-3}.$
$d = h - d' = 450$ mm.
$(1/r)_0 = 1.96 \times 10^{-3}/(0.45\ d) = 9.67 \times 10^{-6}$ mm^{-1}.
$(1/r) = k_r\ k_\varphi\ (1/r)_0 = 1 \cdot 1.04 \cdot 9.67 \cdot 10^{-6} = 10.0 \times 10^{-6}$ mm^{-1}.
$e_2 = (1/r)\ \ell_0{}^2/10 = 10.00 \times 10^{-6} \cdot 256 \times 10^6/10 = 256$ mm.
$M_2 = N_{Ed}\ e_2 = 1000 \cdot 0.256 = 256$ kNm.
$M_{Ed} = M_{0Ed} + M_2 = 628 + 256 = 884$ kNm.

Verification of the Cross-Section at the Restraint

Interaction diagram

The design bending resistance ($\mu = 0.145$) is obtained from the Table U14 corresponding to $f_{ck} = 90$ N/mm^2, with $\omega = 0.28$, in presence of normal force, from which $\nu = \frac{1{,}000{,}000}{500 \cdot 500 \cdot 51} = 0.0784$, and we get

$$M_{Rd} = 0.145 \cdot 500^3 \cdot 51.6 = 924 \text{kNm} > M_{Ed}$$

Analytical calculation

$$(\beta_1 = 0.583;\ \beta_2 = 0.353;\ \varepsilon_{cu2} = 0.0026;\ \varepsilon_{cu2} E_s = 520 \text{N/mm}^2)$$

With reference to Fig. 7.3 of 7.3, we identify the sector where the solution is placed. By the line 0, where the stress of reinforcement A_{s1} is zero, we get:

$$N_{Rd} = -0.583 \cdot 500 \cdot 50 \cdot 51 + 4583 \cdot 391 = -743 + 1792 = +1049$$

By the line no. 1, we get $x = k_1 d' = 4.06 \cdot 50 = 203$ mm and therefore

$$N_{Rd} = -0.353 \cdot 500 \cdot 203 \cdot 51 = -1287 \text{ kN}$$

The solution if $N = -1000$ kN is placed within sector no. 1. Therefore, we get:

$$x^2 - x\left(\frac{4583 \cdot 391 + 1,000,000 - 4583 \cdot 520}{0.583 \cdot 500 \cdot 51}\right) - \frac{4583 \cdot 520 \cdot 50}{0.637 \cdot 400 \cdot 39.6} = 0$$

The solution gives $x = 104.3$ mm from which we get

$$\sigma_{s1} = 520\left(1 - \frac{50}{104.7}\right) = 270.7 \text{ N/mm}^2$$

Finally, through (7.7), we obtain:

$$M_{Rd} = 4583 \cdot 270.7(250 - 50) + 0.583 \cdot 500 \cdot 104.3 \cdot 51(250 - 0.353 \cdot 104.3) + 4583 \cdot 391(250 - 50) = 937\text{kNm} < M_{Ed}$$

Example No. 5—Column 6 m Length Under Normal Force and Biaxial Bending (Fig. 5.14)

Problem Data

Rectangular cross-section with sides.
$h = 800$ mm, $b = 500$ mm.

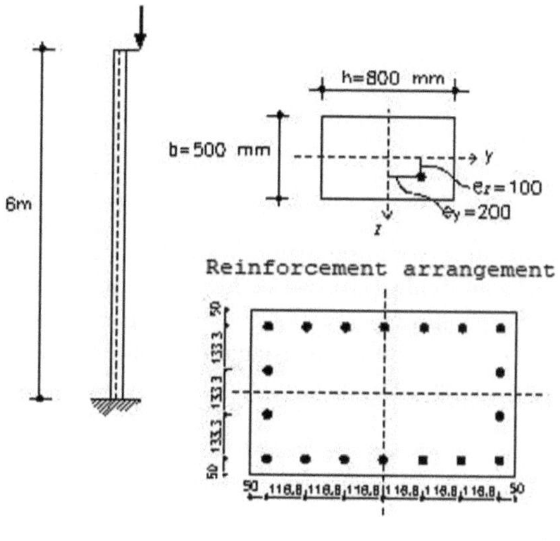

Fig. 5.14 Example no. 5: structural scheme and reinforcement arrangement

Reinforcement.

18 bars 20 mm diameter arranged as follow: 4 on each side parallel to the smaller side ($a_s = 1256$ mm^2); 5 on each side parallel to the greater side ($A_s = 1571$ mm^2). Total area $A_s = 5654$ mm^2 (Fig. 5.14).

Concrete.

C30/37; $f_{cd} = 0.85 f_{ck}/1.5 = 17.0$ N/mm^2.

$E_{cm} = 32{,}000$ N/mm^2.

$E_{cd} = E_{cm}/1.2 = 26{,}666$ N/mm^2.

Effective length

$$\ell_o = 2 \times 6 = 12\, m$$

Biaxial internal force at ULS.

$N_{Ed} = 1000$ kN with eccentricities: $e_z = 0.10$ m; $e_y = 0.20$ m.

Internal forces at SLS, quasi-permanent combination.

$N_{0Eqp} = 500$ kN with the same eccentricities of ULS.

The calculation procedure of [5.8.9(2)] is applied.

First, even if a biaxial moment is present, we perform the verifications of each component with corresponding geometric imperfections.

Formula [5.8.9(3)] says that further verifications are not required if the slenderness ratios fulfil [(5.38a)] and if the corresponding eccentricities fulfil at least one of [(5.38b)]. We calculate therefore separately biaxial bandings along z and y, by applying the method of the nominal stiffness.

Preliminary Calculations

Once reinforcement is assigned, we can deduce:

- geometric ratio

$$\rho = \frac{A_s}{bh} = \frac{5654}{500 \cdot 800} = 0.0141 < 0.04 \quad [9.5.2(3)]$$

- mechanical ratio

$$\omega = \rho \frac{f_{yd}}{f_{cd}} = 0.0141 \cdot \frac{391}{17} = 0.325$$

From Table U16, we deduce the nondimensional design bending resistance μ corresponding to $\omega = 0.325$. We get: $\mu = 0.13$.

Design bending resistances of the cross-section at ULS, employed in the verifications, are the following:

$$M_{Rdz} = \mu \cdot bh^2 f_{cd} = 0.13 \cdot 500 \cdot 800^2 \cdot 17 = 707\,\text{kNm}$$

$$M_{Rdy} = \mu \cdot hb^2 f_{cd} = 0.13 \cdot 800 \cdot 500^2 \cdot 17 = 442\text{kNm}$$

Nondimensional design normal force is:

$$n = v = \frac{N_{Ed}}{A_c f_{cd}} = \frac{1{,}000{,}000}{800 \cdot 500 \cdot 17} = 0.1471$$

Combined Axial Force and Uniaxial Bending Moment with Rotation Around the Axis Z

Limit slenderness can be calculated as follows:

$$\lambda_{\text{lim}}{=}20 \cdot A \cdot B \cdot C/\sqrt{n} \quad [(5.13)]$$

where

$$C = 0.7\text{, as suggested by EC2 for unbraced structures}$$

$$\begin{aligned} A &= 1/(1 + 0.2\varphi_{ef}) = 1/(1 + 0.2 \cdot 1) \\ &= 0.833 \ (\text{for}\varphi_{ef} = 1),\ \text{see creep in the following}) \end{aligned}$$

$$B = \sqrt{1 + 2\omega} = \sqrt{1 + 2 \cdot 0.325} = 1.28$$

$$\sqrt{n} = \sqrt{0.1471} = 0.38$$

We get

$$\lambda_{\text{lim}} = 20 \cdot 0.833 \cdot 1.28 \cdot 0.7/0.38 = 39$$

Slenderness λ_z is equal to

$$\lambda_z = \frac{12}{(0.8/3.46)} = 51.9 > 39$$

Therefore, it is necessary to account for second order effects.
Internal forces at ULS

$$N_{Ed} = 1000\,\text{kN}$$

$$M_{0Edz} = 1000 \cdot 0.20 = 200\,\text{kNm}$$

Internal forces at SLS, quasi-permanent combination

$$N_{0Eqp} = 500 \text{ kN}$$
$$M_{0Eqp} = 100 \text{ kNm}$$

Creep

$$\varphi_{(\infty,t_0)} = 2$$

$$\varphi_{ef} = \varphi_{(\infty,t_0)} \cdot (M_{0Eqp}/M_{0Ed}) = 2 \cdot (100/200) = 1.0 \quad [(5.19)]$$

Internal forces due to constructive imperfections
Verticality imperfection.

$$\theta_i = .0.0052/\sqrt{6} = 0.0045 \quad [(5.1)]$$

$$\theta_i \ell = 25 \text{ mm.}$$

Bending due to imperfections

$$N_{Ed} \cdot (\theta_i \ell) = 25 \text{ kNm} \quad [(5.2)]$$

First order total moment at ULS
$M_{Ed} = 225$ kNm

$$I_{sz} = 2 \cdot 1256 \cdot 350^2 + 2 \cdot 2 \cdot 314 \cdot (116.6^2 + 233.3^2) = 393 \times 10^6 \text{mm}^4$$

$$E_s I_{sz} = 200{,}000 \cdot 393 \times 10^6 = 7.86 \times 10^{13} \text{Nmm}^2$$

$$E_{cd} I_c = 26{,}666 \cdot 800^3 \cdot 500/12 = 5.68 \times 10^{14} \text{Nmm}^2$$

$$\lambda_z = 51.9$$

$$k_1 = \sqrt{\frac{f_{ck}}{20}} = \sqrt{\frac{30}{20}} = 1.22 \quad [(5.23)]$$

$$k_2 = n_{Ed} \cdot \lambda/170 = 0.1471 \cdot 51.9/170 = 0.0449 \quad [(5.24)]$$

$$k_c = k_1 k_2/(1 + \varphi_{ef}) = 1.22 \cdot 0.0449/2 = 0.0274 \quad [(5.22)]$$

$$EI = k_c E_{cd} I_c + E_s I_{sz} = 0.0274 \cdot 5.68 \times 10^{14} + 7.86 \times 10^{13}$$
$$= 9.41 \times 10^{13} \text{Nmm}^2 \quad [(5.21)]$$

$$N_B = \pi^2 EI / \ell_0^2 = 9.86 \cdot \frac{9410 \cdot 10^9}{12,000^2} = 6443 \, \text{kN}$$

Being

$$\beta = \pi^2 / c_0 = 1.23 \text{ (with } c_0 = 8, M_1 \text{uniform diagram)}[(5.29]$$

we get, by applying, slightly rounding up, also to imperfections:

$$M_{Edz} = M_{0Edz} \left[1 + \frac{\beta}{\frac{N_B}{N_{Ed}} - 1} \right] = 225 \left[1 + \frac{1,23}{\frac{6443}{1000} - 1} \right] \quad [(5.28)]$$

In nondimensional terms, combined axial force and uniaxial bending moment are expressed by:

$$\nu = n_{Ed} = \frac{N_{Ed}}{A_c f_{cd}} = \frac{1{,}000{,}000}{800 \cdot 500 \cdot 17} = 0.1471$$
$$\mu = \frac{M_{Edz}}{h^2 b f_{cd}} = \frac{276{,}000{,}000}{800^2 \cdot 500 \cdot 17} = 0.0500$$

The interaction diagram of Table U16 shows that the internal force is suitable with $\omega = 0.325$.

Combined Axial Force and Uniaxial Bending Moment with Rotation Around the y-axis

Limit slenderness, calculated by (5.13N), is given by:

$$\lambda_{\text{lim}} = 39$$

$$\lambda_y = 12/(0.5/3.46) = 83 > 39$$

It is necessary to take into account second order effects.
Internal forces at ULS.
$N_{Ed} = 1000$ kN; $n_{Ed} = N_{Ed}/(A_c f_{cd}) = 0.1471$.
$M_{Ed} = 1000 \cdot 0.10 = 100$ kNm.
Internal forces at SLS, quasi-permanent combination.
$N_{0Eqp} = 500$ kN.
$M_{0Eqp} = 50$ kNm.

Creep

$$\varphi_{(\infty,t_0)} = 2$$

$$\varphi_{ef} = \varphi_{(\infty,t_0)} \cdot (M_{0Eqp}/M_{0Ed}) = 2 \cdot (100/200) = 1.0 \quad [(5.19)].$$

Internal forces due to constructive imperfections.
$M = 25$ kNm.
First order total moment at ULS.

$$M_{0Edy} = 100 + 25 = 125\text{kNm}$$

$$E_{cd}I_c = 26,666 \cdot 5003 \cdot 800/12 = 2.22 \cdot 1014\text{Nmm}^2$$

$$I_{sy} = 2 \cdot 1568 \cdot 200^2 + 2 \cdot 2 \cdot 314 \cdot (66^2 + 200^2) = 181 \cdot 10^6\text{mm}^4$$

$$E_s I_{sy} = 200,000 \cdot 181 \times 10^6 = 3.62 \times 10^{13}\text{Nmm}^2$$

$$\lambda_z = 83.0$$

$$k_1 = \sqrt{\frac{f_{ck}}{20}} = 1.22 \quad [(5.23)]$$

$$k_2 = n_{Ed}\lambda/170 = 0.1471 \cdot 83/170 = 0.0718 \quad [(5.24)]$$

$$k_c = k_1 \cdot k_2/(1 + \varphi_{ef}) = 1.22 \cdot 0.0718/2 = 0.0438 \quad [(5.22)]$$

$$EI = k_c E_{cd} I_c + E_s I_{sy} = 0.0438 \cdot 22.2 \times 10^{13} + 3.62 \times 10^{13} = 4,59 \cdot 10^{13}\text{Nmm}^2$$

$$N_B = \pi^2 EI/\ell_0^2 = 9.86 \cdot \frac{45.9 \times 10^{12}}{12{,}000^2} = 3142\,\text{kN}$$

Being
$\beta = \pi^2/c_0 = 1.23$.
with $c_0 = 8$ and uniform diagram M_1 applied (slightly rounding up) also to imperfections, we get:

$$M_{Edy} = M_{0Ed}\left[1 + \frac{\beta}{\frac{N_B}{N_{Ed}} - 1}\right] = 125\left[1 + \frac{1.23}{\frac{3142}{1000} - 1}\right] = 197\text{ kNm}$$

In nondimensional terms, biaxial bending is given by:

$$\nu = n_{Ed} = \frac{N_{Ed}}{A_c f_{cd}} = \frac{1{,}000{,}000}{800 \cdot 500 \cdot 17} = 0.1471$$
$$\mu = \frac{M_{Edy}}{hb^2 f_{cd}} = \frac{197{,}000{,}000}{800 \cdot 500^2 \cdot 17} = 0.0579$$

The interaction diagram of Table U16 shows that such an internal force is compatible with $\omega = 0.325$.

Verification of the Conditions for the Acceptability of the Separated Verifications

Because of [(5.38a)],

$$\frac{\lambda_y}{\lambda_z} = \frac{83.0}{51.9} = 1.6 \le 2$$
$$\frac{\lambda_z}{\lambda_y} = \frac{51.9}{83.0} = 0.62 \le 2$$

Therefore, equation [5.38a] is fulfilled.

$$\frac{e_y}{h} = \frac{200}{800} = 0.25 \text{ and} \frac{e_z}{b} = \frac{100}{500} = 0.20$$

Being

$$\frac{0.25}{0.20} = 1.25 > 0.2 \text{ and } \frac{0.20}{0.25} = 0.8 > 0.2$$

[(5.38b)] is not fulfilled. We apply therefore equation [(5.39)] of [5.8.9(4)].
Preliminary calculation of (N_{Ed}/N_{Rd}).

$$N_{Rd} = A_c f_{cd} + A_s f_{yd} = 800 \cdot 500 \cdot 17 + 5654 \cdot 391 = 9010\text{kN}$$

$$N_{Ed}/N_{Rd} = 1000/9010 = 0.11$$

By interpolation, we adopt the exponent $a = 1.0083$.
Through [(5.39)],

$$\left(\frac{M_{Edz}}{M_{Rdz}}\right)^{1.0083} + \left(\frac{M_{Edy}}{M_{Rdy}}\right)^{1.0083} = 0.3874 + 0.4427 = 0.8301 < 1.$$

The verification is then fulfilled.

5.5.4.5 Example No. 6—Column 6 m Length Under Normal Force and Biaxial Bending (Fig. 5.15)

Problem Data

Square cross-section with side $h = 500$ mm.
Reinforcement uniformly distributed along the 4 sides ($28\varphi 20$) with

$$d = 50\text{mm}$$

Concrete
C 40/50; $f_{cd} = 22{,}66$ N/mm^2.

$$E_{cm} = \frac{35,000\text{N}}{\text{mm}^2}; \; E_{cd} = \frac{E_{cm}}{1.2} = \frac{29,160\text{N}}{\text{mm}^2} [5.20].$$

Internal forces at ULS
Axial load $N_{Ed} = 900$ kN with $e_{z0} = 0.250$ m, $e_{y0} = 0.250$ m.
Internal forces at SLS, quasi-permanent combination
$N_{0Eqp} = 450$ kN with the same eccentricities of SLU.

Preliminary Calculations

Effective length

$$\ell_0 = 2 \times 6 = 12 \text{ m}$$

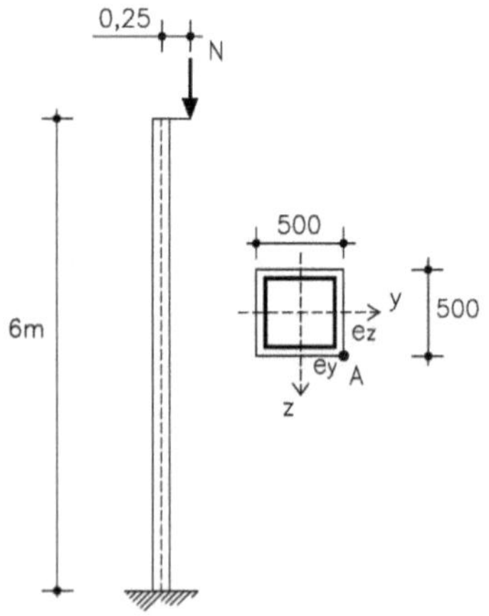

Fig. 5.15 Example no. 6: structural scheme

Slenderness

$$\lambda_y = \lambda_z = \frac{\ell_0}{(h/3.46)} = \frac{12{,}000}{(500/3.46)} = 83$$

First order biaxial internal force at ULS.

$N_{Ed} = 900$ kN with eccentricity: $e_{z0} = e_{y0} = 250$ mm applied in the point A of the cross-section.

$n_{Ed} = N_{Ed}/(A_c f_{cd}) = \frac{900{,}000}{500^2 \cdot 22.66} = 0.1589.$

$M_{Edy} = M_{Edz} = N_{Ed}\, e = 900 \cdot 0.25 = 225$ kNm.

Internal forces due to constructive imperfections.

Verticality imperfections.

$\theta_i = 0.005 \cdot 2/\sqrt{6} = 0.0041.$

$\theta_i \cdot \ell = 0.025$ m.

Moment due to imperfections.

$N_{Ed}\,(\theta_i \cdot \ell) = 900 \cdot 0.025 = 22.5$ kNm.

Total first order monoaxial moment at ULS.

$M_{Ed} = 225.0 + 22.5 = 247.5$ kNm.

Internal forces at SLS, quasi-permanent combination

$N_{0Eqp} = 450$ kN with the same eccentricities.

$M_{0Eqp} = 112.5$ kNm (both along y and z).

Creep

$\varphi_{(\infty,t_0)} = 2.5$

$\varphi_{ef} = \varphi_{(\infty,t_0)}\,(M_{0Eqp}/M_{0Ed}) = (112.5/247.5) \cdot 2.5 = 1.13.$

We remark that relationships [(5.38)] are fulfilled because of the symmetry of cross-section and internal forces, while neither of the [5.38b] is satisfied, in fact, $\frac{e_y/h}{e_z/b} = \frac{0.5}{0.5} = 1 > 0.2$. It is therefore necessary to account for biaxial bending and proceed according to [5.8.9(4)].

Evaluation of Total Monoaxial Internal Forces at ULS

The method of nominal stiffness is applied.

Concrete

$$I_c = 5208.3 \times 106\text{mm}^4$$

$$\lambda_c = 83$$

$$k_1 = \sqrt{\frac{40}{20}} = 1.41$$

$$k_2 = nEd\lambda/170 = 0.1589 \cdot 83/170 = 0.07769$$

$$k_c = k_1 k_2/(1 + ef) = 1.41 \cdot 0.0776/(1 + 1.13) = 0.0514$$

$$k_c E_{cd} I_c = 0.0514 \cdot 29,160 \cdot 5208.3 \cdot 10^6 = 0.78 \cdot 10^{13} \text{Nmm}^2$$

Steel

$$As = 28 \cdot 314 = 8796 \text{mm}^2$$

we indicate by A^*_s the area distributed along each side, i.e. being A_s the total area, we get $A^*_s = A_s/4$. The reinforcement per unit length is $q = A^*s/400$.

We get therefore:

$$A^*s = As/4 = \frac{8796}{400} = 2200\,\text{mm}^2$$

$$q = A^*s/400 = \frac{2200}{400} = 5.5\,\text{mm}^2/\text{mm}$$

The inertial moment I_s of the reinforcement about a barycentric axis parallel to a couple of sides can be written as:

$$I_s = 2 \cdot A_s^* \cdot (\frac{h}{2} - d')^2 + 2q \cdot \frac{(h - 2d')^3}{12} = 2 \cdot 2200 \cdot (200)^2 + 2 \cdot 5.50 \cdot \frac{(400)^3}{12}$$
$$= 234.6 \times 10^6\ \text{mm}^4$$

$$E_s I_s = 234.6 \times 10^6 \cdot 200,000 = 4.69 \times 10^{13}\ \text{Nmm}^2$$

Finally,
$EI = 0.78 \cdot 10^{13} + 4.69 \cdot 10^{13} = 5.47 \cdot 10^{13}$ Nmm^2.
$N_B = \theta^2\ EI/l_0{}^2 = 9.86 \cdot \frac{5.47 \cdot 10^{13}}{12,000^2} = 3748$ kN.
We obtain:

$$M_{Ed} = M_{0Ed}\left[1 + \frac{\beta}{\frac{N_B}{N_{Ed}} - 1}\right] = 247.5 \cdot \left[1 + \frac{1.23}{\frac{3748}{900} - 1}\right] = 344.0\,\text{kNm}$$

$$N_{Rd} = 5002 \cdot 22.66(1 + \omega) = 9064\text{kN}$$

$$\frac{N_{Ed}}{N_{Rd}} = \frac{900}{9064} = 0.1$$

From the interaction diagram (Table U16), design bending resistance μ_{Rd} associated to $\nu = n_{Ed} = 0.1589$ and $\omega = 0.60$ is equal to $\mu_{Rd} = 0.26$.

Therefore, design bending resistance is equal to $M_{Rd} = 0.26 \cdot 500^3 \cdot 22.66 =$ 736.4 kNm.

Biaxial Verification

According to [5.8.9(4)], the value 0.1 of the ratio $\frac{N_{Ed}}{N_{Rd}}$ is corresponding to the unitary value of the exponent in the equation [(5.39)]. This equation, being bending moments about axes y and z equal, gives

$$2 \cdot \left(\frac{344.0}{736.4}\right)^1 = 0.934 < 1$$

Therefore, the verification is fulfilled.

Example No. 7—Verification of a Column in a Structure

All the geometric quantities represented in Fig. 5.3 (together with dimensions and supports) are considered, with the only exception of the storey height, equal to 5.00 m instead of 3.00 m.

Acting normal load on the column: 500 kN.

End-sections moments $M_{02} = 75$ kNm, $M_{01} = 50$ kNm (moments with similar signs tending fibers placed on the same side)

Materials: Concrete C25/30; $f_{cd} = 14.16$ N/mm^2; $E_{cm} = 31{,}000$ kN/mm^2

Steel $f_{yd} = 391$ N/mm^2

Column with square cross-section with side 350 mm.

Column reinforcement: 8 bars 14 mm diameter, (in the 4 corners and in the middle of the side cross-section): $A_s = 1232$ mm^2; we get a reinforcement geometric and mechanical ratio, respectively equal to:

$$\rho = \frac{1232}{350^2} = 0.01; \omega = 0.01 \cdot 391/14.16 = 0.277.$$

For the buckling verification, by taking into account only (3 + 3) bars arranged on the opposite sides, we obtain $\omega = (6/8) \cdot (0.01 \cdot 391/14.16) = 0.208$.

Calculation of the flexibilities k_1 and k_2 corresponding to upper and lower nodes.

We assume the same elastic modulus for beams and the column. Inertial moments are assumed equal to those calculated in 5.4.1.2, here recalled:

- for the column: $I_c = 12,5 \cdot 10^8 mm^4$
- for the beams: $I_{\text{beam}} = 5,20 \cdot 10^8 mm^4$, which takes into account cracking in some ways.

By applying formulas of 5.4.1.1, we get:

$$k_1 = \frac{(12.50 \cdot 10^8/5000) \cdot 2}{3 \cdot 5.20 \times 10^8/4500 + 4 \cdot 5.20 \times 10^8/5000} = 0.65$$

$$k_2 = \frac{(12.50 \times 10^8/5000) \cdot 2}{4 \cdot 5, .20 \times 10^8/4500 + 4 \cdot 5.20 \times 10^8/5000} = 0.57.$$

Using [(5.15)], the effective length is given by:

$$\ell_0 = 0.50 \cdot 5.0 \cdot \sqrt{\left(1 + \frac{0.65}{0.45 + 0.65}\right) \cdot \left(1 + \frac{0.57}{0.45 + 0.57}\right)} = 3.94$$

Column slenderness is given by:

$$\lambda = \frac{\ell_0}{i} = \frac{3940}{350/3.46} = 39.$$

By BSI Tables for braced columns (5.4.1.1.1), under the conditions of top end-section 2 and of base 2, corresponding to $\beta = 0.85$, we get the effective length $\ell_o = 0.85 \cdot 5.00 = 4.25$ m and the slenderness.

$$\lambda = \frac{4250}{350/3.46} = 42.$$

This value will be adopted because more severe for the verification.

Limit slenderness is given by

$$\lambda_{\text{lim}} = 20 \cdot A \cdot B \cdot C/\sqrt{n} \quad [(5.13\text{N})].$$

being:

$$A = 1/(1 + 0.2 \cdot \varphi_{ef}) = 1/(1 + 0.2 \cdot 1) = 0.833$$

$$(\text{for}\varphi_{ef} = 1 \text{ see the following paragraph})$$

$$B = \sqrt{1 + 2 \cdot 0.277} = 1.246.$$

$$C = 1.7 - \frac{50}{75} = 1.033$$

$$n = \frac{500,000}{350 \cdot 350 \cdot 14,16} 02883; \sqrt{0.2883} = 0.5369.$$

We get

$$\lambda_{\text{lim}} = 20 \cdot 0.833 \cdot 1.246. \cdot 1.033/0.5369 = 39.9 < \lambda$$

Second order effects will be taken into account.

Method of the Nominal Curvature

First order internal forces.

By means of [(5.32)], the equivalent moment is given by:

$$M_{0e} = 0.6M_{02} + 0.4M_{01} = 0.6 \cdot 75 + 0.4 \cdot 50 = 65\,\text{kNm} > 0.4 \cdot 75 = 60\,\text{kNm}.$$

Moment due to geometric imperfections:

According to [5.2(9)], for isolated columns in braced structures (as in the studied case), the effect due to imperfections can be taken into account through an eccentricity.

$$e_i = \frac{\ell_0}{400}.$$

Therefore, we get, with $\ell_0 = 4250$ mm.

$$e_i = \frac{4250}{400} = 10.6\text{ mm} = 0.0106\text{ m}$$

and the moment

$$M_{\text{imp}} = 500 \times 0.010 = 5.0\text{kNm}.$$

First order total moment:

$$M_0 = 65.0 + 5.0 = 70.0\text{kNm}$$

Calculation of the second order eccentricity.

$$e_2 = (1/r) \cdot \ell_0^2/c$$

where:

$$\frac{1}{r} = K_r K_\varphi \cdot \frac{1}{r_0}\text{e}(1/\text{r}_0) = \varepsilon_{yd}/(0.45d)$$

$$K_r = 1 \text{ being } n = \frac{N}{A_c f_{cd}} = \frac{500000}{350^2 \cdot 14.16} = 0.288 < n_{bal} = 0.4$$

$$K_\varphi = 1 + \beta\varphi_{ef} \quad [(5.37)]$$

$$\beta = 0.35 + f_{ck}/200 - \lambda/1500 = 0.195$$

$$\varphi_{ef} = \varphi_{(\infty,t_o)} \cdot (M_{Eqp}/M_{0Ed}) = 2 \cdot (0.5) = 1$$

$$K_\varphi = 1 + 0,195 \cdot 1 = 1.195$$

$$(1/r) = 1.195 \cdot \frac{391}{200,000} \cdot \frac{1}{0.45 \cdot 315} = 16.5 \times 10^{-6}\text{mm}^{-1}$$

and, with $c = 10$.

$$e_2 = 16.5 \times 10^{-6} \cdot 4.25^2 \times 10^6/10 = 30\text{mm}$$

Second order moment is therefore given by:

$$M_2 = 500 \times 0.030 = 15.0\text{kNm}$$

Total acting moment is given by.

$$M_{Ed} = 70.0 + 15.0 = 85.0\text{kNm}.$$

Method of the Nominal Stiffness

Stiffness calculation:

$$EI = K_c E_{cd} I_c + K_s I_s \quad [(5.21)]$$

and, using [(5.22)],

$$K_c = k_1 k_2/(1 + \varphi_{ef});\ K_s = 1$$

$$k_1 = \sqrt{25/20} = 1.12$$

$n = 0.288$ (from the nominal curvature method).

$$k_2 = n\frac{\lambda}{170} = 0.288 \cdot \frac{42}{170} = 0.07$$

$$\varphi_{ef} = \varphi_{(\infty,t_o)} \cdot (M_{Eqp}/M_{oEd}) = 2 \times (0.5) = 1$$

$$K_c = 1.12 \cdot 0.07/(1+1) = 0.039$$

$$I_c = \frac{350^4}{12} = 1.25 \times 10^9 \text{mm}^4$$

$$E_{cd} = \frac{E_{cm}}{1.2} = \frac{31,000}{1.2} = 25,800 \text{ N/mm}^2$$

$$I_s = 2 \cdot 462 \cdot 135^2 = 16.8 \times 10^6 \text{mm}^4$$

$$EI = 25,800 \cdot 1.25 \times 10^9 \cdot 0.039 + 200,000 \cdot 16.8 \cdot 10^6 = 4.61 \times 10^{12} \text{Nmm}^2$$

The eulerian critical load is given by:

$$N_b = \frac{9.86 \cdot 4.61 \times 10^{12}}{4250^2} = 2516 \text{ kN}$$

First order total moment is given by: $65.0 + 5.0 = 70.0$ kNm, including geometric imperfections.

Taking into account that diagram M can be considered as uniform, with $\beta = 1.23$, the ultimate acting bending moment is given by:

$$M = 70.0 \cdot \left[1 + \frac{1.23}{\frac{2516}{500} - 1}\right] = 87.0 \text{ kNm.}$$

Verification

Using the two calculation procedures, we obtained bending moments equal, respectively, to 85 and 87 kNm. In nondimensional terms, internal forces for the second calculation are the following:

$$\nu = \frac{500,000}{350^2 \cdot 14.16} = 0.288 = \mu = \frac{87 \times 10^6}{350^3 \cdot 14.16} = 0.143.$$

Considering interaction diagrams of Figure U.10 in the Appendix, the above-calculated couple of internal forces requires $\omega = 0.12$. As recalled in para. 5.5.4.7, the arranged mechanical ratio is 0.208, greater than the required one.

The verification is therefore fulfilled.

Table 5.7 Synthesis of developed examples

Example no	Method of the nominal stiffness			Method of the nominal curvature	General method	Method of the nominal stiffness (biaxial)
	Direct	Differential equation	Displacement of the line			
1	X	X	x	x	x	–
2	X	x	x	x	–	–
3	X	x	x	x	x	–
4	X	x	x	x	–	–
4bis	X	x	x	–	–	–
5	–	–	–	–	–	x
6	–	–	–	–	–	x
7	X	–	–	x	–	–

5.5.5 Synthesis of Developed Examples

Table 5.7 resumes all the calculation procedures adopted in the examples previously developed.

References

Applications de l'Eurocode 2—Calcul des bâtiments en béton—(Sous la direction de J.A.Calgaro et J.Cortade)—Presses de l'Ecole Nationale des Ponts et chaussées—Paris—2005.
Belluzzi, O. Scienza delle costruzioni—Zanichelli, Bologna—1953 (in Italian).
British Standard Institution—BS 8110—1985. Part 1: Code of practice for design and construction.
Westerberg, B. (2004). Second order effects in slender concrete structures. TRITA-BKN. Rapport 77—ISSN 1103-4289—Stockholm.

Chapter 6
Prestressed Concrete

Abstract The chapter deals with the design of prestressed concrete, which presents some peculiar characteristics with respect to reinforced concrete. In EC2, the term "prestress" indicates all the effects of prestressing, which are both internal forces and strains. Today, prestressing is one of the most common techniques for slabs and industrial buildings, road and railway bridges, tanks, towers, antennas, and, in general, in the precast industry. Most of the rules for the design of prestressed concrete are discussed in Sect. 5.10 of EC2, but some rules are dealt with in other sections. Therefore, to help the reader in the identification of all EC2 rules for prestressed concrete, they all are collected in this chapter. The prestress load is applied through high-strength steel wires, strands, or bars. Prestressing steel may be embedded in the concrete or placed out of the structure with discrete contact points at deviators and anchorages. Prestressing steel can be pre-tensioned and bonded or post-tensioned and bonded or unbonded. Following topics are discussed in detail in the chapter: detailing and mechanical properties of prestressing steel, limitations to prestress intensity, losses of prestress, ultimate limit states of prestressed concrete. Many case studies show the application of rules to real cases.

6.1 Introduction

The idea of introducing a compression force into reinforced concrete beams to overcome the low tensile strength of concrete is due to the French engineer Eugène Freyssinet and goes back to 1928. Prestress can be defined as a permanent load applied to RC beams before service loads to improve their structural capacity. Traditional prestressing methods introduce compression forces into reinforced concrete beams through tensioning steel tendons between end-anchorages.

In EC2, the term "prestress" indicates all the effects of prestressing, which are mainly internal forces and strains. Prestress may also be induced by not direct actions, but this aspect is not considered in EC2.

This chapter was authored by Franco Angotti and Maurizio Orlando.

© The Author(s), under exclusive license to Springer Nature Switzerland AG 2022
F. Angotti et al., *Reinforced Concrete with Worked Examples*,
https://doi.org/10.1007/978-3-030-92839-1_6

The design of prestressed concrete structures does not differ from the design of reinforced concrete structures. However, as prestressed concrete structures show some peculiar characteristics, they are dealt with separately in this chapter.

Today, prestressing is one of the most employed techniques for slabs and industrial buildings, road and railway bridges, tanks, towers, antennas, and, in general, in the precast industry.

Most of the features for the design of prestressed members and structures are presented in Section [5.10] of EC2. To help the reader and to simplify the identification of all EC2 rules for prestressed concrete structures, they are collected in Table 6.1.

The prestress load is applied by high-strength steel wires, strands, or bars, whose properties are described in Sect. 6.2. Prestressing steel may be embedded in the concrete or placed out of the structure with discrete contact points at deviators and anchorages.

Prestressing steel can be pre-tensioned and bonded or post-tensioned and bonded or unbonded.

6.2 Stress–Strain Diagram $\sigma - \varepsilon$ for Prestressing Steel

A typical stress–strain diagram from a tensile test on prestressing steel is shown in Fig. 6.1. The most significant values from the test are:

$f_{p0.1}$ 0.1% proof stress (i.e., amount of stress at which the material exhibits 0.1% of plastic deformation; it is determined from the stress-strain curve by drawing a line parallel to the initial straight part of the curve and at a distance from the origin by an amount equal to 0.1% thus determining the stress at which the line cuts the curve),

f_p tensile strength,

ε_u ultimate strain.

For design purposes, design values are required, which are calculated from characteristic values (usually provided by the producers). The characteristic $\sigma - \varepsilon$ curve is labelled with *0* in Fig. 6.2. It fits the experimental curve through a bilinear function and is defined using characteristic values of the *0.1%* proof-stress ($f_{p0.1\,k}$), the tensile strength (f_{pk}), and the ultimate strain ε_{uk} of prestressing steel.

For cross-section design, either of the following two models may be assumed [3.3.6 (7)]:

- *model 1*: an inclined top branch with a strain limit ε_{ud} (curve *1* in Fig. 6.2);
- *model* 2: a horizontal top branch without strain limit (curve *2* in Fig. 6.2).

If more accurate values are not known, EC2 suggests the following values: $\varepsilon_{ud} = 0.02$ and $f_{p0.1k}/f_{pk} = 0.9$ (Table 6.2).

Table 6.1 Sections of EN1992-1-1 with rules for prestressed structures

Definitions		1.5.2.3, 1.5.2.4, 2.3.1.4
Partial coefficients		
	Partial coefficients for prestressing action	2.4.2.2
	Partial coefficients for prestressing steel	2.4.2.4, Table 2.1N
Prestressed members and structures		5.10
Durability		
	Minimum concrete cover for pre-tensioned wires, strands and bars, and for ducts of post-tensioned tendons	4.4.1.2 (3) 4.4.1.2 (4) Table 4.5 (N)
Detailing		8
	Minimum distances between pre-tensioned wire, strands, and bars	8.10
	Prestressing devices: anchoring devices (anchorages) and coupling devices (couplers)	3.4, 8.10.4
Anchorage zones		
	Pre-tensioned wires, strands, and bars	8.10.2
	Post-tensioned tendons	8.10.3
Mechanical properties of pretension steels		3.3
Limitation to stressing force		
	Maximum prestress force at the tensioning	5.10.2.1
	Initial prestress force immediately after the transfer of prestress to the concrete, without initial losses	5.10.3
	Mean stress in prestressing steel under service conditions, after initial and time-dependent losses	7.2(5)
Limitation of concrete stresses		5.10.2.2
Initial losses of prestress for pre-tensioning		5.10.4
Thermal loss		10.5.2
Initial losses of prestress for post-tensioning		5.10.5
	Losses due to the instantaneous elastic deformation of concrete	5.10.5.1
	Losses due to friction	5.10.5.2
	Losses due to anchorage draw-in	5.10.5.3
Time-dependent losses of prestress for pre- and post-tensioning		5.10.6
	Creep of concrete	3.1.4, annex B (B.1)
	Shrinkage of concrete	3.1.4, annex B (B.2)
	Losses due to relaxation of prestressing steel	3.3.2 (7), annex D
	Thermal effects on the relaxation loss	10.3.2

(continued)

Table 6.1 (continued)

Definitions		1.5.2.3, 1.5.2.4, 2.3.1.4
Partial coefficients		
	Partial coefficients for prestressing action	2.4.2.2
	Loss due to heating of prestressing steel during heat curing of precast concrete elements	10.5.2
	Losses due to shrinkage, creep, and relaxation	5.10.6
ULS		
	Prestress effects at ultimate limit state	5.10.8
	Bending with or without axial force	6.1
	Shear	6.2
	Punching shear	6.4
SLS		
Effects of prestressing at serviceability limit state and limit state of fatigue		5.10.9
	Cracking	7.3
	Deflections	7.4
Fatigue		6.8.2
Consideration of prestress in analysis		5.10.7

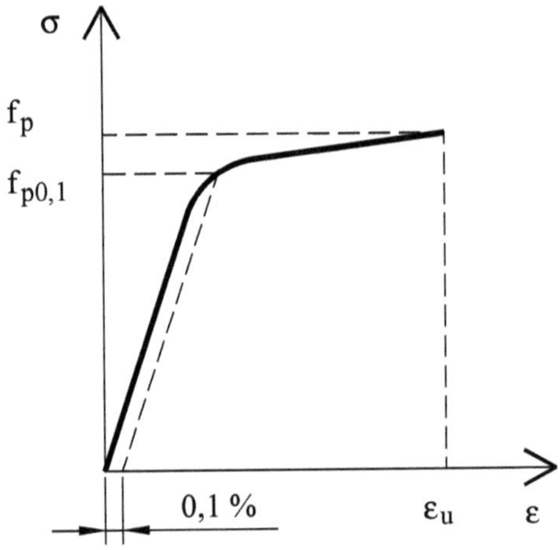

Fig. 6.1 Stress–strain diagram for prestressing steel

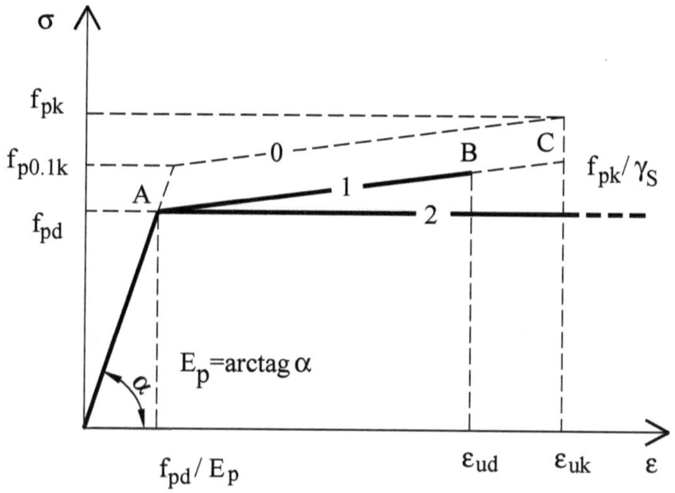

Fig. 6.2 Idealised and design stress–strain diagrams for prestressing steel (*note that segment "1" is not parallel to segment "0"*)

Table 6.2 f_{pk} and $f_{p0.1\,k}$ for different types of prestressing steels

Type of reinforcement	f_{pk} (N/mm^2)	$f_{p0.1k} = 0.9\,f_{pk}$ (N/mm^2)[a]
Hot rolled bars	1030	927
	1230	1107
Wires	1770	1593
	1860	1674[b]

[a]Value recommended by EC2 if more accurate values are not known

[b]In the examples of this chapter, the value of 1600 N/mm^2, provided in catalogues of some producers, is used

6.3 Maximum Prestress Force

6.3.1 Maximum Prestress Force Applied to a Tendon During Tensioning

The force applied to a tendon, P_{max} (i.e. the force at the live end during tensioning), shall not exceed the following value:

$$P_{\max} = A_p \cdot \sigma_{p,\max} \quad [(5.41)]$$

where

A_p is the cross-sectional area of the tendon,

$\sigma_{p,\max} = \min\{k_1 f_{pk};\ k_2 f_{p0.1k}\}$ is the maximum stress applied to the tendon.

The values of $k1$ and $k2$ for use in a Country may be found in its National Annex. The recommended values are $k1 = 0.8$ and $k2 = 0.9$.

6.3.2 Maximum Prestress Force After the Transfer of Prestress to Concrete (After Initial Losses)

At the distance x from the live end of the tendon, the value of the initial prestress force $P_{m0}(x)$ at time $t = t_0$, immediately after the transfer of prestress to concrete,[1] is obtained by subtracting the initial losses $\Delta P_i(x)$ from the force $P_{\max}$ (at tensioning).

According to EC2, $P_{m0}(x)$ should not exceed the following value:

$$P_{m0}(x) = A_p \cdot \sigma_{pm0}(x) \quad [(5.43)]$$

where $\sigma_{pm0}(x) = \min\{k_7 f_{pk};\ k_8 f_{p0.1k}\}$ is the stress in the prestressing steel immediately after the transfer of prestress, with $k_7 = 0.75$ and $k_8 = 0.85$:

$$\sigma_{pm0}(x) = \min\{0.75 f_{pk};\ 0.85 f_{p0.1k}\}$$

6.3.3 Mean Prestress Force in Prestressing Steel at Service Conditions

If more accurate values of $f_{p0.1k}$ are not known, the relationship suggested by EC2 is adopted $(f_{p0.1k}/f_{pk} = 0.9)$ and the maximum stress which can be applied to prestressing steel takes on values collected in Table 6.3.

Table 6.4 collects the maximum values of prestress forces for different nominal bar diameters and the same four different types of steel given in Table 6.3.

[1] In beams with post-stressed bars, the application of global tensioning to concrete occurs when all the operations of tensioning and anchoring of the heads of the tendons are completed; in beams with pre-tensioned steel, prestress is transferred to concrete at the release and/or cut of the strands from the anchorage devices placed at the ends of the casting bed.

Table 6.3 Maximum stress for four different types of prestressing steel ($f_{p0.1k}/f_{pk} = 0.9$ as suggested by EC2, if more accurate values are not known)

f_{pk} (N/mm^2)	$f_{p0.1k}$ (N/mm^2)	Maximum stress during tensioning for pre-tensioned steel $\min\{0.8\, f_{pk};\, 0.9\, f_{p0.1k}\}$	Maximum stress during tensioning for post-tensioned tendons $\min\{0.75\, f_{pk};\; 0.85\, f_{p0.1k}\}$	Maximum stress under service conditions ($\sigma_{p,es} = 0.8\, f_{p0.1k}$)
1030	927	–	772.5 ($= 0.75 f_{pk}$)	741.6
1230	1107	–	922.5 ($= 0.75 f_{pk}$)	885.6
1770	1593	1416 ($= 0.8 f_{pk}$)	1327.5 ($= 0.75 f_{pk}$)	1274.4
1860	1674	1488 ($= 0.8 f_{pk}$)	1395 ($= 0.75 f_{pk}$)	1339.2

6.4 Limitation of Concrete Stress

6.4.1 Concrete Stress at the Transfer of Prestress

The concrete compressive stress resulting from the prestress force and other loads at the time of tensioning (post-tensioned tendons) or release of prestress (pre-tensioned tendons) should be limited to:

$$\sigma_c \leq 0.6\, f_{ck}(t) \quad [(4.1.45)]$$

where $f_{ck}(t)$ is the characteristic compressive strength of concrete at age t when it is subjected to the prestress force.

6.4.2 Limitation of Stresses in Anchorage Zones

At application (post-tensioning) or transfer (pre-tensioning) of prestress, EC2 recommends that the mean compressive strength of the concrete $f_{cm}(t)$ in the anchorage zones does not exceed the minimum value defined in the relevant European Technical Approval (ETA),[2] defined as a function of the adopted prestressing system [5.10.2.2 (3)]. Moreover, the minimum strength should be assured through compressive tests on concrete specimens (at least three specimens whose strengths do not differ more than 5%), cured under the same environmental conditions of the structural element to which prestress is applied.

[2] In Europe, each prestressing system is provided with an ETA *European Technical Approval* which allows the producer to apply the CE certification on the product. The label CE certifies the compliance of the product with the relevant rules.

Table 6.4 Maximum prestress forces at tensioning and maximum prestress forces at service conditions for the same four types of prestressing steel given in Table 6.3

Bar	*Steel type* 1 (f_{pk} = 1030 N/mm^2, $f_{p0.1\,k}$ = 0.9 f_{pk}) ($\sigma_{p,\max}$ = 772.5 N/mm^2 post-tensioning, $\sigma_{p,es}$ = 741.6 N/mm^2)	
	$P_{\max}$ applied by jacks	$P_{m,SLS}$ in SLS (characteristic combination)
d = 26.5 mm (A = 551 mm^2)	425,647 N (≅ 426 kN)	408,622 N (≅ 409 kN)
d = 32 mm (A_p = 804 mm^2)	621,090 N (≅ 621 kN)	596,246 N (≅ 596 kN)
d = 36 mm (A_p = 1020 mm^2)	787,950 N (≅ 788 kN)	756,432 N (≅ 756 kN)
Bar	*Steel type* 2 (f_{pk} = 1230 N/mm^2, $f_{p0.1\,k}$ = 0.9 f_{pk}) ($\sigma_{p,\max}$ = 922.5 N/mm^2 post-tensioning, $\sigma_{p,es}$ = 885.6 N/mm^2)	
	$P_{\max}$ applied by jacks	$P_{m,SLS}$ at SLS (characteristic combination)
d = 26.5 mm (A = 551 mm^2)	508,297 N (≅ 508 kN)	487,966 N (≅ 488 kN)
d = 32 mm (A_p = 804 mm^2)	741,690 N (≅ 742 kN)	712,022 N (≅ 712 kN)
d = 36 mm (A_p = 1020 mm^2)	940,950 N (≅ 941 kN)	903,312 N (≅ 903 kN)
Wire	*Steel type* 3 (f_{pk} = 1770 N/mm^2, $f_{p0.1\,k}$ = 0.9 f_{pk}) ($\sigma_{p,\max}$ = 1416 N/mm^2 pre-tensioning, 1327.5 N/mm^2 post-tensioning, $\sigma_{p,es}$ = 1274.4 N/mm^2)	
	$P_{\max}$ applied by jack	$P_{m,SLS}$ at SLS (characteristic combination)
T13 (A_p = 100 mm^2)	141,600 N (≅ 142 kN) pre-tensioning 132,750 N (≅ 133 kN) post-tensioning	127,440 N (≅ 127 kN)
T15 (A_p = 140 mm^2)	198,240 N (≅ 198 kN) pre-tensioning 185,850 N (≅ 186 kN) post-tensioning	178,416 N (≅ 178 kN)
Wire	*Steel type* 4 (f_{pk} = 1860 N/mm^2, $f_{p0.1\,k}$ = 0.9 f_{pk}) ($\sigma_{p,\max}$ = 1488 N/mm^2 pre-tensioning, 1395 N/mm^2 post-tensioning, $\sigma_{p,es}$ = 1339.2 N/mm^2)	

(continued)

Table 6.4 (continued)

Bar	*Steel type* 1 ($f_{pk} = 1030$ N/mm^2, $f_{p0.1\,k} = 0.9$ f_{pk}) ($\sigma_{p,max} = 772.5$ N/mm^2 post-tensioning, $\sigma_{p,es} = 741.6$ N/mm^2)	
	P_{max} applied by jacks	$P_{m,SLS}$ in SLS (characteristic combination)
	P_{max} applied by jacks	$P_{m,SLS}$ at SLS (characteristic combination)
T13 ($A_p = 100$ mm^2)	148,800 N (≅ 149 kN) pre-tensioning 139,500 N (≅ 140 kN) post-tensioning	133,920 N (≅ 134 kN)
T15 ($A_p = 140$ mm^2)	208,320 N (≅ 208 kN) pre-tensioning 195,300 N (≅ 195 kN) post-tensioning	187,488 N (≅ 187 kN)

It is worth noting that the mean compressive strength required at the transfer of prestress force is dependent on the distance between anchorage devices.

If prestress is applied with multiple stages, the required strength can be reduced as a function of the prestress level reached each step.

For a prestress level not higher than 30% of the full prestress, the mean compressive strength $f_{cm}(t)$ at time t should be at least 50% of the mean compressive strength required by the European Technical Approval for full prestress.

For a prestress level between 30 and 100% of the full prestress, the concrete strength is linearly interpolated between 50 and 100% of the concrete strength required for full prestress [5.10.2.2 (4)]. The linear interpolation of the concrete strength at varying the prestress level gives values collected in Table 6.5.

The linear relationship between the concrete strength ratio K (%) and prestress level G (%) is given by (Fig. 6.3)

$$K = (5G + 200)/7(\text{per } 30\% \leq G \leq 100\%)$$

the inverse relationship is

$$G = (7K - 200)/5(\text{per } 50\% \leq K \leq 100\%)$$

Table 6.5 Ratio K of the mean concrete compressive strength $f_{cm}(t)$ required in anchorage zones as a function of prestress level G

Prestress level (G) (%)	≤30	40	50	60	70	80	90	100
K (%)	50	57	64	71	79	86	93	100

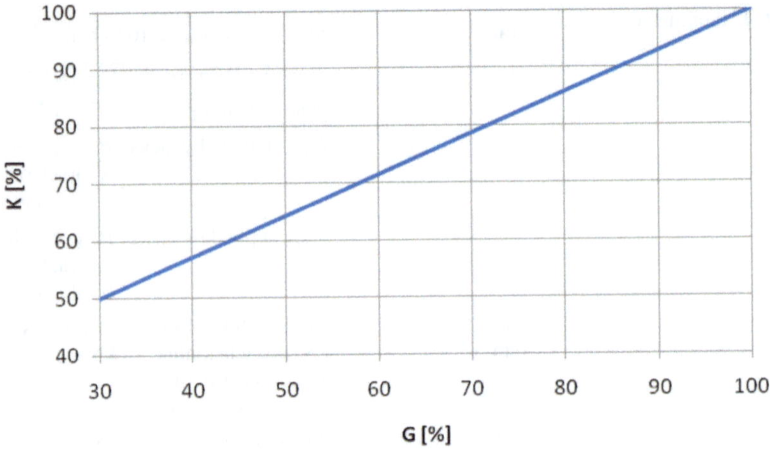

Fig. 6.3 Mean concrete strength ratio K required in anchorage zones vs prestress level G

It is worth reminding that the characteristic strength $f_{ck}(t)$ at time t is then given by [3.1.2 (5)]:

$$\begin{aligned} f_{ck}(t) &= f_{cm}(t) - 8 \text{ for } 3 \le t \le 28 \text{ days} \\ f_{ck}(t) &= f_{ck} \text{ for } t \ge 28 \text{ days} \qquad [(3.1)] \\ f_{cm}(t) &= \beta_{cc}(t) f_{cm} \end{aligned}$$

where

being

$$\beta_{cc}(t) = e^{s\left[1-\sqrt{\frac{28}{t}}\right]} \quad [(3.2)]$$

$s = 0.20$ for Class R cement,
$s = 0.25$ for Class N cement,
$s = 0.38$ for Class S cement.

The three classes of cement R, N and S are corresponding to the early strength of the cement, defined as its mechanical compressive strength at 2 or 7 days.[3] For each class of normalized compressive strength of concrete at 28 days (32.5, 42.5 and 52.5 N/mm^2), there are three classes of early strength: a class with ordinary early strength, indicated by N, a class with high early strength, indicated by R, and a class with low early strength, indicated by S. The strength class S is only defined for CEM III cement (see EN197-1). Table 6.6 collects the values of the coefficient $\beta_{cc}(t)$. Table 6.7 collects values of the characteristic cylinder compressive strength $f_{ck}(t)$ at

[3] The early strength is defined at 7 days for 32.5 S, 32.5 N and 42.5 S classes; at 2 days for all other classes (see EN 197-1).

Table 6.6 β_{cc} coefficients for the three cement classes (R, N, S)

Days	Cement class		
	R	N	S
3	0.66	0.60	0.46
7	0.82	0.78	0.68
14	0.92	0.90	0.85
21	0.97	0.96	0.94

3, 7, 14 and 21 days as a function of the cement class. Table 6.7 also collects values of class S, even if this type of cement is not used for prestressed structures.

Figure 6.4 shows the time evolution of $f_{ck}(t)$ for a concrete C35/45 for all three cement classes (S, N, R).

6.4.2.1 Example 1. Required Mean Compressive Strength at the Application of Prestress (Post-tensioning)

The ETA of a post-tensioning system requires that the mean compressive strength in the anchorage zone at the application of the prestress is not less than 28 N/mm^2 *for anchorages with spacing not less than* 190 mm *and* 36 N/mm^2 *for spacing not less than* 175 mm. *Calculate the maximum prestressing level at 7 days for a C32/40 concrete made with class N cement.*

The mean compressive strength at 28 days is $f_{cm} = f_{ck} + 8 = 40\,\text{N/mm}^2$; for a class N cement, β_{cc} is equal to 0.78 at 7 days (Table 6.6), then:

$$f_{cm}(7) = \beta_{cc}(7) \cdot f_{cm} = 0.78 \cdot 40 = 31.2\,\text{N/mm}^2,$$

$f_{ck}(7) = f_{cm}(7) - 8 = 23.2\,\text{N/mm}^2$.[4]

Being $28\,\text{N/mm}^2 \leq f_{cm}(7) \leq 36\,\text{N/mm}^2$, it is possible to proceed with the full post-tensioning for anchorages with a spacing of 190 mm, while for anchorages arranged at 175 mm spacing, being $f_{cm}(7) = 31.2\,\text{N/mm}^2$ only 86.7% of the required mean strength ($K = 31.2/36 \cong 0.866$), it is necessary to adopt a prestressing level $G < 100\%$. The maximum prestress level which can be used is given by $G = (7\,K - 200)/5$; for $K = 86.7$, $G = (7\,K - 200)/5 \cong 81\%$.

It is worth reminding that the theoretical value of $f_{cm}(7)$ has to be verified through compressive tests performed on concrete specimens cured under the same environmental conditions of the beam.

[4] The calculated value is slightly greater than the value in Table 6.7 (23.15 N/mm^2), being the values of $\beta_{cc}(t)$ of Table 6.6 approximated at the first two decimal places.

Table 6.7 Characteristic cylinder compressive strength $f_{ck}(t)$ in N/mm^2 at 3, 7, 14 and 21 days as a function of the cement class

Concrete	Days	Cement class		
		S	N	R
C28/35 [a]	3	8.49	13.54	15.87
	7	16.62	20.04	21.47
	14	22.76	24.46	25.14
	21	25.94	26.63	26.90
C30/37	3	9.40	14.73	17.19
	7	17.99	21.59	23.11
	14	24.47	26.26	26.98
	21	27.83	28.56	28.84
C32/40 [a]	3	10.32	15.93	18.52
	7	19.35	23.15	24.75
	14	26.17	28.07	28.82
	21	29.72	30.48	30.78
C35/45	3	11.69	17.72	20.51
	7	21.41	25.49	27.21
	14	28.74	30.77	31.58
	21	32.55	33.37	33.69
C40/50	3	13.98	20.72	23.82
	7	24.83	29.38	31.30
	14	33.01	35.28	36.18
	21	37.26	38.18	38.54
C45/55	3	16.27	23.71	27.14
	7	28.24	33.28	35.39
	14	37.28	39.79	40.79
	21	41.97	42.99	43.39
C50/60	3	18.56	26.70	30.45
	7	31.66	37.17	39.49
	14	41.55	44.29	45.39
	21	46.69	47.80	48.23
C55/67	3	20.85	29.69	33.77
	7	35.08	41.06	43.58
	14	45.82	48.80	49.99
	21	51.40	52.61	53.08
C60/75	3	23.14	32.68	37.08
	7	38.50	44.96	47.67

(continued)

Table 6.7 (continued)

Concrete	Days	Cement class		
		S	N	R
	14	50.10	53.31	54.59
	21	56.12	57.42	57.93
C70/85	3	27.72	38.66	43.71
	7	45.34	52.75	55.86
	14	58.64	62.33	63.80
	21	65.55	67.04	67.62
C80/95	3	32.30	44.65	50.34
	7	52.18	60.53	64.05
	14	67.18	71.34	73.00
	21	74.98	76.66	77.32
C90/105	3	36.88	50.63	56.97
	7	59.02	68.32	72.24
	14	75.73	80.36	82.21
	21	84.41	86.28	87.01

[a] The strength classes C28/35 and C32/40 are not listed in [Table 3.1-EC2], but they are provided in some National Standards; therefore, in some figures, tables and examples of this chapter, C28/35 or C32/40 concrete is utilized

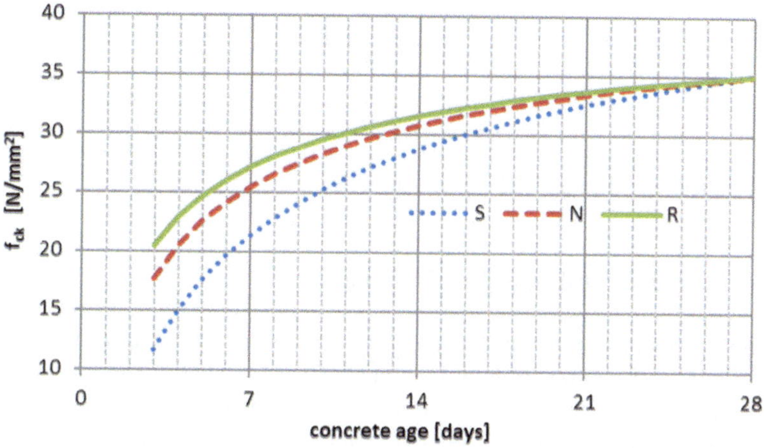

Fig. 6.4 Time evolution $f_{ck}(t)$ for a concrete C35/45 for the three classes of cement (R, N, S)

6.4.3 *Maximum Concrete Stresses at Serviceability Limit State*

Maximum concrete compressive stresses at service conditions, after initial and time-dependent losses, should satisfy the following limitations [7.2(2)]:

$\sigma_c \leq 0.60\, f_{ck}$ for characteristic combination.[5]

$\sigma_c \leq 0.45\, f_{ck}$ for quasi-permanent combination.[6]

6.4.3.1 Characteristic Values of the Prestress Force

For serviceability (and fatigue), possible variations of the prestress should be considered [5.10.9]. At serviceability limit states two characteristic values of the prestress force are calculated as

$$P_{k,\text{sup}} = r_{\text{sup}} P_{m,t}(x) \quad [(5.47)]$$

$$P_{k,\text{inf}} = r_{\text{inf}} P_{m,t}\,(x) \quad [(5.48)]$$

where $P_{k,\text{sup}}$ is the upper characteristic value and $P_{k,\text{inf}}$ is the lower characteristic value.

For post-tensioning with bonded tendons, it is assumed that:

$$r_{\text{sup}} = 1.10$$

$$r_{\text{inf}} = 0.90$$

i.e., a variation equal to $\pm$ 10% is accounted for.

6.4.3.2 Minimum Ordinary Reinforcement

In prestressed members, it is not required any minimum reinforcement if, under the characteristic load combination and the characteristic value of the prestress force, concrete is compressed or the concrete tensile stress is lower than $\sigma_{ct,p}$ [7.3.2 (4)]. The recommended value for $\sigma_{ct,p}$ is $f_{ct,\text{eff}}$, which is corresponding to the mean value of the concrete effective tensile strength at the time of occurrence of first cracks: $f_{ct,\text{eff}} = f_{ctm}$ or $f_{ct,\text{eff}} = f_{ctm}(t)$ if the opening of the first crack is expected before 28 days.

---

[5] According to EC2, the limitation is valid only for exposure classes XD, XF, XS.

[6] Under quasi-permanent loads, EN1992-1-1 also accepts concrete compressive stresses greater than $0.45\, f_{ck}$, if creep nonlinear effects are explicitly calculated.

Table 6.8 Design concrete tensile strength for the evaluation of local effects vs compressive characteristic strength

$f_{ck}\,(\text{N/mm}^2)$	28	30	32	35	40	45	50	55	60	70	80	90
$f_{ck}\,(\text{N/mm}^2)$	1.9	2.0	2.1	2.2	2.5	2.7	2.9	3.0	3.1	3.2	3.4	3.5
$f_{ctd}\,(\text{N/mm}^2)$	1.3	1.3	1.4	1.5	1.7	1.8	1.9	2.0	2.1	2.1	2.3	2.3

6.4.4 Design Prestress Force and Concrete Strength in Anchorage Zones of Post-tensioned Tendons

6.4.4.1 Prestress Force

Local effects of the prestress force in anchorage zones of post-tensioned tendons are evaluated by adopting as design value of the prestress force its mean value[7] multiplied by a partial factor $\gamma_{P,\,\text{unfav}} = 1.2$ [2.4.2.2 (3)]:

$$P_d = \gamma_{P,\text{unfav}}\, P_m(t) = 1.2\, P_m(t)$$

After the application of prestress to concrete, the mean value of the prestress force at time $t = 0$ will be considered, that is after the initial losses. The mean value of the long-term prestress force will be adopted to account for local effects at ULS, after long-term losses (due to shrinkage and creep of concrete and relaxation of steel).

6.4.4.2 Concrete Tensile Stress

The design concrete tensile strength for the analysis of local effects behind the anchorage heads is equal to the lower characteristic value of the concrete tensile strength [8.10.3(2)] divided by the partial factor at ULS (Table 6.8):

$$f_{ctd} = f_{ctk,0.05}\big/\gamma_C = f_{ctk,\,0.05}\big/1.5$$

6.5 Local Effects at Anchorage Devices

Anchorage devices are used to transfer the prestress force from tendons to concrete (Fig. 6.5).

Anchorage devices shall fulfil technical rules given in the ETA of the chosen prestressing system.

---

[7] EN1990—Remark at clause 4.1.2(6): prestressing characteristic value, at a given time t, can take an upper value $P_{k,\text{sup}}(t)$ or a lower value $P_{k,\text{inf}}(t)$. For ULS, medium value $P_m(t)$ can be used.

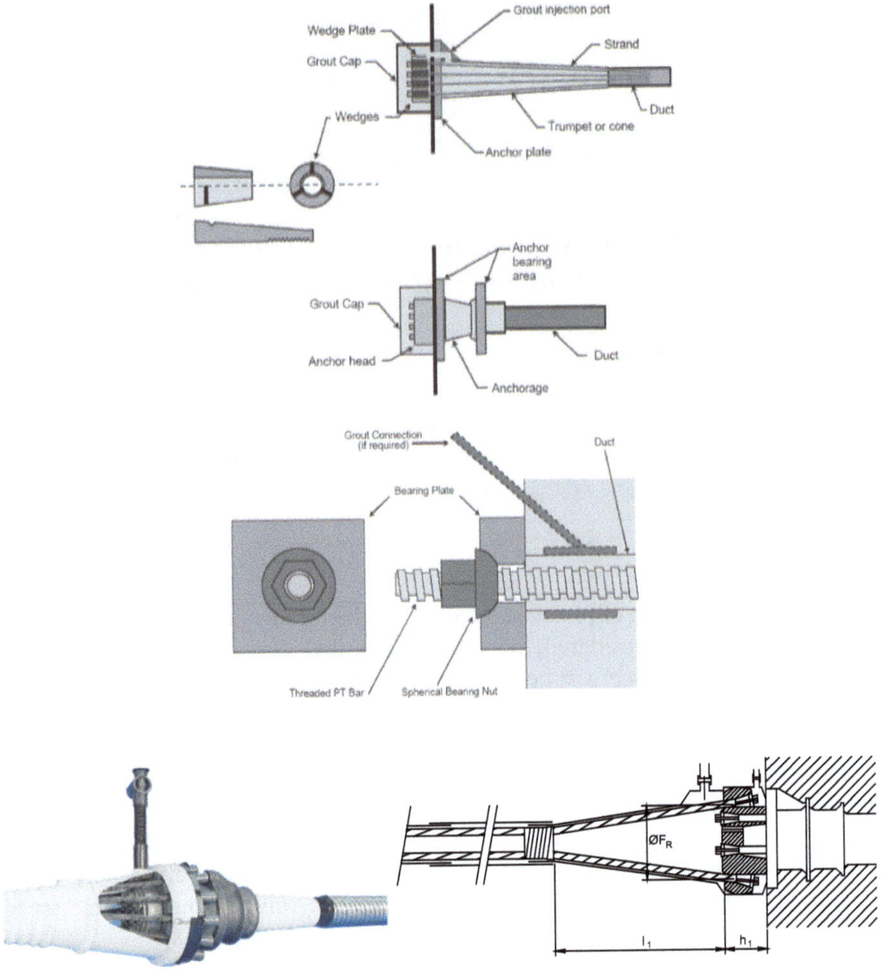

Fig. 6.5 Examples of anchorage devices (at the top) and couplers (at the bottom) for post-tensioning tendons (www.dywidag-systems.com)

The prestress force may be assumed to disperse into the concrete at an angle 2β (Fig. 6.6) from the end of the anchorage device, where β may be assumed equal to arctan (2/3) $\cong 33.7°$.

Couplers shall be placed to not compromise the strength of the structural element: they shall be placed far away from the intermediate supports, avoiding that, in the same cross-section, the ratio of coupled tendons is greater than 50%.

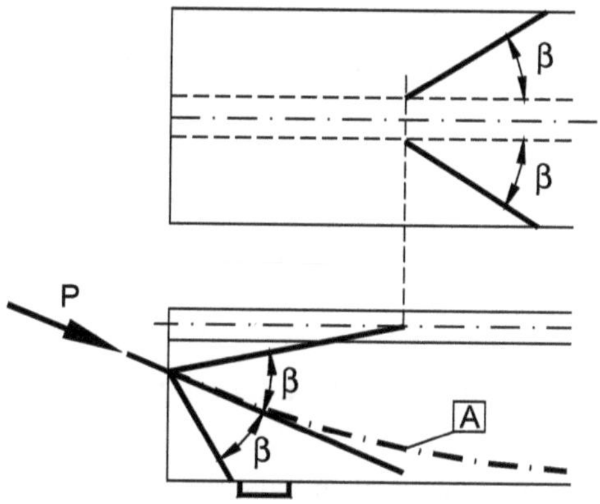

Fig. 6.6 Dispersion of prestress: plan view (at the top) and elevation view (at the bottom) [Fig. 8.18]; (A = prestressing tendon, $\beta = \arctan(2/3) = 33.7^\circ$)

Deviators of prestressing tendons shall be able to withstand both longitudinal and transverse forces that the tendons apply to them and transmit forces to the structure and ensure that the radius of curvature of prestressing tendons does not cause any overstressing or damage to the structure.

The radius of curvature of prestressing tendons shall satisfy the rules given in the ETA. According to clause [8.10.5(4)], angular deviations smaller than 0.01 radiant ($\cong 0^\circ\ 34'\ 23''$) do not require the use of deviators, instead, for greater deviations, it is necessary to insert deviators and to consider, in the design, forces transmitted to the structure.

The transverse tensile forces are resisted by the anti-burst reinforcement, which is designed assuming its tensile stress equal to the steel design strength; if the tensile stress is limited to 300 N/mm^2, the verification of the crack width is not required [8.10.3(4)].

6.6 Minimum Distance of Pre-tensioned Strands or Ducts of Post-tensioned Tendons from Edges

Local concrete crushing or splitting at the end of pre- and post-tensioned members shall be avoided. EC2 does not give any limitation about the minimum distance of prestressing tendons from edges, but it refers to the relevant European Technical Approval [5.10.2.2 (2)].

The only limitation is concerning minimum concrete cover, already given in Table 3.1 in Chap. 3, where minimum values of the concrete cover to fulfil requirements for the bond between concrete and steel are collected, both for ordinary and prestressing steel.

6.7 Minimum Clear Spacing for Pre-tensioned Strands and Ducts of Post-tensioned Tendons

EC2 gives some rules on the minimum spacing between pre-tensioned strands and ducts of post-tensioned tendons so that the concrete can be placed and compacted adequately for the development of bond, particularly in anchorage zones.

6.7.1 Minimum Clear Spacing of Pre-tensioned Strands

For pre-tensioned strands, the clear distance (horizontal e_h and vertical e_v) of individual parallel strands or horizontal layers of parallel strands should fulfil the following limitations (Fig. 6.7):

$$e_v \geq \max\,(2\phi;\ d_g),$$

$$e_h \geq \max\,(2\phi;\ 20\,\text{mm};\ d_g + 5\,\text{mm}),$$

where ϕ is the strand diameter and d_g is the maximum aggregate size.

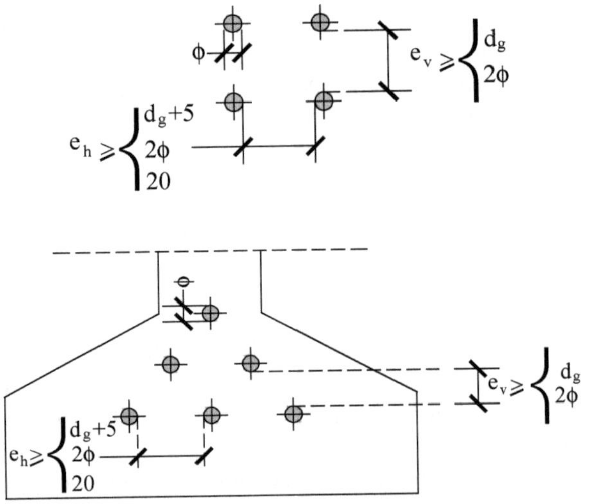

Fig. 6.7 Minimum spacing among pre-tensioned strands (d_g = maximum diameter of aggregates, ϕ = reinforcement diameter) ([Fig. 8.14])

6.7.1.1 Example 2. Calculation of the Minimum Clear Spacing of Pre-tensioned Strands

Verify if clear distances between pre-tensioned strands in the cross-section of Fig. 6.8 satisfy the minimum values. The nominal diameter of strands is 15 mm *and the maximum size of aggregate is* 25 mm.

The minimum vertical clear distance (e_v) and minimum horizontal clear distance (e_h) should not exceed following limits:

$$e_v \geq e_{v,min} = \max(2\phi;\ d_g) = \max(2 \cdot 15;\ 25) = 30\,\text{mm},$$

$$e_h \geq e_{h,min} = \max(2\phi;\ 20\,\text{mm};\ d_g + 5\,\text{mm}) = \max(2 \cdot 15;\ 20;\ 25 + 5) = 30\,\text{mm}.$$

The vertical spacing between the strands indicated in Fig. 6.8 is equal to 50 mm, therefore the corresponding vertical clear distance is $e_v = (50-15) = 35\,\text{mm} > e_{v,\text{min}} = 30\,\text{mm}$.

For the same cross-section of Example 2, Table 6.9 collects the minimum values of the clear distances between strands, for two different values of the strand diameter as a function of the maximum size of aggregate.

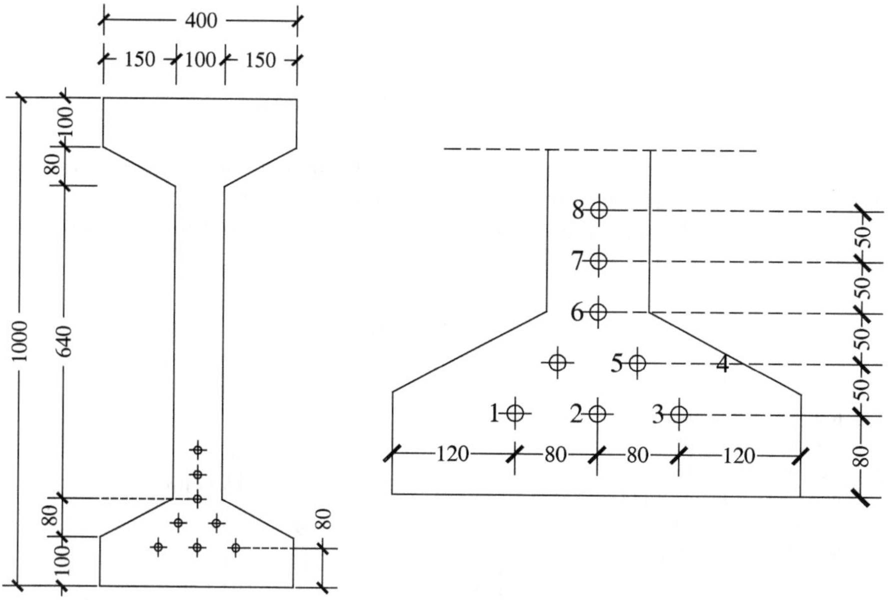

Fig. 6.8 Cross-section of a pre-tensioned beam with the indication of vertical and horizontal spacing between strands

Table 6.9 Minimum clear spacings between strands

Strand diameter ϕ(mm)	Maximum size of aggregate (d_g)	e_h	e_v
13	$d_g \leq 21$ mm	26 mm	26 mm
	21 mm < $d_g \leq 26$ mm	$d_g + 5$ mm	26 mm
	$d_g > 26$ mm	$d_g + 5$ mm	d_g
15	$d_g \leq 25$ mm	30 mm	30 mm
	25 mm < $d_g \leq 30$ mm	$d_g + 5$ mm	30 mm
	$d_g > 30$ mm	$d_g + 5$ mm	d_g

6.7.2 Minimum Clear Spacing Between Post-tensioned Tendons

A minimum clear spacing for ducts of post-tensioned tendons is required to guarantee that the concrete can be adequately cast without damaging ducts and to assure that concrete can resist contact forces arising along with the curved parts of ducts both during tensioning and after.

For this purpose, ducts cannot be bundled, except for a pair of ducts placed vertically one above the other. Following clear spacings should also be satisfied (Fig. 6.9):

$$e_v \geq \max(\phi;\ 40\,\text{mm};\ d_g),$$
$$e_h \geq \max(\phi;\ 50\,\text{mm};\ d_g + 5\,\text{mm}),$$

where e_v is the vertical clear spacing, e_h is the horizontal clear spacing, ϕ is the diameter of the duct and d_g is the maximum size of aggregate.

For not aligned ducts, the vertical clear distance should fulfil the following condition:

$$e_v \geq \max(\phi;\ 40\,\text{mm})$$

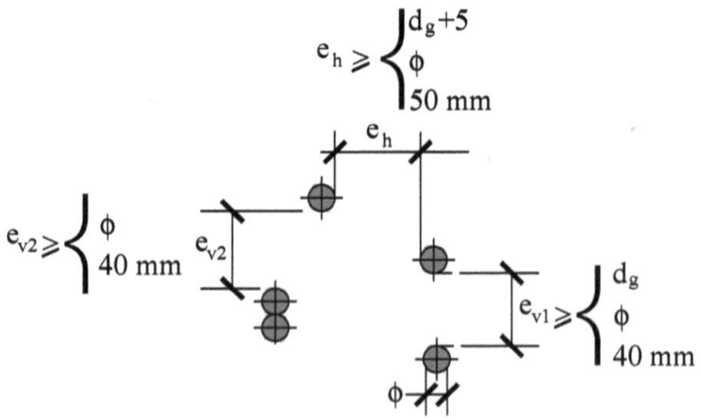

Fig. 6.9 Minimum clear spacings between ducts (e_{v1}, e_{v2} = clear vertical spacing between ducts, aligned or not on the same vertical line, respectively) ([Fig. 8.15])

An adequate concrete cover is required to ensure the protection of ducts and strands against corrosion and to avoid that large diameter ducts can produce abrupt variations of isostatic lines with stress concentrations and stress values greater than the mean stress evaluated on the net cross-section.

Moreover, the adequate compaction of concrete should be assured through the vibration of the cast, to avoid segregation of aggregates and reduce the risk of forming gravel nests.

Problems in casting concrete can also depend on other parameters, as the number of ducts arranged on the same vertical plane and the beam depth.

6.7.2.1 Example 3. Minimum Clear Distance for Post-tensioned Tendons

Verify if the minimum clear distance between the ducts of post-tensioned tendons shown in Fig. 6.10 is fulfilled. Prestressing steel is made of no. 6 *tendons* 12T13 *(twelve 13 mm strands); the duct of each tendon has an internal diameter of* 75 mm *and an external diameter of* 80 mm*; the maximum aggregate size is 30 mm.*

Maximum vertical clear distance (e_v) and horizontal clear distance (e_h) should respect the following limits:

$$e_v \geq \max(\phi;\ 40\,\text{mm};\ d_g) = \max(80;\ 40;\ 30) = 80\,\text{mm},$$

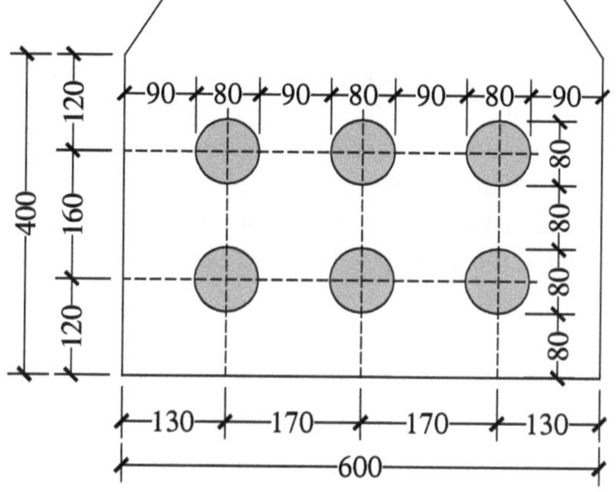

Fig. 6.10 Arrangement of strands within a duct with the external diameter equal to 80 mm; the maximum aggregate size is equal to 30 mm

$$e_h \geq \max(\phi;\, 50\,\text{mm};\ d_g + 5\,\text{mm}) = \max(80;\ 50;\ 30 + 5) = 80\,\text{mm};$$

as shown in the figure, both requirements are fulfilled.

6.8 Initial Losses Occurring in Pre-tensioned Beams

Prestressing losses cause the prestress force to reduce. Prestressing losses can be defined as the difference between the initial force in the prestressing steel at tensioning and the effective prestress force existing in the structural member at time t. When the prestress is transferred to the concrete, initial prestress losses occur: the difference between the prestress force imposed at the jack and the force in the steel immediately after the transfer of the prestress at a particular section represents the immediate losses.

In pre-tensioned beams, the following initial losses occur:

1. *During tensioning*: loss due to the friction at angular deviations and losses due to the return of the anchorage devices at the stressing jack.
2. *Before the transfer of prestress to concrete*: loss due to the relaxation of pre-tensioned steel during the time between the tensioning and the transfer of prestress to concrete; in case of heat curing, it is necessary not only to consider thermal effects on steel relaxation, but also the thermal loss (6.10.4).
3. *During the transfer of prestress to concrete*: loss due to the elastic shortening of concrete produced by the prestress force transmitted by pre-tensioned strands once they are released from anchorage devices.

6.8.1 Losses Due to the Elastic Shortening of Concrete (Pre-tensioned Strands)

In pre-tensioned beams, the release of prestressing steel from anchorages produces an elastic shortening of concrete. Due to the bond between steel and concrete, prestressing steel undergoes the same elastic shortening of adjacent concrete and therefore the prestress force decreases.

If P_0 is the prestress force before it is transferred to concrete and P is the prestress force after the elastic shortening of concrete, the tensile stress in the concrete fibres placed at the same level of the centroid of prestressing steel in the beam of Fig. 6.11 is expressed by

$$\sigma_{c,p} = \frac{P}{A_c} + \frac{(P\, z_{cp})\, z_{cp}}{I_c} = \frac{P}{A_c} \cdot \left(1 + \frac{z_{cp}^2 A_c}{I_c}\right)$$

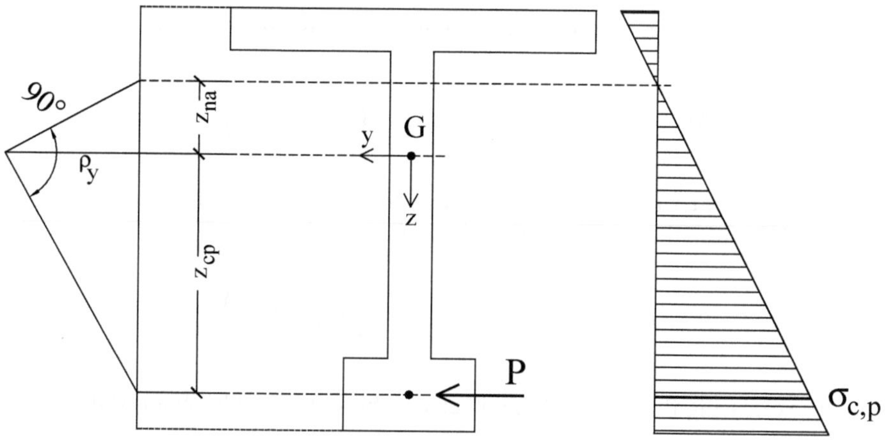

Fig. 6.11 Concrete stress induced by the prestress force P alone ($z_{cp} \cdot z_{a.n.} = \rho_y^2$, where ρ_y is the radius of gyration about to the centroid axis y)

where A_c is the area of the concrete cross-section and I_c is the second moment of area about the y centroid axis, z_{cp} is the distance between the prestress force P and the centre of gravity of the concrete cross-section.

If E_{cm} is the average elasticity modulus of concrete, the axial shortening is equal to

$$\varepsilon_c = \frac{\sigma_{c,p}}{E_{cm}}$$

When the transfer of the prestress occurs, the beam deflects upwards and rises from the formwork, then the beam is subjected to the self-weight (or a portion of it, depending on the stiffness of the formwork). The concrete stress is then reduced by an amount $\sigma'_{c,p} = M_{sw} z_{cp} / I_c$, where M_{sw} is the bending moment produced by the self-weight, while the tensile stress in prestressing steel increases. The increase of the tensile force due to the self-weight does not represent a variation of the loss due to the elastic shortening of concrete (see Sect. 6.8.1.1) and therefore, to evaluate the amount of the loss, the expression of $\sigma_{c,p}$ is used without considering the reduction $\sigma'_{c,p}$.

Under the assumption of a perfect bond between steel and concrete, the concrete compressive strain ε_c is coincident with the reduction of strain $\Delta\varepsilon_p$ in prestressing steel, therefore the normal stress in prestressing steel reduces by

$$\Delta\sigma_p = E_p \cdot \Delta\varepsilon_p = E_p \cdot \varepsilon_c = E_p \cdot \left(\frac{\sigma_{c,p}}{E_{cm}}\right) = \alpha_e \cdot \sigma_{c,p} \text{ where } \alpha_e = \frac{E_p}{E_{cm}}$$

The stress reduction is associated with the following prestress loss

$$\Delta P_{el} = \alpha_e \cdot \sigma_{c,p} \cdot A_p = \alpha_e \frac{P}{A_c} \cdot \left(1 + \frac{z_{cp}^2 A_c}{I_c}\right) \cdot A_p$$

The force P after the transfer of prestress to concrete is obtained by subtracting the loss ΔP_{el} from the initial prestress force P_0

$$P = P_0 - \Delta P_{el} = P_0 - \alpha_e \frac{P}{A_c} \cdot \left(1 + \frac{z_{cp}^2 A_c}{I_c}\right) \cdot A_p$$

and, by expressing P in terms of the other variables, it is finally obtained that

$$P = \frac{P_0}{1 + \alpha_e \frac{A_p}{A_c} \cdot \left(1 + \frac{z_{cp}^2 A_c}{I_c}\right)} = \frac{P_0}{1 + \alpha_e \cdot \left(\frac{A_p}{A_c} + \frac{I_p}{I_c}\right)} \text{ where } I_p = A_p\, z_{cp^2}$$

The expression of P can be also obtained through the following procedure. Under the assumption of a perfect bond between steel and concrete, the elastic shortening of concrete is coincident with the length reduction of strands ΔL_p . ΔL_p is given by the difference between the initial elongation ΔL_0 at the tensioning of steel and the final elongation ΔL_{fin} immediately after the transfer of prestress (Fig. 6.12):

$$\Delta L_p = \Delta L_o - \Delta L_{p,\text{fin}} \rightarrow \Delta \varepsilon_p L_{\text{iniz}} = \left(\varepsilon_{p0} - \varepsilon_p\right) L_{\text{iniz}} \rightarrow \Delta \varepsilon_p = \varepsilon_{p0} - \varepsilon_p$$

where ε_{po} is the initial strain in prestressing steel and ε_p is the strain after the elastic shortening of concrete.

Due to perfect bond, the concrete strain equals the strain reduction in prestressing steel:

$$\varepsilon_c = \Delta \varepsilon_p = \varepsilon_{p0} - \varepsilon_p.$$

In the case of centred prestress, by substituting the concrete and steel stress–strain laws in the compatibility equation

$$\varepsilon_c = \frac{P}{E_{cm} A_c}, \quad \varepsilon_{p0} = \frac{P_0}{E_p A_p}, \quad \varepsilon_p = \frac{P}{E_p A_p}$$

the following expression is obtained

$$\frac{P}{E_{cm} A_c} = \frac{P_0}{E_p A_p} - \frac{P}{E_p A_p}$$

and then

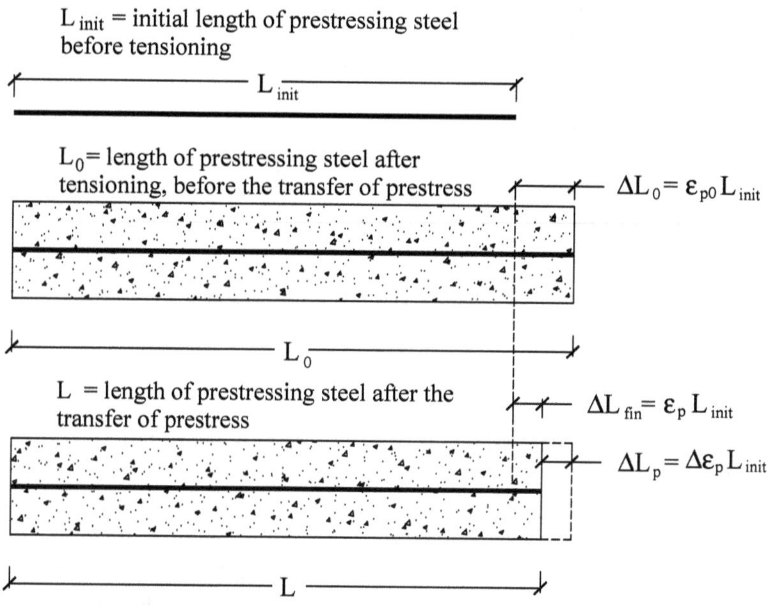

Fig. 6.12 Elongation of prestressing tendons at tensioning (ε_{p0}) and after the transfer of prestress to concrete (ε_p) (centred prestress)

$$P = \frac{A_c}{A_{ci}} P_0$$

where $A_{ci} = A_c + \alpha_e A_p$ is the area of the transformed section.

In the case of eccentric prestress, the constitutive equation of concrete takes the following form

$$\varepsilon_c = \frac{1}{E_{cm}} \left(\frac{P}{A_c} + \frac{P z_{cp}^2}{I_c} \right)$$

while the steel equation is not varying; again, by substituting in the compatibility equation,

$$\frac{1}{E_{cm}} \left(\frac{P}{A_c} + \frac{P z_{cp}^2}{I_c} \right) = \frac{P_0}{E_p A_p} - \frac{P}{E_p A_p}$$

then

$$\frac{P}{E_{cm} A_c} + \frac{P}{E_p A_p} + \frac{P\, z_{cp}^2}{E_{cm} I_c} = \frac{P_0}{E_p A_p} \rightarrow \frac{E_p A_p}{E_{cm} A_c} P + P + \frac{E_p A_p}{E_{cm} I_c} z_{cp}^2 P = P_0$$

$$P = \frac{P_0}{1 + \alpha_e \frac{A_p}{A_c}\left(1 + \frac{A_c}{I_c} z_{cp}^2\right)} = \frac{P_0}{1 + \alpha_e\left(\frac{A_p}{A_c} + \frac{I_p}{I_c}\right)} \text{ where } I_p = A_p z_{cp^2}$$

It can be finally underlined that the loss for elastic shortening of concrete can be automatically accounted for by applying the force P_0, present in the strands before their release, on the transformed section. For example, for centred prestress, it is immediately obtained that

$$\sigma_c = \frac{P_0}{A_{ci}}, \text{ then } P = \sigma_c A_c = \frac{P_0}{A_{ci}} A_c.$$

6.8.1.1 Effect of the Self-weight on the Loss Due to the Elastic Shortening of Concrete

In pre-stressed members, as in post-tensioned members, after ducts are grouted, external loads (and, among them, self-weight) do not modify the prestress force.

This result is clear if one considers effects on a tension zone of a pre-stressed member due to the self-weight and other permanent loads which act at the transfer of prestress.

Under the hypothesis of perfect bond, prestressing steel, and adjacent concrete fibres exhibit the same elongation, and therefore both materials are subjected to tensile forces. Due to elongation, the tensile force in the steel increases, while the compression force in the concrete decreases.

It is then obvious that the increase of the tensile force in the steel due to the self-weight does not represent an increase of the prestress force, because in that case, it would produce an increase of the compression force in the concrete and not a reduction.

To evaluate the loss due to the elastic shortening of the concrete, then it is not significant if the beam deflects upwards at the transfer of prestress or later, because strains due to external loads do not modify the amount of the loss due to elastic shortening of concrete.

6.9 Initial Losses of Prestress for Post-tensioning

Following losses occur at tensioning of tendons for post-tensioned members:

- ΔP_{el} = loss due to the immediate elastic shortening of concrete,
- $\Delta P_{\mu}(x)$ = loss due to friction,
- ΔP_{sl} = loss due to wedge draw-in of anchorage devices.

The total initial loss in a post-tensioned member is therefore given by

$$\Delta P_i(x) = \Delta P_{el} + \Delta P_{\mu}(x) + \Delta P_{sl}$$

6.9.1 Losses Due to the Shortening of Concrete (Post-tensioned Tendons)

In the case of only one post-tensioned tendon, there is not any loss due to immediate elastic shortening of concrete, because it occurs at the tensioning of the tendon. Therefore, when the tendon is anchored, the elastic shortening of the concrete has already completely occurred.

In the case of two or more post-tensioned tendons, tendons are generally tensioned one after the other, therefore each tendon (except for the last one) undergoes the same elastic shortening of the concrete due to the tensioning of subsequent tendons, therefore it undergoes a loss of prestress.

The tensioning of the first tendon produces a shortening of the member and as already observed, it does not suffer any loss of prestress. For the same reason, tensioning and anchoring of the second tendon does not produce any prestress loss in the second tendon itself but reduces the prestress force in the first tendon, due to the elastic shortening produced by the second tendon. When the third tendon is stressed and anchored, prestress forces of the first two tendons decrease, and so on until the tensioning and anchoring of the last tendon, which produces a reduction of the prestressing force in all the other *(n-1)* tendons.

This effect is known as the "mutual effect" and the corresponding total prestress loss is given by

$$\Delta P_{el} = A_p \cdot E_p \cdot \sum \left[\frac{j \cdot \Delta\sigma_c(t)}{E_{cm}(t)} \right] \quad [(5.44)]$$

where

A_p	area of each prestressing tendon,
E_p	modulus of elasticity of prestressing steel,
$E_{cm}(t)$	average modulus of elasticity of concrete at time *t*,
$\Delta\sigma_c(t)$	variation of concrete stress at the centroid of tendons at time *t*,
j	= (*n*–1) / (2*n*) where *n* is the number of identical tendons pre-tensioned successively; as an approximation, *j* may be taken as 1/2,
j	1 for variations produced by permanent loads applied after prestressing.

In the following, the calculation of the loss due to immediate elastic shortening of concrete is performed.

A post-tensioned member is provided with *n* closely placed tendons, so the eccentricity of each tendon can be approximated with the eccentricity of the centroid of the tendons, which is the geometric place where the resultant prestress force is applied along the structural member length. Tendons are pre-tensioned subsequently, one after the other one.

It is assumed that tensioning of a single tendon produces a stress variation equal to the total stress variation divided by the number n of tendons; therefore, at the generic abscissa x

$$\delta\sigma_c(x,t) = \frac{\Delta\sigma_c(x,t)}{n}$$

where $\delta\sigma_c(x,t)$ and $\Delta\sigma_c(x,t)$ are, respectively, the stress variations induced by the single tendon and all tendons.

The stress variation $\delta\sigma_c(x,t)$ is associated with the following concrete strain at the level of the prestressing steel centroid

$$\delta\varepsilon_c(x,t) = \frac{\delta\sigma_c(x,t)}{E_{cm}(t)} = \frac{\Delta\sigma_c(x,t)}{n\,E_{cm}(t)}$$

Under the assumption that tendons are closely spaced, all the already pre-tensioned tendons have the same shortening, but still, they are not bonded to the concrete. Therefore, the equation between the steel and concrete strains at the level of the steel centroid is not valid for each section [$\delta\varepsilon_p(x,t) \neq \delta\varepsilon_c(x,t)$], but only on average along the length of the member [$\delta\varepsilon_p(t) = \delta\varepsilon_c(t)$] (for details, see Sect. 6.9.1.1).

Then each tendon, which has already been stretched, undergoes the following prestress loss

$$\delta\sigma_p(t) = E_p\delta\varepsilon_c(t) = E_p\frac{\Delta\sigma_c(t)}{n\,E_{cm}(t)}$$

where $\delta\varepsilon_c(t)$ indicates the average concrete strain at the level of the steel centroid and is corresponding to the following prestress reduction $\delta P(t)$ in each of (i–1) already pre-tensioned tendons

$$\delta\,P(t) = A_p E_p\delta\varepsilon_c(t) = A_p E_p\frac{\Delta\sigma_c(t)}{n\,E_{cm}(t)}$$

The first pre-tensioned tendon experiences losses due to the tensioning of the other (n – 1) tendons, the second one experiences (n – 2) losses, and so on, until the last tendon, does not suffer any loss. Table 6.10 collects all the losses for each of the n tendons.

By adding all the losses for all the tendons, the total prestress loss is obtained

$$\Delta P_{el}(t) = \sum_{i=1}^{n}\Delta P_i(t) = \sum_{i=1}^{n}(n-i)\delta P(t) = \sum_{i=1}^{n}(n-i)\frac{A_p E_p}{E_{cm}(t)}\frac{\Delta\sigma_c(t)}{n}$$

Table 6.10 Tendon losses for a post-tensioned member due to immediate elastic shortening of concrete

Tendon no	Loss of prestress
1	$\Delta P_1 = (n-1)\,\delta P(t)$[a]
2	$\Delta P_2 = (n-2)\,\delta P(t)$
……	……
i	$\Delta P_i = (n-i)\,\delta P(t)$
……	……
n-1	$\Delta P_{n-1} = \delta P(t)$
n	0

[a] $\delta P(t)$ = loss of prestress for an already tensioned and anchored tendon, after tensioning of a new tendon
n = number of tendons

$$= \frac{n\cdot(n-1)}{2}\frac{A_p E_p}{E_{cm}(t)}\frac{\Delta\sigma_c(t)}{n} = \frac{(n-1)}{2}\frac{A_p E_p}{E_{cm}(t)}\Delta\sigma_c(t)$$

where the following relationship has been used

$$\sum_{i=1}^{n}(n-i) = n^2 - \sum_{i=1}^{n} i = n^2 - \frac{n(n+1)}{2} = \frac{n(n-1)}{2}$$

being $\sum_{i=1}^{n} i = \frac{n(n+1)}{2}$ the well-known Gauss formula.

The loss ΔP_{el} divided by the number of tendons n gives the average loss of prestress for each tendon

$$\Delta P_{mean}(t) = \frac{(n-1)}{2n}\frac{A_p E_p}{E_{cm}(t)}\Delta\sigma_c(t) = j\frac{A_p E_p}{E_{cm}(t)}\Delta\sigma_c(t),\ \text{being}\ j = \frac{n-1}{2n}$$

The total prestress loss can finally be rewritten as

$$\Delta P_{el} = \sum_{i=1}^{n}\Delta P_{media}(t) = A_p E_p \sum_{i=1}^{n} j\frac{\Delta\sigma_c(t)}{E_{cm}(t)}$$

which is coincident with the expression [(5.44)] of EC2.

For a sufficiently high number n of tendons, the ratio $j = (n-1)/(2n)$ tends to 1/2 and the expression above becomes

$$\Delta P_{el} = \frac{1}{2}A_p E_p \sum_{i=1}^{n}\frac{\Delta\sigma_c(t)}{E_{cm}(t)}$$

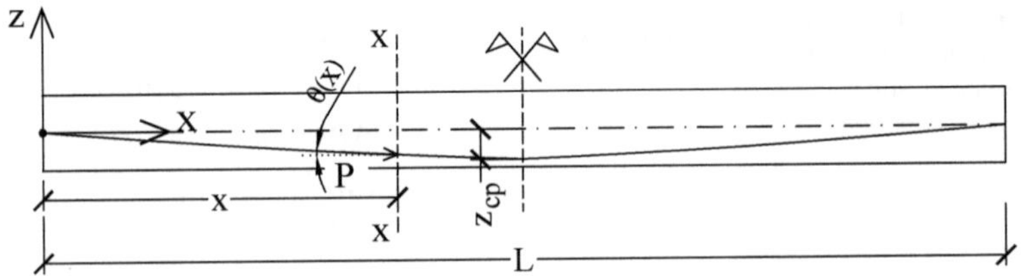

Fig. 6.13 Prestress force P induced by the single tendon at the x cross-section

which gives an overestimation up to 20% for $n = 6$[8] and up to 11% for $n = 10$.

The above calculation does not consider external loads. During tensioning of tendons, the beam deflects upwards and is subjected to its self-weight, which can modify the tensile force of already tensioned tendons but still not grouted. The strict evaluation of this phenomenon depends on the stiffness of the formwork:

- if the formwork is infinitely stiff, the beam bottom detaches from the formwork during tensioning, so the self-weight cannot modify the internal forces of tendons,[9] as the stressing jack imposes a force and not an elongation,
- if the formwork is infinitely deformable, the beam bottom detaches from the formwork only at the end of tensioning, therefore effects of self-weight concern all the tendons except the last one, as they are already anchored but not grouted.

At the cross-section of abscissa x (Fig. 6.13), $\sigma_{c,pp}(x)$ indicates the concrete stress at the level of the steel centroid due to self-weight and, in general, to permanent loads applied during tendon tensioning.

As already seen, the strain of prestressing tendons is coincident only on average with the concrete strain at the level of the tendon centroid, because tendons are still not bonded; therefore, it follows that

$$\varepsilon_{\mathrm{p,pp}}(\mathrm{x}) \neq \varepsilon_{\mathrm{c,pp}}(\mathrm{x})$$

and

$$\overline{\varepsilon}_{p,pp} = \overline{\varepsilon}_{c,pp} = \frac{\overline{\sigma}_{c,pp}}{E_{cm}(t)}.$$

where $\varepsilon_{p,pp}(x)$ and $\varepsilon_{p,pp}(x)$ are the strains of tendons and adjacent concrete in the cross-section of abscissa x, $\overline{\varepsilon}_{p,pp}$ and $\overline{\varepsilon}_{c,pp}$ are the average strains in steel and concrete, $\overline{\sigma}_{c,pp}$ is the concrete average stress, $E_{cm}(t)$ is the average value of the elasticity modulus of concrete at time t when self-weight is applied.

---

[8] The exact expression of j gives $j = 5/12 \cong 0.416$ for $n = 6$ and $j = 0.45$ for $n = 10$.

[9] The beam self-weight does not change the internal forces of tendons if the beam bottom detaches from the formwork during the tensioning of the first tendon, otherwise if the beam bottom detaches from the formwork during the tensioning of tendon *i-th*, the internal forces in the other (*i-1*) tendons, which have already been stretched, change.

The prestress loss corresponding to the strain $\overline{\varepsilon}_{c,pp}$ is equal to

$$\Delta\sigma_p(t) = \overline{\varepsilon}_{c,pp} \cdot E_p = \frac{\overline{\sigma}_{c,pp}}{E_{cm}(t)} \cdot E_p$$

It can be assumed that the self-weight effects during tensioning are intermediate between the two limit cases of infinitely stiff and infinitely deformable formwork. Therefore, the prestress loss is equal to

$$\Delta\sigma_p(t) = \alpha \cdot \overline{\sigma}_{c,pp} \cdot \frac{E_p}{E_{cm}(t)} \quad \text{with } 0 \leq \alpha \leq 1$$

The expression given by EC2 for the loss due to the elastic shortening of concrete considers the variations induced by permanent loads before grouting of ducts through a coefficient $j = 1$.

As already seen for members with pre-tensioned tendons, also in post-tensioned elements, after duct grouting, the increase of tensile force of tendons due to external loads does not represent a reduction of the loss for elastic shortening of the concrete (see Sect. 6.8.1.1).

Finally, for elements with unbonded prestressing steel, the prestress variation due to the variation of applied loads is negligible.

6.9.1.1 Calculation of the Variation of Compressive Stress $\Delta\sigma_c$ to Be Used in [(5.44)]

At the tensioning of the *i-th* tendon, the (*i*–1) stretched tendons have not been grouted and can slide inside their ducts. Therefore, they do not experience the same strain of adjacent concrete at each cross-section but exhibit a constant strain (if friction is neglected) between their ends. For this reason, the calculation of the loss due to immediate elastic shortening of concrete requires considering the mean compressive stress of concrete along the tendon length, and not the compressive stress of each cross-section.

It is supposed that tendons are closely spaced, so their position can be approximated with the position of their centroid.

Under the prestress force P of a single tendon (assumed constant along the length of the tendon), at the cross-section of abscissa x the compression force and bending moment increase respectively by

$$\Delta N(x) = P \cos\theta(x)$$
$$\Delta M(x) = P \cdot z_{cp}(x)$$

where

$\theta(x)$ is the angle between the tendon and the horizontal line at x section,

P is the compression force, assumed positive,
$z_{cp}(x)$ is the eccentricity of the tendon, which is negative if, at the x section, the tendon is beneath the beam axis (Fig. 6.13).

With the sign convention adopted for P and $z_{cp}(x)$, the bending moment $\Delta M(x)$ due to prestress is negative in those cross-sections where the tendon is beneath the axis of the beam.

Being the tendons only slightly inclined to the horizontal, it is possible to assume that $\theta(x) \cong 0$, $\cos\theta(x) \cong 1$ and $\Delta N(x) \cong P$, so the axial force is constant along the whole beam.

In all the cross-sections, the axial force produces a constant increase of the compressive stress, equal to

$$\Delta\sigma_{c,N} = \frac{P}{A_c}$$

while in the x cross-section, the bending moment produces a variation of the compressive stress of the concrete at the level of the tendon centroid, which is variable along the beam and is equal to

$$\Delta\sigma_{c,M}(x) = \frac{\Delta M(x)}{I_c} \cdot z_{cp}(x) = \frac{P}{I_c} \cdot z_{cp}^2(x)$$

where I_c is the second moment of area of concrete about the centroid axis.

The mean compressive stress due to $\Delta M(x)$ is obtained by integrating the function $\Delta\sigma_{c,M}(x)$ along the tendon length and dividing by the tendon length.[10]

$$\Delta\overline{\sigma}_{c,M} = \frac{1}{L} \cdot \int_0^L \frac{P}{I_c} \cdot z_{cp}^2(x) \cdot dx$$

Finally, the total variation of compressive stress to use in [(5.44)] is given by the sum of $\Delta\sigma_{c,N}$ and $\Delta\overline{\sigma}_{c,M}$

$$\Delta\sigma_c = \Delta\sigma_{c,N} + \Delta\overline{\sigma}_{c,M} = \frac{P}{A_c} + \frac{1}{L} \cdot \frac{P}{I_c} \int_0^L z_{cp}^2(x) \cdot dx$$

In the following paragraphs, the mean compressive stress $\Delta\overline{\sigma}_{c,M}$ induced by the prestressing bending moment $\Delta M(x)$ is calculated for two different cases: linear and parabolic CGS profile (CGS = centroid of prestressing steel = centroid of the tendons).

[10] Due to the slight slope of prestressing tendons to the beam axis, the length of each tendon can be approximated with the length of the beam and the curvilinear abscissa along the tendons be assumed coincident with the abscissa x along the axis of the beam.

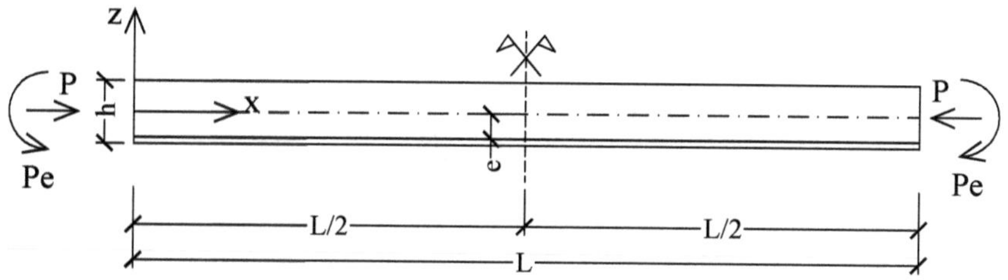

Fig. 6.14 Linear CGS profile with constant eccentricity e

1. Linear CGS profile (Fig. 6.14)

$$\left[z_{cp}(x) = -e = \text{cost.}\right]$$

The bending moment $\Delta M(x)$ is constant, therefore also the variation of concrete compressive stress at the centroid of tendons is constant and equal to:

$$\Delta\sigma_{c,M} = \Delta\overline{\sigma}_{c,M} = \frac{P}{I_c}e^2$$

Alternatively, the same result can be obtained by considering two symmetrical moments $M\,(=Pe)$ applied to end-sections of a simply supported beam, which produce two symmetrical rotations φ of end-sections

$$\varphi = \frac{M}{2R} = \frac{M \cdot L}{2 \cdot E_{cm} \cdot I_c} = \frac{P \cdot e \cdot L}{2 \cdot E_{cm} \cdot I_c}$$

being $R = (E_{cm}\, I_c) / L$.

Therefore, tendons experience an elastic shortening ΔL which is given by (Fig. 6.15)

Fig. 6.15 Rotation of the end section and corresponding shortening of tendons

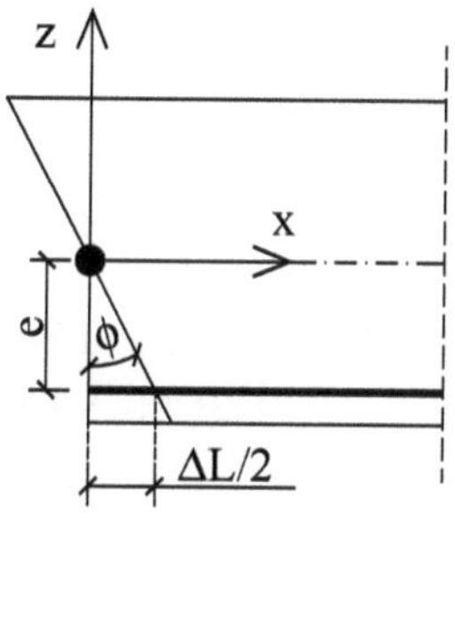

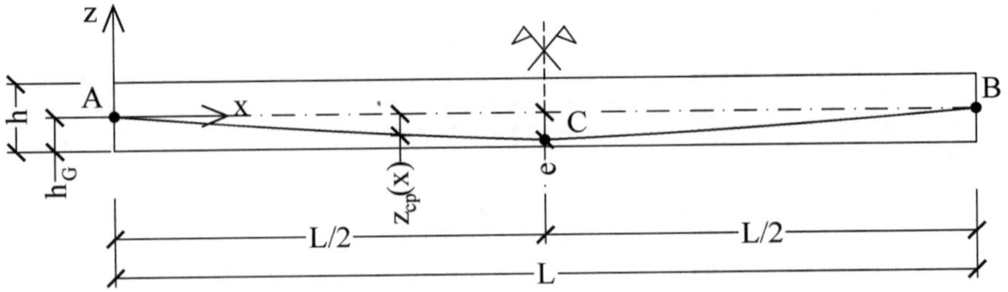

Fig. 6.16 Parabolic CGS profile with end anchorages placed at the centroid of end sections

$$\Delta L = 2 \cdot \varphi \cdot e = \frac{P \cdot e^2 \cdot L}{E_{cm} \cdot I_c}$$

and the corresponding variation of compressive stress in the concrete adjacent to the tendon:

$$\Delta\sigma_{c,M} = \Delta\overline{\sigma}_{c,M} = E_{cm} \cdot \frac{\Delta L}{L} = \frac{P \cdot e^2}{I_c}$$

2. **Parabolic resultant tendon with anchorages placed at the centroids of end-sections (Fig. 6.16)**

[$z_{cp}(x) = \frac{4e}{L^2}\left(x^2 - Lx\right)$ where e is the CGS eccentricity in the midspan]

$$\Delta\overline{\sigma}_{c,M} = \frac{1}{L} \cdot \int_0^L \frac{P}{I_c} \cdot z_{cp}^2(x) \cdot dx = \frac{1}{L} \cdot \frac{P}{I_c} \cdot \int_0^L \left[\frac{4e}{L^2}\left(x^2 - Lx\right)\right]^2 \cdot dx$$

$$= \frac{1}{L} \cdot \frac{P}{I_c}\frac{16\,e^2}{L^4} \int_0^L \left(x^4 - 2Lx^3 + L^2x^2\right) \cdot dx$$

$$= \frac{1}{L} \cdot \frac{P}{I_c}\frac{16\,e^2}{L^4}\left(\frac{x^5}{5} - \frac{Lx^4}{2} + \frac{L^2x^3}{3}\right)\Big|_0^L$$

$$= \frac{8}{15}\frac{P}{I_c}e^2$$

6.9.1.2 Effects of Self-weight and Quasi-Permanent Loads Applied During Tensioning

The variation of the prestress force due to self-weight and quasi-permanent loads acting on the beam during tensioning (when tendons are still not bonded to concrete) are usually negligible; however, for the sake of completeness, it is calculated below.

The tensile stress varies parabolically and it is given by the following expression

$$\Delta\sigma_{c,g}(x) = \frac{M_g(x)}{I_c} \cdot z_{cp}(x)$$

where $M_g(x) = g \cdot \frac{L}{2} \cdot x - g \cdot \frac{x^2}{2} = g \cdot \frac{x \cdot (L-x)}{2}$ being L the length of the beam and g the sum of self-weight and uniformly distributed quasi-permanent loads.

Considered a Cartesian orthogonal system with the x-axis parallel to the beam axis and directed from left to right, and the origin placed in the centroid of the left end section of the beam.

The average increase of the tensile stress due to the load g in concrete fibres adjacent to the tendon is equal to

$$\Delta\overline{\sigma}_{c,g} = \frac{1}{L} \cdot \int_0^L \frac{M_g(x)}{I_c} \cdot z_{cp}(x) \cdot dx = \frac{g}{2\, I_c\, L} \cdot \int_0^L x \cdot (L-x) \cdot z_{cp}(x) \cdot dx$$

It is worth reminding that the variation $\Delta\overline{\sigma}_{c,g}$ induced by the load g should be multiplied by a coefficient α between 0 and 1, as a function of the stiffness of the formwork.

In the following paragraphs, the expression $\Delta\overline{\sigma}_{c,g}(x)$ is evaluated for two relevant cases: constant eccentricity and parabolic eccentricity.

1. **Linear CGS profile**

$$\left[z_{cp}(x) = -e = \text{cost.}\right]$$

$$\Delta\overline{\sigma}_{c,g}(x) = \frac{g}{2\, I_c\, L} \cdot \int_0^L x \cdot (L-x) \cdot z_{cp}(x) \cdot dx = -\frac{g\,\cdot e}{2\, I_c\, L} \cdot \int_0^L x \cdot (L-x) \cdot dx$$

$$= -\frac{g\,\cdot e}{2\, I_c\, L} \cdot \left(\frac{L}{2} x^2 - \frac{x^3}{3}\right)\Bigg|_0^L = -\frac{g\, e\, L^2}{12\, I_c}$$

The negative sign indicates tensile stresses because stresses are assumed as positive when compressive.

2. **Parabolic CGS profile with anchorages placed at the centroid of end-sections**

[$z_{cp}(x) = \frac{4e}{L^2}\left(x^2 - Lx\right)$ where e is the eccentricity of the resultant tendon in the midspan]

$$\Delta\overline{\sigma}_{c,g}(x) = \frac{g}{2\,I_c\,L} \cdot \int\limits_0^L x \cdot (L-x) \cdot z_{cp}(x) \cdot dx = -\frac{4\,g \cdot e}{2\,I_c\,L^3} \cdot \int\limits_0^L x^2 \cdot (L-x)^2 \cdot dx$$

$$= -\frac{2g \cdot e}{I_c\,L} \cdot \left(L^2\frac{x^3}{3} - 2L\frac{x^4}{4} + \frac{x^5}{5}\right)\Bigg|_0^L = -\frac{g\;eL^2}{15\,I_c}$$

6.9.1.3 Example 4. Loss Due to Concrete Elastic Shortening in a Post-tensioned Beam

Calculate the loss due to immediate elastic shortening of concrete for the beam with post-tensioned tendons shown in Fig. 6.17, under the assumption that pre-tensioning is performed at 14 days. *The CGS profile is parabolic in the lateral segments and linear in the central segment and it is anchored at the centroids of end sections.*

Geometrical characteristics of the concrete section (after subtracting the area of duct cross-sections):

$h = 1800$ mm.
$A_c = 907{,}641$ mm^2.
$I_c = 3.5727\text{E}+11\text{mm}^4$.
$z_{cp} \cong 859$ mm (CGS eccentricity from the centroid of the concrete cross-section in the central segment B-B').

Prestressing reinforcement is formed by no. 6 tendons with 7 wires with a nominal diameter of 15.3 mm ($A_p = 7 \times 140 = 980$ mm^2). The total area of prestressing steel is $A_{p,tot} = 5880\,\text{mm}^2$. The internal and external diameter of each duct is 75 mm and 80 mm, respectively.

Materials

Concrete: C35/45 with class R cement:

$$f_{ck} = 35\,\text{N/mm}^2$$

$f_{cm} = f_{ck} + 8 = 43\,\text{N/mm}^2$ ([Table 3.1])
characteristic and mean cylinder compressive strength of concrete at 14 days:

$$f_{ck}(t = 14) = 31.58\,\text{N/mm}^2 \quad \text{(Table 6.7)}$$

$$f_{cm}(t = 14) = f_{ck}(t = 14) + 8 = 39.58\,\text{N/mm}^2$$

average modulus of elasticity of concrete at 28 days ([Table 3.1])[11]:

[11] The modulus of elasticity of concrete is a function of the modulus of elasticity of its components; the values of E_{cm} given in [Table 3.1] are valid for quartzite aggregate, while they are 10% lower

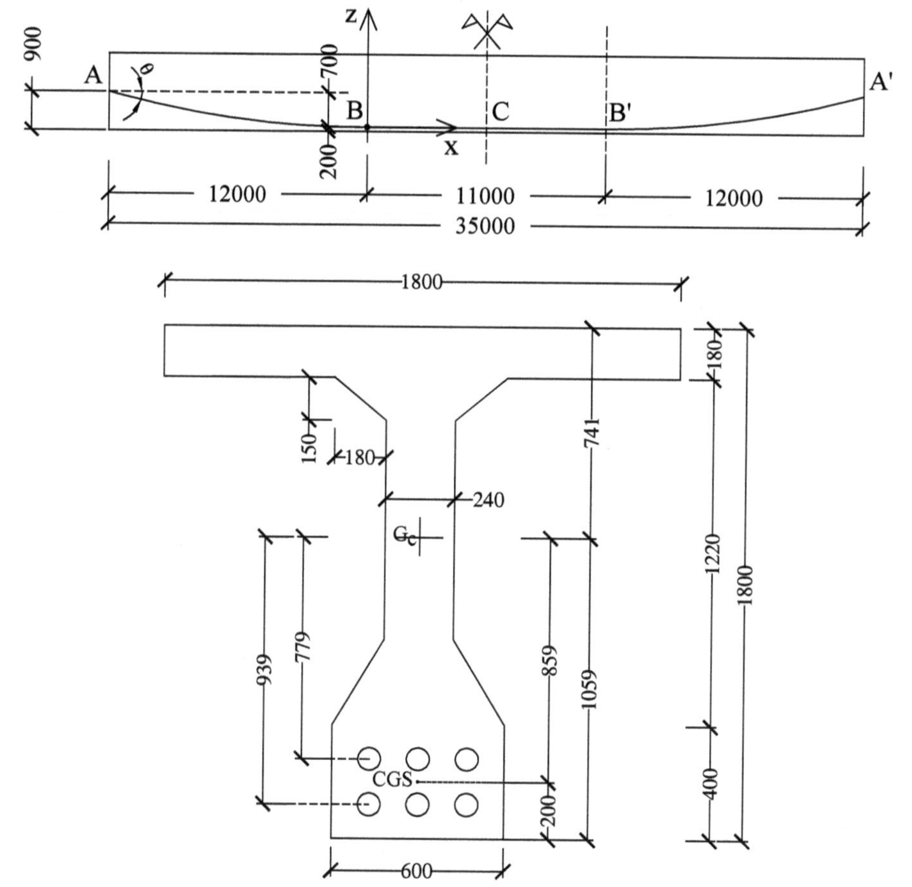

Fig. 6.17 Longitudinal and transverse cross-section of a beam with post-tensioned tendons (unit of length: mm)

$$E_{cm} = 22 \cdot \left(\frac{f_{cm}}{10}\right)^{0.3} = 22 \cdot \left(\frac{43}{10}\right)^{0.3} \cong 34\,GPa = 34000\,\text{N/mm}^2$$

average modulus of elasticity of concrete at 14 days [(3.5)]:

$$E_{cm}(t = 14) = \left[\frac{f_{cm}(t = 14)}{f_{cm}}\right]^{0.3} \cdot E_{cm} = \left(\frac{39.58}{43}\right)^{0.3} \cdot 34000 = 33165\,\text{N/mm}^2$$

characteristics of prestressing steel:

---

for calcareous aggregate, 30% lower for sandstone aggregate and 20% higher for basalt aggregate [3.1.3(2)]. In all the examples of this chapter, the value of E_{cm} given by [Table 3.1] is used under the assumption that concrete is made with quartzite aggregate.

$$E_p = 195{,}000\,\text{N/mm}^2$$
$$f_{pk} = 1860\,\text{N/mm}^2$$
$$f_{p0.1k} = 1600\,\text{N/mm}^2$$

maximum value of the initial prestress force applied at the live end during tensioning:

$$\sigma_{p,\max} \le \min\big(0.75\,f_{pk};\;\; 0.85\,f_{p0.1k}\big) = 1360\,\text{N/mm}^2$$

$$P_{\max} = 5880 \cdot 1360 = 7.996.800\,\text{N} \cong 7.997\,\text{kN}.$$

The variation of the prestress force induced by the self-weight during tensioning and before the grouting of ducts is assumed negligible.

Applying the result of Sect. 6.9.1.1 to the case of a linear tendon with constant eccentricity and the case of a parabolic tendon, the average compressive stress in the concrete at the level of tendon centroid is equal to

$$\Delta\overline{\sigma}_c = \frac{P}{A_c} + \frac{8}{15}\frac{P}{I_c}e^2$$

in the 12 m long lateral segments with parabolic tendons, and is equal to

$$\Delta\overline{\sigma}_c = \frac{P}{A_c} + \frac{P}{I_c}e^2$$

in the 11 m long central segment where tendons are horizontal;

the mean value along the beam takes the following value:

$$\Delta\overline{\sigma}_c = \frac{P}{A_c} + \frac{\left(\frac{8}{15}\frac{P}{I_c}e^2\right)\cdot 24 + \left(\frac{P}{I_c}e^2\right)\cdot 11}{35} = \frac{P}{A_c} + \frac{17}{25}\frac{P}{I_c}e^2 = 20.04\ \text{N/mm}^2$$

and the corresponding total loss of prestress due to immediate elastic shortening of concrete, due to non-contemporary tensioning of tendons, is equal to:

$$\Delta P_{el} = A_p E_p \frac{n-1}{2}\frac{\Delta\sigma_c(t)}{E_{cm}(t)} = 980 \cdot 195.000 \cdot \frac{6-1}{2}\frac{20.04}{33.165}$$
$$= 288.681\ N \cong 289\ \text{kN}$$

which is corresponding to the ratio of 3.6% of the prestress force $P_{\max}$ at the tensioning of tendons.

6.9.2 Loss Due to Friction

The loss due to friction of prestress $\Delta P_\mu(x)$ in post-tensioned tendons may be estimated by:

$$\Delta P_\mu(x) = P_{\max}\left(1 - e^{-\mu(\theta + kx)}\right) \quad [(5.45)]$$

where

- θ(rad) is the sum of the angular deviations over a distance x (irrespective of direction or sign).
- $\mu\left(\text{rad}^{-1}\right)$ is the coefficient of friction between the tendon and its duct.
- k (rad/m) is an unintentional angular deviation for internal tendons (per unit length) (Fig. 6.18),
- x (m) is the distance along the tendon from the point where the prestress force is equal to $P_{\max}$ (the force at the live end applied by the hydraulic jack); in general, the slope of post-tensioned tendons to the horizontal is very slight, therefore the distance x measured along the tendon can be approximated with the distance measured along the axis of the beam.

If the exponential function is approximated with its first-order Taylor series, the loss due to friction can be rewritten in the following form

$$\Delta P_\mu(x) = P_{\max}\left(1 - e^{-\mu(\theta + kx)}\right) \cong P_{\max}\mu(\theta + kx)$$

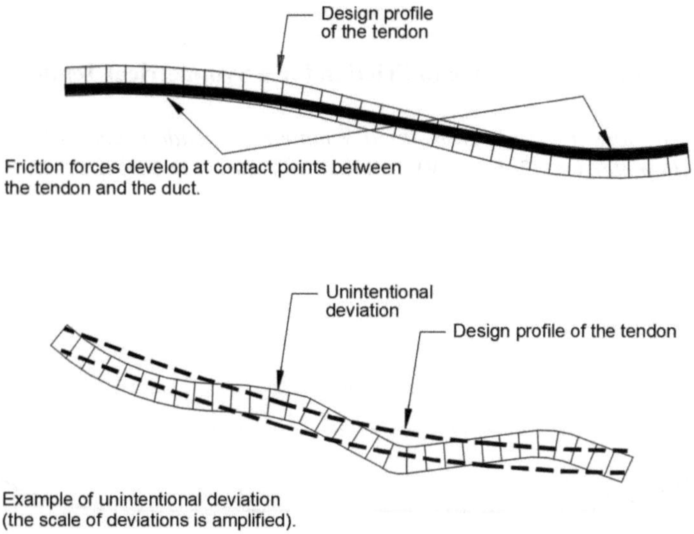

Fig. 6.18 Unintentional deviation of the duct from the design profile

Table 6.11 Coefficient of friction μ of post-tensioned internal and external unbonded tendons [Table 5.1]

	Internal tendons[a]	External unbonded tendons			
Type of reinforcement		Steel duct/unlubricated	HDPE duct/unlubricated	Steel duct/lubricated	HDPE duct/lubricated
Cold drawn wire	0.17	0.25	0.14	0.18	0.12
Strand	0.19	0.24	0.12	0.16	0.10
Deformed bar	0.65	–	–	–	–
Smooth round bar	0.33	–	–	–	–

[a]For tendons that fill about half of the duct
HDPE high-density polyethylene

which can be used for small values of $\mu(\theta + kx)$. Using the linearized formula of $\Delta P_\mu(x)$, the prestress force reduced by the friction loss shows a linear variation along the beam axis.

The values of μ and k are given in the ETA of the prestressing system otherwise μ can take values given in Table 6.11 and the coefficient k values between 0.005 rad/m and 0.01 rad/m, depending on the installation tolerance of tendons. The coefficient k for unintentional angular deviations depends on the quality of workmanship, the distance between tendon supports, the type of duct or sheath, and the degree of vibration during the placing of the concrete.

6.9.2.1 Example 5. Loss Due to Friction for a Symmetrical Tendon

Calculate the loss due to friction for a symmetrical tendon made of 6 *strands of* 15.3 mm *diameter* ($A_p = 6 \times 140 = 840\,\text{mm}^2$) *shown in Fig. 6.19. Consider both*

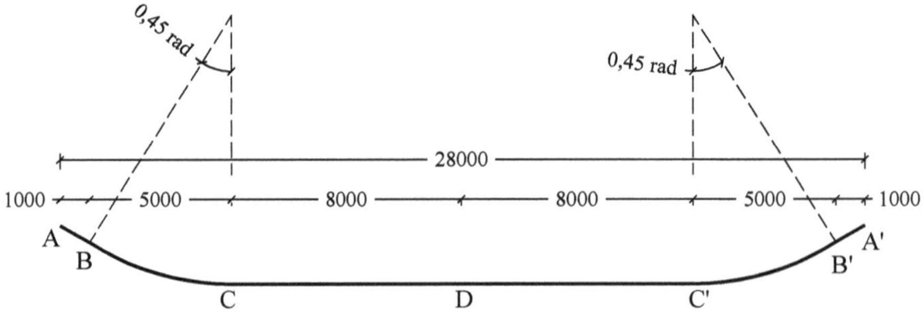

Fig. 6.19 Symmetrical tendon (unit of length: mm)

the case of tensioning from both ends (two live ends A and A′) and from only one live end A.

The following values are assumed:

$P_{max} = 1.000$ kN initial prestress force at the live ends A and A′ applied by the stressing jack; initial stress $\sigma_{pi} = 1.000.000/840 = 1.190\,\text{N/mm}^2$.

$\mu = 0.19\,\text{rad}^{-1}$ coefficient of friction between tendon and duct (Table 6.11).

$k = 0.01$ rad/m unintentional angular deviation for internal tendons per unit length; the value of k, in absence of data from an ETA lies in the range $0.005 \div 0.01$. [5.10.5.2(3)]

$\theta = 0.45$ rad angular deviation between cross-sections B (B′) and C (C′).

In the formula [(5.45)], distances measured along the horizontal line are employed instead of distances measured along the tendon, because of the slight slope of the tendon in segments BC and C′B′. This approximation is always adopted during the calculation of the loss due to friction in all the examples of this Chapter.

By applying the formula [(5.45)], the values collected in Table 6.12 are obtained.

If tensioning is performed only at the live end *A*, the prestress force in cross-sections A, B, C, D takes on values collected in Table 6.13 (Fig. 6.20).

Table 6.12 Loss due to friction for tensioning at both ends (symmetrical tendon)

Section	Abscissa x (m)	θ (rad)	Prestressing loss (kN) [1] $\Delta P_\mu(x) = P_{max}\left(1 - e^{-\mu(\theta+kx)}\right) \cong P_{max}\mu(\theta + kx)$	Prestress force (kN)[a] $P(x) = P_{max}e^{-\mu(\theta+kx)}$	Loss ratio (%)[a]
A	0	0	0	1000	0
B	1	0	$\Delta P\mu(x_B) = 2(2)$	998 (998)	0.2 (0.2)
C	6	0.45	$\Delta P\mu(x_C) = 92\,(97)$	908 (903)	9.2 (9.7)
D	14	0.45	$\Delta P\mu(x_D) = 106(112)$	894 (888)	10.6 (11.2)

[a]Values in round brackets are obtained by applying the simplified formula, which derives from the approximation of the exponential function with its first-order Taylor series

Table 6.13 Loss due to friction for tensioning at one end (symmetrical tendon)

Section	Abscissa x (m)	θ (rad)	Prestressing loss (kN) [1] $\Delta P_{\mu}(x) = P_{max}\left(1 - e^{-\mu(\theta+kx)}\right) \cong P_{max}\mu(\theta + kx)$	Prestress force (kN)[a] $P(x) = P_{max}e^{-\mu(\theta+kx)}$	Loss ratio (%)[a]
A	0	0	0	1000	0
B	1	0.45	$\Delta P\mu(x_B) = 2(2)$	998 (998)	0.2 (0.2)
C	6	0.45	$\Delta P\mu(x_C) = 92\,(97)$	908 (903)	9.2 (9.7)
D	14	0.45	$\Delta P\mu(x_D) = 106(112)$	894 (888)	10.6 (11.2)
C′	22	0.45	$\Delta P\mu(x_{C'}) = 120(127)$	880 (873)	12.0 (12.7)
B′	27	0.90	$\Delta P\mu(x_{B'}) = 199(222)$	801 (778)	19.9 (22.2)
A′	28	0.90	$\Delta P\mu(x_{A'}) = 201(224)$	799 (776)	20.1 (22.4)

[a]Values in round brackets are obtained by applying the simplified formula, which derives from the approximation of the exponential function with its first-order Taylor series

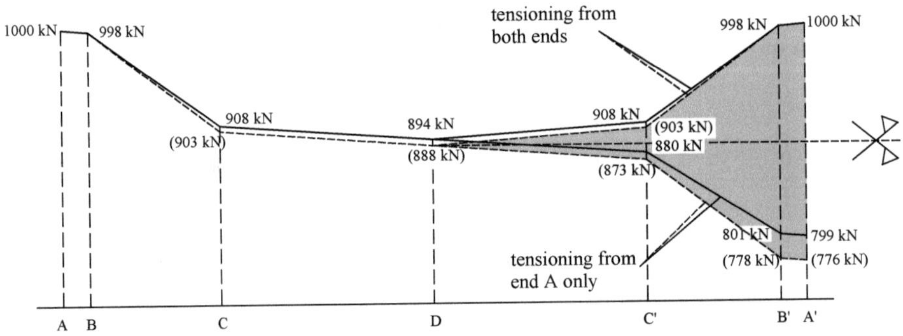

Fig. 6.20 Loss due to friction for tensioning at both ends or only at the left end; both curves are represented, the one obtained by the exponential function (solid lines) and the one obtained by the approximated linear function (dashed lines and values inside round brackets)

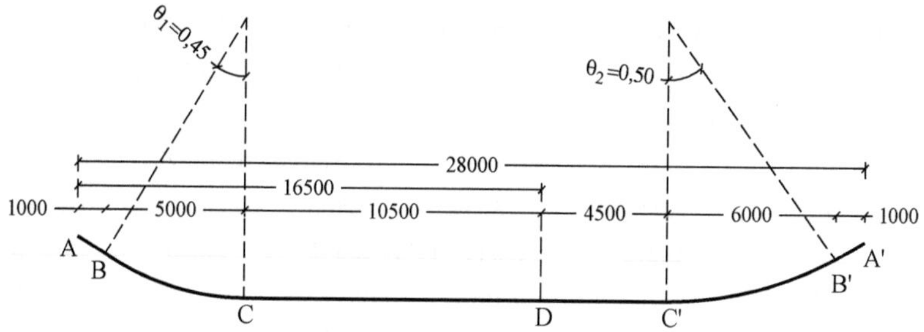

Fig. 6.21 Non-symmetrical tendon (lengths in mm)

6.9.2.2 Example 6. Position of the Fixed Point for a Non-symmetrical Tendon

Calculate the position of the fixed point[12] *of the non-symmetrical tendon shown in Fig. 6.21, under the assumption that tensioning is performed from both ends* (A and A′).

In the case of a non-symmetrical tendon, the fixed-point D along the tendon is not coincident with the midpoint of the tendon. Its abscissa can be determined by imposing that the prestress force in D, calculated from the left end A (from left to right) is equal to the one calculated from the right end A′ (from right to left):

$$P_\mu(x_D) = P_{\max} \cdot \left[1 - e^{-\mu \cdot (\theta_1 + k \cdot x_D)}\right] = P_{\max} \cdot \left[1 - e^{-\mu \cdot (\theta_2 + k \cdot (L - x_D))}\right]$$

$$\theta_1 + k \cdot x_D = \theta_2 + k \cdot (L - x_D) \Rightarrow x_D = \frac{\theta_2 - \theta_1 + kL}{2k}$$

By substituting the numerical values, the following value is obtained:

$$x_D = \frac{0.50 - 0.45 + 0.01 \cdot 28.00}{2 \cdot 0.01} = 16.5 \text{ m}$$

In general, the loss due to friction, in its linear approximation, can be easily divided into two contributions: the first one is due to friction in curvilinear segments (angular deviation) and the second one to unintentional angular deviation, in both curvilinear and linear segments:

$$\Delta P_\mu(x) = P_{max}\mu(\theta + kx) = (P_{max}\mu\theta) + (P_{max}\mu kx) = \Delta P_{\mu,ang}(\theta) + \Delta P_{\mu,rect}(x)$$

where the two contributions are equal to:

---

[12] The fixed point of a tendon tensioned from both ends is the point that does not move along the axis of the tendon during tensioning: all the points placed at the left hand of the fixed point move leftwards and vice versa, those placed at the right hand, move rightwards; the fixed point is also corresponding to the point where the prestress force takes the minimum value. For symmetrical tendons tensioned from both ends, the fixed point is coincident with the midpoint of the tendon.

Table 6.14 Losses due to friction (ratio) in the curve vs θ for $\mu = 0.19$ (strands)

θ (rad)	0.025	0.05	0.075	0.1 (5.73°)	0.125
$\Delta P_{\mu,ang}(\theta)/P_{\max} \cong \mu\,\theta$	0.475%	0.950%	1.425%	1.900%	2.375%
θ (rad)	0.15	0.175	0.2 (11.46°)	0.225	0.25
$\Delta P_{\mu,ang}(\theta)/P_{\max} \cong \mu\,\theta$	2.850%	3.325%	3.800%	4.275%	4.750%
θ (rad)	0.275	0.3 (17.19°)	0.325	0.35	0.375
$\Delta P_{\mu,ang}(\theta)/P_{\max} \cong \mu\,\theta$	5.225%	5.700%	6.175%	6.650%	7.125%
θ (rad)	0.4 (22.92°)	0.425	0.45	0.475	0.5 (28.65°)
$\Delta P_{\mu,ang}(\theta)/P_{\max} \cong \mu\,\theta$	7.600%	8.075%	8.550%	9.025%	9.500%

$$\frac{\Delta P_{\mu,\text{ang}}(\theta)}{P_{\max}} = \mu\,\theta \qquad \frac{\Delta P_{\mu,\,rect}(x)}{P_{max}} = \mu\,k\,x$$

For $\mu = 0.19$ (strands) the friction loss due to unintentional deviation per unit length ($x = 1$ m) varies from 0.095% $P_{\max}$ (for $k = 0.005$ rad/m) to 0.19% $P_{\max}$ (per $k = 0.01$ rad/m).

The variation of $\Delta P_{\mu,ang}(\theta)/P_{\max} \cong \mu\,\theta$ with θ is collected in Table 6.14 for $\mu = 0.19$ (strands).

6.9.2.3 Example 7. Calculation of the Loss Due to Friction Through Table 6.14.

Calculate the loss due to friction in the midspan section D *of Example* 5 *(symmetrical tendon) by using Table 6.14.*

From Example 5, the angular deviation between live end A and midspan section D is equal to 0.45 rad, the distance x between A and D is equal to 14 m and the force at the live end is $P_{\max} = 1000\,\text{kN}$.

From Table 6.14 the loss due to friction in the curvilinear part, for $\theta = 0.45$ rad and $\mu = 0.19$, is equal to 0.0855 $P_{\max}$, while the loss in the linear segment is equal to:

$$\frac{\Delta P_{\mu,\text{rect}}(x)}{P_{max}} = \mu\,k\,x = 0.19 \cdot 0.01 \cdot 14 = 2.66\,\%$$

The total loss due to friction in section D is equal to: $\Delta P_{\mu} = (0.0855 + 0.0266)\,P_{\max} = 0.1121\,P_{\max} = 112.1\,\text{kN}$, which is the value already evaluated in Example 5 (see the value of $\Delta P\mu(x_D)$ between round brackets in the last row of the fourth column of Table 6.12).

6.9.3 Loss Due to Wedge Draw-In of the Anchorage Devices

It is necessary to consider the loss due to the wedge draw-in of the anchorage devices and the loss due to the deformation of the anchorage devices themselves. The values of the wedge draw-in are given in the European Technical Approval of the prestressing system and are around $2 \div 6$ mm.

It is worth noting that, almost every time, the loss due to wedge draw-in has no practical effect because it reduces the prestress only for a small length in proximity of the tendon ends. Moreover, they are often balanced through suitable corrections at tensioning.

After the wedge draw-in of the anchorage device and the deformation of the anchorage device itself, the tendon slides inwards in the duct, by an amount depending on the type of the anchorage device (Fig. 6.22).

The sliding of the tendon inwards occurs until section X at the distance l_p from the end-section A (Fig. 6.23).At the cross-section X, the sum of the friction forces along the segment AB of the tendon balances the loss of prestress due to the sliding of the tendon itself. Therefore, the wedge draw-in of the anchorage does not produce any prestress loss beyond the cross-section X.

If the variation of the prestress force within the tendon is approximated with linear segments, the two lines which represent the variation of the steel prestress before and after the wedge draw-in of the anchorage device are symmetrical about a line parallel to the beam axis (e.g. AB and A′B in Fig. 6.24). This happens because the friction is the same (and therefore the associated losses are varying with the same law) both for the tendon sliding from right to left (at tensioning) and for the tendon sliding from left to right (at the wedge drawing-in).

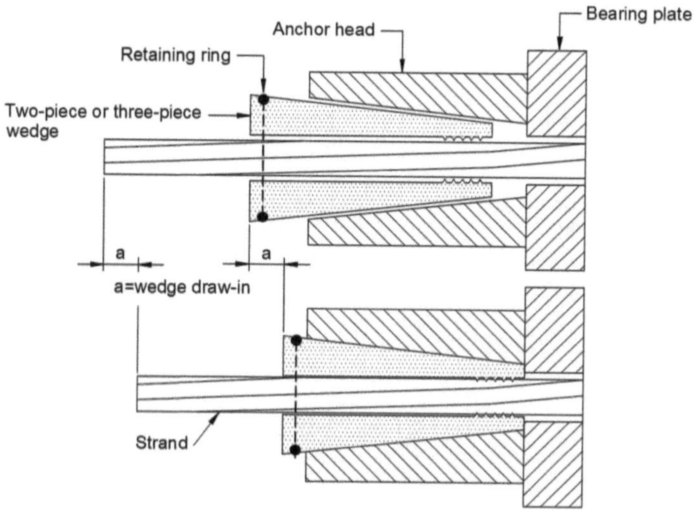

Fig. 6.22 Wedge draw-in a of an anchorage device: $a = 2 \div 6$ mm

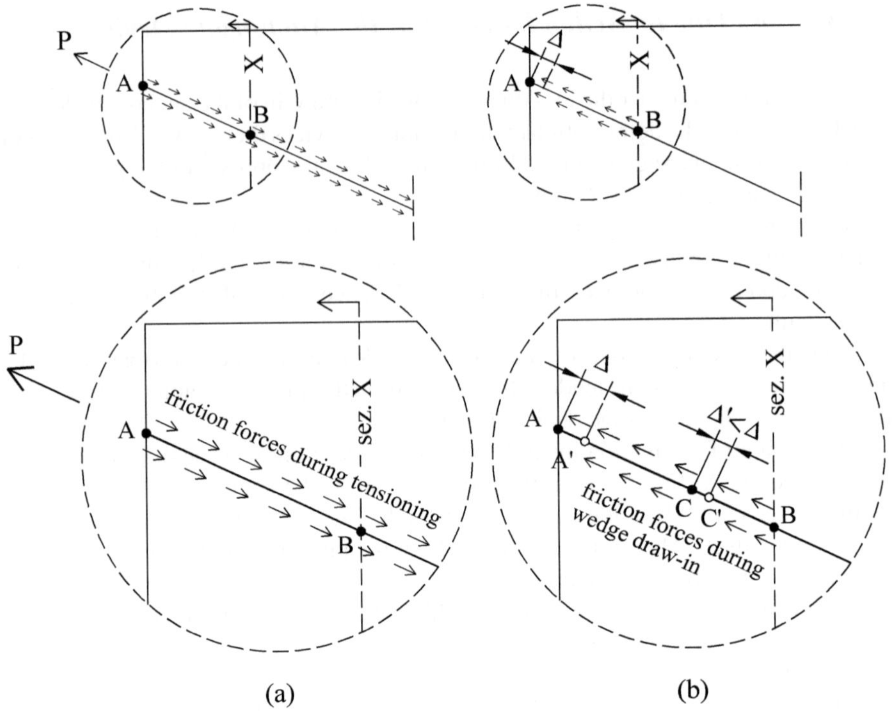

Fig. 6.23 Evolution of friction forces: **a** at tensioning of the tendon **b** at the wedge draw-in Δ of the anchorage device in A (the loss due to the elongation of the tendon is softened by the friction with the duct, therefore point C experiences a displacement $\Delta' < \Delta$). The phenomenon is opposite to the one observed during tensioning, with the difference that it appears only in the initial part AB of the tendon

Fig. 6.24 Segment interested by the wedge draw-in of the anchorage device (lines AB and A′B are symmetrical about the horizontal axis passing through B; $\Delta P'_\mu$ = loss due to friction per unit length)

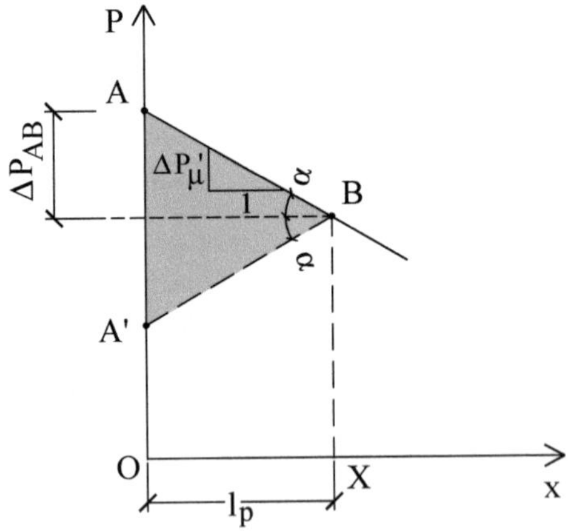

Each elementary segment of length dx, between cross-sections O and X (Fig. 6.24), experiences a shortening of the amount $\delta(dx)$ and produces a reduction of the prestress force of the following amount

$$\delta P = A_p \delta\sigma = A_p E_p \frac{\delta(dx)}{dx}$$

where A_p is the area of the tendon and E_p is the modulus of elasticity of the prestressing steel; by inverting this expression, $\delta(dx)$ assumes the following expression

$$\delta(dx) = \frac{\delta P \cdot dx}{A_p E_p}$$

The total sliding w of the tendon in the duct is obtained by summing the shortening $\delta(dx)$ of all the elementary segments included in the segment A-B with length l_p

$$w = \int_0^{l_p} \delta(dx) = \int_0^{l_p} \frac{\delta P \cdot dx}{A_p E_p}$$

from which $A_p E_p w = \int_0^{l_p} \delta P \cdot dx$. where the integral represents the area of the triangle ABA′ in Fig. 6.24.

It is then possible to conclude that the abscissa l_p and the prestress of the tendon can be calculated accounting for two facts:

- the diagrams of the prestress force before and after the wedge draw-in of the anchorage devices are symmetrical about the horizontal line (AB and A′B in Fig. 6.24),
- the area enclosed between these diagrams (ABA′) is coincident with A_p E_p w.

In general, the profile of a tendon is made of a series of linear and curvilinear segments; the linearized diagram of the prestress shortly after tensioning and shortly before the transfer of the prestress from the jack to the anchorage device is then made of a series of segments with variable slopes. To calculate the loss due to the wedge draw-in of the anchorage device it is then necessary to know the position of the cross-section X, which can be located between A and B or between B and C or beyond section C (Fig. 6.25).

The easiest method to calculate l_p consists in comparing A_p E_p w with the areas ABA′, ABCB′A″, 2etc. For example, if the cross-section X is placed between B and C, the following inequality holds

$$\text{ABA}' < A_p E_p w < \text{ABCB}'\text{A}''$$

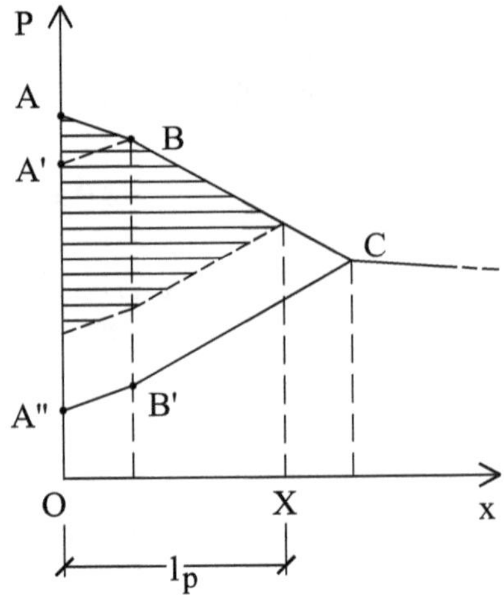

Fig. 6.25 Calculation of areas for evaluating the segment interested by the wedge draw-in of the anchorage

In the special case of a tendon with a parabolic profile, the area ABA′ in Fig. 6.24 is given by the following expression

$$A_{ABA'} = l_p \cdot \Delta P_{AB} = l_p \cdot \left(l_p \cdot \tan\alpha\right) \cong l_p^2 \cdot \alpha$$

which, once it is made equal to $A_p E_p w$, gives

$$l_p = \sqrt{\frac{A_p E_p w}{\alpha}}$$

α angle can be estimated by calculating the loss due to friction per unit length $\Delta P'_\mu$ in the 1.00 m long segment from the live end-section

$$\Delta P'_\mu = \Delta P_\mu(1\,m) = P_{\max} \cdot [1 - e^{-\mu(\theta + k \cdot 1)}] \cong P_{\max}\mu(\theta + k)$$

where θ is the angular deviation of the tendon in the 1.00 m long segment.[13]

It results that $\tan\alpha = \Delta P'_\mu$ (Fig. 6.24) and, by approximating the angle with its tangent, $\alpha = \Delta P'_\mu$; finally, the following expression for l_p is obtained

$$l_p = \sqrt{\frac{A_p E_p}{P_{\max}\mu(\theta + k)} w}$$

[13] Unit angular deviation θ is given by the inverse of the curvature radius R; for a parabola θ is constant and equal to $= \theta 1/R = 8e/L^2$, where L is the span length and e is the sag of the parabola in the midspan section, being null the eccentricity of the tendon at the end-sections.

Table 6.15 Values of ΔP

	A-B	B-C	C-D
Distance (m)	1	5	8
ΔP(kN)	2	90	14

6.9.3.1 Example 8. Loss Due to Wedge Draw-In of the Anchorage Devices

Calculate the loss induced in the tendon of Example 5 (no. 6 *strands of* 15.3 mm *diameter*; $A_p = 840\,\text{mm}^2$) *due to a 3 mm wedge draw-in of the anchorage devices.*

Being the modulus of elasticity of steel strands equal to 195,000 N/mm² [3.3.6(3)], the following expression holds

$$w \cdot A_p \cdot E_p = 3 \cdot 840 \cdot 195{,}000 = 4.914 \times 10^8\,\text{Nmm} = 491.4\,\text{kNm}$$

Table 6.15 shows the distances and the prestress force differences between two adjacent cross-sections after friction losses.

The variation of the prestress force is shown in the following figures, once losses due to friction and draw-in of anchorage devices occurred, under the hypothesis that the cross-section X where effects of the draw-in of the anchorage devices vanish, is coincident with section B (Fig. 6.26), with section C (Fig. 6.27) or with the section D (Fig. 6.28).

Partial areas (Figs. 6.26, 6.27 and 6.28):

(1): $A_1 = 2 \cdot \frac{AB \cdot \Delta P_{AB}}{2} = 1 \cdot 2 = 2\,\text{kNm}$
(2): $A_2 = AB \cdot (2 \cdot \Delta P_{BC}) = 1 \cdot (2 \cdot 90) = 180\,\text{kNm}$
(3): $A_3 = 2 \cdot \frac{BC \cdot \Delta P_{BC}}{2} = 5 \cdot 90 = 450\,\text{kNm}$

(4): $A_4 = BC \cdot (2 \cdot \Delta P_{CD}) = 5 \cdot (2 \cdot 14) = 140\,\text{kNm}$
(5): $A_5 = 2 \cdot \frac{CD \cdot \Delta P_{CD}}{2} = 8 \cdot 14 = 112\,\text{kNm}$

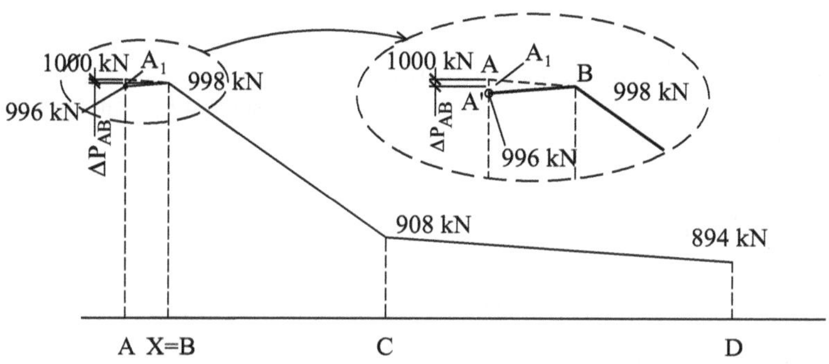

Fig. 6.26 Variation of the prestress force under the hypothesis that the cross-section X is coincident with the cross-section B (dashed line is corresponding to the variation of *P*(*x*) in the segment AB if the wedge draw-in of the anchorage devices is null)

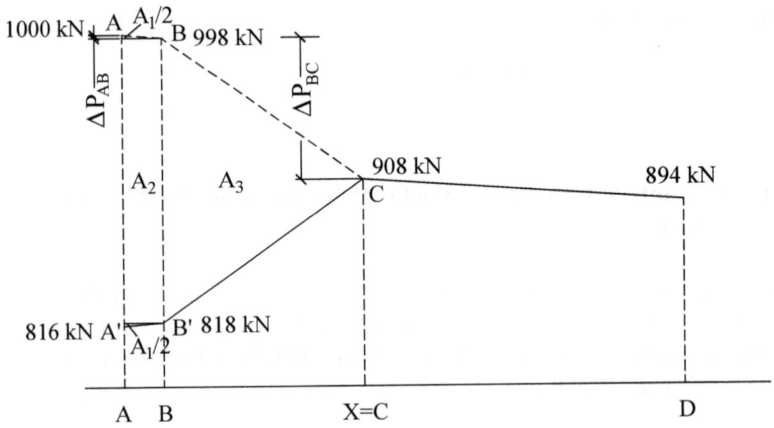

Fig. 6.27 Variation of the prestress force under the hypothesis that the cross-section X is coincident with the cross-section C (dashed lines are corresponding to the variation of *P*(*x*) in the segment AC if the wedge draw-in of the anchorage devices is null)

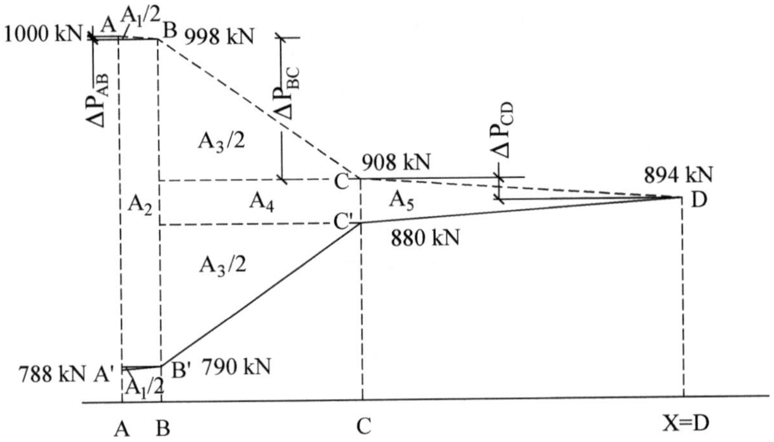

Fig. 6.28 Variation of the prestress force under the hypothesis that the cross-section X is coincident with the cross-section D (dashed lines are corresponding to the variation of *P*(*x*) if the wedge draw-in of the anchorage devices is null)

Area ABA′ (Fig. 6.26):

$$A_{ABA'} = A_1 = 2 \text{ kNm}$$

Area ABCB′A′ (Fig. 6.27):

$$A_{ABCB'A'} = A_1 + A_2 + A_3 = 2 + 180 + 450 = 632 \text{ kNm}$$

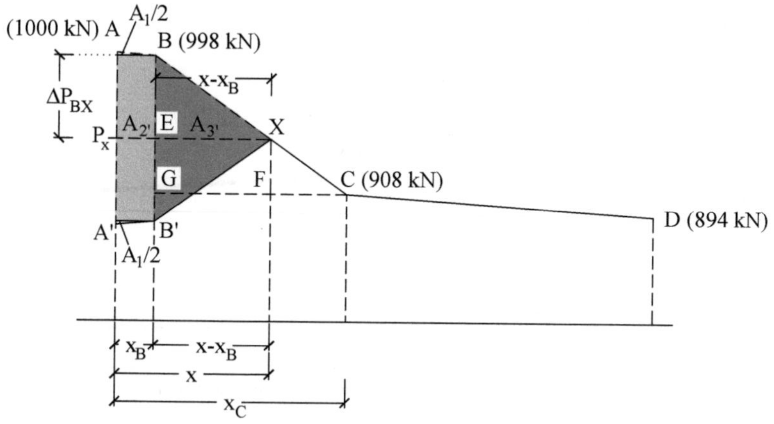

Fig. 6.29 Effective position of the cross-section X

Area ABCDC′B′A′ (Fig. 6.28):

$$A_{ABCDC'B'A'} = A_1 + A_2 + A_3 + A_4 + A_5$$
$$= 2 + 180 + 450 + 140 + 112 = 884 \text{ kNm}$$

The value $w \cdot A_p \cdot E_p = 491.4$ kNm is included between the value of the areas ABA′ and ABCB′A′, therefore section X is placed between B and C.

The similarity of triangles XFC and BGC similar (Fig. 6.29) allows writing:

$$\frac{XF}{BG} = \frac{FC}{GC} \Rightarrow \frac{P_X - P_C}{P_B - P_C} = \frac{x_C - x}{x_C - x_B} \Rightarrow \frac{P_X - 908}{90} = \frac{6 - x}{5} \Rightarrow$$
$$P_X = 90 \cdot \frac{6 - x}{5} + 908 = 108 - 18x + 908 = 1016 - 18x \text{ (KN)}$$

the areas $A_{2'}$ and $A_{3'}$ are equal to:

$$A_{2'} = x_B \cdot 2 \cdot \Delta P_{BX} = 1 \cdot 2 \cdot (998 - P_X) = 1 \cdot 2 \cdot (998 - 1016 + 18x)$$
$$= 2 \cdot (18x - 18) = 36 \cdot (x - 1) \text{ [kNm]}$$

$$A_{3'} = (x - x_B) \cdot \frac{(2 \cdot \Delta P_{BX})}{2} = (x - 1) \cdot (998 - P_X) = (x - 1)$$
$$\cdot (998 - 1016 + 18x) = (x - 1) \cdot (18x - 18) = 18\left(x^2 - 2x + 1\right)\text{[kNm]}$$

and the area ABXB′A′ is given by

$$A = A_1 + A_{2'} + A_{3'} = 2 + 36 \cdot (x - 1) + 18\left(x^2 - 2x + 1\right) = 18\,x^2 - 16 \text{ [kNm]}$$

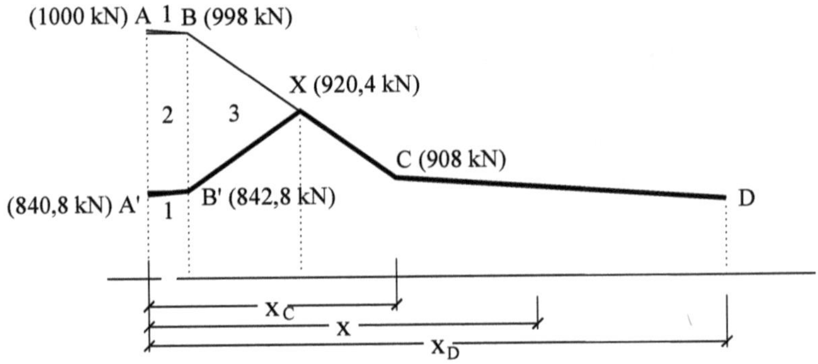

Fig. 6.30 Variation of the prestress force once the losses due to friction and wedge draw-in of the anchorage devices occurred

the x value is obtained imposing that the area ABXB′A′ is equal to $w \cdot A_p \cdot E_p =$ 491.4 kNm:

$$18x^2 - 16 = 491.4 \Rightarrow x = 5.31\ \text{m}$$

therefore (Fig. 6.30):

$$P_X = 1016 - 18 \cdot 5.31 \cong 920.4\ \text{kN}$$

$$\Delta P_{BX} = 998 - 920.4 = 77.6\ \text{kN}$$

$$P_{A'} = P_A - 2 \cdot (\Delta P_{AB} + \Delta P_{BX}) = 1000 - 2 \cdot (2 + 77.6) = 840.8\ \text{kN}$$

$$P_{B'} = P_{A'} + \Delta P_{AB} = 840.8 + 2 = 842.8\ \text{kN}$$

6.10 Time-Dependent Prestress Losses

In both pre-tensioned and post-tensioned beams, the prestress force experiences time-dependent losses, due to the shortening of concrete induced by shrinkage and creep under permanent loads and to the relaxation of prestressing steel.

EC2 gives the expressions to calculate separately effects of the concrete shrinkage and creep and of the "intrinsic" relaxation of steel, whose definition is recalled in Sect. 6.10.2. In prestressed structures, the relaxation of steel is influenced by the shortening of concrete due to shrinkage and creep. This interaction can be approximately accounted for by applying a reduction coefficient equal to 0.8 to the relaxation loss (see [5.10.6 (1)]).

Furthermore, EC2 gives an expression to consider mutual interaction among the three types of long-term prestress loss (Sect. 6.10.7).

6.10.1 Loss Due to Shrinkage of Concrete

Shrinkage is the decrease in either length or volume of concrete resulting from chemical changes or changes in moisture content.

There are three different types of shrinkage: plastic shrinkage, autogenous shrinkage, and drying shrinkage. The plastic shrinkage is concerning with the concrete in the hardening phase, when it is still plastic, while the others are concerning with the hardened concrete.

Plastic shrinkage is due to the evaporation of water from the surface of concrete when it is still plastic and exposed to unsaturated air (relative humidity (RH) less than 95%). This kind of shrinkage cannot occur in form-worked structures, while it can take significant values in projected concretes (*spritz-beton*) and floors.

Autogenous shrinkage occurs in absence of relative humidity variations; it usually occurs after the placing of concrete (that is after that the concrete has lost its initial plasticity) and therefore after plastic shrinkage.

Autogenous shrinkage consists of the auto-drying of the cement paste due to the lack of water in the micropores of the cement matrix, therefore it can be negligible in ordinary concretes with a water-cement ratio over 0.40.

Drying shrinkage occurs in all structures placed in unsaturated air with RH < 95% because the cement paste of the concrete tends to gradually release water to the environment.[14]

The shrinkage occurring between the time of concrete placing and the time of prestress transfer does not have any effect on the prestress itself. The calculation of the prestress loss $\Delta\sigma_p(t)$ at time t due to shrinkage has to be performed by considering the difference of the shrinkage strains in the time interval (t_0, t):

$$\Delta\varepsilon_{cs}(t) = \varepsilon_{cs}(t) - \Delta\varepsilon_{cs}(t_0)$$

where t_0 is the age of the concrete at the transfer of prestress. Under the hypothesis of a perfect bond between steel and concrete, the strain variation $\Delta\varepsilon_p(t)$ in prestressing steel is coincident with $\Delta\varepsilon_{cs}(t)$, and the corresponding stress variation is given by:

[14] Concrete cured in environments with $RH > 95\%$ or protected by impermeable sheets or wetted for long time with water, does not dry and is dimensionally stable. If a concrete block is immersed in water, it tends to swell because absorbs water; finally, if it is placed within an environment with $RH < 95\%$, it tends to dry and to reduce its volume.

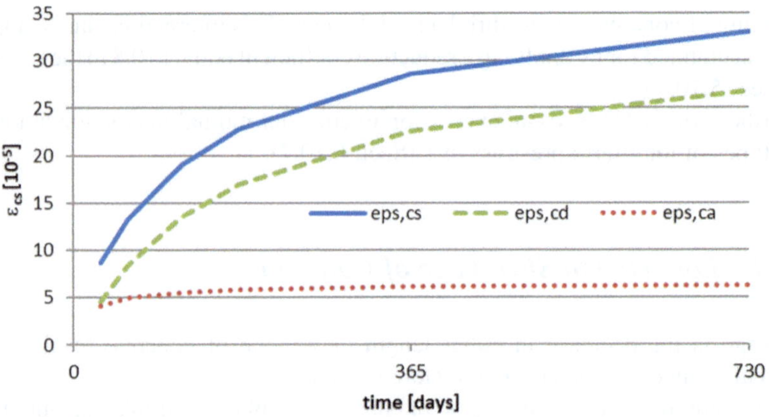

Fig. 6.31 Time evolution of the shrinkage strain: total strain due to shrinkage ε_{cs} (solid line), strain due to drying shrinkage ε_{cd} (dashed line), strain due to autogenous shrinkage ε_{ca} (dotted line) (concrete class C35/45, cement Class N, $h_0 = 261$ mm, RH = 60%, see Example 10)

$$\Delta\sigma_p(t) = E_p \Delta\varepsilon_p(t) = E_p \Delta\varepsilon_{cs}(t)$$

The qualitative variation of strain with time is represented in Fig. 6.31.

The total shrinkage strain is given by the sum of strain due to drying shrinkage (ε_{ca}) and strain due to autogenous shrinkage (ε_{ca})

$$\varepsilon_{cs} = \varepsilon_{cd} + \varepsilon_{ca}. \quad [(3.8)]$$

6.10.1.1 Strain Due to Drying Shrinkage

EC2 gives the following expression for the calculation of strain due to drying shrinkage

$$\varepsilon_{cd}(t) = \beta_{ds}(t, t_s) \cdot k_h \cdot \varepsilon_{cd,0} \quad [(3.9)]$$

where

$$\beta_{ds}(t, t_s) = \frac{(t - t_s)}{(t - t_s) + 0.04 \cdot \sqrt{h_0^3}}$$

with

t age of concrete, in days (for $t = \infty$, $\beta_{ds} = 1$);

Table 6.16 Values of k_h vs h_0 [Table 3.3]

h_0 (mm)	k_h (6.1)
100	1.0
200	0.85
300	0.75
≥500	0.70

(6.1) For values of h_0 different from those provided in the Table, the value of k_h can be obtained by linear interpolation of tabulated values

Table 6.17 Unrestrained drying shrinkage values $\varepsilon_{cd,0}(10^{-5})$ for concrete with Class N cement [Table 3.2]

Concrete strength class	Relative Humidity—RH (%)						
	20	40	60	80	90	100	
C20/25	62	58	49	30	17	0	
C40/50	48	46	38	24	13	0	
C60/75	38	36	30	19	10	0	$\times 10^{-5}$
C80/95	30	28	24	15	8	0	
C90/105	27	25	21	13	7	0	

t_s age of concrete (in days) at the beginning of drying shrinkage, that is generally at the end of hardening in a humid environment ("*curing*") ($t_s = 3 \div 7$ days)[15];

h_0 ($= 2\,A_c\,/\,u$) conventional dimension (in mm) of the transversal cross-section (A_c = area of transversal cross-section of the concrete, u = perimeter of the part of transversal section exposed to drying),

k_h coefficient depending on the conventional dimension h0 (Table 6.16),

$\varepsilon_{cd,0}$ long-term strain due to unrestrained drying shrinkage.

The value of unrestrained drying shrinkage is variable with the cement class. For a Class N cement, $\varepsilon_{cd,0}$ can be directly obtained from Table 3.2 of EC2, which is recalled in the following Table 6.17.

For different classes of concrete strength or relative humidity different from Table 6.17, it is possible to interpolate values provided in the Table. By varying the cement class (S or R), Table 6.17 is not valid anymore, and therefore it is necessary to use the expression [(B.11)], which is also valid for Class N:

$$\varepsilon_{cd,0} = 0.85 \cdot \left[(220 + 110 \cdot \alpha_{ds1}) \cdot \exp\left(-\alpha_{ds2} \cdot \frac{f_{cm}}{f_{cm0}}\right)\right] \cdot 10^{-6} \cdot \beta_{RH} \quad [(B.11)]$$

[15] In the first 3 ÷ 7 days after the placing, the hydration speed of the cement is very high, therefore in this period it is convenient to keep the concrete in a humid environment (by watering or by using suitable sheets) to avoid the stop of hydration and subsequent risks: (a) strength lower than those reachable in concrete with wet curing, (b) very porous concrete.

Table 6.18 Coefficients α_{ds1} and α_{ds2} for the three cement classes

Type of cement	S	N	R
α_{ds1}	3	4	6
α_{ds2}	0.13	0.12	0.11

where $\beta_{RH} = 1.55 \cdot \left[1 - \left(\frac{RH}{RH_0}\right)^3\right]$. with

f_{cm} average compressive strength of concrete,
f_{cm0} $= 10\ \text{N/mm}^2$,

α_{ds1} and α_{ds2} are two coefficients depending on the type of cement (Table 6.18).

RH is the environmental relative humidity (expressed as a ratio),
RH_0 $= 100\ \%$ is the relative humidity of saturated air.

By applying [(B.11)], values collected in Table 6.19 are obtained as concrete compressive strength, cement class and relative humidity are varying. Values of $\varepsilon_{cd,0}$ for Class S cement are equal to 0.76 ÷ 0.81 times those of Class N, while for Class R cements, values vary from 1.37 up to 1.47 times those of Class N.

6.10.1.2 Strain Due to Autogenous Shrinkage

Autogenous shrinkage strain is given by

$$\varepsilon_{ca}(t) = \beta_{as}(t)\,\varepsilon_{ca}(\infty) \quad [(3.11)]$$

where

$$\varepsilon_{ca}(\infty) = 2.5 \cdot (f_{ck} - 10) \cdot 10^{-6} \quad [(3.12)]$$

$$\beta_{as}(t) = 1 - e^{-0.2 \cdot t^{0.5}} \text{ with } t \text{ expressed in days} \quad [(3.13)]$$

In a very short time interval, autogenous shrinkage strain takes on values close to the final long-term value, while drying shrinkage strain develops slowly and takes much more time to take values close to the final long-term one. Table 6.20 collects values of $\varepsilon_{ca}(\infty)$ with the variation of concrete strength class.

Remark. *In precast elements subjected to heat curing, the autogenous shrinkage can be assumed negligible, as indicated in* [10.3.1.2(3)]: $\varepsilon_{ca}(\infty) = 0$.

Figure 6.32 shows the variation of autogenous shrinkage strain ε_{ca} in the first two years after the placing of concrete for four different concrete strength classes.

Table 6.19 Unrestrained drying shrinkage nominal values $\varepsilon_{cd,0}$ for concrete with cement classes S, N [a] and R

Concrete	Cement Class	Relative Humidity – RH (%)								
		20	30	40	50	60	70	80	90	
C30/37	S	43.9	43.0	41.4	38.7	34.7	29.0	21.6	12.0	$\times 10^{-5}$
	N	54.7	53.6	51.6	48.2	43.2	36.2	26.9	14.9	
	R	75.7	74.3	71.4	66.8	59.8	50.1	37.2	20.7	
C35/45	S	41.1	40.3	38.8	36.3	32.5	27.2	20.2	11.2	$\times 10^{-5}$
	N	51.5	50.5	48.6	45.4	40.7	34.1	25.3	14.1	
	R	71.7	70.3	67.6	63.2	56.6	47.5	35.3	19.6	
C40/50	S	38.5	37.8	36.3	34.0	30.4	25.5	18.9	10.5	$\times 10^{-5}$
	N	48.5	47.6	45.8	42.8	38.3	32.1	23.9	13.2	
	R	67.8	66.5	64.0	59.8	53.6	44.9	33.4	18.5	
C45/55	S	36.1	35.4	34.1	31.8	28.5	23.9	17.8	9.9	$\times 10^{-5}$
	N	45.7	44.8	43.1	40.3	36.1	30.2	22.5	12.5	
	R	64.2	63.0	60.6	56.6	50.7	42.5	31.6	17.5	
C50/60	S	33.8	33.2	31.9	29.8	26.7	22.4	16.6	9.2	$\times 10^{-5}$
	N	43.0	42.2	40.6	37.9	34.0	28.5	21.2	11.7	
	R	60.8	59.6	57.3	53.6	48.0	40.2	29.9	16.6	
C55/67	S	31.7	31.1	29.9	28.0	25.0	21.0	15.6	8.7	$\times 10^{-5}$
	N	40.5	39.7	38.2	35.7	32.0	26.8	19.9	11.1	
	R	57.5	56.4	54.3	50.7	45.5	38.1	28.3	15.7	
C60/75	S	29.7	29.1	28.0	26.2	23.5	19.7	14.6	8.1	$\times 10^{-5}$
	N	38.1	37.4	36.0	33.6	30.1	25.3	18.8	10.4	
	R	54.4	53.4	51.4	48.0	43.0	36.1	26.8	14.9	
C70/85	S	26.1	25.6	24.6	23.0	20.6	17.3	12.8	7.1	$\times 10^{-5}$
	N	33.8	33.2	31.9	29.8	26.7	22.4	16.6	9.2	
	R	48.8	47.8	46.0	43.0	38.5	32.3	24.0	13.3	
C80/95	S	22.9	22.5	21.6	20.2	18.1	15.2	11.3	6.3	$\times 10^{-5}$
	N	30.0	29.4	28.3	26.5	23.7	19.9	14.8	8.2	
	R	43.7	42.8	41.2	38.5	34.5	28.9	21.5	11.9	
C90/105	S	20.1	19.7	19.0	17.7	15.9	13.3	9.9	5.5	$\times 10^{-5}$
	N	26.6	26.1	25.1	23.5	21.0	17.6	13.1	7.3	
	R	39.1	38.4	36.9	34.5	30.9	25.9	19.3	10.7	

[a] For Class N cement, values of $\varepsilon_{cd,0}$ for $f_{ck} = 20\text{-}40\text{-}60\text{-}80\text{-}100\ \text{N/mm}^2$ and $RH = 20\text{-}40\text{-}60\text{-}80\text{-}90\%$ are coincident with those collected in Table 3.2 of EC2

Table 6.20 Long-term autogenous shrinkage strain $\varepsilon_{ca}(\infty)$ vs concrete strength class

Concrete strength	C30/37	C35/45	C40/50	C45/55	C50/60
$\varepsilon_{ca}(\infty)$	5×10^{-5}	6.25×10^{-5}	7.5×10^{-5}	8.75×10^{-5}	10×10^{-5}
Concrete strength	C55/67	C60/75	C70/85	C80/95	C90/105
$\varepsilon_{ca}(\infty)$	11.25×10^{-5}	12.5×10^{-5}	15×10^{-5}	17.5×10^{-5}	20×10^{-5}

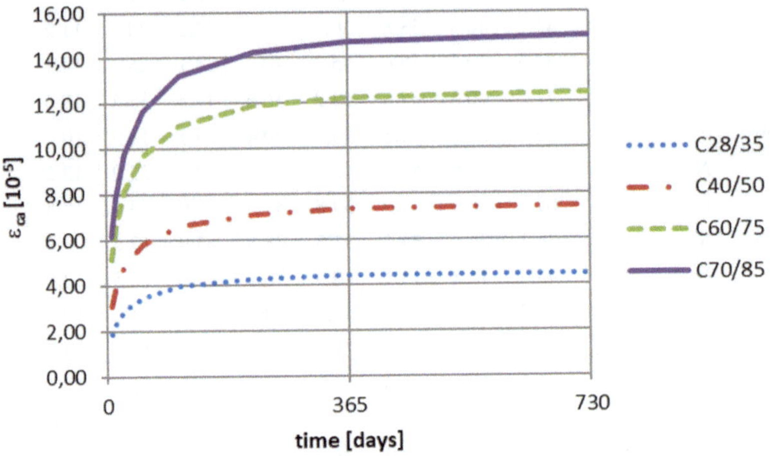

Fig. 6.32 Variation of autogenous shrinkage strain with time

6.10.1.3 Example 9. Calculation of the Autogenous Shrinkage Strain

Calculate the autogenous shrinkage strain for C32/40 *concrete after* 28 days, *one year and two years from the placing of concrete.*

Being:

$$\varepsilon_{ca}(\infty) = 2.5 \cdot (f_{ck} - 10) \cdot 10^{-6} = 2.5 \cdot (32 - 10) \cdot 10^{-6} = 5.5 \times 10^{-5}$$

and

$$\beta_{as}(28) = 1 - e^{-0.2 \cdot 28^{0.5}} = 0.653$$
$$\beta_{as}(365) = 1 - e^{-0.2 \cdot 365^{0.5}} = 0.978$$
$$\beta_{as}(730) = 1 - e^{-0.2 \cdot 730^{0.5}} = 0.9955$$

then

$$\varepsilon_{ca}(28) = \beta_{as}(28) \cdot \varepsilon_{ca}(\infty) = 0.653 \cdot 5.5 \times 10^{-5} = 3.59 \times 10^{-5}$$
$$\varepsilon_{ca}(365) = \beta_{as}(365) \cdot \varepsilon_{ca}(\infty) = 0.978 \cdot 5.5 \times 10^{-5} = 5.38 \times 10^{-5}$$
$$\varepsilon_{ca}(730) = \beta_{as}(730) \cdot \varepsilon_{ca}(\infty) = 0.9955 \cdot 5.5 \times 10^{-5} = 5.48 \times 10^{-5}$$

The example confirms what has already been expressed: autogenous shrinkage reaches quickly its long-term value; after one year it already takes 98% of its long-term value and after two years it is almost equal to $\varepsilon_{ca}(\infty)$.

6.10.1.4 Example 10. Calculation of Drying Shrinkage and Autogenous Shrinkage

Calculate the time variation of drying shrinkage strain and autogenous shrinkage strain for a C35/45 concrete made with Class N cement.

Consider a 200 × 600 mm *rectangular cross-section, with the base and vertical sides exposed for a height of* 360 mm (*Fig. 6.33a*)*; the relative humidity is* 60%.

Calculation of drying shrinkage strain

$$\varepsilon_{cd}(t) = \beta_{ds}(t, t_s) \cdot k_h \cdot \varepsilon_{cd,0}$$

from Table 6.19, for C35/45 concrete $\left(f_{ck} = 35\,\text{N/mm}^2\right)$, Class N cement and $RH = 60\%$,

$$\varepsilon_{cd,0} = 40.7 \times 10^{-5}$$

for a rectangular cross-sections 200 × 600 mm, where the base and a 360 mm long segment of both vertical sides are exposed,

$$h_0 = \frac{2 \cdot A_c}{u} = \frac{2 \cdot 200 \cdot 600}{360 + 200 + 360} = 261 \text{ mm}$$

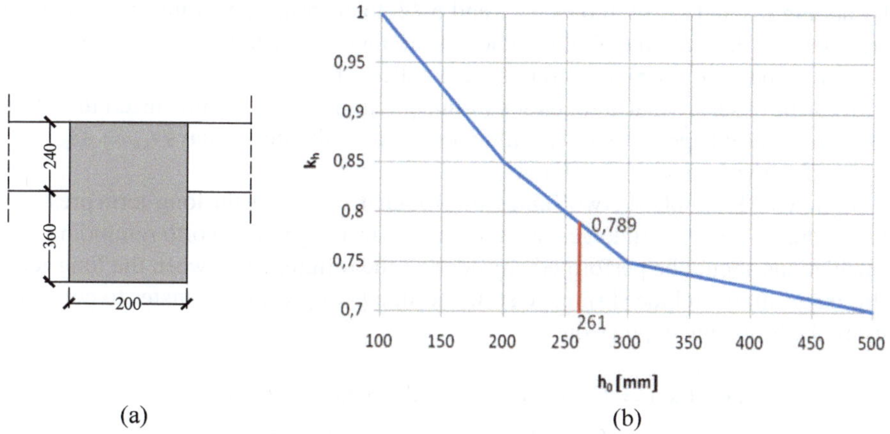

Fig. 6.33 **a** rectangular cross-sections 200 × 600 mm, **b** linear interpolation for the calculation of k_h

from Table 6.16, it results that $k_h = 0.85$ for $h_0 = 200$ mm and $k_h = 0.75$ for $h_0 = 300$ mm; by linearly interpolating, for $h_0 = 261$ mm it results that (Fig. 6.33b):

$$k_h = 0.75 + (0.85 - 0.75) \cdot \frac{300 - 261}{300 - 200} = 0.789$$

Assuming $t_s = 3$ days (see Footnote 16), $\beta_{ds}(t,\ t_s)$ is given by

$$\beta_{ds}(t, t_s) = \frac{(t - t_s)}{(t - t_s) + 0.04 \cdot \sqrt{h_0^3}}$$

for $t = 30, 60, 120, 180, 365, 730$ days and $t = \infty[\varepsilon_{cd}(\infty) = k_h \cdot \varepsilon_{cd,0}]$. Therefore, by multiplying values of $\beta_{ds}(t, t_s)$ by $\varepsilon_{cd,0} = 40.7 \cdot 10^{-5}$, ε_{cd} takes values collected in Table 6.21.

Calculation of autogenous shrinkage strain

$$\varepsilon_{ca}(t) = \beta_{as}(t)\varepsilon_{ca}(\infty)$$

$$\beta_{as}(t) = 1 - e^{-0.2t^{0.5}} \text{ with } t \text{ in days}$$

$$\varepsilon_{ca}(\infty) = 2.5 \cdot (f_{ck} - 10) \cdot 10^{-6} = 2.5 \cdot (35 - 10) \cdot 10^{-6} = 6.25 \times 10^{-5}$$

for $t = 30, 60, 120, 180, 365, 730$ days and $t = \infty$, values of $\beta_{as}(t)$ and $\varepsilon_{ca}(t)$ are collected in Table 6.22.

Calculation of total shrinkage strain

By adding ε_{cd} and ε_{ca} of Tables 6.21 and 6.22, total shrinkage strains of Table 6.23 are obtained. In the same Table, values of the ratio between autogenous shrinkage strain ε_{ca} and total shrinkage strain ε_{cs} are collected.

It can be observed that autogenous shrinkage strain gives a significant contribution to the total shrinkage strain only within a short period: after 30 days $\varepsilon_{ca} \cong 48\,\%\,\varepsilon_{cs}$, while after two years $\varepsilon_{ca} \cong 19\,\%\,\varepsilon_{cs}$.

Following Example 10, two Tables are provided, collecting the long-term prestress losses due to drying shrinkage and autogenous shrinkage. It is worth reminding that calculations should be performed considering the difference between the long-term shrinkage strain and the shrinkage strain at time t_0 of prestress transfer to concrete, using the following expression

$$\begin{aligned}\Delta\sigma_p(\infty) &= E_p\Delta\varepsilon_{cs}(\infty) = E_p[\varepsilon_{cs}(\infty) - \varepsilon_{cs}(t_0)] \\ &= E_p[\varepsilon_{cd}(\infty) - \varepsilon_{cd}(t_0)] + E_p[\varepsilon_{ca}(\infty) - \varepsilon_{ca}(t_0)] \\ &= \Delta\sigma_p(\varepsilon_{cd}) + \Delta\sigma_p(\varepsilon_{ca})\end{aligned}$$

Table 6.21 β_{ds} (for $t_s = 3$ and $h_0 = 261$ mm) and ε_{cd} (for $k_h = 0.789$)

Time t (days)	30	60	120	180	365 (one year)	730(two years)	∞
$\beta_{ds}(t)$	0.138	0.253	0.410	0.512	0.682	0.812	1.000
$\varepsilon_{cd}(t)$	4.43×10^{-5}	8.11×10^{-5}	13.15×10^{-5}	16.44×10^{-5}	21.91×10^{-5}	26.06×10^{-5}	32.11×10^{-5}

Table 6.22 β_{as} and ε_{ca}

Time (days)	30	60	120	180	365(one year)	730 (two years)	∞
β_{as}	0.666	0.788	0.888	0.932	0.978	0.996	1
ε_{ca}.	4.16×10^{-5}	4.92×10^{-5}	5.55×10^{-5}	5.82×10^{-5}	6.11×10^{-5}	6.22×10^{-5}	6.25×10^{-5}

Table 6.23 ε_{cs} and ratio $(\varepsilon_{ca}/\varepsilon_{cs})$

Time s(days)	30	60	120	180	365(one year)	730(two years)	∞
$\varepsilon_{cs} = \varepsilon_{cd} + \varepsilon_{ca}$	8.59×10^{-5}	13.03×10^{-5}	18.70×10^{-5}	22.26×10^{-5}	28.02×10^{-5}	32.28×10^{-5}	38.36×10^{-5}
$(\varepsilon_{ca}/\varepsilon_{cs})$	48.4%	37.8%	29.7%	26.2%	21.8%	19.3%	16.3%

where the first term $(\Delta\sigma_p(\varepsilon_{cd}) = E_p[\varepsilon_{cd}(\infty) - \varepsilon_{cd}(t_0)])$ represents the loss due to drying shrinkage and the second term the loss due to autogenous shrinkage $(\Delta\sigma_p(\varepsilon_{ca}) = E_p[\varepsilon_{ca}(\infty) - \varepsilon_{ca}(t_0)])$.

Table 6.24 collects losses due to long-term drying shrinkage expressed in N/mm^2 for different concrete strength classes and relative humidity.

Table 6.24 has been built by assuming that $t_s = t_0$, i.e., under the hypothesis that the duration of the wet hardening of concrete (see Footnote 29) is coincident with the age of the concrete at the transfer (pre-tensioning) or application (post-tensioning) of prestress. Then $\beta_{ds}(t_0 - t_s) = 0$, $\varepsilon_{cd}(t_0) = 0$ and the expression to calculate the loss due to drying shrinkage takes the following approximated form

$$\Delta\sigma_{\mathrm{p,cd}}(\infty) = E_{\mathrm{p}}\varepsilon_{\mathrm{cd}}(\infty)$$

Moreover, values have been calculated by assuming $E_p = 195{,}000$ N/mm^2 and $k_h = 1.0$. For $k_h < 1$, it is sufficient to multiply tabulated values by k_h, while for $E_p \neq 195{,}000$ N/mm^2 it is enough to multiply tabulated values by the ratio E_p / 195,000, with E_p in N/mm^2.

Finally, if $t_s < t_0$, i.e., if the hardening period is shorter than the age of transfer of prestress, as it usually occurs, tabulated values have to be reduced by the amount $\Delta\sigma_p(t_0) = E_p\varepsilon_{cd}(t_0)$. The value of $\varepsilon_{cd}(t_0)$ is obtained multiplying $\varepsilon_{cd}(\infty)$ by $\beta_{ds}(t, t_s)$; Table 6.21 collects values of $\beta_{ds}(t, t_s)$ for $t_s = 3$ days, t = 30, 60, 120, 180, 365, 730 days and $t = \infty$.

The use of tabulated values also for $t_s < t_0$ involves an overestimation of the loss due to drying shrinkage and therefore the total shrinkage loss. It is then possible to use tabulated values in the design phase and refine the calculation in the verification phase considering the stress reduction $\Delta\sigma_p(t_0) = E_p\varepsilon_{cd}(t_0)$.

Table 6.25 collects values of the prestress loss due to long-term autogenous shrinkage for different values of time t_0 at the transfer of prestress to concrete

Table 6.24 Prestress loss (N/mm^2) due to long-term drying shrinkage (values are calculated for $k_h = 1.0$ and $E_p = 195{,}000\,\text{N/mm}^2$)[a]

Concrete	Cement class	RH50%	RH60%	RH 70%	RH 80%	RH 90%
C30/37	S	75	68	57	42	23
	N	94	84	71	52	29
	R	130	117	98	73	40
C35/45	S	71	63	53	39	22
	N	89	79	66	49	27
	R	123	110	93	69	38
C40/50	S	66	59	50	37	21
	N	83	75	63	47	26
	R	117	105	88	65	36
C45/55	S	62	56	47	35	19
	N	79	70	59	44	24
	R	110	99	83	62	34
C50/60	S	58	52	44	32	18
	N	74	66	56	41	23
	R	105	94	78	58	32
C55/67	S	55	49	41	30	17
	N	70	62	52	39	22
	R	99	89	74	55	31
C60/75	S	51	46	38	28	16
	N	66	59	49	37	20
	R	94	84	70	52	29
C70/85	S	45	40	34	25	14
	N	58	52	44	32	18
	R	84	75	63	47	26
C80/95	S	39	35	30	22	12
	N	52	46	39	29	16
	R	75	67	56	42	23
C90/105	S	35	31	26	19	11
	N	46	41	34	26	14
	R	67	60	51	38	21

[a]Tabulated values have been calculated under the assumption that the strain due to drying shrinkage in the time interval (t_0, ∞), where t_0 is the age of the concrete at the transfer of the prestress, is coincident with the long-term one: $\varepsilon_{cd}(\infty) - \varepsilon_{cd}(t_0) = \varepsilon_{cd}(\infty)$. This hypothesis means that the time t_s at the beginning of drying shrinkage (i.e. the end of the wet hardening of concrete) is coincident with the time t_0 at which concrete is prestressed: $t_s = t_0$, $\beta_{ds} = 0$, $\varepsilon_{cd}(t_0) = 0$.

Table 6.25 Prestress loss (N/mm^2) due to long-term autogenous shrinkage at varying the concrete age at the transfer of prestress to concrete and for $E_p = 195{,}000\,\text{N/mm}^2$

Concrete	Age of the concrete at the transfer of the compression force			
	7 days	14 days	21 days	28 days
C30/37	6	5	4	3
C35/45	7	6	5	4
C40/50	9	7	6	5
C45/55	10	8	7	6
C50/60	11	9	8	7
C55/67	13	10	9	8
C60/75	14	12	10	8
C70/85	17	14	12	10
C80/95	20	16	14	12
C90/105	23	18	16	14

$$\Delta\sigma_p(\varepsilon_{ca}) = E_p[\varepsilon_{ca}(\infty) - \varepsilon_{ca}(\text{t}_0)]$$

as indicated above, the modulus of elasticity of prestressing steel has been taken equal to 195,000 N/mm^2; for $E_p \neq 195{,}000\,\text{N/mm}^2$ it is sufficient to multiply tabulated values by the ratio E_p / 195,000, with E_p in N/mm^2.s

6.10.1.5 Example 11. Loss Due to the Long-Term Concrete Shrinkage for a Beam with Pre-tensioned Strands

Calculate the loss due to shrinkage for a pre-tensioned precast beam subjected to heat curing, whose cross-section is shown in Fig. 6.34. Assume the following input data: area of the geometrical cross-section: 184,000 mm^2, *concrete strength class:* C50/60, *cement class: R, RH:* 70%. *Consider the perimeter fully exposed.*

Autogenous shrinkage strain

For a precast beam subjected to heat treatment, autogenous shrinkage can be assumed negligible, as indicated in [10.3.1.2(3)]

$$\varepsilon_{ca}(\infty) = 0$$

Unrestrained drying shrinkage strain

From Table 6.19, for concrete strength class C50/60, Class R cement and $RH = 70\%$

$$\varepsilon_{cd,0} = 40.2 \times 10^{-5}$$

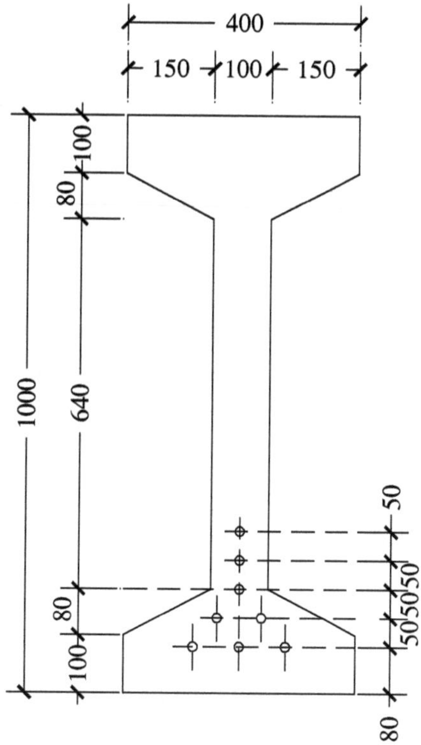

Fig. 6.34 Cross-section of the beam with pre-tensioned strands

Long-term drying shrinkage strain $\varepsilon_{cd}(\infty)$:

$$\varepsilon_{cd}(\infty) = \beta_{ds}(\infty, t_s) \cdot k_h \cdot \varepsilon_{cd,0}$$

where $\beta_{ds}(\infty,\ t_s) = 1.0$ and k_h is obtained by linear interpolation of values collected in Table 6.16: under the hypothesis that the section perimeter is fully exposed

$$h_0 = \frac{2A}{u} = \frac{2 \cdot 184.000}{3.160} = 116.5\ \text{mm}$$

$$k_h = 0.85 + \frac{1.00 - 0.85}{200 - 100} \cdot (200 - 116.5) \cong 0.98$$

and therefore

$$\varepsilon_{\text{cd}}(\infty) = \beta_{\text{ds}}(\infty,\text{t}_\text{s}) \cdot \text{k}_\text{h} \cdot \varepsilon_{\text{cd,0}} = 1.0 \cdot 0.98 \cdot 40.2 \times 10^{-5} = 39.4 \times 10^{-5}$$

assuming $t_s = t_0$, the following values are obtained:

$$\beta_{ds}(t_0, t_s) = 0 \text{ and } \varepsilon_{cd}(t_0) = 0,$$

therefore, the loss due to drying shrinkage is given by the following simplified expression

$$\Delta\sigma_{p,cd}(\infty) = E_p\varepsilon_{cd}(\infty)$$

Loss due to shrinkage

$$\Delta\sigma_{\text{rit}} = \varepsilon_{cd}(\infty) \cdot E_p = 39.4 \times 10^{-5} \cdot 195{,}000 = 76.83\,\text{N/mm}^2 \cong 77\,\text{N/mm}^2$$

Use of values collected in Table 6.24

Calculation is repeated using Table 6.24: for C50/60 concrete, Class R cement and $RH = 70\%$, the loss due to long-term drying shrinkage takes the value 78 N/mm^2, and, being $k_h = 0.98$, $\Delta\sigma_{\text{rit}} = 78 \cdot 0.98 == 76.44\,\text{N/mm}^2 \cong 77\,\text{N/mm}^2$.

6.10.1.6 Example 12. Loss Due to Long-Term Concrete Shrinkage for a Beam with Post-tensioned Tendons

Calculate the loss due to long-term concrete shrinkage for a beam with post-tensioned tendons, whose cross-section is shown in Fig. 6.35. The concrete strength class is C35/45, the *cement class is R and RH is* 70%*. Hardening takes 3 days (*$t_s = 3$ days*) and the transfer of prestress to concrete occurs* 14 days *after the placing. It is assumed that the perimeter of the beam is fully exposed.*

Geometrical characteristics

Area of the concrete cross-section

$$A_c = 937{,}800\,\text{mm}^2$$

Perimeter of the cross-section

$$u = 7465\,\text{mm}$$

Notional size

$$h_0 = \frac{2A}{u} = \frac{2 \cdot 937.800}{7.465} \cong 251\ \text{mm}$$

Unrestrained drying shrinkage strain $\varepsilon_{cd,0}$

From Table 6.19, for concrete strength class C35/45, Class R cement and $RH = 70\%$

$$\varepsilon_{cd,0} = 47.5 \times 10^{-5}$$

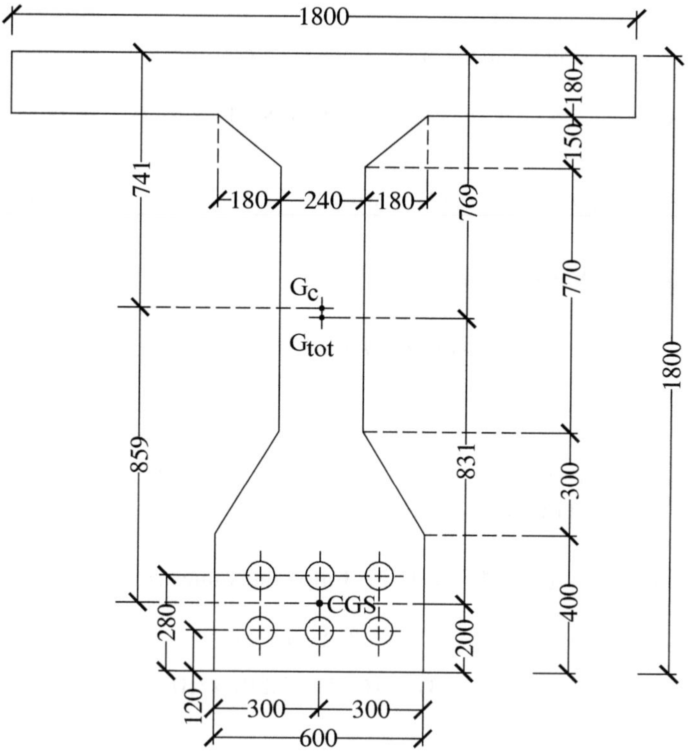

Fig. 6.35 Cross-section of the beam with post-tensioned tendons

Drying shrinkage strain $\varepsilon_{cd}(t_0)$ at time t_0 of prestress transfer:

$$\varepsilon_{cd}(t_0, t_s) = \beta_{ds}(t_0, t_s) \cdot k_h \cdot \varepsilon_{cd,0}$$

where

$$\beta_{ds}(t_0, t_s) = \frac{(t_0 - t_s)}{(t_0 - t_s) + 0.04 \cdot \sqrt{h_0^3}} = \frac{(14 - 3)}{(14 - 3) + 0.04 \cdot \sqrt{251^3}} = 0.065$$

By linearly interpolating values k_h collected in Table 6.16, $k_h = 0.8$ for $h_0 = 251$ mm and, by substituting this value in the expression of $\varepsilon_{cd}(t_0, t_s)$, the following result is obtained

$$\varepsilon_{cd}(t_0, t_s) = \varepsilon_{cd}(14.3) = 0.065 \cdot 0.8 \cdot 47.5 \times 10^{-5} = 2.47 \times 10^{-5}$$

Long-term drying shrinkage strain $\varepsilon_{cd}(\infty)$:

$$\varepsilon_{cd}(\infty) = \beta_{ds}(\infty, t_s) \cdot k_h \cdot \varepsilon_{cd,0}$$

where
$\beta_{ds}(\infty, t_s) = 1$ (independent of the t_s value) therefore:

$$\varepsilon_{cd}(\infty) = \beta_{ds}(\infty, t_s) \cdot k_h \cdot \varepsilon_{cd,0} = 1.0 \cdot 0.8 \cdot 47.5 \times 10^{-5} = 38 \times 10^{-5}$$

$$\Delta\varepsilon_{cd} = \varepsilon_{cd}(\infty) - \varepsilon_{cd}(14) = (38 - 2.47) \times 10^{-5} = 35.53 \times 10^{-5}$$

Autogenous shrinkage strain

$$\varepsilon_{ca}(t) = \beta_{as}(t)\ \varepsilon_{ca}(\infty)$$

where

$$\varepsilon_{ca}(\infty) = 2.5 \cdot (f_{ck} - 10) \times 10^{-6} = 6.25 \times 10^{-5}$$
$$\varepsilon_{ca}(14) = \beta_{as}(14)\ \varepsilon_{ca}(\infty)$$

being $\beta_{as}(14) = 1 - e^{-0.2 \cdot 14^{0.5}} = 0.527$. following values are obtained

$$\varepsilon_{ca}(14) = 0.527 \cdot 6.25 \times 10^{-5} = 3.29 \times 10^{-5}$$
$$\Delta\varepsilon_{ca} = \varepsilon_{ca}(\infty) - \varepsilon_{ca}(14) = (6.25 - 3.29) \times 10^{-5} = 2.96 \times 10^{-5}$$

Total shrinkage strain

$$\varepsilon_{cs}(t_0 = 14) = \varepsilon_{cd}(t_0 = 14, t_s = 3) + \varepsilon_{ca}(t_0 = 14) = 2.47 \times 10^{-5} + 3.29 \times 10^{-5}$$
$$= 5.76 \times 10^{-5}$$
$$\varepsilon_{cs}(\infty) = \varepsilon_{cd}(\infty) + \varepsilon_{ca}(\infty) = 38 \times 10^{-5} + 6.25 \times 10^{-5} = 44.25 \times 10^{-5}$$

Loss due to drying shrinkage

$$\Delta\sigma_{p,cd} = \Delta\varepsilon_{cd} \cdot E_p = 35.53 \times 10^{-5} \cdot 195{,}000 = 69.28\,\mathrm{N/mm^2}$$

Loss due to autogenous shrinkage

$$\Delta\sigma_{p,ca} = \Delta\varepsilon_{ca} \cdot E_p = 2.96 \times 10^{-5} \cdot 195{,}000 = 5.77\,\mathrm{N/mm^2}$$

Total loss due to shrinkage

$$\Delta\sigma_p = 69.28 + 5.77 \cong 75\,\mathrm{N/mm^2}$$

Use of Tables 6.24 and 6.25

The calculation is repeated using Tables 6.24 and 6.25, which collect, respectively, values of the losses due to drying shrinkage and autogenous shrinkage.

From Table 6.24, for a concrete strength class C35/45, Class R cement and $RH = 70\%$, the loss due to long-term drying shrinkage takes the value of 93 N/mm^2; in the present case, being $k_h = 0.8$, the loss is equal to: $\Delta\sigma_{p,cd} = 93 \cdot 0.8 = 74.4\,\text{N/mm}^2$.

This value is greater than the value above (69.28 N/mm^2) because Table 6.24 has been built, as already mentioned, under the hypothesis that $t_s = t_0$.

From Table 6.25, for concrete strength class C35/45 and by assuming the time for the transfer of the prestress equal to 14 days, the loss due to autogenous shrinkage is equal to 6 N/mm^2, slightly greater than the value above (5.77 N/mm^2), because all tabulated values have been obtained after rounding up decimal values to the highest integer.

The total loss due to long-term shrinkage evaluated using Tables 6.24 and 6.25, with the approximation $t_s = t_0$, is equal to:

$$\Delta\sigma_p = 74.4 + 6 = 80.4\,\text{N/mm}^2,$$

around 7% higher than the value above.

Remark. The calculation of $\Delta\sigma_{p,cd}$ performed using Tables can be improved by subtracting to the tabulated value (74.4 N/mm^2) the term $E_p\varepsilon_{cd}(t_0 = 14,\ t_s = 3)$:

$$\begin{aligned}\Delta\sigma_{p,cd} &= 74.4 - E_p \cdot \varepsilon_{cd}(t_0 = 14,\ t_s = 3)\\ &= 74.4 - 4.82 = 69.6\,\text{N/mm}^2 (\cong 69.28\,\text{N/mm}^2),\end{aligned}$$

where

$$\varepsilon_{cd}(t_0 = 14,\ t_s = 3) = \beta_{ds}(t_0 = 14,\ t_s = 3) \cdot k_h \cdot \varepsilon_{cd,0} = 2.47 \cdot^{-5}$$

$$\text{with}\quad \beta_{ds}(t_0 = 14, t_s = 3) = 0.065;\ k_h = 0.8\,\text{ed}\,\varepsilon_{cd,0} = 47.5 \times 10^{-5}.$$

6.10.2 *Loss Due to the Relaxation of Prestressing Steel*

When a steel wire is stretch and maintained at a constant strain between two fixed points, the initial tensile force does not remain constant but decreases with time, because of the steel viscosity. The decrease of stress in steel at constant strain is termed "intrinsic relaxation".

Concerning relaxation, prestressing steel is classified into three classes: *class 1* for wires and strands with high relaxation, *class 2* for wires and strands with low relaxation, and *class 3* for bars.

Table 6.26 Classes of relaxation for prestressing steel and relevant parameters

Relaxation class	Description	$\rho_{1000}(\%)$	k_1	k_2
1	Wires or ordinary strands	8	5.39	6.7
2	Wires or low relaxation strands	2.5	0.66	9.1
3	Hot rolled and processed bars	4	1.98	8

Following EC2, the loss due to intrinsic relaxation of prestressing steel can be evaluated through the following expression:

$$\frac{\Delta\sigma_{pr}}{\sigma_{pi}} = k_1 \cdot \rho_{1000} \cdot e^{k_2 \cdot \mu} \left(\frac{t}{1000}\right)^{0.75 \cdot (1-\mu)} \cdot 10^{-5} \quad [(3.28), (3.29), (3.30)]$$

where:

σ_{pi} for post-tensioned steel $\sigma_{pi} = \sigma_{pm0}$, where σ_{pm0} is the stress in steel after the application of the prestress, once all the initial losses occurred,

for pre-tensioned steel σ_{pi} is the tensile stress applied to the tendon minus the initial loss due to draw-in of anchorages at ends of the precast bed,

ρ_{1000} value of relaxation loss (in %) at 1000 h after tensioning and at a temperature of 20 °C; this value is obtained for initial stress equal to $0.70 f_p$[16] [3.3.2(5)-EC2] and it is expressed as a percentage of the initial stress,

k_1, k_2 are two parameters that depend on the class of relaxation,

$$\mu = \sigma_{\text{pi}} / f_{\text{pk}},$$

t is the time passed after tensioning (in hours).

Table 6.26 collects values of ρ_{1000}, k_1, k_2 for the three classes of relaxation indicated by EC2.

6.10.2.1 Example 13. Loss Due to Steel Relaxation After 1000 h

Calculate the loss due to steel relaxation of a Class 2 strand at 1000 h *after tensioning, assuming an initial prestress of* $0.70\, f_{pk}$. *Steel strengths are the following:* $f_{pk} =$ 1860 N/mm^2, $f_{p0.1\,k} = 1600$ N/mm^2.

Initial prestress: $\sigma_{pi} = 0.70 \cdot 1860 = 1302$ N/mm^2.

The loss due to relaxation is given by the following expression

[16] For strands, the maximum experimental tensile strength f_p is, on average, equal to about 1.15 times the characteristic ultimate strength f_{pk} : $f_p \cong 1.15\, f_{pk}$, therefore $0.7\, f_p \cong 0.7 \cdot 1.15 \cdot f_{pk} = 0.8\, f_{pk}$ i.e. for strands, the loss after 1000 h (ρ_{1000}) is also referable to an initial stress almost equal to $0.80\, f_{pk}$.

$$\frac{\Delta\sigma_{pr}}{\sigma_{pi}} = k_1 \cdot \rho_{1000} \cdot e^{k_2 \cdot \mu} \left(\frac{t}{1000}\right)^{0.75 \cdot (1-\mu)} \cdot 10^{-5}$$

from Table 6.26, for a low relaxation strand (Class 2), the parameters are equal to

$$\rho_{1000} = 2.5;\ k_1 = 0.66;\ k_2 = 9.1$$

which, substituted in the previous expression, give

$$\frac{\Delta\sigma_{pr}}{\sigma_{pi}} = 0.66 \cdot 2.5 \cdot e^{9.1 \cdot 0.7} \cdot 10^{-5} = 0.96\,\%$$

therefore, the loss is equal to: $\Delta\sigma_{pr} = 0.0096 \cdot 1302 = 12.5\,\text{N/mm}^2$.

Remark. Following the calculation above, the percentage loss due to relaxation (0.96%) is not coincident with $\rho_{1000}(= 2.5\,\%)$, but it is much smaller. This result is only apparently in contrast with the definition of ρ_{1000} because the loss due to relaxation after 1000 h is coincident with ρ_{1000} only if the initial prestress is equal to 70% of the real strength of prestressing steel [3.3.2 (5)] and not equal to 70% of the characteristic ultimate strength (see Footnote 17).

6.10.2.2 Example 14. Loss Due to Long-Term Relaxation

Calculate the loss due to long-term relaxation ($t = 500{,}000$ h*) of an ordinary relaxation strand (Class* 1*), a low relaxation strand (Class 2) and a bar (Class* 3*) by assuming an initial prestress of* $0.75 f_{pk}$.[17]

For a strand of Class 1, Table 6.26 gives the following values of parameters to use in the calculation of the loss due to relaxation:

$$\rho_{1000} = 8;\ k_1 = 5.39;\ k_2 = 6.7$$

moreover,

$$\mu = \sigma_{pi}/f_{pk} = 0.75;\ t = 500000\,\text{hours}\ (t = \infty)$$

then

$$\Delta\sigma_{pr}/\sigma_{pi} \cong 0.2104 = 21.04\%.$$

[17] For pre-tensioned strands, the assumed value ($\sigma_{pi} = 0.75\,f_{pk}$) is certainly lower than the limit stress $\left[\sigma_{pi} \leq \min\left(0.8\,f_{pk};\ 0.9\,f_{p0.1k}\right)\right]$; for post-tensioned tendons, the initial prestress shall satisfy the following limitation: $\sigma_{pi} \leq \min\left(0.75\,f_{pk};\ 0.85\,f_{p0.1k}\right)$, therefore the value assumed in the example is acceptable if $0.75\,f_{pk} < 0.85\,f_{p0.1k}$.

Repeating the calculation for an ordinary relaxation strand (Class 2) with diameter of 0.6″ (15 mm), under the same initial prestress, Table 6.26 gives:

$$\rho_{1000} = 2.5;\ k_1 = 0.66;\ k_2 = 9.1$$

moreover, it results

$$\mu = \sigma_{pi} \big/ f_{pk} = 0.75$$

and

$$\Delta\sigma_{pr}/\sigma_{pi} \cong 0.0487 = 4.87\,\%$$

Finally, considering a hot rolled bar (Class 3), again Table 6.26 gives:

$$\rho_{1000} = 4;\ k_1 = 1.98;\ k_2 = 8$$

moreover, it results

$$\mu = \sigma_{pi} \big/ f_{pk} = 0.75$$

and therefore

$$\Delta\sigma_{pr} \big/ \sigma_{pi} \cong 0.1025 = 10.25\,\%.$$

Table 6.27 collects values of the loss due to relaxation after 1000 h and long-term relaxation of all the three relaxation classes for different values of the ratio $\sigma_{pi} \big/ f_{pk}$.

Figure 6.36 shows the variation of the loss due to relaxation 1000 h after tensioning with the variation of the initial prestress.

Finally, Table 6.28 collects values of the ratio between the loss due to relaxation 1, 5, 10, 20, 100, 200, 500, 750 and 500,000 h after tensioning and the loss after 1000 h for different values of the initial prestress. As it is clear from the expression

Table 6.27 Loss (%) due to relaxation after 1000 h and 500,000 h vs initial prestress

Relaxation class	t (hours)	$\sigma_{pi} \big/ f_{pk}$					
		0.50	0.55	0.60	0.65	0.70	0.75
1 = ordinary relaxation strands	1000	1.23	1.72	2.40	3.36	4.69	6.56
	500,000	12.64	13.99	15.50	17.16	19.00	21.04
2 = low relaxation strands	1000	0.16	0.25	0.39	0.61	0.96	1.52
	500,000	1.61	2.00	2.50	3.12	3.90	4.87
3 = hot rolled bars	1000	0.43	0.65	0.96	1.44	2.14	3.20
	500,000	4.45	5.25	6.21	7.34	8.67	10.25

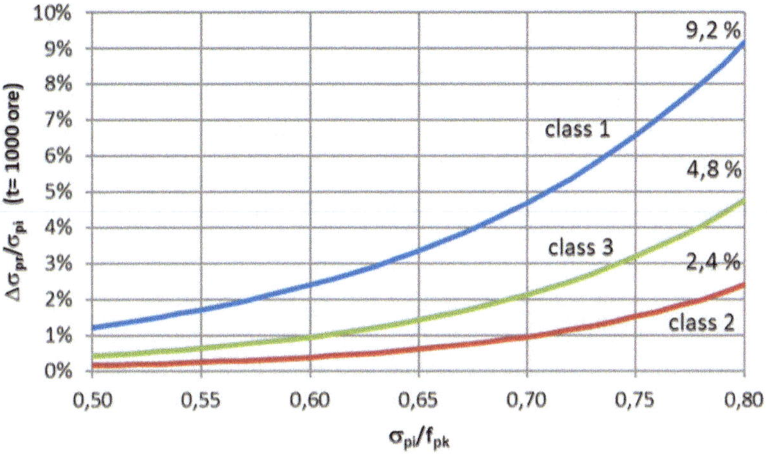

Fig. 6.36 Loss due to relaxation 1000 h after tensioning, as a function of the initial prestress; the values corresponding to $\sigma_{pi} = 0.78 \div 0.80\, f_{pk}$ are very close to ρ_{1000} (see Footnote 17)

Table 6.28 Values of the ratio ρ_t/ρ_{1000} (%) between the loss due to relaxation at time $t(\rho_t)$ and 1000 h (ρ_{1000}) vs σ_{pi}/f_{pk}.

σ_{pi}/f_{pk}	t (hours)											
	1	5	10	20	50	100	200	500	750	1000	500,000	
0.50	7	14	18	23	33	42	55	77	90	100	1028	%
0.60	13	20	25	31	41	50	62	81	92	100	645	
0.70	21	30	35	41	51	60	70	86	94	100	405	
0.75	27	37	42	48	57	65	74	88	95	100	321	

of the relaxation loss, the ratio values do not depend on the relaxation class of the steel.

6.10.3 Effects of the Heat Curing on the Prestress Loss Due to Steel Relaxation

Precast concrete elements with pre-tensioned tendons are often subjected to heat curing to speed up concrete hardening. To evaluate the effects of heat curing at temperatures higher than 20 °C on the steel relaxation, an equivalent time t_{eq} is added to the effective time after tensioning

$$t_{eq} = \frac{1.14^{T_{\max}-20}}{T_{\max}-20} \sum_{i=1}^{n} \left(T_{(\Delta t_i)} - 20\right) \cdot \Delta t_i \quad [(10.2)]$$

where

t_{eq} is the equivalent time (in hours),
$T(\Delta t_i)$ is the temperature (in Celsius degrees) in the time interval Δt_i;
T_{max} is the maximum temperature (in Celsius degrees) reached during the heat curing.

For application purposes, it is useful to rewrite expression [(10.2)] for a heating cycle with constant temperature and a heat cycle with a linear variation of temperature.

Case 1. Constant temperature

If the temperature is maintained constant and equal to T_1 ($T_1 \geq 20$ °C) for a time interval Δt, the equivalent time can be estimated from the expression:

$$t_{eq} = \frac{1.14^{T_{max}-20}}{T_{max} - 20}(T_1 - 20) \cdot \Delta t$$

Case 2. Linearly variable temperature

If the temperature varies linearly from T_1 to T_2 (20 °C $\leq T_1 < T_2 \leq T_{max}$) in the time interval $\Delta t\left[T(t) = T_1 + (T_2 - T_1)/\Delta t\right]$ (Fig. 6.37), [(10.2)] takes the following expression

$$t_{eq} = \frac{1.14^{T_{max}-20}}{T_{max} - 20}\int_0^{\Delta t}[T(t) - 20] \cdot dt = \frac{1.14^{T_{max}-20}}{T_{max} - 20}\int_0^{\Delta t}\left[T_1 + \frac{T_2 - T_1}{\Delta t} \cdot t - 20\right] \cdot dt$$
$$= \frac{1.14^{T_{max}-20}}{T_{max} - 20}\left(\frac{T_1 + T_2}{2} - 20\right) \cdot \Delta t$$

The result shows that the equivalent time t_{eq} for a linear variation of the temperature between T_1 and T_2 in the time interval Δt has the same value of a constant temperature equal to the average temperature $(T_1 + T_2)/2$.

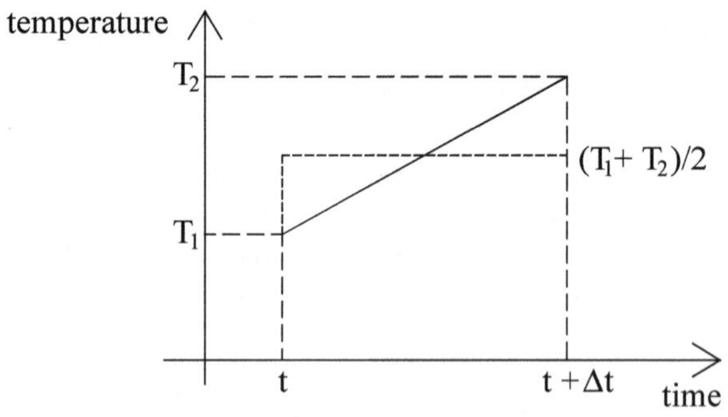

Fig. 6.37 Heat cycle with a linear variation of the temperature between T_1 and T_2 in the time interval Δt

Table 6.29 Values $\left[1.14^{(T_{\max}-20)}/(T_{\max}-20)\right]$ as a function of $T_{\max}$

$T_{\max}$ (°C)	25	30	35	40	45	50
$\frac{1.14^{T_{\max}-20}}{T_{\max}-20}$	0.385	0.371	0.476	0.687	1.058	1.698
$T_{\max}$ (°C)	55	60	65	70	75	80
$\frac{1.14^{T_{\max}-20}}{T_{\max}-20}$	2.803	4.722	8.082	14.005	24.513	43.265

Table 6.29 collects values of the expression $\left[1.14^{(T_{\max}-20)}/(T_{\max}-20)\right]$ as a function of $(T_{\max})$.

6.10.3.1 Example 15. Loss Due to Relaxation in Presence of a Cycle of Heat Curing

Calculate the relaxation loss at the release of a Class 2 *strand for a beam with pre-tensioned strands. Strands are subjected to an initial prestress equal to* $0.8 f_{pk}$ *and are released* 18 h *after tensioning, being the beam subjected to the following 18 h heating cycle (Fig. 6.38):*

- *constant temperature equal to 20 °C (ambient temperature) for 4 h,*
- *ascending ramp until 60 °C, in 4 h (thermal gradient equal to + 10 °C / hour),*
- *constant temperature equal to 60 °C for 6 h,*
- *descending ramp back to 20 °C for 4 h (thermal gradient equal to—10 °C / hour).*

Figure 6.39a shows the effective variation of the temperature and the constant value of temperature in each time interval for the calculation of the equivalent time.

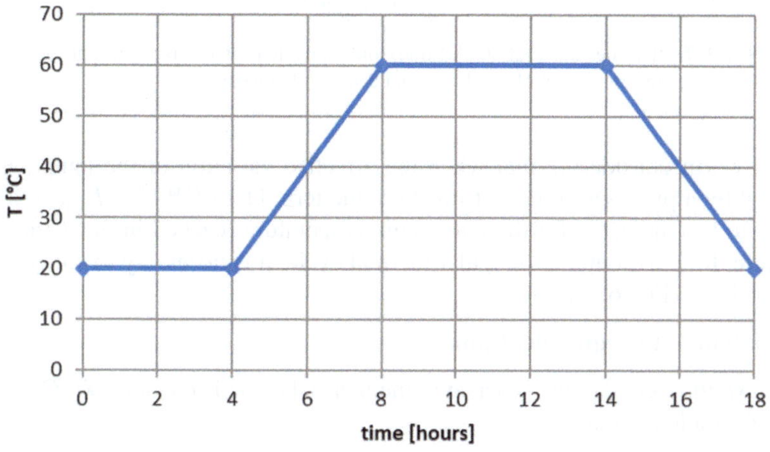

Fig. 6.38 Thermal cycle

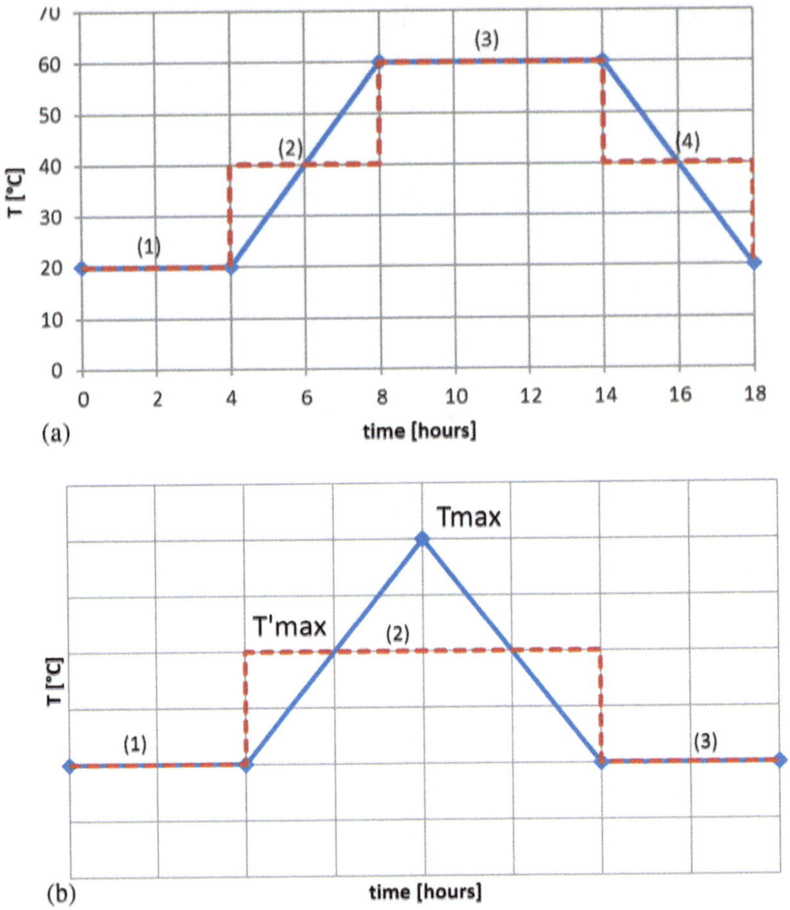

Fig. 6.39 **a** Effective thermal cycle (solid line) and equivalent piecewise constant thermal cycle (dashed line) for the calculation of t_{eq}; **b** thermal cycle with a triangular shape

It is worth considering that when the effective variation of the temperature is replaced by a piecewise constant function, the term $[1.14^{(T_{\max}-20)}/(T_{\max} - 20)]$ has always to be calculated for the maximum temperature reached in the actual cycle; therefore, for a hypothetic triangular thermal cycle, it is necessary to consider $T_{\max}$ and not T'_{max} (Fig. 6.39b).

Calculation of the equivalent time

The maximum temperature reached in the thermal cycle is equal to 60 °C; for $T_{\max} = 60$ °C, Table 6.29 gives

$$\frac{1.14^{T_{\max}-20}}{T_{\max}-20} = 4.722$$

in the segment (1): $T = 20^{\circ}\text{C},\ \Delta t = 4\,h$

$$t_{eq} = 0\ h$$

in the segment (2): linear variation from $T_1 = 20\ ^{\circ}\text{C}$ to $T_2 = 60^{\circ}\text{C},\ \Delta t = 4\,h$

$$t_{eq} = 4.722 \cdot \left(\frac{20 + 60}{2} - 20\right) \cdot 4 = 377.76\ h$$

in the segment (3): $T = 60\ ^{\circ}\text{C},\ \Delta t = 6\,h$

$$t_{eq} = 4.722 \cdot (60 - 20) \cdot 6 = 1133.28\ h$$

in the segment (4): linear variation from $T_1 = 60\ ^{\circ}\text{C}$ to $T_2 = 20\ ^{\circ}\text{C},\ \Delta t = 4\,h$

$$t_{eq} = 4.722 \cdot \left(\frac{60 + 20}{2} - 20\right) \cdot 5 = 377.76\ h$$

adding all the terms, the equivalent time is equal to

$$t_{eq} = 0 + 377.76 + 1133.28 + 377.76 \cong 1889\,h$$

Calculation of the loss due to relaxation

The loss due to relaxation is then evaluated for a time equal to 1889 + 18 = 1907 h, by using the formula.

$$\frac{\Delta\sigma_{pr}}{\sigma_{pi}} = k_1 \cdot \rho_{1000} \cdot e^{k_2 \cdot \mu} \left(\frac{t}{1000}\right)^{0.75 \cdot (1-\mu)} \cdot 10^{-5}$$

with the parameter values given in Table 6.26 for a Class 2 strand

$$\rho_{1000} = 2.5$$
$$k_1 = 0.66$$
$$k_2 = 9.1$$

and with

$$\mu = \sigma_{pi} / f_{pk} = 0.8$$

it results that

$$\Delta\sigma_{pr} / \sigma_{pi} = 2.64\,\%.$$

In absence of the thermal cycle, the loss calculated for t = 18 h would be equal to 1.31%.

6.10.4 Thermal Loss ΔP_θ

In the first part of a heat curing cycle of precast elements, prestressing strands are not bonded to concrete, so that they stretch independently from concrete and undergo a reduction of the prestress force. Later, in the second part of the heating cycle and the successive cooling phase, strands are bonded to concrete, so further strains they undergo are the same as concrete and do not give rise to any variation of the prestress force.

The loss of the prestress force in steel during the initial part of the heating cycle is indicated by EC2 as thermal loss ΔP_θ and is evaluated with the following expression

$$\Delta P_\theta = 0.5\, A_p\, E_p\, \alpha_c\, (T_{\max} - T_0) \quad [(10.3)]$$

where

A_p	is the cross-section of prestressing steel,
E_p	is the elasticity modulus of steel,
$\alpha_c (= 10 \times 10^{-6}\,°\mathrm{C}^{-1})$	is the linear coefficient of thermal expansion for concrete,
$T_{\max} - T_0$	is the difference between the maximum and initial temperature in the concrete adjacent to strands, in Celsius degrees.

The expression [(10.3)] can be obtained with the following reasoning. If one assumes that the perfect bond between strands and concrete occurs instantaneously when the maximum temperature T_{max} is reached, during the heating phase ($T_0 \rightarrow T_{\max}$) strands are not bonded to the concrete and experience the following elongation

$$\varepsilon = \alpha_s (T_{\max} - T_0) = \alpha_c (T_{\max} - T_0)$$

where $\alpha_s (= \alpha_c)$ is the coefficient of linear thermal expansion of steel, and the stress loss $\Delta\sigma_p$, corresponding to the strain ε, is equal to

$$\Delta\sigma_p = E_p \varepsilon = E_p \alpha_c (T_{\max} - T_0)$$

In the successive cooling phase ($T_{\max} \rightarrow T_0$), concrete and strands are bonded, so they experience the same strains without variation of the prestress. If the total area of pre-stressing steel is A_p, at the end of the thermal cycle, the loss of prestress force becomes

$$\Delta P_\theta = A_p \Delta\sigma_p = A_p E_p \alpha_c (T_{\max} - T_0)$$

which is twice the thermal loss given by EC2.

To get the same expression of EC2, one should consider that the bond between prestressing steel and concrete does not occur instantaneously at the temperature $T_{\max}$, as assumed above, but it increases gradually while the temperature varies from the initial value T_0 to the maximum value $T_{\max}$.

The gradual increase of the bond in the interval $(T_0, T_{\max})$ can be equivalently considered by assuming that the bond develops instantaneously as soon as the temperature attains a fictitious value T, between T_0 and $T_{\max}$. Under this assumption, the loss of the prestress occurs in the interval $[T_0, T]$, during which there is not a bond between steel and concrete. The loss of the prestress is therefore given by

$$\Delta P_\theta = A_p \Delta\sigma_p = A_p E_p \alpha_c (T - T_0)$$

Expression [(10.3)] is finally obtained by assuming that the fictitious temperature T is equal to the average value of T_0 and $T_{\max} : T = (T_0 + T_{\max})/2$.

For a more accurate evaluation of the thermal loss, experimental measurements should be performed inside the precast plant: an experimental loss smaller than that calculated with [(10.3)] corresponds to a fictitious temperature lower than the average value $(T_0 + T_{\max})/2$ (i.e., the development of bond is faster), while a higher loss corresponds to a fictitious temperature greater than the mean value (i.e., the development of bond is slower).

6.10.5 Loss Due to Concrete Creep

The word "creep" indicates the strain increase with time while the stress is kept constant.

After an infinite amount of time, creep produces a strain increment equal to $\varphi(\infty, t_0)$ times the instantaneous elastic strain, where $\varphi(\infty, t_0)$ is the long-term creep coefficient and t_0 is the age of concrete at loading time. Prestressing steel, bonded to concrete, experiences with time a strain ε_p equal to the time-dependent concrete strain

$$\varepsilon_p = \varepsilon_{c,creep} = \varphi(\infty, t_0) \cdot \varepsilon_{c,elastic}$$

and shows the following prestress loss

$$\Delta P_c = \varphi(\infty, t_0) \cdot \varepsilon_{c,elastic} \cdot E_p \cdot A_p \cdot = \varphi(\infty, t_0) \cdot \frac{\sigma_c}{E_c} \cdot E_p \cdot A_p$$

According to [3.1.4(2)], creep coefficient is associated with concrete tangent modulus E_c 28 days after the placing, which can be taken as 1.05 E_{cm},[18] where

[18] The expression is valid provided that, at time t_0, concrete is not subjected to a compressive stress greater than $0.45 f_{ck}(t_0)$, where t_0 is the concrete age at loading [3.1.4(2)].

E_{cm} is the secant modulus of elasticity between $\sigma_c = 0$ and $\sigma_c = 0.4\, f_{cm}$. Therefore, the expression of ΔP_c becomes

$$\Delta P_c = \varphi(\infty, t_0) \cdot \frac{\sigma_c}{1.05 \cdot E_{cm}} \cdot E_p \cdot A_p$$

EC2 provides two different methods for the calculation of $\varphi(\infty, t_0)$: a graphic method [3.1.4, Fig. 3.1] (Fig. 6.40) and an analytical method [Annex B].

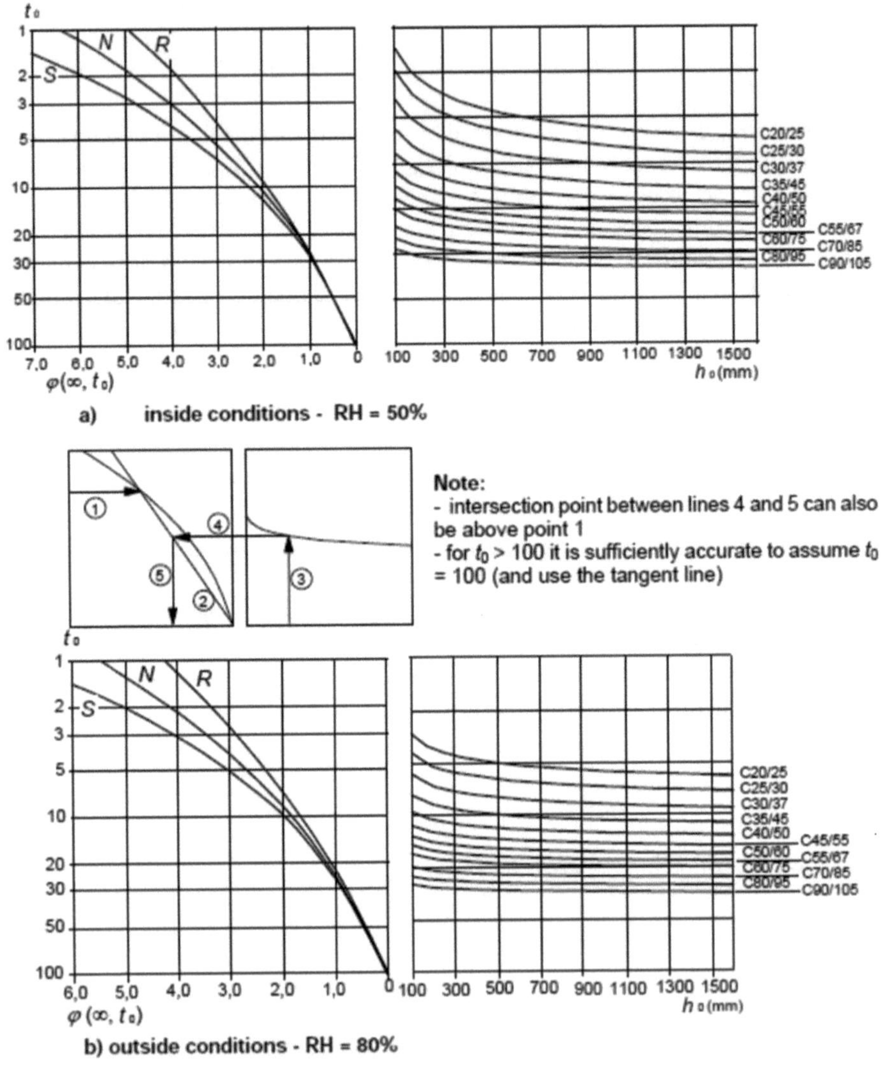

Fig. 6.40 Graphic method for determining the creep coefficient $\varphi(\infty, t_0)$ [Fig. 3.1]

6.10.5.1 Graphic Method (Valid for Temperatures Between −40 °C and + 40 °C and an Average Relative Humidity Between 40 and 100% [3.1.4(5)])

The graphic method for the calculation of $\varphi(\infty, t_0)$ consists of the following procedure (Fig. 6.40):

(a) in the diagram on the left the horizontal line 1 is drawn for $t = t_0$ (t_0 is the concrete age, in days, at the loading time) until the intersection point A with the curve corresponding to the cement Class (S, N, R),
(b) the inclined line 2 passing through A and B (0; 100) is drawn,
(c) in the diagram on the right, the vertical line 3 is drawn, passing through h_0 until the intersection point C with the curve corresponding to the concrete strength class,
(d) starting from point C, the horizontal line 4 is drawn until the intersection point D with line 2,
(e) vertical line passing through D identifies the value of creep coefficient $\varphi(\infty, t_0)$.

6.10.5.2 Analytical Method [Annex B]

For concrete made of Class N cement, EC2 provides the following expression for the calculation of the creep coefficient at time t

$$\varphi(t, t_0) = \varphi_0 \cdot \beta_c(t, t_0) \quad [(\text{B.1})]$$

where

t_0 is the concrete age, in days, at loading,
φ_0 is the nominal creep coefficient, *coincident with the long-term creep coefficient* $[\varphi_0 = \varphi(\infty, t_0)]$, and can be evaluated through the following expression:

$$\varphi_0 = \varphi_{RH} \cdot \beta(f_{cm}) \cdot \beta(t_0)$$

where

φ_{RH} is the coefficient accounting for the effect of relative humidity on the nominal creep coefficient:

$$\varphi_{RH} = 1 + \frac{1 - RH/100}{0.1 \cdot \sqrt[3]{h_0}} \qquad \text{for } f_{cm} \leq 35\,\text{N/mm}^2$$
$$\varphi_{RH} = \left[1 + \frac{1 - RH/100}{0.1 \cdot \sqrt[3]{h_0}} \cdot \alpha_1\right] \cdot \alpha_2 \quad \text{for } f_{cm} > 35\,\text{N/mm}^2$$

with

RH relative humidity of the environment where the element is placed,
$h_0 = 2\,A_c\,/\,u$ notional size (fictitious dimension of the structural element), A_c is the area of transversal cross-section, u is the perimeter of the element in contact with the atmosphere,
α_1, α_2 coefficients accounting for the influence of concrete strength:

$$\alpha_1 = \left[\frac{35}{f_{cm}}\right]^{0.7} \quad \alpha_2 = \left[\frac{35}{f_{cm}}\right]^{0.2} \quad [(\text{B.8c})]$$

where f_{cm} is the average cylinder compressive strength of concrete, expressed in N/mm^2, at 28 days,

$\beta(f_{cm})$ is the coefficient accounting for the effect of the concrete strength:

$$\beta(f_{cm}) = \frac{16.8}{\sqrt{f_{cm}}} \left(f_{cm} \text{in N/mm}^2\right)$$

$\beta(t_0)$ is the coefficient accounting for the concrete age at loading:

$$\beta(t_0) = \frac{1}{0.1 + t_0^{0.20}}$$

$\beta_c(t, t_0)$ is the coefficient describing the variation with time of creep after loading:

$$\beta_c(t, t_0) = \left[\frac{t - t_0}{\beta_H + t - t_0}\right]^{0.3}$$

where

t–t_0 is the non-adjusted loading time, expressed in days,[19]
β_H is a coefficient dependent on the relative humidity (RH in %) and on the notional size of the element (h_0 in mm) and is expressed as:

$$\beta_H = 1.5 \cdot \left[1 + (0.012 \cdot RH)^{18}\right] \cdot h_0 + 250 \le 1500 \qquad \text{for } f_{cm} \le 35\,\text{N/mm}^2$$
$$\beta_H = 1.5 \cdot \left[1 + (0.012 \cdot RH)^{18}\right] \cdot h_0 + 250 \cdot \alpha_3 \le 1500 \cdot \alpha_3 \text{ for } f_{cm} > 35\,\text{N/mm}^2$$

where coefficient α_3 is given by

[19] Low or high temperatures in the range 0  80 °C speed down or up concrete hardening; EC2 considers this effect through the correction of the concrete age as a function of the thermal cycle experimented by the concrete (see Sect. 6.10.6). Non-adjusted age is the concrete age without considering the correction.

$$\alpha_3 = \left[\frac{35}{f_{cm}}\right]^{0.5} \quad (f_{cm}\ \text{in N/mm}^2) \quad [(B.8c)]$$

Effect of the Type of Cement on the Creep Coefficient

If the concrete is not made of Class N cement, it is necessary to consider the effect of a different cement class on the creep coefficient, by modifying the loading age t_0 through the following expression:

$$t_0 = t_{0,T} \cdot \left(\frac{9}{2 + t_{0,T}^{1.2}} + 1\right)^{\alpha} \geq 0.5 \quad [(B.9)]$$

where:

$t_{0,T}$ is the concrete age in days, at loading, adjusted as a function of temperature according to [(B.10)] (see Sect. 6.10.6), to take into account the eventual thermal cycle experienced by the concrete, as it happens in the heat curing within the precast plant,

α is a power depending on the type of cement.

$\alpha = -1$ for Class S cement,
$\alpha = 0$ for Class N cement,
$\alpha = 1$ for Class R cement.

6.10.6 *Effects of a Thermal Cycle on the Concrete Hardening Age*

The effect of low or high temperatures in the range 0 ÷ 80 °C on concrete hardening can be accounted for through a correction of the concrete age t_T [B.1(3)]

$$t_T = \sum_{i=1}^{n} e^{-\{4000/[273+T(\Delta t_i)]-13.65\}} \cdot \Delta t_i \quad [(B.10)]$$

where

$T(\Delta t_i)$ is the temperature, in °C, during the time interval Δt_i,

Δt_i is the time interval, expressed in days, at which the temperature is equal to T.

This value of t_T substitutes t in the calculation of $\varphi(t, t_0)$.

Table 6.30 collects the adjusted concrete age t_T corresponding to a time interval of one day at the temperature T, for T values ranging between 0 and 80 °C. According

Table 6.30 Adjusted concrete age t_T corresponding to one day at temperature T

T (°C)	0	5	10	15	20	25	30	35	40
t_T (days)	0.3671	0.4778	0.6161	0.7875	1	1.2551	1.5662	1.9406	2.3880
t_T (hours)	8.81	11.47	14.79	18.90	24	30.12	37.59	46.57	57.31
T (°C)	45	50	55	60	65	70	75	80	
t_T (days)	2.9194	3.5470	4.2840	5.1448	6.1453	7.3023	8.6343	10.1610	
t_T (hours)	70.07	85.13	102.82	123.48	147.49	175.26	207.22	243.86	

to Table 6.31, for example, one day at 5 °C is corresponding to a time interval of 11.47 h at 20 °C, while one day at 35 °C is corresponding to a time interval of 46.57 h at 20 °C. Values of Table 6.30 are also represented in Fig. 6.41.

Table 6.31 Calculation of t_T according to [(B.10)]

Δt(hours)	t (hours)	Δt(days)	T_{mean} (°C)	Δt_T(days)
4	4	0.1664	20	0.1664
0.4	4.4	0.0167	22	0.0182
0.4	4.8	0.0167	26	0.0219
0.4	5.2	0.0167	30	0.0261
0.4	5.6	0.0167	34	0.0310
0.4	6	0.0167	38	0.0367
0.4	6.4	0.0167	42	0.0432
0.4	6.8	0.0167	46	0.0506
0.4	7.2	0.0167	50	0.0591
0.4	7.6	0.0167	54	0.0688
0.4	8	0.0167	58	0.0797
6	14	0.250	60	1.2862
0.4	14.4	0.0167	58	0.0797
0.4	14.8	0.0167	54	0.0688
0.4	15.2	0.0167	50	0.0591
0.4	15.6	0.0167	46	0.0506
0.4	16	0.0167	42	0.0432
0.4	16.4	0.0167	38	0.0367
0.4	16.8	0.0167	34	0.0310
0.4	17.2	0.0167	30	0.0261
0.4	17.6	0.0167	26	0.0219
0.4	18	0.1067	22	0.0182
			t_T (days) =	2.3232

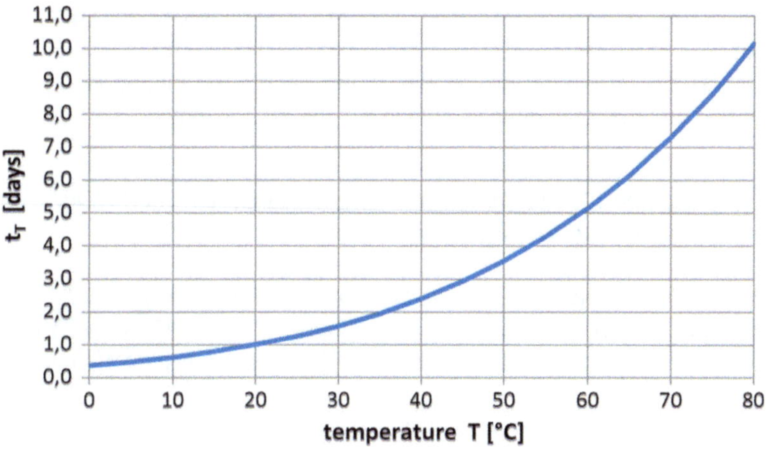

Fig. 6.41 Adjusted concrete age corresponding to one day at temperature T, as a function of T

6.10.6.1 Example 16. Effects of a Thermal Cycle on the Concrete Hardening Age

Consider the beam with pre-tensioned strands of Example 11 *subjected to the thermal cycle of Example* 15 *and calculate the concrete adjusted age accounting for a thermal cycle according to the expression* [(B.10)].

To perform the calculation, the thermal cycle is divided into small time intervals each one characterized by a constant temperature. The time interval where temperature increases from 20 °C to 60 °C is subdivided into ten intervals of 0.4 h each and the same subdivision is adopted for the time interval where temperature decreases from 60 °C to 20 °C (Fig. 6.42).

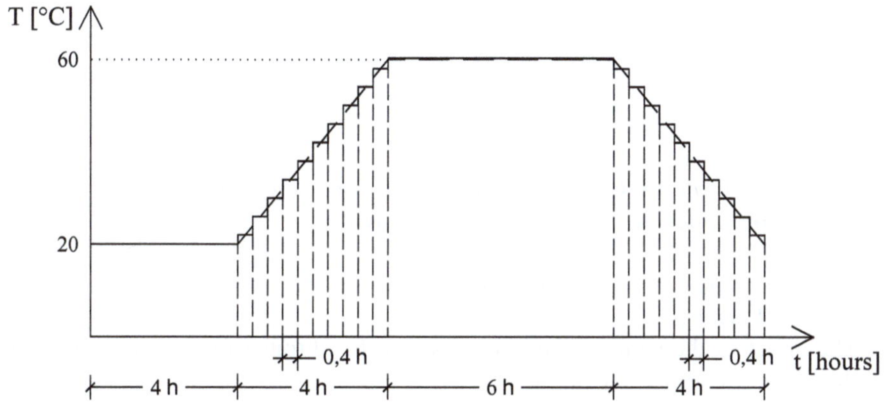

Fig. 6.42 Approximation of inclined lines of the thermal cycle with piecewise constant temperatures

For each time interval, the average value of temperatures in the time interval is assumed.

Table 6.31 collects values of Δt_T (last column) obtained by applying [(B.10)] to each interval and the total value t_T, which is equal to 2.32 days.

6.10.6.2 Example 17. Calculation of Long-Term Creep Coefficient for Concrete Made of Class N Cement

Using the analytical method calculate the long-term creep coefficient for a C50/60 *concrete made of Class* N *cement, for* $t_0 = 28$ days, $RH = 70\%$, $h_0 = 250$ mm.

Being t_0 the loading time, the expression of long-term creep coefficient becomes

$$\varphi(\infty, t_0) = \varphi_0 \text{ with } \varphi_0 = \varphi_{RH} \cdot \beta(f_{cm}) \cdot \beta(t_0)$$

where

$$f_{cm} = f_{ck} + 8 = 58\,\text{N/mm}^2$$

$$\varphi_{RH} = \left[1 + \frac{1 - RH/100}{0.1 \cdot \sqrt[3]{h_0}} \cdot \alpha_1\right] \cdot \alpha_2 = 1.206$$

for $f_{cm} = 58$ N/mm^2 > 35 N/mm^2, $RH = 70\%$ and $h_0 = 250$ mm

$$\alpha_1 = \left[\frac{35}{f_{cm}}\right]^{0.7} = 0.702, \quad \alpha_2 = \left[\frac{35}{f_{cm}}\right]^{0.2} = 0.904$$

$$\beta(f_{cm}) = \frac{16.8}{\sqrt{f_{cm}}} = 2.206, \quad \beta(t_0) = \frac{1}{0.1 + t_0^{0.20}} = 0.488 \; (t_0 = 28 \text{ days})$$

the creep coefficient is equal to

$$\varphi_0 = 1.206 \cdot 2.206 \cdot 0.488 \cong 1.3.$$

6.10.6.3 Example 18. Calculation of Long-Term Creep Coefficient for Concrete Made of Class R Cement

Repeat the calculation of Example 17 under the assumption the concrete is made with Class R *cement.*

The effect of the cement class is accounted for through the adjustment of the loading time according to [(B.9)]

$$t_0 = t_{0,T} \cdot \left(\frac{9}{2 + t_{0,T}^{1.2}} + 1\right)^{\alpha} = 28 \cdot \left(\frac{9}{2 + 28^{1.2}} + 1\right) = 32.46 \geq 0.5 \quad [(\text{B.9})]$$

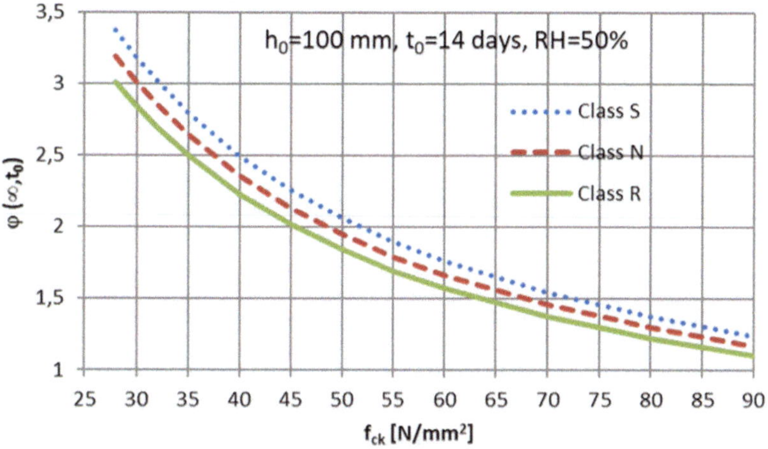

Fig. 6.43 Long-term creep coefficient as a function of concrete strength ($h_0 = 100$ mm, loading time = 14 days, relative humidity = 50%)

where $\alpha = 1$ for a Class R cement,

$$\beta(t_0) = \frac{1}{0.1 + t_0^{0.20}} = \frac{1}{0.1 + 32.46^{0.20}} = 0.475,$$

$$\varphi_0 = 1.206 \cdot 2.206 \cdot 0.475 \cong 1.26$$

It is worth noting that increasing the concrete strength and maintaining unchanged all the other quantities, φ_0 values decrease.

The following figures show the diagrams of the long-term creep coefficient for a C28/35 concrete made with three different classes of cement (S, N, R): the first graph shows the variation of $\varphi(\infty, t_0)$ with the concrete strength, the second one with the relative humidity, the third one with t_0 and the fourth with h_0. In all the diagrams it can be observed that the creep coefficient decreases as the cement class increases (Figs. 6.43 and 6.44).

Tables 6.32, 6.33, 6.34, 6.35, 6.36, 6.37 collect values of the long-term creep coefficient for $t_0 = 7$, 14, 28 days, $h_0 = 100$, 300, 500 mm, $RH = 50\%$ e $RH = 80\%$[20] vs concrete strength for all the three cement classes.

For RH values different from 50 and 80%, it is sufficient to interpolate linearly tabulated values, as $\varphi(\infty, t_0)$ is a linear function of RH.

Analogously, for values of t_0 or h_0 different from tabulated values, it is still possible to use linear interpolation, but the result represents an overestimation of the creep coefficient because both diagrams $\varphi(\infty, t_0)-t_0$ (Fig. 6.45) and $\varphi(\infty, t_0)-h_0$ (Fig. 6.46) have concavity upwards.

---

[20] Adopted RH values are typical for internal ($RH = 50\%$) and external environments ($RH = 80\%$).

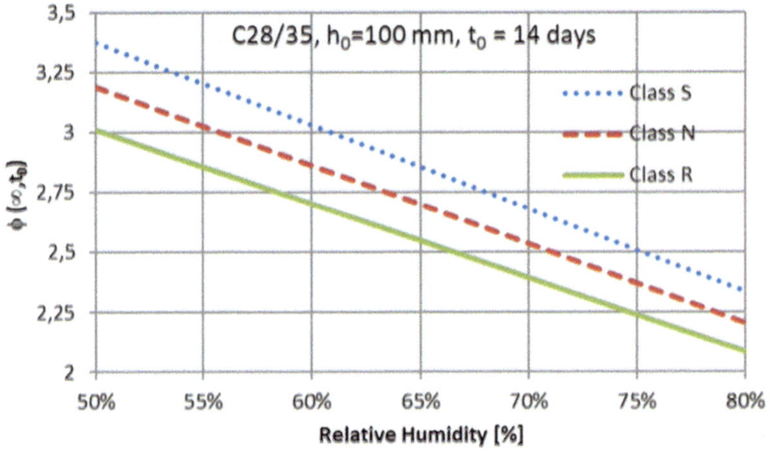

Fig. 6.44 Long-term creep coefficient as a function of relative humidity (concrete strength class C28/35, $h_0 = 100$ mm, loading time = 14 days). The diagram is linear as it can be observed in the expression of φ_{RH}

Table 6.32 Long-term creep coefficient $\varphi(\infty,\ t_0)\varphi(\infty,\ t_0)$

$h_0 = 100$ mm, $RH = 50\%$

Concrete strength class	$t_0 = 7$ days			$t_0 = 14$ days			$t_0 = 28$ days		
	S	N	R	S	N	R	S	N	R
C30/37	3.8	3.4	3.1	3.2	3.0	2.8	2.7	2.6	2.6
C35/45	3.3	3.0	2.7	2.8	2.6	2.5	2.4	2.3	2.3
C40/50	3.0	2.7	2.4	2.5	2.4	2.2	2.1	2.1	2.0
C45/55	2.7	2.4	2.2	2.3	2.1	2.0	1.9	1.9	1.8
C50/60	2.5	2.2	2.0	2.1	2.0	1.8	1.8	1.7	1.7
C55/67	2.3	2.0	1.8	1.9	1.8	1.7	1.6	1.6	1.5
C60/75	2.1	1.9	1.7	1.8	1.7	1.6	1.5	1.5	1.4
C70/85	1.8	1.7	1.5	1.5	1.5	1.4	1.3	1.3	1.2
C80/95	1.6	1.5	1.3	1.4	1.3	1.2	1.2	1.1	1.1
C90/105	1.5	1.3	1.2	1.2	1.2	1.1	1.1	1.0	1.0

6.10.7 Long-Term Losses Due to Concrete Creep and Shrinkage and Steel Relaxation

Intrinsic relaxation of steel is defined as the prestress loss at constant strain (Sect. 6.10.2), therefore it cannot represent the relaxation loss in a prestressed member, where the strain of prestressing steel is not constant but reduces with time due to concrete creep and shrinkage. Therefore, the loss due to relaxation in a tendon stretched at a constant strain between two fixed points is higher than the relaxation

Table 6.33 Long-term creep coefficient $\varphi(\infty, t_0)\varphi(\infty, t_0)$

$h_0 = 100$ mm, $RH = 80\%$

Concrete strength class	$t_0 = 7$ days			$t_0 = 14$ days			$t_0 = 28$ days		
	S	N	R	S	N	R	S	N	R
C30/37	2.7	2.4	2.2	2.2	2.1	2.0	1.9	1.8	1.8
C35/45	2.4	2.1	1.9	2.0	1.9	1.8	1.7	1.6	1.6
C40/50	2.2	1.9	1.8	1.8	1.7	1.6	1.5	1.5	1.5
C45/55	2.0	1.8	1.6	1.7	1.6	1.5	1.4	1.4	1.3
C50/60	1.8	1.6	1.5	1.5	1.4	1.4	1.3	1.3	1.2
C55/67	1.7	1.5	1.4	1.4	1.3	1.3	1.2	1.2	1.1
C60/75	1.6	1.4	1.3	1.3	1.3	1.2	1.1	1.1	1.1
C70/85	1.4	1.3	1.2	1.2	1.1	1.1	1.0	1.0	1.0
C80/95	1.3	1.2	1.0	1.1	1.0	1.0	0.9	0.9	0.9
C90/105	1.2	1.1	1.0	1.0	0.9	0.9	0.8	0.8	0.8

Table 6.34 Long-term creep coefficient $\varphi(\infty, t_0)\varphi(\infty, t_0)$

$h_0 = 300$ mm, $RH = 50\%$

Concrete strength class	$t_0 = 7$ days			$t_0 = 14$ days			$t_0 = 28$ days		
	S	N	R	S	N	R	S	N	R
C30/37	3.2	2.9	2.6	2.7	2.5	2.4	2.3	2.2	2.2
C35/45	2.8	2.6	2.3	2.4	2.3	2.1	2.0	2.0	1.9
C40/50	2.6	2.3	2.1	2.1	2.0	1.9	1.8	1.8	1.7
C45/55	2.3	2.1	1.9	2.0	1.8	1.7	1.7	1.6	1.6
C50/60	2.1	1.9	1.7	1.8	1.7	1.6	1.5	1.5	1.4
C55/67	2.0	1.8	1.6	1.7	1.6	1.5	1.4	1.4	1.3
C60/75	1.8	1.7	1.5	1.5	1.5	1.4	1.3	1.3	1.2
C70/85	1.6	1.5	1.3	1.4	1.3	1.2	1.2	1.1	1.1
C80/95	1.5	1.3	1.2	1.2	1.2	1.1	1.0	1.0	1.0
C90/105	1.3	1.2	1.1	1.1	1.0	1.0	0.9	0.9	0.9

experienced by a second tendon with the same characteristics and initial strain of the first tendon and stretched between the end sections of a prestressed member.

If at time t_1 the stress within the two tendons is the same, at time $t_2 > t_1$ the stress within the second tendon is lower than the first tendon, because of concrete creep and shrinkage. Therefore, the relaxation of the second tendon is lower than the first one. The same relaxation loss is obtained for both tendons only if the first tendon is maintained at a lower constant strain than the initial strain of the second tendon.

The effective relaxation $\Delta P_{r,\text{intr}}$ for a tendon in a prestressed element can be calculated using a reduction coefficient $\chi_r < 1$ to the intrinsic relaxation $\Delta P_{r,\text{eff}}$,

Table 6.35 Long-term creep coefficient $\varphi(\infty,\ t_0)\varphi(\infty,\ t_0)$

$h_0 = 300$ mm, $RH = 80\%$									
Concrete strength class	$t_0 = 7$ days			$t_0 = 14$ days			$t_0 = 28$ days		
	S	N	R	S	N	R	S	N	R
C30/37	2.4	2.2	2.0	2.0	1.9	1.8	1.7	1.7	1.6
C35/45	2.2	2.0	1.8	1.8	1.7	1.6	1.6	1.5	1.5
C40/50	2.0	1.8	1.6	1.7	1.6	1.5	1.4	1.4	1.3
C45/55	1.8	1.6	1.5	1.5	1.4	1.4	1.3	1.3	1.2
C50/60	1.7	1.5	1.4	1.4	1.3	1.3	1.2	1.2	1.1
C55/67	1.6	1.4	1.3	1.3	1.3	1.2	1.1	1.1	1.1
C60/75	1.5	1.3	1.2	1.2	1.2	1.1	1.1	1.0	1.0
C70/85	1.3	1.2	1.1	1.1	1.1	1.0	1.0	0.9	0.9
C80/95	1.2	1.1	1.0	1.0	1.0	0.9	0.9	0.8	0.8
C90/105	1.1	1.0	0.9	0.9	0.9	0.8	0.8	0.8	0.8

Table 6.36 Long-term creep coefficient $\varphi(\infty,\ t_0)\varphi(\infty,\ t_0)$

$h_0 = 500$ mm, $RH = 50\%$									
Concrete strength class	$t_0 = 7$ days			$t_0 = 14$ days			$t_0 = 28$ days		
	S	N	R	S	N	R	S	N	R
C30/37	3.0	2.7	2.4	2.5	2.4	2.3	2.1	2.1	2.0
C35/45	2.7	2.4	2.2	2.2	2.1	2.0	1.9	1.9	1.8
C40/50	2.4	2.2	2.0	2.0	1.9	1.8	1.7	1.7	1.6
C45/55	2.2	2.0	1.8	1.8	1.7	1.6	1.6	1.5	1.5
C50/60	2.0	1.8	1.6	1.7	1.6	1.5	1.4	1.4	1.4
C55/67	1.9	1.7	1.5	1.6	1.5	1.4	1.3	1.3	1.3
C60/75	1.8	1.6	1.4	1.5	1.4	1.3	1.3	1.2	1.2
C70/85	1.5	1.4	1.3	1.3	1.2	1.2	1.1	1.1	1.0
C80/95	1.4	1.3	1.1	1.2	1.1	1.0	1.0	1.0	0.9
C90/105	1.3	1.1	1.0	1.1	1.0	1.0	0.9	0.9	0.9

evaluated for the initial stress which is present immediately after the transfer of the prestress:

$$\Delta P_{r,\text{eff}} = \chi_r \Delta P_{r,\text{intr}}$$

If λ indicates the ratio between the initial stress and the ultimate prestressing steel strength and Ω denotes the ratio between $(\Delta P_{\text{tot}} - \Delta P_r)$ and the initial prestress force, where ΔP_{tot} is corresponding to the total prestressing loss, the coefficient χ_r varies with Ω as shown in Fig. 6.47.

Table 6.37 Long-term creep coefficient $\varphi(\infty,\ t_0)$

$h_0 = 500$ mm, $RH = 80\%$

Concrete strength class	$t_0 = 7$ days			$t_0 = 14$ days			$t_0 = 28$ days		
	S	N	R	S	N	R	S	N	R
C30/37	2.3	2.1	1.9	2.0	1.8	1.7	1.7	1.6	1.6
C35/45	2.1	1.9	1.7	1.8	1.7	1.6	1.5	1.5	1.4
C40/50	1.9	1.7	1.6	1.6	1.5	1.4	1.4	1.3	1.3
C45/55	1.8	1.6	1.4	1.5	1.4	1.3	1.3	1.2	1.2
C50/60	1.7	1.5	1.3	1.4	1.3	1.2	1.2	1.1	1.1
C55/67	1.5	1.4	1.3	1.3	1.2	1.2	1.1	1.1	1.0
C60/75	1.5	1.3	1.2	1.2	1.2	1.1	1.0	1.0	1.0
C70/85	1.3	1.2	1.1	1.1	1.0	1.0	0.9	0.9	0.9
C80/95	1.2	1.1	1.0	1.0	0.9	0.9	0.8	0.8	0.8
C90/105	1.1	1.0	0.9	0.9	0.9	0.8	0.8	0.8	0.7

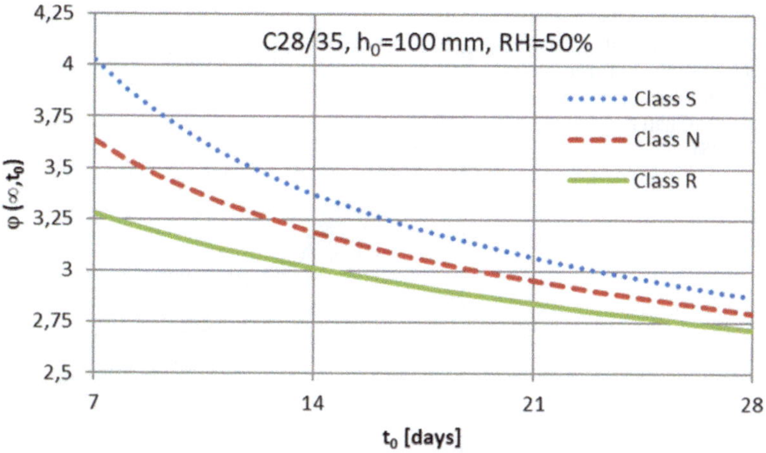

Fig. 6.45 Long-term creep coefficient as a function of the concrete age t_0 at loading (C28/35 concrete, $h_0 = 100$ mm, RH 50%)

To consider the decrease of intrinsic relaxation due to concrete creep and shrinkage, EC2 provides the following relationship

$$\Delta P_{c+s+r} = A_p \Delta\sigma_{p,\,c+s+r}$$

$$= A_p \frac{\overbrace{\varepsilon_{cs} E_p}^{shrinkage} + \overbrace{0.8\Delta\sigma_{pr}}^{steel relaxation} + \overbrace{\frac{E_p}{E_{cm}}\varphi(t,t_0)\sigma_{c,\,QP}}^{creep}}{1 + \frac{E_p}{E_{cm}}\frac{A_p}{A_c}\left(1 + \frac{A_c}{I_c}Z_{cp}^2\right)[1 + 0.8\varphi(t,\ t_0)]} \qquad [(5.46)]$$

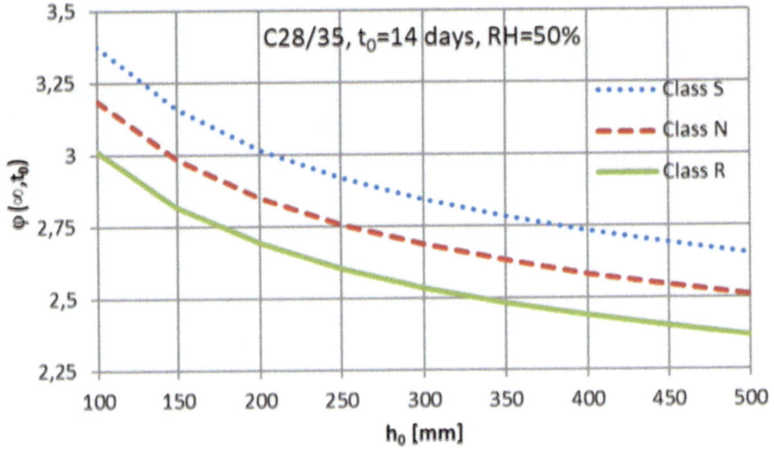

Fig. 6.46 Long-term creep coefficient as a function of notional size h_0 (C28/35 concrete, t_0 = 14 days, $RH = 50\%$)

Fig. 6.47 Variation of the coefficient χ_r of the intrinsic relaxation with the coefficient Ω (Ghali & Trevino, 1985)

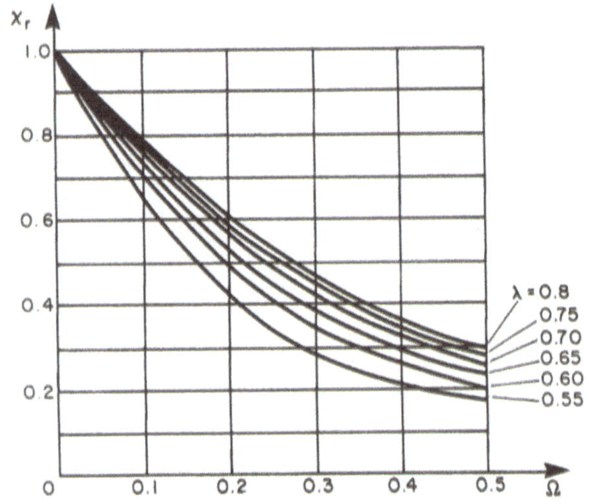

where

$\Delta\sigma_{p,c+s+r}$ is the absolute value of the stress variation due to creep, shrinkage and relaxation at location x, at time t;

ε_{cs} is the shrinkage deformation estimated according to [3.1.4(6)] in absolute value (see Sect. 6.10.1);

E_p is the modulus of elasticity for prestressing steel,

E_{cm} is the mean value of concrete modulus of elasticity,

$\Delta\sigma_{pr}$ is the absolute value of the variation of stress in the tendons at location x, at time t, due to the relaxation of the prestressing steel. This value is calculated for stress $\sigma_p = \sigma_p(G + P_{m0} + \psi_2 Q)$ within the tendons due to prestress force, without the instantaneous losses (P_{m0}) and the combination of quasi-permanent loads ($G + \psi_2 Q$),

$\varphi(t, t_0)$ is the creep coefficient at time t with loading at time t_0,

$\sigma_{c,QP}$ is the stress in the concrete adjacent to tendons, due to self-weight, initial prestress and quasi-permanent loads if present. The value of $\sigma_{c,QP}$ can also be due to part of the self-weight and initial prestress, or the effect of a complete combination of quasi-permanent loads, depending on the analysed constructive phase,

A_p is the area of all the prestressing tendons at the location x,

A_c is the area of the concrete section,

I_c is the second moment of area of the concrete section,

z_{cp} is the distance between the centre of gravity of the concrete section and the tendons.

Equation [(5.46)] can be obtained by imposing the compatibility between the strain reduction experienced by prestressing tendons and strain of adjacent concrete fibres, in the time interval between the transfer of the prestress t_0 and the time t.

If $\Delta\sigma_p$ is the stress decrease in prestressing tendons in the time interval (t_0, t), the following prestress reduction is obtained

$$\Delta P_p = A_p \Delta\sigma_p$$

Due to the equilibrium condition, in the same time interval (t_0, t), the concrete compression force decreases by the same amount

$$\Delta P_c = \Delta P_p = A_p \Delta\sigma_p$$

and the compressive stress reduces by

$$\Delta\sigma_p = \Delta P_c / A_p$$

The stress reduction $\Delta\sigma_p$ in tendons is partially due to their elastic shortening $\Delta\varepsilon_p$ and partially due to the relaxation loss $\Delta\sigma'_{pr}$ at constant strain (intrinsic relaxation); therefore, it follows

$$\Delta\sigma_p = (E_p \Delta\varepsilon_p) + \Delta\sigma'_{pr}$$

and tendon strain is equal to

$$\Delta\varepsilon_p = \frac{\Delta\sigma_p - \Delta\sigma'_{pr}}{E_p}$$

Concrete strain is given by three contributions:

- shrinkage strain $\varepsilon_{cs}(t, t_0)$,

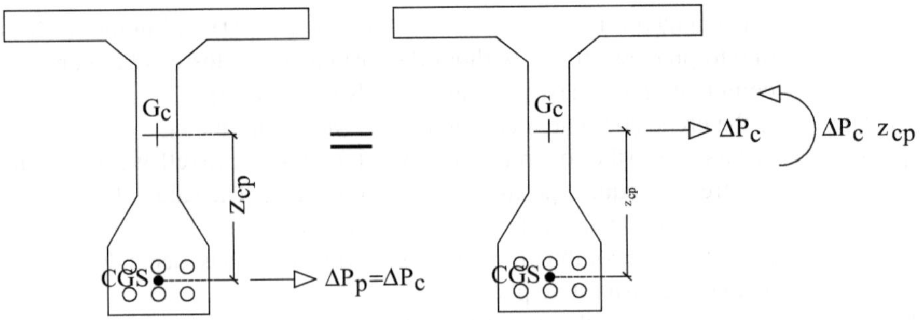

Fig. 6.48 Loss ΔP_c of prestress (the reduction of compressive stress in the concrete at the level of the tendon centroid can be easily calculated once the force ΔP_c is moved to the centroid of the section and the moment $\Delta P_p z_{cp}$ is added)

- creep strain ε_{cc} due to $\sigma_{c,QP}$

  $\varepsilon_{cc} = \frac{\sigma_{c,QP}}{1.05 \cdot E_{cm}} \varphi(t, t_0) \cong \frac{\sigma_{c,QP}}{E_{cm}} \varphi(t, t_0)$[21];
- (elastic + viscous) strain due to the reduction of prestress ΔP_c in the time interval (t_0, t)

$$\varepsilon_{c,\Delta P_c} = \frac{\Delta\sigma_c(\Delta P_c)}{E_{cm}/[1 + 0.8\varphi(t, t_0)]} = \frac{1}{E_{cm}/[1 + 0.8\varphi(t, t_0)]}\left(\frac{\Delta P_c}{A_c} + \frac{\Delta P_c \cdot z_{cp}^2}{I_c}\right)$$

where $\frac{E_{cm}}{1+0.8\varphi(t,t_0)}$ is the age-adjusted elasticity modulus of concrete, to be used for the calculation of the total strain variation (elastic + viscous), due to the stress reduction $\Delta\sigma_c(t)$ in the time interval (t_0, t) and the expression of $\Delta\sigma_c(\Delta P_c)$ considers the eccentricity of the force ΔP_c (Fig. 6.48).

In the expression of $\varepsilon_{c,\Delta Pc}$, the coefficient 0.8 multiplying the creep coefficient $\varphi(t, t_0)$ represents the ageing coefficient χ. This coefficient considers that in the time interval (t_0, t), the reduction of the prestress force P is gradual, therefore the compressive stress σ_{cp} induced by P is also decreasing gradually. It follows that the creep strain associated with σ_{cp} is lower than the strain which would occur if the stress reduction $\Delta\sigma_{cp}$ is immediate. Elementary decrements of the compressive stress $\delta\sigma_{cp}$, obtained by subdivision of the total decrement $\Delta\sigma_{cp}$, are acting on more and more aged concrete.

The ageing coefficient takes generally values between 0.6 and 0.9; if the function $\sigma_c(t)$ is varying with an analogous creep law, i.e., if the variation of σ_c is effectively due to creep, it results $\chi \cong 0.8$.

The variation of concrete total strain at the level of the tendon centroid takes, therefore, the following expression

---

[21] For the calculation of concrete creep strain, formula [(5.46)] uses the secant elasticity modulus E_{cm} of concrete instead of the tangent elasticity modulus (1.05 E_{cm}), differently from formula [3.1.4(2)] (see Sect. 6.10.5).

$$\Delta\varepsilon_c = \varepsilon_{cs}(t, t_0) + \frac{\sigma_{c,QP}}{E_{cm}}\phi(t, t_0) - \frac{1}{E_{cm}}\left(\frac{\Delta P_c}{A_c} + \frac{\Delta P_c \cdot z_{cp}^2}{I_c}\right) \cdot [1 + 0.8\phi(t, t_0)]$$

Under the assumption of a perfect bond between concrete and steel, the strain variation $\Delta\varepsilon_p$ of prestressing steel and the strain variation $\Delta\varepsilon_c$ of concrete at the level of tendon centroid are equal, therefore

$$\begin{aligned}\frac{\Delta\sigma_p - \Delta\sigma'_{pr}}{E_p} &= \frac{\sigma_{c,QP}}{E_{cm}}\phi(t, t_0) + \varepsilon_{cs}(t, t_0)\\ &\quad - \frac{1}{E_{cm}}\left(\frac{\Delta P_c}{A_c} + \frac{\Delta P_c \cdot z_{cp}^2}{I_c}\right) \cdot [1 + 0.8\phi(t, t_0)]\end{aligned}$$

Substituting $\Delta\sigma_p = \Delta P_c / A_p$ in the last expression, the following expression is obtained

$$\begin{aligned}&\frac{\Delta P_c}{A_p E_p} - \frac{\Delta\sigma'_{pr}}{E_p} - \frac{\sigma_{c,QP}\,\phi(t, t_0)}{E_{cm}} - \varepsilon_{cs}(t, t_0)\\ &\quad = -\frac{\Delta P_c}{E_{cm}}\left(\frac{1}{A_c} + \frac{z_{cp}^2}{I_c}\right) \cdot [1 + 0.8\phi(t, t_0)]\end{aligned}$$

therefore, solving respect to ΔP_c, it follows

$$\begin{aligned}&\Delta P_c\left[\frac{1}{A_p E_p} + \frac{1}{E_{cm}}\left(\frac{1}{A_c} + \frac{z_{cp}^2}{I_c}\right)[1 + 0.8\varphi(t, t_0)]\right]\\ &\quad = \frac{\Delta\sigma'_{pr}}{E_p} + \frac{\sigma_{c,QP}\,\varphi(t, t_0)}{E_{cm}} + \varepsilon_{cs}(t, t_0)\end{aligned}$$

And

$$\Delta P_c = A_p \frac{E_p \cdot \varepsilon_{cs}(t, t_0) + \Delta\sigma'_{pr} + \frac{E_p}{E_{cm}} \cdot \sigma_{c,QP}\,\varphi(t, t_0)}{1 + \frac{E_p}{E_{cm}}\frac{A_p}{A_c}\left(1 + \frac{A_c}{I_c}z_{cp}^2\right)[1 + 0.8\varphi(t, t_0)]}.$$

The relaxation loss $\Delta\sigma'_{pr}$ may be expressed as $\Delta\sigma'_{pr} = \chi_r \Delta\sigma_{pr}$, where $\Delta\sigma_{pr}$ is the intrinsic relaxation which would occur in the time interval (t_0, t) if tendons would be stretched between two fixed points. By adopting $\chi_r = 0.8$, as suggested in [5.10.6(1)] and by using the symbol ΔP_{c+s+r} instead of ΔP_c, the following final expression is found [(5.46)]:

$$\Delta P_{c+s+r} = A_p \frac{E_p \cdot \varepsilon_{cs}(t, t_0) + 0.8 \cdot \Delta\sigma_{pr} + \frac{E_p}{E_{cm}} \cdot \sigma_{c,QP}\,\varphi(t, t_0)}{1 + \frac{E_p}{E_{cm}}\frac{A_p}{A_c}\left(1 + \frac{A_c}{I_c}z_{cp}^2\right)[1 + 0.8\varphi(t, t_0)]} \qquad [(5.46)]$$

The quantity in the denominator of [(5.46)] can be approximately assumed equal to 1.0, obtaining the following overestimation of long-term losses

$$\Delta P_{c+s+r} = A_p \Delta\sigma_{p,c+s+r} \cong A_p \left[\varepsilon_{cs} E_p + 0.8 \Delta\sigma_{pr} + \frac{E_p}{E_{cm}} \varphi(t, t_0)\, \sigma_{c,QP} \right]$$

6.10.7.1 Example 19. Calculation of Prestress Losses of a Pre-tensioned Beam Subjected to Steam Curing

Calculate prestress losses of a pre-tensioned simply supported beam subjected to steam curing, whose section is shown in Fig. 6.49. Assume the following design data.

Materials

Concrete strength class C50/60, Class R cement

$$f_{ck} = 50\,\mathrm{N/mm^2}$$

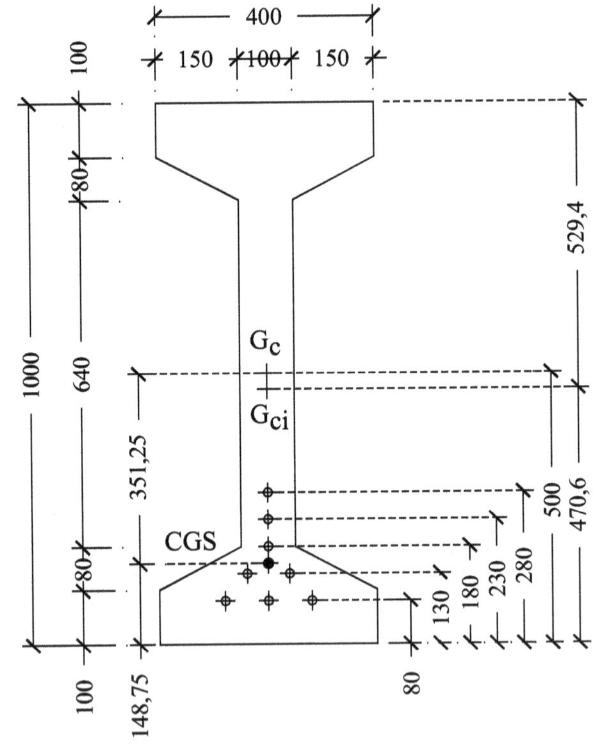

Fig. 6.49 Section of the beam with pre-stressed tendons of Example 19 (G_c: geometrical centroid, G_{ci}: centroid of the transformed section)

$$E_{cm} = 37{,}000\,\mathrm{N/mm^2}$$

Steel for ordinary reinforcement B450C

$$f_{yk} = 450\,\mathrm{N/mm^2}$$

Prestressing steel.
0.6" strands (d = 15.3 mm)

$$\begin{aligned}
A_p &= 140\,\mathrm{mm^2}(\text{area of each strand})\\
A_{p,\text{tot}} &= 8\,A_p = 1120\,\mathrm{mm^2}(\text{total area})\\
E_p &= 195{,}000\,\mathrm{N/mm^2}[3.3.6\,(3)]\\
f_{pk} &= 1860\,\mathrm{N/mm^2}\\
f_{p0.1k} &= 1600\,\mathrm{N/mm^2}
\end{aligned}$$

Relaxation class: Class 2 (wires or low relaxation strands)
$\rho_{1000} = 2.5\,\%$ (recommended value at [3.3.2 (6)] for Class 6.2)

$$f_{pk}/f_{p0.1k} = 1.11 \geq 1.1\,[3.3.4\,(5)]$$

Wedge draw-in of anchorage devices: $w = 3$ mm on each side (data from the ETA given by the producer).

Geometrical data

Length of casting bed $L_{\text{bed}} = 80.00$ m.
Beam span $L = 20.00$ m.
Area of concrete cross-section (approximated by the. geometrical cross-section without eliminating strand area) $A_c = 184{,}000\ \mathrm{mm^2}$.
Cross-section perimeter $u = 3160$ mm.
Second moment of area of the geometric section $I_c = 2.3887 \times 10^{10}\ \mathrm{mm^4}$.
Distance of the geometrical centroid G_c from bottom $z_{inf} = 500$ mm.
Distance of the geometrical centroid G_c from top $z_{sup} = 500$ mm.
Strand number and corresponding distance from the bottom no. 3 strands at 80 mm.
no. 2 strands at 130 mm.
no. 1 strand at 180 mm.
no. 1 strand at 230 mm.
no. 1 strand at 280 mm.
Distance of the centroid of strands CGS from bottom $z_{G,str} = 148{,}75$ mm.
Eccentricity between *CGS* and G_c $z_{cp} = 500 - 148.75 = 351.25$ mm.

Loads

Beam self-weight $G_{k1} = 4.60$ kN/m.
Permanent load $G_{k2} = 5.20$ kN/m.
Variable load (snow) $Q_{k1} = 7.20$ kN/m.

(ψ_2 coefficient for snow load is null)

Execution phases

1. Casting time $t = 0$
2. Thermal cycle duration of 18 h
3. Release of reinforcement at time $t = 18$ h
4. Application of permanent load at 90 days from the transfer of prestress
5. Design life of the beam: 500000 hours ($\cong$ 57 years) [3.3.2(8)]

Limit stress in prestressing steel

Limit stress at tensioning

$$\sigma_{p,\max} = \min\left\{0.8 f_{pk};\ 0.9 f_{p0.1k}\right\} = \min\{0.8 \cdot 1860;\ 0.9 \cdot 1600\}$$
$$= \min\{1488;\ 1440\} = 1440 \text{ N/mm}^2 \qquad [5.10.2(1)\text{P}]$$

Limit stress immediately after the transfer of prestress to concrete

$$\sigma_{pm0} = \min\left\{0.75 f_{pk};\ 0.85 f_{p0.1k}\right\} = \min\{0.75 \cdot 1860;\ 0.85 \cdot 1600\}$$
$$= \min\{1395;\ 1360\} = 1360 \text{ N/mm}^2 \qquad [(5.10.3(2))]$$

Average prestress under service conditions

$$\sigma_{p,es} = 0.8\ f_{p0.1k} = 0.8 \cdot 1600 = 1280\,\text{N/mm}^2.$$

Thermal cycle

To speed up the concrete hardening, the beam is subjected to the thermal cycle discussed in Example 15, whose total duration is 18 h; in the following, effects of the thermal cycle are considered on the following aspects:

- loss due to steel relaxation,
- elasticity modulus of concrete at the transfer of the prestress,
- concrete creep coefficient.

Calculation of initial losses for a single strand

Initial losses are grouped in the following way:

(a) **Losses during tensioning**

ΔP_1: loss due to wedge draw-in of anchorages (in this example it is assumed that strands are not overstressed to compensate the loss ΔP_1, as it often occurs in precast plants),

(b) **Losses before the transfer of prestress to concrete**

ΔP_2: loss due to steel relaxation during concrete steam curing.
ΔP_3: thermal losses.

(c) **Losses during the transfer of prestress to concrete**

ΔP_4: loss due to immediate elastic shortening of concrete.
The calculation of losses is performed for a single strand.

(a) **Loss during tensioning**

Loss due to wedge draw-in of anchorage devices (ΔP_1)

Length of casting bed: $L_{\text{bed}} = 80{,}000$ mm.
Wedge draw-in of anchorage devices: $\Delta L_{\text{bed}} = 2w = 3 + 3 = 6\,\text{mm}$.
Loss due to wedge draw-in: $\Delta P_1 = (\Delta L_{\text{bed}}/L_{\text{bed}})E_p A_p = 2047\,\text{N}$.
Initial prestress applied to each strand: $P_0 = 140 \cdot 1440 = 201.600\,\text{N}$.
Prestress force after wedge draw-in: $P_1 = P_0 - \Delta P_1 = 199553\,\text{N}$.
Steel stress after wedge draw-in:

$$\sigma_{pi} = 199553/140 = 1425\,\text{N/mm}^2$$

Ratio between steel stress and characteristic strength after wedge draw-in:

$$\mu = \sigma_{pi}/f_{pk} = 1425/1860 = 0.766$$

(b) **Losses before transfer of prestress to concrete (in the time interval between tensioning of strands and their release)**

Loss due to steel relaxation during steam curing of concrete **(ΔP_2)**
According to Example 15, it results that $\Delta\sigma_{pr}/\sigma_{pi} = 2.64\,\%$, and therefore

$$\Delta\sigma_{pr} = 0.0264 \cdot \sigma_{pi} = 0.0264 \cdot 1425 = 37.62\,\text{N/mm}^2$$
$$\Delta P_2 = 37.62 \cdot 140 = 5267\,\text{N}$$

$$P_2 = P_1 - DP_2 = 199{,}553 - 5267 = 194{,}286\,\text{N},$$

Thermal loss (ΔP_3)

$$\Delta P_3 = 0.5\, A_p E_p \alpha_c (T_{\max} - T_0) \quad [(10.3)]$$

where

$$A_p = 140\,\text{mm}^2$$
$$E_p = 195{,}000\,\text{N/mm}^2$$
$$\alpha_c = 10^{-5}\,{}^\circ\text{C}^{-1}$$
$$T_{\max} - T_0 = 60\text{–}20 = 40\,{}^\circ\text{C}$$

with the previous data, the thermal loss is equal to

$$\Delta P_3 = 0.5\, A_p E_p \alpha_c (T_{\max} - T_0) = 5460\,\text{N} = 5.46\,\text{kN}$$

At the release of strands, the prestress force in each strand is given by:

$$P_3 = P_2 - \Delta P_3 = 194286 - 5460 = 188826\,\text{N}$$

i.e., just before the release of strands, the prestress reduced by 6.3% [=(2047 + 5267 + 5460) / 201600] compared to the initial prestress at the stressing jack (Fig. 6.50).

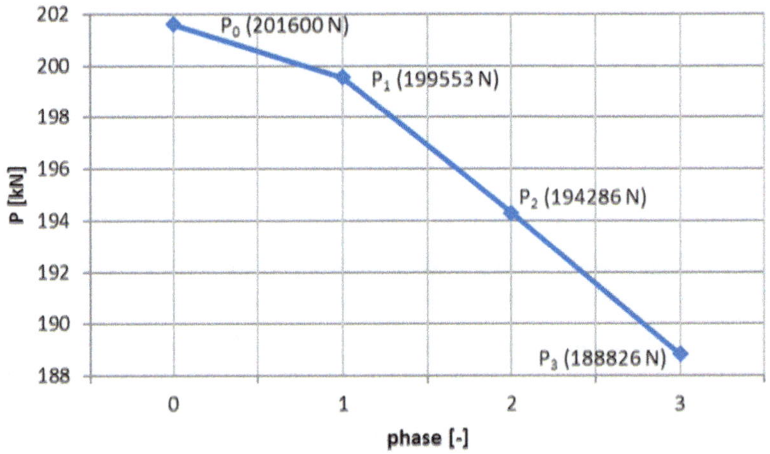

Fig. 6.50 Prestress force in a single strand before the transfer of prestress to concrete: **a** at tensioning, **b** after wedge draw-in of anchorages at the heads of the casting bed, **c** after steel relaxation during the thermal cycle, **d** after thermal losses

(c) **Losses during the transfer of prestress to concrete**

Loss due to concrete initial elastic shortening (ΔP_4)

The total prestress force $P_{4,\text{tot}}$ immediately after the transfer of prestress to concrete is evaluated through the following formula, where the total prestress force $P_{3,\text{tot}}$ at the time of strand release is used:

$$P_{4,\text{tot}} = \frac{P_{3,\text{tot}}}{1 + \frac{E_p}{E_{cm}(t)} \frac{A_{p,\text{tot}}}{A_c} \cdot \left(1 + \frac{z_{cp}^2 A_c}{I_c}\right)}$$

$$= \frac{8 \cdot 188826}{1 + \frac{195{,}000}{31{,}895} \frac{8 \cdot 140}{184{,}000} \cdot \left(1 + \frac{351.25^2 \cdot 184{,}000}{2.3887 \times 10^{10}}\right)} = 1{,}408{,}385 \text{ N}$$

the prestress force in each strand is equal to

$$P_4 = P_{4,\text{tot}}/8 = 176{,}048 \text{ N}$$

and the corresponding normal stress is equal to:

$$\sigma_p = 176{,}048/140 = 1257 \text{ N/mm}^2 \ (\leq \sigma_{pm0} = 1360 \text{ N/mm}^2).$$

In the calculation of $P_{4,\text{tot}}$ above, the value of concrete elasticity modulus at the release of strands $E_{cm}(t)$ has been used. In the present case, by adjusting the concrete age to account for the thermal cycle ($t \cong 2.32$ days from Example 16), $f_{cm}(t)$ holds

$$f_{cm}(t) = e^{s \cdot \left[1 - \sqrt{\frac{28}{t}}\right]} \cdot f_{cm} = e^{0.2 \cdot \left[1 - \sqrt{\frac{28}{2.32}}\right]} \cdot 58 = 35.36 \text{ N/mm}^2 \quad [(3.1), (3.2)]$$

where

$$f_{cm} = f_{ck} + 8 = 58 \text{ N/mm}^2,$$

$s = 0.2$ for Class R cement,
therefore

$$E_{cm}(t) = \left(\frac{f_{cm}(t)}{f_{cm}}\right)^{0.3} \cdot E_{cm} = \left(\frac{35.36}{58}\right)^{0.3} \cdot 37{,}000 = 31{,}895 \text{ N/mm}^2$$

The prestress loss in each strand is equal to

$$\Delta P_4 = P_3 - P_4 = 188826 - 176048 = 12778 \text{ N}$$

Total initial loss in the midspan section

Initial losses give a total reduction of the prestress force in each strand equal to:

$$\Delta P_{\text{init.}} = \Delta P_1 + \Delta P_2 + \Delta P_3 + \Delta P_4 = 2047 + 5267 + 5460 + 12778 = 25552 \text{ N}$$

which represents 12.7% of the initial prestress at the stressing jack ($P_0 = 201{,}600$ N).

Calculation of time-dependent losses for a single strand

Time-dependent losses in each strand are identified as follows:

- ΔP_5 : loss due to steel relaxation,
- ΔP_6 : loss due to concrete shrinkage,
- ΔP_7 : loss due to concrete creep.

Their calculation can be developed as follows.

***Loss due to steel relaxation* (ΔP$_5$)**

The calculation is performed by considering that the steel stress is constant[22] and equal to the normal stress immediately after the wedge draw-in of anchorage devices ($\sigma_{pi} = 1425\,\text{N/mm}^2$—see the calculation of σ_{pi} after the wedge draw-in of anchorages), therefore it results that $\mu = \sigma_{pi} / f_{pk} = 1425/1860 = 0.766$,

and, for $t = 500{,}000$ h,

$$\Delta\sigma'_{p5} = \Delta\sigma_{pr} = k_1 \cdot 2.5 \cdot e^{k_2 \cdot \mu} \left(\frac{t}{1000}\right)^{0.75\cdot(1-\mu)} \cdot 10^{-5} \cdot \sigma_{pi}$$

$$= 0.66 \cdot 2.5 \cdot e^{9.1 \cdot 0.766} \left(\frac{500000}{1000}\right)^{0.75\cdot(1-0.766)} \cdot 10^{-5} \cdot 1425 \cong 74.52\,\text{N/mm}^2$$

The loss due to steel relaxation starting from the moment in which strands are anchored at the stressing heads of the casting bed is equal to:

$$\Delta P'_5 = \Delta\sigma'_{p5} \cdot A_p = 74.52 \cdot 140 = 10433\,\text{N}$$

The loss due to steel relaxation measured since the strand release is given by the difference between $\Delta P'_5$ and ΔP_2, being ΔP_2 the loss due to relaxation occurring between the anchorage of strands at the heads of the casting bed and their release (Fig. 6.51):

$$\Delta P_5 = \Delta P'_5 - \Delta P_2 = 10433 - 5267 = 5166\,\text{N}$$

corresponding to a stress loss equal to:

$$\Delta\sigma_{p5} = \Delta P_5 / A_p = 5166/140 = 36.9\,\text{N/mm}^2$$

[22] Steel stress decreases with time because of concrete creep and shrinkage; the decrease of steel stress is accounted for multiplying the relaxation loss by a reduction coefficient equal to 0.8 (see Sect. 6.10.7).

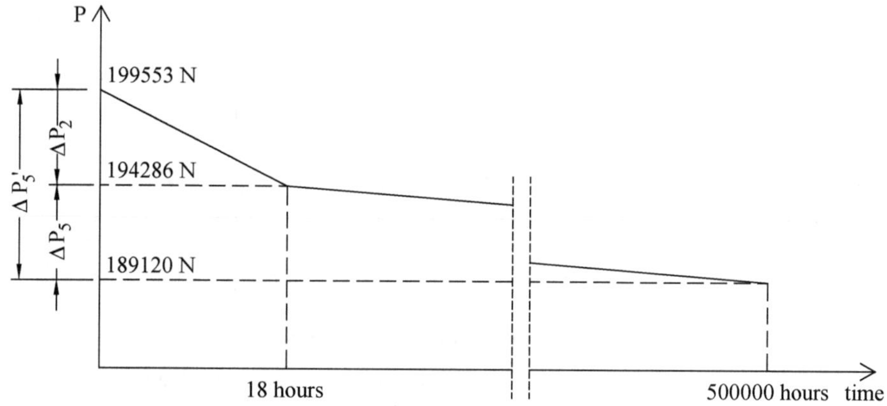

Fig. 6.51 Loss due to relaxation for each strand at 18 h and 500,000 h

Loss due to concrete autogenous shrinkage

In a precast concrete beam subjected to steam-hardening, autogenous shrinkage can be assumed negligible, as indicated in [10.3.1.2(3)]:

$$\varepsilon_{ca}(\infty) = 0$$

***Loss due to concrete drying shrinkage* (ΔP₆)**

From Example 11

$$\Delta\sigma_{p6} = 77\,\mathrm{N/mm^2}$$

corresponding to a loss of prestress in each strand equal to

$$\Delta P_6 = \Delta\sigma_{p6} \cdot A_p = 77 \cdot 140$$
$$= 10780\,\mathrm{N}\ (\cong 5.35\,\%\ \text{of the initial prestress of}\ 201{,}600\,\mathrm{N})$$

***Loss due to concrete creep* (ΔP$_7$)**

Preliminarily it is verified that under quasi-permanent loads the concrete stress is lower than $0.45 f_{ck}$ (see Sect. 6.4.3). The prestress force to be considered in evaluating creep effects is P_4, i.e., the force acting immediately after the loss due to the concrete immediate elastic shortening.

Quasi-permanent combination of loads:

$$q_{QP} = G_{k1} + G_{k2} = 4.60 + 5.20 = 9.80\,\mathrm{kN/m}$$

(snow load coefficient $\psi_{2,\,\text{snow}}$ is assumed null)

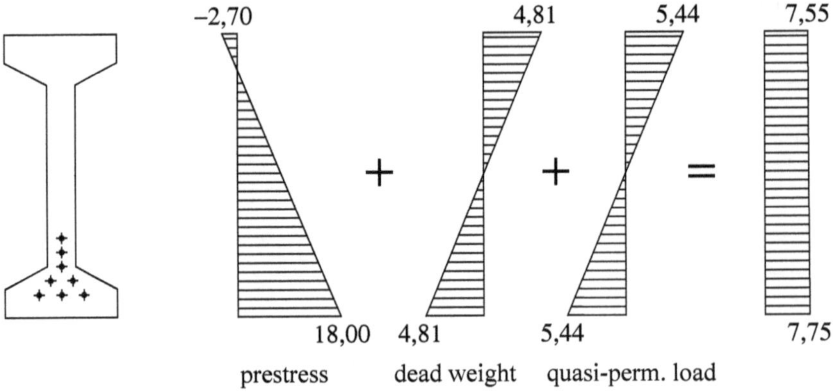

Fig. 6.52 Concrete normal stresses (positive values = compressive stresses)

Bending moment due to quasi-permanent uniformly distributed load is equal to:

$$M_{QP} = q_{QP}L^2/8 = 9.80 \cdot 20^2/8 = 490\,\text{kNm}$$

(230 kNm is the bending moment due to the beam self-weight and 260 kNm is the bending moment due to other permanent loads).

Assuming as positive the concrete compressive stresses, it follows (Fig. 6.52):

$$\begin{aligned}\sigma_{c,\text{sup}} &= \frac{8 \cdot P_4}{A_c} - \frac{8 \cdot P_4 \cdot z_{cp}}{I_c} \cdot z_{\text{sup}} + \frac{M}{I_c} z_{\text{sup}}\\ &= \frac{8 \cdot 176{,}048}{184{,}000} - \frac{8 \cdot 176{,}048 \cdot 351.25}{2.3887 \times 10^{10}} \cdot 500 + \frac{490.000.000}{2{,}3887 \times 10^{10}} \cdot 500\\ &= 7.5\,\text{N/mm}^2 \le 0.45 \cdot 50 = 22.5\ \text{N/mm}^2\end{aligned}$$

$$\begin{aligned}\sigma_{c,\text{inf}} &= \frac{8 \cdot P_4}{A_c} - \frac{8 \cdot P_4 \cdot z_{cp}}{I_c} \cdot z_{\text{sup}} + \frac{M}{I_c} z_{\text{sup}}\\ &= \frac{8 \cdot 176{,}048}{184{,}000} + \frac{8 \cdot 176{,}048 \cdot 351{,}25}{2.3887 \times 10^{10}} \cdot 500 - \frac{490.000.000}{2.3887 \times 10^{10}} \cdot 500\\ &= 7.75\,\text{N/mm}^2 \le 0.45 \cdot 50 = 22.5\ \text{N/mm}^2\end{aligned}$$

where stresses are evaluated considering the concrete section alone (approximated with the geometrical cross-section) because the calculation is performed using the value of the prestress force after the concrete elastic shortening. Moreover, stresses induced by external loads are also calculated approximately, considering the concrete section alone, even if they are acting on the transformed section, where steel is considered as an equivalent concrete area. If the calculation is repeated on the transformed section, the prestress force $P_{3,tot}$ before the transfer of the prestress should be used (see the last paragraph of Sect. 6.8.1).

Calculation of creep coefficient

Long-term creep

$$\varphi(\infty, t_0) = \varphi_0 \text{ with } \varphi_0 = \varphi_{RH} \cdot \beta(f_{cm}) \cdot \beta(t_0)$$

where: $\varphi_{RH} = [1 + \frac{1-RH/100}{0.1 \cdot \sqrt[3]{h_0}} \cdot \alpha_1] \cdot \alpha_2 = 1.294$ for $f_{cm} > 35\,\text{N/mm}^2$. with $RH = 70\%$, $h_0 = 116.5$ mm.

$$\alpha_1 = \left[\frac{35}{f_{cm}}\right]^{0.7} = 0.702, \alpha_2 = \left[\frac{35}{f_{cm}}\right]^{0.2} = 0.904$$

$$\beta(f_{cm}) = \frac{16.8}{\sqrt{f_{cm}}} = 2.206$$

For the calculation of $\beta(t_0)$, prestress and self-weight are applied at the time of strand release, while other permanent loads are applied 90 days after the transfer of prestress (see the design data at the beginning of the Example). Two different values of the creep coefficient are therefore calculated: the first one corresponding to prestress and self-weight and the second one corresponding to the other permanent loads.

The application time of self-weight $t_{0,a}$ is obtained by adjusting the value of Example 16, which accounts for the thermal cycle,

$$t_{0,T} = \sum_{i=1}^{n} e^{-\{4000/[273+T(\Delta_i)]-13.65\}} \cdot \Delta t_i = 2.32 \text{ days}$$

to consider effects of the cement class through [(B.9)]

$$t_{0,a} = t_{0,T} \cdot \left(\frac{9}{2 + t_{0,T}^{1.2}} + 1\right)^{\alpha} = 2.32 \cdot \left(\frac{9}{2 + 2.32^{1.2}} + 1\right) = 6.72 \geq 0.5$$

Application time $t_{0,b}$ of other permanent loads is obtained by adding 90 days to $t_{0,T}$ and applying again [(B.9)] to account for the cement class:

$$t'_{0,T} = t_{0,T} + 90 = 92.32 \text{ days}$$

$$t_{0,b} = t'_{0,T} \cdot \left(\frac{9}{2 + t'^{\,1.2}_{0,T}} + 1\right)^{\alpha} = 92.32 \cdot \left(\frac{9}{2 + 92.32^{1.2}} + 1\right) = 95.93 \geq 0.5$$

Once the values of $t_{0,a}$ and $t_{0,b}$ are known, it is possible to calculate the corresponding values of $\beta(t_{0,a})$ and $\beta(t_{0,b})$

$$\beta(t_{0,a}) = \frac{1}{0.1 + t_{0,a}^{0.20}} = \frac{1}{0.1 + 6.72^{0.20}} = 0.639$$
$$\beta(t_{0,b}) = \frac{1}{0.1 + t_{0,b}^{0.20}} = \frac{1}{0.1 + 95.93^{0.20}} = 0.386$$

and the values of the creep coefficient

$$\varphi(\infty, t_{0,a}) = 1.294 \cdot 2.206 \cdot 0.639 \cong 1.82$$
$$\varphi(\infty, t_{0,b}) = 1.294 \cdot 2.206 \cdot 0.386 \cong 1.10$$

Concrete stress

The concrete stress at the level of strand centroid is given by the following contributions.

- stress due to prestress:

$$\sigma_{cp} = \frac{8 \cdot P_4}{A_c} + \frac{8 \cdot P_4 \cdot z_{cp}^2}{I_c} = \frac{8 \cdot 176{,}048}{184{,}000} + \frac{8 \cdot 176{,}048 \cdot 351.25^2}{2.3887 \times 10^{10}} = 14.93\ \text{N/mm}^2$$

- stress due to self-weight (tensile and therefore negative)

$$\sigma_{c,pp} = \frac{M_{pp}}{I_c} \cdot z_{cp} = -\frac{(4.6 \cdot 20^2/8) \cdot 10^6}{2.3887 \cdot 10^{10}} \cdot 351.25 = -3.38\ \text{N/mm}^2$$

- stress due to quasi-permanent loads ($\psi_{2,\text{snow}} = 0$; tensile and therefore negative)

$$\sigma_{c,QP} = \frac{M_{pp}}{I_c} \cdot z_{cp} = -\frac{(5.2 \cdot 20^2/8) \cdot 10^6}{2.3887 \times 10^{10}} \cdot 351.25 = -3.82\ \text{N/mm}^2$$

Once the concrete stress at the strand centroid is known, the long-term creep strain can be calculated: the immediate shortening induced by initial prestress and self-weight is multiplied by the coefficient $\varphi(\infty, t_{0,a})$ and the one due to quasi-permanent loads by $\varphi(\infty, t_{0,b})$:

$\varepsilon_c(\infty, t_{0,a}, t_{0,b}) = \varphi(\infty, t_{0,a}) \cdot \frac{\sigma_{cp}+\sigma_{c,pp}}{E_{cm}} + \varphi(\infty, t_{0,b}) \cdot \frac{\sigma_{c,QP}}{E_{cm}}$[23]

therefore, each strand experiences a loss due to concrete creep equal to $\Delta\sigma_{p7} = \varepsilon_c(\infty, t_{0,a}, t_{0,b}) \cdot E_p = \frac{1.82 \cdot (14.93-3.38) - 1.10 \cdot 3.82}{37000} \cdot 195{,}000 = 88.6\ \text{N/mm}^2$[24] corresponding to the following loss of the prestress force in each strand

[23] The secant modulus E_{cm} has been used instead of the tangent modulus (see Footnote 22).

[24] If the tangent modulus ($1.05\ E_{cm}$) is used, the result is $\Delta\sigma_{p7} = 84.4\ \text{N/mm}^2$.

$$\Delta P_7 = \Delta\sigma_7 \cdot A_p = 88.6 \cdot 140 = 12404 \text{ N}$$

which is equal to 6.15% of the initial prestress force of 201,600 N.

Loss due to shrinkage and creep of concrete and steel relaxation (ΔP_{5+6+7})

Summing losses due to concrete shrinkage and creep and losses due to steel relaxation, the total time-dependent loss is equal to

$$\Delta\sigma_{p,5+6+7} = 36.9 + 77 + 88.6 \cong 202.5 \, \text{N/mm}^2$$

corresponding to the following loss of the prestress force in each strand:

$$\Delta P_{5+6+7} = \Delta\sigma_{p,5+6+7} \cdot A_p = 202.5 \cdot 140 = 28350 \, \text{N}$$

which is equal to 14% of the initial prestress force.

A more accurate estimation of the total time-dependent loss can be obtained by [(5.46)] (see Sect. 6.10.7):

$$\begin{aligned}\Delta P_{c+s+r} &= A_p \Delta\sigma_{p,c+s+r} = A_p \cdot \frac{\varepsilon_{cs} E_p + 0.8 \cdot \Delta\sigma_{pr} + \frac{E_p}{E_{cm}} \varphi(\infty, t_0) \cdot \sigma_{c,QP}}{1 + \frac{E_p}{E_{cm}} \cdot \frac{A_{p,\text{tot}}}{A_c} \left(1 + \frac{A_c}{I_c} z_{cp}^2\right)\left[1 + 0.8\varphi(\infty, t_{0,a})\right]} \\ &= 140 \cdot \frac{77 + 0.8 \cdot 36.9 + 88.6}{1 + \frac{195.000}{37.000} \cdot \frac{1.120}{184.000}\left(1 + \frac{184.000}{2.3887 \times 10^{10}} \cdot 351.25^2\right)[1 + 0.8 \cdot 1.82]} \\ &= 140 \cdot 169 = 23660 \text{ N}\end{aligned}$$

In the denominator of the formula the creep coefficient $\varphi\,(\infty, t_{0,a})$, corresponding to the application time of the prestress to concrete, is used.

Time-dependent losses are corresponding to 11.7% of the initial prestress force.

Total prestress loss

Finally, after an infinite amount of time, considering all the losses, the prestress force in each strand is given by

$$\begin{aligned}P &= P_0 - (\Delta P_1 + \Delta P_2 + \Delta P_3 + \Delta P_4) - \Delta P_{5+6+7} \\ &= 201600 - (2047 + 5267 + 5460 + 12778) - 23660 = 152388 \, \text{N}\end{aligned}$$

i.e., it is 24.4% lower than the initial prestress force ($P_0 = 201{,}600$ N).

6.10.7.2 Example 20. Post-tensioned Beam

Calculate the prestress losses (initial and time-dependent) of the simply supported beam of Example 4 (Fig. 6.53) *with six post-tensioned tendons; tensioning is performed from both ends* 14 days *after the placing of concrete. Other design data are*

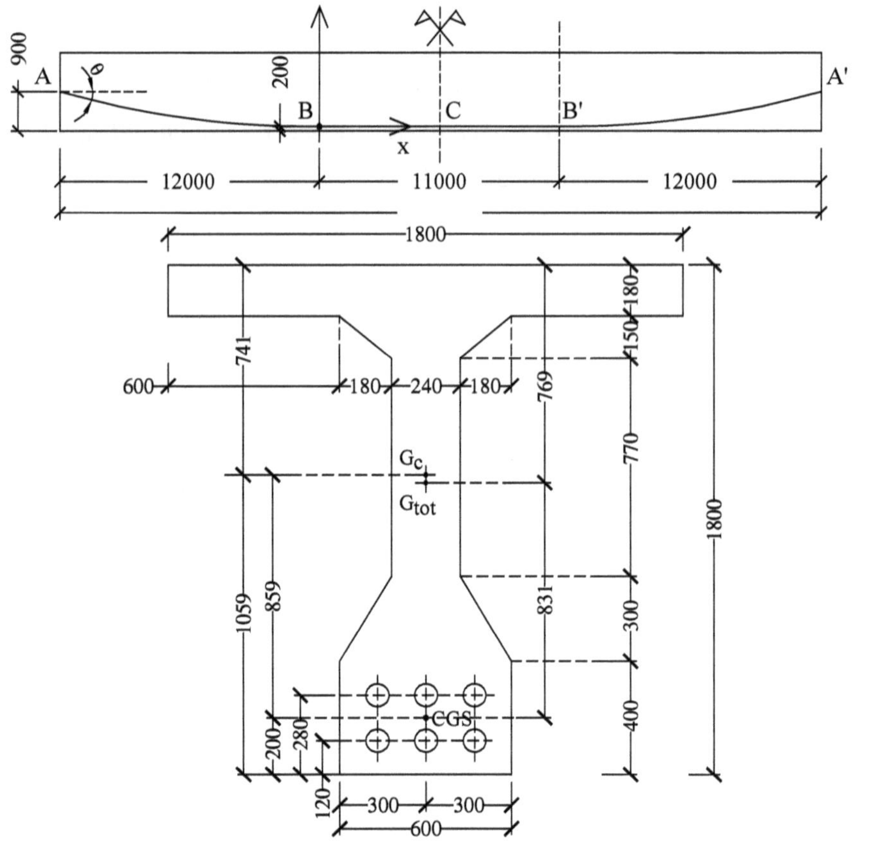

Fig. 6.53 Beam with post-tensioned tendons: CGS profile and midspan cross-section, with the identification of the centroids of geometrical cross-section (G_{tot}), concrete section without the area of ducts (G_c) and tendons (CGS)

friction coefficient $\mu = 0.19$ rad^{-1}, *unintentional angular deviation* $k = 0.005$ rad/m, *wedge draw-in of anchorage devices* $w = 3$ mm *(on each side), RH* $= 70\%$, *concrete strength class* C35/45, *strengths of prestressing steel* $f_{pk} = 1860$ N/mm^2 *and* $f_{p0.1\,k} = 1600$ N/mm^2.

Materials

Concrete strength class C35/45, made of Class R cement

$$f_{ck} = 35\,\text{N/mm}^2$$
$$E_{cm} = 34{,}000\,\text{N/mm}^2$$

Ordinary steel reinforcement of Class C

$$f_{yk} = 450\,\text{N/mm}^2$$

$$f_{yd} = f_{yk}/1.15 = 391\,\text{N/mm}^2$$

$$E_s = 200{,}000\,\text{N/mm}^2$$

Prestressing steel.
no. 6 tendons each made of no. seven 0.6" (d = 15.3 mm) strands

$$A_p = 7 \cdot 140 = 980\,\text{mm}^2 (\text{area of each tendon cross - section})$$

$$A_{p,\text{tot}} = 6 \cdot A_p = 5880\,\text{mm}^2 (\text{total area of tendon cross - sections})$$

$$E_p = 195{,}000\,\text{N/mm}^2 \quad [3.3.6(3)]$$

$$f_{pk} = 1860\,\text{N/mm}^2$$

$$f_{p0.1k} = 1600\,\text{N/mm}^2$$

Relaxation Class: 2 (wires or low relaxation strands)
$\rho_{1000} = 2.5\,\%$ (recommended value in [3.3.2 (6)] for Class 2)

$$f_{pk} \big/ f_{p0.1k} = 1.11 \geq 1.1 \quad [3.3.4(5)]$$

Wedge draw-in of anchorage devices: $w = 3$ mm on each side.

Geometrical data

Beam span $L = 35{,}000$ mm
Characteristics of the midspan cross-section (Table 6.38)
Perimeter $u = 7465$ mm.
Number and distance of tendons from the bottom: no. 3 tendons at 120 mm; no. 3 tendons at 280 mm
Distance of the centroid of tendons (CGS) from bottom $z_{G,\text{tendons}} = 200$ mm.
Eccentricity between CGS and G_c (Fig. 6.53) $z_{cp} = 859$ mm.

Loads

Self-weight: $G_{k1} = 23.50$ kN/m
Permanent load: $G_{k2} = 7.00$ kN/m
Quasi-permanent load: $G_{QP} = 30.50$ kN/m

Concrete mechanical characteristics

Average compressive strength at 14 days:

Table 6.38 Geometrical characteristics of midspan cross-section

Type of section	Geometrical	Concrete (without area of ducts)	Transformed section
u (mm)	7465	7465	7465
A_c (mm^2)	937,800	907,641	995,841
I_c (mm^4)	3.7901×10^{11}	3.5727×10^{11}	4.1715×10^{11}

$$f_{cm}(t) = e^{s\cdot\left[1-\sqrt{\frac{28}{t}}\right]} \cdot f_{cm} = e^{0.2\cdot\left[1-\sqrt{\frac{28}{14}}\right]} \cdot 43 = 39.58\,\text{N/mm}^2 \quad [(3.1), (3.2)]$$

where

$$f_{cm} = f_{ck} + 8 = 43\,\text{N/mm}^2$$

$s = 0.2$ for Class R cement (see Sect. 6.4.2)
Elasticity modulus at 14 days

$$E_{cm}(t) = \left(\frac{f_{cm}(t)}{f_{cm}}\right)^{0.3} \cdot E_{cm} = \left(\frac{39.58}{43}\right)^{0.3} \cdot 34{,}000 = 33{,}165\ \text{N/mm}^2$$

Maximum steel stresses

Maximum stress at the live end at tensioning [5.10.2.1][25]

$$\sigma_{p,\max} \leq \min\left(0.75\, f_{pk};\ 0.85\, f_{p0.1k}\right) = \min(0.75 \cdot 1860;\ 0,85 \cdot 1600)$$
$$= 1360\,\text{N/mm}^2$$

Maximum stress under service conditions after initial and long-term losses:

$$\sigma_{p,es} \leq 0.8\, f_{p0.1k} = 0.8 \cdot 1600 = 1280\ \text{N/mm}^2$$

Maximum prestress force

Maximum force at the live end at tensioning:

$$P_0 = P_{\max} = \sigma_{p,\max}\, A_{p,\text{tot}} = 1360 \cdot 5880 = 7{,}996{,}800\,\text{N} \cong 7997\,\text{kN}$$

Maximum stress under service conditions after initial and time-dependent losses:

$$P_{es} = \sigma_{p,\text{es}}\, A_{p,\text{tot}} = 1280 \cdot 5880 = 7{,}526{,}400\,\text{N} \cong 7526\,\text{kN}$$

[25] EC2 also provides a limit for the maximum stress once initial losses have occurred; for post-tensioned tendons, this limit is coincident with the limit valid for $\sigma_{p,\max}$ (see Sect. 6.3.2).

Calculation of initial losses

Initial losses can be identified as follows:

ΔP_μ: loss due to friction,
ΔP_{sl}: loss due to wedge draw-in of anchorage devices,
ΔP_{el}: loss due to concrete elastic shortening.

Loss due to friction (ΔP_μ)

The rigorous calculation would require calculating separately the friction loss of each tendon along its profile and adding together friction losses of all the tendons. Approximately, the calculation is here performed considering the CGS profile, whose end sections coincide with the centroids of end cross-sections of the beam. The CGS profile is parabolic in the two 12 m long end-segments (AB and B′A′) and is linear in the remaining 11 m long central part BB′ (Fig. 6.53). If a rectangular Cartesian coordinate system x–z with the origin in the CGS at section B is considered, the equation of the parabolic part of the CGS profile is given by

$z = 4.861 \times 10^{-3} x^2$(with x and z expressed in m)[26]

In the end-section A, the CGS profile is inclined to the horizontal of the following angle[27]:

$$|\theta_A| \cong 0.116\,\text{rad} \cong 6.654^\circ$$

which also represents the angular deviation of the tendon between the end-section A and section B, as in section B the tangent to the tendon is horizontal; the loss due to friction between sections A and B is therefore given by

$$\begin{aligned}\Delta P_\mu(12\,\text{m}) &= P_{\max}\left(1 - e^{-\mu(\theta+kx)}\right) = 7997 \cdot \left[1 - e^{-0.19\cdot(0.116+0.005\cdot 12.00)}\right]\\ &= 263\ \text{kN}\ (\cong 3.3\% P_{\max})\end{aligned}$$

while the loss between section A and the midspan section C[28] is given by

$$\begin{aligned}\Delta P_\mu(17.50\,\text{m}) &= P_{\max}\left(1 - e^{-\mu(\theta+kx)}\right) = 7997 \cdot \left[1 - e^{-0.19\cdot(0.116+0.005\cdot 17.50)}\right]\\ &\cong 303\ \text{kN}\ (\cong 3.8\ \%\ P_{\max})\end{aligned}$$

---

26 The equation of the parabola AB in the reference system x–z with origin in B (Fig. 6.53) is given by $z = f x^2 / L^2$, where L is the length of the beam segment AB and f is the ordinate of the point A in the reference system x–z; if x and z are expressed in m, the following equation is obtained: $z = f\, x^2 / L^2 = 0.70\, x^2 / 12{,}00^2 = 4.861 \times 10^{-3}\, x^2$.

27 The tangent to θ angle is given by the derivative of z ($z' = 2 f x / L^2$), evaluated in $x = -L/2$: $\tan\theta = z'_{(x=-L/2)} = -2\, f/L = -2 \cdot (0.90–0.20)/12.00 = -0.116$; therefore, $\theta = \arctan(-0.116) \cong -0.116\,\text{rad} \cong -6.654^\circ$ (negative sign means that the tangent is rotated clockwise respect to the horizontal).

28 The calculation of the friction loss between sections B and C gives: $\Delta P_\mu(x_C - x_B) = 7734 \cdot \left[1 - e^{0.19\cdot(0.005\cdot 5.50)}\right] \cong 40\,\text{kN}$, therefore the loss between A and C is equal to $\Delta P_\mu(x_C - x_A) = \Delta P_\mu(x_B) + 40\,\text{kN} = 303\,\text{kN}$. The result does not change if the friction loss between sections A and C is calculated at once.

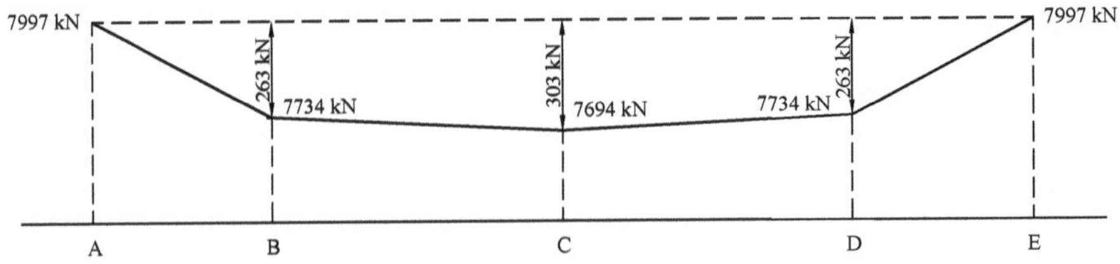

Fig. 6.54 Broken-line approximation of the diagram of the prestress force after the friction loss for tensioning from both end sections

Therefore, once the prestress loss due to friction has occurred, the prestress force in sections A, B and C takes the following values:

$$P_{\mathrm{A}} = 7997\,\mathrm{kN} = P_{\max}$$

$$P_{\mathrm{B}} = 7997 - 263 = 7734\,\mathrm{kN}$$

$$P_{\mathrm{C}} = 7997 - 303 = 7694\,\mathrm{kN}$$

In Fig. 6.54 the diagram of the prestress force along the beam axis is approximated with a broken line.

If the linearized expression of the friction loss is used, losses due to friction assume slightly higher values:

$$\Delta P_{\mu}(12\,\mathrm{m}) = P_{\max}\mu(\theta + kx) = 7997 \cdot 0.19 \cdot (0.116 + 0.005 \cdot 12.00) \cong 267\,\mathrm{kN}\ (> 263\,\mathrm{kN})$$

$$\Delta P_{\mu}(17.50\,\mathrm{m}) = P_{\max}\mu(\theta + kx) = 7997 \cdot 0.19 \cdot (0.116 + 0.005 \cdot 17.50) \cong 309\,\mathrm{kN}\ (> 303\,\mathrm{kN})$$

Loss due to wedge draw-in of the anchorage devices (ΔP_{sl})

The length of the tendon interested by the wedge draw-in is calculated assuming that $w = 3$ mm. According to the procedure indicated in 6.9.3, preliminarily the following quantity is evaluated:

$$w \cdot A_{p,\mathrm{tot}} \cdot E_p = 3 \cdot 5880 \cdot 195{,}000 = 3.4398 \times 10^9\,\mathrm{Nmm} = 3439.8\,\mathrm{kNm}$$

Therefore, areas between the diagrams of the prestress force after the friction loss and after the wedge draw-in are calculated.

The length of the segment AX, interested by the wedge draw-in, is not known a priori, therefore two limit positions are supposed: section X coincident with section

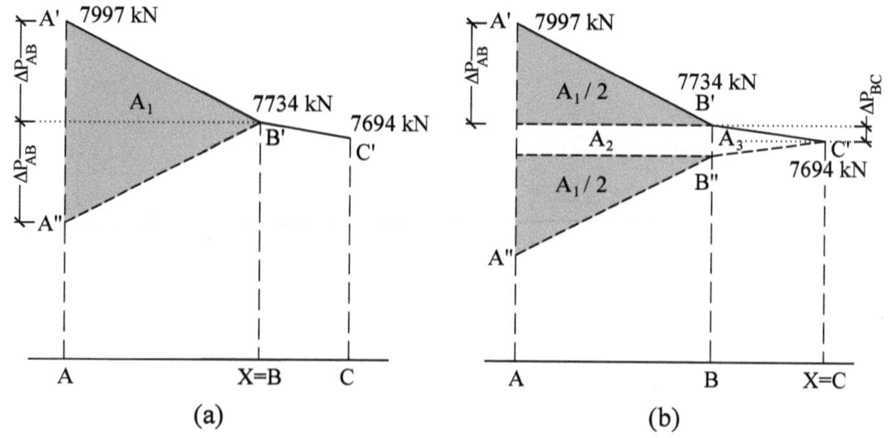

Fig. 6.55 Calculation of areas between the diagrams of the prestress force after the friction loss and after the wedge draw-in: **a** section X coincident with section B, **b** section X coincident with section C

B (Fig. 6.55a) and section X coincident with midspan section C (Fig. 6.55b); this second hypothesis is equivalent to assume that wedge draw-in of the anchorage devices affects the full length of the beam.

The calculation of areas shown in Fig. 6.55 gives:

(1): $A_1 = 2 \cdot \frac{AB \cdot \Delta P_{AB}}{2} = 12 \cdot 263 = 3156$ kNm
(2): $A_2 = AB \cdot (2 \cdot \Delta P_{BC}) = 12 \cdot (2 \cdot 40) = 960$ kNm
(3): $A_3 = 2 \cdot \frac{BC \cdot (\Delta P_{BC})}{2} = 5.50 \cdot 40 = 220$ kNm

Hypothesis 1: section X coincident with section B

$$\mathrm{A_{A'B'A''}} = \mathrm{A_1} = 3156\,\mathrm{kNm}$$

Hypothesis 2: section X coincident with section C

$$\mathrm{A_{A'B'C'B''A''}} = \mathrm{A_1} + \mathrm{A_2} + \mathrm{A_3} = 3156 + 960 + 220 = 4336\,\mathrm{kNm}$$

The value of $w\, A_{p,\mathrm{tot}}\, E_p$ (= 3439.8 kNm) is included between the values of areas A′B′A″ and A′B′C′B″A″, therefore section X is located between sections B and C.

Considering the similarity of triangles B′X′$\overline{\mathrm{B}}$ and B′C′B″ (Fig. 6.56), the following expression is obtained

$$\frac{B'\overline{B}}{B'B''} = \frac{BX}{BC} \Rightarrow \frac{2 \cdot \Delta P_{BX}}{2 \cdot \Delta P_{BC}} = \frac{x - x_B}{x_C - x_B} \Rightarrow \Delta P_{BX} = \Delta P_{BC} \frac{x - x_B}{x_C - x_B} = 40 \frac{x - 12}{5.50}$$

the value of x is evaluated by imposing that area A′B′X′$\overline{\mathrm{BA}}$ (in solid grey in Fig. 6.56) is equal to $w\, A_{p,\mathrm{tot}}\, E_p = 3439.8$ kNm:

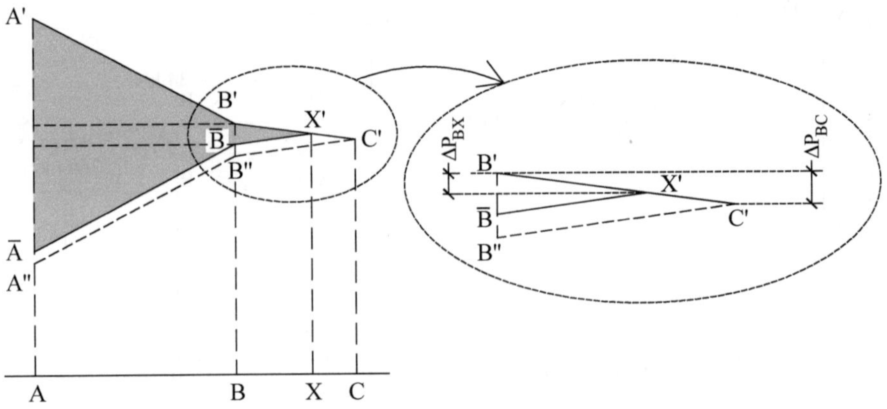

Fig. 6.56 Calculation of areas corresponding to the actual position of section X

$$(x - x_B) \cdot (2 \cdot \Delta P_{BX})/2 + A_1 + A_2 = 3439.8\, kNm$$

introducing numerical values of $x_B - x_A (= 12\,\text{m})$ and $\Delta P_{AB} (= 263\,\text{kN})$ in the last equation and using the above expression of ΔP_{BX}, it follows that

$$x \cong 13.53\,\text{m}$$

so, the following values due to wedge draw-in of anchorage devices are obtained:

$$\Delta P_{BX} = 40 \frac{13.53 - 12}{5.50} \cong 11\ \text{kN}$$

$$P_X = P_B - \Delta P_{BX} = 7734 - 11 = 7723\,\text{kN}$$

$$P_{\overline{A}} = P_A - 2 \cdot (\Delta P_{AB} + \Delta P_{BX}) = 7997 - 2 \cdot (263 + 11) = 7449\ \text{kN}$$

$$P_{\overline{B}} = P_B - 2 \cdot \Delta P_{BX} = 7734 - 2 \cdot 11 = 7712\ kN$$

In all the sections between X and X_1 (Fig. 6.57), including obviously the midspan section, there is not any prestress loss due to the wedge draw-in of anchorage devices.

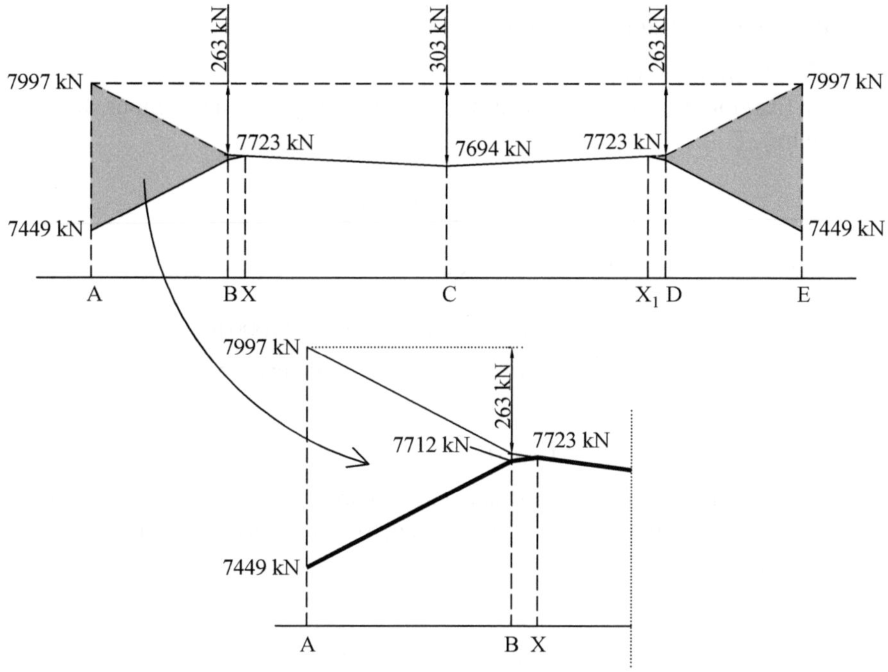

Fig. 6.57 Prestress force after losses due to friction and wedge draw-in of anchorage devices

Loss due to concrete elastic shortening (ΔP_{el})

From Example 4

$$\Delta P_{el} = 288.681\ N \cong 289\ \text{kN}$$

corresponding to 3.6% of the initial prestress ($P_{\max} = 7997$ kN).

Total initial losses at the midspan section

The total loss of prestress in the midspan section due to initial losses is given by

$$\Delta P_i = \Delta P_\mu + \Delta P_{sl} + \Delta P_{el} = 303 + 0 + 289 = 592\ kN$$

equal to 7.4% of the initial prestress force ($P_{\max} = 7997$ kN).

After initial losses, the prestress force in the midspan section is given by

$$P_i = 7997 - 592 = 7405\ \text{kN}$$

corresponding to the following steel stress.

$\sigma_{pi} = \frac{P_i}{A_{p,tot}} = \frac{7{,}405{,}000}{5880} = 1259\ \text{N/mm}^2 \leq \sigma_{pm0} = 1360\ \text{N/mm}^2.$

Calculation of time-dependent losses

Concrete shrinkage loss **($\Delta P_{shrinkage}$)**

From Example 12, the total loss due to drying shrinkage and autogenous shrinkage is given by

$$\Delta\sigma_p \cong 75\,\text{N/mm}^2$$

corresponding to a total prestress loss equal to

$$\begin{aligned}\Delta P_{shrinkage} &= \Delta\sigma_p \cdot A_{p,\text{tot}} = 75 \cdot 5880 = 441000\,\text{N}\\ &= 441\,\text{kN}\ (\cong 5.5\%\ \text{of the initial prestress force}).\end{aligned}$$

Loss due to concrete creep **(ΔP_{creep})**

Preliminarily, it is required to verify that in the midspan section the concrete compressive stress due to prestress and quasi-permanent loads G_{QP} is lower than $0.45 f_{ck}$ (see Sect. 6.4.3);

being

$$G_{QP} = 30.50\,\text{kNm}$$

the bending moment at midspan due to quasi-permanent loads is equal to

$$M_{QP} = G_{QP}L^2/8 = 30.50 \cdot 35^2/8 = 4670\,\text{kNm}$$

- Stress due to quasi-permanent loads (positive if compressive)

$$\sigma_{c,\text{sup}} = \frac{M_{G_{QP}}}{I_c} \cdot z_{\text{sup}} = \frac{4.67 \times 10^9}{3.5727 \times 10^{11}} \cdot 741 = 9.7\,\text{N/mm}^2$$

$$\begin{aligned}\sigma_{c,\text{inf}} &= -\frac{M_{G_{QP}}}{I'_c} \cdot z_{\text{inf}} = -\frac{4.67 \times 10^9}{3.5727 \times 10^{11}} \cdot 1059\\ &= -13.8\ \text{N/mm}^2\ (\text{tensile stress})\end{aligned}$$

$$\sigma_{c,\text{cavi}} = -\frac{M_{G_{QP}}}{I'_c} \cdot z_{cp} = -\frac{4.67 \times 10^9}{3.5727 \times 10^{11}} \cdot 859 = -11.2\ \text{N/mm}^2$$

(concrete stress at the level of tendon centroid).

- Stresses due to prestress

$$\sigma_{c,\text{sup}} = \frac{P_c}{A_c} - \frac{P_c \cdot z_{cp}}{I_c} \cdot z_{\text{sup}} = \frac{7{,}405{,}000}{907{,}641} - \frac{7{,}405{,}000 \cdot 859}{3.5727 \times 10^{11}} \cdot 741$$
$$= -5 \text{ N/mm}^2$$

$$\sigma_{c,\text{inf}} = \frac{P_c}{A_c} + \frac{P_c \cdot z_{cp}}{I_c} \cdot z_{\text{inf}} = \frac{7{,}405{,}000}{907{,}641} + \frac{7{,}405{,}000 \cdot 859}{3.5727 \times 10^{11}} \cdot 1059$$
$$= 27 \text{ N/mm}^2$$

$$\sigma_{c,\text{cavi}} = \frac{P_c}{A_c} + \frac{P_c \cdot z_{cp}^2}{I_c} = \frac{7{,}405{,}000}{907{,}641} + \frac{7{,}405{,}000 \cdot 859^2}{3.5727 \times 10^{11}}$$
$$= 23.5 \text{ N/mm}^2$$

- Total stresses

$$\sigma_{c,\text{sup}} = 9.7 - 5 = 4.7 \text{ N/mm}^2$$

$$\sigma_{c,\text{inf}} = -13.8 + 27 = 13.2 \text{ N/mm}^2 \leq 0.45 \cdot 35 = 15.75 \text{ N/mm}^2$$

$$\sigma_{c,\text{cavi}} = -11.2 + 23.5 = 12.3 \text{ N/mm}^2$$

For the sake of simplicity, the calculation has been performed considering the net section (i.e. without the area of duct cross-sections), but, the calculation is more complex because the resistant section is varying with the grouting of tendon ducts. The rigorous calculation requires two steps. In the first step, the elastic deformation of the beam is prevented by imaginary horizontal supports at both ends, so the concrete creep strain is also prevented, and the beam is subjected to two outward horizontal reaction forces and tensile stresses. In the second step, imaginary supports are removed, that is their reaction forces are applied to the transformed section in the inward direction.

Then, the creep coefficient is calculated

$$\varphi(\infty, \text{t}_0) = \varphi_0 \text{ with } \varphi_0 = \varphi_{RH} \cdot \beta(f_{cm}) \cdot \beta(t_0)$$

where $\varphi_{RH} = 1 + \frac{1-RH/100}{0.1\cdot\sqrt[3]{h_0}} = 1.475$ for $f_{cm} \leq 35\,\text{N/mm}^2$. with $RH = 70\%$, $h_0 = \frac{2A}{u} = \frac{2\cdot 937.800}{7.465} \cong 251$ mm.

$\beta(f_{cm}) = \frac{16.8}{\sqrt{f_{cm}}} = 2.562$ where $f_{cm} = f_{ck} + 8 = 43$ N/mm^2.

The concrete age at the time of loading is adjusted using the expression [(B.9)] to consider the cement Class (R)

$$t_a = t_{0,T} \cdot \left(\frac{9}{2 + t_{0,T}^{1.2}} + 1\right)^{\alpha} = 14 \cdot \left(\frac{9}{2 + 14^{1.2}} + 1\right) = 18.9 \geq 0.5$$

where $\alpha = 1$ for the Class R cement. Once t_a is known, it is possible to calculate $\beta(\mathrm{t_0})$

$$\beta(t_0) = \frac{1}{0.1 + t_0^{0.20}} = \frac{1}{0.1 + 18.9^{0.20}} = 0.526$$

and finally, the creep coefficient

$$\varphi_0 = 1.475 \cdot 2.562 \cdot 0.526 = 1.988 \cong 2$$

The corresponding loss of the prestress is given by

$$\begin{aligned} \Delta P_{\text{visc}} &= A_{p,\text{tot}} \cdot \left(\varphi_0 \cdot \sigma_{QP} \cdot \frac{E_p}{E_{cm}}\right) = 5880 \cdot \left(2 \cdot 12.3 \cdot \frac{195{,}000}{34{,}000}\right) \\ &= 829{,}599\,\text{N} \cong 830\,\text{kN} \end{aligned}$$

corresponding to the stress of 141 N/mm^2(= 10.4% of $\sigma_{\text{p,max}}$).

Loss due to relaxation **($\Delta \mathbf{P_r}$)**

The ratio between the steel stress once initial losses have occurred[29] and steel strength is given by

$$\mu = \frac{\sigma_{pi}}{f_{pk}} = \frac{1259}{1860} = 0.677$$

therefore, the loss due to relaxation, with the parameters k_1, k_2 and ρ_{1000} of Class 2 (low relaxation strands), is given by:

$$\Delta\sigma_{pr} = 1259 \cdot 0.66 \cdot 2.5 \cdot e^{9.1 \cdot 0.677} \cdot \left(\frac{500000}{1000}\right)^{0.75 \cdot (1-0.677)} \cdot 10^{-5} \cong 44\ \text{N/mm}^2$$

corresponding to the following loss of the prestress force

$$\Delta P_{pr} = \Delta\sigma_{pr} \cdot A_{p,\text{tot}} = 44 \cdot 5880 = 258720\,\text{N} \cong 259\,\text{kN}$$

[29] Once initial losses have occurred, the prestress is varying along the beam axis, but to evaluate the loss due to relaxation, a constant stress equal to the stress in the midspan section is considered. It is worth to note that for symmetrical tendons pre-tensioned from only one end-section, the force P in the midspan section is about the average prestress force (see Fig. 6.20).

Total time-dependent losses

Adding losses due to shrinkage and creep of concrete to the steel relaxation loss, the total time-dependent loss is equal to

$$\Delta\sigma_{p,s+c+r} = 75 + 141 + 44 \cong 260\,\text{N/mm}^2 (= 19\%\sigma_{\text{p,max}})$$

corresponding to the following loss of the prestress force

$$\Delta P_{s+c+r} = \Delta\sigma_{p,s+c+r} \cdot A_{p,\text{tot}} = 260 \cdot 5880 = 1528800\,\text{N} \cong 1529\,\text{kN}$$

A more accurate estimation of the total time-dependent loss is given by ([(5.46)]—see Sect. 6.10.7),

$$\begin{aligned}\Delta P_{c+s+r} &= A_p \Delta\sigma_{p,c+s+r} = A_p \cdot \frac{\varepsilon_{cs} E_p + 0.8 \cdot \Delta\sigma_{pr} + \frac{E_p}{E_{cm}}\varphi(t,t_0) \cdot \sigma_{c,Qp}}{1 + \frac{E_p}{E_{cm}} \cdot \frac{A_p}{A_c}\left(1 + \frac{A_c}{I_c} z_{cp}^2\right)[1 + 0.8\varphi(t,t_0)]} \\ &= 5880 \cdot \frac{75 + 0.8 \cdot 44 + 141}{1 + \frac{195{,}000}{34{,}000} \cdot \frac{5880}{907{,}641}\left(1 + \frac{907{,}641}{3.5727 \times 10^{11}} \cdot 859^2\right)[1 + 0.8 \cdot 2]} \\ &= 5880 \cdot 197 \cong 1{,}156{,}000\, N = 1156\,\text{kN}\end{aligned}$$

equal to 75.6% of the simple algebraic sum of all the time-dependent losses.

Total prestress loss

After an infinite amount of time the prestress force is therefore given by:

$$\begin{aligned}P &= P_0 - (\Delta P_\mu + \Delta P_{\text{el}}) - \Delta P_{c+s+r} = 7997 - (303 + 289) - 1156 \\ &= 6249\,\text{kN} \quad (< P_{es} = 7526\,\text{kN})\end{aligned}$$

i.e., it is 22% of the initial one at the stressing jack (branch a in Fig. 6.58). If the algebraic sum would be used, (branch b in Fig. 6.58) the result would be:

$$P = P_0 - (\Delta P_\mu + \Delta P_{\text{el}}) - (\Delta P_c + \Delta P_s + \Delta P_r) = 5875\,\text{kN}$$

with a reduction of 26.5% respect to P_0.

6.11 ULS in Flexure[30]

Figure 6.59b shows the ultimate strain diagram of a prestressed cross-section with

[30] For the ultimate limit state of RC sections under bending and axial force (M and N) see Chap. 7.

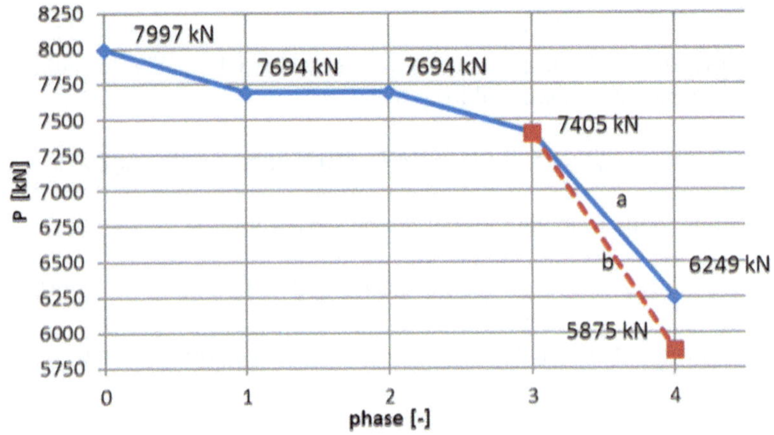

Fig. 6.58 Variation of the prestress force in the midspan section after initial and time-dependent losses: (0) after tensioning, (1) after the friction loss, (2) after the loss due to wedge draw-in of anchorage devices, (3) after the loss due to concrete elastic shortening (mutual effect), (4) after time-dependent losses: a—formula [(5.46)], b—algebraic sum)

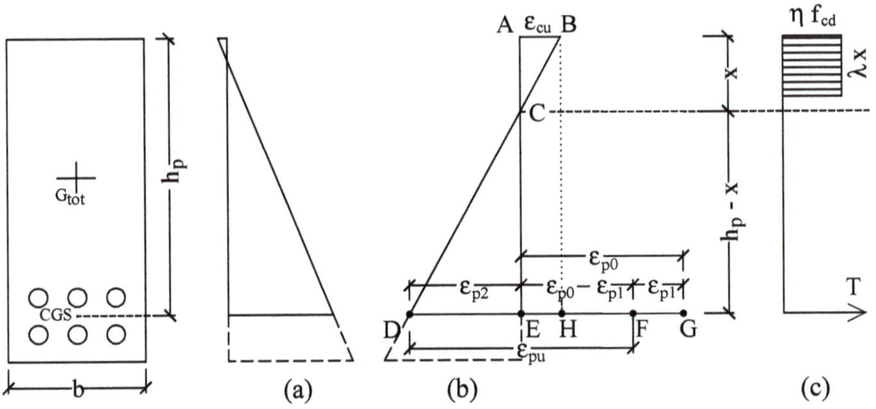

Fig. 6.59 Prestressed cross-section with bonded prestressing steel: **a** Strain due to prestress (ε_{p0}) and permanent loads (ε_{p1}), **b** Strain at failure (ε_{p2}), **c** Concrete compression diagram ("*stress-block*") and resultant tendon tensile force (T)

bonded prestressing steel when concrete strain is equal to ε_{cu} (= 3.5 ‰ for concrete grade not greater than C50/60) and steel strain is equal to ε_{p2}.

As triangles ABC and CDE in Fig. 6.59b are similar, the following compatibility equation is obtained

$$\frac{\varepsilon_{p2}}{\varepsilon_{cu}} = \frac{h_p - x}{x} \Rightarrow \varepsilon_{p2} = \frac{h_p - x}{x}\varepsilon_{cu}$$

where h_p is the distance of the tendon centroid measured from the top compressed edge and x is the depth of the neutral axis.

The normal stress in prestressing tendons is calculated considering that immediately after the transfer of prestress to concrete, tendons are subjected to the following strain:

$$\varepsilon_{p0} - \varepsilon_{p1}$$

where

ε_{p0} steel strain at tensioning,
ε_{p1} steel strain after the prestress transfer, once initial and time-dependent losses have occurred.[31]

Therefore, at failure (BD segment shown in Fig. 6.59), the total steel strain is equal to

$$\varepsilon_{pu} = (\varepsilon_{p0} - \varepsilon_{p1}) + \varepsilon_{p2}$$

The translational equilibrium equation is obtained by making equal the concrete resultant compression force C and the tendon resultant tensile force T: $C = T$.

The resultant compression force in the concrete is expressed as (Fig. 6.59)

$$C = \eta\, f_{cd} \cdot \lambda\, x \cdot b$$

with

$$\begin{aligned}
&\eta = 1.0 && \text{for } \mathrm{f_{ck}} \leq 50\,\mathrm{N/mm^2},\\
&\eta = 1.0 - (\mathrm{f_{ck}} - 50)/200 && \text{for } 50\,\mathrm{N/mm^2} < \mathrm{f_{ck}} \leq 90\,\mathrm{N/mm^2}\\
&\lambda = 0.8 && \text{for } \mathrm{f_{ck}} \leq 50\,\mathrm{N/mm^2}\\
&\lambda = 0.8 - (\mathrm{f_{ck}} - 50)/400 && \text{for } 50\,\mathrm{N/mm^2} < \mathrm{f_{ck}} \leq 90\,\mathrm{N/mm^2}
\end{aligned}$$

while the tensile resultant force in prestressing steel, if the contribution of ordinary reinforcement is neglected, is given by

$$T = f_{pu} \cdot A_{p,tot}$$

where f_{pu} is the steel stress corresponding to ε_{pu} and $A_{p,tot}$ the total area of prestressing steel.

Equating C and T, the following expression of the neutral axis depth is obtained

$$x = \frac{f_{pu} \cdot A_{p,tot}}{\eta\, f_{cd} \lambda\, b}$$

[31] Strain ε_{p1} is corresponding to the shortening of the prestressing steel.

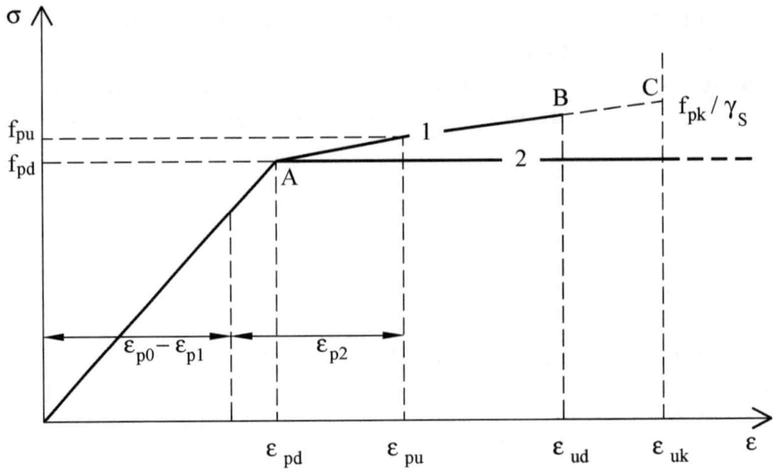

Fig. 6.60 σ–ε diagram with identification of strains $(\varepsilon_{p0} - \varepsilon_{p1})$ and ε_{p2}

In the following paragraphs, the discussion is focused on the special case of elastic perfectly plastic stress/strain law for steel (branch 2 of Fig. 6.60).

If $\varepsilon_{pu} = (\varepsilon_{p0} - \varepsilon_{p1}) + \varepsilon_{p2} \geq \varepsilon_{pd}$, steel is yielded, and the depth of the neutral axis is obtained by substituting f_{pu} with f_{pd} in the previous expression:

$$x = \frac{f_{pd} \cdot A_{p,tot}}{\eta\, f_{cd} \lambda\, b}$$

vice versa if $\varepsilon_{pu} = (\varepsilon_{p0} - \varepsilon_{p1}) + \varepsilon_{p2} < \varepsilon_{pd}$, steel is still in elastic state ($f_{pu} < f_{pd}$) with $f_{pu} = E_p \cdot [(\varepsilon_{p0} - \varepsilon_{p1}) + \varepsilon_{p2}]$, where $\varepsilon_{p2} = (h_p - x)\varepsilon_{cu}/x$ and the tensile resultant force T is rewritten in the following form

$$T = f_{pu} \cdot A_{p,tot} = E_p \cdot \left(\varepsilon_{p0} - \varepsilon_{p1} + \frac{h_p - x}{x} \varepsilon_{cu} \right) \cdot A_{p,tot}$$

so, the equilibrium equation becomes

$$E_p \cdot \left(\varepsilon_{p0} - \varepsilon_{p1} + \frac{h_p - x}{x} \varepsilon_{cu} \right) \cdot A_{p,tot} = \eta\, f_{cd} \cdot \lambda\, x \cdot b$$

and, by rearranging the terms,

$$\eta\, f_{cd} \lambda\, b\, x^2 - E_p A_{p,tot} (\varepsilon_{p0} - \varepsilon_{p1} - \varepsilon_{cu})\, x - E_p h_p \varepsilon_{cu} A_{p,tot} = 0$$

Once the depth of the neutral axis x is identified, the resistant moment M_{Rd} can be written down as

$M_{Rd} = f_{pd} \cdot A_{p,tot} \cdot \left(h_p - \frac{\lambda x}{2}\right)$ if the steel is yielded,
$M_{Rd} = f_{pu}(\varepsilon_{pu}) \cdot A_{p,tot} \cdot \left(h_p - \frac{\lambda x}{2}\right)$ if the steel is in the elastic state.

6.11.1 Dimensionless Calculation of the Failure Type of a Pre-stressed Rectangular Cross-Section Under the Hypothesis of Elastic-Perfectly Plastic Stress–Strain Law for Prestressing Steel

Following quantities are defined

$\rho_p = \frac{A_{p,tot}}{b \cdot h_p}$ prestressing steel geometrical ratio
$\omega_p = \frac{f_{pd} \cdot A_{p,tot}}{\eta f_{cd} \cdot b \cdot h_p}$ prestressing steel geometrical ratio

where, in addition to the symbols already introduced,

$A_{p,tot}$ is the total area of prestressing steel,
b is the width of the cross-section,
h_p is the effective depth measured from the centroid of prestressing steel.

By applying the definition of ω_p, the translational equilibrium equation can be rewritten as follows

$$f_{pu} = \frac{\eta f_{cd} \cdot \lambda x \cdot b}{A_{p,tot}} = \frac{\lambda x}{\omega_p \cdot h_p} f_{pd}$$

which, for failure with prestressing steel in the plastic state, (i.e. for $f_{pu} = f_{pd}$) becomes

$$\omega_p \cdot h_p = \lambda x$$

This expression is valid only for failure with prestressing steel in the plastic state and defines a relationship between the mechanical ratio of prestressing steel and the effective depth of the neutral axis.

It allows to easily identify the failure type with steel in the elastic state or the plastic state (see Example 21).

6.11.2 Dimensionless Calculation of the Failure Type for a T or I Pre-stressed Cross-Section Under the Hypothesis of an Elastic-Perfectly Plastic Stress–Strain Law for Prestressing Steel

For T or I cross-sections, the geometrical ratio ρ_p and mechanical ratio ω_p of prestressing steel take the following expressions

$$\rho_p = \frac{A_{p,tot}}{b_w \cdot h_p}$$

$$\omega_p = \frac{f_{pd} \cdot A_{p,tot}}{\eta f_{cd} \cdot b_w \cdot h_p}$$

where (Fig. 6.61a).

$A_{p,tot}$ is the total area of prestressing steel,
b_w is the width of the beam web,
h_p is the effective depth measured from the centroid of prestressing steel.

If the "*stress-block*" extends until the beam web ($0.8\,x \geq s$ i.e. $x \geq 1.25$ s, where s is the thickness of the compressed flange), the translational equilibrium condition can be expressed as follows

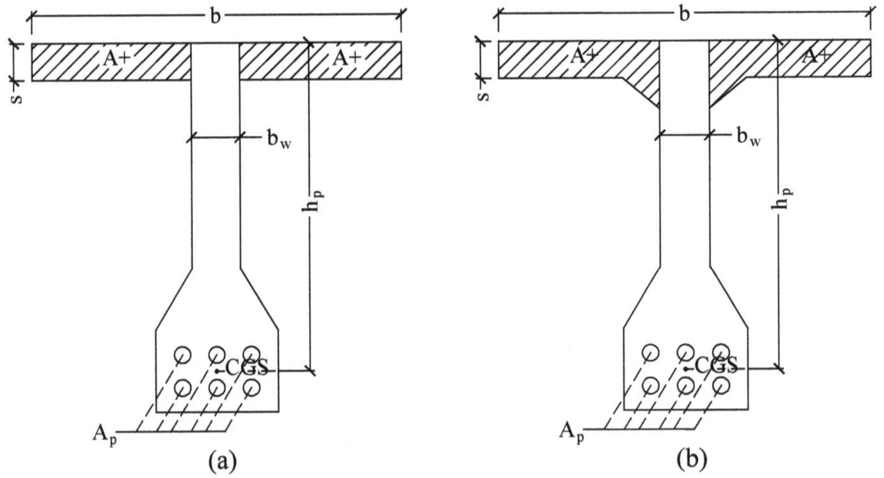

Fig. 6.61 Prestressed T cross-section; dashed lines identify the area A^+ to consider for the calculation of ω_p^*: **a** without fillets between web and flange, **b** with fillets

$$f_{pu} = \frac{\eta f_{cd} \cdot [(b - b_w) \cdot s + \lambda x \cdot b_w]}{A_{p,tot}} = \frac{\eta f_{cd} b_w \left[\left(\frac{b}{b_w} - 1\right) \cdot s + \lambda x\right]}{A_{p,tot}}$$
$$= \frac{(b/b_w - 1) \cdot s + \lambda x}{\omega_p h_p} f_{pd}$$

(b is the width of the compressed flange). For failure with prestressing steel in the plastic state ($f_{pu} = f_{pd}$), this expression becomes

$$\omega_p h_p = (b/b_w - 1) \cdot s + \lambda x$$

The similarity of triangles ABC and BDH shown in Fig. 6.59 gives

$$x = \frac{\varepsilon_{cu}}{\varepsilon_{p2} + \varepsilon_{cu}} h_p$$

and, by substituting it in the equilibrium condition,

$$f_{pu} = \frac{\left(\frac{b}{b_w} - 1\right) \cdot s + \lambda x}{\omega_p h_p} f_{pd} = \frac{\left(\frac{b}{b_w} - 1\right) \cdot s + \lambda \frac{\varepsilon_{cu}}{\varepsilon_{p2} + \varepsilon_{cu}} h_p}{\omega_p h_p} f_{pd}$$
$$= \frac{\left(\frac{b}{b_w} - 1\right) \cdot \frac{s}{h_p} + \lambda \frac{\varepsilon_{cu}}{\varepsilon_{p2} + \varepsilon_{cu}}}{\omega_p} f_{pd}$$

For failure with steel in the plastic state [$f_{pu} = f_{pd}$ and $(\varepsilon_{p0} - \varepsilon_{p1} + \varepsilon_{p2}) \geq f_{pd}/E_p$], the equilibrium equation becomes

$$\omega_p = \left(\frac{b}{b_w} - 1\right) \cdot \frac{s}{h_p} + \lambda \frac{\varepsilon_{cu}}{\varepsilon_{p2} + \varepsilon_{cu}}$$

The ω_p expression suggests defining a new quantity ω_p^* as the difference between ω_p and the ratio between the area A$^+$ (dashed area in Fig. 6.61) and the area (b_w h_p) $\omega_p^* = \omega_p - \left(\frac{b}{b_w} - 1\right) \cdot \frac{s}{h_p} = \frac{f_{pd} \cdot A_{p,tot}}{\eta f_{cd} \cdot b_w \cdot h_p} - \left(\frac{b}{b_w} - 1\right) \cdot \frac{s}{h_p}$[32]and the equilibrium equation, rewritten as a function of ω_p^* takes the same expression of the equation valid for the rectangular cross-section

$$\omega_p^* = \lambda \frac{\varepsilon_{cu}}{\varepsilon_{p2} + \varepsilon_{cu}}$$

therefore, the maximum values of ω_p^*, when the failure occurs with steel in the plastic state, are the same maximum values assumed by ω_p for rectangular cross-sections.

[32] Quantity $\omega{*}_p$ is coincident with ω_p when $b_w = b$, i.e., for rectangular cross-sections.

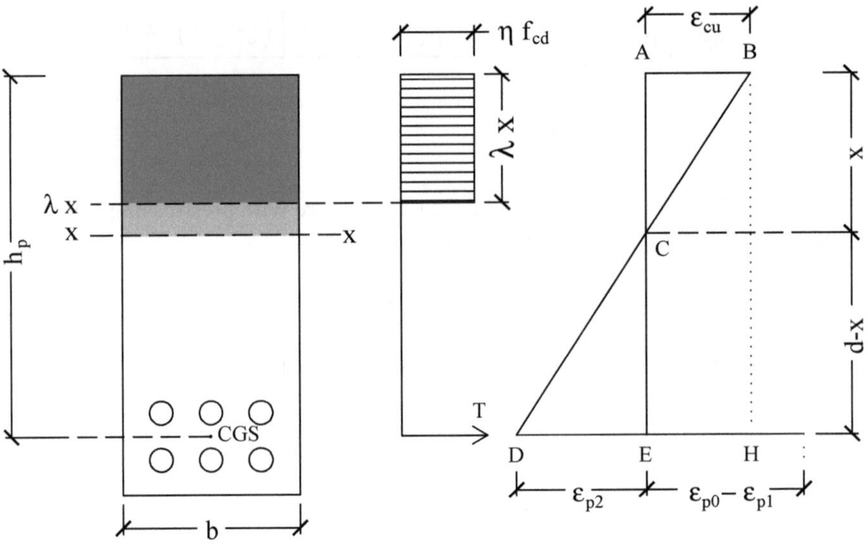

Fig. 6.62 Stress–strain diagram for a prestressed rectangular cross-section at failure

More in general, for a T or I cross-section with fillets between web and flange, the expression of ω_p^* becomes

$\omega_p^* = \omega_p - \frac{A^+}{b_w \cdot h_p}$[33]

where A^+ is corresponding to the total area of the cross-section included fillets and the two parts of slabs out of the web.

6.11.3 Example 21. Calculation of the Failure Type of a Pre-stressed Rectangular Cross-Section (Fig. 6.62) Under the Hypothesis of Elastic-Perfectly Plastic Stress–Strain Law for Prestressing Steel

By applying the definition of ω_p^*, *it is possible to evaluate the failure type by varying the steel mechanical ratio, if the following data are known:* $f_{ck} \leq 50$ N/mm², $f_{pk} = 1860$ N/mm², $f_{p0.1k} = 1600$ N/mm², $f_{pd} = 1391$ N/mm², $E_p = 195{,}000$ N/mm².

[33] This expression is valid only when the concrete rectangular diagram of compressive stresses ("*stress block*") is extended beneath the fillets between web and slab, i.e. for $x \geq 1.25\,(s + h_r)$, where s is the slab depth and h_r is the height of fillets; without fillets, the same condition becomes: $x \geq 1.25\,s$.

It is assumed that, once losses have occurred, the prestressing steel strain is around 5‰[34] ($\varepsilon_{p0} - \varepsilon_{p1} \cong 5$ ‰),

therefore the value $\varepsilon_{p2,\,\mathrm{el}}$ of the strain ε_{p2} due to external loads, which, together with the initial one ($\varepsilon_{p0} - \varepsilon_{p1}$) takes the prestressing steel to the yield strain, is equal to

$$\varepsilon_{p2,\,\mathrm{el}} = f_{pd}/E_p - 0.005 = 1391/195{,}000 - 0.005 = 0.00213 = 2.13‰$$

If the steel is yielded, the translational equilibrium equation takes the following form

$$\omega_p = \frac{\lambda\, x}{h_p}$$

moreover, to ensure compatibility (according to triangle similarity for triangles ABC and BDH shown in Fig. 6.62), the neutral axis depth is given by

$$x = \frac{\varepsilon_{cu}}{\varepsilon_{cu} + \varepsilon_{p2}} h_p$$

and, by combining the two equations,

$$\omega_p = \lambda \frac{\varepsilon_{cu}}{\varepsilon_{cu} + \varepsilon_{p2}}$$

Finally, by putting $\varepsilon_{p2} = \varepsilon_{p2,\,\mathrm{el}} = 2.13‰$, the value of the steel mechanical ratio ω_p at steel yielding is obtained.

$$\omega_{p,el} = 0.8 \frac{3.5}{3.5 + 2.13} = 0.497$$

where it has been assumed that $\lambda = 0.8$, being $f_{ck} \leq 50$ N/mm^2.

From the equilibrium equation, it results that $x < x_{\mathrm{el}}$ for $\omega_p < \omega_{p,\,\mathrm{el}}$, where x_{el} is the depth of the neutral axis corresponding to yielding of steel; therefore, for $\omega_p < \omega_{p,\,\mathrm{el}}$, the steel strain is greater than the yield strain and steel is in the plastic state (Fig. 6.63).

For concrete C55/67, the ultimate concrete strain is equal to

$$\varepsilon_{cu} = 3.2‰$$

and

[34] In prestressing steel, under service conditions, a strain of approximately 5‰ is already present; this value is obtained starting from the strain associated to the initial prestress ($\varepsilon_{p0} = 0.85\, f_{p0.1k}/E_p \cong 7‰$), and by reducing it by 30% to take into account initial and time-dependent losses.

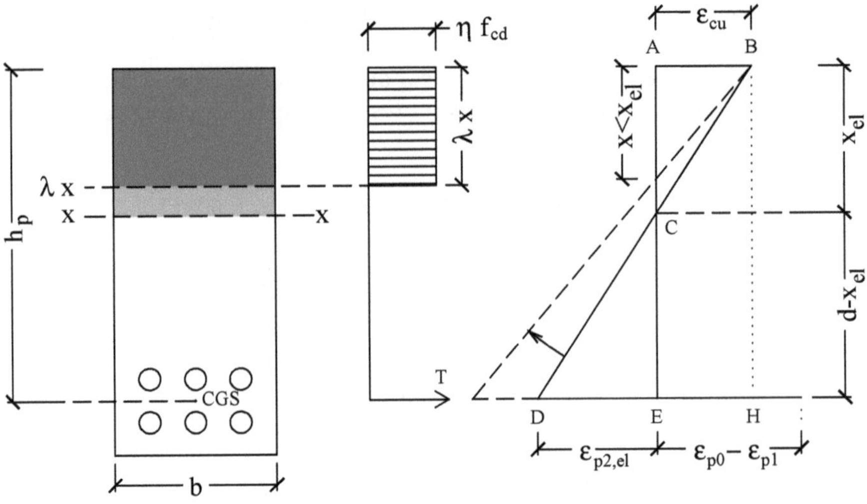

Fig. 6.63 Depth of the neutral axis x_{el} corresponding to the yield strain of prestressing steel and depth of the neutral axis x for $\omega_p < \omega_{p,\text{el}}$

$$\lambda = 0.8 - (f_{ck} - 50)/400 = 0.7875$$

so the mechanical steel ratio corresponding to yielding of prestressing steel takes the following value.

$\omega_p = \frac{\lambda\, \varepsilon_{cu}}{(\varepsilon_{p2} + \varepsilon_{cu})} = \frac{0.7875 \cdot 3.2}{(2.13+3.2)} \cong 0.472.$

Table 6.39 collects the maximum value of the steel mechanical ratio beneath which failure occurs with steel in the plastic state, for four different types of steel and different concrete strength classes.

Values of Table 6.39 have been calculated by assuming an initial strain $(\varepsilon_{p0} - \varepsilon_{p1})$ equal to 3‰ for steel type 1 and 2 (laminated bars) and 5‰ for steel type 3 and 4 (strands).

Table 6.39 Maximum values of the steel mechanical ratio when prestressing steel is in the plastic state at failure ($\varepsilon_{p0} - \varepsilon_{p1} = 3‰$ for bars—steel type 1 and 2, $\varepsilon_{p0} - \varepsilon_{p1} = 5‰$ for strands—steel type 3 and 4, by neglecting the contribution of the ordinary reinforcement)

Steel	f_{pk} (N/mm^2)	$f_{p0.1\,k}$[a] (N/mm^2)	E_p (N/mm^2)	f_{ck} (N/mm^2)					
				≤ 50	55	60	70	80	90
1	1030	835	205,000	0.693	0.673	0.656	0.628	0.607	0.586
2	1230	1080	205,000	0.551	0.527	0.508	0.479	0.463	0.447
3	1770	1520	195,000	0.530	0.506	0.487	0.459	0.443	0.428
4	1860	1600	195,000	0.497	0.472	0.453	0.426	0.411	0.397

[a]Values taken from the catalogues of producers and therefore different from those collected in Table 6.2, where $f_{p0.1\,k} = 0.9 f_{pk}$

6.11.4 Example 22. Calculation of the Resistant Moment for a Prestressed T Cross-Section (Under the Hypothesis of an Elastic-Perfectly Plastic Stress–Strain Law for Prestressing Steel)

Calculate the resistant moment for the midspan section of the beam of Example 20. *For prestressing steel, it is assumed an elastic-perfectly plastic stress–strain law (without any upper limit for strain).*

Prestressing steel

Design strength for prestressing steel [Sect. 3.3.6 (6)]:

$$f_{pd} = \frac{f_{p0.1k}}{\gamma_S} = \frac{1600}{1.15} = 1391 \text{ N/mm}^2$$

strain at the design yield strain:

$$\varepsilon_{pd} = \frac{f_{pd}}{E_p} = \frac{1.391}{195.000} = 7.13‰$$

An elastic-perfectly plastic stress–strain law is assumed (branch 2 of Fig. 6.60).

Ordinary reinforcement

Ordinary reinforcement is neglected.

Concrete

The "*stress-block*" diagram is assumed over a depth equal to λx (x = depth of the neutral axis measured from the compressed edge, $\lambda = 0.8$ for $f_{ck} \leq 50$ N/mm^2). The concrete design compressive strength takes the value

$$f_{cd} = \alpha_{cc} \cdot f_{ck}/\gamma_C = 0.85 \cdot 35/1.5 = 19.83 \,\text{N/mm}^2$$

Prestressing steel strain immediately after the transfer of prestress to concrete

Once initial and time-dependent losses have occurred, the prestressing steel strain is equal to[35]

$$\varepsilon_{p0} - \varepsilon_{p1} = \frac{P_m}{A_{p,tot} \cdot E_p} = \frac{6,249,000}{5880 \cdot 195,000} = 5.45‰$$

where $P_m = 6249$ kN is the mean value of long-term prestress of Example 20, obtained by calculating time-dependent losses through expression [(5.46)].

[35] The mean value of prestress force is used, as indicated in [5.10.8(1)], with $\gamma_p = 1.0$ according to [2.4.2.2(1)].

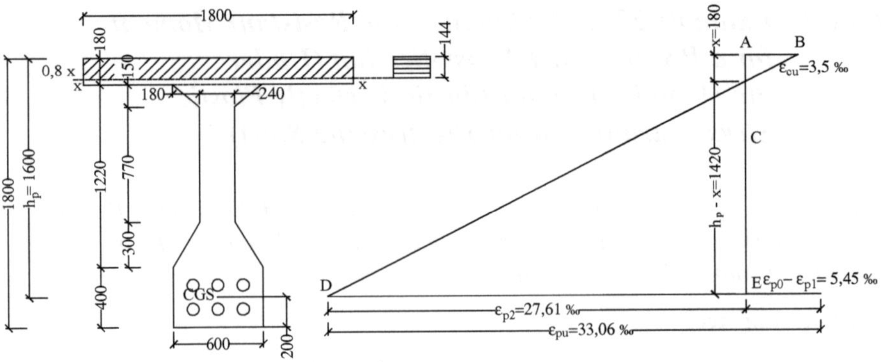

Fig. 6.64 First attempt depth of the neutral axis within the compressed flange

Calculation of the depth of the neutral axis

At first, the neutral axis is assumed as coincident with the lower edge of the flange ($x = s = 180$ mm), and prestressing steel is assumed yielded; under these hypotheses, following values are calculated:

$$\lambda x = 0.8\,\mathrm{s} = 144\,\mathrm{mm}\ \left(\text{depth of the}\ ''stress - block''\ \text{diagram}\right)$$

$\varepsilon_{p2} = \frac{h_p - x}{x} \cdot \varepsilon_{cu} = \frac{1420}{180} \cdot 3.5‰ = 27.61‰$ (Fig. 6.64)
Concrete resultant compression force

$$C = f_{cd} \cdot B \cdot x = 19.83 \cdot 1800 \cdot 144 = 5139936\ N \cong 5140\ \text{kN}$$

Strand resultant tensile force

$$T = A_{p,tot} \cdot f_{pd} = 5880 \cdot 1391 = 8.179.080\ N \cong 8.179\,\text{kN}$$

Being $C < T$, the neutral axis depth is greater than the slab depth.
It is then assumed that $x \geq 1.25 \cdot (180 + 150) = 412{,}5$ mm, i.e. the “*stress-block*” diagram cuts the web beneath the fillets between web and flange, therefore (Fig. 6.65)

$$\begin{aligned} C &= f_{cd} \cdot [A_1 + A_2 + A_3] \\ &= \frac{0.85 \cdot 35}{1.5} \cdot \left[(1800 \cdot 180) + 2 \cdot \frac{180 \cdot 150}{2} + 240 \cdot (0.8 \cdot x - 180)\right] \\ &= (6104700 + 3808 \cdot x)\ \text{N} \end{aligned}$$

and, by imposing $C = T$,

$$\begin{aligned} &(6104700 + 3808 \cdot x) = 8179080 \rightarrow x \cong 545\,\text{mm}\ > 1.25 \cdot (180 + 150) \\ &= 412.5\,\text{mm} \end{aligned}$$

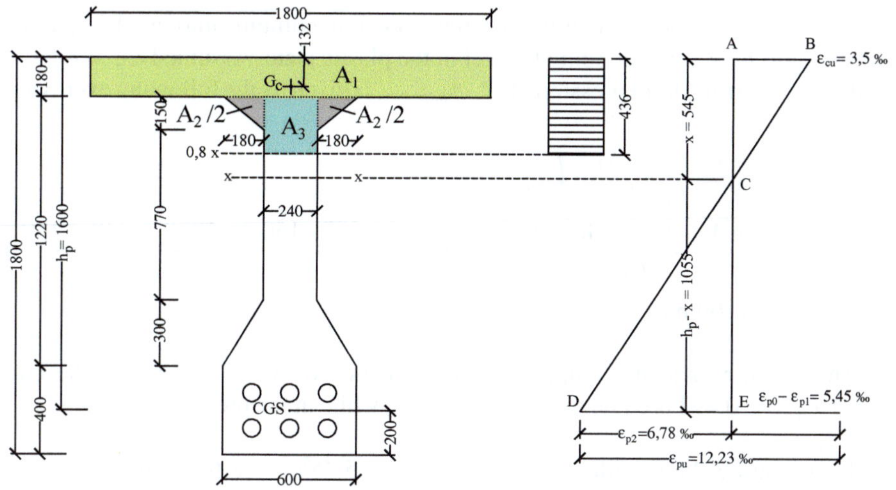

Fig. 6.65 Actual depth of the neutral axis

therefore, the hypothesis that the "*stress-block*" diagram cuts the web beneath the fillets between the web and the flange is satisfied.

It is then required to verify if steel is in the plastic state, as already assumed, or in the elastic state. To this purpose, it is sufficient to calculate strain ε_{p2} (accounting for similarity of triangles ABC and CDE of Fig. 6.65) and total strain ε_{pu}

$$\varepsilon_{p2} = \varepsilon_{cu} \cdot (h_p - x)/x = 3.5 \times 1055/545 = 6.78 ‰$$

the ultimate steel strain is therefore equal to

$$\varepsilon_{pu} = (\varepsilon_{p0} - \varepsilon_{p1}) + \varepsilon_{p2} = 5.45 + 6.78 = 12.23 ‰ >> \varepsilon_{pd}$$

i.e., steel is in the plastic state.

Use of Table 6.39 to identify the type of failure

$$\omega_p = \frac{f_{pd} \cdot A_{p,tot}}{\eta\, f_{cd} \cdot b_{an} \cdot h_p} = \frac{1391 \cdot 5880}{1.0 \cdot 19.83 \cdot 240 \cdot 1600} = 1.074$$

and

$$\omega_p^* = \omega_p - \frac{A^+}{b_n \cdot h_p} = 1.074 - \frac{180 \cdot 150 + (1800 - 240) \cdot 180}{240 \cdot 1600} = 0.272 \leq 0.497$$

being ω_p^* smaller than 0.497 (maximum value of ω_p^* beneath which steel is in the plastic state according to Table 6.39 for type 4 steel and $f_{ck} \leq 50$ N/mm^2) failure occurs with steel in the plastic state.

It is also possible to identify the theoretical maximum amount $A_{p,tot,\max}$ for prestressing steel corresponding to steel in the plastic state. Assumed $\omega_p^* = \omega_{p,el}^* = 0.497$, by inverting the expression of ω_p^*, $A_{p,tot,\max}$ is given by the following expression

$$A_{p,tot,\max} = \frac{\eta\, f_{cd} \cdot b_{an} \cdot h_p}{f_{pd} \cdot} \cdot \left(\omega_p^* + \frac{A^+}{b_n \cdot h_p}\right)$$
$$= \frac{1.0 \cdot 19.83 \cdot 240 \cdot 1600}{1391} \cdot \left(0.497 + \frac{180 \cdot 150 + (1800 - 240) \cdot 180}{240 \cdot 1600}\right)$$
$$\cong 7109 \text{ mm}^2$$

This result represents an approximate evaluation of $A_{p,tot,\max}$, because the limit value $\omega_{p,el}^* = 0.497$ is equal to $\varepsilon_{p0} - \varepsilon_{p1} = 5‰$, while in the present case $\varepsilon_{p0} - \varepsilon_{p1} = 5.45‰$.

By repeating the calculation for $\varepsilon_{p0} - \varepsilon_{p1} = 5.45‰$, the following results are obtained

$$\varepsilon_{p2,\text{el}} = f_{pd}/E_p - (\varepsilon_{p0} - \varepsilon_{p1}) = 1391/195{,}000 - 0.00545 = 0.00168 = 1.68‰$$

$$\omega_{p,el}^* = 0.8 \cdot 3.5/(1.68 + 3.5) = 0.54$$

$$A_{p,tot,\max} = \frac{\eta\, f_{cd} \cdot b_{an} \cdot h_p}{f_{pd} \cdot} \cdot \left(\omega_p^* + \frac{A^+}{b_n \cdot h_p}\right)$$
$$= \frac{1.0 \cdot 19.83 \cdot 240 \cdot 1600}{1391} \cdot \left(0.54 + \frac{180 \cdot 150 + (1800 - 240) \cdot 180}{240 \cdot 1600}\right)^{36}$$
$$\cong 7344 \text{ mm}^2$$

Ultimate resistant moment

The distance of the compression force centroid measured from the compressed edge is equal to 132 mm[37] and the ultimate resistant moment takes the following value:

$$M_{Rd} = T \cdot (1600 - 132) = 8.179.080 \cdot 1468$$
$$= 12.006.889.440 \text{ Nmm} \cong 12.007 \text{ kNm}$$

[36] The use of the expression of ω_p^*, valid for $x \geq 1.25\,(s + h_r)$, is correct also in the calculation of $A_{p,tot,max}$. Under the hypothesis that, at failure, steel remains in the plastic state even if an area $A_{p,tot,max}$ is adopted and being $A_{p,tot,max} \geq A_{p,tot}$, the translational equilibrium equation gives the following inequality $x_{max} \geq x \geq 1.25(s + h_r)$, where x_{max} is the depth of the neutral axis in the cross-section reinforced with $A_{p,tot,max}$.

[37] For the sake of brevity, the calculation of centroid G_c of the area $A_1 + A_2 + A_3$ of Fig. 6.65 is omitted.

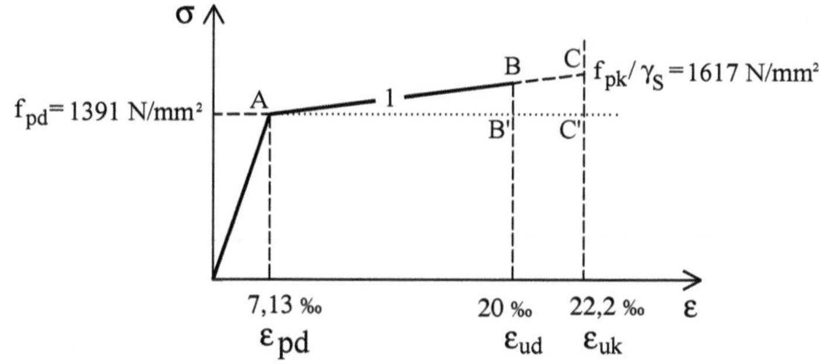

Fig. 6.66 Elastic/strain hardening diagram for prestressing steel

6.11.5 Example 23. Calculation of the Resistant Moment for a Prestressed T Cross-Section Under the Hypothesis of an Elastic/Strain Hardening Stress–Strain Law for Prestressing Steel

Calculate the resistant moment for the midspan section of the beam of Example 22, under the assumption that the stress–strain diagram for prestressing steel is elastic/strain hardening.

Elastic / strain-hardening law for prestressing steel

The similarity of triangles ABB′ and ACC′ of Fig. 6.66 gives

$$\frac{BB'}{AB'} = \frac{CC'}{AC'}$$

$$\frac{f_B - f_{pd}}{\varepsilon_{ud} - \varepsilon_{pd}} = \frac{f_C - f_{pd}}{\varepsilon_{uk} - \varepsilon_{pd}} \quad \rightarrow$$

and, by substituting the numerical values in the expression of f_B,

$$f_B = 1391 + (1617-1391) \cdot (20-7.13)/(22.2-7.13) = 1584\,\text{N/mm}^2$$

taking also into account the relationship[38]

$$\varepsilon_{uk} = \varepsilon_{ud}/0.9 = 20\text{‰}/0.9 = 22.2\text{‰}$$

---

[38] Value $\varepsilon_{ud} = 20\text{‰}$ is recommended by EC2 in [3.3.6(7)], where more accurate values are not known; moreover, EC2 suggests assuming $\varepsilon_{ud} = 0.9\,\varepsilon_{uk}$.

Calculation of $\varepsilon_{p0}-\varepsilon_{p1}$

$$\varepsilon_{p0}-\varepsilon_{p1}=\frac{P_m}{A_{p,tot}\cdot E_p}=\frac{6249000}{5880\cdot 195{,}000}=5.45\text{‰}$$

as already seen in the previous Example.

Ultimate limit strain for prestressing steel

The ultimate limit strain for strands ε_{p2} is obtained by subtracting the strain $(\varepsilon_{p0}-\varepsilon_{p1})$ after the transfer of prestress to concrete, to the ultimate design strain ε_{ud}, after all the initial and time-dependent losses:

$$\varepsilon_{p2}=\varepsilon_{ud}-\left(\varepsilon_{p0}-\varepsilon_{p1}\right)=20-5.45=14.55\text{ ‰}$$

Identification of the type of failure

To identify the type of failure, firstly it is assumed that both materials reach contemporarily their ultimate strains; under this hypothesis, the "stress-block" diagram and the strain diagram at failure are shown in Fig. 6.67.

For $\varepsilon_c=\varepsilon_{cu}$ and $\varepsilon_p=\varepsilon_{p2}$, the distance of the neutral axis measured from the compressed edge is given by

$$x=h_p\cdot\frac{\varepsilon_{cu}}{\varepsilon_{cu}+\varepsilon_{p2}}=1600\cdot\frac{3.5}{3.5+14.55}\cong 310\text{ mm}$$

The depth of the "*stress-block*" diagram is therefore $x' = 0.8\,x = 248$ mm.

The tendon tensile force is given by

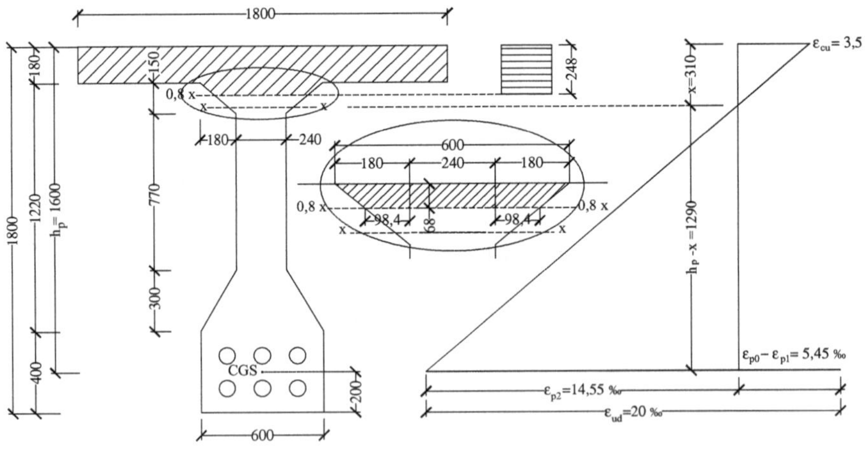

Fig. 6.67 Depth of the neutral axis if concrete and prestressing steel reach contemporarily their ultimate strains

$$T = A_{p,tot} \cdot f_B = 5880 \cdot 1584 = 9313920\,\text{N} \cong 9314\,\text{kN}$$

while the concrete resultant compression force takes the following value (see the detail of Fig. 6.67)

$$C = 19.83 \cdot \{1.800 \cdot 180 + [600 + (240 + 2 \cdot 98.4)] \cdot 68/2\}$$
$$= 712{,}3951\,\text{N} \cong 7124\,\text{kN}$$

which is smaller than T, so the translational equilibrium is not fulfilled.

Being the modulus of T greater than C, failure occurs with concrete at its ultimate limit strain before the steel attains its ultimate strain.

Then it is required to find out if, at failure, prestressing steel has exceeded or not the design yield strain ε_{pd}.

To this purpose, it is required to calculate the depth of the neutral axis corresponding to concrete ultimate strain ε_{cu} at the top and steel strain ε_{p2}, where ε_{p2}, once added to $(\varepsilon_{p0} - \varepsilon_{p1})$, gives the design yield strain ε_{pd} for prestressing steel.

Under this hypothesis, the strain diagram, and the concrete "stress-block" diagram at failure are those shown in Fig. 6.68.

The value of ε_{p2} corresponding to strain ε_{pd} is given by

$$\varepsilon_{p2} = \varepsilon_{pd} - (\varepsilon_{p0} - \varepsilon_{p1}) = 7.13 - 5.45 = 1.68\ ‰$$

the depth x of the neutral axis is equal to

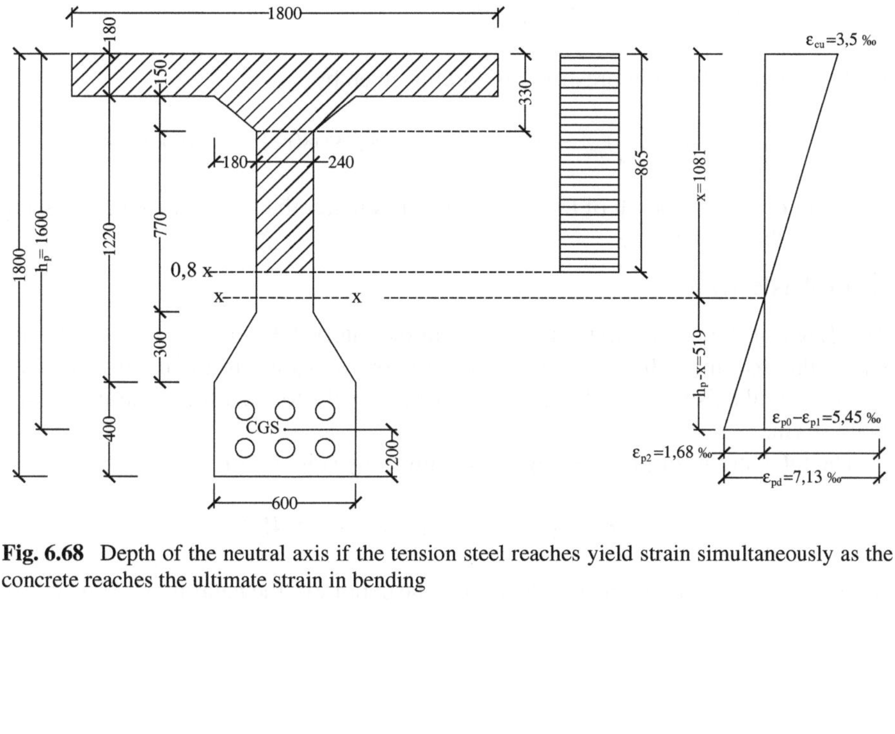

Fig. 6.68 Depth of the neutral axis if the tension steel reaches yield strain simultaneously as the concrete reaches the ultimate strain in bending

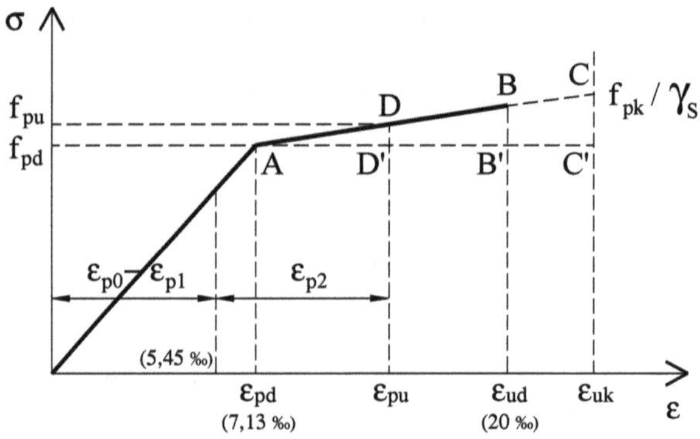

Fig. 6.69 $\sigma - \varepsilon$ diagram with identification of the strain ε_{pu} and of the associated stress f_p

$$x_{el} = h_p \cdot \frac{\varepsilon_{cu}}{\varepsilon_{cu} + \varepsilon_{p2}} = 1600 \cdot \frac{3.5}{3.5 + 1.68} = 1081 \text{ mm}$$

and the depth *x*' of the rectangular "*stress-block*" diagram is given by

$$x' = 0.8\, x_{\text{el}} \cong 865 \,\text{mm}$$

The concrete compression force is given by (Fig. 6.68)

$$\begin{aligned} C &= 19.83 \cdot [1800 \cdot 180 + (600 + 240) \cdot 150/2 + 240 \cdot (865 - 330)] \\ &= 10.220.382 \,\text{N} = 10.220 \text{ kN} \end{aligned}$$

while the tensile force in prestressing steel is given by

$$T = A_{p,tot} \cdot f_{pd} = 5880 \cdot 1391 = 8179080 \text{ N} \cong 8179 \,\text{kN}$$

Being $C > T$, the depth of the neutral axis is smaller than x_{el} and the prestressing steel is in the plastic state.

Neutral axis depth

The depth of the neutral axis is searched in the interval 310 mm $\leq x \leq$ 1081 mm, where the extreme values of the interval are corresponding to the depths assumed by the neutral axis for a steel strain equal, respectively, to the ultimate strain and the yield strain.

The total strain for prestressing steel is given by (Fig. 6.69)

$$\varepsilon_{pu} = \varepsilon_{p2} + (\varepsilon_{p0} - \varepsilon_{p1}) = \varepsilon_{p2} + 5.45\%_{0}$$

where ε_{p2} can be expressed as a function of the depth *x* of the neutral axis (Fig. 6.70).

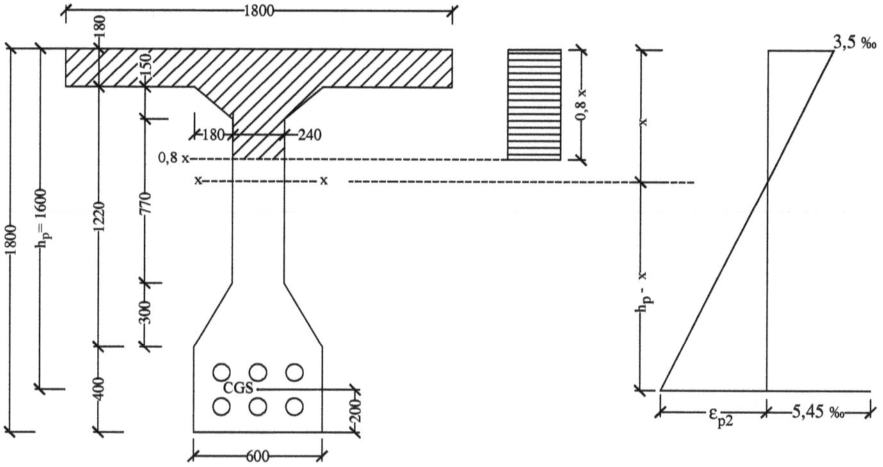

Fig. 6.70 Depth of the neutral axis at failure

$$\varepsilon_{p2} = \frac{h_p - x}{x} 3.5\text{‰}$$

The stress f_p, associated with strain ε_{pu}, can be obtained using the similarity of triangles ADD′ and ABB′ of Fig. 6.69

$$\frac{DD'}{AD'} = \frac{BB'}{AB'} \rightarrow \frac{f_{pu} - f_{pd}}{\varepsilon_{pu} - \varepsilon_{pd}} = \frac{f_B - f_{pd}}{\varepsilon_{ud} - \varepsilon_{pd}}$$

from which it follows that

$$\begin{aligned} f_{pu} &= f_{pd} + \frac{\varepsilon_{pu} - \varepsilon_{pd}}{\varepsilon_{ud} - \varepsilon_{pd}}(f_B - f_{pd}) \\ &= 1391 + \frac{(\varepsilon_{p2} + 5.45\text{‰}) - 7.13\text{‰}}{20\text{‰} - 7.13\text{‰}}(1584 - 1391) \end{aligned}$$

By inserting in the expression of f_{pu}, the relationship which expresses ε_{p2} as a function of $x[\varepsilon_{p2} = (h_p - x) \cdot 0.0035/x]$, the tendon tensile force is also expressed as a function of x $[T(x) = A_{p,tot} \cdot f_{pu}(x)]$; for the sake of brevity, the complete expression of $T(x)$ is omitted.

Under the hypothesis that $x \geq 1.25 \cdot 330 = 412.5$ mm, i.e. the "*stress-block*" diagram is extended beneath the fillets between the web and the flange, the concrete compression force is given by:

$$C = f_{cd} \cdot [1800 \cdot 180 + (600 + 240) \cdot 150/2 + 240 \cdot (0.8\,x - 330)]$$

By making equal T and C, the equation for the calculation of x is obtained, which, iteratively solved, gives

$$x \cong 631\,\text{mm} > 412.5\,\text{mm}$$

therefore, the hypothesis that $x \geq 412.5$ mm, assumed in writing the expression of C, is confirmed.

By substituting the calculated value of x in the expression of C or T, their common value is finally obtained

$$T = C \cong 8505\,\text{kN}$$

Prestressing steel strain ε_{p2} is given by

$$\varepsilon_{p2} = \frac{(h_p - x)}{x} 3.5^0/_{00} = \frac{(1600 - 631)}{631} 3.5^0/_{00} = 5.37^0/_{00}$$

and the total steel strain is equal to

$$\varepsilon_{pu} = \varepsilon_{p2} + (\varepsilon_{p0} - \varepsilon_{p1}) = 5.37\,^0/_{00} + 5.45^0/_{00} = 10.82^0/_{00}$$

Calculation of the ultimate resistant moment

The distance x_G of the compression force C measured from the compressed edge is equal to $x_G = 145$ mm and the ultimate resistant moment takes the following value

$$\begin{aligned} M_{Rdu} &= T \cdot (h_p - x_G) = 8{,}505{,}000 \cdot (1600 - 145) \\ &= 12{,}374{,}775{,}000\ \text{Nmm} \cong 12{,}375\,\text{kNm} \end{aligned}$$

Remark. The difference with the value of M_{Rdu} obtained by assuming an elastic-perfectly plastic material law (see Example 22) is slightly greater than 3%.

6.12 Anchorage Length for Pre-tensioned Tendons

In the anchorage of pre-tensioned tendons, it is necessary to consider two different situations characterized by different bond conditions due to transversal deformation of prestressing steel due to the Poisson effect.

At the transfer of the prestress, tendons experience a stress decrease and a diameter increase, carrying out a radial pressure on the concrete around (Hoyer effect), while, at ULS, following the stress increase due to ultimate loads, their diameter decreases and, contemporarily, bond strength decreases.

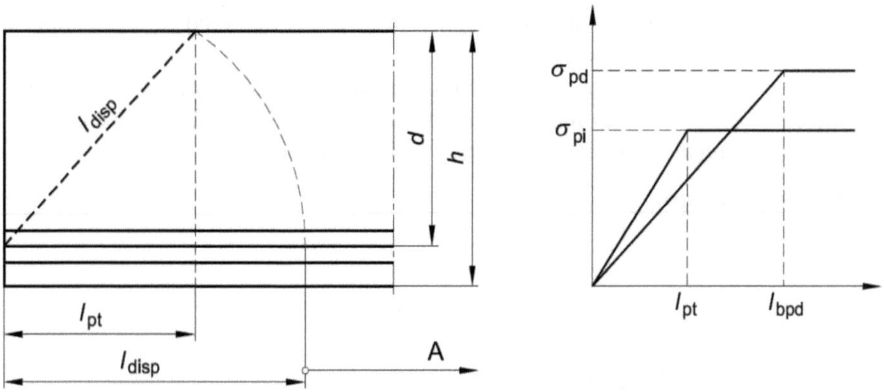

Fig. 6.71 Transfer of prestress in pre-tensioned tendons; length parameters for pre-tensioned steel (d is the distance of the centroid of pre-tensioned tendons measured from the upper edge, A: linear stress distribution in member cross-section) [Fig. 8.16]

Different values of the transmission length of the prestress force to the concrete are corresponding to these two different situations, lower at the transfer of the prestress and higher at ULS.

For the anchorage zone of pre-tensioned tendons, EC2 defines the following quantities (Fig. 6.71):

- transmission length, l_{pt}, required to the full transfer of prestress to concrete.
- dispersion length, l_{disp}, required to disperse concrete stresses gradually to a linear distribution across the cross-section,
- anchorage length at ULS, l_{bpd}, required to fully anchor to concrete the tendon force F_{pd} at the ultimate limit state.

6.12.1 Transfer of Prestress

Bond strength at the transfer of prestress

At the cutting or release of tendons, the prestress may be assumed to be transferred to concrete by a constant bond stress f_{bpt}, defined as

$$f_{bpt} = \eta_{p1}\eta_1 f_{ctd}(t) \text{ (bond strength)} \quad [(8.15)]$$

where η_{p1} is a coefficient that considers the type of tendon and the bond situation at release.

η_{p1} = 2.7 for indented wires,
= 3.2 for 3 and 7-wire strands,

η_1 = 1.0 for good bond conditions (see Sect. 12.1.4—Chap. 12),

= 0.7 otherwise, unless a higher value can be justified regarding special circumstances during the execution

$f_{ctd}(t) = \alpha_{ct}\, 0.7\, f_{ctm}(t)/\gamma_C$ (design tensile stress at time of prestress transfer).

Values of η_{p1} other than those given above may only be used if indicated in the European Technical Approval of the prestressing system.

Transmission length, l_{pt}

The basic value of the transmission length l_{pt} is given by

$$l_{pt} = \alpha_1 \alpha_2 \phi\, \sigma_{pm0}/f_{bpt} \quad [(8.16)]$$

where α_1 = 1.0 for gradual release,
= 1.25 for sudden release,
α_2 = 0.25 for tendons with circular cross-section,
= 0.19 for 3 and 7-wires strands,
ϕ = nominal diameter of tendon,
σ_{pm0} = tendon stress just after release,
f_{bpt} = bond strength.

The design value of the transmission length should be taken as the less favourable of two values, depending on the design situation

$$l_{pt1} = 0.8\, l_{pt}$$
$$l_{pt2} = 1.2\, l_{pt}$$

where normally the lower value is used for verifications of local transversal stresses at release, in the area of the anchorage, while the higher value is used for ULS (e.g. shear), as already mentioned at the beginning of the Section.

EC2 does not give any further indication for the calculation of transversal tensile stresses in the anchorage zone of pre-tensioned members. The behaviour is like post-tensioned tendons, but the length of the local zone (defined in Sect. 6.13) is greater.

Dispersion length

The dispersion length may be calculated with the following formula (Fig. 6.71)

$$l_{\text{disp}} = \sqrt{l_{pt}^2 + d^2}$$

As already mentioned, out of the dispersion length concrete stresses are assumed to have a linear distribution over the cross-section.

6.12.2 Anchorage of Pre-tensioned Tendons at ULS

The verification is required only in those cross-sections where the concrete tensile stress σ_{ct} exceeds the value $f_{ctk,0.05}$. Therefore, this check is usually concerning tendons not bonded to concrete at the end sections of the beam, but only at an assigned distance measured from the support.

This situation can occur, for example, in the case of bonded pre-tensioned tendons with a linear profile. If the prestress force should be modulated along the beam axis, at the end segments, debonding sleeves are placed around tendons; ducts can be maintained for a segment Δx_1 for the first group of strands, for a segment Δx_2 for the second group of strands, and so on (Fig. 6.72).

The anchorage of strands of the first group is therefore starting in the cross-section where the bending moment M_1 acts, while the anchorage of the section group is starting in the cross-section where the bending moment is equal to M_2, and so on. Bending moments M_1 and M_2 may induce tensile stresses greater than $f_{ctk,0.05}$, depending on the distances Δx_1 and Δx_2 measured from the support and on the value of the external loads at the transfer of prestress.

Bond strength at ULS

The bond strength at ULS is given by

$$f_{bpd} = \eta_{p2}\eta_1 f_{ctd} \quad [(8.20)]$$

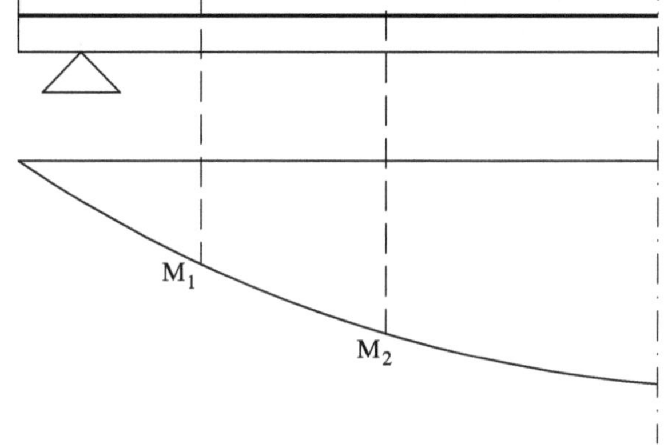

Fig. 6.72 Length Δx_1 and Δx_2 of debonding sleeves placed around some tendons

where η_{p2} = 1.4 for indented wires,
η_{p2} = 1.2 for 7-wire strands,
η_1 = 1.0 in good bond conditions (see Sect. 12.1.4 of Chap. 12),
= 0.7 otherwise, unless a higher value can be justified regarding special circumstances during the execution,

$$f_{ctd} = \alpha_{ct} \cdot f_{ctk,0.05}/\gamma_C \quad [(3.16)]$$

$f_{ctk,0.05}$ value shall be limited to the value of concrete C60/75, for which $f_{ctk,0.05}$ = 3.1 N/mm^2, to consider the possibility of brittle fracture for high strength concretes.

Anchorage length at ULS

The total anchorage length for a strand under the design stress σ_{pd} is given by

$$l_{bpd} = l_{pt2} + \alpha_2 \cdot \phi \cdot (\sigma_{pd} - \sigma_{pm\infty}), \quad [(8.21)]$$

where

$$l_{pt2} = 1.2\, l_{pt},$$

α_2 is the same coefficient used in the definition of the transmission length,
σ_{pd} is the strand stress corresponding to the force calculated for a cracked cross-section, including the shear effect according to [6.2.3 (7)],
$\sigma_{pm\infty}$ is the long-term stress in strands after all the losses have occurred.

The stress variation along the anchorage zone is shown in Fig. 6.73.

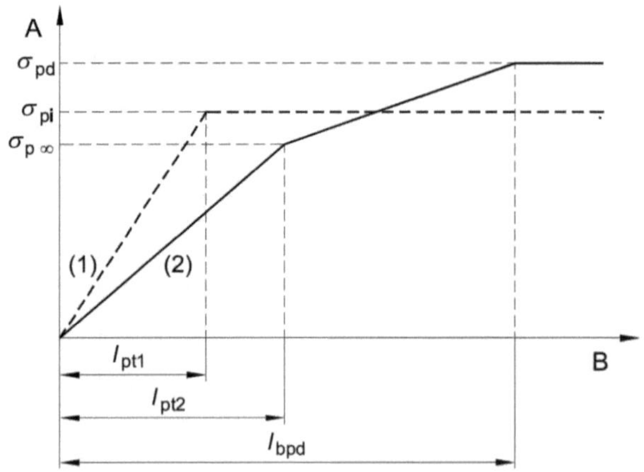

Fig. 6.73 Stress variation in the anchorage zone for pre-tensioned members: (1) at the release of tendons, (2) at ULS (A = stress in pre-tensioned tendons, B = distance measured from the end-section) [Fig. 8.17]

6.12.3 Example 24. Calculation of Transmission Length, Dispersion Length and Anchorage Length for Pre-tensioned Tendons

Calculate the transmission length for the pre-tensioned beam of Example 19, under the hypothesis of a sudden release of tendons, assuming good bond conditions.

Materials

Concrete C50/60, made with Class R cement

$$f_{ck} = 50\,\text{N/mm}^2$$

$$E_{cm} = 37000\,\text{N/mm}^2$$

Reinforcing steel Class C

$$f_{yk} = 450\,\text{N/mm}^2$$

Span L of the beam 20.00 m.

Loads

$$\begin{array}{lll} \text{Self}-\text{weight of the beam} & G_{k1} & = 4.60\,\text{kN/m} \\ \text{Permanent loads} & G_{k2} & = 5.20\,\text{kN/m} \\ \text{Variable load (snow)} & Q_{k1} & = 7.20\,\text{kN/m} \end{array}$$

$$q_{Ed} = 1.3 \cdot (4.60 + 5.20) + 1.5 \cdot 7.20 = 23.54\,\text{kN/m}$$

$$R_{Ed} = q_{Ed} \cdot L/2 = 235.4\,\text{kN (support reaction)}$$

Following quantities are calculated

- bond strength
- transmission length
- dispersion length
- anchorage length at ULS.

Bond strength at prestress transfer

$$f_{bpt} = \eta_{p1}\eta_1 f_{ctd}(t)\ \text{(bond strength)} \quad [(8.15)]$$

with

η_{p1} = 3.2 for 3 and 7-wire strands,
η_1 = 1.0 in good bond conditions,
$f_{ctd}(t)$ $= \alpha_{ct} 0.7\, f_{ctm}(t)/\gamma_C$, is the design tensile strength at time of release, where

$$f_{ctm}(t) = [\beta_{cc}(t)]^{\alpha} \cdot f_{ctm}$$

$$\alpha = 1 \text{ for } t < 28 \text{ days},$$

$$\beta_{cc}(t) = \exp\left\{s\left[1 - \left(\frac{28}{t}\right)^{1/2}\right]\right\} = \exp\left\{0.20\left[1 - \left(\frac{28}{2.32}\right)^{1/2}\right]\right\} \cong 0.61 \quad [(3.2)]$$

being

s = 0.20 for Class R cement.
t = 2.32 days (from Example 16).
f_{ctm} = 4.1 N /mm^2 from [Table 3.1] for concrete C50/60.

following values are obtained

$$f_{ctm}(t) = [\beta_{cc}(t)]^{\alpha} \cdot f_{ctm} = 0.61 \cdot 4.1 = 2.5 \text{ N/mm}^2$$

$$\begin{aligned} f_{ctd}(t) &= \alpha_{ct} 0.7 f_{ctm}(t)/\gamma_C \\ &= 1 \cdot 0.7 \cdot 2.5/1.5 \cong 1.17 \text{ N/mm}^2 \quad [3.1.6(2)\text{P}] \end{aligned}$$

and therefore

$$f_{bpt} = \eta_{p1}\eta_1 f_{ctd}(t) = 3.2 \cdot 1.0 \cdot 1.17 = 3.74 \text{ N/mm}^2$$

Transmission length

The basic transmission length l_{pt} is given by

$$l_{pt} = \alpha_1 \alpha_2 \phi \, \sigma_{pm0}/f_{bpt} \quad [(8.16)]$$

where, in the present case, coefficients take following values:

α_1 = 1.25, for sudden release,
α_2 = 0.19, for 7-wire strands,
ϕ = 15 mm,

$\sigma_{pm0} = \frac{P_{m0}}{A_p} = \frac{176048}{140} = 1257$ N/mm^2 ($P_{m0} = P_4$ from Example 19).

Substituting the numerical values of all parameters, the following value of l_{pt} is obtained:

$l_{pt} = \alpha_1\alpha_2\phi \, \sigma_{pm0}/f_{bpt} = 1.25 \cdot 0.19 \cdot 15 \cdot 1257/3.74 = 1197$ mm.

The design value of the transmission length is equal to the most disadvantageous of the following values:

$$l_{pt1} = 0.8 \cdot l_{pt} = 958 \text{ mm}$$

$$l_{pt2} = 1.2 \cdot l_{pt} = 1436 \text{ mm}$$

where the first value is used for the local verification at release (better bond conditions) and the second value is used for the ULS verifications (worse bond conditions).

Dispersion length

$$l_{\text{disp}} = \sqrt{l_{pt}^2 + d^2} = \sqrt{1197^2 + (1000 - 148.75)^2} = 1469 \text{ mm}$$

calculation of l_{disp} has been performed at the level of the tendon centroid, at a distance equal to 148.75 mm measured from the lower edge (see Fig. 6.49).

Bond strength at ULS

The bond strength at ULS is given by

$$f_{bpd} = \eta_{p2} \cdot \eta_1 \cdot f_{ctd} = 1.2 \cdot 1.0 \cdot 1.93 = 2.316 \text{ N/mm}^2 \quad [(8.20)]$$

where, in the present case,

η_{p2} = 1.2 for 7-wire strands,
η_1 = 1.0 for good bond conditions,

$$f_{ctd} = \alpha_{ct} \cdot f_{ctk,0.05}/\gamma_C = 1 \cdot 2.9/1.5 = 1.93 \text{ N/mm}^2 \quad [(3.16)]$$

($f_{ctk,0.05} = 2.9$ N/mm^2 from [Table 3.1] for concrete C50/60)

Anchorage length at ULS

The total anchorage length for a strand with stress σ_{pd} is given by

$$l_{bpd} = l_{pt2} + \alpha_2 \cdot \phi \cdot \left(\sigma_{pd} - \sigma_{pm\infty}\right)/f_{bpd}$$

where:

l_{pt2} $= 1436$ mm, $\alpha_2 = 0.19$ for 7-wire strands,
σ_{pd} is the tendon stress corresponding to the load calculated in a cracked cross-section, including the shear effect according to [6.2.3 (7)],
$\sigma_{pm\infty}$ $= P/A_p = 152388/140 = 1088$ N/mm^2 (where $P = 152{,}388$ N is the value of prestress force after an infinite amount of time calculated at the end of Example 19).

Internal forces in the cross-section placed at the end of transmission length l_{pt2} are equal to:

$$V_{Ed} = R_{Ed} - q_{Ed} \cdot l_{pt2} = 235.4 - 23.54 \cdot 1.436 = 201.6 \text{ kN}$$

$$M_{Ed} = R_{Ed} \cdot l_{pt2} - q_{Ed} \cdot l_{pt2}^2/2$$
$$= 235.4 \cdot 1.436 - 23.54 \cdot 1.436^2/2 = 313.76\,\text{kNm}$$

and the concrete compressive stress at the bottom is given by.

$$\sigma_c = \frac{P_{tot}}{A_c} + \frac{P_{tot} \cdot z_{cp}}{I_c} \cdot z_{\inf} - \frac{M_{Ed}}{I_c} \cdot z_{\inf}$$
$$= \frac{1{,}219{,}104}{184{,}000} + \frac{1{,}219{,}104 \cdot 351.25}{2.3887 \times 10^{10}} \cdot 500 - \frac{313.76 \times 10^6}{2.3887 \times 10^{10}} \cdot 500 \text{; where } P_{tot}$$
$$= 9.02\,\text{N/mm}^2$$

$= 8\,P$.

By repeating the calculation in other cross-sections placed at distances smaller than l_{pt2} from the end-section, the compressive stress is never greater than $f_{ctk,0.05}$, therefore the verification of the anchorage is not required [8.10.2.3 (1)].

6.13 Anchorage Zones of Post-tensioned Members

In the anchorage zone of post-tensioned members, the concrete behind anchorage plates is subjected to high compressive stresses.

Moreover, the dispersion of the prestress inside the member induces high tensile transversal forces both in a localized zone ("local zone") immediately behind each anchorage and in a larger zone ("general zone"), where load spreads over the whole cross-section, generating a linear stress distribution (Fig. 6.74).

Concrete in the local zone shall support high compression stresses due to anchorage and transfer those stresses to the general zone. The local zone is defined as the concrete volume placed immediately around and behind the anchorage plate with transverse dimensions a and a', equal to twice the minimum distance of the tendon axis measured from the free edge or equal to the distance of the tendon axis measured from an adjacent tendon. The length l of the local zone in the tendon direction is equal to the maximum transversal dimension [$l = \max(a; a')$].

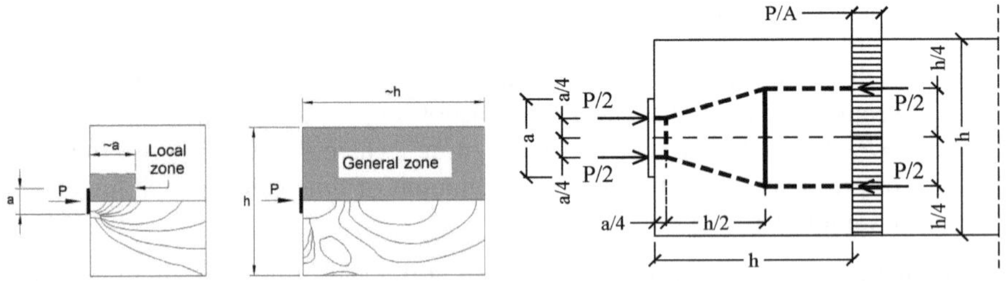

(a) local zone (b) general zone (c) truss scheme for the general zone

Fig. 6.74 Local and general zones for a post-tensioned anchorage

According to Annex J of EN1992-2 (Concrete bridges), the transversal dimensions of the local zone c and c' in the plane of the anchorage plate shall fulfil the following condition

$$P_{\max}/(c \cdot c') \leq 0.6\, f_{ck}(t)$$

where $P_{\max}$ is the maximum tendon prestress and $f_{ck}(t)$ is the concrete characteristic compressive strength at time t of tensioning. It is worth noting that, because of the presence of the duct hole, the actual compressive stress is greater than the calculated stress. It is also required that c and c' fulfil a geometrical condition, i.e., the shapes of the cross-section of the local zone and the plate are homothetic.

The behaviour of the local zone is strongly influenced by the characteristics of anchorage devices and the layout of confining reinforcement, which is usually provided in the shape of helical spirals. For cross-sections with small heights, as slabs, local and general zones can have the same depth in the vertical plane.

In general, the designer does not have to perform verifications in the local zone, where the producer of the prestressing system provides special transverse reinforcement and indicates the minimum design distances measured from the edge and between adjacent anchorage devices.

According to Saint Venant's principle, the general zone is extended for a length equal to the cross-section depth. Within this length, the three-dimensional stress field can be assessed through the discretization of internal forces with a truss made of tensile and compressed members ("strut-and-tie model"[39]).

There are also further tensile stresses in the proximity of the end-section, where anchorage plates are fixed; those stresses are indicated as surface stresses or *spalling* stresses. In the case of a single central tendon, they are not correlated to equilibrium conditions, but only to compatibility conditions of strains between the loaded zone and adjacent zones, therefore they take small values. This result is confirmed by the distribution of stresses which is obtained if lateral zones adjacent to the anchorage plate are eliminated: the stress field is almost the same with or without those zones (Fig. 6.75b, c).

For more than one tendon, compressed areas between two adjacent anchorages are interested by *spalling* stresses greater than those present in the case of a single anchorage; moreover, they increase as the distance between anchorage plates increases.

To prevent the *spalling* phenomenon in the proximity of anchorage plates, it is also possible to adopt a suitable reinforcement capable to resist a tensile force equal to 3% of the maximum prestress load $P_{\max}$, multiplied by the partial coefficient $\gamma_{P,\text{unfav}} = 1.2$,

$A_{s,\text{spalling}} = 0.03 \cdot (1.2 \cdot P_{\max})/f_{yd}$[40]

---

[39] See Chap. 10 for the general discussion about design of reinforced concrete elements using strut-and tie ("*S&T*") models.

[40] The minimum area of surface reinforcement is given in Annex J to EN1992-2 (Concrete bridges), which also provides the minimum area of reinforcement to prevent bursting and spalling of the

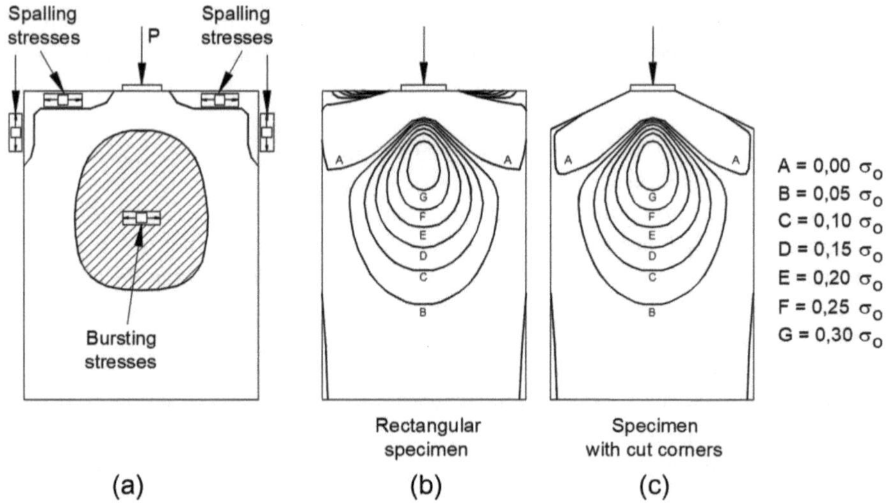

Fig. 6.75 *Spalling* stresses for one centred prestressing tendon (Burdet, 1990): **a** identification of the areas with *spalling* stresses or "bursting" stresses, **b** evolution of σ_0 stress orthogonal to the tendon, in presence of concrete lateral zones, **c** evolution of $\sigma_{0\ \mathrm{stress}}$ in absence of concrete lateral zones

Very high values of *spalling* stresses are also registered when the prestress force has a high eccentricity (Fig. 6.76). In this case, differently from the case of a single central tendon, tensile *spalling* stresses are needed to satisfy equilibrium conditions and can be obtained through the scheme shown in Fig. 6.77, taken from Model Code 1990.[41]

Figure 6.76 shows a member end-section where prestress is applied with very high eccentricity; at the end of transmission length, a linear diagram of normal stresses on the cross-section is assumed. For a very high eccentricity of the prestress force, this diagram has a non-symmetrical butterfly shape, therefore it exists a cross-section 1-2 where shear stresses are zero. The resultant spalling force can be evaluated by writing the equilibrium conditions for element 1234, placed above section 1-2 and between the end-section and section 2-4 at the end of transmission length. The translational equilibrium in the horizontal direction implies that horizontal forces applied on the face 2-4 have to be self-balanced, being the only horizontal forces acting on the element 1234. Therefore, on face 2-4 normal stresses only give a non-zero bending moment M. Face 1-2 is unloaded, and face 3-4 is also unloaded if eventual external loads applied on the top of the beam are neglected. Therefore,

concrete in each regularization prism (or local zone), to be distributed in each direction over the length of the prism: $A_s = 0.15\cdot(1.2\cdot P_{max})/f_{yd}$.

[41] For small depth elements, as hollow-core slabs, this model overestimates the value of "spalling" stresses; for these elements, Model Code 1990 provides a stress diagram of spalling stresses variable with eccentricity and transmission length, based on the linear analysis of elements with depth not greater than 400 mm.

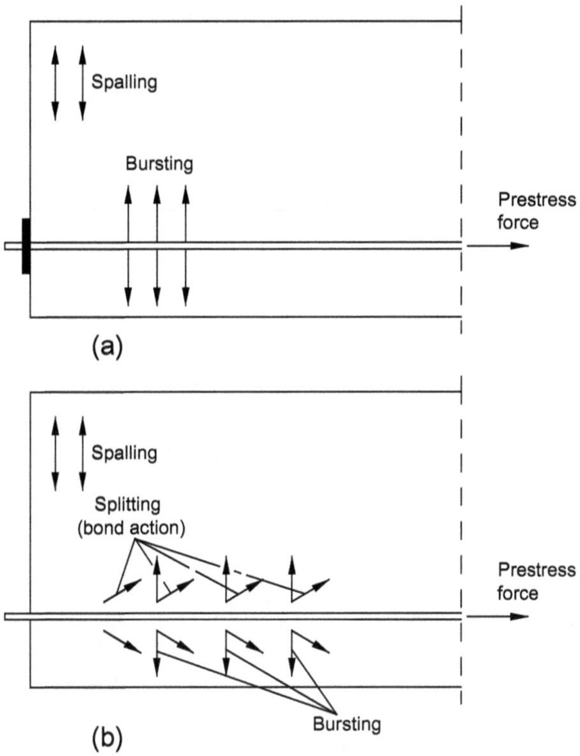

Fig. 6.76 Spalling, splitting, and bursting forces for an eccentric prestressing tendon (Model Code 1990)

Fig. 6.77 Evaluation of “spalling” forces due to an eccentric prestressing tendon (Model Code 1990)

the rotational equilibrium of element 1234 is fulfilled only if the lower face 1-2 is subjected to a moment M equal and opposite to the moment acting on the face 2-4. Finally, under the assumption that the internal lever arm equals half the transmission length, the resultant *spalling* force is given by the following ratio

$$F_{sp} = \frac{M}{0.5\,l}$$

6.13.1 "Bursting" Stresses and Anti-burst Reinforcement

Transverse tensile stresses due to the spread of prestress are known as *bursting* stresses and they are present both in the vertical and horizontal plane.

The analysis of *bursting* stresses can be performed separately within the local prisms of each anchorage and then in the general zone; in small depth elements like slabs, the two zones tend to overlap themselves.

The reinforcement provided to absorb *bursting* stresses, also known as anti-burst reinforcement, is made of spiral helices and/or closed stirrups in the local zone and by stirrups and/or bent bars in the general zone (Fig. 6.78a). The helical reinforcement for the local zone is provided directly by the producer of the prestressing system, who indicates the diameter, spacing, and length of reinforcement. In some situations, as in small depth slabs, limitations on the minimum cover do not leave enough space to place spiral helices. In this case, a conventional reinforcement is used in both directions: it enhances the concrete compressive strength by confinement and absorbs transverse tensile stresses (Fig. 6.78b).

In the following, the main steps for the design of anti-burst reinforcement in the general zone are described.

The prestress dispersion within the general zone can be analysed through a strut-and-tie model S&T (*Strut and Tie*) (see Chap. 10), whose geometry depends on the number and position of anchorage devices. For a given value of the prestress force, tensile stresses due to prestress dispersion decreases as the number of anchorage devices increases. In the theoretical case of an infinite number of anchorage devices, the prestress force would be applied uniformly to the cross-section, so "bursting" stresses would be zero and no anti-burst reinforcement would be required.

Section 6.13.1.2 deals with the method of the equivalent deep beam, introduced by Magnel (1954), which is less approximated than the strut-and-tie model.

Remark. Prestress dispersion happens both in the vertical and horizontal planes. All the following schemes concern the dispersion in the vertical plane; the dispersion in the horizontal plane is accounted for by considering, for each tendon, the S&T model sketched in Sect. 10.6.1.1 of Chap. 10.

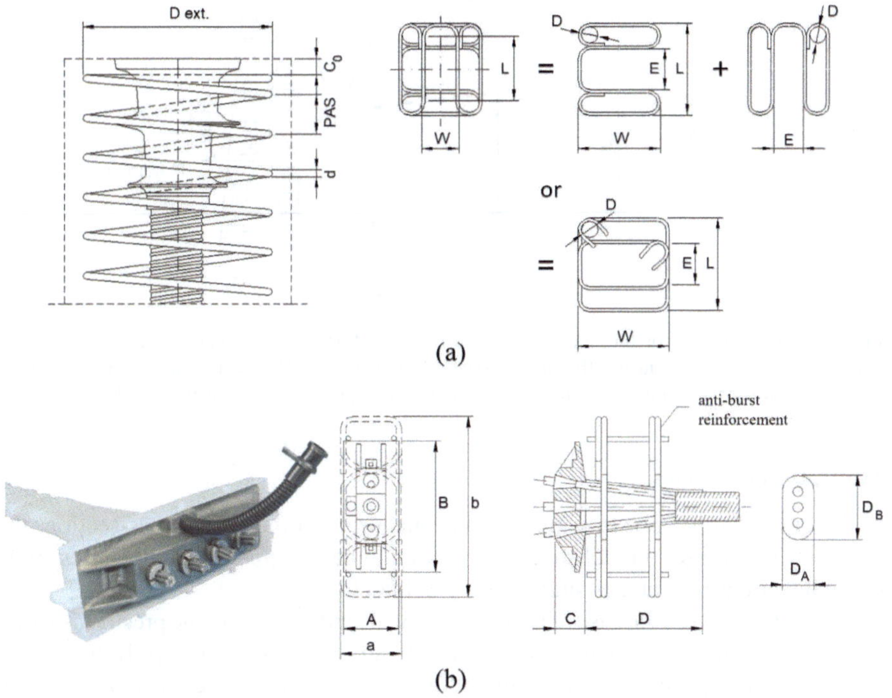

Fig. 6.78 Examples of anti-burst reinforcement in the local zone of a post-tensioned tendon for **a** beam and **b** slab

6.13.1.1 Design of Anti-burst Reinforcement Through S&T Model

Single anchorage

For the anchorage of a single centred tendon parallel to the member axis, it is possible to use the S&T model of Sect. 10.6.1.1 of Chap. 10. To this purpose, it is worth noting that EC2 suggests the adoption of a dispersion angle of 33.7°. By adopting the model described in Chap. 10, the spread angle is varying with the ratio between the heights of the anchorage plate and cross-section. If the anchorage plate is placed eccentrically, it is possible to refer to the S&T model of Sect. 10.6.2.

If the tendon is centred but inclined at an angle α to the horizontal (Fig. 6.79), both a shear force $P{\cdot}\sin\alpha$ and a bending moment $P{\cdot}\sin\alpha{\cdot}h$ are present on the cross-section placed at the distance h measured from the end-section, where h is the beam height. It follows that the normal stress distribution on this cross-section is not constant anymore, but trapezoidal, therefore the geometry of the S&T model can be identified by subdividing the stress diagram into two parts, each of them with resultant force $P/2$, and by applying the same procedure described in the following section for multiple anchorages.

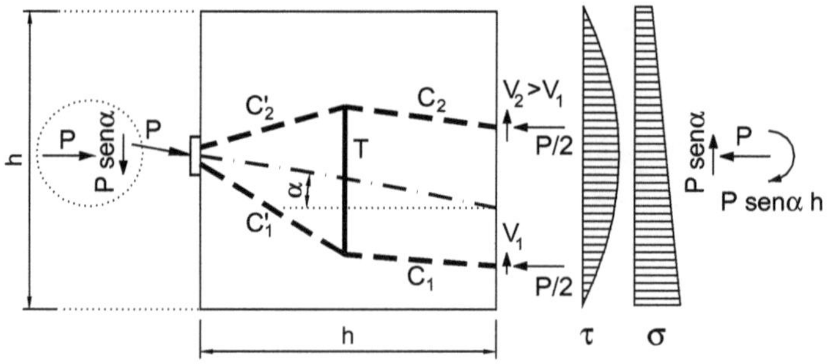

Fig. 6.79 S&T model for the anchorage zone of an inclined centred tendon; struts C_1 and C_2 have different slopes because different transverse forces V_1 and V_2 are assigned to the two struts depending on the parabolic distribution of shear stresses (dashed lines = struts, solid lines = ties)

Multiple anchorages

In the case of more prestressing tendons anchored on the same cross-section, the geometry of the S&T model should be chosen to better approximate the stress field in the linear elastic state, as suggested by Schlaich.[42]

Figure 6.80 shows the case of two symmetrical tendons, where the prestress dispersion can be analysed using the S&T model already described in Chap. 10 for a short deep beam. The position of the resultant dispersion force changes with the distance of the two prestress forces: if the spacing between the two forces is greater than half the cross-section height, the tie is placed immediately behind the anchorage plates (Fig. 6.80b), otherwise, the tie is placed far away from the end-section (Fig. 6.80a).

In the case of n anchorages, the S&T model can be obtained through the following procedure, valid also for deep beams, as already described in Chap. 10 (Fig. 6.81):

- on the cross-section placed at distance h measured from the end-section (h = height of the section), the diagram of normal stresses is subdivided into n parts, so each part is in equilibrium with the prestress force of the corresponding tendon;
- resultant forces are drawn for each part until the intersection in n points (A_1, A_2, ..., A_n) with the cross-section placed at the distance 0.5 h from the cross-section where the centroids G_i of local zones are placed;
- each points A_i is connected with the centroid G_i of the corresponding local zone;
- normal forces in struts and ties are calculated, and the anti-burst reinforcement is designed, to be distributed over a 0.6 h long segment, centred on the tie of the S&T model (at the distance $a/4 + h/2$, where a is the width of anchorage plates); normally, the anti-burst reinforcement is extended to the whole general zone, with total length h.

The values of the prestress force and concrete tensile strength to adopt for local verification of anchorage zones are discussed in Sect. 6.4.2.

[42] See Chap. 10 for the Strut & Tie method for the design of non-slender reinforced concrete elements.

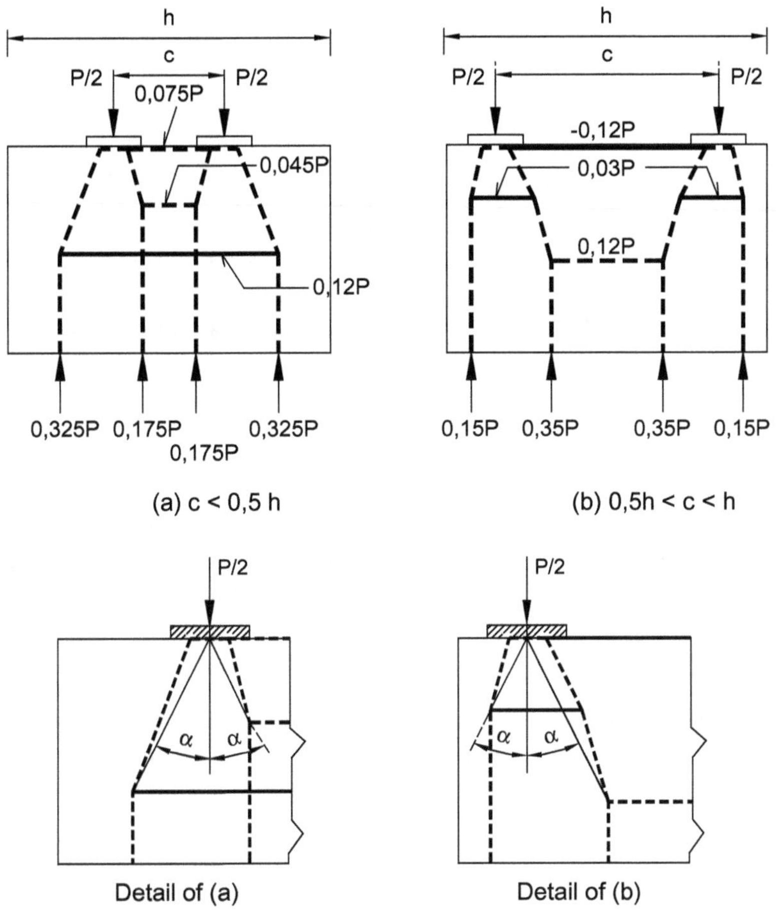

Fig. 6.80 Identification of S&T models for two symmetrical anchorages following isostatic lines obtained through the elastic analysis in the uncracked state: **a** $c < 0.5\ h$, **b** $0.5\ h < c < h$, where c is the distance between anchorages and h is the height of the cross-section (dashed lines = struts, solid lines = ties)

6.13.1.2 Design of Anti-burst Reinforcement Through the Method of the Equivalent Deep Beam

As already mentioned, an alternative method, but less approximated, to evaluate tensile bursting stresses in the dispersion zone consists of considering the anchorage zone as a deep beam (with the axis in the vertical direction) subjected to balanced horizontal forces, which consist of prestress forces transmitted by anchorage plates on one side and a linearly distributed load on the other side.

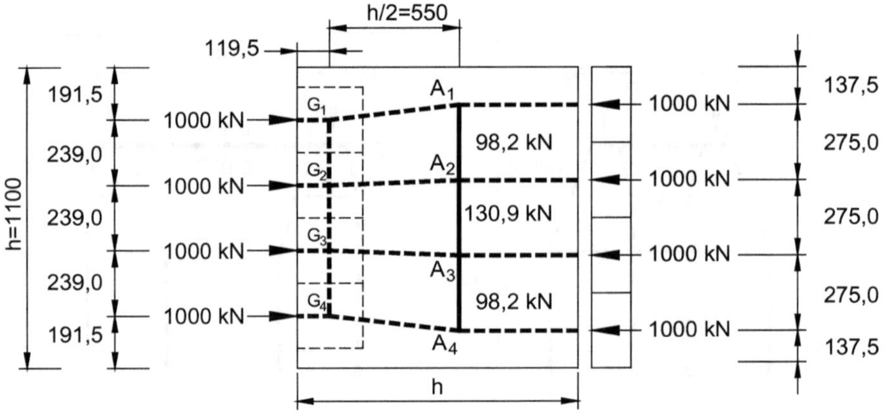

Fig. 6.81 Example of S&T model for multiple anchorages (no. 4 tendons) (dashed lines = struts, solid lines = ties)

The axis of the deep beam is orthogonal to the prestressed beam axis; its height is coincident with the length of the general zone, i.e. with the dispersion length, while its length is equal to the section height.

Single anchorage

Figure 6.82 shows the equivalent deep beam for the anchorage of a single centred tendon. The bending moment required for the rotational equilibrium is given by (Fig. 6.83):

$$M_b = \frac{P}{2} \cdot \left(\frac{h}{4} - \frac{a}{4}\right) = \frac{P}{8} \cdot (h - a)$$

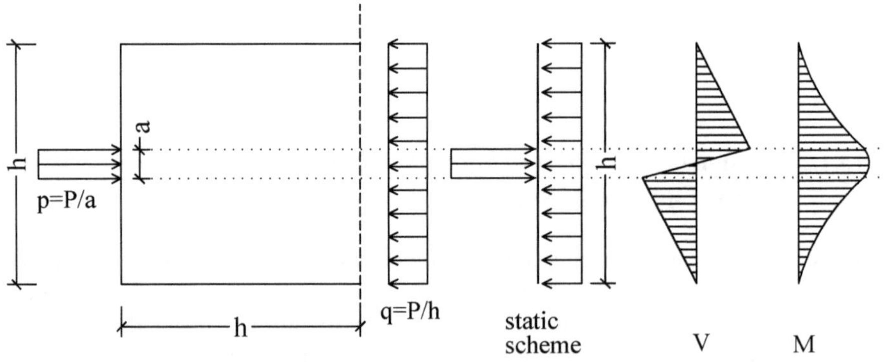

Fig. 6.82 Equivalent deep beam for a single centred anchorage

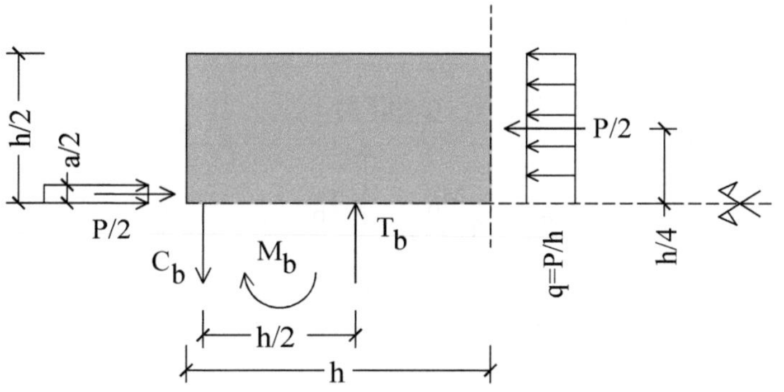

Fig. 6.83 Bending moment M_b in the midspan of the equivalent deep beam for a single centred anchorage

In the elastic uncracked state, the lever arm between C_b and T_b is equal to 0.5 h (Fig. 6.83), which is a good approximation of the lever arm also in the cracked state, therefore.

$$T_b = \frac{M_b}{h/2} = \frac{\frac{P}{8} \cdot (h-a)}{h/2} = \frac{P}{4} \cdot \left(1 - \frac{a}{h}\right)$$

which is the same result of the S&T model (Fig. 6.74c) (formula [(6.58)] of EC2) (see also Sect. 10.6.1.1).

Two symmetrical anchorages

Considering the equivalent deep beam loaded by two prestress forces P placed symmetrically at spacing c, the bending moment diagram is different for values of c, smaller, equal or greater than 0.5 h, being h the height of the cross-section (Fig. 6.84).

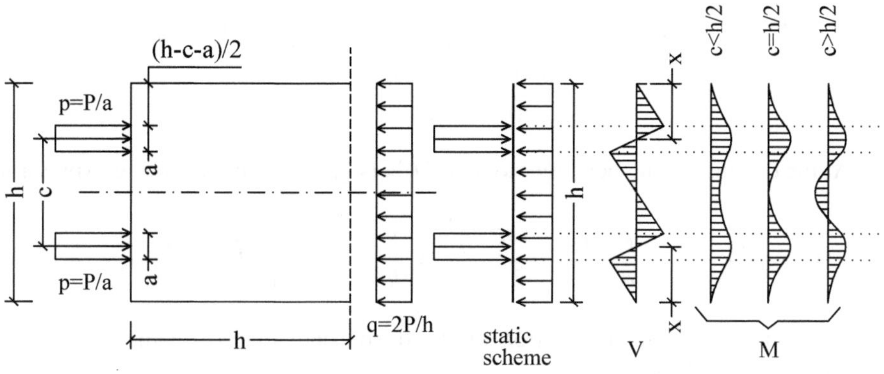

Fig. 6.84 Equivalent deep beam for two symmetrical anchorages

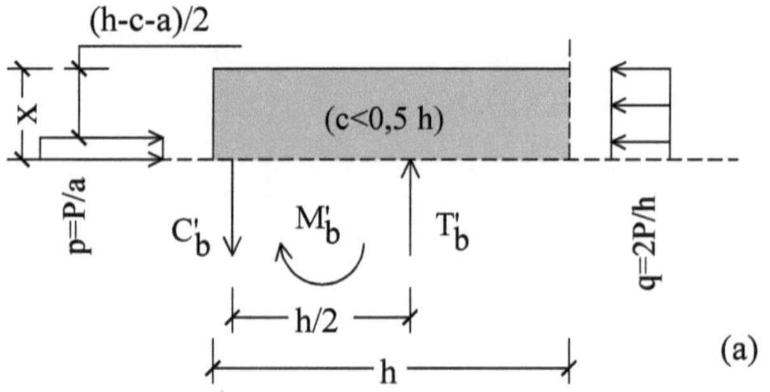

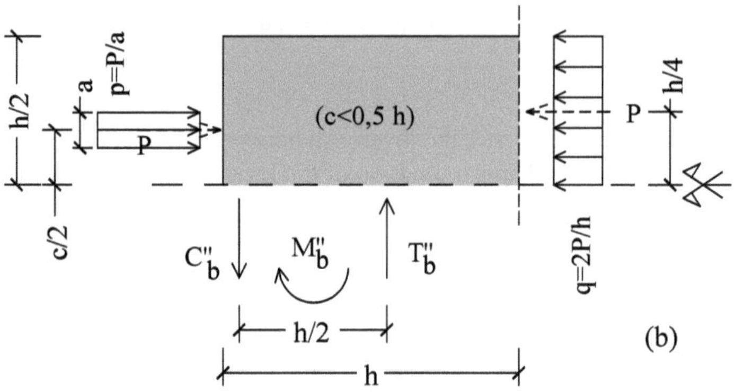

Fig. 6.85 "Bursting" bending moments for the equivalent deep beam of two concentric symmetric anchorages with $c < 0.5$ h: **a** in the cross-section placed on the axis of upper anchorage, **b** in the cross-section placed on the axis of the cross-section

In the deep beam, the shear force vanishes at the following distance x measured from the top of the prestressed beam (Fig. 6.85a):

$$p \cdot \left(x - \frac{h-c-a}{2}\right) = q \cdot x \rightarrow \frac{P}{a} \cdot \left(x - \frac{h-c-a}{2}\right) = \frac{2P}{h} \cdot x \rightarrow x = \frac{h \cdot (h-c-a)}{2(h-2 \cdot a)}$$

At the distance x, the bending moment M'_b is given by the following expression

$$M'_b = \frac{2P}{h} \cdot \frac{x^2}{2} - \frac{P}{a} \cdot \frac{\left(x - \frac{h-c-a}{2}\right)^2}{2} = \frac{P}{h} \cdot x^2 - \frac{P}{2a} \cdot \left(x - \frac{h-c-a}{2}\right)^2$$

while moment M''_b in the section of the deep beam placed on the axis of the prestressed beam is given by

$$M''_b = P \cdot \left(\frac{h}{4} - \frac{c}{2}\right)$$

Finally, the resultant force of *bursting* stresses is obtained by dividing M''_b by the lever arm *b*, which can be assumed equal to 0.5 h.

$$T'_b = \frac{M'_b}{b} \text{ and } T''_b = \frac{M''_b}{b}$$

Multiple anchorages

As in the other cases, the equivalent deep beam is assumed as loaded by all the prestress forces directed from left to right and by the linear load distribution in equilibrium with them, directed from right to left (Fig. 6.86).

Considering the same Example of Fig. 6.81, with centred prestress forces transmitted by four equal anchorages, the bending moment in the equivalent deep beam is equal to 98,301 kNcm, and, by assuming the internal lever arm equal to 0.5 h, the force in the anti-burst reinforcement is given by $N = 98{,}301 / (0.5 \cdot 1100) = 179$ kN (through the S&T model, a value around 131 kN is obtained—see Fig. 6.81).

6.14 Shear Strength

In this Section, the shear capacity of prestressed members is analysed, while the shear capacity of reinforced concrete elements is discussed in Chap. 8.

6.14.1 Beneficial Effects of Prestressing on the Shear Strength

In a prestressed member, shear strength experiences some beneficial effects which are not present in reinforced concrete members:

- longitudinal prestress reduces tensile internal forces due to concrete shear; this reduction is easily recognizable in the displacement of the Mohr circle of the stress field in the cross-section centroid (Fig. 6.87),
- in the calculation of shear-compression strength, an increase of the strength of compressed struts is registered for stress values of concrete prestress non greater than $0.6 f_{cd}$. EC2 considers this effect through the coefficient α_{cw} in the expression of the shear-compression strength; the variation of α_{cw} with the ratio σ_{cp}/f_{cd} is shown in Fig. 6.88,
- in post-tensioned members with inclined tendons, the vertical component of the prestress force reduces the shear force induced by external loads.

Fig. 6.86 Static scheme and bending moments for the equivalent deep beam of the multiple anchorages shown in Fig. 6.81

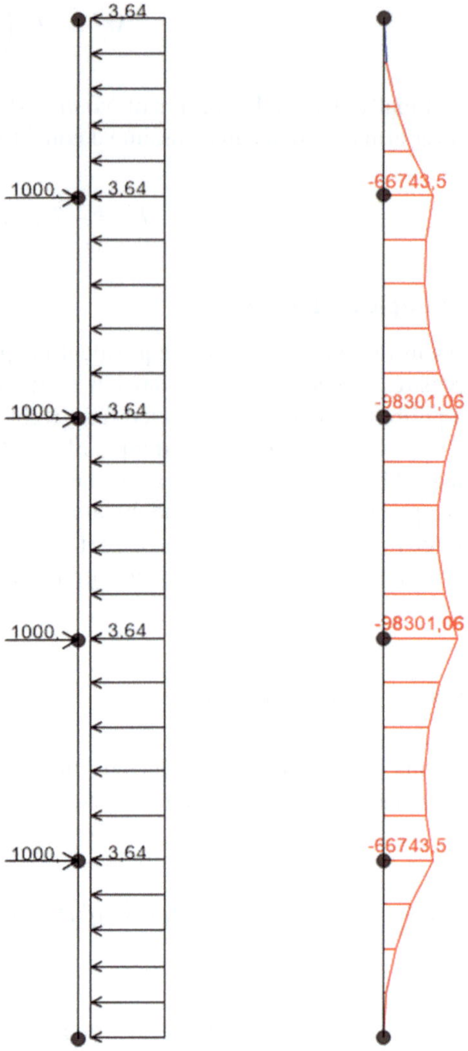

If prestress gives favourable effects, the partial coefficient to adopt for prestress is $\gamma_{P,\text{fav}} = 1.0$ (Remark to [2.4.2.2 (1)]), while if prestress effects are not favourable, $\gamma_{P,\text{unfav}} = 1.3$ (Remark to [2.4.2.2 (2)]).

6.14.1.1 Example 25. Shear Capacity of the Prestressed Beam of Example 19

For the sake of simplicity, in the calculation of design shear, the segment of length c between the end-section and the support axis (Fig. 6.89) is neglected; moreover, all

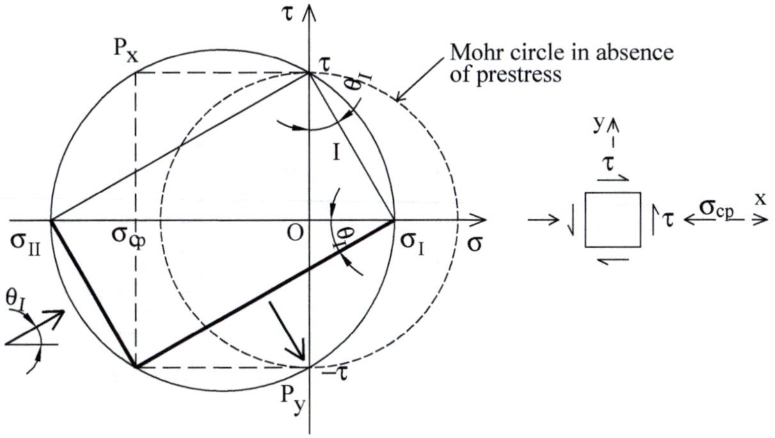

Fig. 6.87 Mohr circle in presence of prestress with identification of principal planes and principal trajectories

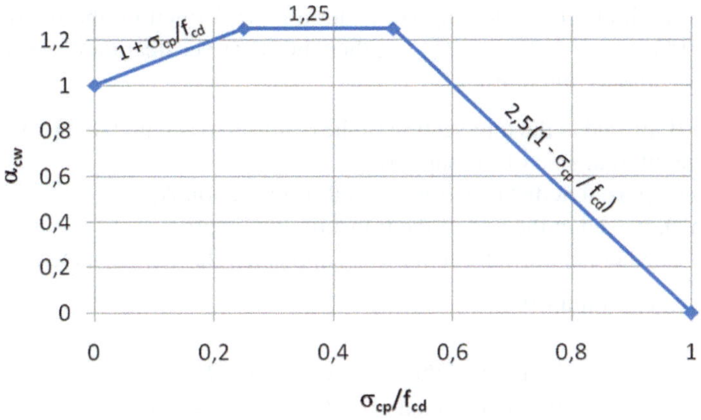

Fig. 6.88 α_{cw} coefficient at varying the compressive stress σ_{cp}

normal stresses are calculated approximately considering the geometric section.

The calculation is performed after an infinite amount of time, because, in general, the short-term calculation is less heavy. Following [6.2.1(8)], for elements mainly subjected to uniformly distributed loads, the shear ULS shall not be verified at distances smaller than d (effective height) from the support. Moreover, [6.2.2 (3)] specifies that the shear verification in uncracked zones is not required for cross-sections placed at distances from the support smaller than the distance of intersection point between the centroid axis and the line drawn at 45° from the internal side of the support.

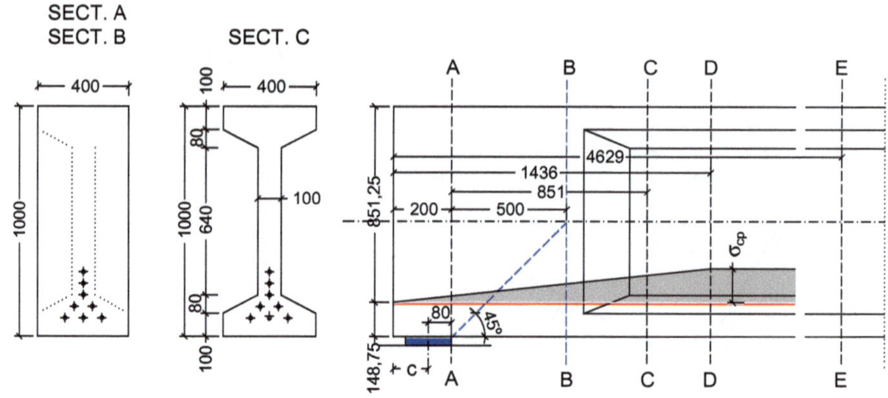

Fig. 6.89 Transverse and longitudinal sections

However, it is necessary to verify that the design shear, in the proximity of the support, is not exceeding $V_{Rd,\max}$, i.e. the shear-compression strength.

Therefore, firstly the following verification is performed on the internal side of the support (section A): $V_{Ed} \leq V_{Rd,\max}$, then the shear verification is executed in the following sections (Fig. 6.89):

- section B, placed at the intersection of the centroid axis with the line drawn at 45° from the internal side of the support,
- section C, placed at distance d measured from section A,
- section D, placed at the end of the transmission length,
- section E, where at the bottom of the section concrete reaches its tensile strength.

Loads (see Example 19):

$$\begin{array}{lll} \text{Self}-\text{weight} & G_{k1} & = 4.60\,\text{kN/m} \\ \text{Permanent load} & G_{k2} & = 5.20\,\text{kN/m} \\ \text{Variable load (snow)} & Q_{k1} & = 7.20\,\text{kN/m} \end{array}$$

Distributed load at ULS is given by

$$q_{Ed} = 1.3 \cdot (4.60 + 5.20) + 1.5 \cdot 7.20 = 23.54\,\text{kN/m}$$

Verification of section A on the internal side of the support

Section A is rectangular with dimensions 400 × 1000 mm. Shear on the internal side of the support shall not exceed the design value of the maximum shear force which can be sustained by the element ($V_{Rd,\max}$ [6.2.1 (8)]), limited by the cracking of the concrete compressed struts. $V_{Rd,\max}$ is given by [(6.5)][43]

$$V_{Rd,\max} = 0.5\ b_w\, d\ \ v\ f_{cd}$$

where

b_w =400 mm
d =1000−148.75 ≅ 851 mm
v =0.5
f_{cd} $=\frac{0.85\cdot f_{ck}}{\gamma_C} = 28.33\ \text{N/mm}^2$

then

$$V_{Rd.\max} = 2410883 \cong 2411\,\text{kN} >> V_{Ed} = 235.4\,\text{kN}$$

Verification of section B

Section B is rectangular with dimensions 400 × 1000 mm.

Design shear

$$d = 1000{-}148.75 \cong 851\,\text{mm}$$

$$q_{Ed} = 23.54\,\text{kN/m}$$

$$V_{Ed} = q_{Ed}\cdot[L/2{-}(0.50+0.08)] = 23.54\cdot(10.00{-}0.58) \cong 221.75\,\text{kN}$$

Normal stress at the neutral axis level

From Example 19 it follows that once initial and time-dependent losses have occurred, the prestress force of each strand is equal to

$$P = 152{,}388\,\text{N}$$

and the total prestress force is given by

$$P_{\text{tot}} = 8\,\text{P} = 1{,}219{,}104\,\text{N}$$

while, from Example 24, the following transmission length is obtained

$$l_{pt2} = 1436\,\text{mm} \cong 1.44\,\text{m}$$

In cross-section B, the total prestress force is equal to

[43] Expression [(6.5)] of EC2 is coincident (unless the coefficient α_{cw}) with [(6.9)], if the angle θ is assumed equal to 45° in [(6.9)].

$$\alpha_l P_{\text{tot}} = [(0.20 + 0.50)/1.44] \cdot 1{,}219{,}104 = 592{,}620\,\text{N}$$

where 0.20 m is the distance of section A from the end section and 0.50 m is the distance of section B from section A.

The normal stress on the neutral axis is given by

$$\alpha_l \sigma_{cp} = (\alpha_l P_{\text{tot}})/A_c = 592{,}620/(1000 \cdot 400) \cong 1.48\,\text{N/mm}^2$$

Compressive stress at the bottom due to prestress force

Prestress force induces at the bottom of section B the following compressive stress:

$$\sigma_{cp,\inf} = \frac{P_{tot}}{A_c} + \frac{P_{tot} \cdot z_{cp}}{I_c} \cdot z_{\inf} = \frac{592.620}{1000 \cdot 400} + \frac{592.620 \cdot 351.25}{\frac{400 \times 1000^3}{12}} \cdot 500$$
$$= 4.6\ \text{N/mm}^2$$

Tensile stress at the bottom due to external load

Bending moment due to ultimate load q_{Ed} is given by

$$M_{Ed} = \frac{q_u \cdot L}{2} \cdot 0.50 - \frac{q_u \cdot 0.50^2}{2} = 235.4 \cdot 0.50 - \frac{23.54 \cdot 0.50^2}{2} = 114.76\,\text{kNm}$$

and tensile stress induced at the bottom is given by

$$\sigma_{cq,\inf} = \frac{M_{Ed}}{I} z_{\inf} = \frac{114.76 \times 10^6}{\frac{400 \times 1000^3}{12}} \cdot 500 = 1.7\ \text{N/mm}^2$$

Total normal stress at the bottom

The total stress induced by the prestress and ultimate load at the bottom is compressive and it is given by

$$\sigma_{c,\inf} = \sigma_{cp,\inf} - \sigma_{cq,\inf} = 4.6 - 1.7 = 2.9\ \text{N/mm}^2$$

therefore, section B is uncracked.

Shear strength

Shear strength is evaluated through the formula [(6.4)]

$$V_{Rd,c} = \frac{I \cdot b_w}{S} \sqrt{f_{ctd}^2 + \alpha_l \sigma_{cp} f_{ctd}} \cong 0.7 b_w d \sqrt{f_{ctd}^2 + \alpha_l \sigma_{cp} f_{ctd}}$$

in the present case

$$V_{Rd,c} = (0.7 \cdot 851) \cdot 400 \cdot \sqrt{1.93^2 + 1.48 \cdot 1.93} = 611.284\,N$$
$$\cong 611\,kN > V_{Ed} = 221.75\text{ kN}$$

where it has been assumed

$$f_{ctd} = \alpha_{ct}\frac{f_{ctk,0.05}}{\gamma_C} = 1.0\frac{2.9}{1.5} = 1.93\text{ N/mm}^2$$

Verification of section C placed at distance *d* from section A

Section C has the following characteristics:

I-shaped profile with area $A_c = 184{,}000\text{ mm}^2$ and second moment of area $I_c = 2.3887 \times 10^{10}\text{ mm}^4$.

Design shear

$$d = 1000 - 148.75 \cong 851\text{ mm}$$

$$q_{Ed} = 23.54\,\text{kN/m}$$

$$V_{Ed} = q_{Ed} \cdot [L/2 - (d + 0.08)] = 23.54 \cdot [10.00 - (0.851 + 0.08)] \cong 213.5\,\text{kN}$$

Normal stress along the neutral axis

Total prestress force transferred to concrete is given by

$$\alpha_l P_{\text{tot}} = [(0.20 + 0.851)/1.44] \cdot 1.219.104 = 889.777\,\text{N}$$

where 0.20 m is the distance of section A measured from the end-section and 0.851 m is the distance of section C measured from the section A.

Normal stress along the neutral axis is given by

$$\alpha_l \sigma_{cp} = (\alpha_l P_{\text{tot}})/A_c = 889{,}777/184{,}000 \cong 4.83\,\text{N/mm}^2$$

Compressive stress at the bottom due to prestress

Prestress force induces at the bottom of section C the following compressive stress

$$\sigma_{cp,\text{inf}} = \frac{P_{\text{tot}}}{A_c} + \frac{P_{\text{tot}} \cdot z_{cp}}{I_c} \cdot \frac{h}{2} = \frac{889{,}777}{184{,}000} + \frac{889{,}777 \cdot 351.25}{2.3887 \times 10^{10}} \cdot 500$$
$$= 11.4\text{ N/mm}^2$$

Tensile stress at the bottom due to external load

Bending moment due to ultimate load q_{Ed} is given by

$$M_{Ed} = \frac{q_u \cdot L}{2} \cdot d - \frac{q_u \cdot d^2}{2} = 235.4 \cdot 0.851 - \frac{23.54 \cdot 0.851^2}{2} = 191.8\,\text{kNm}$$

and tensile stress induced by q_{Ed} at the bottom is given by

$$\sigma_{cq,\inf} = \frac{M_{Ed}}{I} z_{\inf} = \frac{191.8 \times 10^6}{2.3887 \times 10^{10}} \cdot 500 = 4\ \text{N/mm}^2$$

Total normal stress at the bottom

At the bottom the total stress induced by the prestress and ultimate load is compressive, and it is given by

$$\sigma_{c,\inf} = \sigma_{cp,\inf} - \sigma_{cq,\inf} = 11.4 - 4 = 7.4\,\text{N/mm}^2$$

therefore, section C is not cracked.

Shear strength

Shear strength is evaluated through the formula [(6.4)]:

$$V_{Rd,c} = \frac{I \cdot b_w}{S}\sqrt{f_{ctd}^2 + \alpha_1 \sigma_{cp} f_{ctd}} \cong 0.7 b_w d \sqrt{f_{ctd}^2 + \alpha_1 \sigma_{cp} f_{ctd}}$$

in the present case

$$V_{Rd,c} = (0.7 \cdot 851) \times 100 \cdot \sqrt{1.93^2 + 4.83 \cdot 1.93} = 215.169\,N$$
$$\cong 215\ kN > V_{Ed} = 213.5\ \text{kN}$$

where

$$f_{ctd} = \alpha_{ct} \frac{f_{ctk,0.05}}{\gamma_c} = 1.0 \frac{2.9}{1.5} = 1.93\,\text{N/mm}^2$$

Verification of section D placed at the end of the transmission zone

Section D is I-shaped

with area $A_c = 184{,}000\,\text{mm}^2$. and moment of inertia of concrete cross-section $I_c = 2.3887 \cdot 1010\ \text{mm}^4$.

For the sake of brevity, only the value of the shear strength is reported, after performing the same calculations of section C, because section D is also uncracked. The shear capacity is equal to 242 kN and it is greater than the design shear by about 203 kN.

Verification of section E where concrete attains its tensile strength

Calculation of the distance x of section E from the support axis

The position of section E is obtained by equating the total stress at the bottom due to prestress and external load to concrete tensile strength; the distance x is equal to

$$x_E \cong 4629\,\text{mm},$$

as the following stress values are obtained at the bottom compressive stress at the bottom due to P_{tot}

$$\sigma_{cp,\inf} = \frac{P_{\text{tot}}}{A_c} + \frac{P_{\text{tot}} \cdot z_{cp}}{I_c} \cdot z_{\inf} = \frac{1.219.104}{184{,}000} + \frac{1.219.104 \cdot 351.25}{2.3887 \times 10^{10}} \cdot 500$$
$$= 15.6\,\text{N/mm}^2 \text{ (compression)}$$

Bending moment induced by the external load

$$M_{Ed} = \frac{q_u \cdot L}{2} \cdot x_E - \frac{q_u \cdot x_E^2}{2} = 235.4 \cdot 4.629 - \frac{23.54 \cdot 4.629^2}{2} = 837 \text{ kN}$$

Tensile stress at the bottom due to M_{Ed}

$$\sigma_{cq,\inf} = \frac{M_{Ed}}{I} z_{\inf} = \frac{837 \times 10^6}{2.3887 \times 10^{10}} \cdot 500 = 17.52 \text{ N/mm}^2 \text{ (tensile stress)}$$

Total stress at the bottom (tensile stress):

$$\sigma_{c,\inf} = \sigma_{cp,\inf} - \sigma_{cq,\inf} = 15.6 - 17.52 = -1.92 \text{ N/mm}^2 \cong f_{ctd}$$

Design shear

$$V_{Ed} = 235.40 - 23.54 \cdot 4.629 \cong 126\,\text{kN}$$

Shear strength

Shear strength without reinforcement is calculated by [(6.2)]

$$V_{Rd,c} = \left[C_{Rd,c} k \; (100 \; \rho_l \; f_{ck})^{1/3} + 0.15 \cdot \sigma_{cp}\right] \cdot b_w d$$
$$\geq \left(v_{\min} + 0.15 \cdot \sigma_{cp}\right) \cdot b_w d \quad [(6.2 \text{ a} - \text{b})]$$

where

$$C_{Rd,c} = \frac{0.18}{\gamma_c} = \frac{0.18}{1.5} = 0.12$$

$$k = 1 + \sqrt{\frac{200}{d}} = 1.485 \leq 2$$

$$\rho_l = A_{sl}/(b_w d) = 1120/(100 \cdot 851) = 0.013 \leq 0.02$$

$$\sigma_{cp} = \frac{P_{\text{tot}}}{A_c} = \frac{1.219.104}{184,000} = 6.62\,\text{N/mm}^2$$

$$v_{Rd,c} = 0.12 \cdot 1.485\ (100 \cdot 0.013 \cdot 50)^{1/3} + 0.15 \cdot 6.62 = 1.71\ \text{N/mm}^2$$

$$\begin{aligned} v_{Rd,c,\min} &= 0.035 \cdot k^{1.5} \cdot f_{ck}^{0.5} + 0.15 \cdot \sigma_{cp} = 0.035 \cdot 1.485^{1.5} \cdot 50^{0.5} + 0.15 \cdot 6.62 \\ &= 1.44\ \text{N/mm}^2 \end{aligned}$$

then

$$V_{Rd,c} = v_{Rd,c} \cdot b_w \cdot d = 1.71 \cdot 100 \cdot 851 = 145521\,\text{N} \cong 145\,\text{kN}$$

being $V_{Rd,c} > V_{Ed}$, it is not required to provide any shear reinforcement.

Transverse reinforcement.

In all sections, the minimum amount of reinforcement is provided according to [(9.4), (9.5N), (9.6N)].

6.14.1.2 Shear Verification of a Post-tensioned Beam

Shear verification of a simply supported beam with post-tensioned tendons has to be performed under both short-term conditions, immediately after the transfer of prestress to concrete, and long-term conditions, even if long-term verification is usually heavier. In the first case, it is necessary to consider initial losses, while in the second case both initial and time-dependent losses must be accounted for. In both cases, verification can be performed according to the following procedure:

1. Design shear is calculated including the vertical component of internal prestress force in inclined tendons [6.2.1 (3)]; if prestress effect is favourable, partial coefficients $\gamma_{G,\text{unfav}} = 1.3$ for permanent loads and $\gamma_{P,\text{fav}} = 1.0$ for prestress are adopted ($\gamma_{P,\text{fav}} = 1.0$ according to Remark to [2.4.2.2 (1)]), otherwise the partial coefficient $\gamma_{G,\text{fav}} = 1.0$ and $\gamma_{P,\text{unfav}} = 1.3$ are adopted ($\gamma_{P,\text{unfav}} = 1.3$ according to the Remark to 2.4.2.2 (2) of EC2).
2. Design shear at support $V_{Ed(\text{supp.})}$ should not exceed $V_{Rd,\max}$, according to [6.2.1 (8)], otherwise, the support section should be re-designed:

$$V_{Ed(\text{app.})} \leq V_{Rd,\max},\ \text{where}\ V_{Rd,\max} = 0.5 \cdot b_{w,nom} \cdot d \cdot \nu \cdot f_{cd} \quad [(6.5)]$$

being $b_{w,\text{nom}}$ the nominal width of the web beam.

If the web beam contains injected ducts with diameter $\phi > b_w/8$, it is assumed that

$$b_{w,\text{nom}} = b_w - 0.5\Sigma\phi \quad [(6.16)];$$

(for injected metallic ducts with $\phi \leq b_w/8$, $b_{w,\text{nom}} = b_w$, while for non-injected ducts, injected plastic ducts and unbonded reinforcement,

$$b_{w,\text{nom}} = b_w - 1.2\Sigma\phi \quad [(6.17)].$$

3. Shear strength is calculated without transverse reinforcement, in the section identified by the intersection between the centroid axis and the line drawn at 45° from the internal side of the support [6.2.2 (3)]; expression [(6.4)] is used, being valid for uncracked sections (see the discussion and the meaning of symbols in Chap. 8):

$$V_{Rd,c} = \frac{I \cdot b_w}{S} \cdot \sqrt{f_{ctd}^2 + \alpha_l \sigma_{cp} f_{ctd}} \quad [(6.4)]$$

where σ_{cp} is the concrete compressive stress along the centroid axis induced by prestress force ($\sigma_{cp} = \gamma_{P,\text{fav}} P/A_c$, with $\gamma_{P,\text{fav}} = 1$) and the width b_w is the nominal width $b_{w,\text{nom}}$, which takes into account the presence of ducts (see previous point 2 for the definition of $b_{w,\text{nom}}$).

In I-shaped or T-shaped sections, where the web width varies with depth, the principal tensile stress can occur along with a fibre different from the centroid axis, therefore, it is necessary to calculate $V_{Rd,c}$ at different levels to identify the minimum value, and it is necessary to account also for the normal stress σ_{ext} induced by bending external loads, therefore $V_{Rd,c}$ expression becomes:

$$V_{Rd,c} = \frac{I \cdot b_w}{S} \cdot \sqrt{f_{ctd}^2 + \left(\alpha_l \sigma_{cp} + \sigma_{ext}\right) f_{ctd}}$$

It is worth noting that long-term shear strength is different from short-term one because both prestress σ_{cp} and nominal web width $b_{w,\text{nom}}$ are different.

4. On each side, the uncracked length of the beam is identified, where normal stress σ_{inf} at the bottom does not exceed the design tensile strength:$\sigma_{\text{inf}} \leq f_{ctd} = f_{ctk,0.05}/1.5$.
5. The shear strength without transverse reinforcement is calculated in the section where the transition from the uncracked end segment to the cracked central part happens. The relationship [(6.2a)], valid for cracked sections (see Chap. 8 for the meaning of the symbols), is used:

$$V_{Rd,c} = \left[C_{Rd,c} \cdot k \cdot (100 \cdot \rho_l \cdot f_{ck})^{1/3} + k_1 \cdot \sigma_{cp}\right] \cdot b_w \cdot d \quad [(6.2a)]$$

where $\sigma_{cp} = \gamma_{P,\text{fav}} P/A_c \leq 0.2\, f_{cd}$ is the compressive stress induced by the prestress and b_w is the web width without reduction due to ducts.

6. The minimum number of stirrups is adopted ($\sin\alpha = 1$) where the design shear is lower than shear strength without reinforcement (see Chap. 8 for the meaning of symbols):

$$\frac{A_{sw}}{s \cdot b_w} \geq 0.08 \cdot \frac{\sqrt{f_{ck}}}{f_{yk}} \quad [(9.5\ \text{N})]$$

7. Stirrups are designed in sections where shear strength without transverse reinforcement is lower than short-term or long-term design shear; to this purpose, [(6.13)] and [(6.14)] are utilized (see Chap. 8 for details of calculation and meaning of the symbols):

$$V_{Rd,s} = \frac{A_{sw}}{s} \cdot z \cdot f_{ywd} \cdot (\cot\theta + \cot\alpha)\ \sin\alpha, \quad [(6.13)]$$

$$V_{Rd,\max} = \alpha_{cw} b_w z (\nu_1 f_{cd})(\cot\theta + \cot\alpha)/(1 + \cot^2\theta), \quad [(6.14)]$$

where [(6.14)] b_w has to be assumed equal to $b_{w,\text{nom}}$, to account for ducts (see previous point 2).

Stirrups designed for the section described at the above point 3 would be extended until the end section of the beam.

8. The additional internal tensile force induced by shear is calculated through [(6.18)] and the need for additional longitudinal reinforcement is verified.

For T-shaped or I-shaped cross-sections, it is also required to verify the shear transfer from the web to flanges (see Chap. 8).

Reference

Ghali, A., & Trevino, J. (1985). Relaxation of steel in prestressed concrete. *PCI Journal, September-October, 1985*, 82–93.

Chapter 7
Ultimate Limit State for Bending with or Without Axial Force

Abstract The chapter illustrates the calculation methods (design and verification) at ULS of sections subject to bending with and without axial force. The case study is very extensive, both for the number of the strength classes of the concrete, the variety of sections (rectangular, T, circular) and the stress distribution models. The chapter is completed by a series of calculation tools: ULS tables and interaction diagrams.

7.1 Introduction

Paragraph [6.1] of EC2 "*Bending with or without axial force*" reports as Principles (P) the main hypotheses of the problem and strain limits of concrete and reinforcements. In the present Chapter, the concepts of EC2 are considered, developed, and applied to the most common cases: rectangular, circular and T-shaped cross-sections, strength calculation of assigned cross-sections, reinforcement design.

7.2 Main Hypotheses

The main hypotheses reported in [6.1(2)P] and [6.1(3)P] are the following:

- plane sections remain plane (Fig. 7.1);
- the strain in bonded reinforcement or bonded prestressing tendons, whether in tension or compression, is the same as that one in the surrounding concrete;
- the tensile concrete strength is ignored;
- stresses in compressive concrete are obtained from the design stress/strain diagram and relationships given in [3.17] or (2.1.5) of Chap. 2;
- the maximum strain of compressive concrete is obtained from Table [3.1] or Table 2.2, or Fig. 2.4 of Chap. 2;
- stresses in reinforcing steel are derived from the design diagram in [3.2.7] or in Fig. 2.4 of Chap. 2;

This chapter was authored by Piero Marro and Matteo Guiglia.

© The Author(s), under exclusive license to Springer Nature Switzerland AG 2022
F. Angotti et al., *Reinforced Concrete with Worked Examples*,
https://doi.org/10.1007/978-3-030-92839-1_7

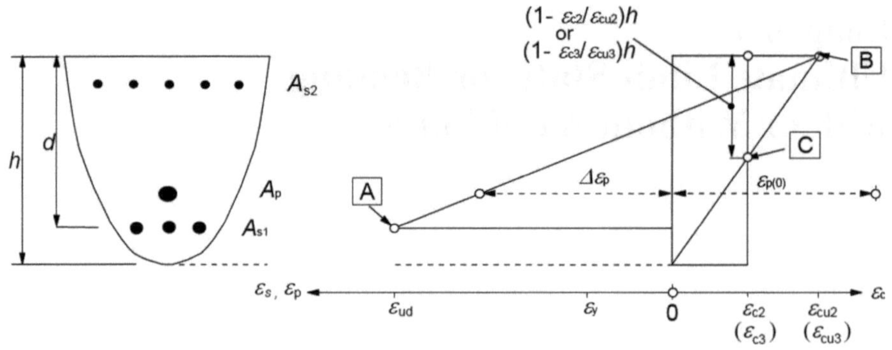

Fig. 7.1 Possible strain distributions at the ULS [Figure 6.1]

- stresses in prestressing steel are derived from the design diagram in [3.3.6] or in Fig. 2.8 of Chap. 2;
- the maximum design strain of reinforcing steel is indicated by ε_{ud}. Values are reported in the following comments.

Moreover, point [6.1(4)] specifies that in case of compression not due to prestressing, a minimum eccentricity has to be considered ($e_0 = h/30$ but not less than 20 mm, where h is the depth of the section), to take into account the uncertainty of the load application point.

- *Comments to Fig. 7.1 ([(6.1)] of EC2)*

Figure 7.1 shows all the possible deformed configurations of the cross-section at ULS. The most significant points are:

- point A: strain ε_{uk} of steel Class C is 0,075. Considering [3.2.7(2)], the abscissa of A is given by $\varepsilon_{ud} = 0.9 \cdot \varepsilon_{uk} = 0.067$.

  For steel Class A, being $\varepsilon_{uk} = 0.025$, we obtain $\varepsilon_{ud} = 0.022$.

- point B has abscissa corresponding to the limit strain of the concrete in compression ε_{cu2} (considering the generalized parabola-rectangle diagram) or ε_{cu3} (bilinear diagram). ε_{cu2} and ε_{cu3} take the same values as reported in Table [3.1] or Table 2.2.
- abscissa ε_y: for operational purposes, strain ε_{yd} corresponding to design elastic limit $\varepsilon_{yd} = \frac{f_{yd}}{E_s}$ takes the following value for steels B450C and B450A:

$$\varepsilon_{yd} = \frac{f_{yk}}{\gamma_S} \cdot \frac{1}{E_s} = \frac{450}{1.15} \cdot \frac{1}{200{,}000} = \frac{1.96}{1000}$$

- ε_{c2}, ε_{c3} are the strains corresponding to the concrete design strength, respectively according to the generalized parabola-rectangle diagram and according to the bilinear model. They are a function of f_{ck}.

- ε_{cu2}, ε_{cu3} are the limit strains of concrete in compression, according to the two models already mentioned, always a function of f_{ck}.
- point C: the distance of this point from the most compressed side depends on the concrete class and the stress–strain diagram. For $f_{ck} \leq 50$ N/mm^2 and parabola-rectangle diagram, it takes the value (3/7)h. Formulas shown in the Figure can be applied in any other case.

Once the transverse cross-section is assigned, by associating to each deformation line the stress–strain diagrams of concrete and steel, internal forces can be deduced and combined to originate the resultant force and moment at ULS. This approach is valid for the "undisturbed" regions of beams and columns, i.e., for members not locally stressed by concentrated forces and without geometric discontinuities. Discontinuity regions shall be treated according to the rules of Chap. 10 (struts-and-ties), i.e. [6.5].

7.3 Resultant Compressive Force in Case of Rectangular Cross-Section

The calculation of the resultant compressive force and its position in the case of a rectangular cross-section is discussed here adopting the generalized parabola-rectangle and bilinear diagrams.

It is necessary to distinguish two cases:

- real neutral axis ($x \leq h$);
- virtual neutral axis ($x > h$).

Real neutral axis

The resultant compressive force F_C of the stress-block related to a rectangle with width b and depth x is given by $F_C = \beta_1 f_{cd} b$ x. Its position, measured from the most compressed side, is defined by $\beta_2 x$.

β_1 and β_2 can be expressed as follows, as a function of the strain ε_c:

- in the case of the generalized parabola-rectangle diagram (EC2),

$$\beta_1(\varepsilon_{cu2}) = \frac{\int_0^{\varepsilon_{cu2}} \sigma_c(\varepsilon) \cdot d\varepsilon}{f_{cd} \cdot \varepsilon_{cu2}}$$

$$\beta_2(\varepsilon_{cu2}) = 1 - \frac{\int_0^{\varepsilon_{cu2}} \sigma_c(\varepsilon) \cdot \varepsilon . d\varepsilon}{\varepsilon_{cu2} . \int_0^{\varepsilon_{cu2}} \sigma_c(\varepsilon) \cdot d\varepsilon}$$

where the expressions of stresses σ_c are given by (2.4) and (2.5) of Chap. 2 or [3.17] and [3.18];

- in the case of the bilinear diagram

$$\beta_1(\varepsilon_{cu3}) = \frac{\int_0^{\varepsilon_{cu3}} \sigma_c(\varepsilon) \cdot d\varepsilon}{f_{cd} \cdot \varepsilon_{cu3}}$$

$$\beta_2(\varepsilon_{cu3}) = 1 - \frac{\int_0^{\varepsilon_{cu3}} \sigma_c(\varepsilon) \cdot \varepsilon \cdot d\varepsilon}{\varepsilon_{cu3} \cdot \int_0^{\varepsilon_{cu3}} \sigma_c(\varepsilon) \cdot d\varepsilon}$$

with $\sigma_c = f_{cd} \cdot \frac{\varepsilon_c}{\varepsilon_{c3}}$ in the first segment and $\sigma_c = f_{cd}$ in the second segment.

The integral calculation is simple in the case of the parabola–rectangle diagram (exponent = 2 for $f_{ck} \leq 50$ N/mm^2), but it can be more difficult in the case of the generalized parabola–rectangle diagrams ($f_{ck} > 50$ N/mm^2), being the exponents not an integer. To facilitate the solution of practical problems, β_1 and β_2 formulas have been developed as follows. Finally, calculated numerical values are reported in Tables 7.1, 7.2 and 7.3 as a function of f_{ck} for the three constitutive laws.

Solving formulas for β_1 and β_2

(a) Generalized parabola–rectangle ($2 \geq n > 1$)

$$\beta_1 = 1 - \frac{\varepsilon_{c2}}{(n+1) \cdot \varepsilon_{cu2}}$$

$$\beta_2 = 1 - \frac{\frac{n \cdot (n+3)}{(n+1) \cdot (n+2)} \cdot \varepsilon_{c2}^2 + \left(\varepsilon_{cu2}^2 - \varepsilon_{c2}^2\right)}{2 \cdot \beta_1 \cdot \varepsilon_{cu2}^2}$$

Table 7.1 β_1 and β_2—Generalized parabola–rectangle diagram (n = 2 and n < 2 in [(3.17)])

f_{ck} (N/mm^2)	≤50	55	60	70	80	90
β_1	0.8095	0.7419	0.6950	0.6372	0.5994	0.5833
β_2	0.4160	0.3919	0.3772	0.3620	0.3548	0.3529

Table 7.2 β_1 and β_2—Parabola–rectangle diagram (n = 2 in [(3.17)])

f_{ck} (N/mm^2)	≤50	55	60	70	80	90
β_1	0.8095	–	–	–	–	–
β_2	0.4160	–	–	–	–	–

Table 7.3 β_1 and β_2—Bilinear diagram

f_{ck} (N/mm^2)	≤50	55	60	70	80	90
β_1	0.7500	0.7097	0.6724	0.6296	0.5769	0.5577
β_2	0.3888	0.3746	0.3628	0.3511	0.3401	0.3373

(b) Parabola–rectangle ($n = 2$)

$$\beta_1 = 1 - \frac{\varepsilon_{c2}}{3 \cdot \varepsilon_{cu2}}$$

$$\beta_2 = 1 - \frac{1 - \frac{1}{6} \cdot \left(\frac{\varepsilon_{c2}}{\varepsilon_{cu2}}\right)^2}{2 \cdot \beta_1}$$

(c) Bilinear ($n = 1$)

$$\beta_1 = 1 - \frac{\varepsilon_{c3}}{2 \cdot \varepsilon_{cu3}}$$

$$\beta_2 = 1 - \frac{1 - \frac{1}{3} \cdot \left(\frac{\varepsilon_{c3}}{\varepsilon_{cu3}}\right)^2}{2 \cdot \beta_1}$$

Virtual neutral axis. Development with generalized parabola–rectangle diagrams.

With reference to Fig. 7.2, assigning $\xi' = x/h$, maximum and minimum strains at the cross-section sides, ε_t and ε_b respectively (top and bottom), are given by:

$$\varepsilon_t = \frac{\xi' \cdot \varepsilon_{c2}}{\frac{\varepsilon_{c2}}{\varepsilon_{cu2}} + \xi' - 1}$$

$$\varepsilon_b = \left(1 - \frac{1}{\xi'}\right) \cdot \varepsilon_t$$

Defining β_{1t} and β_{2t} as the coefficients corresponding respectively to the resultant force and its position with reference to all the extension x, and β_{1b} and β_{2b} as the analogous quantities corresponding to the portion x–h, the resultant force β_3 and its position β_4 respect to the most compressed side with reference to the depth h are

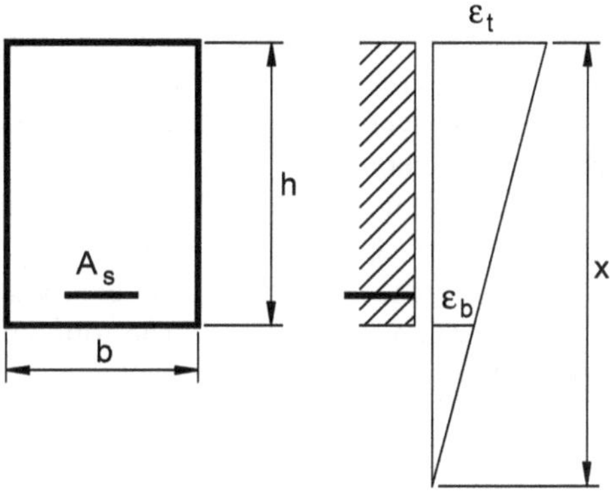

Fig. 7.2 Rectangular cross-section with virtual neutral axis

Table 7.4 β_3 and β_4—Generalized parabola–rectangle diagram

x/h	$f_{ck} \leq 50$ N/mm^2		$f_{ck} =$ 60 N/mm^2		$f_{ck} =$ 70 N/mm^2		$f_{ck} =$ 80 N/mm^2		$f_{ck} =$ 90 N/mm^2	
	β_3	β_4	β_3	β_4	β_3	β_4	β_3	β_4	β_3	β_4
1.00	0.8095	0.4160	0.6950	0.3772	0.6372	0.3620	0.5994	0.3548	0.5833	0.3529
1.20	0.8955	0.4583	0.7871	0.4244	0.7297	0.4102	0.6925	0.4035	0.6772	0.4019
1.40	0.9341	0.4748	0.8413	0.4472	0.7883	0.4349	0.7538	0.4291	0.7399	0.4276
1.60	0.9547	0.4830	0.8761	0.4605	0.8283	0.4497	0.7968	0.4446	0.7842	0.4433
1.80	0.9669	0.4878	0.9001	0.4689	0.8569	0.4595	0.8283	0.4550	0.8170	0.4539
2.00	0.9748	0.4908	0.9173	0.4748	0.8784	0.4664	0.8523	0.4624	0.8421	0.4614
2.50	0.9855	0.4947	0.9442	0.4835	0.9135	0.4770	0.8925	0.4738	0.8845	0.4731
5.00	0.9970	0.4989	0.9828	0.4951	0.9694	0.4923	0.9597	0.4909	0.9562	0.4906

given by:

$$\beta_3 = \xi' \cdot \beta_{1t} - (\xi' - 1) \cdot \beta_{1b}$$

$$\beta_4 = \frac{\xi'^2 \cdot \beta_{1t} \cdot \beta_{2t} - (\xi' - 1) \cdot \beta_{1b} \cdot ((\xi' - 1) \cdot \beta_{2b} + 1)}{\beta_3}$$

The coefficients β_{1t}, β_{2t}, β_{1b} and β_{2b} shall be calculated using formulas developed in Marro-Ferretti and Ferretti (1995) (see [7.10]) and not reported here; the values of β_3 and β_4 as a function of x/h are reported in Table 7.4.

7.4 Equilibrium Configurations of a Rectangular Cross-Section Under Axial Force Combined with Bending

The analysis of the equilibrium configurations at the ULS of a rectangular cross-section under axial force combined with bending is studied here by following the generalized parabola–rectangle diagram.

The cross-section has dimensions b, h, d, d', and reinforcements are A_{s1} at the top and A_{s2} at the bottom (Fig. 7.3).

In the case of steel B450C, a bilateral stress–strain diagram is assumed, with the horizontal branch without strain limit (Fig. 2.8 of Chap. 2).

With reference to Fig. 7.1, the configurations of the strain line passing through points B or C are considered. The following six sectors are obtained (Fig. 7.3 and Table 7.5):

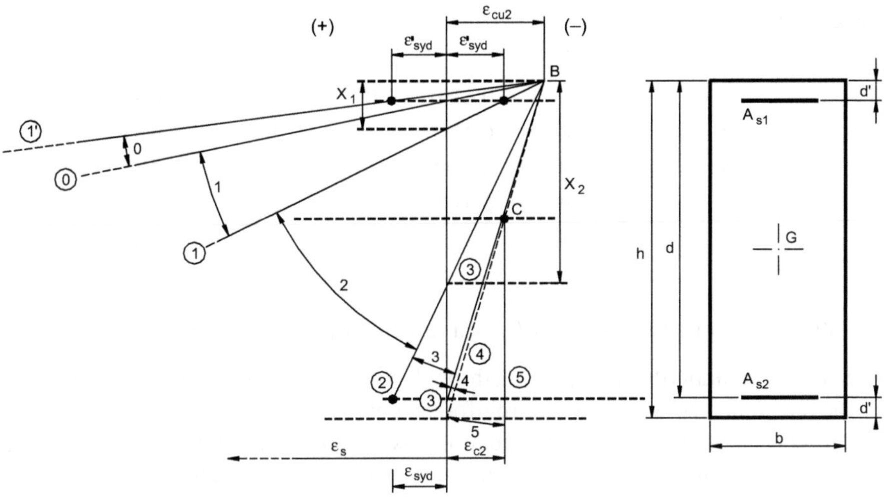

Fig. 7.3 Deformed configurations for a rectangular cross-section

Table 7.5 Summary of lines and sectors, with reference to Fig. 7.3

Sectors	Lines	Neutral axis x
0	1′–0	$x \le d'$
1	0–1	$d' \le x \le k_1 d'$
2	1–2	$k_1 d' \le x \le k_2 d$
3	2–3	$k_2 d \le x \le d$
4	3–4	$d \le x \le h$
5	4–5	$h \le \infty$

- **Sector 0** contains the lines passing through point B from the one indicated with 1′ corresponding to the point where reinforcement A_{s1} takes its positive strain $\varepsilon_{syd} = f_{yd}/E_s$, to the one indicated with 0, where stress in A_{s1} is null. In this Sector, the depth of the neutral axis x is variable between $x_0 = k_0\, d'$ and $d'(k_0\, d' < x < d')$. It is depending on the distance d' of the reinforcement to the top side of the cross-section and from the ultimate concrete strain ε_{cu2}, therefore on its class. As a result

$$k_0 = \frac{\varepsilon_{cu2}}{\varepsilon_{cu2} + \varepsilon_{syd}}$$

For example, with
$f_{ck} \le 50\,\text{N/mm}^2$,
$\varepsilon_{cu2} = 0.s0035$,
$\varepsilon_{syd} = \frac{450}{1.15} \cdot \frac{1}{200{,}000} = 0.00196$, we get:
$k_0 = \frac{0.0035}{0.0035+0.00196} = 0.64$.

Table 7.6 Concrete parameters

f_{ck} (N/mm^2)	k_0	k_1	k_2	k_5	$1000\varepsilon_{c2}$	$1000\varepsilon_{cu2}$	β_1	β_2
12–50	0.64	2.27	0.64	0.4286	2.0	3.5	0.809	0.416
60	0.60	3.08	0.60	0.2069	2.3	2.9	0.695	0.377
70	0.58	3.65	0.58	0.1111	2.4	2.7	0.637	0.362
80	0.57	4.06	0.57	0.0385	2.5	2.6	0.599	0.355
90	0.57	4.06	0.57	0.00	2.6	2.6	0.583	0.377

Table 7.6 reports the values of k_0 as a function of f_{ck}.

- **Sector 1** contains the lines passing through point B from the one indicated with 0 to the one indicated with 1, where reinforcement A_{s1} takes the strain $\varepsilon'_{syd} = f'_{yd}/E_s$. The corresponding depth of the neutral axis $x_1 = k_1\, d'$ is dependent on d' and the ultimate concrete strain ε_{cu2}, therefore on its class. As a result

  $k_1 = \frac{\varepsilon_{cu2}}{\varepsilon_{cu2}-\varepsilon_{syd}}$ with strain ε expressed in absolute value.
  Table 7.6 reports the values of k_1 as a function of f_{ck}.

- **Sector 2** contains the lines passing through point B, from 1 to 2 where the strain of the reinforcement A_{s2} takes the design elastic limit ε_{syd}. The corresponding depth of the neutral axis $x_2 = k_2 d$ is a function of f_{ck}. Values of k_2 are reported in Table 7.6.
- **Sector 3** contains the lines passing through point B, from 2 to 3 where the strain of A_{s2} is zero.
- **Sector 4** contains the lines passing through point B, from 3 to 4 where $x = h$.
- **Sector 5** contains the lines passing through point C, from line 4 to the vertical one (line 5). The position of point C from the top side is depending on f_{ck}, as shown in Fig. 7.1.

The values of the ratio "vertical projection of BC/height $h'' = (1 - \frac{\varepsilon_{c2}}{\varepsilon_{cu2}}) = k_5$ are reported in Table 7.6.

Analysis of the 6 sectors

The translational equilibrium equation gives the value of N_{Rd} on the 7 deformation lines which separate the sectors. The rotational equilibrium equation gives the value of M_{Rd} to the centroid of the concrete cross-section. In all the expressions, ε_{cu2} is assumed as positive.

Equations relating to lines

Line 1'

$$N_{Rd} = A_{s1} f_{yd} - \beta_1 b \overline{x} f_{cd} + A_{s2} f_{yd} \quad (\overline{x} = \frac{\varepsilon_{cu2}}{\varepsilon_{cu2} + \varepsilon_{syd}} d')$$

$$M_{Rd} = -A_{s1} f_{yd}\left(\frac{h}{2} - d'\right) + \beta_1 b \overline{x} f_{cd}\left(\frac{h}{2} - \beta_2 \overline{x}\right) + A_{s2} f_{yd}\left(\frac{h}{2} - d'\right)$$

Line 0

$$N_{Rd} = -\beta_1 b d' f_{cd} + A_{s2} f_{yd}$$

$$M_{Rd} = +\beta_1 b d' f_{cd}\left(\frac{h}{2} - \beta_2 d'\right) + A_{s2} f_{yd}\left(\frac{h}{2} - d'\right)$$

Line 1

$$N_{Rd} = -A_{s1} f_{yd} - \beta_1 b (k_1 d') f_{cd} + A_{s2} f_{yd}$$

$$M_{Rd} = A_{s1} f_{yd}\left(\frac{h}{2} - d'\right) + \beta_1 b (k_1 d') f_{cd}\left(\frac{h}{2} - \beta_2 k_1 d'\right) + A_{s2} f_{yd}\left(\frac{h}{2} - d'\right)$$

Line 2

$$N_{Rd} = -A_{s1} f_{yd} - \beta_1 b k_2 d f_{cd} + A_{s2} f_{yd}$$

$$M_{Rd} = A_{s1} f_{yd}\left(\frac{h}{2} - d'\right) + \beta_1 b k_2 d f_{cd}\left(\frac{h}{2} - \beta_2 k_2 d\right) + A_{s2} f_{yd}\left(\frac{h}{2} - d'\right)$$

Line 3

$$N_{Rd} = -A_{s1} f_{yd} - \beta_1 b d f_{cd}$$

$$M_{Rd} = A_{s1} f_{yd}\left(\frac{h}{2} - d'\right) + \beta_1 b d f_{cd}\left(\frac{h}{2} - \beta_2 d\right)$$

Line 4

$$N_{Rd} = -A_{s1} f_{yd} - \beta_1 b h f_{cd} - A_{s2}\sigma_{s2}; \quad \left(\sigma_{s2} = E_s \frac{d'}{h} \cdot \varepsilon_{cu2}\right)$$

$$M_{Rd} = A_{s1} f_{yd}\left(\frac{h}{2} - d'\right) + \beta_1 b h f_{cd}\left(\frac{h}{2} - \beta_2 h\right) - A_{s2}\sigma_{s2}\left(\frac{h}{2} - d'\right)$$

Line 5

$$N_{Rd} = -A_{s1} f_{yd} - \beta_3 b h f_{cd} - A_{s2} f_{yd}$$

$$M_{Rd} = A_{s1} f_{yd}\left(\frac{h}{2} - d'\right) - A_{s2} f_{yd}\left(\frac{h}{2} - d'\right)$$

If within each sector a compatible value of N_{Ed} is considered, i.e., between the values of N_{Rd} corresponding to the two lines which identify the Sector, the corresponding value of x can be determined by the translational equilibrium equation. The rotational equilibrium equation gives the value of M_{Rd} associated with the proposed value of N_{Ed}.

Sector 0 ($x < d'$)

Reinforcement A_{s1} is in tension in the elastic range. Stress is given by

$$\sigma_{1s} = \varepsilon_s E_s = E_s|\varepsilon_{cu2}|\left(\frac{d'}{x} - 1\right) \tag{7.1}$$

Stress in the reinforcement A_{s2} is equal to f_{yd}.

Translational equilibrium equation in presence of the assigned axial force N_{Ed} (positive if tensile, negative if compressive, which shall be compatible with the studied sector), can be written as

$$A_{s1}\sigma_s - \beta_1 bxf_{cd} + A_{s2}f_{yd} = N_{Ed} \tag{7.2}$$

which, by the suitable substitutions, becomes the second-order equation in x

$$x^2 - x\left(\frac{A_{s2}f_{yd} - N_{Ed} - A_{s1}E_s\varepsilon_{cu2}}{\beta_1 bf_{cd}}\right) - \frac{A_{s1}E_s\varepsilon_{cu2}d'}{\beta_1 bf_{cd}} = 0 \tag{7.3}$$

which gives the following solution

$$x = \frac{1}{2}\cdot\left(\frac{A_{s2}f_{yd} - N_{Ed} - A_{s1}E_s\varepsilon_{cu2}}{\beta_1 bf_{cd}}\right) + \sqrt{\frac{1}{4}\cdot\left(\frac{A_{s2}f_{yd} - N_{Ed} - A_{s1}E_s\varepsilon_{cu2}}{\beta_1 bf_{cd}}\right)^2 + \left(\frac{A_{s1}E_s\varepsilon_{cu2}d'}{\beta_1 bf_{cd}}\right)}$$

Once x is known, σ_{1s} is also known by applying Eq. (7.1). The associated design bending resistance, calculated to the centroid axis of the concrete cross-section, is given by

$$M_{Rd} = -A_{s1}\sigma_{1s}\left(\frac{h}{2} - d'\right) + \beta_1 xbf_{cd}\left(\frac{h}{2} - \beta_2 x\right) + A_{s2}f_{yd}\left(\frac{h}{2} - d'\right) \tag{7.4}$$

Sector 1

Reinforcement A_{s1} is in compression in the elastic range. The stress is given by

$$\sigma_{1s} = E_s \varepsilon_{cu2}(1 - \frac{d'}{x}) \tag{7.5}$$

Translational equilibrium equation in presence of an assigned axial force N_{Ed} (compression), can be written as follows

$$-A_{1s} E_s \varepsilon_{cu2}\left(1 - \frac{d'}{x}\right) - \beta_1 b x f_{cd} + A_{2s} f_{yd} = N_{Ed} \tag{7.6}$$

It is worth noting that the value of N_{Ed} shall be compatible with the examined sector, as we will see in the following discussion.

The application of (7.6) gives the second-order equation in x

$$x^2 - x\left(\frac{A_{s2} f_{yd} + N_{Ed} - A_{s1} E_s \varepsilon_{cu2}}{\beta_1 b f_{cd}}\right) - \frac{A_{s1} E_s \varepsilon_{cu2} d'}{\beta_1 b f_{cd}} = 0$$

Once x is known, σ_{1s} is defined by Eq. (7.5). Therefore, the associated design bending resistance can be determined, evaluated to the centroid axis of the concrete cross-section:

$$M_{Rd} = A_{s1}\sigma_{1s}\left(\frac{h}{2} - d'\right) + A_{s2} f_{yd}\left(\frac{h}{2} - d'\right) + \beta_1 b x f_{cd}\left(\frac{h}{2} - \beta_2 x\right) \tag{7.7}$$

β_1 and β_2 are a function of f_{ck} and are reported in Table 7.1.

Sector 2 ($x_1 < x < x_2$)

Both reinforcements have stress f_{yd} (A_{S1} in compression, A_{S2} in tension).

Translational equilibrium equation can be written as:

$$-A_{1s} f_{yd} - \beta_1 b x f_{cd} + A_{2s} f_{yd} = N_{Ed} \tag{7.8}$$

The solution is given by

$$x = \frac{(A_{s2} - A_{s1}) f_{yd} - N_{Ed}}{\beta_1 b f_{cd}}$$

The design bending resistance is given by

$$M_{Rd} = (A_{s1} + A_{s2}) f_{yd}\left(\frac{h}{2} - d'\right) + \beta_1 b x f_{cd}\left(\frac{h}{2} - \beta_2 x\right) \tag{7.9}$$

Sector 3: ($d > x > x_2$)

A_{s1} has stress f_{yd}. A_{s2} is in the elastic range with stress given by

$$\sigma_s = E_s \varepsilon_{cu2} \left(\frac{d}{x} - 1 \right) \tag{7.10}$$

Translational equilibrium equation can be written as

$$-A_{s1} f_{yd} - \beta_1 b x f_{cd} + A_{s2} E_s \varepsilon_{cu2} \left(\frac{d}{x} - 1 \right) = N_{Ed} \tag{7.11}$$

which leads to the following equation

$$x^2 - x \frac{(-N_{Ed} - A_{s1} f_{yd} - A_{s2} E_s \varepsilon_{cu2})}{\beta_1 b f_{cd}} - \frac{A_{s2} E_s \varepsilon_{cu2} d}{\beta_1 b f_{cd}} = 0$$

Once this equation is solved, as already described for Sector 1, the design bending resistance is given by

$$\begin{aligned} M_{Rd} = {} & A_{s1} f_{yd} \left(\frac{h}{2} - d' \right) + \beta_1 b x f_{cd} \left(\frac{h}{2} - \beta_2 x \right) \\ & + A_{s2} E_s \varepsilon_{cu2} \left(\frac{d}{x} - 1 \right) \cdot \left(\frac{h}{2} - d' \right) \end{aligned} \tag{7.12}$$

Sector 4: ($d < x < h$)

A_{s1} has stress f_{yd}. A_{s2} is in the elastic range with stress given by

$$\sigma_s = -E_s \varepsilon_{cu2} \cdot \left(1 - \frac{d}{x} \right) \; (\varepsilon_{cu2} \text{ in absolute value}) \tag{7.13}$$

Translational equilibrium equation can be written as

$$-\beta_1 x b f_{cd} - A_{s1} f_{yd} - A_{s2} E_s \varepsilon_{cu2} \left(1 - \frac{d}{x} \right) = N_{Ed} \tag{7.14}$$

which leads to the following equation

$$x^2 - x \frac{(N_{Ed} - A_{s1} f_{yd} - A_{s2} E_s \varepsilon_{cu2})}{\beta_1 b f_{cd}} - \frac{A_{s2} E_s \varepsilon_{cu2} d}{\beta_1 b f_{cd}} = 0$$

Once the equation is solved, σ_{s2} is also known. The design bending resistance is given by

$$M_{Rd} = \beta_1 x b f_{cd} \left(\frac{h}{2} - \beta_2 x \right) + A_{s1} f_{yd} \left(\frac{h}{2} - d' \right)$$

$$- A_{s2}E_s\varepsilon_{cu2}\left(1 - \frac{d}{x}\right)\cdot\left(\frac{h}{2} - d'\right) \tag{7.15}$$

Sector 5: $(h \leq x \leq \infty)$

In this case, it is necessary to introduce β_3 and β_4 coefficients which are functions of the ratio x/h (see Table 7.4). The translational equilibrium equation can be written for assigned values of x/h, corresponding to the values of the two mentioned coefficients. It is necessary to proceed tentatively and by approximation, until the design value N_{Ed} equals the value of N_{Rd} given by

$$N_{Rd} = -bhf_{cd}\beta_3 - A_{s1}f_{yd} - A_{s2}E_s\varepsilon_{c2}\cdot\frac{\left(\frac{x}{h} - \frac{d}{h}\right)}{\left(\frac{x}{h} - k_5\right)} \tag{7.16}$$

(the third term in the second member cannot exceed $A_{s2}f_{yd}$).

k_5 is given by the expression $k_5 = (1 - \frac{\varepsilon_{c2}}{\varepsilon_{cu2}})$.

ε_{c2} and ε_{cu2} are reported in Table [3.1] and Table 2.2 of Chap. 2.

Once x/h is defined, design bending resistance is given by

$$M_{Rd} = bhf_{cd}\beta_3\left(\frac{h}{2} - \beta_4 h\right) + A_{s1}f_{yd}\left(\frac{h}{2} - d'\right)$$
$$- A_{s2}E_s\varepsilon_{c2}\frac{\left(\frac{x}{h} - \frac{d}{h}\right)}{\left(\frac{x}{h} - k_5\right)}\cdot\left(\frac{h}{2} - d'\right) \tag{7.17}$$

(the observation proposed above for A_{s2} is valuable also in this case).

- *Operating procedure*

To place N_{Ed} in the right sector, the value of N_{Rd} has to be calculated for all the possible configurations corresponding to lines 1′, 0, 1, 2, 3, etc., by using the given expressions.

By applying this procedure, the sector where the proposed N_{Ed} is placed can be identified. The corresponding translational equilibrium equation gives the depth of the neutral axis x; therefore, the design bending resistance of the cross-section associated with $N_{Ed} = N_{Rd}$ can be calculated.

It is worth noting that the couples of values N_{Rd}–M_{Rd} corresponding to lines are points of the interaction diagram *N–M* of the cross-section at ULS.

Table 7.6 reports the values of the parameters for each concrete strength.

β_1 and β_2 are reported in Table 7.1 and β_3 and β_4 in Table 7.4.

- Example

In the following, three examples are discussed for a rectangular cross-section $b = 400$ mm; $h = 700$ mm; $d = 650$ mm; $d' = 50$ mm for both sides. $A_{s1} = 2000$ mm^2;

Fig. 7.4 Cross-section geometry and reinforcement

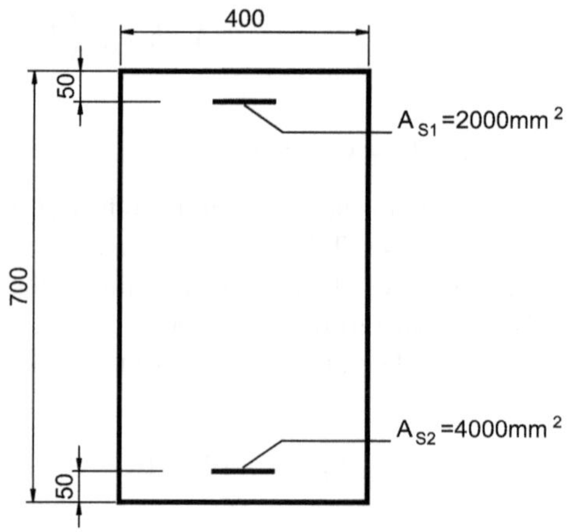

$A_{s2} = 4000$ mm^2; $f_{yk} = 450$ N/mm^2, $f_{yd} = 391$ N/mm^2. Examples are referred to concretes: $f_{ck} = 30, 60, 90$ N/mm^2 (Fig. 7.4).

Preliminarily neutral axis depth is calculated for lines 1′, 0, 1, etc. and, therefore, the corresponding values of N_{Rd} and M_{Rd}. Results are reported in Table 7.7.

(1) The line corresponding to $N_{Rd} = 0$ and the corresponding design bending resistance are calculated for the three concrete classes.

- Concrete $f_{ck} = 30$ N/mm^2 ($f_{cd} = 17$ N/mm^2).

According to Table 7.7, the searched line is placed within Sector 2 between lines 1 and 2, being the two values of N_{Rd} with opposite sign. The neutral axis is obtained through (7.8) by assigning $N_{Ed} = 0$.

We obtain:

$-2000 \cdot 391 - 0.809 \cdot 400 \times 17 + 4000 \cdot 391 = 0$, from which $x = 142$ mm.

The design bending resistance, obtained through (7.9), is given by:

$$\begin{aligned} M_{Rd} &= (2000 + 4000) \cdot 391 \cdot (350 - 50) \\ &\quad + 0.809 \cdot 142 \cdot 400 \cdot 17(350 - 0.416 \cdot 142) \\ &= 931\,\text{kNm} \end{aligned}$$

- Concrete $f_{ck} = 60$ N/mm^2 ($f_{cd} = 34$ N/mm^2)

According to Table 7.7, the neutral axis is placed in Sector 1 between lines 0 and 1. Reinforcement A_{s1} is in compression in the elastic range. By Eq. (7.6) we get:

$$x^2 - x\left(\frac{4000 \cdot 391 - 2000 \cdot 580}{0.695 \cdot 400 \cdot 34}\right) - \frac{2000 \cdot 580 \cdot 50}{0.695 \cdot 400 \cdot 34} = 0$$

from which $x = 102$ mm

Table 7.7 N_{Rd} and M_{Rd} strength

Line	$f_{ck} = 30$ N/mm^2			$f_{ck} = 60$ N/mm^2			$f_{ck} = 90$ N/mm^2		
	x (mm)	N_{Rd} (kN)	M_{Rd} (kNm)	x (mm)	N_{Rd} (kN)	M_{Rd} (kNm)	x (mm)	N_{Rd} (kN)	M_{Rd} (kNm)
1'	32	+2170	+294	30	+2063	+330	28,5	+2007	+350
0	50	+1289	+560	50	+1091	+626	50	+969	+667
1	113	+157	+892	154	− 673	+1128	203	− 1632	+1376
2	416	− 1511	+1109	390	− 2904	+1451	370	− 3624	+1670
3	650	− 4360	+519	650	− 6926	+879	650	− 8512	+1166
4	700	− 4835	+401	700	− 7564	+755	700	− 9255	+1046
5	∞	− 7106	− 234	∞	− 11,866	− 234	∞	− 16,626	− 234

$$\sigma_{1s} = 200 \cdot 2.9\left(1 - \frac{50}{102}\right) = 296\,\text{N/mm}^2$$

and therefore:

$$M_{Rd} = 296 \cdot 2000\,(350 - 50) + 4000 \cdot 391\,(350 - 50) \\ + 0.695 \cdot 400 \cdot 102 \cdot 34\,(350 - 0.377 \cdot 102) = 947\,\text{kNm}$$

- Concrete $f_{ck} = 90$ N/mm^2 ($f_{cd} = 51$ N/mm^2)

According to Table 7.7, the neutral axis is placed in Sectors 1 between lines 0 and 1. Reinforcement A_{s1} is in compression in the elastic range. By applying the Eq. (7.6) we obtain:

$$x^2 - x\frac{4000 \cdot 391 - 2000 \cdot 520}{0.583 \cdot 400 \cdot 51} - \frac{2000 \cdot 520 \cdot 50}{0.583 \cdot 400} = 0$$

Solution is $x = 91.7$ mm

$$\sigma_{1s} = -520(1 - \frac{0.50}{91.7}) = -283\,\text{N/mm}^2$$

and, by (7.7), we obtain

$$M_{Rd} = 283 \cdot 2000 \cdot 300 + 4000 \cdot 391 \cdot 300 \\ + 0.583 \cdot 400 \cdot 91.7 \cdot 51(350 - 0.353 \cdot 91.7) \\ = 985\text{ kNm}$$

Figure 7.5 shows the diagrams M_{Rd} - N_{Rd} for the three concrete classes.

(2) Evaluation of the design bending resistance associated with the assigned N_{Rd}.

- Concrete $f_{ck} = 30$ N/mm^2

$N_{Ed} = +\,1500$ kN $= +\,1{,}500{,}000$ N
The solution is placed in Sector 0.
Neutral axis position is obtained by (7.3)

$$x^2 - x\left(\frac{4000 \cdot 391 - 1{,}500{,}000 - 2000 \cdot 700}{0.809 \cdot 400 \cdot 17}\right) - \frac{2000 \cdot 700 \cdot 50}{0.809 \cdot 400 \cdot 17} = 0$$

from which we obtain $x = 44.3$ mm. Therefore, it results

$$\sigma_{1s} = 700 \cdot (\frac{50}{44.3} - 1) = 90\,\text{N/mm}^2$$

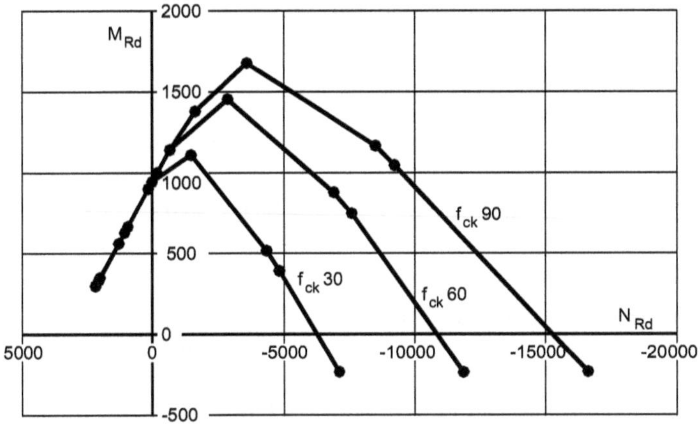

Fig. 7.5 Interaction diagrams $N_{Rd} - M_{Rd}$

and by (7.4)

$$M_{Rd} = -2000 \cdot 90 \cdot (350 - 50) + 0.809 \cdot 44.3 \cdot 400 \cdot 17 \cdot (350 - 0.416 \cdot 44.3) + 4000 \cdot 391 \cdot (350 - 50) = 496 \text{ kNm}.$$

- Concrete $f_{ck} = 60$ N/mm^2

  $N_{Ed} = -7200$ kN
  The solution is placed in Sector 4. By (7.14) we obtain:

$$x^2 - x\left(\frac{7{,}200{,}000 - 2000 \cdot 391 - 4000 \cdot 580}{0.695 \cdot 400 \cdot 34}\right) - \frac{4000 \cdot 580 \cdot 50}{0.695 \cdot 400} = 0$$

  It is obtained that $x = 671$ mm

$$\sigma_{1s} = -580 \cdot (1 - \frac{650}{671}) = -18\,\text{N/mm}^2$$

and by (7.15)

$$M_{Rd} = 0.695 \cdot 671 \cdot 400 \cdot 400 \cdot 34 \cdot (350 - 0.377 \cdot 671) + 2000 \cdot 391 \cdot (350 - 50) = 828 \text{ kNm}$$

- Concrete $f_{ck} = 90$ N/mm^2

  $N_{Ed} = -11000$ kN
  Solution is placed in Sector 5.
  In this case, with $x/h > 1$, it is necessary to proceed by attempts.

It is assigned $x/h = 1.2$ corresponding to:
$\beta_3 = 0.677$ and $\beta_4 = 0.402$; k_5 is equal to zero for $f_{ck} = 90$.
Using (7.16) we obtain:

$$N_{Rd} = -400 \cdot 700 \cdot 51 \cdot 0.677 - 2000 \cdot 391 - 4000 \cdot 520 \cdot \left(\frac{1.2 - \frac{650}{700}}{1.2 - 0} \right) = -10917\,\text{kN}$$

which is a value similar to the required one.

The associated design bending resistance, evaluated by (7.17), is given by:

$$M_{Rd} = 400 \cdot 700 \cdot 51 \cdot 0.677 \cdot (350 - 0.402 \cdot 700) + 2000 \cdot 391 \cdot (350 - 50) = 757\,\text{kNm}.$$

(3) For the sake of comparison, the following reinforcement arrangement is assumed: $A_{s1} = 4000\ \text{mm}^2$, $A_{s2} = 2000\ \text{mm}^2$ (opposite to the previous case). Only for $f_{ck} = 30$, calculations of N_{Rd} and M_{Rd} corresponding to lines from 1′ to 5 are repeated. Table 7.8 reports relevant results.

The configuration corresponding to the pure axial force is placed between lines 1′ and 0 (neutral axis $x = 37$ mm). N_{Rd} is equal to +1562 kN.

The configuration corresponding to uniaxial bending is placed between lines 0 and 1 (neutral axis $x = 59.6$ mm). M_{Rd} is equal to +476 kNm.

Figure 7.6 shows the diagram of the calculated values, also reported in Table 7.8: in this way, it is possible to represent the entire strength domain in the plane $M_{Rd} - N_{Rd}$ (Tables 7.9 and 7.10).

Comments

(1) If reinforcements A_{s1} and A_{s2} are equal, the diagram of Fig. 7.6 ($f_{ck} = 30\ \text{N/mm}^2$) is symmetrical to the axis N and consequently the extreme values of N, positive and negative, are on the axis itself. Therefore, it is sufficient to represent only one portion of the diagram.

Table 7.8 N_{Rd} and M_{Rd} strength

Line	$f_{ck} = 30\ \text{N/mm}^2$		
	x (mm)	N_{Rd} (kN)	M_{Rd} (kNm)
1′	32	+2170	– 175
0	50	+507	+325
1	113	– 1406	+892
2	416	– 3070	+1114
3	650	– 5140	+754
4	700	– 5415	+696
5	∞	– 7106	+234

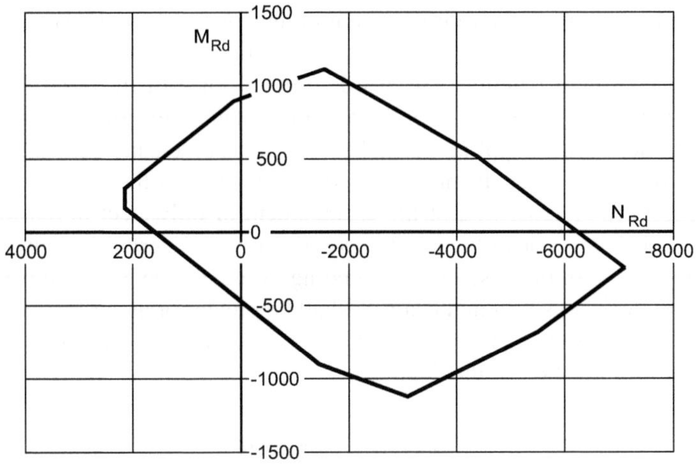

Fig. 7.6 Interaction diagram $N_{Rd} - M_{Rd}$

(2) The polygonal curves of Figs. 7.5 and 7.6 are enclosed in the continuous curves which could be obtained by developing the calculations for close points, i.e., for close deformation lines. In the Appendix, continuous diagrams of this type are shown and can be useful for the cross-section design.

7.5 Reinforcement Design for Uniaxial Bending and Axial Force Combined with Bending

A transverse cross-section symmetrical to the y axis is considered (Fig. 7.7). The applied force is contained in the symmetry plane. Once design internal forces at the ULS are assigned (N_{Ed} applied to the centroid of the concrete cross-section and

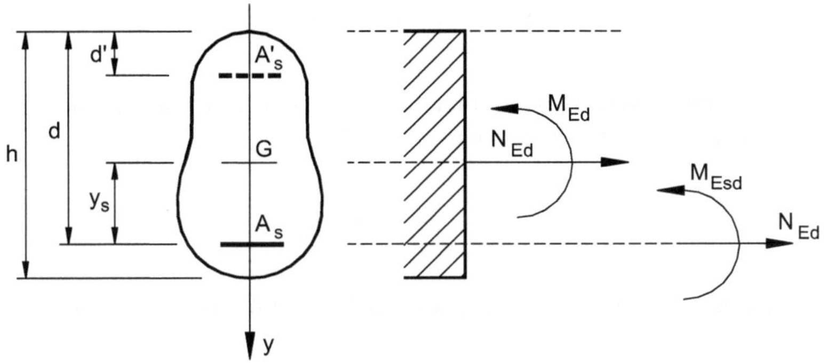

Fig. 7.7 Uniaxial bending and axial loads

M_{Ed}), the moment about the reinforcement in tension can be calculated as $M_{Esd} = M_{Ed} - N_{Ed}\, y_s$ and the reinforcement area A_s required by M_{Esd} can be determined as in the case of pure bending Afterwards, the presence of the axial force N_{Ed} is considered correcting the strength of the reinforcement in tension with an equivalent quantity. In these cases, applied normal force shall be considered with its sign: if N_{Ed} is a compressive force, its presence reduces the area of the steel in tension.

The general rule, consisting of the arrangement of only steel in tension (A_s), is adopted and, if this is not sufficient, it is possible to add a certain amount of steel in compression (A'_s). In the case of pure bending, to guarantee a ductile behaviour to the structure, the steel strain in tension ε_s shall be greater or at least equal to the one corresponding to the design elastic limit, i.e. $\varepsilon_s \geq \varepsilon_{yd} = f_{yd}/E_s$. In this way, the neutral axis does not exceed the depth

$$x_{\lim} = \frac{\varepsilon_{cu}}{\varepsilon_{cu} + \varepsilon_{yd}} \cdot d$$

being ε_{cu} the deformation of the side in compression. This limit, as it will be shown in the following, is also a useful operating reference for the design of axial load combined with bending.

ε_{cu} value depends only on the concrete class, being the values ε_{cu2} (for the generalized parabola-rectangle diagram) and ε_{cu3} (bilinear diagram) identical each other (Table 2.2 or [Table 3.1]). ε_{yd} is depending on the design strength $f_{yd} = f_{yk}/\gamma_s$. For example, for concrete $f_{ck} \leq 50$ N/mm^2, $\varepsilon_{cu2} = 0.0035$; steel B450C, $f_{yd} = 391$ N/mm^2, $\varepsilon_{yd} = 0.00196$, we get:

$$x_{\lim} = \frac{0.0035}{0.0035 + 0.00196} \cdot d = 0.641d.$$

If M_{Esd} is greater than $M_{\lim}$, corresponding to $x_{\lim}$ with only reinforcement in tension, it is necessary to arrange a certain amount of steel in compression. It is then possible to calculate the difference

$$\Delta M_{Esd} = M_{Esd} - M_{lim}$$

which can be supported by two reinforcements, in compression and tension, working at the design elastic limit, whose cross-section is

$$A_s^{'} = \frac{\Delta M_{\mathrm{Esd}}}{f_{yd} \cdot (d - d')}$$

The amount A'_s in tension can be added to the area A_s corresponding to the limit moment.

- *Remark*

Limitations on x more severe than the one above mentioned ($x \leq x_{\text{lim}}$) are required to fulfil ductility needs for bending in statically indeterminate structures when the moment redistribution is performed (Chap. 4). Moreover, both in statically indeterminate and determinate cases, the stress limitation at SLS ([7.2(2)] and Chap. 11) requires implicitly a limitation of the depth of the neutral axis at ULS.

In the design of slender members, point [9.2.1.1(1)] requires a minimum amount of reinforcement in tension to avoid brittleness on the steel side. In other words, the design bending resistance at ULS must be greater than the moment causing cracking.

Special cases

Axial force combined with bending—Reinforcement design

If N_{Ed} is applied to a point of the symmetry axis far away from the centroid at the distance e, on the opposite side to the reinforcement in tension, with distance y_s, we get:

$$M_{Esd} = N_{Ed}(y_s + e)$$

If $M_{Esd} \leq M_{lim}$, the reinforcement A_s required by M_{Esd} is calculated and the amount N_{Ed}/f_{yd}. is subtracted to this.

If $M_{Esd} > M_{lim}$, the reinforcement corresponding to M_{lim} is calculated, the couple of reinforcements corresponding to $\Delta M = M_{Esd} - M_{lim}$ is added and the amount N_{Ed}/f_{yd} is subtracted to the reinforcement in tension.

7.6 Rectangular Cross-Section in Uniaxial Bending and Axial Force Combined with Bending

7.6.1 Rectangular Cross-Section: Generalized Parabola-Rectangle Diagrams; Bilinear Diagram

The calculation is performed through equilibrium and compatibility equations (plane sections) using coefficients β_1, β_2, β_3 and β_4, already recalled in Sect. 7.3.

If the distribution of compressive stress of the portion of the cross-section experiencing the shortening is associated with the lines of Fig. 7.1, it is possible to calculate the value of the resultant compressive force $F_c = bx\,\beta_1 f_{cd}$ and the distance $\beta_2 x$ of the force itself from the compressed side. The force is a function of the depth x of the neutral axis and the coefficients β_1 and β_2.

In bending case, if the deformation of steel in tension is greater or at least equal to the deformation corresponding to the design elastic limit $\varepsilon_{syd} = \frac{f_{yd}}{E_s}$ (this happens if $x \leq x_{\text{lim}}$), translational equilibrium equation can be written as

$$A_s f_{yd} = bx\beta_1 f_{cd} \tag{7.18}$$

and rotational equilibrium equation referred to the level of tensile reinforcement is given by

$$M_{Rd} = bx\beta_1 f_{cd}(d - \beta_2 x) \tag{7.19}$$

By introducing the dimensionless notations:

- relative depth of the neutral axis

$$\xi = \frac{x}{d} \tag{7.20}$$

- reduced moment

$$\mu = \frac{M_{Rd}}{bd^2 f_{cd}} \tag{7.21}$$

- reinforcement mechanical ratio

$$\omega = \frac{A_s f_{yd}}{bdf_{cd}} \tag{7.22}$$

the previous relationships become

$$\beta_1\xi = \omega \tag{7.23}$$

$$\mu = \beta_1\xi(1 - \beta_2\xi) \text{ or } \mu = \omega(1 - \frac{\beta_2}{\beta_1}\omega) \tag{7.24}$$

These expressions show that, in the domain of $\xi < \xi_{\text{lim}}$, μ and ω are univocal functions of ξ. Based on this observation, tables μ–ω–ξ have been built for cross-sections with reinforcement only in tension ω and for cross-sections with reinforcement in tension ω and with reinforcement in compression ω' for different ratios ω'/ω, and, as a constitutive law, the generalized parabola-rectangle diagram (Tables U1–U5 in the Appendix).

The value of $\xi_{\text{lim}} = \frac{x_{\text{lim}}}{d} = \frac{\varepsilon_{cu}}{\varepsilon_{cu}+\varepsilon_{syd}}$ is depending on f_{ck} through ε_{cu}, and on $\varepsilon_{syd} = \frac{f_{yd}}{E_s}$ which is given by $\frac{391}{200{,}000} = 1.96/1000$ for steels B450C and B450A.

ξ_{lim} corresponds to values of ω_{lim} and μ_{lim} reported in Tables 7.9 and 7.10, one for each constitutive law. These values are also reported in the last row of Tables from U1 to U5 in the Appendix.

Table 7.9 ξ_{lim}, ω_{lim} and μ_{lim} as a function of f_{ck}—generalized parabola—rectangle diagram

f_{ck} (N/mm^2)	ξ_{lim},	ω_{lim}	μ_{lim}
≤50	0.641	0.519	0.380
55	0.613	0.455	0.345
60	0.598	0.415	0.321
70	0.580	0.369	0.292
80	0.570	0.342	0.273
90	0.570	0.332	0.266

Table 7.10 ξ_{lim}, ω_{lim} and μ_{lim} as a function of f_{ck}—bilinear diagram

f_{ck} (N/mm^2)	ξ_{lim}	ω_{lim}	μ_{lim}
≤50	0.641	0.481	0.360
55	0.613	0.435	0.335
60	0.598	0.402	0.315
70	0.580	0.365	0.291
80	0.570	0.329	0.265
90	0.570	0.318	0.257

7.6.1.1 Solving Formulas for Uniaxial Bending and Axial Force Combined with Bending

Uniaxial Bending

If $M_{Ed} < M_{\lim}$, it is not necessary to arrange reinforcement in compression. In this case, neutral axis depth is given by

$$x = \frac{d}{2\beta_2}\left[1 - \sqrt{1 - \frac{4M_{Rd}\beta_2}{\beta_1 bd^2 f_{cd}}}\right] \tag{7.25}$$

obtained by Eq. (7.19), by imposing $M_{Ed} = M_{Rd}$. Using (7.18), we get the following reinforcement A_s in tension

$$A_s = \frac{bx\beta_1 f_{cd}}{f_{yd}}$$

Using dimensionless notations, previous formulas can be written as

$$\xi = \frac{1}{2\beta_2}\left[1 - \sqrt{1 - 4\mu\frac{\beta_2}{\beta_1}}\right] \tag{7.26}$$

$$\omega = \beta_1\xi \tag{7.23}$$

If A_s is assigned, such as $\omega < \omega_{\lim}$, from the previous equations we obtain

$$x = \frac{A_s f_{yd}}{b\beta_1 f_{cd}};\ M_{Rd} = b\,x\,\beta_1 f_{cd}(d - \beta_2 x)$$

With dimensionless notations, by Eq. (7.22), we obtain ω; then, through (7.23), we obtain:

$$\xi = \frac{\omega}{\beta_1}$$

$$\mu = \omega\left(1 - \frac{\beta_2}{\beta_1}\omega\right) \tag{7.27}$$

If $M_{Ed} > M_{lim}$, we can write: $M_{Ed} = M_{lim} + \Delta M$.

The reinforcement in tension $A_{s,lim} = \frac{b \cdot x_{\lim} \cdot \beta_1 \cdot f_{cd}}{f_{yd}}$ is corresponding to M_{lim} ($A_{s,lim} = \frac{M_{\lim}}{z_{\lim} \cdot f_{yd}}$, being $z_{\lim} = d - \beta_2 \cdot x_{\lim}$ the lever arm of the internal couple corresponding to $x_{\lim}$). The reinforcement area $A'_s = \frac{\Delta M}{(d-d') \cdot f_{yd}}$ (in compression and in tension) is corresponding to ΔM.

With dimensionless notations, once assigned $\Delta\mu = \frac{\Delta M}{bd^2 f_{cd}}$ and $\delta' = \frac{d'}{d}$ (for d and d' see Fig. 7.7), the reinforcement in compression is given by

$$\omega' = \frac{\Delta\mu}{1 - \delta'} \tag{7.28}$$

and the reinforcement in tension is given by

$$\omega = \omega_{\lim} + \omega' \tag{7.29}$$

Axial Force and Bending Moment

The total moment is calculated about the reinforcement in tension (Fig. 7.7)

$$M_{Esd} = M_{Ed} - N_{Ed}\, y_s$$

If $M_{Esd} > M_{\lim}$, we can write: $M_{Esd} = M_{\lim} + \Delta M$ and tensile reinforcement is given by

$$A_s = \frac{1}{f_{yd}}\left(\frac{M_{\lim}}{z_{\lim}} + \frac{\Delta M}{(d - d')} + N_{Ed}\right)\ (N_{Ed}\ \text{negative if compressive}) \tag{7.30}$$

being $z_{\lim} = d - x_{\lim}\beta_2$.

Compressive reinforcement is given by

$$A'_s = \frac{\Delta M}{(d - d')f_{yd}} \tag{7.31}$$

With dimensionless notations, once it is assigned that

$$\nu = \frac{N_{Ed}}{bdf_{cd}} \tag{7.32}$$

we obtain

$$\omega = \omega_{\lim} + \frac{\Delta\mu}{1 - \delta'} + \nu \tag{7.33}$$

$$\omega' = \frac{\Delta\mu}{1 - \delta'} \tag{7.34}$$

If $M_{Esd} < M_{lim}$, reinforcement in compression is not required. x is calculated as a function of M_{Esd} and the reinforcement is given by

$$A_s = \frac{b \cdot x \cdot \beta_1 \cdot f_{cd}}{f_{yd}} + \frac{N_{Ed}}{f_{yd}} \tag{7.35}$$

In dimensionless terms, if $\mu < \mu_{lim}$
, we obtain.$\xi = \frac{1}{2\beta_2}\left[1 - \sqrt{1 - 4\mu\frac{\beta_2}{\beta_1}}\right]$
which is (7.26)

and

$$\omega = \beta_1\xi + \nu \tag{7.36}$$

(for axial force combined with bending, N_{Ed} and ν are negative).

7.6.2 Rectangular Cross-Section: Rectangular Stress Diagram

Analogous formulations can be obtained by assuming a uniform stress diagram according to [3.1.7(3)] or to (2.1.5) $\sigma_c = \eta f_{cd}$ on the height $y = \lambda x$. η and λ values given by EC2 in 3.1.7(3) are here reported in Table 7.11.

Table 7.11 A λ and η as a function of f_{ck}

f_{ck} (N/mm^2)	λ	η
≤50	0.8000	1.0000
55	0.7875	0.9750
60	0.7750	0.9500
70	0.7500	0.9000
80	0.7250	0.8500
90	0.7000	0.8000

7.6.2.1 Uniaxial Bending

Dimensional formulas for bending with $M_{Ed} \leq M_{lim}$:

Translational equilibrium equation is given by

$$b\, y\, \eta\, f_{cd} = A_s f_{yd} \tag{7.37}$$

Rotational equilibrium equation is given by

$$b\, y\, \eta\, f_{cd}\left(d - \frac{y}{2}\right) = M_{Rd} \tag{7.38}$$

If $\frac{y}{d} = k$, equations become

$$b\, k\, d\, \eta\, f_{cd} = A_s f_{yd} \tag{7.39}$$

$$b\, d^2\, k\, \eta\, f_{cd}\left(1 - \frac{k}{2}\right) = M_{Rd} \tag{7.40}$$

When geometry, A_s and materials are assigned, the value of y can be obtained with Eq. (7.37) and then M_{Rd} can be calculated applying formula (7.38). Analogously, it is possible to calculate k using (7.39) and (7.40).

If we want to design A_s for a given M_{Ed}, k is obtained by (7.40) and it can be substituted in (7.39). In such a case we obtain the equation

$$k^2 - 2k + \frac{2M_{Ed}}{bd^2\eta f_{cd}} = 0$$

which can be solved, placing

$$k = 1 - \sqrt{1 - \frac{2M_{Ed}}{bd^2\eta f_{cd}}} \tag{7.41}$$

In dimensionless terms, being

$$\mu = \frac{M_{Rd}}{bd^2 f_{cd}}, \quad \omega = \frac{A_s f_{yd}}{bd f_{cd}}$$

Equations (7.37) and (7.38) can be written as

$$\omega = k\eta \tag{7.42}$$

$$\mu = k\eta\left(1 - \frac{k}{2}\right) = \omega\left(1 - \frac{\omega}{2\eta}\right) \tag{7.43}$$

and the equation which allows obtaining k is the following

$$k = 1 - \sqrt{1 - \frac{2\mu}{\eta}} \tag{7.44}$$

Finally, we get

$$\omega = \eta - \sqrt{\eta^2 - 2\eta\mu} \tag{7.45}$$

If $f_{ck} \leq 50$ N/mm^2, being $\eta = 1$, Eqs. (7.42–7.44) allow to obtain the particular relationships

$$\omega = k = 1 - \sqrt{1 - 2\mu} \tag{7.46}$$

$$\mu = \omega(1 - \frac{\omega}{2}) \tag{7.47}$$

Expressions from (7.37) to (7.47) are valuable for reinforcements with stress f_{yd}, for $y \leq y_{\lim} = \lambda x_{\lim}$, i.e. for $k \leq k_{\lim} = \lambda \xi_{\lim}$.

$k_{\lim}$, $\omega_{\lim}$ and $\mu_{\lim}$ are reported in Table 7.12.

If $M > M_{\lim}$, the procedure is analogous to the one already discussed in Sect. 7.6.1.1.

Table 7.12 $k_{\lim}$, $\omega_{\lim}$ and $\mu_{\lim}$ as a function of f_{ck}

f_{ck} (N/mm^2)	$k_{\lim}$	$\omega_{\lim}$	$\mu_{\lim}$
≤50	0.513	0.513	0.381
55	0.483	0.471	0.357
60	0.463	0.394	0.312
70	0.434	0.391	0.306
80	0.413	0.351	0.278
90	0.400	0.320	0.256

7.6.3 *Examples of Application for Rectangular Cross-Sections in Uniaxial Bending (7.6.1.1.1 and 7.6.2)*

In the following Examples, steel is B450C with $f_{yk} = 450$ N/mm^2 and $f_{yd} = 391$ N/mm^2.

7.6.3.1 Example 1—Bending, Reinforcement Design

Problem data (Fig. 7.8).

Cross-section: $b = 300$ mm; $h = 550$ mm; $d = 500$ mm.

Concrete: C25/30; $f_{cd} = 14.16$ N/mm^2.

$M_{Ed} = 200$ kNm $= 200 \times 10^6$ Nmm.

Preliminarily, we calculate M_{lim}: according to Table 7.9, we get $\mu_{\text{lim}} = 0.380$, from which $M_{lim} = 0.380 \cdot 300 \cdot 500^2 \cdot 14.16 = 403$ kNm.

Dimensional Procedure

Parabola-rectangular diagram:

$\beta_1 = 0.8095$; $\beta_2 = 0.4160$ (Table 7.1)

Reference to the expressions introduced in 7.6.1

The neutral axis according to (7.25) is given by

$$x = \frac{500}{2 \cdot 0.416}\left[1 - \sqrt{1 - \frac{4 \cdot 200 \times 10^6 \cdot 0.416}{0.8095 \cdot 300 \cdot 500^2 \cdot 14.16}}\right] = 130.5\,\text{mm}$$

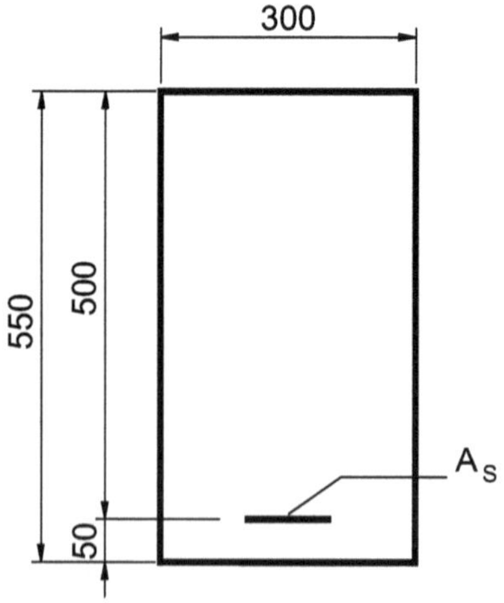

Fig. 7.8 Bending—Example 1: cross-section and reinforcement arrangement

$$\frac{x}{d} = \frac{130.5}{500} = 0.261 < \left(\frac{x}{d}\right)_{\lim} = 0.64$$

$$A_s = \frac{300 \cdot 130.5 \cdot 0.8095 \cdot 14.16}{391} = 1148\,\mathrm{mm}^2\ (3\phi 22)$$

Bilinear diagram
$\beta_1 = 0.750$; $\beta_2 = 0.3888$ (Table 7.2)
Neutral axis:

$$x = \frac{500}{2 \cdot 0.3888}\left[1 - \sqrt{1 - \frac{4 \cdot 200 \times 10^6 \cdot 0.3888}{0.75 \cdot 300 \cdot 500^2 \cdot 14.16}}\right] = 141.0\,\mathrm{mm}$$

$$A_s = \frac{300 \cdot 141 \cdot 0.75 \cdot 14.10}{391} = 1144\,\mathrm{mm}^2$$

With Dimensionless Notations and Parabola–Rectangle Diagram

Using (7.21) and (7.26) we get:

$$\mu = \frac{200 \times 10^6}{300 \cdot 500^2 \cdot 14.16} = 0.1883$$

$$\xi = \frac{1}{2 \cdot 0.416}\left[1 - \sqrt{1 - 4 \cdot 0.1883 \cdot \frac{0.4160}{0.8095}}\right] = 0.261$$

Through (7.23) we get:
$\omega = 0.8095 \cdot 0.261 = 0.2113$, from which:

$$A_s = 0.2113 \cdot 300 \cdot 500 \cdot \frac{14.16}{391} = 1148\,\mathrm{mm}^2$$

With Dimensionless Notations and Rectangular Stress Diagram

Being $f_{ck} = 25$ N/mm^2, $\eta = 1$ and $\lambda = 0.80$.
From the previous calculation, $\mu = 0.1883$, and, by (7.46), we get:
$\omega = 1 - \sqrt{1 - 2 \cdot 0.1883} = 0.2104$ and, therefore, $k = 0.2104$

$$\xi = \frac{x}{d} = \frac{y}{\lambda d} = \frac{k}{\lambda} = \frac{0.2104}{0.8} = 0.2631$$

These values are practically coincident with the ones of the previous paragraph.

With Dimensionless Notations and Use of Table U1 (Appendix)

We enter with $\mu = 0.19$ (slightly rounded up). We get: $\omega = 0.2134$; $\xi = 0.2636$.
Then, we obtain $\frac{\omega b d f_{cd}}{f_{yd}} = \frac{0.2134 \cdot 300 \cdot 500 \cdot 14.16}{391} = 1159\,\text{mm}^2$.

7.6.3.2 Example 2—Bending, Calculation of the Design Bending Resistance

Problem data (Fig. 7.9).
Cross-section: $b = 300$ mm; $h = 550$ mm; $d = 500$ mm; $A_s = 1886$ mm^2 ($6\phi 20$ mm).
Concrete: C30/37; $f_{cd} = 17$ N/mm^2.

Procedure with Dimensional Notations—Parabola–Rectangle Diagram ($\beta_1 = 0.8095,\ \beta_2 = 0.4160$)

Neutral axis

$$x = \frac{1886 \cdot 391}{300 \cdot 0.8095 \cdot 17} = 178.6\,\text{mm}\left(\frac{x}{d} = 0.357 < 0.64\right)$$

$$\begin{aligned} M_{Rd} &= 300 \cdot 178.6 \cdot 0.8095 \cdot 17 \cdot (500 - 0.4160 \cdot 178.6) \\ &= 313.9\,\text{kNm} \end{aligned}$$

With Dimensionless Notations and Table U1 (Appendix)

$\omega = \frac{1886 \cdot 391}{300 \cdot 500 \cdot 17} = 0.289$ corresponding, by interpolation, to $\mu = 0.245$.

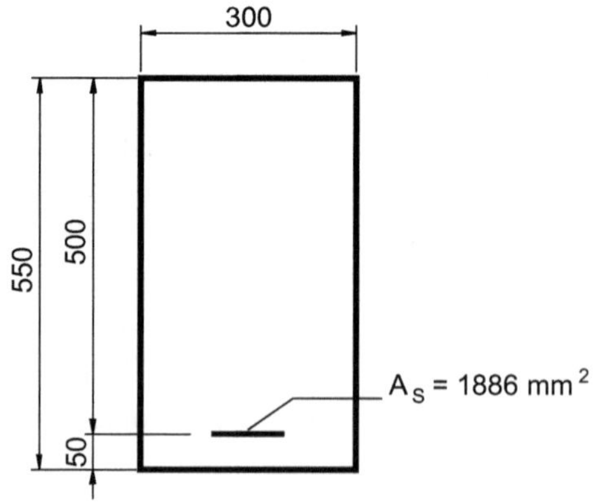

Fig. 7.9 Bending—Example 2: cross-section and reinforcement

We get:

$$M_{Rd} = 0.245 \cdot 300 \cdot 500^2 \cdot 17 = 312.3\,\text{kNm}$$

With Dimensionless Notations and Rectangular Stress Diagram ($\eta = 1$ Being $f_{ck} < 50$ N/mm^2)

Being $\omega = 0.289$, according to the previous point we get:

$$\xi = \frac{0.289}{0.8} = 0.361$$

$\mu = 0.289\left(1 - \frac{0.289}{2}\right) = 0.247$, which is a value analogous to that one already obtained.

7.6.3.3 Example 3—Bending, Calculation of Reinforcements

Problem data (Fig. 7.10).

Cross-section: $b = 300$ mm; $h = 550$ mm; $d = 500$ mm.

Concrete: C70/85; $f_{cd} = 39.66$ N/mm^2.

$M_{Ed} = 500$ kNm.

Preliminarily M_{lim} is calculated: from Table 7.9, $\mu_{\text{lim}} = 0.292$, from which

$$M_{\text{lim}} = 0.292 \cdot 300 \cdot 500^2 \cdot 39.66 = 868\,\text{kNm}$$

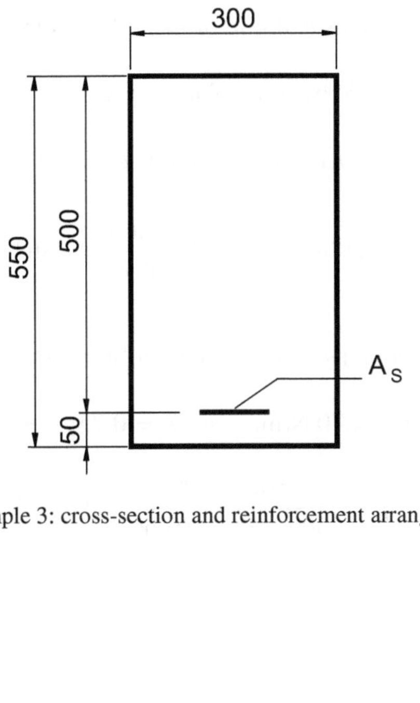

Fig. 7.10 Bending—Example 3: cross-section and reinforcement arrangement

With Dimensional Notations—Generalized Parabola–Rectangle Diagram ($\beta_1 = 0.6372$; $\beta_2 = 0.3620$)

Neutral axis

$$x = \frac{500}{2 \cdot 0.362}\left[1 - \sqrt{1 - \frac{4 \cdot 500 \times 10^6 \cdot 0.362}{0.637 \cdot 300 \cdot 500^2 \cdot 39.66}}\right] = 147.7\,\text{mm}$$

$$\frac{x}{d} = \frac{147.7}{500} = 0.295 < \frac{x_{\text{lim}}}{d} = 0.58$$

$$A_s = \frac{bx\beta_1 f_{cd}}{f_{yd}} = \frac{300 \cdot 147.7 \cdot 0.637 \cdot 39.66}{391} = 2864\,\text{mm}^2$$

With Dimensionless Notations—Generalized Parabola–rectangle Diagram

$$\mu = \frac{500 \cdot 10^6}{300 \cdot 500^2 \cdot 39.66} = 0.1681$$

$$\xi = \frac{1}{2 \cdot 0.362}\left[1 - \sqrt{1 - \frac{4 \cdot 0.1681 \cdot 0.362}{0.637}}\right] = 0.295$$

$$\omega = \beta_1 \xi = 0.637 \cdot 0.295 = 0.1882$$

$$A_s = 0.1882 \cdot 300 \cdot 500 \cdot \frac{39.66}{391} = 2864\,\text{mm}^2$$

With Dimensionless Notations and Use of Table U3 (Appendix)

Being $\mu = 0.1681$, Table U3, by interpolation, gives:

$$\omega = 0.1780 + 0.8 \cdot 0.0127 = 0.1882$$

$$A_s = 0.1882 \cdot 300 \cdot 500\frac{39.66}{391} = 2864\,\text{mm}^2$$

With Dimensionless Notations and Rectangular Stress Diagram

Coefficients related to $f_{ck} = 70\,\text{N/mm}^2$ are $\lambda = 0.75$; $\eta = 0.90$

$$\mu = \frac{500 \times 10^6}{300 \cdot 500^2 \cdot 39.66} = 0.1681$$

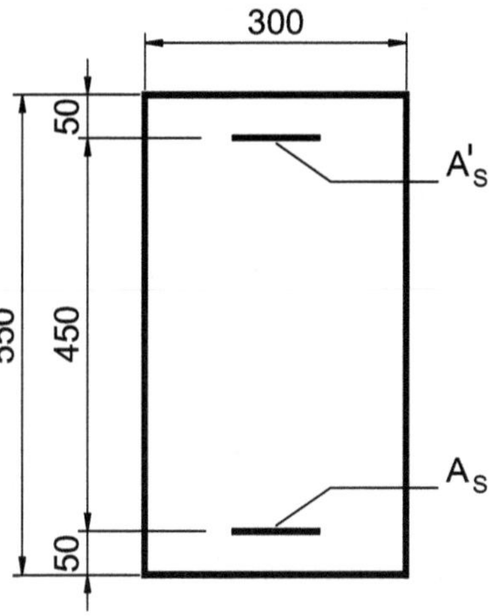

Fig. 7.11 Bending—Example 4: cross-section and reinforcement arrangement

with (7.45)

$$\omega = \eta - \sqrt{\eta^2 - 2\mu\eta} = 0.90 - \sqrt{0.90^2 - 2 \cdot 0.1681 \cdot 0.90} = 0.1877$$

$$A_s = 0.1877 \cdot 300 \cdot 500 \cdot \frac{39.66}{391} = 2855\,\text{mm}^2$$

7.6.3.4 Example 4—Bending, Reinforcement Design

Problem data
Cross-section: $b = 300$ mm; $h = 550$ mm; $d = 500$ mm (Fig. 7.11)
Concrete: C25/30; $f_{cd} = 14.16$ N/mm^2
$M_{Ed} = 450$ kNm

7.6.3.5 With Dimensional Notations—Parabola–Rectangle Diagram ($\beta_1 = 0.8095$, $\beta_2 = 0.4160$)

Neutral axis

$$x = \frac{500}{2 \cdot 0.416}\left[1 - \sqrt{1 - \frac{4 \cdot 450 \cdot 10^6 \cdot 0.416}{0.8095 \cdot 300 \cdot 500^2 \cdot 14.16}}\right] = 385\,\text{mm}$$

$\frac{x}{d} = \frac{385}{500} = 0.77 > \left(\frac{x}{d}\right)_{\text{lim}} = 0.64$: compressive reinforcement is required.

Limit value of x is given by:

$$x_{lim} = 0.64 \cdot 500 = 320\,\text{mm}$$

The limit moment is given by:

$$M_{\text{lim}} = 300 \cdot 320 \cdot 0.8095 \cdot 14.16 \cdot (500 - 0.416 \cdot 320) = 404\,\text{kNm}$$

and the corresponding reinforcement in tension is given by

$$A_s = \frac{300 \cdot 320 \cdot 0.8095 \cdot 14.16}{391} = 2814\,\text{mm}^2$$

The complementary moment that reinforcements A'_s, in tension and in compression, have to equilibrate is given by:

$$\Delta M = M_{Ed} - M_{\text{lim}} = 450 - 404 = 46\,\text{kNm}$$

Being the distance between reinforcements equal to 450 mm, we get:

$$A'_s = \frac{46 \cdot 10^6}{391 \cdot 450} = 261\,\text{mm}^2$$

Finally, we get:

$$A_s = 2814 + 261 = 3075\,\text{mm}^2$$

$$A'_s = 261\,\text{mm}^2$$

With Dimensionless Notations

According to Table 7.9, we obtain:

$$\mu_{\text{lim}} = 0.38$$

$\mu = \frac{450 \times 10^6}{300 \cdot 500^2 \cdot 14.16} = 0.424 > 0.38$ and, therefore, $\Delta\mu = 0.044$.
According to Table 7.9, we get $\mu_{\text{lim}} = 0.38$, corresponding to $\omega_{\text{lim}} = 0.519$

$$\omega' = \frac{\Delta\mu}{(1 - \delta')} = \frac{0.044}{(1 - 0.1)} = 0.049$$

Therefore, we get:

$$\omega = 0.519 + 0.049 = 0.568$$

$$\omega' = 0.049$$

Corresponding tensile reinforcement is given by

$$A_s = \frac{0.568 \cdot 300 \cdot 500 \cdot 14.16}{391} = 3086\,\text{mm}^2$$

and reinforcement in compression is given by

$$A_s = \frac{0.049 \cdot 300 \cdot 500 \cdot 14.16}{391} = 266\,\text{mm}^2$$

According to Table U1, column $\omega'/\omega = 0.1$, we get:
$\mu = 0.424$ which can be satisfied with $\omega = 0.552$, $\xi = 0.61$.
Reinforcements are therefore given by:
$A_s = 2998\,\text{mm}^2$; $A_s' = 300\,\text{mm}^2$.

7.6.4 Examples of Application for Rectangular Cross-Section with Axial Force Combined with Uniaxial Bending (7.6.1.1.2)

In the following examples steel is B450C with $f_{yk} = 450$ N/mm^2 and $f_{yd} = 391$ N/mm^2.

7.6.4.1 Example 1—Reinforcement Design

Problem data

Rectangular cross-section: $b = 300$ mm; $h = 550$ mm; $d = 500$ mm (Fig. 7.12).

Concrete: C40/50; $f_{cd} = 22.66$ N/mm^2

$N_{Ed} = -600$ kN with eccentricity 1.0 m upwards to the centroid of the concrete cross-section.

Solution in Dimensional Terms—Parabola–Rectangle Diagram
($\beta_1 = 0.8095,\ \beta_2 = 0.4160$)

The problem can be solved by applying the procedure shown in Sects. 7.5 and 7.6.1.1.2.

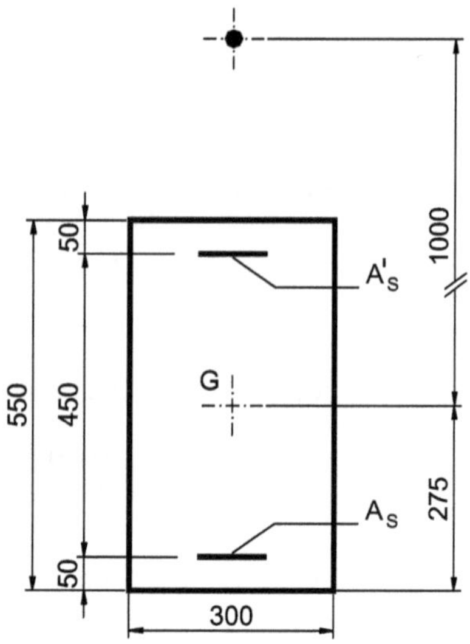

Fig. 7.12 Bending with axial force—Example 1: cross-section and pressure centre

The distance between the application point of N_{Ed} and reinforcement in tension is given by $(e + y_s) = (1000 + 225) = 1225$ mm. Therefore, the moment calculated about the reinforcement in tension is given by

$$M_{\text{Esd}} = 600 \cdot (1.0 + 0.225) = 735\,\text{kNm}$$

Limit moment for ductility (according to Table 7.9, $\mu_{\text{lim}} = 0.38$; $\xi_{\text{lim}} = 0.64$)

$$M_{\text{lim}} = \mu_{\text{lim}} \cdot b \cdot d^2 \cdot f_{cd} = 0.38 \cdot 300 \cdot 500^2 \cdot 22.66 = 646\,\text{kNm}$$

$$\Delta M = 735 - 646 = 89\,\text{kNm}$$

$$x_{\text{lim}} = 0.64 \cdot 500 = 320\,\text{mm}$$

$$z_{\text{lim}} = d - 0.416 \cdot x_{\text{lim}} = 500 - 0.416 \cdot 320 = 366.8\,\text{mm}$$

Therefore, according to (7.30) and (7.31), we get:

$$A_s = \left(\frac{M_{\text{lim}}}{z_{\text{lim}}} + \frac{\Delta M}{d - d'} + N_{Ed}\right) \cdot \frac{1}{f_{yd}} = \left(\frac{646 \times 10^6}{366.8} + \frac{89 \times 10^6}{(500 - 50)} - 600 \times 10^3\right) \cdot \frac{1}{391} = 3474\,\text{mm}^2$$

$$A_s^{'} = \frac{89 \cdot 10^6}{(500 - 50) \cdot 391} = 505\,\text{mm}^2$$

The sum of reinforcements is equal to 3979 mm^2.

Procedure Using Dimensionless Notations

According to (7.32) we get

$$\nu = \frac{600,000}{300 \cdot 500 \cdot 22.66} = 0.1765$$

$\frac{e}{d} = \frac{1225}{500} = 2.45$ eccentricity to the reinforcement in tension.
The reduced moment evaluated about the reinforcement in tension is given by:

$$\mu = 0.1765 \cdot 2.45 = 0.4324 > \mu_{\text{lim}} = 0.38$$

$$\Delta\mu = 0,4324 - 0,38 = 0,0524$$

According to Table 7.9 with $\omega' = 0$, $\omega_{\text{lim}} = 0,519$ is corresponding to $\mu_{\text{lim}} = 0.38$.

For the reinforcement couple: if $\delta' = \frac{d'}{d} = 0.1$ then $\omega' = \frac{\Delta\mu}{1-\delta'} = \frac{0.052}{0.9} = 0.058$.
On the tension side: $\omega = \omega_{\text{lim}} + \omega' - \nu = 0.519 + 0.058 - 0.176 = 0.401$.
On the compression side: $\omega' = 0.058$.
Corresponding reinforcements are given by

$$A_s = 0.401 \cdot 300 \cdot 500 \cdot \frac{22.66}{391} = 3483\,\text{mm}^2$$

$$A_s' = 0.058 \cdot 300 \cdot 500 \cdot \frac{22.66}{391} = 505\,\text{mm}^2$$

Solution in Dimensionless Terms by Use of Tables (Appendix)

According to Table U1 $\left(\omega'/\omega = 0.2\right)$.

The value corresponding to $\mu = 0.4324$ (calculated above), by interpolation, is given by

$$\omega = 0.5400;\ \omega' = 0.108$$

On the side in tension, it is necessary to arrange $\omega = 0.5400 - 0.1765 = 0.3635$, corresponding to 3150 mm^2.

On the side in compression: $\omega' = 0.108$, corresponding to 937 mm^2.

The sum of the two reinforcements, 4087 mm^2, is practically equal to the sum obtained by the first procedure.

Solution with Rectangular Stress Diagram (7.6.2)

Limit moment

According to Table 7.12: $\mu_{\text{lim}} = 0.381$. Therefore,

$$M_{\text{lim}} = 0.381 \cdot 300 \cdot 500^2 \cdot 22.66 = 646\,\text{kNm}$$

Moreover: $k_{\text{lim}} = 0.513$

$$y_{\text{lim}} = 0.513 \cdot 500 = 256\,\text{mm}$$

$$z_{\text{lim}} = d - \frac{y_{\text{lim}}}{2} = 372\,\text{mm}$$

$$A_s = \left(\frac{646 \times 10^6}{372} + \frac{89 \times 10^6}{(500 - 50)} - 600 \times 10^3\right) \cdot \frac{1}{391} = 3412\,\text{mm}^2$$

$$A_s' = \frac{89 \times 10^6}{(500 - 50) \cdot 391} = 505\,\text{mm}^2.$$

7.6.4.2 Example 2—Reinforcement Design

Problem data (Fig. 7.13)

Rectangular cross-section: $b = 400$ mm; $h = 700$ mm; $d = 650$ mm.

Concrete: C20/25; $f_{cd} = 11.33$ N/mm^2.

$N_{Ed} = -700$ kN with eccentricity 1.0 m from the centroid of the concrete cross-section.

Procedure with Dimensional Notations—Parabola–Rectangle Diagram ($\beta_1 = 0.8095,\ \beta_2 = 0.4160$)

The procedure is analogous to the one already conducted in the previous case.

The moment calculated about the tensile reinforcement is given by

$$M_{\text{Esd}} = 700 \cdot (1.0 + 0.30) = 910\,\text{kNm}$$

The limit moment for ductility (according to Table 7.9, $\mu_{\text{lim}} = 0.38;\ \xi_{\text{lim}} = 0.64$) is given by

$$M_{\text{lim}} = 0.38 \cdot 400 \cdot 650^2 \cdot 11.33 = 727\,\text{kNm}$$

Fig. 7.13 Axial force combined with uniaxial bending—Example 2: cross-section and pressure center

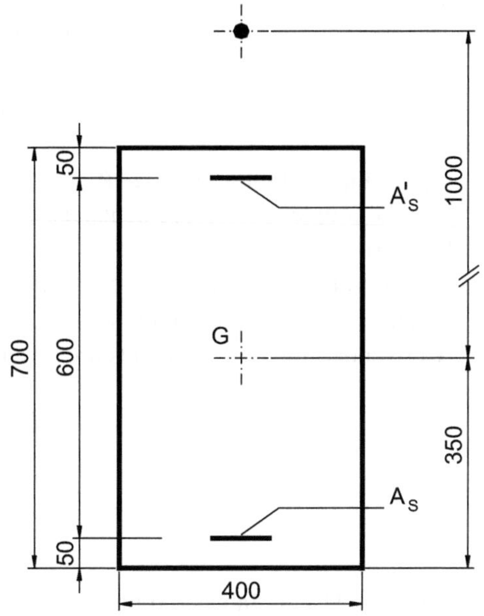

$$\Delta M = 910 - 727 = 183\,\text{kNm}$$

$$x_{\lim} = 0.64 \cdot 650 = 416\,\text{mm}$$

$$z_{\lim} = d - 0.416 \cdot x_{\lim} = 650 - 0.416 \cdot 416 = 477\,\text{mm}$$

$$\begin{aligned} A_s &= \left(\frac{M_{\lim}}{z_{\lim}} + \frac{\Delta M}{d - d'} + N_{Ed}\right) \cdot \frac{1}{f_{yd}} \\ &= \left(\frac{727 \cdot 10^6}{477} + \frac{183 \cdot 10^6}{(650 - 50)} - 700 \cdot 10^3\right) \cdot \frac{1}{391} \\ &= 2888\,\text{mm}^2 \end{aligned}$$

$$A_s^{'} = \frac{183 \cdot 10^6}{(650 - 50) \cdot 391} = 780\,\text{mm}^2$$

Solution in Dimensionless Terms—Parabola–Rectangle Diagram

$\nu = \frac{700{,}000}{400 \cdot 650 \cdot 11.33} = 0.238$ is the relative eccentricity from the tensile reinforcement

$$\frac{e}{d} = \frac{1300}{650} = 2.0$$

$$\mu = 0.2376 \cdot 2.0 = 0.4752 > \mu_{\lim} = 0.38$$

$$\Delta\mu = 0.4752 - 0.38 = 0.0952$$

According to Table U1 ($\omega' = 0$),$\omega_{\lim} = 0.518$ is corresponding to $\mu_{\lim} = 0.38$.

For the couple of reinforcements: if $\delta' = \frac{d'}{d} = 0.1$ is assigned, we get $\omega' = \frac{\Delta\mu}{1-\delta'} = \frac{0.0952}{0.9} = 0.106$.

On the side in tension: $\omega = \omega_{\lim} + \omega' - \nu = 0.518 + 0.106 - 0.238 = 0.386$.

On the side in compression: $\omega' = 0.106$.

Corresponding reinforcements are given by:

$$A_s = 2908\,\text{mm}^2$$

$$A_s' = 797\,\text{mm}^2$$

Solution by Use of Tables (Appendix)

According to Table U1 (assigned ratio $\omega'/\omega = 0.3$).

$\omega = 0.575$; $\omega' = 0.172$ are corresponding to the value $\mu = 0.4752$,

On the tension side, it is necessary to arrange $\omega = 0.575 - 0.238 = 0.337$, corresponding to 2538 mm^2.

On the side in compression, it is necessary to arrange $\omega' = 0.172$, corresponding to 1300 mm^2.

It is easy to verify that the sums of the reinforcements required by the two dimensionless procedures are slightly different from each other.

Solution in Dimensional Terms—Bilinear Diagram

According to Table 7.2: $\beta_2 = 0.3888$.

According to Table 7.10: $\xi_{\lim} = 0.641$; $\omega_{\lim} = 0.481$; $\mu_{\lim} = 0.360$.

Being $M_{Esd} = 910$ kNm, we get:

$$M_{\lim} = 0.360 \cdot 400 \cdot 650^2 \cdot 11.33 = 689\,\text{kNm}$$

$$\Delta M = 910 - 689 = 221\,\text{kNm}$$

$$x_{\lim} = 0.641 \cdot 650 = 416\,\text{mm}$$

$$z_{\lim} = d - \beta_2 x_{\lim} = 650 - 0.388 \cdot 416 = 488\,\text{mm}$$

$$A_s = \frac{689 \times 10^6}{488 \cdot 391} + \frac{221 \times 10^6}{600 \cdot 391} - \frac{700 \times 10^3}{391}$$
$$= 3611 + 942 - 1790 = 2763\,\text{mm}^2$$

$$A'_s = 942\,\text{mm}^2$$

7.7 T-shaped Cross-Section in Bending with or Without Axial Force

7.7.1 T-shaped Cross-Section in Bending

T-shaped cross-sections are the most rational solutions to resist positive bending moment: the extended flange is appropriate to support compressive force, while the reinforcement, placed on the bottom side of the web, supports tensile forces.

To perform the design, two problems can arise: calculate reinforcement once the cross-section, the materials and the force are assigned; otherwise, if the cross-section and the reinforcement are assigned, calculate strength. Solutions can be obtained quite easily by assuming a uniform stress distribution within the concrete (rectangular diagram).

In the case of bending, three sectors can be identified for the solution:

- Sector 1: the depth y of the portion of cross-section under uniform compressive force is contained in the flange ($y \leq h_f$). Reinforcement is working at stress f_{yd};
- Sector 2: $h_f < y \leq \lambda\, x_{\lim}$, the area in uniform compression is extended to a portion of the web and the reinforcement in tension is working as in the previous case at a value corresponding to the design elastic limit;
- Sector 3: if the bending requires $y > \lambda\, x_{\lim}$ for the compliance of ductility, the procedure is the one already shown at the beginning of the Chapter. The moment is subdivided into two terms: the first one is the limit moment and the second one is a complementary moment that can be absorbed by two equal amounts of reinforcement, one in compression and another one in tension. However, it should be observed that, for this type of cross-section (extended flange), the last case is not easy to find in bending, mainly in the case of high-resistance concretes. It can arise, on the contrary, in the case of axial force combined with bending.

7.7.1.1 Development of Calculation Procedures

Basic notations

With reference to Fig. 7.14, we set:

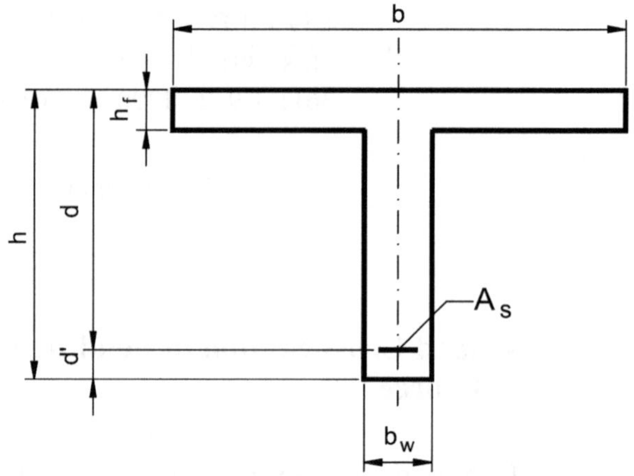

Fig. 7.14 T-shaped cross-section: geometry and notations

$$\frac{b}{b_w} = n; \ \frac{h_f}{d} = w; \ \omega = \frac{A_s f_{yd}}{b d f_{cd}}; \ \mu = \frac{M}{b d^2 f_{cd}}$$

A rectangular stress diagram is adopted (see 7.6.2). The depth of the portion of the section in uniform compression has extension $y = \lambda x$ and the stresses are given by $\sigma_c = \eta f_{cd}$. Coefficients λ and η are reported in Table 7.11.

Sector 1

It is defined by

$$y = \lambda x \leq h_f$$

i.e., once $k = \frac{y}{d}$ is assumed, $k \leq \frac{h_f}{d} = k_1$.

The depth y of the portion of the cross-section under uniform compressive force is contained in the flange.

The expressions used here are those of 7.6.2, from (7.37) to (7.47), established for rectangular cross-sections, both in dimensional and dimensionless terms.

Table 7.13 reports ω_1 and μ_1 values at the limit of Sector 1 ($y = h_f$) as a function of $w = h_f/d$, i.e. for different relative depths of the slab and for different values of f_{ck} until 70 N/mm^2. Values greater than 70 N/mm^2 have no practical interest.

Sector 2

It is defined by $h_f < y \leq \lambda x_{\lim}$.

i.e. $w < k \leq \lambda \xi_{\lim} = k_2$,

where $\xi_{\lim} = \frac{\varepsilon_{cu}}{\varepsilon_{cu} + \varepsilon_{syd}}$.

Table 7.14 reports ε_{cu}, $\xi_{\lim}$ and k_2 values (λ in Table 7.11).

Table 7.13 μ_1 and ω_1 as a function of w for different values of f_{ck}

w	$f_{ck} \leq 50$ N/mm^2		$f_{ck} = 60$ N/mm^2		$f_{ck} = 70$ N/mm^2	
	μ_1	ω_1	μ_1	ω_1	μ_1	ω_1
0.10	0.0950	0.1000	0.0903	0.0950	0.0855	0.0900
0.15	0.1388	0.1500	0.1318	0.1425	0.1249	0.1350
0.20	0.1800	0.2000	0.1710	0.1900	0.1620	0.1800
0.25	0.2188	0.2500	0.2078	0.2375	0.1969	0.2250
0.30	0.2550	0.3000	0.2423	0.2850	0.2295	0.2700
0.35	0.2888	0.3500	0.2743	0.3325	0.2599	0.3150
0.40	0.3200	0.4000	0.3040	0.3800	0.2880	0.3600

Table 7.14 ε_{cu}, ξ_{lim} and $k_2 = \lambda\, \xi_{lim}$ as a function of f_{ck}

f_{ck} (N/mm^2)	ε_{cu}	$\xi_{\lim}$	κ_2
≤50	0.0035	0.6410	0.5128
60	0.0029	0.5967	0.4624
70	0.0027	0.5794	0.4346

Relationships of Sector 2

The area A_c corresponding to the depth y is given by

$$A_c = bh_f + b_w(y - h_f) = bwd + \frac{b}{n}(kd - wd)$$

Translational equilibrium equation is given by

$$bd\left(w + \frac{k}{n} - \frac{w}{n}\right) \cdot \eta f_{cd} = A_s f_{yd}$$

From this, it results

$$\omega = \eta \cdot \left(w + \frac{k - w}{n}\right) \tag{7.48}$$

The rotational equilibrium equation about the reinforcement in tension is given by:

$$M = \eta\, f_{cd}\left[b_w \cdot y \cdot \left(d - \frac{y}{2}\right) + (b - b_w) \cdot h_f \cdot \left(d - \frac{h_f}{2}\right)\right] \tag{7.49}$$

By developing the previous equation, we get

$$y^2 - (2d) \cdot y - \frac{2}{b_w}\left[(b - b_w) \cdot h_f \cdot \left(d - \frac{h_f}{2}\right) - \frac{M}{\eta \cdot f_{cd}}\right] = 0 \tag{7.50}$$

from which

$$y = d - \sqrt{d^2 + \frac{2}{b_w}\left[(b - b_w) \cdot h_f \cdot \left(d - \frac{h_f}{2}\right) - \frac{M}{\eta \cdot f_{cd}}\right]} \tag{7.51}$$

In dimensionless terms, the three previous equations can be written as

$$\mu = \frac{\eta}{2n} \cdot \left[2k - k^2 + 2(n-1)(w - 0.5w^2)\right] \tag{7.52}$$

$$k^2 - 2k - 2(n-1)(w - 0.5w^2) + \frac{2n\mu}{\eta} = 0 \tag{7.53}$$

$$k = 1 - \sqrt{1 + 2(n-1)(w - 0.5w^2) - \frac{2n\mu}{\eta}} \tag{7.54}$$

Tables 7.15, 7.16 and 7.17 report μ_2 and ω_2 values at the superior side of the sector ($k_2 = \lambda\xi_{\text{lim}}$), for concretes with $f_{ck} \leq 50$ N/mm^2, $f_{ck} = 60$ N/mm^2 and $f_{ck} = 70$ N/mm^2, respectively. λ and η values are reported in Table 7.11.

Tables from U6 to U9 (Appendix) report $\mu - \omega - \xi$ correlated values in Sector 2, for $f_{ck} \leq 50$ N/mm^2.

Remark: solutions in Sectors 1 and 2 guarantee ductility ($\varepsilon_s \geq \frac{f_{yd}}{E_s}$).

7.7.1.2 Approximated Method for Bending Verification

If the cross-section and the moment M_{Ed} are assigned, the internal couple is built by the uniform compression on the slab and by reinforcement in tension. In such a case, the lever arm of the couple is given by $z = d - h_f/2$, and the required reinforcement in tension is obtained by the formula $A_s = M_{Ed}/(z f_{yd})$.

It is necessary to verify that compressive stress is allowable, which means that:

Table 7.15 Limit values of μ_2 and ω_2 as a function of w and n for $f_{ck} \leq 50$ N/mm^2

w	$n = 2$		$n = 3$		$n = 4$		$n = 5$	
	μ_2	ω_2	μ_2	ω_2	μ_2	ω_2	μ_2	ω_2
0.10	0.2380	0.3060	0.1903	0.2373	0.1665	0.2030	0.1522	0.1824
0.15	0.2598	0.3310	0.2195	0.2707	0.1993	0.2405	0.1872	0.2224
0.20	0.2805	0.3560	0.2470	0.3040	0.2302	0.2780	0.2202	0.2624
0.25	0.2998	0.3810	0.2728	0.3373	0.2593	0.3155	0.2512	0.3024
0.30	0.3180	0.4060	0.2970	0.3707	0.2865	0.3530	0.2802	0.3424
0.35	0.3348	0.4310	0.3195	0.4040	0.3118	0.3905	0.3072	0.3824
0.40	0.3505	0.4560	0.3403	0.4373	0.3352	0.4280	0.3322	0.4224

Table 7.16 Limit values of μ_2 and ω_2 as a function of w and n for $f_{ck} = 60$ N/mm^2

w	$n = 2$		$n = 3$		$n = 4$		$n = 5$	
	μ_2	ω_2	μ_2	ω_2	μ_2	ω_2	μ_2	ω_2
0.10	0.2140	0.2671	0.1727	0.2098	0.1521	0.1811	0.1397	0.1639
0.15	0.2348	0.2909	0.2005	0.2414	0.1833	0.2167	0.1730	0.2019
0.20	0.2544	0.3146	0.2266	0.2731	0.2127	0.2523	0.2043	0.2399
0.25	0.2728	0.3384	0.2511	0.3048	0.2403	0.2879	0.2338	0.2779
0.30	0.2900	0.3621	0.2741	0.3364	0.2661	0.3236	0.2613	0.3159
0.35	0.3060	0.3859	0.2955	0.3681	0.2902	0.3592	0.2870	0.3539
0.40	0.3209	0.4096	0.3152	0.3998	0.3124	0.3948	0.3107	0.3919

Table 7.17 Limit values of μ_2 and ω_2 as a function of w and n for $f_{ck} = 70$ N/mm^2

w	$n = 2$		$n = 3$		$n = 4$		$n = 5$	
	μ_2	ω_2	μ_2	ω_2	μ_2	ω_2	μ_2	ω_2
0.10	0.1958	0.2405	0.1590	0.1904	0.1406	0.1653	0.1296	0.1502
0.15	0.2155	0.2630	0.1853	0.2204	0.1702	0.1990	0.1611	0.1862
0.20	0.2340	0.2855	0.2100	0.2504	0.1980	0.2328	0.1908	0.2222
0.25	0.2515	0.3080	0.2333	0.2804	0.2242	0.2665	0.2187	0.2582
0.30	0.2678	0.3305	0.2550	0.3104	0.2486	0.3003	0.2448	0.2942
0.35	0.2830	0.3530	0.2753	0.3404	0.2714	0.3340	0.2691	0.3302
0.40	0.2970	0.3755	0.2940	0.3704	0.2925	0.3678	0.2916	0.3662

$$\sigma_c = \frac{M_{Ed}}{z \cdot h_f \cdot b} \leq f_{cd} \cdot \eta$$

If $\mu \leq \mu_1$, verification is fulfilled implicitly.

7.7.2 *Examples of Application for T-shaped Cross-Section in Bending*

In the following examples, steel is B450C with $f_{yk} = 450$ N/mm^2 and $f_{yd} = 391$ N/mm^2.

7.7.2.1 Example 1—Ribbed Slab in Bending—Reinforcement Calculation

Problem data (Fig. 7.15).

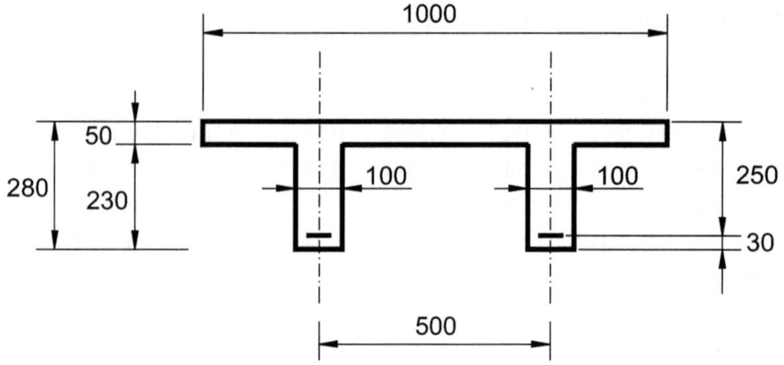

Fig. 7.15 Bending—Example 1: extended slab cross-section

Slab with ribs at distance 500 mm.
$h = 280$ mm; $d = 250$ mm; $h_f = 50$ mm; $b_w = 100$ mm.
Dimensionless parameters: $w = h_f/d = 0.20$; $n = b/b_w = 5$.
Concrete: C25/30; $f_{cd} = 14.16$ N/mm^2.
Load at ULS: $M_{Ed} = 60$ kNm for 1 m strip.

Calculation with Dimensionless Notations

The reduced moment is given by:

$$\mu = \frac{60 \times 10^6}{1000 \cdot 250^2 \cdot 14.16} = 0.0678$$

As $0.0678 < \mu_1 = 0.18$ (Table 7.13), the configuration is placed in Sector 1, i.e. the height of the portion in uniform compression is smaller than h_f.

From the (7.44), valid for rectangular cross-sections, we get:
$k = 1 - \sqrt{1 - \frac{2\mu}{\eta}}$; and being $\eta = 1$:

$$k = 1 - \sqrt{1 - 2 \cdot 0.0678} = 0.0703$$

The depth of the portion under uniform compression is given by:

$$y = kd = 0.0703 \cdot 250 = 18\,\text{mm}$$

$\omega = k = 0.0703$. Required reinforcement is given by:
$A_s = \frac{0.0703 \cdot 1000 \cdot 250 \cdot 14.16}{391} = 636\,\text{mm}^2/\text{m}$ i.e. 318 mm^2 for each rib.

Calculation by the Approximated Method of 7.7.1.2

With the given problem data, we obtain:

$$z = d - \frac{h_f}{2} = 250 - \frac{50}{2} = 225\,\text{mm}$$

$$A_s = \frac{M_{Ed}}{z \cdot f_{yd}} = \frac{60 \times 10^6}{225 \cdot 391} = 682\,\text{mm}^2$$

The verification of the slab is unnecessary because $\mu < \mu_1$.

7.7.2.2 Example 2—T-shaped Cross-Section in Bending—Reinforcement Design

Problem data

T-shaped cross-section with the following mechanical and geometric characteristics (Fig. 7.16).

$b = 800$ mm; $b_w = 200$ mm; $h = 850$ mm; $d = 800$ mm; $h_f = 120$ mm.

$$n = \frac{800}{200} = 4; \; w = \frac{120}{800} = 0.15$$

Concrete: C16/20; $f_{cd} = 9.00$ N/mm^2; $\eta = 1$; $\lambda = 0, 8$.
Design bending moment at ULS: $M_{Ed} = 800$ kNm.
Preliminarily, we identify the sector where the solution has to be searched:

$$\mu = \frac{800 \times 10^6}{800 \cdot 800^2 \cdot 9} = 0.1736$$

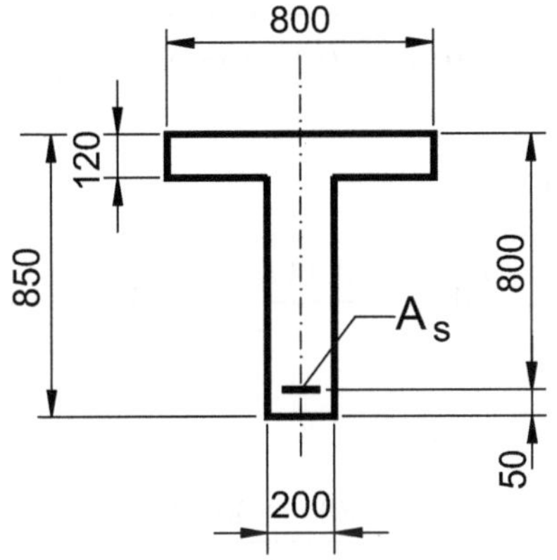

Fig. 7.16 Bending—Example 2: T-shaped cross-section

Being $w = 0.15$ and $f_{ck} = 16\ \mathrm{N/mm^2}$, from Tables 7.13 and 7.15, we get:

$$0.1388 = \mu_1 < 0.1736 < \mu_2 = 0.1993.$$

The solution is then placed in Sector 2.

Dimensional Discussion

Through (7.51), we get:

$$y = 800 - \sqrt{800^2 + \frac{2}{100}\left[(800 - 200) \cdot 120 \cdot (800 - 60) - \frac{800 \times 10^6}{9}\right]} = 267\ \mathrm{mm}$$

$$x = 267/0.8 = 333.9\ \mathrm{mm}$$

$$F_c = [(267-120) \cdot 200 + 800 \cdot 120] \cdot 9 = 1{,}128{,}600\ \mathrm{N}$$

$$\mathrm{F_c = F_s\ then\ A_s} = 1{,}128{,}600/391 = 2886\ \mathrm{mm^2}$$

(solved by 5 bars 28 mm diameter arranged on two layers)

Calculation by Dimensionless Formulas

$$\mu = \frac{800 \times 10^6}{800 \cdot 800^2 \cdot 9} = 0.1736 > \mu_1 = 0.1388$$

By means of (7.54), we can calculate k. We get:

$$k = 1 - \sqrt{1 + 2 \cdot (4 - 1) \cdot \left(0.15 - 0.5 \cdot 0.15^2\right) - 2 \cdot \frac{4}{1} \cdot 0.1736} = 0.334 > 0.15$$

therefore, by means of (7.48)

$$\omega = \left(0.15 + \frac{0.334 - 0.15}{4}\right) = 0.1960$$

and finally:

$$A_s = 0.1960 \cdot \frac{800 \cdot 800 \cdot 9}{391} = 2887\ \mathrm{mm^2}$$

Calculation by Tables (Appendix)

The problem can be easily solved by Table U6.

Being $n = 4$ and $w = 0.15$, the table valid for $\mu = 0.17$ (slightly rounded) gives:

$$\mathrm{x} = 0.39; \ \omega = 0.19.$$

From these values, we get:

$$x = \xi \cdot d = 0.39 \cdot 800 = 312\,\mathrm{mm}$$

$$A_s = 0.19 \cdot \frac{800 \cdot 800 \cdot 9}{391} = 2800\,\mathrm{mm}^2$$

7.7.2.3 Example 3—T-shaped Cross-Section in Bending—Reinforcement Design

Problem data

T-shaped cross-section with the following geometrical and mechanical data (Fig. 7.17).

$b = 400$ mm; $b_w = 200$ mm; $h = 900$ mm; $d = 800$ mm; $h_f = 120$ mm.

$n = 2$; $w = 0.15$.

Concrete: C60/75; $f_{cd} = 34$ N/mm^2; $\eta = 0.95$; $\lambda = 0.775$ (Table 7.11).

Moment at the ULS: $M_{Ed} = 1200$ kNm.

Preliminarily, we identify the sector where the solution has to be searched for:

$$\mu = \frac{1200 \times 10^6}{400 \cdot 800^2 \cdot 34} = 0.1378$$

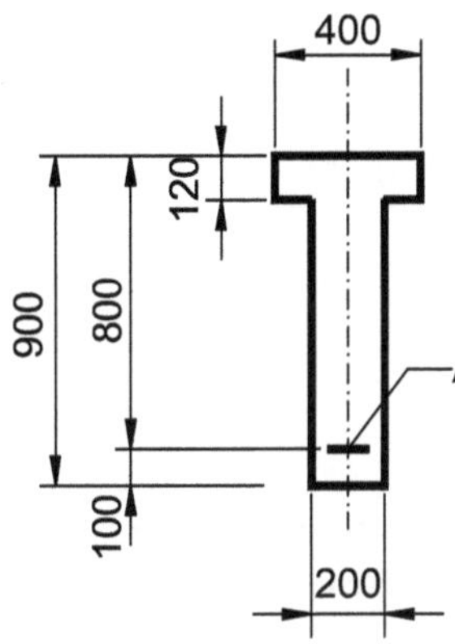

Fig. 7.17 Bending—Example 3: T-shaped cross-section

Being $w = 0.15$ and $f_{ck} = 60\ \text{N/mm}^2$, from Table 7.13 we get $\mu > \mu_1 = 0.1318$. The solution is then placed in Sector 2.

Dimensional Procedure

By means of (7.51), we get:

$$y = 800 - \sqrt{800^2 + \frac{2}{200} \cdot \left[(400 - 200) \cdot 120 \cdot \left(800 - \frac{120}{2}\right) - \frac{1200 \times 10^6}{0.95 \cdot 34}\right]}$$
$$= 132\ \text{mm}$$

$$x = 132/0.775 = 170\ \text{mm}$$

$$F_c = [132 \cdot 200 + 200 \cdot 120] \cdot 34 \cdot 0.95 = 1{,}628{,}000\ \text{N}$$

$$A_s = \frac{1{,}628{,}000}{391} = 4165\ \text{mm}^2$$

Calculation by Dimensionless Formulas

$$\mu = \frac{1200 \times 10^6}{400 \cdot 800^2 \cdot 34} = 0.1378$$

using (7.54)

$$k = 1 - \sqrt{1 + 2 \cdot \left(0.15 - \frac{0.15^2}{2}\right) - \frac{2 \cdot 2}{0.95} \cdot 0.1378} = 0.165 > 0.15$$

through (7.48)

$$\omega = (0,15 + \frac{0.165 - 0.15}{2}) \cdot 0.95 = 0.1496$$
$$A_s = 0.1496 \cdot \frac{400 \cdot 800 \cdot 34}{391} = 4163\ \text{mm}^2$$

7.7.2.4 Example 4—Ribbed Slab in Bending—Calculation of the Design Bending Resistance

It is the opposite problem to the discussed one in Example 1 (Fig. 7.15 of 7.7.2.1).

Here we calculate the design bending resistance M_{Rd} of a slab with geometric characteristics analogous to those of Example 1, but made of concrete C40/45, $f_{cd} = 22.66$ N/mm², $A_s = 400$ mm² for each rib:

$$w = 0.2;\ \eta = 1;\ \lambda = 0.8.$$

Reinforcement mechanical ratio is given by
$\omega = \frac{2 \cdot 400 \cdot 391}{1000 \cdot 250 \cdot 22.66} = 0.0552 < \omega_1 = 0.20$ (Table 7.13)
The solution is then placed in Sector 1.
From (7.43), we get:

$$\mu = 0.0552 \cdot \left(1 - \frac{0.0552}{2}\right) = 0.0537$$

and, therefore, design bending resistance is given by:

$$M_{Rd} = 0.0537 \cdot 1000 \cdot 250^2 \cdot 22.66 = 76\,\text{kNm}$$

7.7.2.5 Example 5—T-shaped Cross-Section in Bending. Calculation of the Design Bending Resistance

Problem data
$b = 400$ mm; $h = 900$ mm; $d = 800$ mm.
$b_w = 200$ mm; $h_f = 120$ mm (Fig. 7.18).
$w = 0.15$; $n = \frac{400}{200} = 2$.
Concrete: C30/37; $f_{cd} = 17$ N/mm²; reinforcement $A_s = 3180$ mm² (6ϕ26)

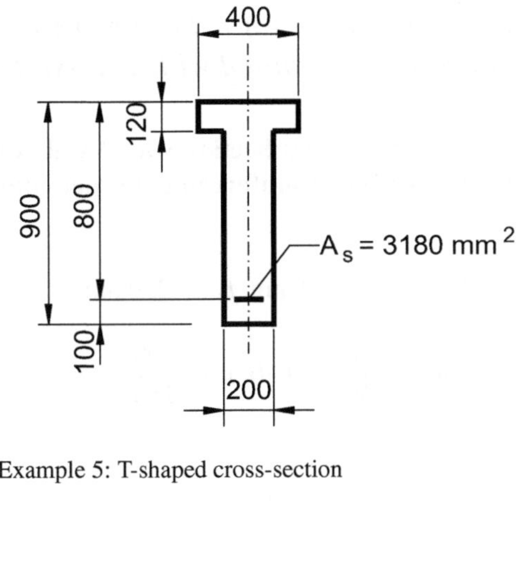

Fig. 7.18 Bending—Example 5: T-shaped cross-section

Reinforcement mechanical ratio is given by:

$\omega = \frac{3180 \cdot 391}{400 \cdot 800 \cdot 17} = 0.2286 > \omega_1 = 0.15$ but also $< \omega_2 = 0{,}331$ (see Tables 7.13 and 7.15).

The solution is then placed within Sector 2.

Once ω is known, through (7.48) we calculate k:

$0.2286 = \left(0.15 + \frac{k-0.15}{2}\right)$, from which $k = 0.3072$.

Therefore, we get: $y = 0.3072 \cdot 800 = 246$ mm.

By means of (7.52), we get:

$$\mu = \frac{1}{2 \cdot 2}\left[2 \cdot 0.3072 - 0.3072^2 + 2 \cdot \left(0.15 - \frac{0.15^2}{2}\right)\right] = 0.1994$$

and therefore:

$$M_{Rd} = 0.1994 \cdot 400 \cdot 800^2 \cdot 17 = 867.7\,\text{kNm}$$

The solution can be immediately found by Table U6: $\omega = 0.2295$ is corresponding to $\mu = 0.200$.

7.7.3 T-shaped Cross-Section with Axial Force and Bending

The procedure is the same as the presented one for rectangular cross-sections, that is considering the moment about the reinforcement in tension.

In the following examples, steel is B450C with $f_{yk} = 450$ N/mm^2 and $f_{yd} = 391$ N/mm^2.

7.7.4 Examples of Application for a T-shaped Cross-Section Under Axial Force Combined with Uniaxial Bending

7.7.4.1 Example 1—T-shaped Cross-Section with Axial Force Combined with Bending. Reinforcement Calculation

Problem data

$b = 500$ mm; $h = 800$ mm; $d = 750$ mm; $b_w = 250$ mm; $h_f = 150$ mm (Fig. 7.19)

$$w = \frac{150}{750} = 0.20;\ n = \frac{500}{250} = 2$$

Concrete: C30/37; $f_{cd} = 17$ N/mm^2.

$N_{Ed} = -500$ kN at the upper edge of the cross-section

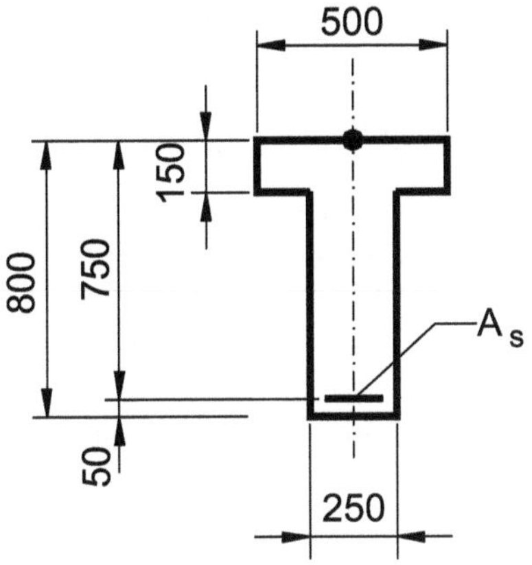

Fig. 7.19 Axial force combined with bending—Example 1: T-shaped cross-section

$M_{Ed} = 500$ kNm
The total moment about the reinforcement in tension is given by:

$$M_{Esd} = 500 + 500 \cdot 0.75 = 875\,\text{kNm}$$

Solution in Dimensional Terms

To identify the sector where the solution is placed, the reduced moment μ is calculated:

$$\mu = \frac{875 \times 10^6}{500 \cdot 750^2 \cdot 17} = 0.183$$

This value is within Sector 2, between $\mu_1 = 0.18$ (Table 7.13) and $\mu_2 = 0.2085$ (Table 7.15) corresponding to $w = 0.20$, $n = 2, f_{ck} \leq 50$ N/mm^2 (Tables 7.16 and 7.17).

Therefore using (7.51) we get:

$$y = 750 - \sqrt{750^2 + \frac{2}{250} \cdot \left[(500 - 250) \cdot 150 \cdot \left(750 - \frac{150}{2}\right) - \frac{875 \times 10^6}{17}\right]}$$
$$= 155\,\text{mm}$$

$$A_c = 155 \cdot 250 + 250 \cdot 150 = 76250\,\text{mm}^2$$

The resultant compressive force is given by $F_c = 76250 \cdot 17 = 1296\,\text{kN}$.
The tensile force is given by $1296 - 500 = 796$ kN.
The required reinforcement in tension is given by $A_s = \frac{796{,}000}{391} = 2036\,\text{mm}^2$.
The reinforcement in compression is not necessary.

Solution in Dimensionless Terms

The reduced moment is given by
$\mu = \frac{875\times10^6}{500\cdot750^2\cdot17} = 0.183 > \mu_1 = 0.1800$ (Table 7.13)
but also $0.183 < \mu_2 = 0.2805$ (Table 7.15).

The solution is then placed in Sector 2. By means of (7.54) we get

$$k = 1 - \sqrt{1 + 2\cdot1\cdot(0.20 - 0.5\cdot0.20^2) - 2\cdot2\cdot0.1830} = 0.2075$$

By means of (7.48) we get:

$$\omega = \left(0.20 + \frac{0.2075 - 0.20}{2}\right) = 0.2038;$$
$$A_s = \frac{0.2038\cdot500\cdot750\cdot17}{391} = 3322\,\text{mm}^2.$$

It is then necessary to subtract to these values the amount corresponding to the normal force transferred to the reinforcements, i.e. $\frac{N_{Ed}}{f_{yd}} = \frac{500{,}000}{391} = 1278\,\text{mm}^2$.

Finally, the following amount is required: $3322 - 1278 = 2044$ mm^2.

By Table U7, approximately but rapidly, we found out that $\omega = 0.2062$ is corresponding to $\mu = 0.1850$ and, therefore, to $A_s = 3363$ mm^2 and $3363 - 1278 = 2085$ mm^2.

7.7.4.2 Example 2—T-shaped Cross-Section with Axial Force and Bending. Calculation of the Moment Which Can Be Associated with an Assigned Normal Force

Problem data.
$b = 500$ mm; $b_w = 250$ mm; $h = 1100$ mm; $d = 1000$ mm; $h_f = 200$ mm (Fig. 7.20)

$$w = \frac{200}{1000} = 0.20; \quad n = \frac{500}{250} = 2;$$

Concrete: C30/37; $f_{cd} = 17$ N/mm^2; $A_s = 3000$ mm^2

The load on the cross-section is $N_{Ed} = -750$ kN, applied at point B.

The problem is to evaluate the design bending resistance which can be associated with N_{Ed}.

The problem can be solved by transporting N_{Ed} on the reinforcement, i.e. by introducing the fictitious moment $M^* = N_{Ed}(d-h_f) = 750\cdot0.80 = 600$ kNm and by assuming simple bending as if the reinforcement is $A_s^* = 3000 + \frac{N_{Ed}}{f_{yd}} = 3000 + \frac{750{,}000}{391} = 4918\,\text{mm}^2$.

From $A_s^* = 4918\,\text{mm}^2$, we get $\omega = \frac{4918\cdot391}{500\cdot1000\cdot17} = 0.2262 > \omega_1$.

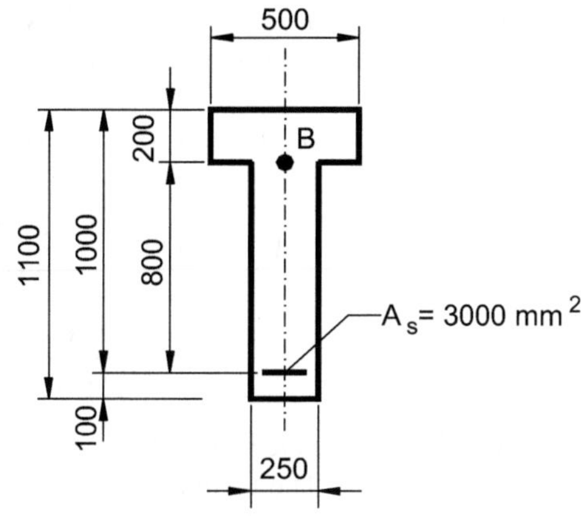

Fig. 7.20 Axial force combined with bendi—Examples 2 and 3: T-shaped cross-section

Recalling Tables 7.15 and 7.13, for $w = 0.20$, $n = 2$ and $f_{ck} \leq 50$ N/mm^2, it results:

$$\omega_2 = 0.356 > 0.2262 > \omega_1 = 0,20$$

the solution is then placed in Sector 2.

Therefore, by obtaining k from (7.48), being $\eta = 1$, we get:

$$k = n \cdot \left(\omega - w + \frac{w}{n}\right) = 2 \cdot \left(0.2262 - 0.2 + \frac{0.2}{2}\right) = 0.2524$$

The depth of the compression block is given by $y = 0.2524 \cdot 1000 = 252\,\text{mm}$.

The reduced moment is obtained by (7.52):

$$\mu = \frac{1}{2 \cdot 2}\left[2 \cdot 0.2524 - 0.2524^2 + 2 \cdot 1 \cdot \left(0.2 - 0.5 \cdot 0.2^2\right)\right] = 0.20$$

The corresponding moment is $M = 0.20 \cdot 500 \cdot 1000^2 \cdot 17 = 1700\,\text{kNm}$.

By subtracting the fictitious moment, we obtain the searched design bending resistance:

$$M_{Rd} = 1700–600 = 1100\,\text{kNm}.$$

7.7.4.3 Example 3—T-shaped Cross-Section with Axial Force and Bending. Calculation of the Moment Which Can Be Associated with an Assigned Axial Force

We consider the same geometry and the same problem of the previous Example (Fig. 7.20), with the following data:

Concrete: C60/75; $f_{cd} = 34$ N/mm^2; $A_s = 3000$ mm^2
$N_{Ed} = -1500$ kN, applied at point B, on the vertical axis of the cross-section.

$$M^* = 1500 \cdot 0.8 = 1200\,\text{kNm (moment about the reinforcement)}$$

$$A_s^* = 3000 + \frac{1,500,000}{391} = 6836\ mm^2$$

$$\omega^* = \frac{6836 \cdot 391}{500 \cdot 1000 \cdot 34} = 0.1572$$

The solution is placed in Sector 1 (neutral axis cutting the slab, being $\omega^* < 0.19$—Table 7.13). The formulas are those of the rectangular cross-Sect. (7.6.2).
$\omega^* = k\eta$, with $\eta = 0.95$ (Table 7.11).

$$k = \frac{0.1572}{0.95} = 0.1655;\ y = 0.1655 \cdot 1000 = 165.5\,\text{mm}$$

Using (7.43), we get:

$$\mu^* = \omega^*\left(1 - \frac{\omega^*}{2\eta}\right) = 0.1572 \cdot \left(1 - \frac{0.1572}{2 \cdot 0.95}\right) = 0.1442$$

and therefore:

$$M^* = 0.1442 \cdot 500 \cdot 1000^2 \cdot 34 = 2451\,\text{kNm}$$

By subtracting the quantity $N_{Ed}\ e = 1500 \cdot 0.8 = 1200\,\text{kNm}$ from this moment, we get
$M = 1251$ kNm

7.8 Interaction Diagrams for Axial Force and Bending at ULS

Interaction diagrams ν - μ describe the strength of cross-sections under the action of axial loads combined with bending in a plane of symmetry of the cross-section itself. In these diagrams, reinforcements are arranged in a symmetrical way about the two axes, both to absorb moments with opposite signs and to simplify the design. Diagrams are built by using the reinforcement mechanical ratio ω as a parameter. With a given ω, a point is identified on a curve, and its ν and μ coordinates are corresponding to the reduced axial force and reduced design bending resistance

$$\omega = \frac{A_{s,tot} \cdot f_{yd}}{A_c \cdot f_{cd}}$$

$$\nu = \frac{N_d}{A_c \cdot f_{cd}}$$

$$\mu = \frac{M_d}{A_c \cdot h \cdot f_{cd}}$$

These diagrams are valuable for cross-sections with all dimensions (being equal the shape, the arrangement of reinforcements, and the position of the reinforcement to the sides of the section), but they are dependent on the quality of the employed steel and on the constitutive law $\sigma - \varepsilon$ of the concrete (parabola-rectangle with $\varepsilon_{c2} = 0.0020$ and $\varepsilon_{cu2} = 0.0035$ for f_{ck} until 50 N/mm^2; generalized parabola-rectangle for f_{ck} > 50 N/mm^2 with exponents lower than 2 and ε_{c2} and ε_{cu2} smaller of the above-indicated values). This circumstance involves that, to cover all the possible cases, it is necessary to have at disposal a lot of diagrams.

In the present case, to provide a series of diagrams suitable for the discussion, the following choices have been made:

- rectangular cross-section: only one type of reinforcement (f_{yk} = 450 N/mm^2); symmetrical arrangement of the bars in layers parallel to the opposite sides (U10 – U15); two different concrete covers (d'/d = 0.05 and 0.10) (U16 – U17); continuous arrangement on four sides; reinforcement at the corner of the section (U23 – U27);
- circular cross-section: reinforcement arranged on a circumference (U18 – U22).

For both groups of cross-sections, the following concrete classes have been considered: until 50 N/mm^2 and f_{ck} = 60, 70, 80, 90 N/mm^2.

The diagrams are reported in the Appendix.

The application of the interaction diagrams is very easy.

For the design purpose, once the cross-section is assigned, it is simply required to enter the diagram with the values of the reduced axial force and bending moment. It is sufficient to read the value ω corresponding to the curve passing through the point defined by the couple of values ν and μ.

For the verification of a cross-section, it is possible to define the maximum axial force that the cross-section can support with an assigned bending moment (in this case, generally, we have two solutions), or the maximum bending moment acceptable with an assigned axial load.

7.8.1 Examples of Application of the Interaction Diagrams $\nu - \mu$

In the following examples steel is B450C with f_{yk} = 450 N/mm^2 and f_{yd} = 391 N/mm^2.

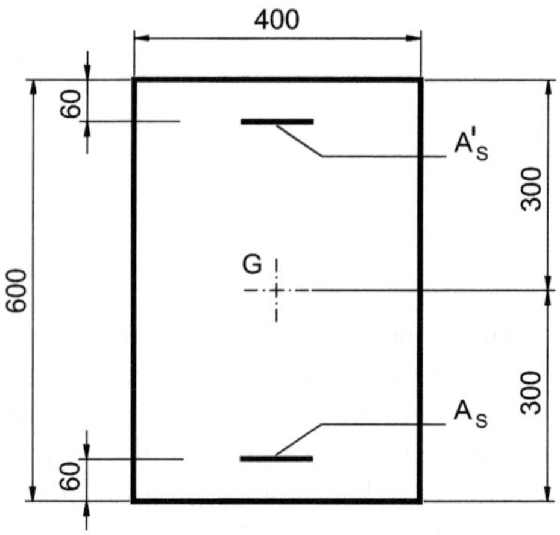

Fig. 7.21 Axial force combined with uniaxial bending with small eccentricity—Example 1: rectangular cross-section

7.8.1.1 Example 1—Rectangular Cross-Section Under Axial Force Combined with Bending. Reinforcement Calculation

Problem data.

Rectangular cross-section: $b = 400$ mm; $h = 600$ mm; symmetrical reinforcements at the sides (Fig. 7.21).

Concrete: C30/37; $f_{cd} = 17$ N/mm^2

Axial force combined with uniaxial bending: $N_{Ed} = -1000$ kN; $M_{Ed} = 500$ kNm (to the centroid)

Eccentricity $e = \frac{500}{1000} = 0.5$ m.

Diagrams of Table U10 (Appendix) and relevant formulas are employed.

Dimensionless parameters:

$$\nu = \frac{1{,}000{,}000}{400 \cdot 600 \cdot 17} = 0.2451$$

$$\mu = \frac{500 \times 10^6}{400 \cdot 600^2 \cdot 17} = 0.2042$$

Based on the above data, the Table gives $\omega = 0.30$.

Total reinforcement is $A_{s,\text{tot}} = 0.30 \cdot 400 \cdot 600 \cdot \frac{17}{391} = 3130\,\text{mm}^2$, on each edge, $A_s = 3130/2 = 1565$ mm^2 corresponding, for example, to $5\phi20$ mm.

7.8.1.2 Example 2—Circular Cross-Section Under Axial Force Combined with Bending. Reinforcement Calculation

Problem data (Fig. 7.22).

Circular cross-section with diameter 1000 mm; reinforcement uniformly arranged along the circumference with diameter 800 mm, $d' = 100$ mm.

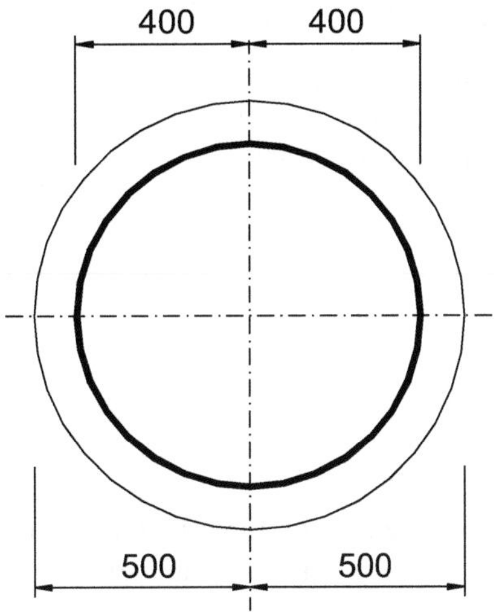

Fig. 7.22 Axial force combined with uniaxial bending with small eccentricity—Example 2: circular cross-section

Concrete: C40/50; $f_{cd} = 22.66$ N/mm^2
Axial force: $N_{Ed} = -10{,}000$ kN; bending moment $M_{Ed} = 3500$ kNm.
Eccentricity $\frac{3500 \times 10^3}{10000} = 350$ mm.
Diagrams of Table U18 (Appendix) and relevant formulas are employed.
Dimensionless parameters:
Being $A_c = 785{,}000$ mm^2, we get:

$$\nu = \frac{10{,}000{,}000}{785{,}000 \cdot 22.66} = 0.5622$$

$$\mu = \frac{3500 \times 10^6}{785{,}000 \cdot 1000 \cdot 22.66} = 0.1968$$

Based on the above data, the Table gives $\omega = 0.48$.

Total reinforcement is $A_s = 0.48 \cdot 785{,}000 \cdot \frac{22.66}{391} = 21837$ mm^2. A possible arrangement is made of 49 ϕ 24 mm bars, with a distance between the bar axes equal to 50 mm.

7.8.1.3 Example 3—Square Cross-Section Under Axial Force Combined with Bending. Reinforcement Calculation

Problem data

Square cross-section with side 500 mm; uniformly distributed reinforcement along a square 400 mm side (Fig. 7.23).

Concrete: C30/37; $f_{cd} = 17$ N/mm^2.

Axial force: $N_{Ed} = -3000$ kN with eccentricity $e = 150$ mm.

$M_{Ed} = 3000 \cdot 0.15 = 450$ kNm.

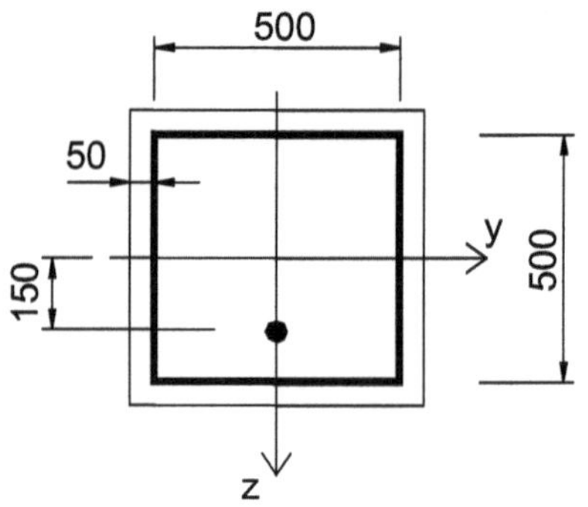

Fig. 7.23 Axial force combined with uniaxial bending with small eccentricity—Example 3: square cross-section

Diagrams of Table U16 (Appendix) and relevant formulas are employed.
Dimensionless parameters:

$$\nu = \frac{3{,}000{,}000}{500^2 \cdot 17} = 0.7059; \ \mu = \frac{450 \times 10^6}{500^3 \cdot 17} = 0.2118$$

Based on the above data, the Table gives $\omega = 0.42$.

Total reinforcement is $A_s = 0.42 \cdot 500^2 \cdot \frac{17}{391} = 4565\ \text{mm}^2$. A possible arrangement is made of 12 ϕ 22 mm bars, with a distance between the bar axes equal to 133 mm.

7.9 "Rose" Shaped Diagrams for Axial Force Combined with Uniaxial or Biaxial Bending

For axial force combined with uniaxial or biaxial bending, interaction diagrams are available, collecting ω values (total reinforcement) required for all the three internal forces ν, μ_x, μ_y.

For rectangular cross-sections with symmetric reinforcement, placed in the four corners of the cross-section, dimensionless interaction diagrams have the "rose" shape, because each diagram is subdivided into eight sectors and each sector is corresponding to a special value of the reduced axial force.

A series of those diagrams is reported in Appendix, Tables U23, U24, U25, U26 and U27.

7.9.1 Example 1: Axial Force Combined with Biaxial Bending

Problem data

Rectangular cross-section: $b = 400$ mm; $a = 600$ mm; reinforcement bars placed in the corners of the cross-sections.

Concrete: C30/37; $f_{cd} = 17\ \text{N/mm}^2$
Steel: B450C with $f_{yk} = 450\ \text{N/mm}^2$ and $f_{yd} = 391\ \text{N/mm}^2$.
Axial force combined with biaxial bending: $N_{Ed} = -1600$ kN; $M_a = 395$ kNm; $M_b = 150$ kNm (Fig. 7.24).

Reinforcement design

Diagrams of Table U23 and relevant formulas are employed:

$$\nu = \frac{1,600,000}{600 \cdot 400 \cdot 17} = 0.39$$

$$\mu_a = \frac{395 \times 10^6}{600^2 \cdot 400 \cdot 17} = 0.1614$$

$$\mu_b = \frac{150 \times 10^6}{600 \cdot 400^2 \cdot 17} = 0.0919$$

For the given data, the table gives $\omega = 0.27$.
Total reinforcement is given by

$$A_s = 0.27 \cdot 400 \cdot 600 \cdot \frac{17}{391} = 2817\,\text{mm}^2$$

4 bars ϕ 30 mm, one for each corner of the cross-section, fulfil the requirement of the calculation.

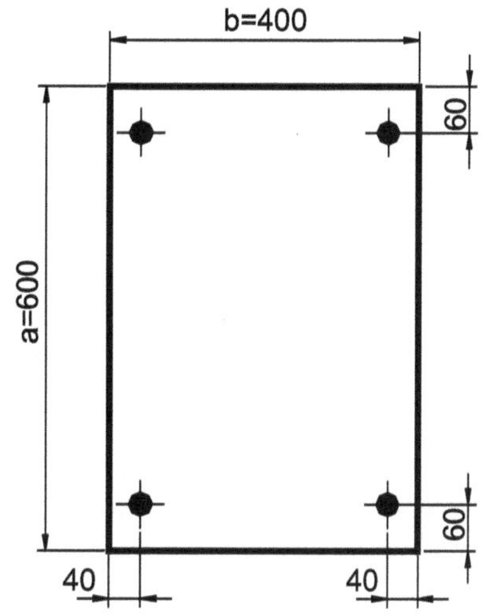

Fig. 7.24 Axial force combined with biaxial bending—Cross-section

Reference

Marro, P., & Ferretti, D. (1995), ULS in Bending and/or normal force—Extension of Model Code 90 to HSC. Atti del Dipartimento di Ingegneria Strutturale n. 52 – Politecnico di Torino (in Italian)

Chapter 8
Shear and Torsion at Ultimate Limit State

Abstract The chapter examines the calculation rules (design and verification) at ULS of structures subjected to shear and torsion forces, with possible presence of bending and axial force. Rectangular, T, tubular sections are analysed.

8.1 Shear

8.1.1 Symbols and Definitions

For the verification of the shear resistance, [6.2.1(1)P] defines the following symbols:

$V_{Rd,c}$ is the design shear resistance of the member without shear reinforcement;

$V_{Rd,s}$ is the design value of the shear force which can be sustained by the yielding shear reinforcement of a member provided with transverse reinforcements and limited by the strength of the reinforcement itself;

$V_{Rd,max}$ is the design value of the maximum shear force which can be sustained by a member provided with transverse reinforcements, limited by crushing of the compression struts.

Here some assumptions of EC2 are recalled:

- [6.2.1(4)]: when, based on the design shear calculation, no shear reinforcement is required, minimum shear reinforcement should nevertheless be provided according to 9.2.2. The minimum shear reinforcement may be omitted in members such as slabs (solid, ribbed or hollow-core slabs) where transverse redistribution of loads is possible. Minimum reinforcement may also be omitted in members of minor importance (e.g., lintels with span ≤ 2 m) which do not contribute significantly to the overall resistance and stability of the structure;
- [6.2.1(8)]: for members subject to predominantly uniformly distributed loading the design shear force need not be verified at a distance less than d from the face of the support. Any shear reinforcement required should continue to the support.

This chapter was authored by Piero Marro and Matteo Guiglia.

© The Author(s), under exclusive license to Springer Nature Switzerland AG 2022
F. Angotti et al., *Reinforced Concrete with Worked Examples*,
https://doi.org/10.1007/978-3-030-92839-1_8

In addition, it should be verified that the shear at the support does not exceed $V_{Rd,\max}$ (see also 6.2.2 (6) and 6.2.3 (8));

- [6.2.1(9)]: where a load is applied near the bottom of a section, sufficient vertical reinforcement to carry the load to the top of the section should be provided in addition to any reinforcement required to resist shear.

8.1.2 Members Without Transverse Reinforcements

8.1.2.1 Evolution of the Behaviour and Calculation of the Strength

The beam shown in Fig. 8.1, provided with only linear longitudinal reinforce to bending and subjected to two loads in symmetric positions, has a central part with a uniform bending moment and null shear and two lateral parts with a linearly variable bending moment and constant shear. Under service loads, beam cracks under bending, not only in the central part but also in the lateral parts, where the bending moment is greater than the cracking moment. In these portions of the member, shear transmission occurs with three different mechanisms acting in a parallel way:

- shear strength of the compressed zone, above the cracks;
- aggregate interlock along the cracks;
- dowel effect, provided by the longitudinal reinforcement at cracks.

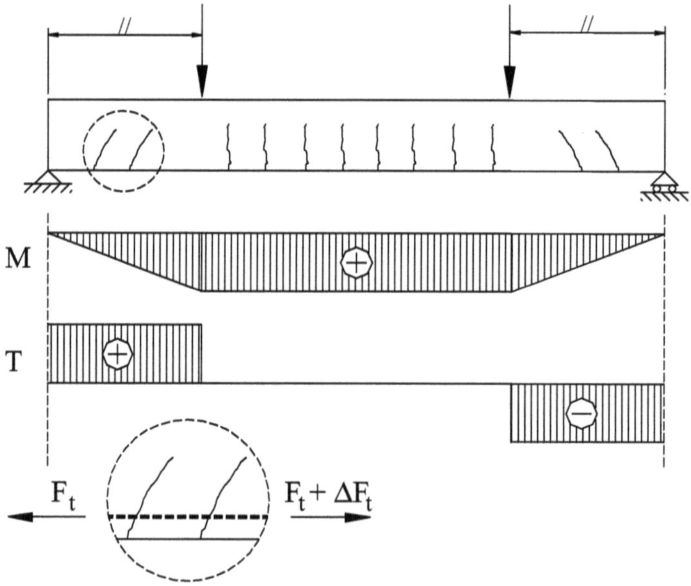

Fig. 8.1 Beam without transverse reinforcements

A concrete portion between two subsequent cracks can be considered as a sub-vertical cantilever joined in the compressed strut and loaded by the force ΔF_t given by the difference of the tensile forces of the reinforcement at the two cracks (see the detail of Fig. 8.1). For a certain value of the load, the bending moment at the fixed support reaches the strength value and the aggregate interlock goes to zero: the other two mechanisms are not able, except that in exceptional cases, to sustain additional loads. Shear can be therefore transferred to the support passing over the ineffective cantilever. If the longitudinal reinforcements are adequate and well anchored, the failure occurs gradually; otherwise, the failure occurs suddenly due to loss of bond between concrete and bars.

The strength of the beam can be therefore calculated by an expression considering all the contributions described above:

$$V_{Rd,c} = \left[C_{Rd,c}k(100\,\rho_1\,f_{ck})^{1/3} + k_1\sigma_{cp}\right]b_w\,d \quad [(6.2a)]$$

$C_{Rd,c} = 0.18/\gamma_c$ is the basic resistant tangential stress. The value of $C_{Rd,c}$ is modified through the coefficient $k = 1 + \sqrt{\frac{200}{d}} \leq 2.0$ (with d in mm), which highlights the low efficiency of the aggregate interlock with the increase of the effective depth d, by the quantity between brackets, which is a function of the concrete strength expressed in N/mm^2 and by the ratio of the longitudinal reinforcement $\rho_1 = \frac{A_{sl}}{b_w d} \leq 0.02$, where A_{sl} is the area of the longitudinal tensile reinforcement out of the verification cross-section for a length at least equal to $(d + l_{bd})$, being l_{bd} the anchorage length [8.4.4 (1)]. For the definition of A_{sl}, Fig. 8.2 is a useful reference.

Table 8.1 gives some of the values of the corrective factor between brackets.

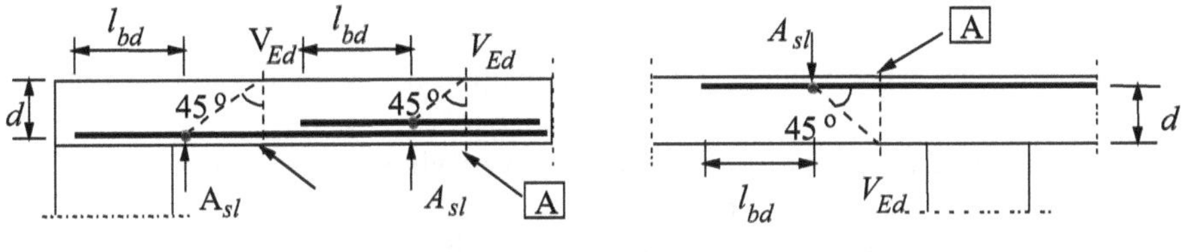

Fig. 8.2 Definition of A_{sl} in the expression [6.2a] [Fig. 6.3]

Table 8.1 Values of $(100\,\rho_l\,f_{ck})^{1/3}$

	f_{ck}(N/mm^2)			
	20	30	50	70
$100\rho_1 = 0.5$	2.15	2.46	2.92	3.27
$100\rho_1 = 1.0$	2.71	3.10	3.68	4.12
$100\rho_1 = 1.5$	3.10	3.55	4.21	4.71
$100\rho_1 = 2.0$	3.42	3.91	4.64	5.19

Table 8.2 Values of $\nu_{\min}$

$f_{ck}\,(\mathrm{N/mm^2})$	20	30	40	50	60	70	80	90
$\nu_{\min}\,(\mathrm{N/mm^2})$	0.443	0.542	0.626	0.700	0.767	0.828	0.885	0.939

The analysis of the Table allows noting that the influence of f_{ck} for each value of ρ_l is very low: passing from 20 to 70, an increment of 50% is obtained. Slightly bigger is the influence of ρ_l for each value of f_{ck}.

If the beam is prestressed, the strength is increased by the quantity $k_1\sigma_{cp}$ with

$$k_1 = 0.15$$

$$\sigma_{cp} = \frac{N_{Ed}}{A_c} \le 0.20 f_{cd}$$

being N_{Ed} the prestressing force.

The term $k_1\sigma_{cp}$ takes into account the favourable effect of prestressing because it delays the crack opening.

Formula [(6.2a)] has a lower limit

$$V_{Rd,\mathrm{c}} = \left[\nu_{\min} + k_1\sigma_{cp}\right] b_w d \quad [(6.2\mathrm{b})]$$

where $\nu_{\min}$ is given by $\nu_{\min} = 0.035\,k^{3/2}\,f_{ck}^{1/2}$ [(6.3N)]

$$k_1 = 0.15$$

Table 8.2 gives some values of $\nu_{\min}$ as a function of f_{ck} under the assumption $k = 2$.

Expression [6.2.2(5)] underlines the need of verifying longitudinal reinforcement in cracked zones under bending by shifting the diagram M_{Ed} over a distance equal to the effective depth d in the unfavourable direction, by recalling also, to this purpose, [9.2.1.3 (2)].

[6.2.2 (6)] underlines that, for loads applied on the upper side within a distance a_v between $0.5d$ and $2d$ from the anchorage limit if this is rigid, or from the support axis if a soft mechanism is provided, the contribution of such loads to the shear V_{Ed} may be reduced by the amount $\beta = a_v/2d$.

In case of a_v is smaller than $0.5d$, the value $a_v = 0.5d$ is adopted, because in the proximity of the support, the load is transferred to the same support basically by compressed struts. The reduction is allowed if the longitudinal reinforcement is completely anchored on the support. Moreover, shear V_{Ed}, calculated without the reduction β, shall be smaller or equal to the quantity $0.5\, b_w\, d\, \nu\, f_{cd}$, where

b_w is the minimum with of the portion of the cross-section in traction (mm);
d is the effective depth;

ν is a coefficient for the reduction of the strength of the shear-cracked concrete.

8.1.2.2 Simply Supported Prestressed Beams Without Transverse Reinforcements

Formula [(6.2a)] can be used in the beam segments cracked under bending. Where tensile stress due to bending is smaller than $f_{ctk,0-05}/\gamma_C$, shear strength is conditioned by concrete tensile strength at the centroid level (i.e., where the main tensile stress σ_{c1} is equal to the design tensile strength $f_{ctd} = f_{ctk,0.05}/\gamma_C$). In this case, EC2 gives the following expression for resistant shear.

$$V_{Rd,c} = \frac{Ib_w}{S}\sqrt{(f_{ctd})^2 + \alpha_I \sigma_{cp} f_{ctd}} \quad [(6.4)]$$

where

- I the inertial moment;
- b_w the width of the cross-section at the centroid axis, allowing for the presence of ducts according to [(6.16)] and [(6.17)];
- S the first moment of area to the centroid axis for the portion of cross-section placed above or below;
- $\alpha_I = \ell_x/\ell_{pt2}$ smaller than 1 in case of pre-tensioned tendons; $\alpha_I = 1.0$ for other prestressing types;
- ℓ_x the distance of the cross-section considered at the point where the transmission length starts;
- ℓ_{pt2} the upper bond value of the transmission length for the prestressed element according to [(8.18)];
- σ_{cp} the concrete compressive stress at the centroidal axis due to axial loading and/or to prestressing ($\sigma_{cp} = N_{Ed}/A_c$ in N/mm^2, with N_{Ed} positive).

The meaning of the square root (tangential stress τ), when it is assumed $\alpha_I = 1$, is the following (Fig. 8.3): τ is the proportional mean between σ_{c1} and σ_{c2}, which are the principal stress respectively in tension and compression, i.e., in absolute value

$$\tau^2 = \sigma_{c1} \cdot \sigma_{c2} = f_{ctd} \cdot \left(f_{ctd} + \sigma_{cp}\right) = (f_{ctd})^2 + f_{ctd} \cdot \sigma_{cp}$$

Coefficient $\alpha_I < 1$ limits the stress σ_{cp} when the verification is performed on a cross-section within the pre-tensioning transmission length. For this detail, which concerns prestressing through pre-tensioned tendons, [8.10.2.2(3)] should be referred to.

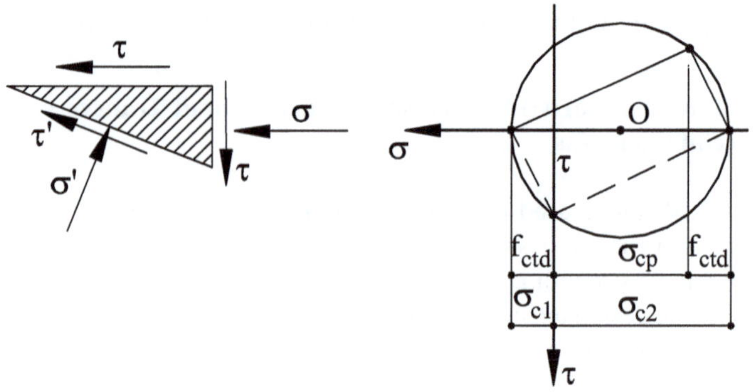

Fig. 8.3 Mohr circle for a prestressed beam without transverse reinforcement

8.1.3 Members with Transverse Reinforcements

8.1.3.1 Resistant Mechanism

The configuration at ULS of a beam provided with transverse reinforcements under bending and shear forces is described by an isostatic scheme to which the static plasticity theorem is applied. Theoretical results are combined with experimental results (Nielsen, 1990).

Figure 8.4 shows a portion of a beam with a rectangular or T-shaped cross-section without loads (also without self-weight) but under the action of positive shear and moment. Under this hypothesis, shear is the same as the two end sections and it is assumed applied with uniform distribution of tangential stresses.

Figure 8.4 shows the following resistant elements, identified by capital letters of the alphabet:

(B) concrete struts in compression with slope θ to the beam axis: theoretically, these struts are connected by hinges to upper and lower chords;

(D) transverse bars in tension with slope α to the beam axis: they are also assumed as elements connected to upper and lower chords through hinges.

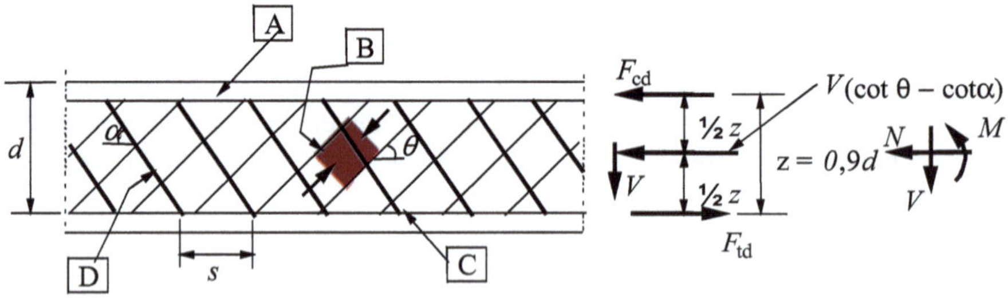

Fig. 8.4 Truss scheme for members provided with traverse reinforcements [6.5]

Struts and ties, mutually intersecting, draw a resistant reticulum.

Moreover:

(A) upper chord made basically of concrete in compression;
(C) lower chord made of longitudinal reinforcements.

Following limits are given by EC2 for angles θ and α:

- for struts in compression: $1 \leq \cot\theta \leq 2.5$, i.e., $45° \geq \theta \geq 21.8°$ [(6.7 N)]. Because of the plasticity theory, the angle θ can be assumed within the assigned limits, chosen by the designer. The limit $1 < \cot\theta$ would not be theoretically acceptable. However, experimental tests have shown that strengths related to angles bigger than 45° do not exist (see Model Code 1990). The limit $\cot\theta \leq 2.5$, applying at the ULS, intends to guarantee that, at serviceability, stresses in transverse reinforcements do not reach design elastic limit f_{yd} and therefore cracks have small acceptable width;
- for transverse reinforcements: $90° \geq \alpha \geq 45°$ [(9.2.2(1)].

In the general case of transverse reinforcements with slope α, shear strength of concrete struts in compression is given by

$$V_{Rd,\max} = \alpha_{cw} b_w z(\nu_1 f_{cd})(\cot\theta + \cot\alpha)/(1 + \cot^2\theta) \quad [(6.14)]$$

and the relevant strength of transverse reinforcements is given by

$$V_{Rd,s} = \frac{A_{sw}}{s} \cdot z \cdot f_{ywd} \cdot (\cot\theta + \cot\alpha)\ \text{sen}\alpha \quad [(6.13)]$$

In the special case with $\alpha = 90^{\circ}$, the two previous formulas can be simplified respectively as follows.

$$V_{Rd,\max} = \alpha_{cw} b_w z(\nu_1 f_{cd})/(\cot\theta + \tan\theta) \quad [(6.9)]$$

$$V_{Rd,s} = \frac{A_{sw}}{s} \cdot z \cdot f_{ywd} \cdot \cot\theta \quad [(6.8)]$$

In [(6.9)] and [(6.14)], α_{cw} is a coefficient referred to the stress of the concrete struts in compression, which, without prestressing, takes the value of 1. In the same formulas, coefficient ν_1, which is assumed equal to ν (smaller than 1), reduces the design strength f_{cd} as the struts in compression are weakened by the inclined cracks and they are subjected to bending because they are partially fixed to trusses. Theoretically (Nielsen 1990), ν is a decreasing function of characteristic strength f_{ck}, as expressed by [(6.6N)].

If we consider again Fig. 8.4, we can observe that:

- the upper chord is subjected to force F_{cd} which can be expressed as a combination of bending and shear $F_{cd} = (M_{Ed}/z) + 0.5V_{Ed}(\cot\theta - \cot\alpha)$. The bending component is negative, i.e. it represents a compressive force if M is positive.
  The shear term is positive if V_{Ed} is positive because the quantity $(\cot\theta - \cot\alpha)$ is positive or at least zero, due to the combined effect of [6.7N] and [9.2.2(1)];
- lower chord is subjected to force $F_{td} = (M_{Ed}/z) + 0.5\,V_{Ed}(\cot\theta - \cot\alpha)$. Under the assumptions sketched above, both terms are positive.

Finally, the effect of the shear on trusses is always tensile because it represents the horizontal component of the oblique force with slope θ, mitigated by the component $\cot\alpha$ when inclined reinforcements with a slope different from 90° are present.

The shear effect on the bending-compressed chords is favourable, while the situation gets worst in the tensile chord. This fact is considered directly by applying the formula

$$\Delta F_{td} = 0.5\,V_{Ed}(\cot\theta - \cot\alpha) \quad [(6.18)]$$

under the assumption that $(M_{Ed}/z) + \Delta F_{td}$ shall be assumed not bigger than $\left(M_{Ed,\max}/z\right)$. Alternatively, it is possible to translate the diagram M_{Ed} of the amount

$$a_\ell = (z/2)\,(\cot\theta - \cot\alpha)$$

in the less favourable direction.

Figure [9.2] shows a case with positive and negative bending moments.

Web beam is compressed by the horizontal force $H = V_{Ed}(\cot\theta - \cot\alpha)$, because of the equilibrium to the two tensile forces of trusses $0.5\,V_{Ed}(\cot\theta - \cot\alpha)$. This circumstance, which derives from the limits of [(6.7N)] and [9.2.2(1)], has a favourable effect on structural durability.

At the end-section supports of the members, where $M_{Ed} = 0$, longitudinal reinforcements at the lower side shall be calculated and suitably anchored to stand tensile force F_t given by [6.18], i.e.: $F_t = 0.5 \cdot V_{Ed} \cdot (\cot\theta - \cot\alpha)$.

Moreover, it is necessary to consider that, [9.2.1.4(1)] of EC2 recommends that bottom reinforcement at the end-section supports is at least equal to the 25% of the reinforcement arranged on the span length. Therefore, the heaviest request between the two has to be satisfied. Effectively, by [6.18], it can be obtained that:

- when vertical stirrups are arranged ($\cot\alpha = 0$), if $\cot\theta = 1$, $F_t = 0.5\,V_{Ed}$; if $\cot\theta = 2$, $F_t = V_{Ed}$; if $\cot\theta = 2.5$, $F_t = 1.25\,V_{Ed}$;
- when stirrups inclined by 45° are arranged ($\cot\alpha = 1$), if $\cot\theta = 1$, $F_t = 0$; if $\cot\theta = 2$, $F_t = 0.5\,V_{Ed}$; if $\cot\theta = 2.5$, $F_t = 0.75\,V_{Ed}$.

It is clear that, in some cases, the second prescription is prevailing.

[9.2.1.5(1)] extends to supports of continuous beams the prescription of 25% of the reinforcement along the span length.

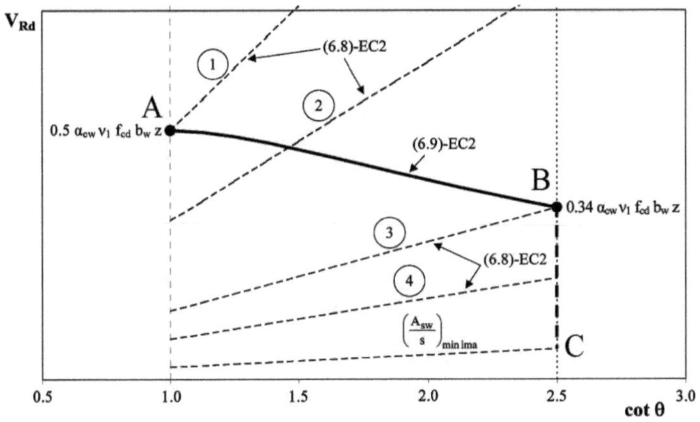

Fig. 8.5 Resistant shear V_{Rd} versus cotθ

Accounting for expressions [(6.8)], [(6.9)] and [(6.18)], Fig. 8.5 shows the representation of the resistant mechanism of a beam segment with rectangular or T-shaped cross-section, provided with vertical stirrups. Along the horizontal axis, we found the acceptable values of cotθ (from 1 to 2.50). Along the vertical axis, we found the resistant shear V_{Rd}. The curved line A–B, drawn according to [(6.9)], is the set of the representative points of the resistant shear of compressed struts with the variation of the slope θ. The value 0.5 in A, corresponding to $\theta = 45^{\circ}$, reduces itself to 0.34 in B where the slope θ is equal to 21.8° ($\cot\theta = 2.5$). By arranging suitable transverse reinforcements (vertical stirrups with cross-section $\frac{A_{sw}}{s}$), each of the previous points represents a configuration of ULS with simultaneous collapse of struts in compression and stirrups in tension. The lines expressed by (6.8) have a slope $\frac{A_{sw}}{s} \cdot z \cdot f_{yd}$. From B to C, where the slope θ is not varying and it is equal to the minimum value allowed by EC2, the limit state is reached with intact concrete struts and yielded stirrups. The ordinate of C is dependent on the minimum value of the geometrical ratio of transverse reinforcement.

$$\rho_{w,\min} = \frac{0.08 \cdot \sqrt{f_{ck}}}{f_{yk}} \quad [9.5N]$$

being $\rho_w = \frac{A_{sw}}{s \cdot b_w \cdot \sin\alpha}$ [9.4N]

The vertical value of C is therefore a function of the strength of both materials.

Figure 8.5 allows obtaining various quantitative aspects of the problem.

A point of line A–B is considered (Fig. 8.6). The ratio between the ordinate and the abscissa, $V_{Rd}/\cot\theta$, based on the [(6.8)], is proportional to the area of transverse reinforcement A_{sw}/s. We get $\frac{V_{Rd}}{\cot\theta} = \frac{A_{sw}}{s} \cdot z \cdot f_{ywd}$. Such a relationship is also valid in the vertical segment below B, until C where we have the minimum transverse reinforcement [9.5N]. Figure 8.7 shows the diagram of A_{sw}/s, normalized by the maximum value, vs cotθ. It can be observed that passing from A to B, the required reinforcement passes from 1 to 0.276, reducing itself further in the segment B–C

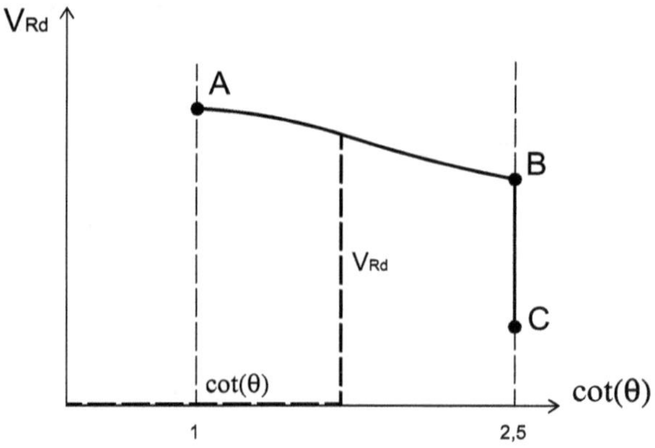

Fig. 8.6 Resistant shear V_{Rd} versus cotθ

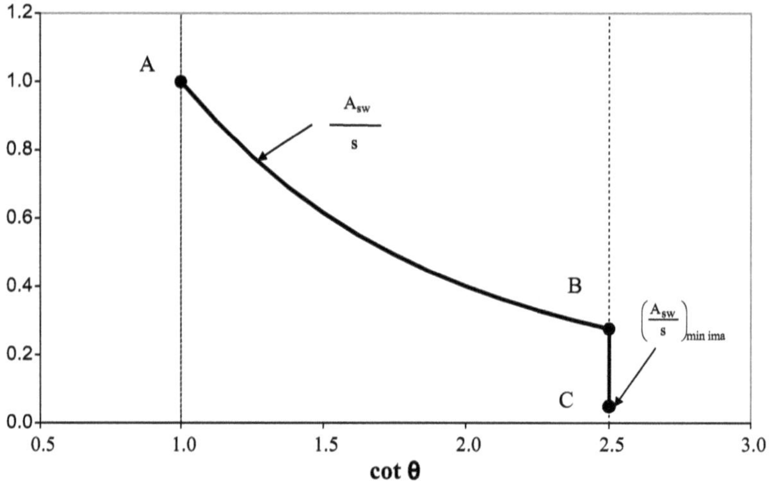

Fig. 8.7 Normalized transverse reinforcement versus cotθ

until the minimum reinforcement. This is a function of f_{ck} and f_{yk}, by the [(9.5N)]. In case of $f_{ck} = 30\,\text{N/mm}^2$ and $f_{yk} = 450\,\text{N/mm}^2$, the vertical value of C is equal to 0.09 (Fig. 8.5).

Analogously, if we consider the product of the ordinate of a point of the lines A–B and B–C through the corresponding abscissa cotθ, this is proportional to the number of longitudinal bars required by shear $\Delta A_{s\ell}$. In facts, from [(6.18)] we get:

$$\Delta A_{s\ell} = 0.5 \cdot V_{Ed} \cot\theta / f_{y\ell d}$$

Figure 8.8 shows the requirement of this reinforcement, normalized to the maximum value. It can be noted that from the value in A equal to 0,58, we pass

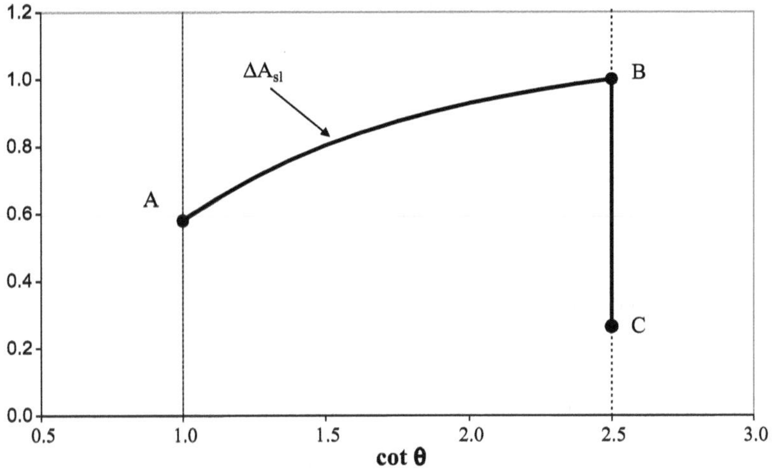

Fig. 8.8 Normalized longitudinal reinforcement versus cotθ

to increasing values until 1 in B; the values are then decreasing in the segment B–C, according to the decreasing values of shear until the minimum transverse reinforcement.

In the case of $f_{ck} = 30\,\text{N/mm}^2$ of Fig. 8.8, the high of C is equal to 0.27.

As an example, we can calculate the normalized ordinates of points A, B and C of Figs. 8.6, 8.7 and 8.8, for a rectangular cross-section $b_w = 200\,\text{mm}$, $z = 675\,\text{mm}$, C40/50 and then $f_{cd} = 22.66\,\text{N/mm}^2$, $\nu f_{cd} = 11.33\,\text{N/mm}^2$, reinforced with stirrups $\alpha = 90^\circ$ and $f_{yd} = 391\,\text{N/mm}^2$.

- For point A (cot$\theta = 1$).

  By equation [(6.9)] we calculate $V_{Rd,\max}$ with cotθ = 1:

$$V_{Rd,\max} = \frac{200 \cdot 675 \cdot 11.33}{(1+1)} = 765\,\text{kN}$$

and, by [(6.8)], we get

$$\frac{A_{sw}}{s} = \frac{V}{z \cdot f_{yd} \cdot \cot\theta} = \frac{765{,}000}{675 \cdot 391 \cdot 1} = 2.90\,\text{mm}^2/\text{mm}$$

  Finally, by [(6.18)], we obtain

$$A_{s\ell} = \frac{0.5 \cdot V \cdot \cot\theta}{f_{yd}} = \frac{0.5 \cdot 765{,}000 \cdot 1}{391} = 978\,\text{mm}^2$$

- For point B (cot$\theta = 2.5$)

$$V_{Rd,\max} = \frac{200 \cdot 675 \cdot 11.33}{(2.0 + 0.4)} = 527\,\text{kN}$$

$$\frac{A_{sw}}{s} = \frac{527{,}000}{675 \cdot 391 \cdot 2.5} = 0.80\,\text{mm}^2/\text{mm}$$

$$A_{s\ell} = \frac{0.5 \cdot 527{,}000 \cdot 2.5}{391} = 1685\,\text{mm}^2$$

- For point C, preliminarily the minimum transverse reinforcement has to be calculated using equation [(9.5N)]:

$$\frac{A_{sw,\min}}{s \cdot b_w} = (0.08 \cdot \sqrt{f_{ck}})/f_{yk} = (0.08 \cdot \sqrt{40})/450 = 0.0011$$

$$\frac{A_{sw,\min}}{s} = 0.0011 \cdot 200 = 0.225\,\text{mm}^2/\text{mm}$$

Therefore, by [(6.8)] we get $V_{Rd} = 0.225 \cdot 675 \cdot 391 \cdot 2.5 = 148\,\text{kN}$

$$A_{s\ell} = \frac{0.5 \cdot 148{,}000 \cdot 2.5}{391} = 473\,\text{mm}^2$$

Normalized ordinates are the following:
Figure 8.6 (V_{Rd}): A = 1.0; B = $\frac{527}{765} = 0.69$; C = $\frac{148}{765} = 0.19$
Figure 8.7 ($\frac{A_{sw}}{s}$): A = 1.0; B = $\frac{0.80}{2.90} = 0.27$; C = $\frac{0.225}{2.90} = 0.08$
Figure 8.8 ($A_{s\ell}$): A = $\frac{978}{1685} = 0.58$; B = 1.0; C = $\frac{473}{1685} = 0.28$

8.1.3.2 Verification and Design Procedures

- Theme 1: Being assigned geometry, materials and reinforcements, calculate the resistant shear V_{Rd} (verification)
  - Preliminarily, in presence of vertical stirrups, it is necessary to verify that

$$\frac{A_{sw,\max} f_{ywd}}{b_w s} \leq \frac{1}{2} \alpha_{cw} \nu_1 f_{cd} \quad [(6.12)]$$

or that in presence of inclined stirrups by angle α to the beam axis

$$\frac{A_{sw,\max} f_{ywd}}{b_w s} \leq \frac{\alpha_{cw} \nu_1 f_{cd}}{2 \sin \alpha} \quad [(6.15)]$$

that shear transverse reinforcement is not greater than the compatible one with the concrete strut strength. Equation [(6.12)] is obtained by equating the second members of [(6.8)] and [(6.9)] with the assumption $\cot\theta = 1$, i.e., by imposing the simultaneous crisis of compressed struts in the configuration of their maximum strength. Equation [(6.15)] is obtained analogously, by the second members of [(6.13)] and [(6.14)].

If the verification is fulfilled, transverse reinforcements are yielded, θ is smaller or at least equal to 45° and the limit state configuration is ductile.

- If ductility is verified, θ can be evaluated (see formulas expressed in the following) by imposing the equality of the second members [(6.8)] = [(6.9)] or [(6.13)] = [(6.14)]:

  if $\theta > 21.8^{\circ}$ (i.e. $\cot\theta < 2.50$), simultaneous collapse of concrete struts and stirrups occurs. In such a case, V_{Rd} can be calculated by writing equations [(6.8)] or [(6.9)] or, in the other case, equations [(6.13)] or [(6.14)] with the angle θ already calculated;
  if $\theta < 21.8^{\circ}$, concrete compressed struts do not collapse; $\cot\theta = 2.5$ is assumed and V_{Rd} is calculated by equations [(6.8)] or [(6.13)]. This is the case of a ductile crisis due to the stirrup failure.

- If [(6.12)] or [(6.15)] are not verified, failure occurs when transverse reinforcements are still in the elastic range. In such a case, strength is dependent on compressed strut strength according to [(6.9)] or [(6.14)] with $\cot\theta = 1$. Therefore, after the calculation of $V_{Rd,}$ it is possible to calculate stress σ_s by [(6.8)] or [(6.13)], obtaining it instead of f_{ywd}. Collapse is brittle.
- In all the cases, it is necessary to verify that tensile reinforcement in the truss in compression is adequate to the local bending moment increased by the amount required by [(6.18)].

- *Theme 2: Being assigned geometry, materials and* $V_{Ed,}$ *calculate the reinforcements (design)*

  (a) We verify that V_{Ed} is not greater than $V_{Rd,\max}$ evaluated with $\cot\theta = 1$ (strength of compressed struts). In such a case, failure is ductile. If $V_{Ed} > V_{Rd,\max}$ concrete cross-section has to be changed by increasing its width or depth or both until the equation is satisfied.
  (b) When condition a) is fulfilled, we calculate θ by equation [(6.9)] or [(6.14)], by placing V_{Ed} at the first member. Therefore:

  - if $\theta > 21.8^{\circ}$, equation [(6.8)] or [(6.13)] give directly (A_{sw}/s);
  - if $\theta < 21.8^{\circ}$, failure occurs with intact compressed struts. It is assumed $\cot\theta = 2.5$ and (A_{sw}/s) is calculated by [(6.8)] or [(6.13)].

In both cases, once θ is known, equation [(6.18)] allows evaluating ΔF_{td} and therefore longitudinal reinforcement required for shear.

Solving formulas

Theme 1 (verification)

Formulas for vertical stirrups

If A_{sw}, b_w, s, z, f_{yd}, f_{cd} are assigned, we calculate the resistant shear V_{Rd}.

First, we verify if the reinforcement, associated with all the other problem data, gives rise to a ductile ULS, by fulfilment of

$$\frac{A_{sw} f_{ywd}}{b_w s(\nu f_{cd})} \leq \frac{1}{2} \quad [(6.12)]$$

If [(6.12)] is not satisfied, ULS is brittle because concrete struts failure occurs as long as stirrups are in the elastic range. In such a case, resistant shear is given by [(6.9)], with $\cot\theta = 1$

$$V_{Rd,\max} = \frac{b_w z \nu f_{cd}}{2}$$

If [(6.12)] is satisfied, ULS is ductile. In such a case, by equating the second members of [(6.8)] and [(6.9)], which means the simultaneous collapse of concrete struts and stirrups, we obtain

$$\frac{A_{sw} f_{ywd}}{b_w s(\nu f_{cd})} = \text{sen}^2\theta$$

Given $\frac{A_{sw} f_{ywd}}{b_w s(\nu f_{cd})} = \omega_t$ (being ω_t the mechanical ratio of transverse reinforcement), by substituting, we get $\omega_t = \text{sen}^2\theta$.

From this equation, we obtain the strut slope θ at failure

$$\theta = \text{arcsen}\sqrt{\omega_t}$$

According to [(6.7N)], it should be

$$45^\circ \geq \theta \geq 21.8^\circ$$

i.e. $0.5 \geq \omega_t \geq 0.14\left(0.14 = \sin^2 21.8^\circ\right)$

If ω_t is between the two limits, resistant shear is the one corresponding to reinforcement, given by

$$V_{Rd,s} = \frac{A_{sw}}{s} \cdot z \cdot f_{yd} \cdot \cot\theta \quad [(6.8)]$$

If $\omega_t < 0.14$, the limit state is reached when the failure occurs with the yielding of stirrups and intact concrete struts. In such a case, according to [(6.7N)], resistant shear is always calculated by equation [(6.8)], but with $\cot\theta \leq 2.5$.

Formulas for inclined stirrups

The procedure already shown for stirrups with $\alpha = \pi/2$ is modified as follows.
Ductility verification:

$$\frac{A_{sw} f_{ywd}}{bs(\nu f_{cd})} \cdot \text{sen}\alpha \leq \frac{1}{2} \quad [(6.15)]$$

If equation [(6.15)] is not fulfilled, resistant shear is given by [(6.14)] with $\cot\theta = 1$

$$V_{Rd,\max} = \frac{b_w z \nu f_{cd}}{2} \cdot (1 + \cot\alpha)$$

If [(6.15)] is fulfilled, by equating the second members of [(6.13)] and [(6.14)], we get

$$\frac{A_{sw} f_{ywd}}{b_w s(\nu f_{cd})} \cdot \text{sen}\alpha = \text{sen}^2\theta$$

and, by considering that $\frac{A_{sw} f_{ywd}}{b_w s(\nu f_{cd})} = \omega_t$, we get

$$\omega_t \cdot \text{sen}\alpha = \text{sen}^2\theta$$

from which

$$\theta = \text{arcsen}\sqrt{\omega_t \text{sen}\alpha}$$

If $0.5 \geq \omega_t \text{sen}\alpha \geq 0.14$ the simultaneous collapse of struts and stirrups occurs, and the resistant shear is given by

$$V_{Rd,s} = \frac{A_{sw}}{s} \cdot z \cdot f_{yd} \cdot (\cot\theta + \cot\alpha) \cdot \text{sen}\alpha \quad [(6.13)]$$

If $\omega_t \text{sen}\alpha < 0.14$, equation [(6.13)] is valid, with $\cot\theta \leq 2.5$.

Theme 2 (design)

Formulas for vertical stirrups

Design reinforcements, given the following data: V_{Ed}, b_w, s, z, νf_{cd}.
Preliminarily, we check if the concrete cross-section is adequate by comparing V_{Ed} with a shear strength of compressed struts [(6.9)] with $\cot\theta = 1$

$$V_{Rd,\max} = \frac{b_w z \nu f_{cd}}{2}$$

If $V_{Ed} > \frac{b_w z v f_{cd}}{2}$, the cross-section is not adequate.
If $V_{Ed} \leq \frac{b_w z v f_{cd}}{2}$, the cross-section is adequate.
In such a case, by writing the equation [(6.9)] and imposing $V_{Ed} = V_{Rd,\max}$, we get

$$V_{Ed} = \frac{1}{2} b_w z v f_{cd} \cdot \operatorname{sen} 2\theta$$

Introducing

$$k = \frac{2V_{Ed}}{b_w z v f_{cd}}$$

from equation [(6.9)] we get

$$k = \operatorname{sen} 2\theta$$

and, from this relationship, we obtain

$$\theta = \frac{1}{2} \operatorname{arcsen} k$$

According to [(6.7N)], it has to be:
$1 \geq k \geq 0.69$, being $0.69 = \operatorname{sen} 2 \cdot 21.8°$.
In such a case, transverse reinforcement is given by [(6.8)]

$$\frac{A_{sw}}{s} = \frac{V_{Ed}}{z f_{yd} \cot\theta}$$

If $k < 0.69$, the same formula is valid, with $\cot\theta \leq 2.5$.
Finally, according to [(6.18)], we get

$$\Delta F_{td} = \frac{V_{Ed}}{2} \cdot \cot\theta$$

Formulas for inclined stirrups

Verification of the concrete cross-section adequacy.
By applying equation [(6.14)] with $\cot\theta = 1$, we get

$$V_{Rd,\max} = \frac{b_w z v f_{cd}}{2} \cdot (1 + \cot\alpha)$$

If $V_{Ed} > \frac{b_w z v f_{cd}}{2} \cdot (1 + \cot\alpha)$, the cross-section is not adequate.
If $V_{Ed} \leq \frac{b_w z v f_{cd}}{2} \cdot (1 + \cot\alpha)$, the cross-section is adequate.

In such a case, we rewrite [(6.14)]: $V_{Ed} = \alpha_{cw} b_w z(\nu_1 f_{cd})(\cot\theta + \cot\alpha)/(1 + \cot^2\theta)$.

and, given that $\frac{2V_{Ed}}{\alpha_{cw} b_w z(\nu_1 f_{cd})} = k$ with $\alpha_{cw} = 1$ e $\nu_1 = \nu$, the following equation is obtained

$$\cot^2\theta - \frac{2}{k}\cot\theta - \left(\frac{2}{k}\cot\alpha - 1\right) = 0$$

whose solution is $\cot\theta = \left[\frac{1}{k} + \sqrt{\left(\frac{1}{k}\right)^2 + \frac{2\cot\alpha}{k} - 1}\right]$.

According to [(6.7N)], imposing that $\cot\theta \geq 1$, it should be $k \leq (1 + \cot\alpha)$.

From equation [(6.13)], we get

$$\frac{A_{sw}}{s} = \frac{V_{Ed}}{z \cdot f_{yd} \cdot (\cot\theta + \cot\alpha) \cdot \text{sen}\alpha}$$

and, by [(6.18)] $\Delta F_{td} = \frac{V_{Ed}}{2} \cdot (\cot\theta - \cot\alpha)$.

8.1.4 *Examples of Verification of Beams Provided with Transverse Reinforcements (Theme 1)*

Materials

Concrete strength $f_{ck} = 20$–40–$60\,\text{N/mm}^2$ and relevant characteristics are reported in Table 8.3.

Reinforcements $f_{yk} = 450\,\text{N/mm}^2$; $f_{ywd} = 391\,\text{N/mm}^2$.

Examples are referred to rectangular or T-shaped cross-sections with:

$h = 600$ mm; $d = 550$ mm; $z = 500$ mm; $b_w = 150$ mm.

8.1.4.1 Example 1: 2-arms Vertical Stirrups 6 mm Diameter, s = 200 mm

Concrete: C20/25; $f_{cd} = 11.33\,\text{N/mm}^2$

Table 8.3 f_{cd} and νf_{cd}.

f_{ck} (N/mm^2)	f_{cd} (N/mm^2)	νf_{cd} (N/mm^2)
20	11.33	5.66
40	22.66	11.33
60	34.00	17.00

$$A_{sw}/s = 56.5/200 = 0.28\,\text{mm}^2/\text{mm}$$

$$\rho_w = \frac{A_{sw}}{s \cdot b_w} = \frac{56.5}{150 \cdot 200} = 0.0019 > \rho_{w,\min} = \frac{\left(0.08\sqrt{f_{ck}}\right)}{f_{yk}} = 0.00079$$

Ductility verification according to [(6.12)].
$\omega_t = \frac{A_{sw} \cdot f_{yd}}{b_w \cdot s \cdot v f_{cd}} = \frac{56.5 \cdot 391}{150 \cdot 200 \cdot 5.66} = 0.130 < 0.5$, verified

$$\theta = \text{arcsen}\sqrt{0.130} = 21.13^\circ < 21.8^\circ$$

At ULS, concrete struts do not collapse.
By imposing $\cot\theta = 2.50$, equation [(6.8)] gives:

$$V_{Rd} = \frac{A_{sw}}{s} z \cdot f_{yd} \cdot \cot\theta = 0.28 \cdot 500 \cdot 391 \cdot 2.50 = 136.8\,\text{kN}$$

8.1.4.2 Example 2: 2 Arms—Vertical Stirrups 6 mm Diameter, s = 150 mm

Concrete: C20/25; $f_{cd} = 11.33\,\text{N/mm}^2$

$$A_{sw}/s = 56.5/150 = 0.37\,\text{mm}^2/\text{mm}$$

$$\rho_w = 0.0025 > \rho_{w,\min} = 0.00079$$

Ductility verification using [(6.12)].
$\omega_t = \frac{56.5 \cdot 391}{150 \cdot 150 \cdot 5.66} = 0.1735 < 0.5$, verified.
$\theta = \text{arcsen}\sqrt{0.1735} = 24.61^\circ$; $\cot\theta = 2.18$

$$V_{Rd} = 0.37 \cdot 500 \cdot 391 \cdot 2.18 = 158\,\text{kN}$$

8.1.4.3 Example 3: 2 Arms—Vertical Stirrups 8 mm Diameter, s = 150 mm

Concrete: C20/25; $f_{cd} = 11.33\,\text{N/mm}^2$

$$A_{sw}/s = 100/150 = 0.66\,\text{mm}^2/\text{mm}$$

$$\rho_w = 0.0044 > \rho_{w,\min} = 0.00079$$

By the application of [(6.12)], we get:
$\omega_t = \frac{100 \cdot 391}{150 \cdot 150 \cdot 5.66} = 0.307 < 0.5$, verified

$$\theta = \text{arcsen}\sqrt{0.307} = 33.6^\circ > 21.8^\circ;\ \cot\theta = 1.50$$

$$V_{Rd} = 0.66 \cdot 500 \cdot 391 \cdot 1.50 = 193\,\text{kN}$$

8.1.4.4 Example 4: 2 Arms—Vertical Stirrups 10 mm Diameter, s = 150 mm

Concrete: C40/50; $f_{cd} = 22.66\,\text{N/mm}^2$

$$A_{sw}/s = 157/150 = 1.04\,\text{mm}^2/\text{mm}$$

$$\rho_w = \frac{157}{150 \cdot 150} = 0.0069 > \rho_{w,\min} = 0.00112$$

By the application of [(6.12)], we get:
$\omega_t = \frac{157 \cdot 391}{150 \cdot 150 \cdot 11.33} = 0.241 < 0.5$, verified

$$\theta = \text{arcsen}\sqrt{0.241} = 29.38^\circ > 21.8^\circ;\ \cot\theta = 1.77$$

$$V_{Rd} = 1.04 \cdot 500 \cdot 391 \cdot 1.77 = 360\,\text{kN}$$

8.1.4.5 Example 5: 2 Arms—Vertical Stirrups 12 mm Diameter, s = 150 mm

Concrete: C20/25; $f_{cd} = 11.33\,\text{N/mm}^2$

$$A_{sw}/s = 226/150 = 1.50\,\text{mm}^2/\text{mm}$$

$$\rho_w = 0.010 > \rho_{w,\min} = 0.00079$$

By the application of [(6.12)], we get:

$$\omega_t = \frac{226 \cdot 391}{150 \cdot 150 \cdot 5.66} = 0.69 > 0.5$$

which means insufficient ductility. Failure occurs side concrete with stirrups in the elastic range (fragile collapse).

By [(6.9)] with $\cot\theta = 1.0$, we get:

$$V_{Rd} = 150 \cdot 500 \cdot 5.66/(1 + 1) = 212.25\,\text{kN}$$

From [(6.8)] by substituting f_{yd} with σ_s we get:

$$\sigma_s = \frac{212{,}250}{1.50 \cdot 500 \cdot 1.0} = 283\,\text{N/mm}^2$$

8.1.4.6 Example 6: 2 Arms—Vertical Stirrups 12 mm Diameter, s = 150 mm

Concrete: C60/75; $f_{cd} = 34.0\,\text{N/mm}^2$

$$A_{sw}/s = 226/150 = 1.50\,\text{mm}^2/\text{mm}$$

$$\rho_w = 0.010 > \rho_{w,\min} = \left(0.08\sqrt{60}\right)/450 = 0.0014$$

By the application of [(6.12)], we get:

$$\omega_t = \frac{226 \cdot 391}{150 \cdot 150 \cdot 17.0} = 0.231 < 0.5$$

$$\theta = \text{arcsen}\sqrt{0.231} = 28.7° > 21.8°;\ \cot\theta = 1.82$$

$$V_{Rd} = 1.50 \cdot 500 \cdot 391 \cdot 1.82 = 534\,\text{kN}$$

8.1.4.7 Example 7: 2-arms Inclined Stirrups 12 mm diameter, $\alpha = \pi/4$ ($\sin\alpha = 0.707;\ \cot\alpha = 1.0$), s = 150 mm

Concrete: C40/50; $f_{cd} = 22.66\,\text{N/mm}^2$

$$A_{sw}/s = 226/150 = 1.50\,\text{mm}^2/\text{mm}$$

$\rho_w = \frac{226}{150 \cdot 150 \cdot 0.707} = 0.0141 > \rho_{w,\min} = 0.00112$ (see 8.1.4.4)

By [(6.15)] we get:

Table 8.4 Results' summary–Exercise 1 (verification)

	f_{ck}	A_{sw}/s	$\cot\theta$	V_{Rd}(kN)	Note
Example 1 ($\alpha = \pi/2$)	20	56.5/200	2.50	136.8	Intact struts
Example 2 ($\alpha = \pi/2$)	20	56.5/150	2.18	158.0	
Example 3 ($\alpha = \pi/2$)	20	100/150	1.50	193.0	
Example 4 ($\alpha = \pi/2$)	40	157/150	1.77	360.0	
Example 5 ($\alpha = \pi/2$)	20	226/150	1.00	212.2	Brittle failure
Example 6 ($\alpha = \pi/2$)	60	226/150	1.82	534.0	
Example 7 ($\alpha = \pi/4$)	40	226/150	1.75	570.0	

$$\omega_t = \frac{226 \cdot 391}{150 \cdot 150 \cdot 11.33} = 0.3466$$

$$\omega_t \text{sen}\alpha = 0.3466 \cdot 0.707 = 0.245 < 0.5$$

$$\theta = \text{arcsen}\sqrt{0.245} = 29.7^\circ > 21.8^\circ;\ \cot\theta = 1.75$$

And, by applying equation [(6.13)]:

$$V_{Rd} = 1.50 \cdot 500 \cdot 391 \cdot (1.75 + 1.0) \cdot 0.707 = 570\,\text{kN}$$

8.1.4.8 Results Summary of Theme 1

See Table 8.4.

8.1.5 Examples of Reinforcement Design (Theme 2)

Examples concern rectangular or T-shaped cross-sections, with: $b_w = 200\,\text{mm}$; $h = 800\,\text{mm}$; $d = 750\,\text{mm}$; $z = 675\,\text{mm}$.

Concretes with various $f_{ck}.f_{cd}$ and νf_{cd} values are reported in Table 8.3.

8.1.5.1 Example 1

Problem data:

$V_{Ed} = 400\,\text{kN}$; vertical stirrups.

C20/25; $f_{cd} = 11.33\,\text{N/mm}^2$

$$k = \frac{2 \cdot 400{,}000}{200 \cdot 675 \cdot 5.66} = 1.05 > 1$$

$k > 1$ means $V_{Ed} > V_{Rd,\max}$.
The resistant cross-section has to be modified increasing the width b_w to 220 mm. Therefore, we get:

$$k = \frac{2 \cdot 400{,}000}{220 \cdot 675 \cdot 5.66} = 0.95 < 1$$

$$\theta = \frac{1}{2}\text{arcsen}\ k = \frac{1}{2}\text{arcsen}\ 0.95 = 35.9^{\circ} > 21.8^{\circ};\ \cot\theta = 1.38$$

From [(6.8)], we get

$$\frac{A_{sw}}{s} = \frac{V_{Ed}}{z \cdot f_{ywd} \cdot \cot\theta} = \frac{400{,}000}{675 \cdot 391 \cdot 1.38} = 1.10\,\text{mm}^2/\text{mm}$$

fulfilled, for example, 10 mm diameter 2-arms stirrups, $s = 140$ mm. Moreover, we get:

$$\Delta F_{td} = \frac{V_{Ed}}{2} \cdot \cot\theta = \frac{400}{2} \cdot 1.38 = 276\,\text{kN}$$

8.1.5.2 Example 2

Problem data:
$V_{Ed} = 500\,\text{kN}$; vertical stirrups.
Concrete: C40/50

$$k = \frac{2 \cdot 500{,}000}{200 \cdot 675 \cdot 11.33} = 0.654 < 1$$

if $k < 0.69$ = crisis on steel side

$$\theta = \frac{1}{2}\text{arcsen}\ k = \frac{1}{2}\text{arcsen}\ 0.654 = 20.4^{\circ} < 21.8^{\circ}$$

It is assumed $\cot\theta = 2.50$

$$\frac{A_{sw}}{s} = \frac{V_{Ed}}{z \cdot f_{ywd} \cdot \cot\theta} = \frac{500{,}000}{675 \cdot 391 \cdot 2.50} = 0.76\,\text{mm}^2/\text{mm}$$

By arranging 12 mm diameter 2-arms stirrups, $s = 300$ mm, solution satisfies equation [(9.5N)]. We obtain

$$\rho_w = \frac{A_{sw}}{s \cdot b_w} = \frac{113 \cdot 2}{300 \cdot 200} = 0.0038 > \frac{\left(0.08\sqrt{40}\right)}{450} = 0.0011$$

Solution fulfils also [(9.6N)], being

$$s = 300 < 0.75 \cdot d(1 + \cot\alpha) = 0.75 \cdot 750 = 562.5\,\text{mm}$$

$$\Delta F_{td} = \frac{V_{Ed}}{2} \cdot \cot\theta = \frac{500}{2} \cdot 2.50 = 625.0\,\text{kN}$$

8.1.5.3 Example 3

Problem data:
$V_{Ed} = 750\,\text{kN}$; vertical stirrups.
Concrete: C60/75

$$k = \frac{2 \cdot 750{,}000}{200 \cdot 675 \cdot 17.0} = 0.654 < 1$$

Being $k < 0.69$, failure occurs only side steel.
$\theta = \frac{1}{2}\text{arcsen}\ k = 20.4° < 21.8°$; it is assumed that $\cot\theta = 2.5$
$\frac{A_{sw}}{s} = \frac{V_{Ed}}{z \cdot f_{ywd} \cdot \cot\theta} = \frac{750{,}000}{675 \cdot 391 \cdot 2.5} = 1.14\ \text{mm}^2/\text{mm}$;

$$\rho_w = 0.0057 > \rho_{w,\min} = 0.00138$$

satisfied by 12 mm diameter 2-arms stirrups, $s = 180$ mm. Finally:

$$\Delta F_{td} = \frac{V_{Ed}}{2} \cdot \cot\theta = \frac{750}{2} \cdot 2.5 = 937.5\ \text{kN}$$

8.1.5.4 Example 4

Problem data:
$V_{Ed} = 750\,\text{kN}$; 45° inclined stirrups ($\sin\alpha = 0.707$; $\cot\alpha = 1.0$).
Concrete: C40/50

$$k = \frac{2 \cdot 750{,}000}{200 \cdot 675 \cdot 11.33} = 0.98 < 1 + \cot\alpha = 2$$

$\cot\theta = \left[\frac{1}{0.98} + \sqrt{\left(\frac{1}{0.98}\right)^2 + 2 \cdot \frac{1}{0.98} - 1}\right] = 2.46$ from which $\theta = 22.1^\circ > 21.8^\circ$.

From equation [(6.13)], we get

$$\frac{A_{sw}}{s} = \frac{750{,}000}{675 \cdot 391 \cdot (2.46 + 1) \cdot 0.707} = 1.16\ \text{mm}^2/\text{mm}$$

By arranging 12 mm diameter stirrups each 200 mm ($\frac{A_{sw}}{s} = 1.13 \cong 1.16\,\text{mm}^2/\text{mm}$), the solution satisfies [(9.5N)]. In facts,

$$\rho_w = \frac{A_{sw}}{s \cdot b_w \cdot \text{sen}\alpha} = \frac{1.16}{200 \cdot 0.707} = 0.0082 > \frac{\left(0.08\sqrt{40}\right)}{450} = 0.0011$$

$$\Delta F_{td} = \frac{1}{2} V_{Ed}(\cot\theta - \cot\alpha) = \frac{1}{2} 750 \cdot (2.46 - 1.0) = 547\ \text{kN}$$

which requires 1400 mm^2.

8.1.5.5 Results Summary of Theme 2

See Table 8.5.

8.1.6 Shear Strength in Case of Loads Near to Supports

For beams not requiring shear resistant reinforcements $\left(V_{Ed} < V_{Rc,d}\right)$, [6.2.2(6)] says that for loads applied at the top, at a distance $0.5d \le a_v \le 2d$ from the edge of the anchorage disposal or the centre of the support disposal (if this is flexible), the shear contribution V_{Ed} can be reduced by $\beta = \frac{a_v}{2d}$. These loads, in a certain manner, transfer themselves directly to the support through compressed struts (under the assumption that longitudinal reinforcements are well anchored at the support—see Figure [6.4]). If $a_v < 0.5d$, the value $a_v = 0.5d$ has to be adopted.

Table 8.5 Results summary—Theme 1 (design)

α	f_{ck}	V_{Ed}(kN)	$\cot\theta$	A_{sw}/s	ΔF_{td}(kN)	Remark
Example 1 ($\alpha = \pi/2$)	20	400	1.38	1.10	276.0	$b_w = 220$ mm
Example 2 ($\alpha = \pi/2$)	40	500	2.50	0.76	625.0	Intact struts
Example 3 ($\alpha = \pi/2$)	60	750	2.50	1.14	937.5	Intact struts
Example 4 ($\alpha = \pi/4$)	40	750	2.46	1.16	547.0	

For beams requiring transverse reinforcements ($V_{Ed} > V_{Rd,c}$), equation [6.2.3(8)] for the calculation of transverse reinforcements allows a reduction of the design shear due to loads applied at the top in the proximity of supports, by the application of the coefficient β defined above.

However, it is possible to account only for transverse reinforcements arranged on a length of 0.75 a_v. Shear force V_{Ed} calculated in this way has to satisfy the following relationship.

$$V_{Ed} \leq A_{sw} f_{ywd} \cdot \sin\alpha \quad [(6.19)]$$

where A_{sw} is the area of reinforcement bars in the length 0.75 a_v central over the full length a_v.

Moreover following conditions have to be satisfied: longitudinal reinforcement has to be fully anchored at the support; shear V_{Ed} calculated without reduction β has not to be greater than $V_{Rd,\max}$ (formulas [(6.9)] or [(6.14)]) evaluated once the angle θ has been chosen.

8.1.6.1 Application Example

The cantilever of Fig. 8.9 has rectangular cross-section with $b_w = 200$ mm, $h = 300$ mm, $d = 270$ mm².

Materials: Concrete $f_{ck} = 30\,\text{N/mm}^2$; steel $f_{yk} = 450\,\text{N/mm}^2$.

First case: $V_{Ed} < V_{Rd,c}$.

Load $P = 100$ kN (ULS) being $a_v = 300$ mm. Longitudinal reinforcement: $A_s = 400$ mm

$$\rho_1 = \frac{400}{200 \cdot 270} = 0.0074;\ \text{k} = 1 + \sqrt{\frac{200}{270}} = 1.86$$

Fig. 8.9 Cantilever scheme

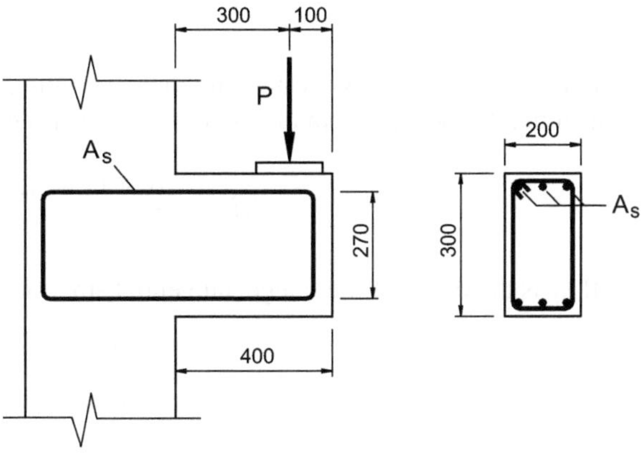

Applying [(6.2.a)], we get: $V_{Rd,c} = \frac{0.18}{1.5} \cdot 1.86 \cdot (100 \cdot 0.0074 \cdot 30)^{1/3} 200 \cdot 270 =$ 76 kN.

Being $\beta = \frac{a_v}{2d} = \frac{300}{2 \cdot 270} = 0.55$, design shear can therefore be taken equal to 100 × 0.55 = 55 kN, suitable with the strength 76 kN.

Second case: $V_{Ed} > V_{Rd,c}$.

Load $P = 150$ kN (ULS) being $a_v = 300$ mm. Longitudinal reinforcement: $A_s = 600\ \text{mm}^2$.

In this case, transverse reinforcements are required. We arrange 2-arms stirrups diameter 8 mm, $s = 100$ mm. Over the length 0.75 $a_v = 225$ mm we have 3 stirrups.

Design shear is given by $V_{Ed} = 150 \times 0.55 = 82.5$ kN.

Stirrup strength is given by $3 \cdot 2 \cdot 50 \cdot 391 = 117.5\ \text{kN} > V_{Ed}$.

Verification of the concrete strength, under the assumption $\cot\theta = \tan\theta = 1$

$$V_{Rd,\max} = b_w z \nu f_{cd}/(\cot\theta + \tan\theta) = 200 \cdot (0.9 \cdot 270) \cdot 8.5/(1+1)$$
$$= 206\ \text{kN} > 150\ \text{kN}$$

Remark: the cantilever in the example is characterized by $a_v = 1.11\,d$ *and, for this reason, we can follow the procedure already discussed. For cantilever with* $a_v < 0.5\,d$*, the reference model is described in Appendix [J3] of EC2.*

8.1.7 Shear Between Web and Flanges for T-beams

For a T-beam under the action of positive bending moment M, the internal couple is given by two forces F, the compression one passing through the centre of the flange, the tensile one at the reinforcement level, at a distance z, with modulus $F = M/z$. If $\eta = b_w/b_{\text{eff}}$, the force F_d of Fig. 8.10 zone 1, relative to each of the two parts of the flange protruding from the web is given by

$$F_d = \frac{M}{z}\frac{(1-\eta)}{2}$$

For each segment Δx where moment variation is ΔM, the tangential force transmitted to the web from one half, or the other half of the flange is given by

$$\Delta F_d = \frac{\Delta M}{z}\frac{(1-\eta)}{2}$$

If h_f is the depth of the flange, tangential stress is given by

$$v_{Ed} = \frac{\Delta F_d}{h_f \Delta x}$$

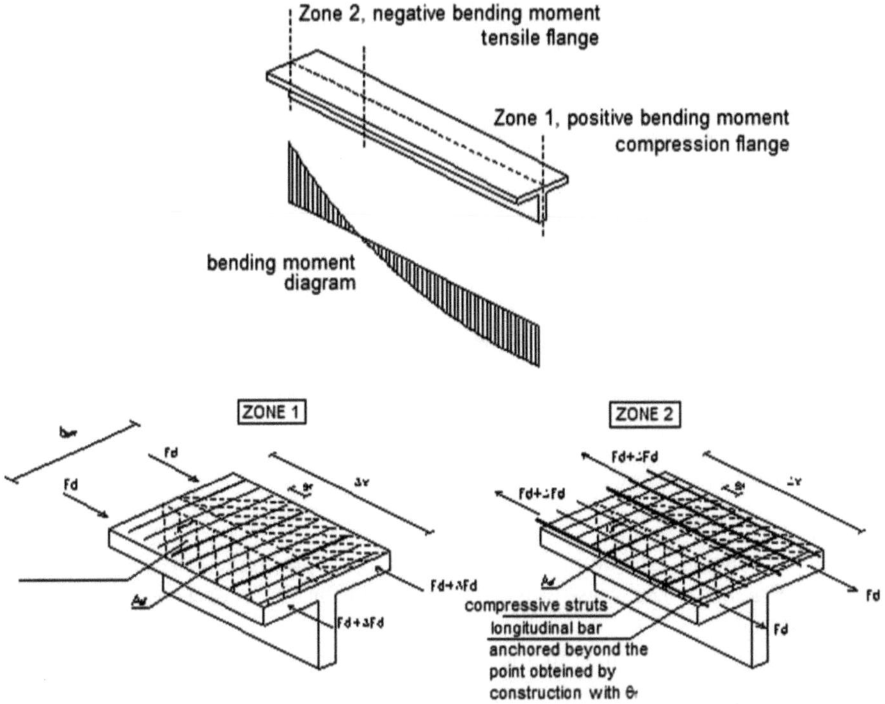

Fig. 8.10 Connection between flange and web for a T-beam

The transfer mechanism of ΔF_d can be explained as a system of compressive struts with slope θ_f to the axis of the beam combined with ties in the form of tensile reinforcement, at a distance s_f.

Let consider now the equilibrium of a node "reinforcement—strut—web". With reference to Fig. 8.11, the tangential force transmitted over a length s_f has to be in equilibrium with the longitudinal component of the strut force

$$v_{Ed} h_f s_f = \sigma_c h_f s_f \sin\theta_f \cos\theta_f$$

and the transverse component has to be in equilibrium with the reinforcement bar

$$\sigma_c h_f s_f \sin^2\theta = A_s \sigma_s$$

By imposing, at ULS, that $\sigma_c \leq \nu f_{cd}$, $\sigma_s \leq f_{yd}$, we obtain the following formulas.

$$v_{Ed} \leq \nu f_{cd} \sin\theta_f \cos\theta_f \quad [(6.22)]$$

$$\frac{A_{sf} f_{yd}}{s_f} \geq v_{Ed} h_f \tan\theta_f \quad [(6.21)]$$

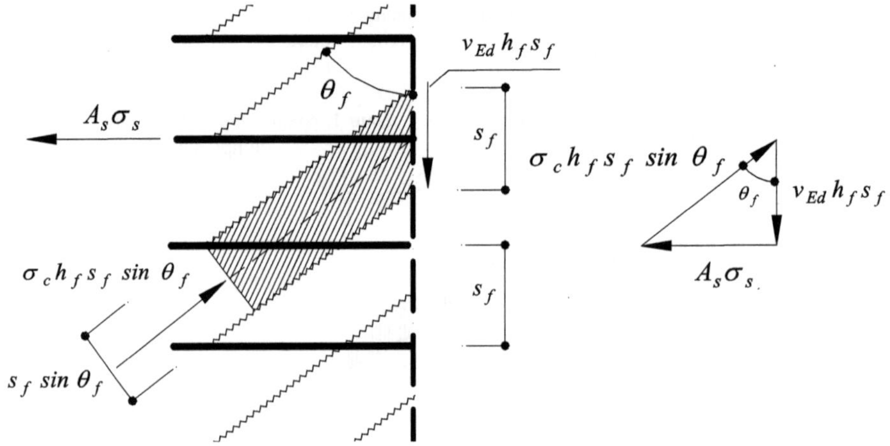

Fig. 8.11 Web-flange connection—node equilibrium

The link can concern compressed slabs (positive M) o in traction (negative M), as shown in Fig. 8.10. The stress transfer mechanism for the second case is analogous to the first one.

Concerning the moment sign, the following limitations apply for the angle θ_f.

- slab in compression:$26.5^\circ \le \theta_f \le 45^\circ$
- slab in tension: $38.6^\circ \le \theta_f \le 45^\circ$

Concerning the segment Δx, this shall be not greater than half of the length between the cross-section with the null bending moment and the one with the maximum (or minimum) bending moment. In the case of concentrated loads, length shall not exceed the distance between two subsequent force application points.

Finally, it should be remarked that the width b_{eff} shall be calculated through [(5.7)] adopting the distance l_0 of [Figure 5.2] between two sections with null moment.

For an application of the described calculation procedures, see the example of the continuous beam in Chap. 4, para. 4.3.7.5.

8.2 Torsion

8.2.1 General

Torsion can be present in structural members with two different levels of relevance:

- compatibility torsion, not necessary to the equilibrium condition, emerging because of the displacement compatibility: for example, the edge beam of a floor calculated under the assumption of simple supports receives from the floor transversal bending moments which becomes torsion on the beam. In general, this torsion can be neglected at the ULS, being sufficient to arrange a minimum amount of reinforcement as indicated in [7.3 and 9.2] to avoid too much torsional cracking at SLS;

- equilibrium torsion, necessary for the equilibrium of a structural element, which requires verifications, both at ULS and SLS.

Reinforced concrete beams in torsion have a behaviour much different in the two cases, uncracked and cracked: the main difference is the strong reduction of the torsional stiffness (until 1/5) after cracking.

In the first phase, stresses are calculated with the common formulas of structural Mechanics. In the second phase, the behaviour mode is a thin wall tubular structure built as a special truss of concrete struts in compression inclined to the axis and ties in tension represented by transverse and longitudinal reinforcements.

Compact cross-sections can be modelled as an equivalent tubular cross-section. Complex cross-sections, as T-shaped cross-sections, can be decomposed into rectangular sub-sections, each of them modelled by a thin wall tubular cross-section. Total torsion strength can be assumed as the sum of the strength of each sub-section. The distribution of torsion moment among the sub-sections can be based on the stiffness of the various parts under the assumption of null cracking. In the case of hollow cross-sections, the thickness of the equivalent thin wall cannot be greater than the effective thickness.

8.2.2 Calculation Procedure

A convex compact cross-section in torsion can be studied considering an inscribed ring-shaped cross-section with the same perimeter and uniform thickness $t_{\mathrm{ef}} = A/u$, where A is the area of the cross-section inside perimeter u, without accounting for possible holes. It is recommended that this is at least equal to $2c$, being c the distance between the edge of the cross-section and the axis of the longitudinal reinforcement. Bredt method is applied.

With reference to Fig. 8.12, we consider a ring-shaped cross-section with convex polygonal perimeter and sides with length z. According to the Bredt method, for the ith side of the polygon, we have

$$\tau_{t,i} \cdot t_{ef,i} = \frac{T_{Ed}}{2A_k} \quad [(6.26)]$$

where T_{Ed} is the torsion moment and A_k is the area inside the mean line.

Because all the sides of the polygon have thickness A/u, expression [(6.26)] can be written

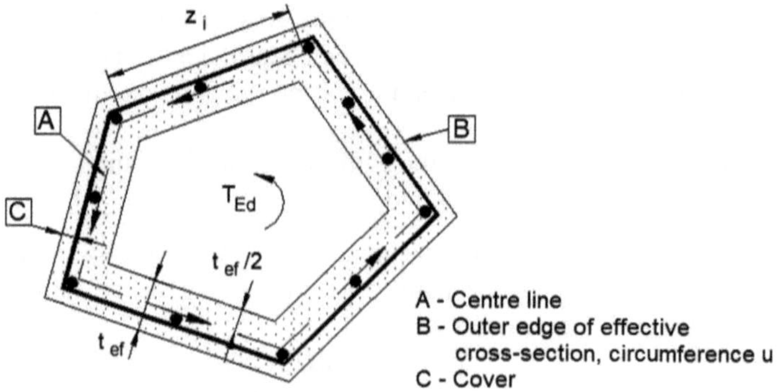

Fig. 8.12 Torsion: notations and definitions [6.11]

$$\tau \cdot t = \frac{T_{Ed}}{2A_k}$$

with the index *i*, if necessary, to distinguish the parameters of the *i*th side.

If we extend to a side z_i of the polygon the flux τt, we get the shear force

$$V_i = \tau\, t\, z_i$$

which can be studied through the shear formulas given in [6.2].

Shear on the *i*th side of the polygonal cross-section, originated by the torsion, is

$$V_{Edi} = (T_{Ed} \cdot z_i)/2A_k$$

Formulas have been obtained from the shear ones.

Recalling [6.9] we get:

$$\begin{aligned} V_{Rd,\max} &= (\alpha_{cw}\ \nu\ \mathrm{f_{cd}})\mathrm{b_w}\ \mathrm{z}/(\cot\theta + \tan\theta) \\ &= (\alpha_{cw}\ \nu\ \mathrm{f_{cd}})\mathrm{b_w}\ \mathrm{z}\ \sin\theta\cos\theta \qquad [(6.9)] \end{aligned}$$

by substituting to b_w the thickness *t* and by distinguishing the sides of the polygon by the index *i*, it results

$$V_{Rdi,\max} = (\alpha_{cw}\nu f_{cd})t\, z_i \sin\theta\, \cos\theta$$

We write now the moment given by the various V_i about the point *P* of the plane. If r_i is the distance of the action line of V_i (i.e., the distance of the axis of the *i*th side) from the point *P*, the maximum torsion of concrete compressive struts is obtained

$$T_{Rd,\max} = \Sigma(\alpha_{cw}\nu f_{cd})t\, z_i r_i \sin\theta\, \cos\theta.$$

But being $\Sigma z_i r_i = 2A_k$, we get

$$T_{Rd,\max} = 2(\alpha_{cw} \nu f_{cd})t\ A_k \sin\theta\ \cos\theta \quad [(6.30)]$$

which is the maximum torsion of transverse reinforcements.

Recalling relationship [(6.8)].

$$V_{Rd,s} = (A_{sw}/s)z\ f_{ywd} \cot\theta \quad [(6.8)]$$

and repeating the previous calculation, we obtain

$$T_{Rd,s} = (A_{sw}/s) f_{ywd} \cot\theta\ \Sigma z_i r_i = (A_{sw}/s) f_{ywd} 2\ A_k \cot\theta \tag{8.1}$$

For longitudinal reinforcements, the total tensile force can be obtained by adding all the contributions of [(6.8)], considering that each compressive strut transmits the force $\frac{V}{2} \cot\theta$ to both the ends, i.e.

$$F_{td} = 2 \sum \frac{V}{2} \cot\theta$$

and with V expressed by $V_{Rd,s}$, we obtain

$$F_{td} = (A_{sw}/s) f_{ywd} \cot\theta\ \ \Sigma z_i \cot\theta$$

But recalling expression [6.30] and by substituting Σz_i with u_k, we get

$$F_{td} = A_{s1} f_{yld} = T_{Rs,d} \frac{u_k}{2A_k} \cot\theta \tag{8.2}$$

For safety, it is necessary that

$$T_{Ed} \le T_{Rd,\max}$$

Transverse reinforcements are defined by

$$\frac{A_{sw}}{s} = \frac{T_{Ed}}{2A_k \cot\theta f_{ywd}} \tag{8.3}$$

while longitudinal ones are given by

$$A_{sl} = \frac{T_{Ed} u_k \cot\theta}{2A_k f_{yld}} \quad [(6.28)]$$

8.2.3 General and Practical Rules of EC2

- Slope of compressed struts should be limited: being the formulas derived from the shear ones, also for torsion the limitation $1 \leq \cot\theta \leq 2.5$ is valid.
- Stirrups must be closed and forming an angle of 90° with the member axis [9.2.3(1)]. The minimum amount of stirrups is the same as required for shear ([9.2.2(5) and 9.2.2(6)]). The distance between stirrups must be not greater than 1/8 of the cross-section perimeter [9.2.3(3)].
- Longitudinal reinforcements shall be arranged inside the stirrups with a bar in each corner and the spacing of the other bars is not more than 350 mm [9.2.3(4)].
- $\alpha_{cw} = 1$.

8.2.4 Verification and Design in Case of Pure Torsion

Verification

Design torsion resistance is calculated with assigned geometry, materials and reinforcement.

Design torsion resistance provided by stirrups is equated to design torsion resistance provided by longitudinal reinforcements

$$\frac{A_{sw}}{s} \cdot f_{ywd} \cdot 2A_k \cot\theta = \frac{A_{s\ell}}{u_k} \cdot 2A_k f_{y\ell d} \cdot \frac{1}{\cot\theta}$$

Solving, we obtain

$$\cot^2\theta = \left(\frac{A_{s\ell}}{u_k} \cdot f_{y\ell d}\right) \cdot \left(\frac{s}{A_{sw} f_{ywd}}\right) \tag{8.4}$$

from which we get cotθ.

If $1 \leq \cot\theta \leq 2.5$, with the obtained value of cotθ, design torsion resistance provided by reinforcements is calculated

$$T_{Rd,s} = \frac{A_{sw}}{s} \cdot f_{ywd} \cdot 2A_k \cot\theta$$

Therefore, with the same value of cotθ, we calculate

$$T_{Rd,\max} = 2\nu f_{cd} A_k t \sin\theta \cos\theta \quad [(6.30)]$$

Design torsion resistance is the smaller of the two.

If cot$\theta > 2.50$, the calculation above is performed with cot$\theta = 2.50$.

Design

Given $T_{Ed,}$ geometry and materials, the adequacy of the cross-section is verified, and reinforcement is calculated.

Preliminarily θ angle is calculated by making equal T_{Ed} and $T_{Rd,\max}$

$$T_{Ed} = (\nu f_{cd}) \cdot t \cdot A_k \sin 2\theta$$

from which

$$\theta = \frac{1}{2} \arcsin \frac{T_{Ed}}{(\nu f_{cd}) \cdot t \cdot A_k} \tag{8.5}$$

If $45^\circ > \theta > 21.8^\circ$ (that is to say if $1 \geq \frac{T_{Ed}}{(\upsilon f_{cd}) t A_k} \geq 0.69$), the concrete cross-section is adequate, and $\frac{A_{sw}}{s}$ and $A_{s\ell}$ can be calculated.

If $\theta < 21.8^\circ$, i.e., if $\frac{T_{Ed}}{(\upsilon f_{cd}) t A_k} < 0.69$, strength on the concrete side is exuberant and, for the calculation of reinforcements, we set $45^\circ \leq \theta \leq 21.8^\circ$.

If $\frac{T_{Ed}}{(\upsilon f_{cd}) t A_k} > 1$, i.e., if θ is greater than 45°, the concrete cross-section is inadequate to support the torsion T_{Ed}.

8.2.4.1 Example 1—Evaluation of Design Torsion Resistance

The rectangular cross-section of Fig. 8.13 is assigned, with $b = 400$ mm, $h = 500$ mm, concrete C25/30, with 8 mm diameter stirrups spacing 200 mm, and longitudinal reinforcement available for torsion $A_{s\ell} = 1800\ \text{mm}^2$ and, subsequently, $A_{s\ell} = 3000\ \text{mm}^2$. Design torsion resistance is calculated in both cases.

Steel B450C for longitudinal and web reinforcement ($f_{y\ell d} = f_{ywd}$).

Geometrical and mechanical parameters

$$A = 400 \cdot 500 = 200{,}000\ \text{mm}^2$$

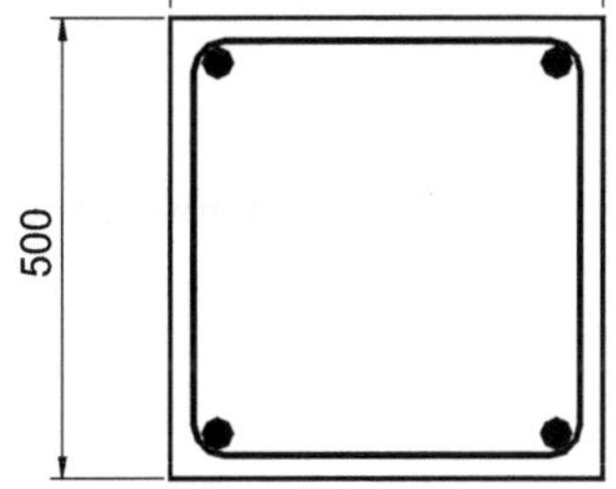

Fig. 8.13 Torsion—Example 1, beam cross-section

$$u = 2 \cdot (400 + 500) = 1800\,\text{mm}$$

$t = \frac{A}{u} = \frac{200{,}000}{1800} = 111$ mm: for the sake of simplicity, it is assumed 110 mm

$$u_k = 2 \cdot [(500 - 110) + (400 - 110)] = 1360\,\text{mm}$$

$$A_k = (500 - 110) \cdot (400 - 110) = 113{,}100\,\text{mm}^2$$

$$\begin{aligned} \nu f_{cd} &= 0.5 \cdot 14.16 = 7.08\text{N/mm}^2\text{N/mm}^2; \\ f_{yd} &= 391\,\text{N/mm}^2 \end{aligned}$$

$$f_{ywd} = f_{y\ell d} = 391\,\text{N/mm}^2$$

$$A_{sw} = 50\,\text{mm}^2;\ s = 200\,\text{mm}$$

Evaluation of cotθ.
If $A_{s\ell} = 1800\,\text{mm}^2$.
$\cot^2\theta = \frac{1800}{1360} \cdot \frac{200}{50} = 5.2941$ from which cot$\theta = 2.30$ and $\theta = 23.50^\circ$

$$\begin{aligned} T_{Rd,s} &= \frac{A_{sw}}{s} \cdot f_{yd} \cdot 2A_k \cdot \cot\theta \\ &= \frac{50}{200} \cdot 391 \cdot 2 \cdot 113{,}100 \cdot 2.30 = 50.85\,\text{kNm} \end{aligned}$$

$$\begin{aligned} T_{Rd,\max} &= 2 \cdot (\nu f_{cd}) \cdot t \cdot A_k \cdot \sin\theta \cdot \cos\theta \\ &= 2 \cdot 7.08 \cdot 110 \cdot 113{,}100 \cdot 0.3987 \cdot 0.9171 \\ &= 64.41\ \text{kNm} \end{aligned}$$

Design torsion resistance is the smaller of the two values, i.e. 50.85 kNm.

If $A_{s\ell} = 3000\,\text{mm}^2$ (if all the other parameters are equal), we get cot$\theta = 2.97$. In such a case, assuming that cot$\theta = 2.50$, we get:

$$T_{Rd,s} = 55.27\,\text{kNm}$$

$$T_{Rd,\max} = 60.74\,\text{kNm}$$

Design torsion resistance is 55.27 kNm.

8.2.4.2 Example 2—Design of Torsion Reinforcements

Considering the cross-section of Example 1, with identical materials, we design the reinforcements necessary for three values of torsion T_{Ed}.

(a) $\mathbf{T_{Ed} = 75\ kNm}$

Applying formula $\theta = \frac{1}{2}\arcsin\frac{T_{Ed}}{(\nu f_{cd})\cdot\alpha_{cw}\cdot t\cdot A_k}$, we get:

$$\theta = \frac{1}{2}\arcsin\frac{75{,}000{,}000}{7.08\cdot 1.0\cdot 110\cdot 113{,}100} = 29.18^{\circ}$$

which is the collapse angle of compressed struts. Therefore, we get

$$\cot\theta = 1.79$$

We obtain the following reinforcements:

$$\frac{A_{sw}}{s} = \frac{T_{Ed}}{2\cdot A_k\cdot f_{ywd}\cdot\cot\theta} = \frac{75{,}000{,}000}{2\cdot 113{,}100\cdot 391\cdot 1.79}$$
$$= 0.47\ \text{mm}^2/\text{mm}\ (1\phi 10/165\ \text{mm})$$

$$A_{s\ell} = \frac{T_{Ed}\cdot u_k\cdot\cot\theta}{2\cdot A_k\cdot f_{y\ell d}} = \frac{75{,}000{,}000\cdot 1360\cdot 1.79}{2\cdot 113{,}100\cdot 391} = 2064\ \text{mm}^2$$

(b) $\mathbf{T_{Ed} = 100\ kNm}$

In this case

$$\frac{T_{Ed}}{(\nu f_{cd})\cdot t\cdot A_k} = \frac{100{,}000{,}000}{7.08\cdot 1.0\cdot 110\cdot 113{,}100} = 1.13 > 1$$

therefore, the concrete cross-section is not sufficient. It is not possible to proceed further.

(c) $\mathbf{T_{Ed} = 40\ kNm}$

With the proposed value, the angle θ is given by

$$\theta = \frac{1}{2}\arcsin\frac{40{,}000{,}000}{7.08\cdot 110\cdot 113{,}100}$$
$$= 13.50^{\circ}\left(\frac{T_{Ed}}{(\nu f_{cd})\alpha_{cw} t A_k} = 0.45 < 0.69\right)$$

Collapse occurs on the steel side.
We assume $\theta = 21.8^{\circ}$ and therefore $\cot\theta = 2.50$.

We get:

$$\frac{A_{sw}}{s} = \frac{T_{Ed}}{2 \cdot A_k \cdot f_{ywd} \cdot \cot\theta} = \frac{4{,}000{,}000}{2 \cdot 113{,}100 \cdot 391 \cdot 2.50}$$
$$= 0.18 \text{ mm}^2/\text{mm } (1\phi 8/250 \text{ mm})$$

$$A_{s\ell} = \frac{T_{Ed} \cdot u_k \cdot \cot\theta}{2 \cdot A_k \cdot f_{y\ell d}} = \frac{40{,}000{,}000 \cdot 1360 \cdot 2.50}{2 \cdot 113{,}100 \cdot 391} = 1537 \text{ mm}^2$$

Alternatively, we can assume, for example, $\theta = 30^{\circ}$ and therefore $\cot\theta = 1.7321$. We get:

$$\frac{A_{sw}}{s} = \frac{T_{Ed}}{2 \cdot A_k \cdot f_{ywd} \cdot \cot\theta} = \frac{4{,}000{,}000}{2 \cdot 113{,}100 \cdot 391 \cdot 1.7321}$$
$$= 0.259 \text{ mm}^2/\text{mm } (1\phi 8/190 \text{ mm})$$

$$A_{s\ell} = \frac{T_{Ed} \cdot u_k \cdot \cot\theta}{2 \cdot A_k \cdot f_{y\ell d}} = \frac{40{,}000{,}000 \cdot 1360 \cdot 1.7321}{2 \cdot 113{,}100 \cdot 391} = 1064 \text{ mm}^2$$

i.e., increase of stirrups and reduction of the longitudinal reinforcement.

8.2.5 Shear-Torsion Interaction Diagrams

If torsion is generated by eccentric forces to the beam axis, the same beam is also solicited by bending and shear. For the couple shear-torsion, EC2 provides calculation rules at ULS for compact cross-sections strongly solicited and for approximately rectangular cross-sections under small loads not inducing cracking.

In both cases, calculation rules are represented by interaction diagrams V-T.

- *First case: strong loads*

The formula for the verification at ULS is given by (compressed struts).

$$T_{Ed}/T_{Rd,\max} + V_{Ed}/V_{Rd,\max} \leq 1 \quad [(6.29)]$$

being, for [(6.30)]

$$T_{Rd,\max} = 2\nu\, f_{cd}\, A_k\, t_{\text{ef}} \sin\theta \cos\theta$$

and for [(6.9)]

$$V_{Rd,\max} = (\nu\,f_{cd}) \cdot b_w \cdot z\,/\,(\cot\theta + \tan\theta) = (\nu\,f_{cd}) \cdot b_w \cdot z \cdot \sin\theta \cdot \cos\theta$$

- *Second case: weak loads*

The formula is the following (without cracking).

$$T_{Ed}/T_{Rd,c} + V_{Ed}/V_{Rd,c} \le 1 \quad [(6.31)]$$

being $T_{Rd,c}$ the torsion cracking moment evaluated by imposing in the expression $[(6.26)]\tau = f_{ctd} = f_{ctk}\gamma_c$

$$T_{Rd,c} = 2 f_{ctd} t_{ef} A_k$$

and $V_{Rd,c}$ being calculated by means of [(6.2a)].

The fulfilment of the relationship [(6.31)] allows arranging only minimum transversal reinforcement.

8.2.5.1 Example 1—Beam Under Shear, Torsion and Bending

Beam with span 6 m fully restrained at the end-sections, with a load P in the mean cross-section with three different eccentricities (Fig. 8.14).

Concrete: C60/75; $f_{cd} = 34$ N/mm^2. Steel: B450 C.

Static exam and reinforcement design for the load $P = 300$ kN at ULS, with eccentricities: 0.80, 0.60, 0.40 m. The beam self-weight is neglected.

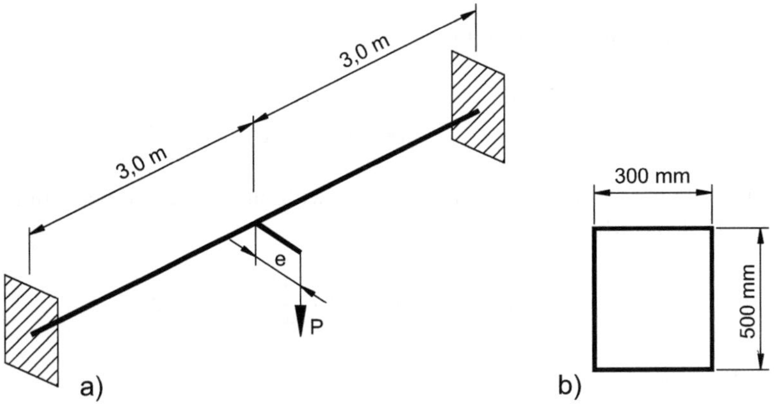

Fig. 8.14 Interaction shear-torsion—**a** static scheme; **b** beam cross-section

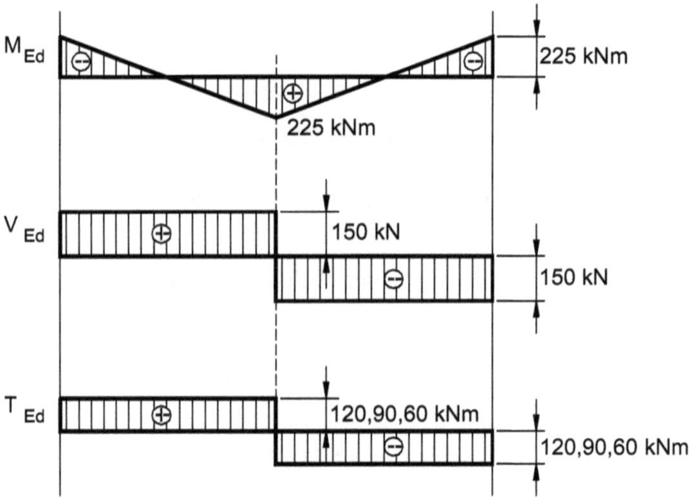

Fig. 8.15 Internal forces M_{Ed}, V_{Ed}, T_{Ed} diagrams

Static exam and reinforcement design

Bending

Being the beam rigidly supported at the end-sections, moments in the mean cross-section and at the rigid supports have the same absolute value (Fig. 8.15)

$$M_{Ed} = \frac{P\ell}{8} = \frac{300 \cdot 6}{8} = 225\,\text{kNm}$$

The reduced moment is given by

$$\mu_{Ed} = \frac{M_{Ed}}{bd^2 f_{cd}} = \frac{225 \times 10^6}{300 \cdot 450^2 \cdot 34} = 0.1090$$

From Table U2 we deduce, for the solution with only reinforcements in tension, $\omega = 0.1175$, from which

$$\rho = \frac{A_s}{bd} = \omega \frac{f_{cd}}{f_{yd}} = 0.1175 \cdot \frac{34}{391} = 0.0102$$

and, therefore,

$$A_s = 0.0102 \cdot 300 \cdot 450 = 1380\,\text{mm}^2$$

Table 8.6 Internal forces of shear and torsion

P(kN)	e(m)	V_{Ed}(kN)	T_{Ed}(kNm)
300	0.80	150	120
300	0.60	150	90
300	0.40	150	60

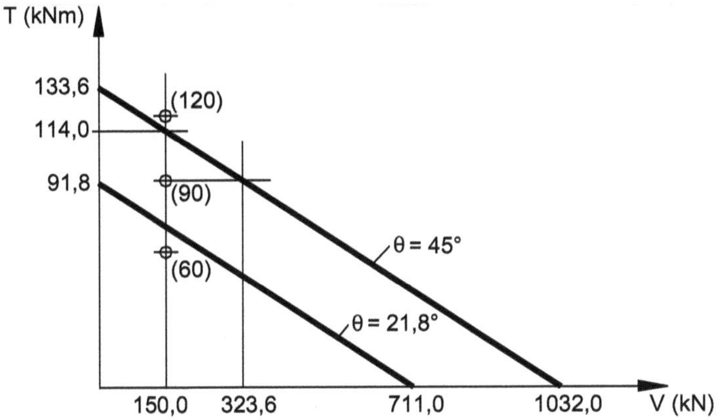

Fig. 8.16 Interaction diagrams $V_{Rd,\max} - T_{Rd,\max}$

Shear and torsion

Internal forces V and T have uniform diagrams, antisymmetric to the mean cross-section of the beam (Fig. 8.15). Relevant values are collected in Table 8.6.

The following procedure is proposed to analyse internal forces associated with the beam strength.

We sketch (Fig. 8.16) the two interaction diagrams $V_{Rd,\max} - T_{Rd,\max}$ referred to the concrete struts strength with slope $\theta = 45^\circ$ and $\theta = 21.8^\circ$ respectively, by applying the relationship:

$$T_{Ed}/T_{Rd,\max} + V_{Ed}/V_{Rd,\max} = 1 \quad [(6.29)]$$

In the same Figure, we design the points representative of the couples $V_{Ed} - T_{Ed}$ of Table 8.6. Based on the comparison, we can design reinforcements.

$T_{Rd,\max}$ and $V_{Rd,\max}$ are calculated by means of

$$T_{Rd,\max} = 2\nu f_{cd} A_k t_{ef} \sin\theta \cos\theta \quad [(6.30)]$$

$$V_{Rd,\max} = b_w z \nu f_{cd} \sin\theta \cos\theta \quad [(6.9)]$$

Figure 8.17a shows the torsion-resistant tubular cross-section. The values of the geometrical terms of [(6.30)] are given by:

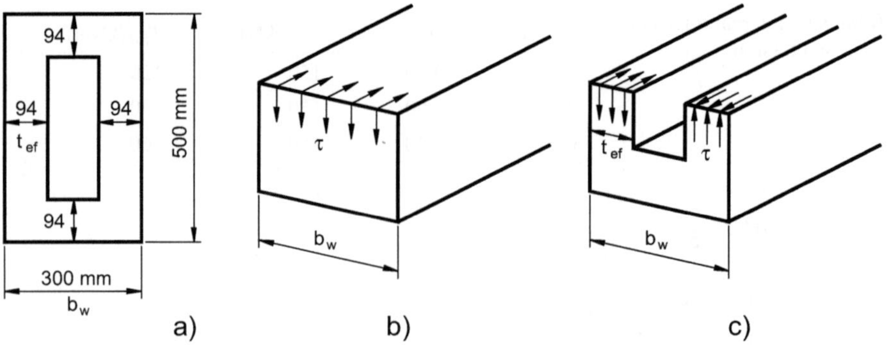

Fig. 8.17 **a** Resistant cross-section; **b** τ shear stress; **c** τ torsion stress

$$A = 300 \cdot 500 = 15000\,\text{mm}^2$$

$$u = (300 + 500) \cdot 2 = 1600\,\text{mm}$$

$$t_{ef} = A/u = 94\,\text{mm}$$

$$u_k = 2 \cdot (500 - 94 + 300 - 94) = 1224\,\text{mm}$$

$$A_k = (500 - 94) \cdot (300 - 94) = 83636\,\text{mm}^2$$

$$d = 450\,\text{mm}$$

$$z = 0.9d = 405\,\text{mm}$$

The values of the terms referred to in equation [(6.9)] are contained in the problem data.

Calculation of the terms referred to [(6.30)] and [(6.9)].

Assuming $\theta = 45^\circ$, we get

$$T_{Rd,\max} = 2 \cdot 0.50 \cdot 34 \cdot 83636 \cdot 94 \cdot 0.707 \cdot 0.707 = 133.6\,\text{kN m}$$

$$V_{Rd,\max} = 300 \cdot 405 \cdot 0.50 \cdot 34 \cdot 0.707 \cdot 0.707 = 1032\,\text{kN}$$

Assuming $\theta = 21.8^\circ$

$$T_{Rd,\max} = 2 \cdot 0.50 \cdot 34 \cdot 83636 \cdot 94 \cdot 0.3714 \cdot 0.9285 = 91.8\,\text{kN m}$$

$$V_{Rd,\max} = 300 \cdot 405 \cdot 0.50 \cdot 34 \cdot 0.3714 \cdot 0.9285 = 711\,\text{kN}$$

Interaction diagrams are made of two parallel lines (Fig. 8.16). If on the same Figure we design the points with coordinates $V_{Ed} = 150\,\text{kN}$ and T_{Ed} of Table 8.6, we observe that:

- $T_{Ed} = 120\,\text{kN m}$ is not acceptable, because the strength is exceeded;
- $T_{Ed} = 90\,\text{kN m}$ is the range between the two lines, therefore it is acceptable. The corresponding angle θ shall be determined;
- $T_{Ed} = 60\,\text{kN m}$ is under the two lines: it is acceptable and ULS occurs because of the reinforcement collapse.

Evaluation of slope θ of compressed struts

Preliminarily we observe that shear causes on the rectangular cross-section, the following tangential stress distribution

$$\tau_v = \frac{V_{Ed}}{b_w \cdot z}$$

which is uniform on the width b_w and depth z (Fig. 8.17b).

Torsion instead causes on a flux tube with width t_{ef} and height z, a uniform distribution of stress τ_t with a sign, and on the parallel tube the same distribution with opposite sign (Fig. 8.17c)

$$\tau_t = \frac{T_{Ed}}{2A_k \cdot t_{ef}}$$

The most solicited flux tube is the one where shear and torsion stresses add and give the resultant force

$$V^* = (\tau_t + \tau_v) \cdot t_{ef} \cdot z \tag{8.6}$$

By substituting, we get

$$V^* = \frac{T_{Ed} \cdot z}{2A_k} + V_{Ed} \cdot \frac{t_{ef}}{b_w} \tag{8.7}$$

V^* is therefore a known term.

If we recall the relationship between tangential stresses τ, both due to shear and torsion, to stresses σ_c of the compressed struts with slope θ

$$\tau = \sigma_c \sin\theta \cdot \cos\theta$$

within the tube flux where these stresses add each other, we can write

$$(\tau_t + \tau_v) = \sigma_c \cdot \sin\theta \cdot \cos\theta$$

and at ULS

$$(\tau_t + \tau_v) = (\nu f_{cd}) \cdot \sin\theta \cdot \cos\theta$$

By substituting in expression (8.6) we get

$$V^* = (\nu f_{cd}) \cdot z \cdot t_{ef} \cdot \sin\theta \cdot \cos\theta = \frac{1}{2}(\nu f_{cd}) \cdot z \cdot t_{ef} \cdot \sin 2\theta$$

which is a relationship that allows calculating θ. Therefore, we get

$$\theta^* = \frac{1}{2} \arcsin \frac{2V^*}{t_{ef} z \nu f_{cd}} \tag{8.8}$$

In the second case ($T_{Ed} = 90\,\text{kN m}$), by applying (8.7) and (8.8), we get:

$$V^* = \frac{90 \times 10^6 \cdot 405}{2.83636} + 150 \times 10^3 \cdot \frac{94}{300} = 218 + 47 = 265\ \text{KN}$$

$$\theta^* = \frac{1}{2} \arcsin \frac{2 \cdot 265{,}000}{94 \cdot 405 \cdot 0.5 \cdot 34} = 27.5^{\circ}$$

In the third case ($T_{Ed} = 60\,\text{kN m}$) we get

$$V^* = \frac{60 \times 10^6 \cdot 405}{2.83636} + 150 \times 10^3 \cdot \frac{94}{300} = 145 + 47 = 192\ \text{KN}$$

$$\theta^* = \frac{1}{2} \arcsin \frac{2 \cdot 192{,}000}{94 \cdot 405 \cdot 0.5 \cdot 34} = 18.20^{\circ}$$

In the second case, θ^* is the lower limit value for the possible solutions. These solutions are limited in the range θ^*–45° (and not 21.8°–45°). If the value θ^* is adopted, we get the minimum amount of reinforcement. If the value 45° is adopted with $V_{Rd} = 150\,\text{kN}$, we get $T_{Rd} = 114\,\text{kNm}$; with $T_{Rd} = 90\,\text{kNm}$ we get $V_{Rd} = 323\,\text{kN}$, in each case with strength greater than the required one. These values are graphically obtained by intersecting the line of the superior diagram with two lines, one vertical and another one horizontal, passing through the point with coordinates $V_{Ed} = 150\,\text{kN}$, $T_{Ed} = 90\,\text{kNm}$.

In the third case, being $\theta^* = 18.20^{\circ} < 21.80^{\circ}$, because of [(6.7N)] it is necessary to adopt at least 21.80°.

Calculation of reinforcement

Stirrups

As a rule, reinforcement is calculated by summing reinforcement required for torsion and shear, calculated using the same angle θ.

Under the assumption of 2-arms stirrups, required reinforcement for torsion (one arm), can be deduced by (8.3)

$$\frac{A_{sw}}{s} = \frac{T_{Ed}}{2A_k f_{yd} \cot\theta}$$

and reinforcement for shear is obtained by [(6.8)]

$$\frac{A_{sw}}{s} = \frac{V_{Ed}}{z \cdot f_{yd} \cot\theta}$$

This last one is instead the total reinforcement required for shear, i.e. in the present case 2-arm stirrups. To add the two results, the one obtained for shear shall be divided by two. Therefore, we get

$$\frac{A_{sw}}{s} = \frac{T_{Ed}}{2A_k f_{yd} \cot\theta} + \frac{V_{Ed}}{2 \cdot z \cdot f_{yd} \cot\theta}$$

In the second case ($T_{Ed} = 90$ kNm), being $\theta^* = 27.5^\circ$ and $\cot\theta^* = 1.92$, we get, for one stirrup arm:

$$\begin{aligned}\frac{A_{sw}}{s} &= \frac{90 \times 10^6}{2 \cdot 83{,}636 \cdot 391 \cdot 1.92} + \frac{150{,}000}{2 \cdot 405 \cdot 391 \cdot 1.92} \\ &= 0.7167 + 0.2461 = 0.9628\,\text{mm}^2/\text{mm}\end{aligned}$$

which are required, for example, with 1 ϕ 12/110 mm.

In the third case ($T_{Ed} = 60$ kNm), being $\theta^* < 21.8^\circ$ to fulfil [(6.7N)], we assume $\cot\theta = 2, 5$, corresponding to 21.8°. For one stirrup arm we get:

$$\begin{aligned}\frac{A_{sw}}{s} &= \frac{60 \times 10^6}{2 \cdot 83{,}636 \cdot 391 \cdot 2.50} + \frac{150{,}000}{2 \cdot 405 \cdot 391 \cdot 2.50} \\ &= 0.3670 + 0.1890 = 0.5560\,\text{mm}^2/\text{mm}\end{aligned}$$

which are fulfilled, for example, by 1 ϕ 10/140 mm.

Longitudinal reinforcement

Required reinforcement for torsion is given by

$$A_{s\ell} = \frac{T_{Ed} u_k}{2 \cdot A_k \cdot f_{yd}} \cdot \cot\theta$$

and it shall be distributed along the perimeter, within stirrups.

Shear reinforcing is given by

$$A_{s\ell} = \frac{V_{Ed}}{2 \cdot f_{yd}} \cdot \cot\theta$$

and shall be arranged at the bottom.

In the second case ($T_{Ed} = 90\,\text{kNm}$), we get:

$$\begin{aligned} A_{s\ell} &= \frac{90 \times 10^6 \cdot 1224}{2 \cdot 83{,}636 \cdot 391} \cdot 1.92 + \frac{150{,}000}{2 \cdot 391} \cdot 1.92 \\ &= (3234 + 368)\,\text{mm}^2 \end{aligned}$$

In the third case ($T_{Ed} = 60\,\text{kN m}$) we get:

$$\begin{aligned} A_{s\ell} &= \frac{60 \times 10^6 \cdot 1224}{2 \cdot 83{,}636 \cdot 391} \cdot 2.50 + \frac{150{,}000}{2 \cdot 391} \cdot 2.50 \\ &= (2807 + 480)\,\text{mm}^2 \end{aligned}$$

The first value within brackets is referred to torsion, the second one is referred to shear.

Reference

Nielsen, M. P. (1990). Commentaries on Shear and Torsion. Eurocode 2 Editorial Group—1st Draft—October 1990.

Chapter 9
Punching Shear

Abstract Punching shear is a local failure mechanism of an RC structural element, typically a slab, caused by a concentrated load. It can occur when the structural element is too thin to support the applied load and/or the size of the loaded area is too small. This chapter deals with the verification of punching shear for slabs and footings subjected to vertical loads only, while effects due to the cyclic inversion of bending moments at column-slab joints during seismic events are not considered. In literature, various models are available to calculate the punching shear strength: some models are based on the plasticity theory, others on strut and tie (S&T) models or fracture mechanics, but an analytical complete formulation is still missing. Formulas adopted in the international codes, like EC2, are substantially experimental. To evaluate the punching shear strength of a structural element, EC2 defines a conventional control perimeter along which the load is distributed. For slabs, the control perimeter is placed at a distance equal to twice the effective depth from the edge of the column or from the loaded area. For foundations, the control perimeter is placed at a distance not higher than twice the effective depth and it must be identified through an iterative procedure. All the verifications are described in detail and many case studies show their application to real cases.

9.1 Introduction

Punching shear is a local failure mechanism of an RC structural element, typically a slab, caused by a concentrated load. It can occur when the structural element is too thin to support the applied load and/or the size of the loaded area is too small (Fig. 9.1).

This chapter deals with the verification of punching shear for slabs and footings subjected to vertical loads only, while effects due to the cyclic inversion of bending moments at column-slab joints during seismic events are not considered.

This chapter was authored by Franco Angotti and Maurizio Orlando.

© The Author(s), under exclusive license to Springer Nature Switzerland AG 2022
F. Angotti et al., *Reinforced Concrete with Worked Examples*,
https://doi.org/10.1007/978-3-030-92839-1_9

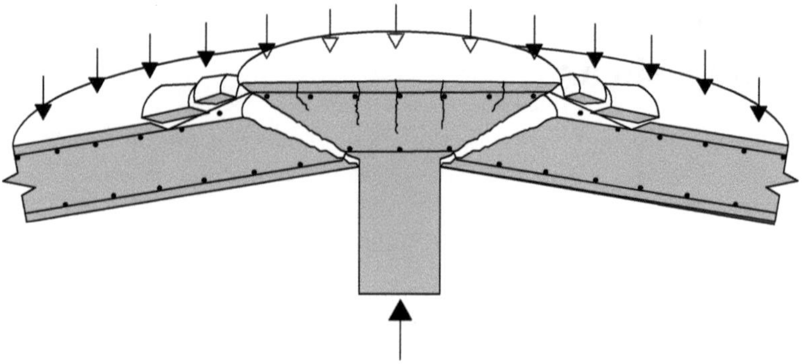

Fig. 9.1 Punching shear of a slab at an internal circular column

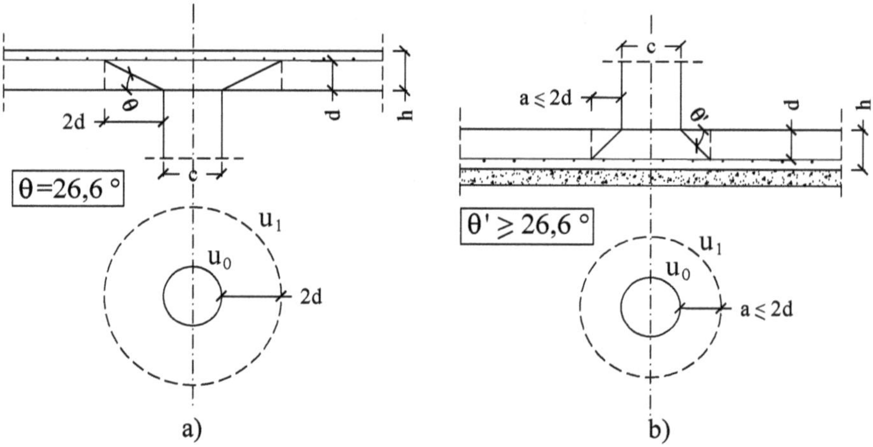

Fig. 9.2 Control perimeter u_1 according to EC2 for a circular column: **a** slab, **b** footing (u_0 is the perimeter of the column)

In literature, various models are available to calculate the punching shear strength: some models are based on the plasticity theory, others on strut and tie (S&T) models or fracture mechanics, but an analytical complete formulation is still missing. Formulas adopted in the international codes, like EC2, are substantially experimental.

To evaluate the punching shear strength of a structural element, EC2 defines a conventional control perimeter along which the load is distributed. This perimeter is denoted as u_1.

For slabs, the control perimeter is placed at a distance equal to $2d$ from the edge of the column (where d is the effective depth) or from the loaded area (Fig. 9.2a). For foundations, the control perimeter is placed at a distance $a \leq 2d$ (Fig. 9.2b), which must be identified through an iterative procedure (see 9.15).

The angle of the failure surface with the horizontal is $\theta = 26.6°$ for slabs and $\theta' \geq 26.6°$ for footings.

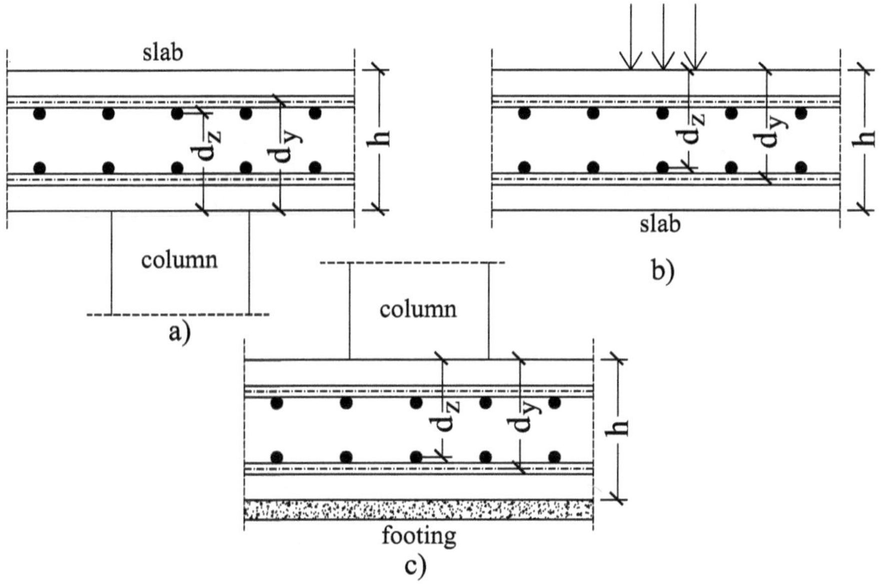

Fig. 9.3 Effective depths in y and z directions: **a** slab on a column, **b** concentrated load on a slab, **c** footing

For circular columns, the verification is performed on the lateral surface S_1 of the cylinder, whose circumference of the base is given by the perimeter u_1 and whose height is equal to the effective depth of the slab: $S_1 = u_1 \cdot d$). The effective depth of the slab can be calculated as the average of the effective depths in the two orthogonal directions (Fig. 9.3).

For polygonal columns, the control perimeter can be obtained by moving outwardly of *2d* all the sides of the column perimeter and by filleting the translated sides with circumference arcs of radius *2d*, centred on the cross-section corners (Fig. 9.4).

In special cases, e.g. when footings are subjected to high contact pressure or when reactions act at a distance less than *2d* from the loaded area (Fig. 9.5), it is required to consider a control perimeter located at a distance lower than 2d.

For loaded areas near to openings, if the minimum distance between the loaded area and the opening is not higher than *6d*, the segment PQ of the perimeter u_1 between the two tangent lines from the centre of the loaded area to the perimeter of the opening should be considered as ineffective (Fig. 9.6).

In every case, further verification is always required on the perimeter u_0 of the column or loaded area; on this perimeter, the shear force should not exceed a limit value related to the compressive concrete strength.

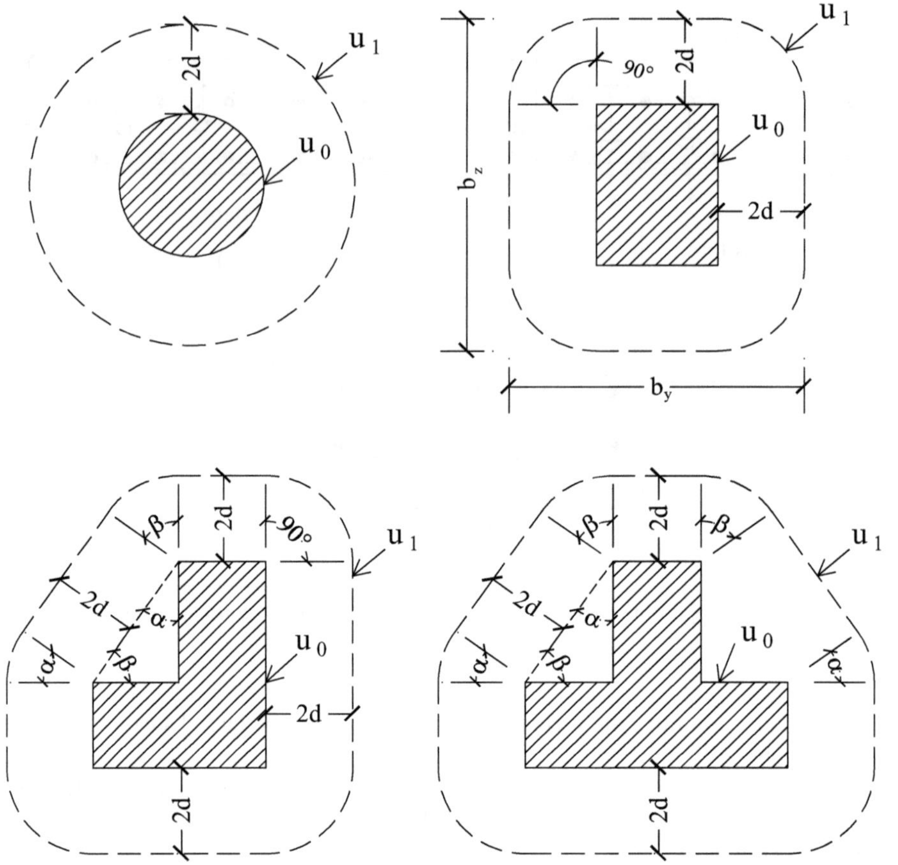

Fig. 9.4 Control perimeter for different shapes of the cross-section or loaded area [Fig. 6.13]

Because of the experimental nature of the formulas, both the shape and the distance of the control perimeter from the loaded area are not univocally defined, and international codes provide different expressions for both design force and punching strength.

Figure 9.7 shows the control perimeter according to various international codes; its distance from the column perimeter varies from *0.5d* (ACI code) to *2d* (EC2 and Model Code 1990).

9.2 The Failure Mechanism Due to Punching Shear

Punching shear failure is characterized by inclined cracks developing from the perimeter of the column or loaded area, which define a failure surface with the shape of a truncated pyramid (rectangular perimeter) or truncated cone (circular perimeter). This mechanism is typical of a brittle failure, where the flexural reinforcement can remain in the elastic field until failure.

The punching shear failure at an internal column can be described considering the experimental results for a slab on a circular column, subjected to a linearly distributed load on a circumference centred on the column axis.

As the load increases, the following phases can be easily identified:

1. the slab deforms elastically without cracking
2. some tangential cracks appear on the top face close to the column (Fig. 9.8a)
3. some radial cracks open, due to circumferential bending moments; these cracks draw circular sectors on the slab (Fig. 9.8b)
4. in the proximity of the column bending moments remain approximately constant, while new circumferential cracks appear at a higher distance from the column (Fig. 9.8c)
5. new cracks open by creating a separation surface between the slab and the column with the shape of a truncated cone; this is the typical shape of the punching failure surface (Fig. 9.8d)
6. existing cracks widen until the slab fails: the failure occurs suddenly, at least for slabs with a normal amount of longitudinal reinforcement
7. after the slab failure, on the top face of the slab, a big circumferential crack can be observed, with a radius higher than the radius of the punching shear cone because of the yielding and plastic deformation of the tension top reinforcement.

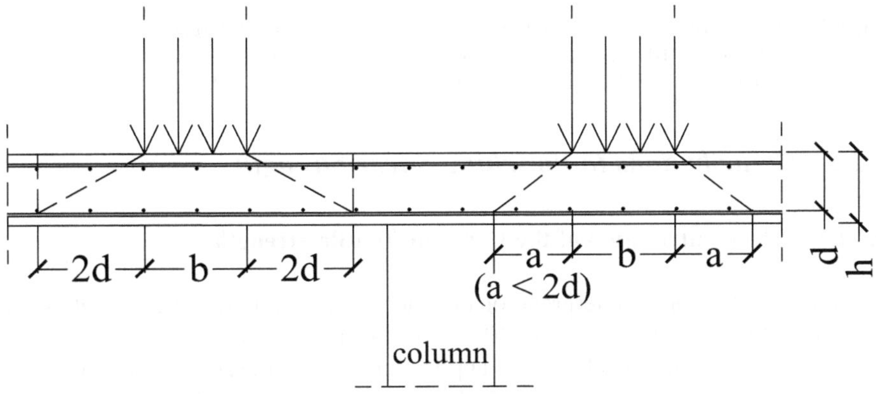

Fig. 9.5 Loaded area located at a distance greater (left) or less (right) than $2d$ from the column

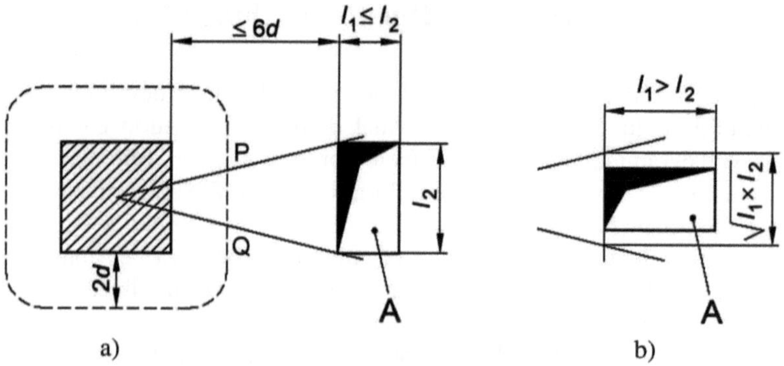

Fig. 9.6 Control perimeter in presence of a rectangular opening [Fig. 6.14]: **a** the PQ segment is ineffective; **b** if the opening turns its smallest size towards the loaded column or area, the equivalent size $\sqrt{l_1 \cdot l_2}$ is considered to draw the tangents. *Legend* A = opening

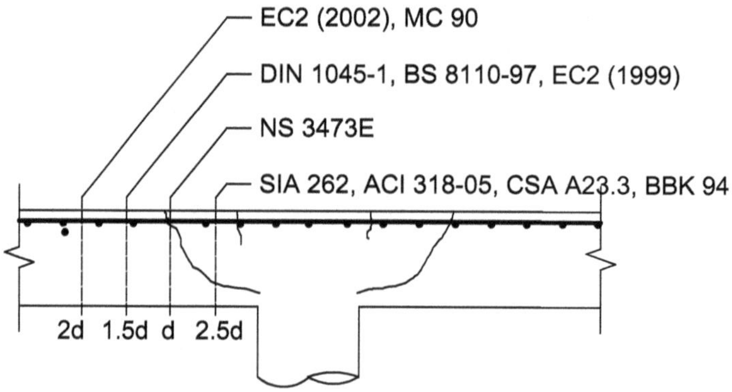

Fig. 9.7 Control perimeter for punching shear verification according to different international codes (ACI—United States, BBK—Sweden, BS—Great Britain, CSA—Canada, MC90—Model Code 1990, DIN—Germany, NS—Norway, SIA—Switzerland) (Guandalini, 2005)

9.2.1 Contributions to Punching Shear Strength

9.2.1.1 The Contribution of the Concrete Tensile Strength

Literature models for the interpretation of punching shear failure are almost all based on strut and tie models (see Chap. 10). Strut and tie models identify a mechanism where the load is transferred from its application point to the column perimeter.

One of the proposed models extends the strut and tie model within the slab portion placed on the column, considering the presence of a concrete tie (Fig. 9.9). In this model, the punching shear failure is coincident with the tension failure of the concrete tie: the ultimate load can be evaluated by considering tensile stresses on the truncated conical punching surface and by integrating their vertical components.

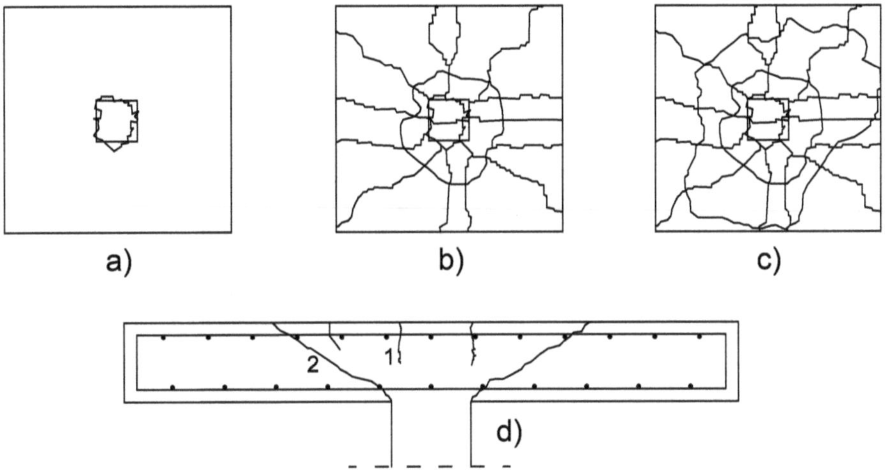

Fig. 9.8 Cracking pattern evolution on the tension top face of a slab at increasing the punching load (**a–c**) sectional view of the final cracking pattern with 1-circumferential and 2-radial cracks (**d**)

The contribution due to tension in the concrete tie should be added to the contribution of steel reinforcement passing through the punching surface (longitudinal reinforcement and transverse reinforcement, if any) and the contribution of prestressing reinforcement if any.

9.2.1.2 The Contribution of the Flexural Reinforcement ("Dowel Action")

The contribution to shear transfer given by the flexural reinforcement through the inclined punching shear surface is called "dowel action". It is like the transfer shear mechanism that occurs in beams through flexural cracks.

Some experimental tests have highlighted that in slabs with orthogonal flexural reinforcements, the dowel action can contribute up to 34% to the punching shear strength (CEB-FIP, 2001).

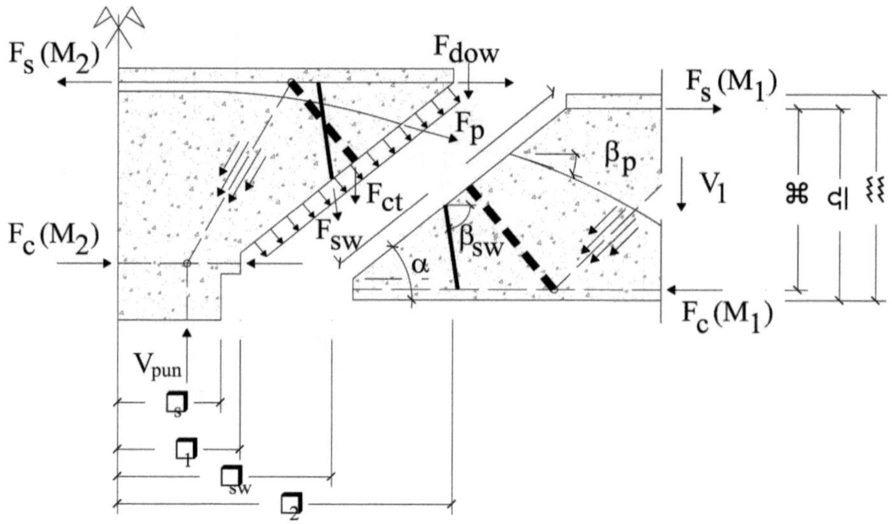

Fig. 9.9 Analytical model of the shear-punching resistant mechanism: in bold the concrete tie (which has the shape of a truncated cone); the base punching resistance of the slab is corresponding to the resistance of the truncated conical concrete tie (Menetréy, 1996). *Legend* r_1, r_2 = distances at the bottom and top of the inclined crack from the column axis; r_{sw} = distance from the column axis of the intersection point of the transverse reinforcement with the crack; β_{sw}, β_p = inclination of the transverse steel and prestressing reinforcement, F_{dow} = force due to the dowel effect in the top reinforcement, F_{ct} = vertical component of the resultant tension force on the truncated conical surface of the crack, F_p = prestressing force, F_{sw} = tension force in the transverse reinforcement

The flexural reinforcement ratio can strongly influence the punching shear strength; the load–displacement curves obtained from tests on slabs with different reinforcement ratios ($\rho_l = 0.2 - 0.4 - 0.8 - 1.2 - 2\%$) show that, after an initial linear elastic branch, higher values of ρ lead to higher values of the punching shear strength, while the ultimate displacement (and then ductility) decreases (Fig. 9.10).

The flexural reinforcement ratio also influences the geometry of the mechanical model. Experimental tests on slabs with transverse reinforcement have highlighted that the shape of the compression field is different as ρ_l is varying.

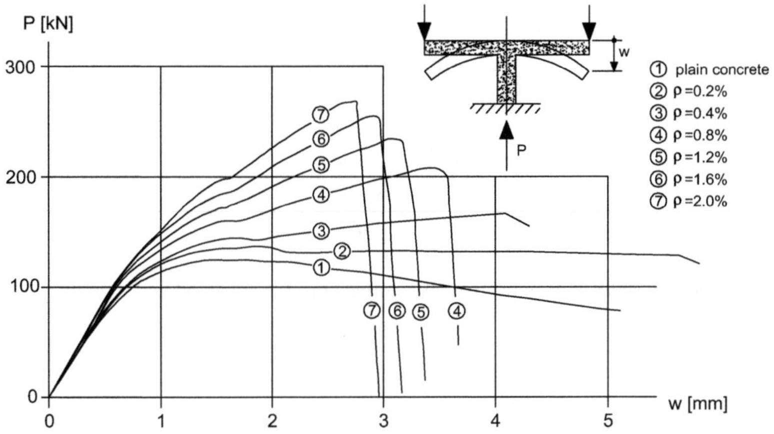

Fig. 9.10 Influence of the flexural reinforcement ratio on the punching shear strength (Menétrey, 2002)

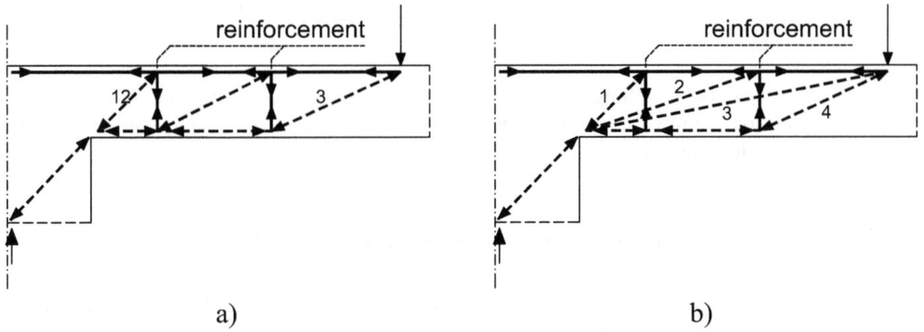

Fig. 9.11 S&T models for slabs with low (**a**) or high (**b**) flexural reinforcement ratio (CEB-FIP, 2001)

If the reinforcement ratio is low, the compression field is characterized by very high stresses between adjacent transverse reinforcement bars. The resistant mechanism can be well described by a strut and tie model with struts between adjacent reinforcement perimeters and ties represented by the transverse reinforcement (Fig. 9.11a).

For high values of the flexural reinforcement ratio, compression struts develop (Fig. 9.11b) from both the load perimeter and the top of each reinforcement perimeter up to the column perimeter. This mechanical model is also confirmed by very high tensile stresses experimentally measured in flexural reinforcement.

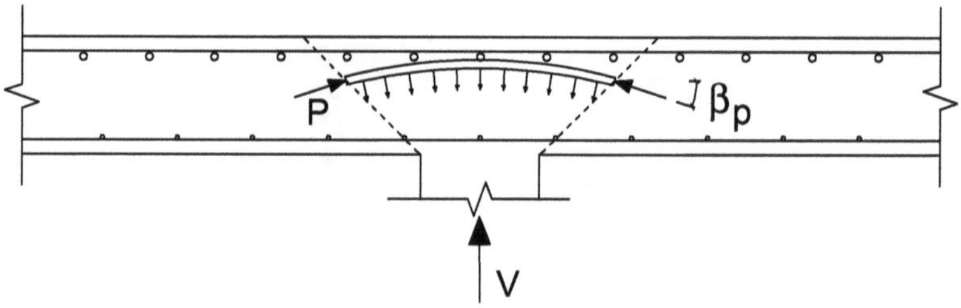

Fig. 9.12 Punching shear resistance given by inclined prestressing tendons

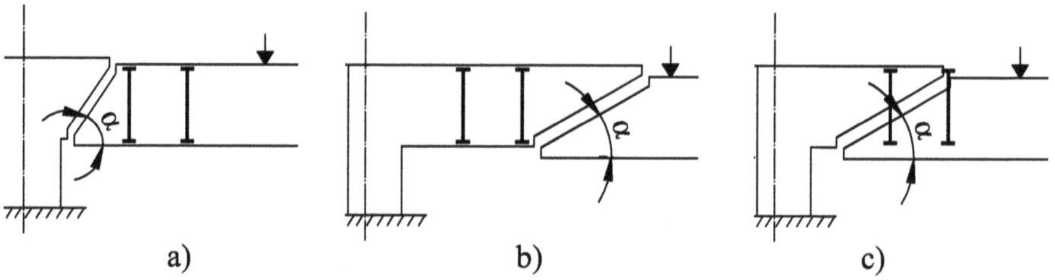

Fig. 9.13 Position of the punching shear surface of slabs with transverse reinforcement: **a** between the column and first reinforcement perimeter, **b** beyond the last reinforcement perimeter, **c** in the region equipped with transverse reinforcement (CEB-FIP, 2001)

9.2.1.3 The Contribution of the Prestressing Reinforcement

The vertical component of the prestressing force in inclined tendons gives a further contribution to the slab punching shear strength (Fig. 9.12).

To calculate this contribution, the actual tensile stress in the tendons should be considered and not the steel yield strength, which cannot be reached because punching shear failure occurs with small deformations.

Prestress also gives a dowel action, then the prestressing steel crossing the failure surface can be added to the flexural reinforcement.

9.2.1.4 The Contribution of the Transverse Reinforcement (Links, Stirrups, Bent-Up Bars)

The transversal reinforcement gives a contribution to the punching shear strength because it guarantees a link action orthogonally to the failure surface.

For slabs with punching reinforcement (like vertical links, stirrups, bent-up bars, etc.), failure can occur in one of the following three regions (Fig. 9.13).

1. between the column and the first reinforcement perimeter,
2. beyond the last reinforcement perimeter,
3. within the region equipped with transverse reinforcement with the formation of a cracked surface crossing transverse reinforcement.

To avoid failures shown in Fig. 9.13a, b, EC2 gives some geometrical rules for the position of the first and last reinforcement perimeter, which are collected in paragraph 9.13.

Failure shown in Fig. 9.13c can be avoided by adopting a minimum area of transverse reinforcement.

Some researchers have also evaluated the variation of the punching shear strength if the position of the first reinforcement perimeter is changed, while the position of the other reinforcement perimeters is kept unchanged.

The highest values of the punching shear strength are obtained for distances of the first perimeter from the column perimeter in the range 0.5d ÷ 1.0d when the strut between the top of the first perimeter and the perimeter of the column (strut 1 in Fig. 9.11) has a slope between 45° and 60° to the horizontal.

If the distance between the first reinforcement perimeter and the column decreases, the intersection between the punching shear surface and the first reinforcement perimeter moves to the bottom. For small values of the distance, the length of the bar under the failure surface could be so short that the bond strength is lower than reinforcement yielding strength.

Vice versa, by increasing the distance between the first perimeter and the column, the compression force in the strut increases and early concrete crushing occurs.

9.2.2 Size Effect

Experimental tests have shown a "size effect" in the punching shear failure: if the slab thickness increases, the unit punching shear strength (namely the resistant shear divided by the area of the control surface) decreases.

Some tests on slabs with transverse reinforcement have shown that the unit punching shear strength decreases also if the distance of the control perimeter from the column increases. For example, failures occurring beyond the last reinforcement perimeter in slabs with transverse reinforcement, are characterized by values of the unit punching shear strength lower than non-reinforced slabs, whose failure surfaces arise at a lower distance from the column or the loaded area.

9.2.3 Types of Punching Shear Reinforcement

Figures 9.14, 9.15 and 9.16 show different types of punching shear reinforcement, which can be adopted in slabs and foundations.

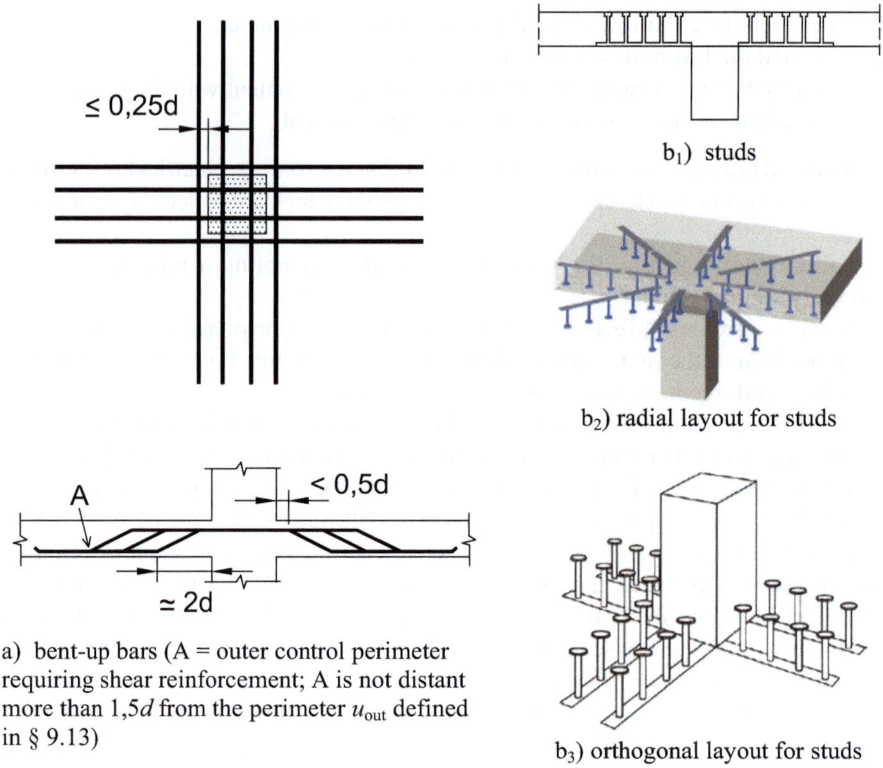

a) bent-up bars (A = outer control perimeter requiring shear reinforcement; A is not distant more than 1,5d from the perimeter u_{out} defined in § 9.13)

b_1) studs

b_2) radial layout for studs

b_3) orthogonal layout for studs

Fig. 9.14 Types of punching shear reinforcement (part I)

Experimental tests have proved that stirrups give ductility to the mechanism of punching shear failure, even if the flexural reinforcement is not involved; to activate this mechanism, it is sufficient that stirrups are anchored at the same level of the flexural reinforcement (Fig. 9.15c).

The calculation of vertical links and bent-up bars and their layouts are described in paragraph 9.13.

Prefabricated reinforcement shown in Fig. 9.16 goes usually with a "European Technical Agreement" containing all the information for design and assembly.

9.3 Phases of Punching Shear Verification

The punching shear verification should be performed:

1. along the perimeter u_0 of the column or loaded area,
2. along the control perimeter u_1.

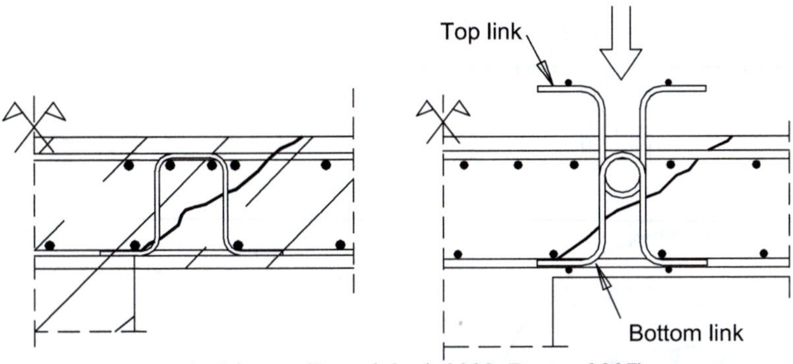

a) Stirrups (Beutel & al. 2002, Broms 2007)

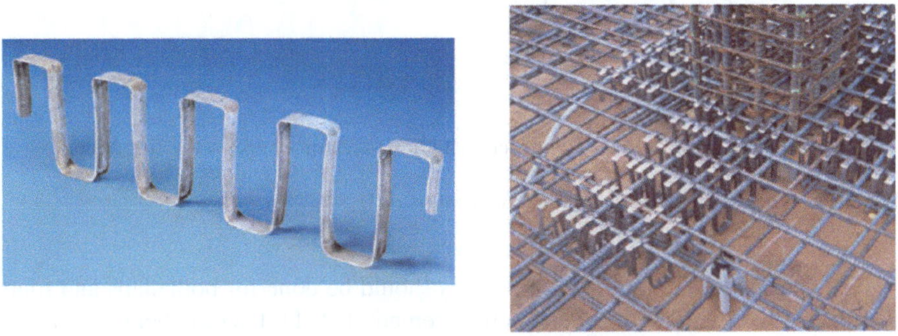

b) Links made with omega-shaped steel strips
(LENTON system by ERICO – http://www.erico.com)

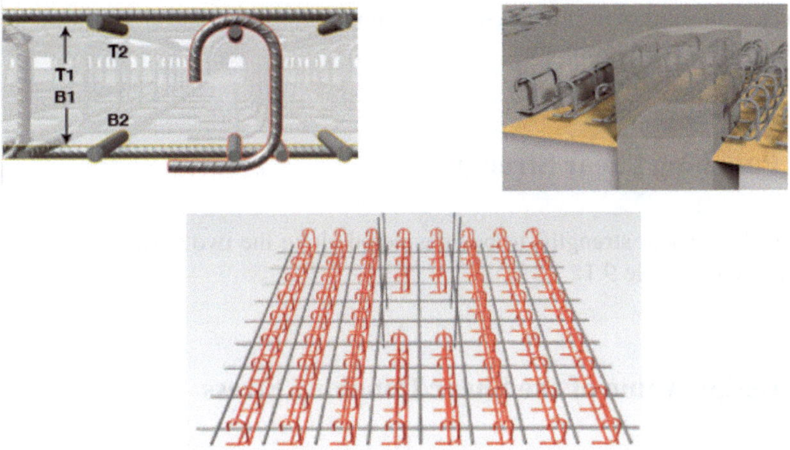

c) Prefabricated punching shear ladders (shear ladder system by ROM, Ltd. UK)

Fig. 9.15 Types of punching shear reinforcement (part I)

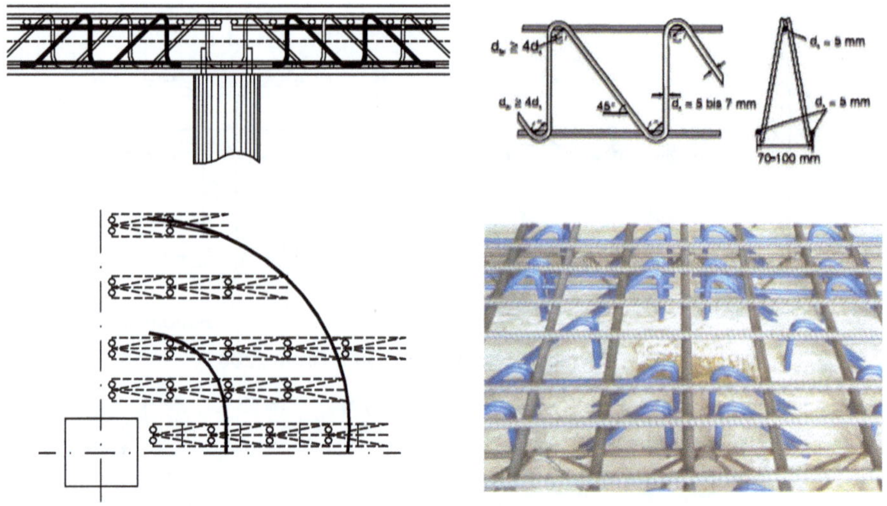

d) Prefabricated lattice-girders (FILIGRAN system – http://www.filigran.de)

Fig. 9.16 Types of punching shear reinforcement (part II)

The verification along the perimeter u_0 should be done for both slabs and foundations, with or without transverse reinforcement. In 9.11, the ratio between the two punching shear strengths along u_0 and u_1 for slabs without transverse reinforcement is evaluated for different effective depth, concrete grade, and cross-section dimensions. For thin slabs, the minimum strength is obtained along the perimeter u_1, while beyond a limit value of the thickness, the minimum strength is that along the perimeter u_0 of the column or loaded area.

9.4 Punching Shear Strength

The punching shear strengths to be considered along the two perimeters u_0 and u_1 are collected in Table 9.1.

9.5 Design Value of the Shear/Punching Stress

The design punching shear stress v_{Ed} along the control perimeter u_i is calculated with the following formula:

$$v_{Ed} = \frac{\beta \cdot V_{Ed}}{u_i \cdot d}$$

Table 9.1 Punching shear strength and control perimeter

Control perimeter	Design strength	Meaning
u_0	$v_{Rd,max}$	design value of the maximum punching shear resistance along the perimeter u_0 [(6.53)]: it represents the maximum value of the unit punching shear resistance (per unit area) in the proximity of a column or loaded area
u_1	$v_{Rd,c}$	design value of the punching shear resistance of a slab without punching shear reinforcement along the control perimeter u_1 at distance *2d* for slabs or $a \leq 2d$ for foundations [(6.47)]
	$v_{Rd,cs}$	design value of the punching shear resistance of a slab with punching shear reinforcement along the control perimeter u_1 at distance *2d* [(6.52)]

where

$\beta = 1$ if the column reaction (or load for loaded areas) passes through the centroid of the control perimeter

$\beta > 1$ if the column reaction (or load for loaded areas) does not pass through the centroid of the control perimeter

V_{Ed} punching shear design force.

u_{i} u_0 = column or loaded area perimeter.

u_1 = control perimeter at distance $2d$ from the column perimeter for slabs (Fig. 9.2a) and $a \leq 2d$ for foundations (Fig. 9.2b).

EC2 uses the same symbol v_{Ed} for punching shear stress calculated along the two perimeters u_0 and u_1; in this chapter, a different subscript is added to the symbol v_{Ed} to avoid misunderstanding:

$v_{\mathrm{Ed},0}$ punching shear stress along the perimeter u_0 of the column or loaded area.

$v_{\mathrm{Ed},1}$ punching shear stress along the control perimeter u_1.

EC2 gives formulas to calculate the β coefficient where the column reaction is eccentric, considering axial load combined with uniaxial or biaxial bending and different column positions (internal, edge or corner column) (Fig. 9.17). Expressions and tables to calculate β are provided in paragraph 9.7.

9.6 Perimeters u_0 and u_1 for Rectangular Columns

The calculation of u_0 and u_1 perimeters for rectangular columns varies with the position of the column.

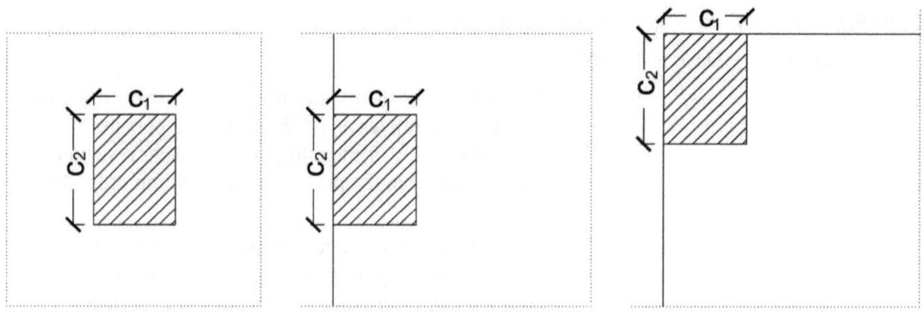

a) b) c)

Fig. 9.17 Rectangular column; **a** internal column, **b** edge column, **c** corner column

9.6.1 *Internal Column*

$$u_0 = 2(c_1 + c_2)$$
$$u_1 = 2(c_1 + c_2) + 4\pi\, d$$

9.6.2 *Edge Column*

c_1 = dimension orthogonal to the edge
c_2 = dimension parallel to the edge

$$u_0 = c_2 + 3d \le c_2 + 2c_1$$
$$u_1 = 2c_1 + c_2 + 2\pi\, d$$

In addition to u_1, the reduced control perimeter u_1* can be defined as follows (Fig. 9.18):

$$u_1^* = \min(0.5c_1;\ 1.5\, d) + c_2 + 2\pi d$$

The reduced control perimeter u_1* can be used as described at point *e* of paragraph 9.7.2.

9.6.3 *Corner Column*

$$u_0 = 3d \le c_1 + c_2$$
$$u_1 = c_1 + c_2 + \pi\, d$$

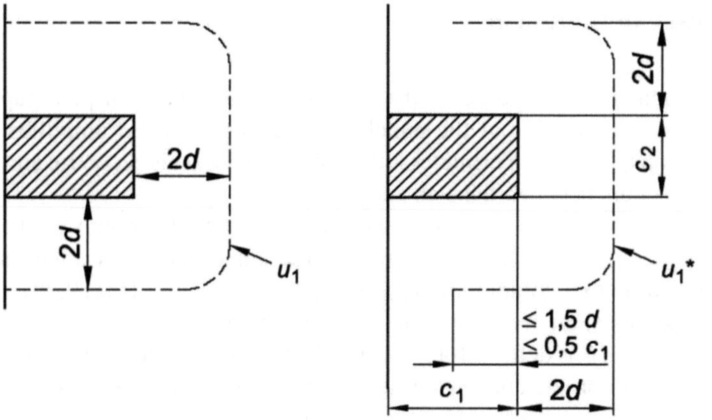

Fig. 9.18 Control perimeter u_1 and reduced control perimeter u_1* for rectangular edge column [Figures 6.15 and Figure 6.20]

In addition to u_1, the reduced control perimeter u_1* can be defined as follows (Fig. 9.19)

$$u_1^* = \min(0.5c_1;\ 1.5\,d) + \min(0.5\,c_2;\ 1.5\,d) + \pi d$$

The reduced control perimeter u_1* can be used as shown at point *f* of paragraph 9.7.2.

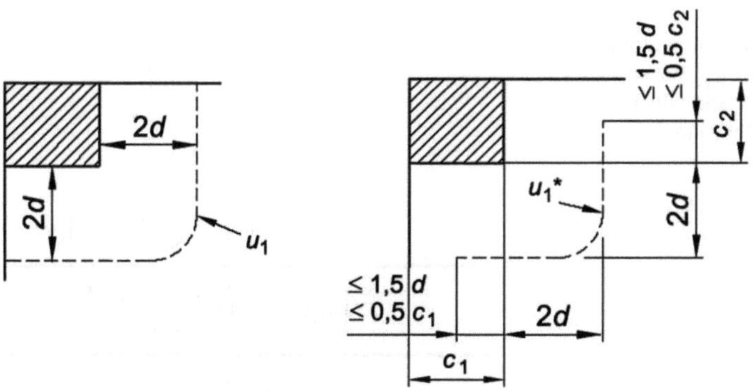

Fig. 9.19 Control perimeter u_1 and reduced control perimeter u_1* for rectangular corner column [Figure 6.15 and Figure 6.20]

9.7 The Coefficient β

The coefficient $\beta \geq 1$ is used in the expression introduced in paragraph 9.15 to calculate the design value of the punching shear stress v_{Ed}. β amplifies the mean shear stress on the control perimeter if the column reaction is eccentric and a transfer of moment occurs between the slab and column.

The bending moment M_{Ed} (which should be considered in both y e z directions of the slab, namely $M_{Ed,y}$ and $M_{Ed,z}$) is given by the sum of moments at the end sections of the top and bottom column: $M_{Ed,y(z)} = M'_{Ed,y(z)} + M''_{Ed,y(z)}$ (Fig. 9.20).

For foundations, the moment is only transferred by the column.

The coefficient β is defined for the control perimeter u_1 and its meaning is evident if the shear stress on the control surface S_1 ($S_1 = u_1 \cdot d$) is expressed as the sum of the shear stress $v_{Ed(V)}$ produced by V_{Ed} and the shear stress $v_{Ed(M)}$ produced by M_{Ed}

$$v_{Ed} = v_{Ed(V)} + v_{Ed(M)} = V_{Ed}/(u_1 d) + v_{Ed(M)}$$

Comparing this expression with the previous one, the following relationship is obtained

$$\beta = 1 + v_{Ed(M)} u_1 d / V_{Ed}$$

The shear stress $v_{Ed(M)}$ can be calculated by equating the moment M_{Ed} exchanged between the slab and column with the moment produced by the shear stress $v_{Ed(M)}$ distributed along the perimeter u_1

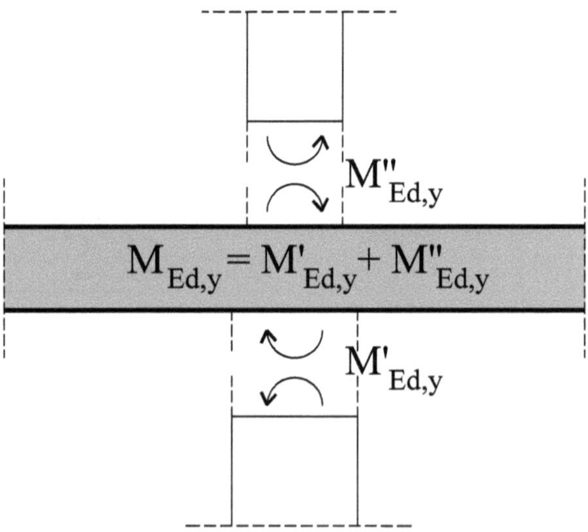

Fig. 9.20 Bending moments $M'_{Ed,y}$ and $M''_{Ed\,y}$ transmitted by the lower and upper column to the slab

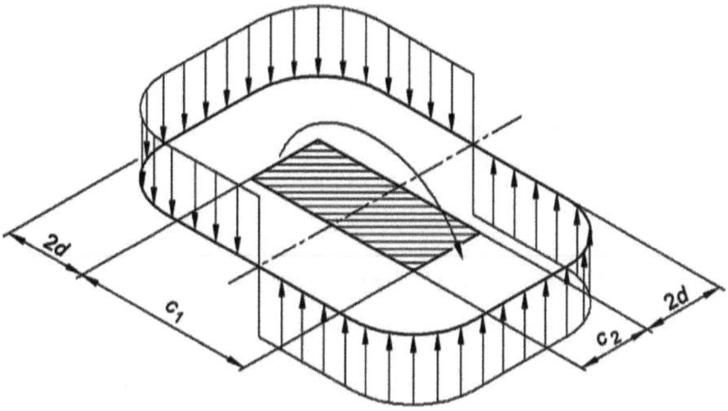

Fig. 9.21 Plastic distribution of shear stresses on the perimeter u_I due to bending moment at an internal slab-column joint [Figure 6.19]

$$M_{Ed} = \int_0^{u_1} \left(v_{Ed(M)} d\right) e \, \mathrm{d}u$$

where

$\mathrm{d}u$ infinitesimal segment of the control perimeter
e distance of $\mathrm{d}u$ to the axis of the moment M_{Ed}.

If shear stresses have a plastic distribution ($v_{\mathrm{Ed(M)}}$ = constant) (Fig. 9.21), the rotational equilibrium can be written as follows

$$M_{Ed} = v_{Ed(M)}\, d \int_0^{u_1} |e|\, du = v_{Ed(M)}\, d\, W_1$$

where $d\cdot$ is the bending moment given by a uniform distribution (complete plasticity) of unit shear stresses ($v_{\mathrm{Ed(M)}} = 1$) on the control surface S_1 with effective depth d and perimeter u_1.

From the previous expression, the following expression is obtained:

$$v_{\mathrm{Ed(M)}} = M_{\mathrm{Ed}}/(d W_1)$$

which is substituted in the β formula to give

$$\begin{aligned} \beta &= 1 + v_{\mathrm{Ed(M)}} u_1 d / V_{\mathrm{Ed}} \\ &= 1 + M_{\mathrm{Ed}} u_1 d / (d W_1 V_{\mathrm{Ed}}) \\ &= 1 + (M_{\mathrm{Ed}} / V_{\mathrm{Ed}})(u_1 / W_1) \end{aligned}$$

An additional coefficient k is introduced because the bending moment M_{Ed} is not only balanced by shear stresses, but also by bending moments in the slab strips parallel to the bending plane and torsion moments in the orthogonal strips.

The following expression is then obtained:
$\beta = 1 + k(M_{Ed}/V_{Ed})(u_1/W_1)$ [(6.39)]
The coefficient k also accounts for the non-uniform distribution of shear stresses, and it depends on the ratio c_1/c_2 (see paragraph 9.7.1).

9.7.1 Values of the Coefficient k

In the following paragraphs, the values of the coefficient k, to calculate β by using the formula [(6.39)], are provided.

Rectangular columns

The coefficient k takes different values with the variation of the ratio c_1/c_2 between dimensions of the column section (Tables 9.2 and 9.3), where c_1 is the dimension parallel to the stress plane.

For c_1/c_2 values different from those listed in Table 9.2, linear interpolation can be performed

$$k = 0.45 + 0.3 \cdot (c_1/c_2 - 0.5) \geq 0.45 \quad \text{if}\, c_1/c_2 \leq 1 \tag{9.1}$$

$$k = 0.6 + 0.1 \cdot (c_1/c_2 - 1) \leq 0.8 \quad \text{if}\, c_1/c_2 > 1 \tag{9.2}$$

Table 9.2 Values of the coefficient k for rectangular loaded areas [Table 6.1]

c_1/c_2	≤ 0.5	1.0	2.0	≥ 3.0
k	0.45	0.60	0.70	0.80

Table 9.3 Values of the coefficient k versus c_1/c_2

c_1/c_2	≤ 0.50	0.55	0.60	0.65	0.70	0.75	0.80	0.85	0.90	0.95	1.00
k	0.450	0.465	0.480	0.495	0.510	0.525	0.540	0.555	0.570	0.585	0.600
c_1/c_2		1.05	1.10	1.15	1.20	1.25	1.3	1.35	1.40	1.45	1.50
k		0.605	0.610	0.615	0.620	0.625	0.630	0.635	0.640	0.645	0.650
c_1/c_2		1.55	1.60	1.65	1.70	1.75	1.80	1.85	1.90	1.95	2.00
k		0.655	0.660	0.665	0.670	0.675	0.680	0.685	0.690	0.695	0.700
c_1/c_2		2.05	2.10	2.15	2.20	2.25	2.30	2.35	2.40	2.45	2.50
k		0.705	0.710	0.715	0.720	0.725	0.730	0.735	0.740	0.745	0.750
c_1/c_2		2.55	2.60	2.65	2.70	2.75	2.80	2.85	2.90	2.95	≥ 3.00
k		0.755	0.760	0.765	0.770	0.775	0.780	0.785	0.790	0.795	0.800

k values obtained by linear interpolation are shown in Table 9.3 for c_1/c_2 values multiples of 0.05.

Circular columns

For circular columns, the coefficient k takes the same value of the square section ($c_1/c_2 = 1$), namely $k = 0.6$.

9.7.2 Calculation of the Coefficient β for Rectangular or Circular Columns

The expression to evaluate β varies with the shape of the cross-section, the position of the column (internal, edge or corner column), and the type of eccentricity (uniaxial or biaxial).

Below the coefficient β is evaluated for all cases listed in Table 9.4.

(a) *Internal rectangular column and uniaxial eccentricity*

The coefficient β is evaluated as follows

$$\beta = 1 + k \frac{M_{Ed}}{V_{Ed}} \frac{u_1}{W_1}$$

By applying the definition of W_1 ($W_1 = \int_0^{u_1} |e|\, du$), the following expression is obtained (Fig. 9.22)

$$W_1 = \int_0^{u_1} |e|\, du = 2 \cdot \left\{ \int_{AB} |e|\, du + \int_{BC} |e|\, du + \int_{CC'} |e|\, du + \int_{C'B'} |e|\, du + \int_{B'A'} |e|\, du \right\}$$

Table 9.4 List of evaluation cases for the calculation of coefficient β

Internal rectangular column	
(a)	Uniaxial eccentricity
(b)	Biaxial eccentricity
Edge rectangular column	
(c)	Eccentricity orthogonal to the edge
(d)	Eccentricity parallel to the edge
(e)	Biaxial eccentricity
Corner rectangular column	
(f)	Inward eccentricity
(g)	Outward eccentricity
Internal circular column	
(h)	Uniaxial eccentricity

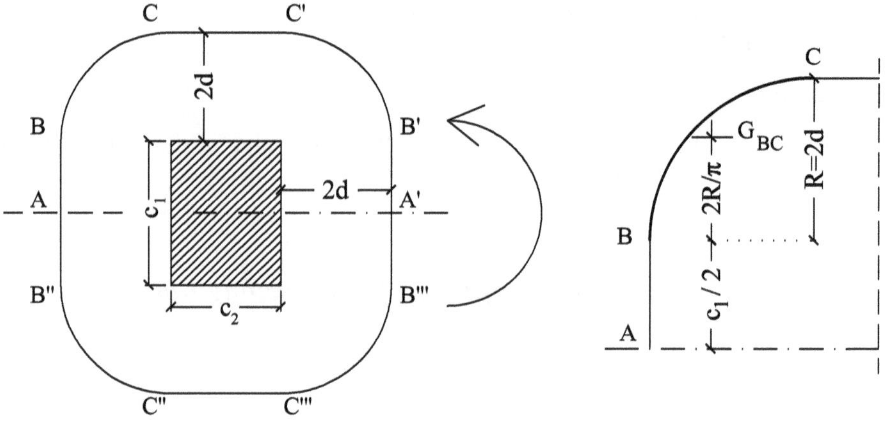

Fig. 9.22 Scheme for the calculation of W_1 for an internal rectangular column (the distance of the centroid of the circular arc BC of radius R = $2d$ from AA' axis is shown on the right)

$$= 2 \cdot \left\{ 2 \cdot \int_{AB} |e|\, du + 2 \cdot \int_{BC} |e|\, du + \int_{CC'} |e|\, du \right\}$$

$$= 2 \cdot \left\{ 2 \cdot \underbrace{\frac{c_1}{2}}_{AB} \cdot \frac{c_1}{4} + 2 \cdot \underbrace{\frac{\pi\,(2d)}{2}}_{BC} \cdot \left[\frac{2\,(2d)}{\pi} + \frac{c_1}{2} \right] + \underbrace{c_2}_{CC'} \cdot \left(2d + \frac{c_1}{2} \right) \right\}$$

$$= \frac{c_1^2}{2} + 16d^2 + 2\pi d c_1 + 4c_2 d + c_1 c_2 \qquad [(6.41)]$$

where

c_1 is the column dimension parallel to the load eccentricity
c_2 is the column dimension orthogonal to the load eccentricity.

In the case of a square column ($c_1 = c_2$), W_1 becomes

$$W_1 = \frac{3}{2} c_1^2 + (4 + 2\pi)\, c_1\, d + 16\, d^2$$

The coefficient β for an internal rectangular column can also be expressed in the following form

$$\beta = 1 + k \frac{M_{Ed}}{V_{Ed}} \frac{u_1}{W_1} = 1 + k \cdot \frac{M_{Ed}}{V_{Ed}} \cdot \frac{2(c_1 + c_2) + 4\pi\, d}{\frac{c_1^2}{2} + 16d^2 + 2\pi\, d c_1 + 4c_2 d + c_1 c_2} = 1 + \frac{e}{\alpha}$$

where

$$e = M_{\mathrm{Ed}}/V_{\mathrm{Ed}}$$

$$\alpha = \frac{1}{k} \cdot \frac{\frac{c_1^2}{2} + 16d^2 + 2\pi dc_1 + 4c_2 d + c_1 c_2}{2(c_1 + c_2) + 4\pi d}$$

and α has the dimension of a length.

The values of α are listed in Table 9.5 for different values of the ratio c_1/c_2 and the depth d; they allow the immediate calculation of the coefficient β.

Table 9.5 Values of α (mm) to be used for the calculation of the coefficient β ($\beta = 1 + e/\alpha$), with e in mm, for an internal rectangular column with varying c_1, c_2 and d (c_1 is the dimension parallel to the bending plane, namely orthogonal to the moment axis)

c_1 (mm)	d (mm)	c_2 (mm)							
		250	300	350	400	450	500	550	600
250	200	592	654	708	756	798	836	844	852
	300	805	888	959	1021	1076	1125	1134	1143
	500	1231	1352	1456	1546	1626	1696	1707	1717
300	200	597	625	681	731	776	816	853	887
	300	803	839	912	977	1034	1086	1134	1177
	500	1215	1264	1371	1465	1548	1622	1689	1751
350	200	601	632	657	709	756	799	838	873
	300	801	840	872	938	998	1052	1102	1148
	500	1200	1254	1298	1394	1479	1556	1626	1690
400	200	605	639	666	690	738	783	823	861
	300	799	841	875	904	966	1022	1073	1121
	500	1186	1245	1292	1331	1418	1498	1570	1636
450	200	609	645	675	700	722	768	810	849
	300	797	842	879	910	937	995	1048	1097
	500	1173	1235	1286	1328	1364	1445	1519	1587
500	200	612	650	682	710	734	755	798	838
	300	795	843	882	916	944	970	1024	1074
	500	1160	1226	1280	1325	1364	1398	1473	1542
550	200	615	656	690	719	744	767	787	828
	300	793	844	885	921	951	978	1002	1054
	500	1148	1218	1275	1322	1363	1399	1431	1501
600	200	617	660	697	727	754	778	800	819
	300	791	844	888	925	958	986	1012	1035
	500	1137	1210	1269	1320	1363	1400	1434	1464

For values of c_1, c_2 and d different from those listed in the tables, linear interpolation can be performed.

9.7.3 Example No. 1—Evaluation of the Coefficient β for an Internal Rectangular Column

Calculate the coefficient β *for An internal rectangular column with dimensions* $c_1 \times c_2 = 500 \times 300$ mm *supporting a slab with an effective depth of* 250 *mm. Design data are:* $V_{Ed} = 1000$ kN, $M_{Ed} = 500$ kNm*; the eccentricity is parallel to the dimension* c_1.

The coefficient β is given by the expression

$$\beta = 1 + k\frac{M_{Ed}}{V_{Ed}}\frac{u_1}{W_1}$$

where

$$\begin{aligned} W_1 &= \frac{c_1^2}{2} + 16d^2 + 2\pi\, dc_1 + 4c_2d + c_1c_2 \\ &= \frac{500^2}{2} + 16\cdot 250^2 + 2\pi\cdot 250\cdot 500 \\ &\quad + 4\cdot 300\cdot 250 + 500\cdot 300 = 236{,}0398\ \text{mm}^2 \\ u_1 &= 2\cdot(c_1 + c_2) + 4\pi\, d = 2\cdot(500 + 300) \\ &\quad + 4\cdot\pi\cdot 250 = 4742\ \text{mm} \end{aligned}$$

$k = 0.6 + 0.1 \cdot (c_1/c_2 - 1) = 0.6 + 0.1 \cdot (500/300 - 1) = 0.666$ (from formula (9.2) because the ratio c_1/c_2 is equal to $1.66 > 1$)

Substituting these values in the above expression, β assumes the following value

$$\beta = 1 + k\frac{M_{Ed}}{V_{Ed}}\frac{u_1}{W_1} = 1 + 0.666\frac{500{,}000}{1000}\cdot\frac{4742}{2{,}360{,}398} = 1.67$$

Now it is possible to calculate again the coefficient β, by using Table 9.5

$$\beta = 1 + e/\alpha$$

where $e = M_{\text{Ed}}/N_{\text{Ed}} = 500{,}000/1000 = 500$ mm.

for $c_1 = 500$ mm and $c_2 = 300$ mm, Table 9.5 gives the following α values:

$$\alpha = 650\,\text{mm} \qquad \text{for}\, d = 200\,\text{mm}$$

$$\alpha = 843\,\text{mm} \qquad \text{for } d = 300\,\text{mm}$$

Using linear interpolation for $d = 250$ mm, the coefficient α is expressed as

$$\alpha = 650 + (843 - 650) \cdot (250 - 200)/(300 - 200) = 747\,\text{mm}$$

Thus, the coefficient β is equal to:

$$\beta = 1 + e/\alpha = \beta = 1 + 500/747 = 1.67$$

(b) *Internal rectangular column and biaxial eccentricity*

For an internal rectangular column, if the eccentricity is biaxial (Fig. 9.23), the expression for the calculation of β is the following:

$$\beta = 1 + k_1 \frac{M_{Ed1}}{V_{Ed}} \frac{u_1}{W_1} + k_2 \frac{M_{Ed2}}{V_{Ed}} \frac{u_1}{W_2}$$

where

$M_{\text{Ed1}}(M_{\text{Ed2}})$ is the moment with bending plane parallel to $c_1(c_2)$;
$k_1(k_2)$ is the coefficient k evaluated using expressions or tables already given in 9.7.1, by assuming $c_1(c_2)$ as the dimension parallel to the bending plane;

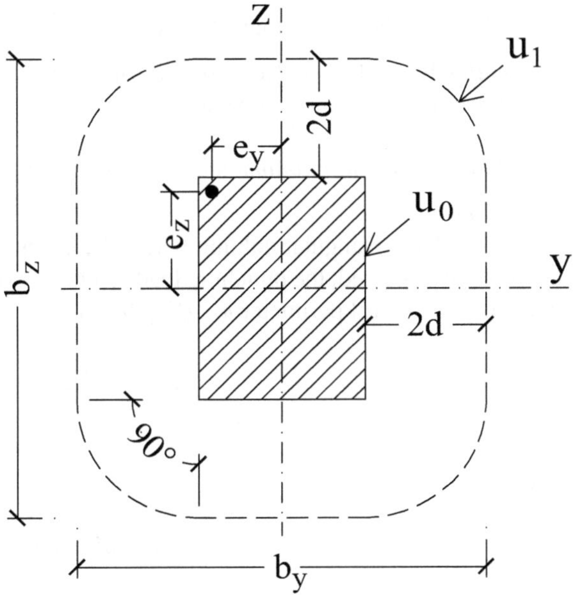

Fig. 9.23 External dimensions (b_y and b_z) of the control perimeter of an internal rectangular column and eccentricities (e_y and e_z) in both directions y and z

Table 9.6 Values of the coefficient β versus ratios e_y/b_z and e_z/b_y

β		e_z/b_y							
		0.01	0.02	0.05	0.10	0.20	0.30	0.40	0.50
e_y/b_z	0.01	1.025	1.040	1.092	1.181	1.360	1.540	1.720	1.900
	0.02	1.040	1.051	1.097	1.184	1.362	1.541	1.721	1.901
	0.05	1.092	1.097	1.127	1.201	1.371	1.547	1.726	1.904
	0.10	1.181	1.184	1.201	1.255	1.402	1.569	1.742	1.918
	0.20	1.360	1.362	1.371	1.402	1.509	1.649	1.805	1.969
	0.30	1.540	1.541	1.547	1.569	1.649	1.764	1.900	2.050
	0.40	1.720	1.721	1.726	1.742	1.805	1.900	2.018	2.153
	0.50	1.900	1.901	1.904	1.918	1.969	2.050	2.153	2.273

$W_1(W_2)$ is the moment produced by a plastic distribution of unit shear stresses ($v_{\mathrm{Ed(M)}} = 1$) on the control surface S_1 with the bending plane parallel to the dimension $c_1(c_2)$.

EC2 also gives the following approximated expression for β.

$$\beta = 1 + 1,8\sqrt{\left(\frac{e_y}{b_z}\right)^2 + \left(\frac{e_z}{b_y}\right)^2} \qquad [(6.43)]$$

where

e_y and e_z are the eccentricities in the y and z-axis, respectively;
b_y and b_z are the dimensions of the control perimeter (Fig. 9.23).

Table 9.6 lists values of β as e_y/b_z and e_z/b_y vary; for values different from tabulated ones, linear interpolation can be used.

(c) *Edge rectangular column and eccentricity orthogonal to the edge*

Inward eccentricity

For an edge rectangular column where the eccentricity is orthogonal to the edge and inwardly directed, the punching load can be considered as uniformly distributed on a reduced control perimeter u_1^* (Fig. 9.24)

$$v_{Ed} = \frac{V_{Ed}}{u_1^* \cdot d}$$

where

$$u_1^* = 2 \cdot \min(1.5d;\ 0.5\,c_1) + c_2 + 2\pi d;$$

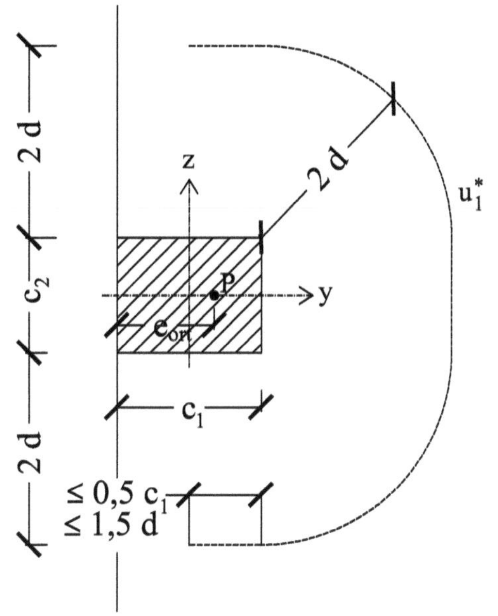

Fig. 9.24 Reduced control perimeter u_1* for edge rectangular column with inward eccentricity orthogonal to the edge [Figure 6.20] (P = centre of pressure)

c_1 is the dimension orthogonal to the edge.

Using this expression for v_{Ed} is equivalent to assume $\beta = u_1/u_1{}^*$ in the general expression $v_{Ed} = \beta\, V_{Ed}/(u_1\, d)$.

Outward eccentricity

For an edge column where the eccentricity orthogonal to the edge is outwardly directed and there is not eccentricity parallel to the edge, the coefficient β is given by the following expression

$$\beta = 1 + k\frac{M_{Ed}}{V_{Ed}}\frac{u_1}{W_1}$$

where

M_{Ed} is the moment evaluated about the z' axis, parallel to the edge and passing through the centroid G of the control perimeter (Fig. 9.25)

$$M_{\mathrm{Ed}} = M_z + V_{\mathrm{Ed}} \cdot (e_g - c_1/2)$$

e_g is the distance of the centroid G of the control perimeter from the slab edge

$$e_g = \frac{c_1^2 + c_2(c_1 + 2d) + 2\pi\, dc_1 + 8d^2}{2c_1 + c_2 + 2\pi\, d}$$

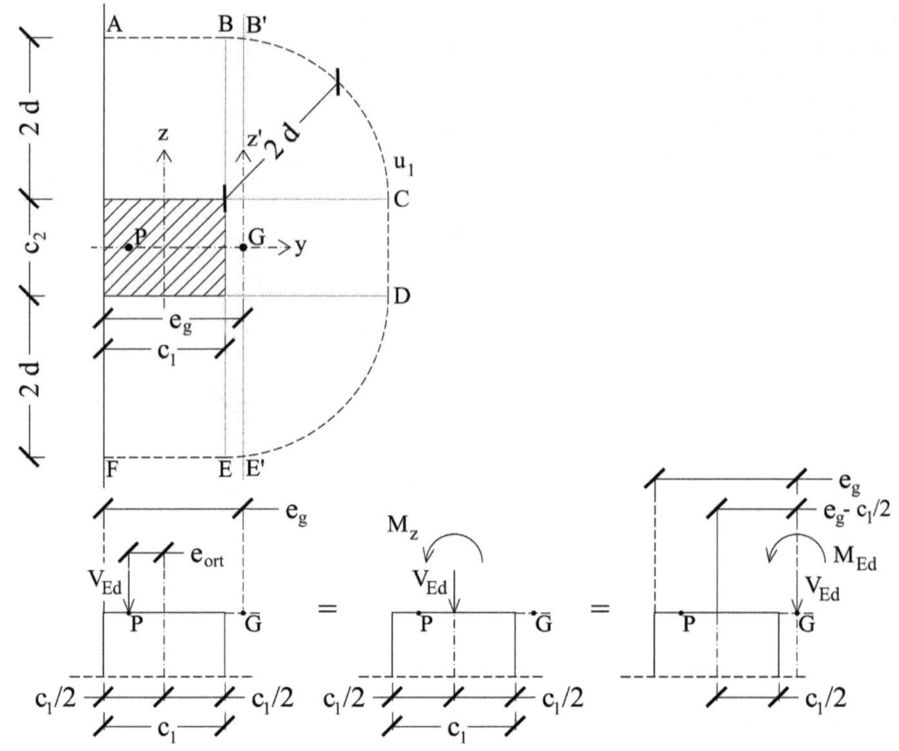

Fig. 9.25 Edge column with outward eccentricity orthogonal to the slab edge (P = centre of pressure, G = centroid of the control perimeter u_1)

W_1 is calculated about the z' axis

$$W_1 = 2c_1(e_g - c_1/2) + c_2(c_1 + 2d - e_g) + 2BB' \cdot (e_g - e_{BB'}) + 2B'C \cdot (e_{B'C} - e_g)$$

($e_{BB'}$ and $e_{B'C}$ are the distances from the edge of centroids of the circular arcs BB′ and B′C).

k is obtained from Tables 9.2 and 9.3 with ratio c_1/c_2 replaced by $c_1/(2c_2)$.

(d) *Edge rectangular column with eccentricity parallel to the edge*

For an edge rectangular column, with uniaxial eccentricity parallel to the edge (Fig. 9.26), the coefficient β is given by the following expression

$$\beta = 1 + k\frac{u_1}{W_1}e_{\text{par}}$$

Fig. 9.26 Edge column with eccentricity e_{par} parallel to the edge (P = centre of pressure, G = centroid of the control perimeter u_1)

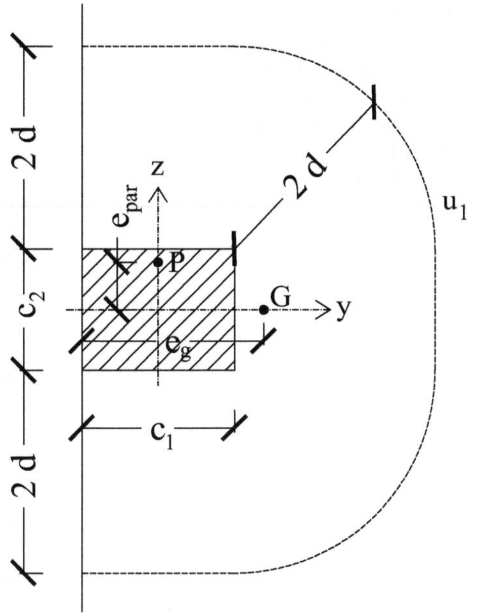

where

u_1 is the control perimeter
e_{par} is the eccentricity parallel to the slab edge
k is the coefficient obtained from Tables 9.2 and 9.3 being c_1/c_2 replaced by $c_2/(2c_1)$
W_1 is calculated about the y-axis, orthogonal to the edge

$$W_1 = \frac{c_2^2}{4} + c_1\, c_2 + 4\, c_1\, d + 8\, d^2 + \pi\, d\, c_2 \qquad [(6.45)]$$

(e) *Edge rectangular column with biaxial eccentricity*

Outward eccentricity orthogonal to the edge

For an edge rectangular column, if the eccentricity is biaxial and the eccentricity orthogonal to the edge is inwardly directed (Fig. 9.27), the coefficient β is given by

$$\beta = \frac{u_1}{u_1^*} + k \frac{u_1}{W_1} e_{par} \qquad [(6.44)]$$

where

u_1 is the control perimeter

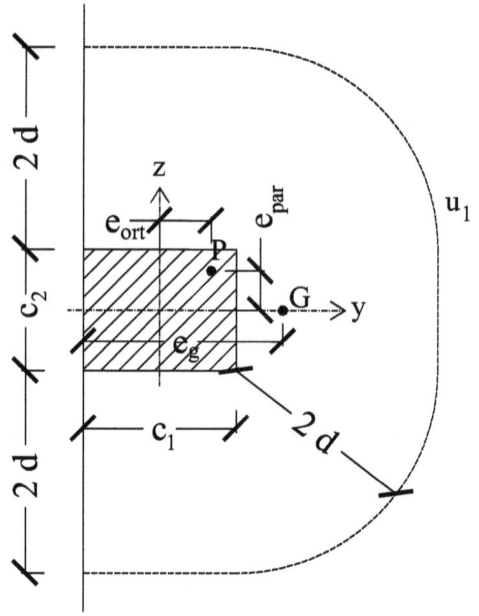

Fig. 9.27 Edge column with biaxial eccentricity, where the eccentricity orthogonal to the edge is inwardly directed (P = centre of pressure, G = centroid of the control perimeter u_1)

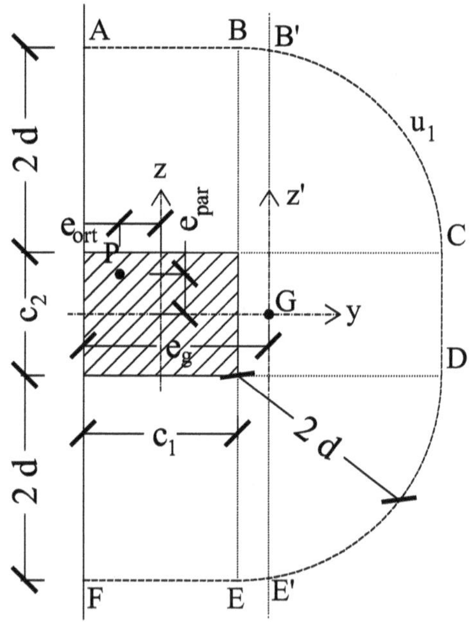

Fig. 9.28 Edge column with biaxial eccentricity, where the eccentricity orthogonal to the edge is outwardly directed (P = centre of pressure, G = centroid of the control perimeter u_1)

$u_1{}^*$ is the reduced control perimeter (Fig. 9.24)
e_{par} is the eccentricity parallel to the slab edge
k is the coefficient obtained from Tables 9.2 and 9.3 with ratio c_1/c_2 replaced by the ratio $c_2/(2c_1)$
W_1 has the same expression of case d

$$W_1 = \frac{c_2^2}{4} + c_1\, c_2 + 4\, c_1\, d + 8\, d^2 + \pi\, d\, c_2 \qquad [(6.45)]$$

Outward eccentricity orthogonal to the edge

If the eccentricity orthogonal to the edge is outwardly directed, the expression of β, which considers the biaxial eccentricity, becomes

$$\beta = 1 + k_z \frac{M_{Ed,z'}}{V_{Ed}} \frac{u_1}{W_{1,z'}} + k_y \frac{M_{Ed,y}}{V_{Ed}} \frac{u_1}{W_{1,y}}$$

where

$W_{1,z'}$ is calculated about the z' axis, parallel to the edge and passing through the centroid G of the control perimeter (Fig. 9.28)

$$W_{1,z'} = 2c_1\left(e_g - c_1/22\right) + c_2\left(c_1 + 2d - e_g\right) + 2BB' \cdot \left(e_g - e_{BB'}\right) + 2B'C \cdot \left(e_{B'C} - e_g\right)$$

($e_{\mathrm{BB'}}$ and $e_{\mathrm{B'C}}$ are the distances from the edge of centroids of the circular arcs BB′ and B′C)

$W_{1,y}$ is calculated about the y axis, orthogonal to the edge

$$W_{1,y} = \frac{c_2^2}{4} + c_1\, c_2 + 4\, c_1\, d + 8\, d^2 + \pi\, d\, c_2$$

e_g is the distance from the slab edge of the centroid G of the control perimeter

$$e_g = \frac{c_1^2 + c_2(c_1 + 2d) + 2\pi\, dc_1 + 8d^2}{2c_1 + c_2 + 2\pi\, d}$$

$M_{\mathrm{Ed},z'}$ is the moment which bends strips orthogonal to the edge, calculated about the z' axis parallel to the edge and passing through the centroid G of the control perimeter

$$M_{\mathrm{Ed},z'} = M_z + V_{\mathrm{Ed}} \cdot (e_g - c_1/2)$$

$M_{\mathrm{Ed,y}}$ is the moment about the y axis, which bends slab strips parallel to the edge

k_z is obtained from Tables 9.2 and 9.3 with c_1/c_2 replaced by $c_1/(2c_2)$
k_y is obtained from Tables 9.2 and 9.3 with c_1/c_2 replaced by $c_2/(2c_1)$.

(f) *Corner rectangular column with inward eccentricity*

For a corner rectangular column with inward eccentricity, the punching shear stress can be considered uniformly distributed on the reduced control perimeter u_1^* (Fig. 9.29)

$$v_{Ed} = \frac{V_{Ed}}{u_1^* \cdot d}$$

where $u_1^* = \min(1.5\,d; 0.5\,c_1) + \min(1.5\,d; 0.5\,c_2) + \pi\,d$.

As already shown for case *c)*, the use of this expression for v_{Ed} is equivalent to assume $\beta = u_1/u_1^*$ in the general expression $v_{Ed} = \beta\, V_{Ed}/(u_1\, d)$.

(g) *Corner rectangular column with outward eccentricity*

If the eccentricity is outwardly directed, the expression of β becomes

$$\beta = 1 + k_z \frac{M_{Ed,z'}}{V_{Ed}} \frac{u_1}{W_{1,z'}} + k_y \frac{M_{Ed,y'}}{V_{Ed}} \frac{u_1}{W_{1,y'}}$$

where

$W_{1,z'}$ is calculated about z' axis and $W_{1,y'}$ about y' axis
e_{gy} is the distance from the slab edge of the centroid G of the control perimeter, in the y direction (Fig. 9.30)

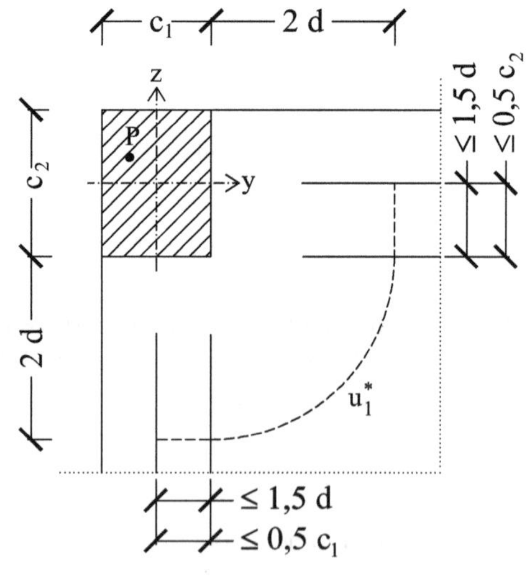

Fig. 9.29 Reduced control perimeter u_1* with distances from the slab edges for a corner column [Figure 6.20b]

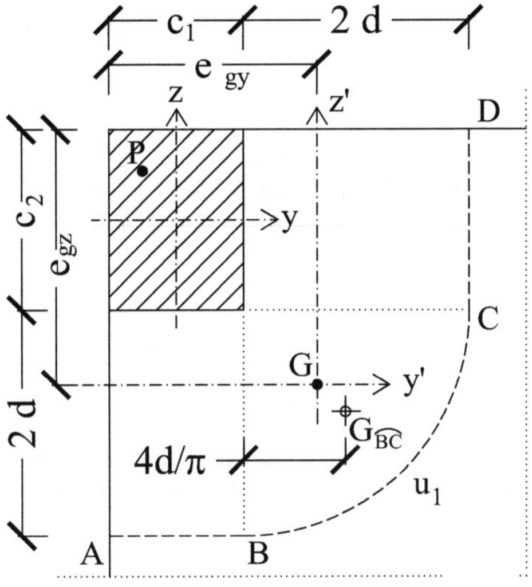

Fig. 9.30 Corner column with outward eccentricity (P = centre of pressure, G = centroid of the control perimeter u_1, G_{BC} = centroid of the arc BC)

$$e_{gy} = \frac{\frac{c_1^2}{2} + c_2(c_1 + 2d) + \pi\, d\, c_1 + 4\, d^2}{c_1 + c_2 + \pi\, d}$$

e_{gz} is the distance from the slab edge of the centroid G of the control perimeter, in the z direction (Fig. 9.30)

$$e_{gz} = \frac{\frac{c_2^2}{2} + c_1(c_2 + 2d) + \pi\, d\, c_2 + 4\, d^2}{c_1 + c_2 + \pi\, d}$$

$M_{\mathrm{Ed,z'}}$ is the bending moment about z' axis (bending plane parallel to c_1)
$M_{\mathrm{Ed,y'}}$ is the bending moment about y' axis (bending plane parallel to c_2)
k_z is obtained from Tables 9.2 and 9.3
k_y is obtained from Tables 9.2 and 9.3 with the ratio c_1/c_2 replaced by c_2/c_1.

(h) *Internal circular column*

For an internal circular column of diameter c, W_1 is expressed as follows

$$W_1 = \int_0^{u_1} |e|\, dl = 2 \cdot \int_{ABA'} |e| dl = 2 \cdot \left(\pi\, R \cdot \frac{2R}{\pi}\right)$$
$$= 2 \cdot \pi(c/2 + 2d) \cdot \frac{2 \cdot (c/2 + 2d)}{\pi} = (c + 4d)^2$$

where $R = c/2 + 2d$ (Fig. 9.31).

Putting W_1 and u_1 into the expression of the coefficient β, the following expression is obtained

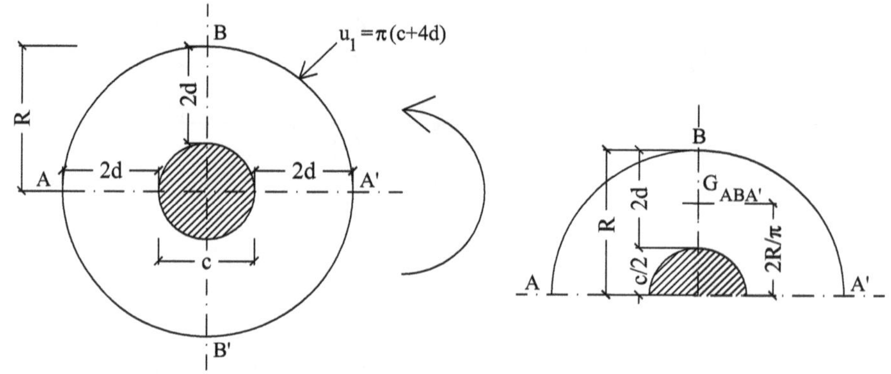

Fig. 9.31 Scheme for the calculation of W_1 for an internal circular column (the centroid $G_{ABA'}$ of the semi-circle perimeter ABA' is shown on the right)

$$\beta = 1 + k\frac{M_{Ed}}{V_{Ed}}\frac{u_1}{W_1} = 1 + k \cdot e \cdot \frac{\pi(c+4d)}{(c+4d)^2} = 1 + k \cdot e \cdot \frac{\pi}{c+4d}$$

The coefficient k assumes the same value as the square section (for $c_1/c_2 = 1$ from Table 9.2 k = 0.6), then the final expression of β is.

$$\beta = 1 + 0.6\pi\frac{e}{c+4d} \qquad [(6.42)]$$

9.7.3.1 *Approximate Values of β*

For structures whose lateral stability does not depend on the frame action between slabs and columns, and where adjacent spans do not differ more than 25%, approximate values may be used for the coefficient β (Fig. 9.32).

For concentrated loads close to a slab-column joint, it is not possible to reduce the shear force as described in EC2 paragraphs [6.2.2(6)] and [6.2.3(8)].

9.8 Punching Shear Calculation on the Perimeter of the Column or Loaded Area

The verification of the punching shear resistance of a slab or column base, with or without shear reinforcement, requires the fulfilment of the following condition

$$v_{Ed,0} \leq v_{Rd,\max}$$

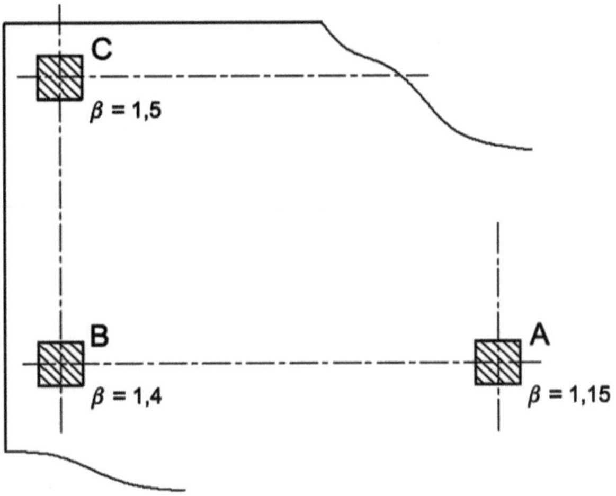

Fig. 9.32 EC2 recommended values for β: A = internal column, B = edge column, C = corner column [Figure 6.21 N].

where

$v_{Rd,\max}$ is the maximum punching shear resistance (9.8.1), related to the concrete compressive strength on the perimeter of the column or loaded area.

$v_{Ed,0}$ is the design value of the punching shear resistance (9.8.2).

If the verification is not satisfied, it is necessary to adopt one of the following strategies:

- to increase dimensions of the column cross-section or loaded area
- to increase the slab thickness
- to enlarge the column head, which is equivalent to increase the thickness of the slab in the proximity of the column
- to adopt a higher strength class of the concrete.

If the verification on the perimeter u_0 is satisfied, then the punching shear verification on the control perimeter u_1 can be performed.

9.8.1 Maximum Punching Shear Resistance $v_{Rd,max}$

The maximum punching shear resistance on the perimeter of the column or loaded area is given by

$$v_{\mathrm{Rd,max}} = 0.4\, \nu\, f_{\mathrm{cd}}$$

where

$\nu = 0.5$ for concrete class $\leq$ C70/85

$f_{cd} = \alpha_{cc} \cdot f_{ck} \,/\, \gamma_C =$ design value of concrete compressive strength.

Table 9.7 Values of $v_{Rd,\max}$ versus f_{ck}

f_{ck} (N/mm^2)	20	25	28	30	32	35
$v_{Rd,\max}$ (N/mm^2)	2.267	2.833	3.173	3.400	3.627	3.967
f_{ck} (N/mm^2)	40	45	50	55	60	70
$v_{Rd,\max}$ (N/mm^2)	4.533	5.100	5.667	6.233	6.800	7.933

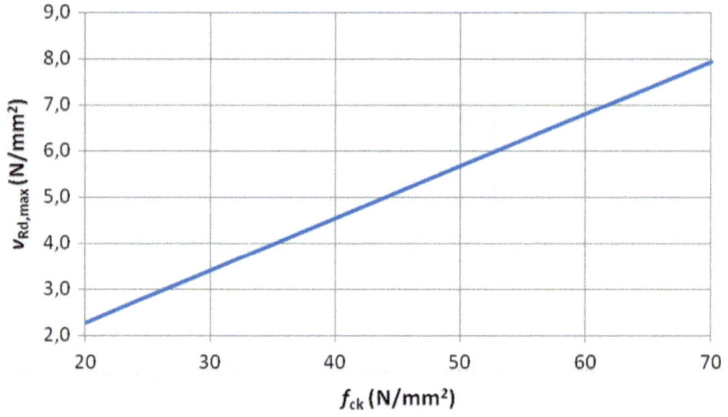

Fig. 9.33 Maximum punching shear resistance $v_{Rd,\max}$ (N/mm^2) with varying f_{ck}

The former expression proposed by EC2 for the evaluation of $v_{Rd,\max}$ was given by

$$v_{Rd,\max} = 0.5V f_{cd}$$

Afterwards, the expression has been modified by decreasing the maximum punching shear resistance of 20% to go beyond the inconsistency related to the use of the coefficient β, which is defined considering the control perimeter u_1, also for the verification on the perimeter u_0 of the column or loaded area.

Values of $v_{Rd,\max}$ are collected in Table 9.7 and the diagram of Fig. 9.33 for f_{ck} values between 20 and 70 N/mm^2.

9.8.2 Design Value of the Punching Shear Resistance

The design value of the punching shear resistance $v_{Ed,0}$ on the perimeter of the column or loaded area u_0 is given by

$$v_{Ed,0} = \frac{\beta \cdot V_{Ed}}{u_0 \cdot d}$$

where

V_{Ed} is the design value of the shear force

u_0 is the perimeter of the column or loaded area

d $= (d_y + d_z)/2$ is the slab effective depth (average of the effective depths in the two directions y and z)

$\beta \geq 1$ is the coefficient that considers the eccentricity of the column reaction or concentrated load, already given in 9.7.

9.8.3 Maximum Punching Shear Resistance for Slabs on Circular Columns

For a slab on a circular column with diameter c, the punching shear verification on the perimeter u_0 is expressed by

$$v_{Ed,0} = \frac{\beta\, V_{Ed}}{u_0 d} = \frac{\beta\, V_{Ed}}{\pi\, c d} \leq v_{Rd,\max} = 0.2\ f_{cd}$$

Then, for given slab effective depth d and concrete compressive strength, the shear force should satisfy the following expression

$V_{Ed} \leq V_{Rd,\max} = 0.2\ f_{cd} \frac{\pi\, c\, d}{\beta}$.

Maximum values of the maximum shear force $V_{\mathrm{Rd,max}}$ for a slab on an internal circular column are collected in Table 9.8 at varying the slab effective depth d and the column diameter c.

9.8.4 Maximum Punching Shear Force for Slabs on Rectangular Columns

For a slab on a rectangular column with dimensions c_1 and c_2, the punching shear stress on the perimeter u_0 is expressed by

$$v_{Ed,0} = \frac{\beta\, V_{Ed}}{u_0 d} = \frac{\beta\, V_{Ed}}{2\,(c_1 + c_2)\, d} \leq v_{Rd,\max} = 0.2\ f_{cd}$$

The maximum punching shear force for assigned effective depth d and concrete compressive strength f_{cd} should be limited as follows

$$V_{Ed} \leq V_{Rd,\max} = 0.2\ f_{cd} \frac{2\,(c_1 + c_2)\, d}{\beta}$$

Table 9.8 Values of $V_{Rd,max}$ (kN) versus effective depth d and diameter c of internal circular column ($f_{ck} = 25$ N/mm^2; $f_{cd} = 14.17$ N/mm^2 and $\beta = 1$)

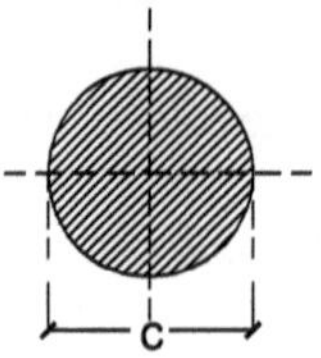

Slab made of C25/30 concrete
on an internal circular column
(for concrete class different from C25/30 and β≠1 see notes[a, b])

c	d (mm)							
(mm)	150	200	250	300	350	400	450	500
200	267	356	445	534	623	712	801	890
250	334	445	556	668	779	890	1 001	1 113
300	401	534	668	801	935	1 068	1 202	1 335
350	467	623	779	935	1 090	1 246	1 402	1 558
400	534	712	890	1 068	1 246	1 424	1 602	1 780
450	601	801	1 001	1 202	1 402	1 602	1 802	2 003
500	668	890	1 113	1 335	1 558	1 780	2 003	2 225
550	734	979	1 224	1 469	1 713	1 958	2 203	2 448
600	801	1 068	1 335	1 602	1 869	2 136	2 403	2 670
650	868	1 157	1 446	1 736	2 025	2 314	2 604	2 893
700	935	1 246	1 558	1 869	2 181	2 492	2 804	3 115
750	1 001	1 335	1 669	2 003	2 337	2 670	3 004	3 338
800	1 068	1 424	1 780	2 136	2 492	2 848	3 204	3 560
850	1 135	1 513	1 892	2 270	2 648	3 026	3 405	3 783
900	1 202	1 602	2 003	2 403	2 804	3 204	3 605	4 006
950	1 268	1 691	2 114	2 537	2 960	3 382	3 805	4 228
1000	1 335	1 780	2 225	2 670	3 115	3 560	4 006	4 451
1050	1 402	1 869	2 337	2 804	3 271	3 738	4 206	4 673
1100	1 469	1 958	2 448	2 937	3 427	3 917	4 406	4 896
1150	1 535	2 047	2 559	3 071	3 583	4 095	4 606	5 118
1200	1 602	2 136	2 670	3 204	3 738	4 273	4 807	5 341

[a] The grey boxes indicate greater values than $f_{cd}\, A_c$, under the hypothesis that also the column is made of C25/30 concrete.
[b] If the characteristic cylinder compressive strength f_{ck} of the concrete slab is different from C25/30, $V_{Rd,max}$ values have to be multiplied by the ratio $f_{ck}/25$.
[c] Table 9.8 can also be used for the case of the uniaxial or biaxial bending moment combined with axial load by dividing tabulated values by the coefficient β.

$V_{Rd,max}$ values of slabs on internal rectangular columns are listed in Table 9.9 for varying effective depth d and $(c_1 + c_2)$, c_1 and c_2 being the two dimensions of the cross-section.

Table 9.9 Values of $V_{Rd,max}$ (kN) versus slab effective depth d and sum $(c_1 + c_2)$ of the cross-section dimensions of an internal rectangular column ($f_{ck} = 25$ N/mm^2, $f_{cd} = 14{,}17$ N/mm^2 and $\beta = 1$)

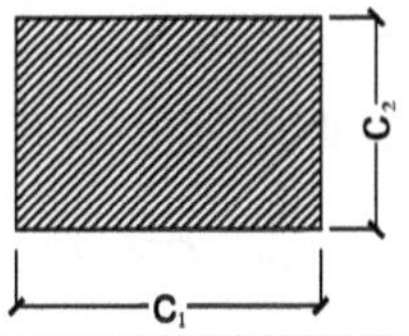

Slab made of C25/30 concrete on an internal rectangular column with dimensions $c_1 \times c_2$
(for concrete class different from C25/30 and $\beta \neq 1$ see notes[a, b])

c_1+c_2 (mm)	d (mm)							
	150	200	250	300	350	400	450	500
500	425	567	708	850	992	1 133	1 275	1 417
550	468	623	779	935	1 091	1 247	1 403	1 558
600	510	680	850	1 020	1 190	1 360	1 530	1 700
650	553	737	921	1 105	1 289	1 473	1 658	1 842
700	595	793	992	1 190	1 388	1 587	1 785	1 983
750	638	850	1 063	1 275	1 488	1 700	1 913	2 125
800	680	907	1 133	1 360	1 587	1 813	2 040	2 267
850	723	963	1 204	1 445	1 686	1 927	2 168	2 408
900	765	1 020	1 275	1 530	1 785	2 040	2 295	2 550
950	808	1 077	1 346	1 615	1 884	2 153	2 423	2 692
1000	850	1 133	1 417	1 700	1 983	2 267	2 550	2 833
1050	893	1 190	1 488	1 785	2 083	2 380	2 678	2 975
1100	935	1 247	1 558	1 870	2 182	2 493	2 805	3 117
1150	978	1 303	1 629	1 955	2 281	2 607	2 933	3 258
1200	1 020	1 360	1 700	2 040	2 380	2 720	3 060	3 400
1250	1 063	1 417	1 771	2 125	2 479	2 833	3 188	3 542
1300	1 105	1 473	1 842	2 210	2 578	2 947	3 315	3 683
1350	1 148	1 530	1 913	2 295	2 678	3 060	3 443	3 825
1400	1 190	1 587	1 983	2 380	2 777	3 173	3 570	3 967
1450	1 233	1 643	2 054	2 465	2 876	3 287	3 698	4 108
1500	1 275	1 700	2 125	2 550	2 975	3 400	3 825	4 250

[a] For columns with square cross-section the grey boxes give values of $V_{Rd,max}$ greater than $f_{cd}\,A_c$, under the hypothesis that also the column is made with concrete C25/30

[b] If the characteristic compressive cylinder strength f_{ck} of the slab concrete is different from C25/30, values of $V_{Rd,max}$ have to be multiplied for $f_{ck}/25$

[c] Table 9.9 can be employed also for the case of bending moment in one or two directions combined with compression/tension, by dividing the values of the Table by the coefficient β

9.8.4.1 Example No. 2—Calculation of the Maximum Punching Shear Resistance

Calculate the maximum punching shear resistance $V_{Rd,max}$ *for a slab made of* C30/37 *concrete, with effective depth* 250 mm, *on a circular column with a diameter of* 250 mm *and for a slab with the same effective depth, made of* C25/30 *concrete, on a column with* 300 mm *square cross-section.*

Column with circular cross-section

For $c = 250$ mm, $d = 250$ mm and $f_{ck} = 25$ N/mm^2, Table 9.8 gives $V_{Rd,\max} = 556$ kN.

To calculate the maximum punching shear resistance for a C30/37 concrete, it is sufficient to multiply the tabulated value by $f_{ck}/25 = 30/25 = 1.2$

$$V_{Rd,\max} = 556 \cdot 1.2 = 667\,\text{kN}$$

Column with a square cross-section

For $c_1 + c_2 = 600$ mm, $d = 250$ mm and $f_{ck} = 25$ N/mm^2, Table 9.9 gives

$$V_{Rd,\max} = 850\,\text{kN}$$

9.8.5 Minimum Value of the Slab Effective Depth

From the punching shear verification on the perimeter u_0 of the column or loaded area, it results that the minimum value of the slab effective depth to avoid the concrete crushing is given by the following expression

$$d_{\min} = \frac{\beta\, V_{Ed}}{0.2\ f_{cd} u_0}$$

Tables 9.10 and 9.11, concerning columns with circular or rectangular cross-sections respectively, give values of $d_{\min}/V_{Ed}$ for $\beta = 1$, with the variation of the sum $(c_1 + c_2)$, c_1 and c_2 being the cross-section dimensions.

If it is not possible to increase the slab thickness, Tables 9.10 and 9.11 can be used to calculate the thickness of an enlarged head to add to the column. Alternatively, a higher concrete strength class can be used for the whole slab or the slab portion close to the column (Fig. 9.34).

Table 9.10 Slab on a circular column: ratio d_{min}/V_{Ed} between the minimum effective depth d_{min} (mm) and the shear force V_{Ed} (kN) (d in mm, V_{Ed} in kN) with $\beta = 1$ versus diameter of the cross-section and f_{ck}

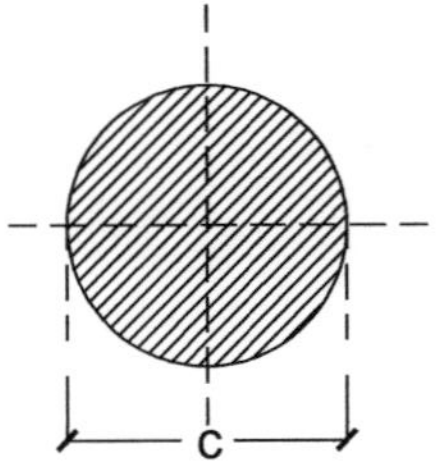

d in mm
V_{Ed} in kN

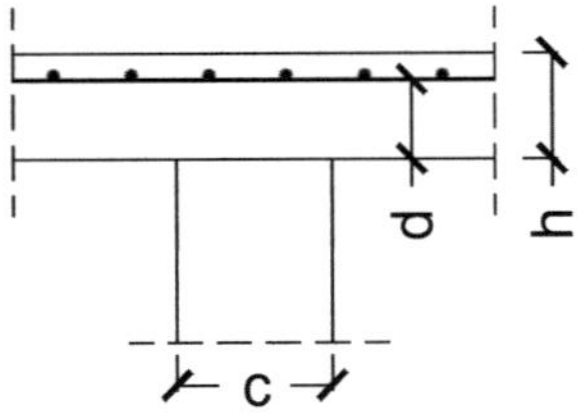

c (mm)	f_{ck} (N/mm^2)											
	20	25	28	30	32	35	40	45	50	55	60	70
200	0.702	0.562	0.502	0.468	0.439	0.401	0.351	0.312	0.281	0.255	0.234	0.201
250	0.562	0.449	0.401	0.374	0.351	0.321	0.281	0.250	0.225	0.204	0.187	0.160
300	0.468	0.374	0.334	0.312	0.293	0.267	0.234	0.208	0.187	0.170	0.156	0.134
350	0.401	0.321	0.287	0.267	0.251	0.229	0.201	0.178	0.160	0.146	0.134	0.115
400	0.351	0.281	0.251	0.234	0.219	0.201	0.176	0.156	0.140	0.128	0.117	0.100
450	0.312	0.250	0.223	0.208	0.195	0.178	0.156	0.139	0.125	0.113	0.104	0.089
500	0.281	0.225	0.201	0.187	0.176	0.160	0.140	0.125	0.112	0.102	0.094	0.080
550	0.255	0.204	0.182	0.170	0.160	0.146	0.128	0.113	0.102	0.093	0.085	0.073
600	0.234	0.187	0.167	0.156	0.146	0.134	0.117	0.104	0.094	0.085	0.078	0.067
650	0.216	0.173	0.154	0.144	0.135	0.123	0.108	0.096	0.086	0.079	0.072	0.062
700	0.201	0.160	0.143	0.134	0.125	0.115	0.100	0.089	0.080	0.073	0.067	0.057
750	0.187	0.150	0.134	0.125	0.117	0.107	0.094	0.083	0.075	0.068	0.062	0.053

(continued)

Table 9.10 (continued)

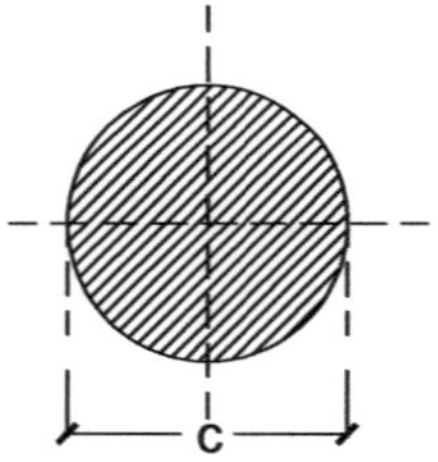

d in mm
V_{Ed} in kN

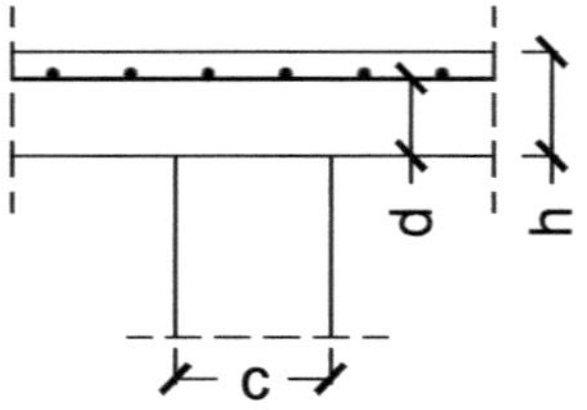

c (mm)	f_{ck} (N/mm^2)											
	20	25	28	30	32	35	40	45	50	55	60	70
800	0.176	0.140	0.125	0.117	0.110	0.100	0.088	0.078	0.070	0.064	0.059	0.050
850	0.165	0.132	0.118	0.110	0.103	0.094	0.083	0.073	0.066	0.060	0.055	0.047
900	0.156	0.125	0.111	0.104	0.098	0.089	0.078	0.069	0.062	0.057	0.052	0.045
950	0.148	0.118	0.106	0.099	0.092	0.084	0.074	0.066	0.059	0.054	0.049	0.042
1000	0.140	0.112	0.100	0.094	0.088	0.080	0.070	0.062	0.056	0.051	0.047	0.040

Table 9.11 Slab on a rectangular column: ratio d_{min}/V_{Ed} between the minimum effective depth d_{min} (mm) and the shear force V_{Ed} (kN) (d in mm, V_{Ed} in kN) with $\beta = 1$ vs the sum $(c_1 + c_2)$ of the cross-section dimensions and f_{ck}

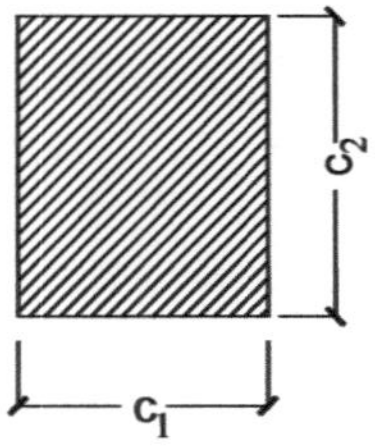

d in mm
V_{Ed} in kN

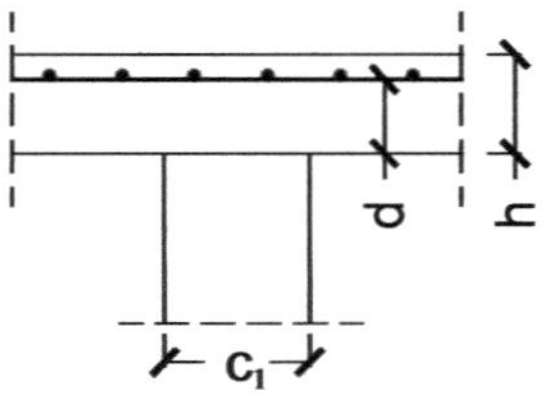

c_1+c_2 (mm)	f_{ck} (N/mm^2)											
	20	25	28	30	32	35	40	45	50	55	60	70
500	0.441	0.353	0.315	0.294	0.276	0.252	0.221	0.196	0.176	0.160	0.147	0.126
550	0.401	0.321	0.286	0.267	0.251	0.229	0.201	0.178	0.160	0.146	0.134	0.115
600	0.368	0.294	0.263	0.245	0.230	0.210	0.184	0.163	0.147	0.134	0.123	0.105
650	0.339	0.271	0.242	0.226	0.212	0.194	0.170	0.151	0.136	0.123	0.113	0.097
700	0.315	0.252	0.225	0.210	0.197	0.180	0.158	0.140	0.126	0.115	0.105	0.090
750	0.294	0.235	0.210	0.196	0.184	0.168	0.147	0.131	0.118	0.107	0.098	0.084
800	0.276	0.221	0.197	0.184	0.172	0.158	0.138	0.123	0.110	0.100	0.092	0.079
850	0.260	0.208	0.185	0.173	0.162	0.148	0.130	0.115	0.104	0.094	0.087	0.074
900	0.245	0.196	0.175	0.163	0.153	0.140	0.123	0.109	0.098	0.089	0.082	0.070
950	0.232	0.186	0.166	0.155	0.145	0.133	0.116	0.103	0.093	0.084	0.077	0.066
1000	0.221	0.176	0.158	0.147	0.138	0.126	0.110	0.098	0.088	0.080	0.074	0.063
1050	0.210	0.168	0.150	0.140	0.131	0.120	0.105	0.093	0.084	0.076	0.070	0.060
1100	0.201	0.160	0.143	0.134	0.125	0.115	0.100	0.089	0.080	0.073	0.067	0.057

(continued)

Table 9.11 (continued)

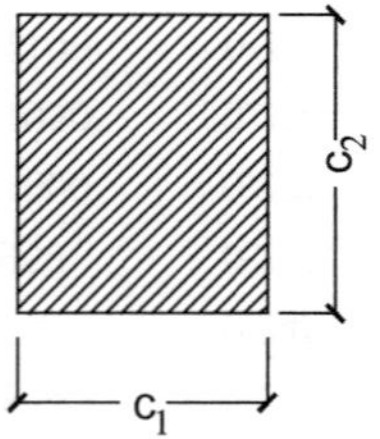

d in mm
V_{Ed} in kN

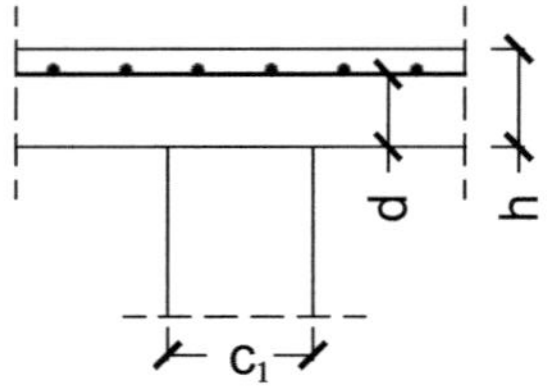

$c_{1+}c_2$ (mm)	f_{ck} (N/mm^2)											
	20	25	28	30	32	35	40	45	50	55	60	70
1150	0.192	0.153	0.137	0.128	0.120	0.110	0.096	0.085	0.077	0.070	0.064	0.055
1200	0.184	0.147	0.131	0.123	0.115	0.105	0.092	0.082	0.074	0.067	0.061	0.053
1250	0.176	0.141	0.126	0.118	0.110	0.101	0.088	0.078	0.071	0.064	0.059	0.050
1300	0.170	0.136	0.121	0.113	0.106	0.097	0.085	0.075	0.068	0.062	0.057	0.048
1350	0.163	0.131	0.117	0.109	0.102	0.093	0.082	0.073	0.065	0.059	0.054	0.047
1400	0.158	0.126	0.113	0.105	0.098	0.090	0.079	0.070	0.063	0.057	0.053	0.045
1450	0.152	0.122	0.109	0.101	0.095	0.087	0.076	0.068	0.061	0.055	0.051	0.043
1500	0.147	0.118	0.105	0.098	0.092	0.084	0.074	0.065	0.059	0.053	0.049	0.042

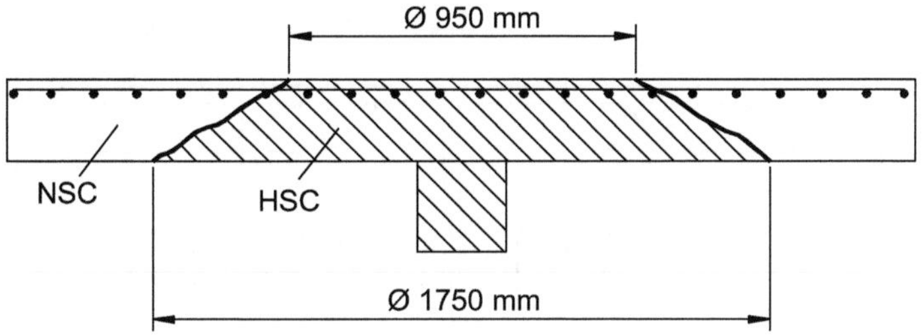

Fig. 9.34 Slab portion on an internal column made of high strength concrete (HSC) (CEB-FIP, 2001)

9.8.5.1 Example No. 3—Calculation of the Minimum Effective Depth of a Slab

Calculate the effective depth of a slab made of C25/30 *concrete, placed on an internal rectangular column with dimensions* 300 × 400 mm *under a centred axial load of* 850 kN or *the same axial load combined with bending moment, assuming for* β *the approximated values of* 1.15 *indicated in* Fig. 9.32.

Centred axial load

For $c_1 + c_2 = 300 + 400 = 700$ mm and $f_{ck} = 25$ N/mm^2, Table 9.11 gives the values 0.252 for $d_{\min}/V_{Ed}$.

The required value is therefore

$$d_{\min} = 0.252 \times 850 = 214\,\text{mm}.$$

Axial load combined with bending moment

By assuming $\beta = 1.15$, it is obtained $d_{\min} = 214 \times 1.15 \cong 246$ mm.

9.9 Columns with Enlarged Heads

The punching shear resistance can be increased by increasing the effective depth of the slab or by enlarging the head of the column itself.

EC2 gives the expressions of the punching shear resistance for circular and rectangular columns with enlarged heads.

The geometry of the enlarged head is defined by the projection l_H beyond the column perimeter and the height h_H (Figs. 9.35 and 9.38). Depending on the ratio l_H/h_H, two different cases are possible

(a) $l_H \leq 2\,h_H$, (Fig. 9.35)
(b) $l_H > 2\,h_H$, (Fig. 9.38).

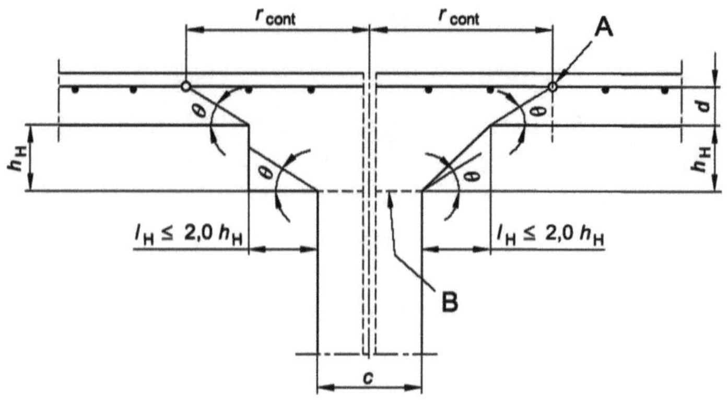

Fig. 9.35 Enlarged column head with $l_H \leq 2.0\ h_H$ (A: control section, B: loaded area A_{load}, $\theta = \arctan(1/2) = 26.6°$) [Figure 6.17]

The two cases are examined separately in the following paragraphs.

9.9.1 Enlarged Column Head with $l_H \leq 2h_H$

9.9.1.1 Circular Column

If the projection l_H of the enlarged head is not higher than twice its height h_H ($l_H \leq 2.0\ h_H$, Fig. 9.35), the verification of the punching shear stress has to be done only on the control perimeter beyond the enlarged head. The distance of the control perimeter to the column centre, named r_{cont}, is

$r_{cont} = 2d + (l_H + 0.5\,c)$ [(6.33)]

where

l_H is the distance between the perimeter of the enlarged head and the perimeter of the column

c is the diameter of the circular column.

The control perimeter, that is the circumference with radius r_{cont} (Fig. 9.36), is given by

$$u_1 = 2\pi r_{cont} = 2\pi(0.5c + l_H + 2d)$$

which is coincident with the perimeter of a circular column with diameter $c + 2l_H$, without an enlarged head.

It is then possible to conclude that, focusing on the verification on the control perimeter u_1, the enlargement of the column head with $l_H \leq 2h_H$ is like to increase the radius of the column of the amount l_H.

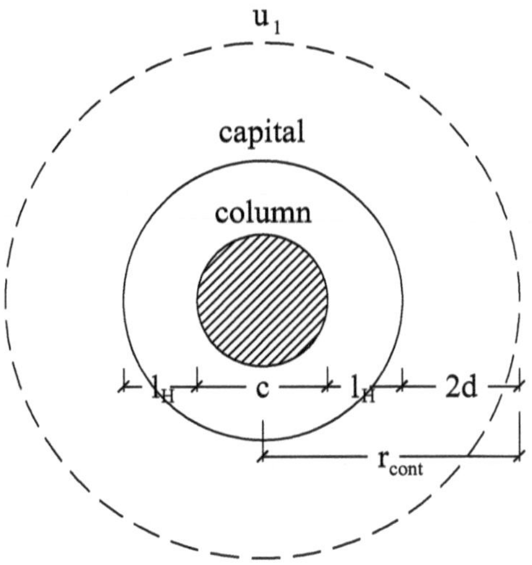

Fig. 9.36 Enlarged column head and control perimeter for a circular column

9.9.1.2 Rectangular Column

For a rectangular column with enlarged head having $l_H \leq 2d$ and total dimensions l_1 and l_2 ($l_1 = c_1 + 2l_{H1}$, $l_2 = c_2 + 2l_{H2}$, $l_1 \leq l_2$) (Fig. 9.37), the radius r_{cont} of the perimeter u_1 can be taken as the lesser of

$$r_{cont} = 2d + 0.56\sqrt{(l_1\, l_2)} \qquad [(6.34)]$$

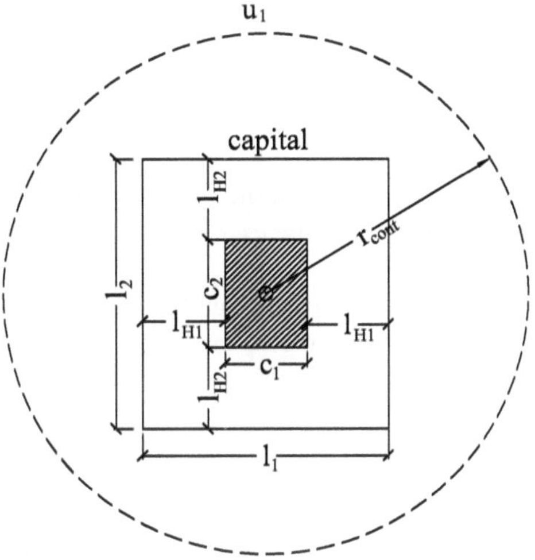

Fig. 9.37 Enlarged column head for a rectangular column

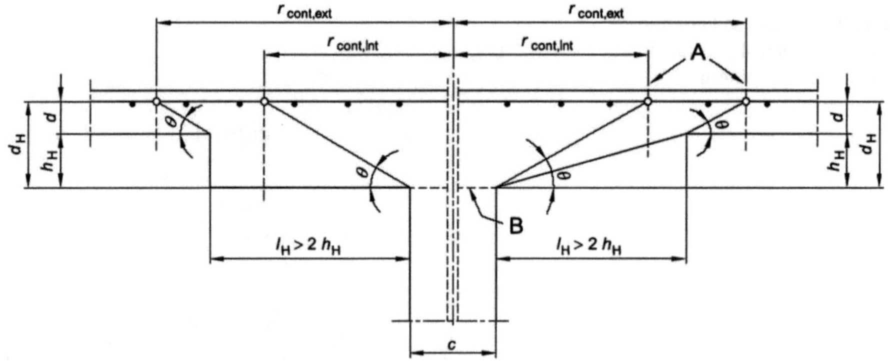

Fig. 9.38 Slab on enlarged circular column head with $l_H > 2\ h_H$ (A = control section, B = loaded area A_{load}, $\theta = \arctan(1/2) = 26.6°$) [Figure 6.18]

$$r_{\text{cont}} = 2d + 0.69\, l_1 \qquad [(6.35)]$$

Expression [(6.34)] is obtained by considering an equivalent circular enlarged head of radius r_{eq} with the same area of the rectangular enlarged head

$$\pi r_{\text{eq}}^2 = l_1 l_2 \rightarrow r_{\text{eq}} = (l_1 l_2/\pi)^{0.5} \cong 0.56(l_1 l_2)^{0.5}$$

from which

$$r_{\text{cont}} = r_{\text{eq}} + 2d = 0.56(l_1 l_2)^{0.5} + 2d$$

By comparing [(6.34)] with [(6.35)], following results are obtained

- for $l_2 = 1.5\ l_1$, [(6.34)] becomes [(6.35)]
- for $l_2 < 1.5\ l_1$ the value given by [(6.35)] is smaller than the value calculated with [(6.34)].

It is then possible to conclude that the choice of the smallest value given by [(6.34)] and [(6.35)] is equivalent to use the same expression [(6.34)] with $l_2 = 1.5\ l_1$ if $l_2 > 1.5\ l_1$.

Therefore, it is not suitable to adopt enlarged heads with $l_2 > 1.5\ l_1$ because the portion of l_2 exceeding $1.5\ l_1$ does not give any contribution to the punching shear resistance.

9.9.2 Enlarged Column Head with $l_H > 2h_H$

9.9.2.1 Circular Column

For enlarged heads with $l_H > 2h_H$, it is necessary to verify both the section of the enlarged head and the slab section.

Rules given for the calculation of the control perimeter [6.4.2] and punching shear stress [6.4.3] can be also applied to the punching shear verification within the enlarged head, by assuming the effective depth d equal to d_H ($d_H = d + h_H$).

For circular columns, the distances of control surfaces ($r_{\text{cont,ext}}$ and $r_{\text{cont,int}}$ in Fig. 9.38) can be taken as

$$r_{\text{cont,ext}} = l_H + 2d + 0.5c \qquad [(6.36)]$$

$$r_{\text{cont,int}} = 2(d + h_H) + 0.5c \qquad [(6.37)]$$

9.9.2.2 Rectangular Column

EC2 does not give any suggestion for rectangular columns with an enlarged head and $l_H > 2h_H$. Like circular columns, the radius $r_{\text{cont,ext}}$ can be calculated using the same expression of r_{cont} for $l_H \leq 2\,h_H$,

$$r_{\text{cont,ext}} = \min\left[2d + 0.56\sqrt{l_1\,l_2};\;\; 2d + 0.69\,l_1\right]$$

where l_1 and l_2 are the dimensions of the enlarged head ($l_1 = c_1 + 2l_{H1}$, $l_2 = c_2 + 2l_{H2}$, $l_1 \leq l_2$); for the verification within the enlarged head, it is possible to use the control perimeter u_1 valid for rectangular columns ($u_1 = 2\,c_1 + 2\,c_2 + 4\,\pi\,d$) and effective depth d_H (Fig. 9.39).

9.9.2.3 Example No. 4—Circular Column with $l_H > 2\,h_H$

Calculate $r_{cont,int}$ and $r_{cont,ext}$ for a circular column with diameter of 300 mm *and enlarged head of* $l_H = 750$ mm *and* $h_H = 150$ mm. *The slab effective depth is* 200 mm.

Being $l_H > 2\,h_H$, $r_{\text{cont,ext}}$ and $r_{\text{cont,int}}$ are given by

$$r_{\text{cont,ext}} = l_H + 2d + 0.5c = 750 + 2 \cdot 200 + 0.5 \cdot 300 = 1300\,\text{mm}$$
$$r_{\text{cont,int}} = 2(d + h_H) + 0.5c = 2(200 + 150) + 0.5 \cdot 300 = 850\,\text{mm}$$

Fig. 9.39 Enlarged head of a rectangular column with $l_H > 2h_H$

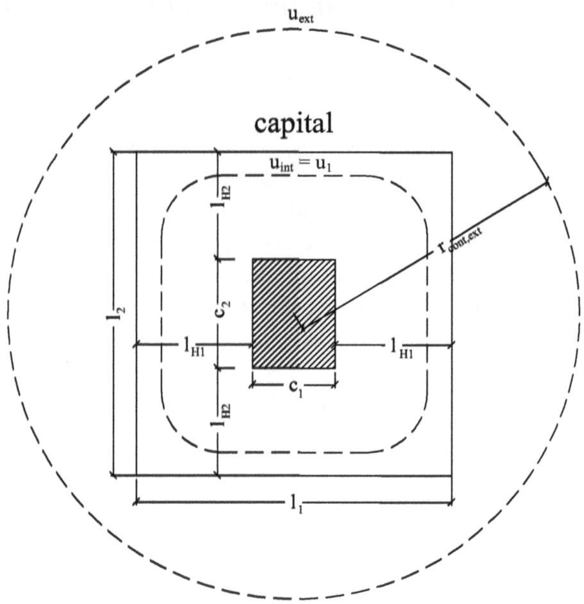

9.9.2.4 Example No. 5—Rectangular Column with an Enlarged Head

Calculate the control perimeter of a rectangular cross-section with dimensions $c_1 \times c_2 = 300 \times 400$ mm *and* 250 mm *high enlarged head; considering* $l_H = 400$ mm *or* $l_H = 600$ mm. *The slab effective depth is equal to* 250 mm.

Case $l_H = 400$ **mm**

Being $l_H \leq 2\,h_H$, the distance r_{cont} is given by

$$r_{\text{cont}} = \min\begin{cases} 2d + 0.56\sqrt{l_1\,l_2} \\ 2d + 0.69\,l_1 \end{cases}$$

If

$$l_1 = c_1 + 2\,l_H = 300 + 2 \cdot 400 = 1100\,\text{mm}$$
$$l_2 = c_2 + 2\,l_H = 400 + 2 \cdot 400 = 1200\,\text{mm}$$

r_{cont} becomes

$$r_{\text{cont}} = \min\begin{cases} 1143\ \text{mm} \\ 1259\ \text{mm} \end{cases} = 1143\ \text{mm}$$

and the control perimeter is

$$u_1 = 2\pi\, r_{\text{cont}} = 2\pi \cdot 1143 = 7182 \text{ mm}$$

Case $l_H = 600$ **mm**

Being $l_H > 2\, h_H$, $r_{\text{cont,ext}}$ is

$$r_{\text{cont,ext}} = \min\left[2d + 0.56\sqrt{l_1\, l_2};\ 2d + 0.69\, l_1\right] = 1367 \text{ mm}$$

where

$$l_1 = c_1 + 2\, l_H = 300 + 2 \cdot 600 = 1500\,\text{mm}$$
$$l_2 = c_2 + 2\, l_H = 400 + 2 \cdot 600 = 1600\,\text{mm}$$

and the control perimeters are

$$u_{\text{ext}} = 2\pi\, r_{\text{cont,ext}} = 2\pi \cdot 1367 = 8589\,\text{mm}$$
$$u_{\text{int}} = 2(c_1 + c_2) + 4\pi(d + h_H)$$
$$= 2(300 + 400) + 4\pi 500 = 7683\,\text{mm}$$

9.10 Punching Shear Verification Along the Control Perimeter u₁

The punching shear verification along the control perimeter u_1 of slabs without transverse reinforcement is expressed by

$$v_{Ed,1} \le v_{Rd,c}$$

where

$v_{Rd,c}$ is the punching shear resistance of a slab without punching shear reinforcement (see the expression in 9.10.1).

If the verification is not satisfied, it is possible:

- to adopt the same solutions listed in 9.8
- to introduce punching shear reinforcement, without changing the geometrical dimensions of the problem.

For slabs with punching shear reinforcement, the verification is expressed by

$$v_{Ed,1} \le v_{Rd,cs}$$

where $v_{Rd,cs}$ is the punching shear resistance of a slab with punching shear reinforcement (the expression is given in 9.12).

9.10.1 Punching Shear Resistance of Slabs Without Shear Reinforcement $v_{Rd,c}$

The punching shear resistance of slabs without shear reinforcement is calculated using the following expression.[1]

$$v_{Rd,c} = C_{Rd,c}\, k\, (100\, \rho_l\, f_{ck})^{1/3} + k_1\, \sigma_{cp} \geq \left(v_{\min} + k_1\, \sigma_{cp}\right) \qquad [(6.47)]$$

where

$C_{Rd,c}$	is equal to $0.18/\gamma_C$, being $\gamma_C = 1.5$ for persistent and transient loads and $\gamma_C = 1.0$ for exceptional loads
$k = 1 + \sqrt{\frac{200}{d}} \leq 2.0$	is the scale factor (see 9.2.2), with $d = (d_y + d_z)/2$ in mm, being d_y and d_z the slab effective depths in y and z directions.
$\rho_l = \sqrt{\rho_{ly}\, \rho_{lz}} \leq 0.02$	is the geometrical ratio of flexural steel reinforcement, also including bonded prestressing steel, calculated as the geometric mean of ratios in both directions of the slab
f_{ck}	is the characteristic cylinder compressive strength of concrete in N/mm^2.
k_1	= 0.1
$\sigma_{cp} = (\sigma_{cy} + \sigma_{cz})/2$	where σ_{cy} e σ_{cz} are design normal stresses (in N/mm^2, positive for compression) applied on the concrete control section in y and z directions:

$$\sigma_{cy} = N_{Ed,y}/A_{cy} \text{ e } \sigma_{cz} = N_{Ed,z}/A_{cz}.$$

with $N_{Ed,y}$ and $N_{Ed,z}$ axial forces across the full bay for internal columns (Fig. 9.40) and across the control section for edge columns (Fig. 9.41); they can be produced by loads or prestressing;

A_{cy} (A_{cz}) concrete area according to the definition of $N_{Ed,y}$ ($N_{Ed,z}$) (Figs. 9.40 and 9.41).

In the expression of $v_{Rd,c}$, the size effect, which was mentioned in 9.2.2, is considered through the parameter $k = 1 + \sqrt{(200/d)}$. This parameter, however, only

[1] The expression is identical to [(6.2.a)] that provides the design value of the shear resistance $V_{Rd,c}$ of a beam without shear reinforcement except that the term k_1, which here takes the value 0.10 instead of 0.15.

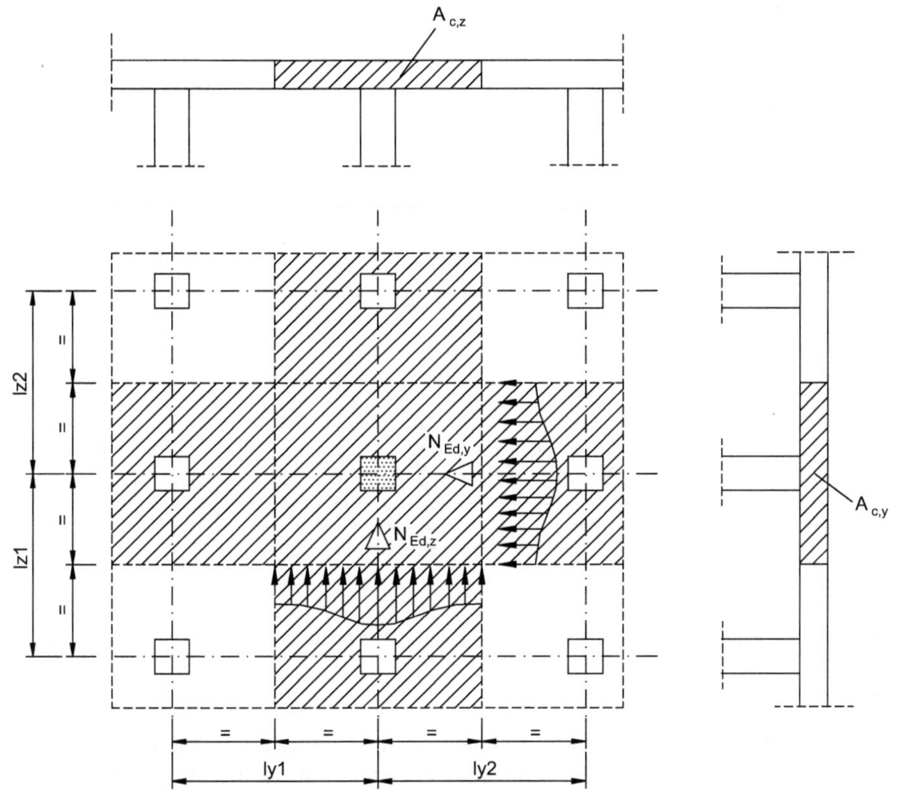

Fig. 9.40 Scheme for the calculation of $N_{Ed,y}$ and $N_{Ed,z}$ and corresponding concrete sections for an internal column

considers the variation of the shear punching resistance at varying the slab effective depth, but not at varying the radius of the punching surface.

Some researchers have proposed to replace the constant value of $C_{Rd,c}$ with a variable value, which changes with the size of the column cross-section and the effective depth of the slab.[2]

The values ρ_{ly} e ρ_{lz} are calculated as average values on a strip of width equal to the column width increased by *3d* on each side (Fig. 9.42).

$$\rho_{ly} = \frac{A_{ly}}{(c_2 + 6d)\,d}, \; \rho_{lz} = \frac{A_{lz}}{(c_1 + 6d)\,d}$$

[2] Leskelä (2008): $C_{Rd,c} = 0.3\,(c + 1.5\,d)/(c + 4d)$, where $c = \sqrt{c_1 c_2}$ for rectangular columns with dimensions c_1 and c_2 and c is equal to the diameter of the circular column.

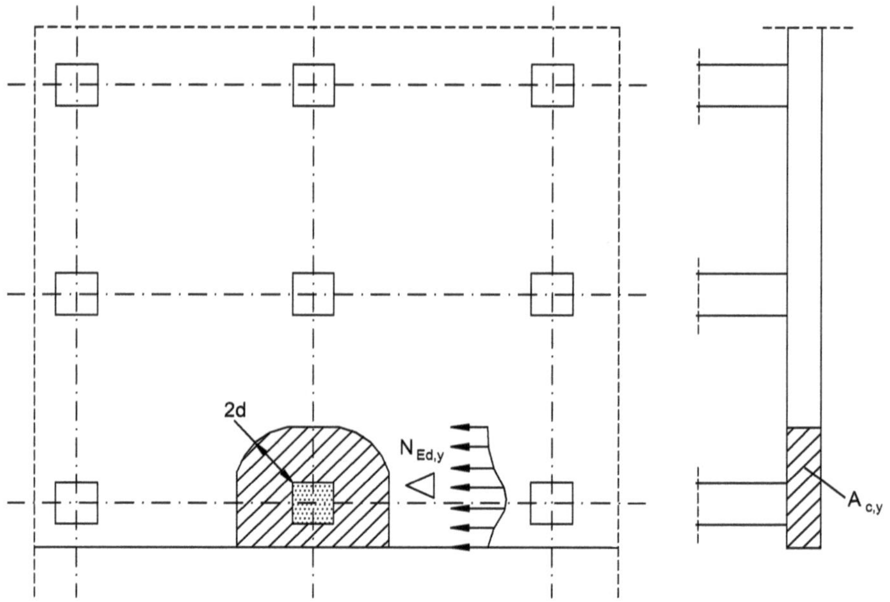

Fig. 9.41 Scheme for the calculation of $N_{Ed,y}$ and corresponding concrete sections for an edge column

where A_{ly} is the area of the reinforcement in the y direction in the portion having a width $c_2 + 6d$ and A_{lz} is the reinforcement area in the z direction in the portion having a width $c_1 + 6d$.

If the spacing of the reinforcement in the two directions is constant, the reinforcement geometric ratios are simply obtained with the following expressions

$$\rho_{ly} = \frac{A_{1,ly}}{i_y\, d}, \ \rho_{lz} = \frac{A_{1,lz}}{i_z\, d}$$

where: $A_{1,ly}$ ($A_{1,lz}$) is the area of a bar in the y (z) direction; i_y (i_z) is the spacing of bars placed in the y (z) direction.

The shear stress $v_{\min}$ is the lower limit of the punching resistance in the absence of normal stresses. It is only a function of the concrete tensile strength. The expression for the calculation of $v_{\min}$ is the following (Table 9.12).

$$v_{\min} = 0.035\ k^{3/2}\ f_{ck}^{1/2} \qquad [(6.3.\text{N})]$$

Linear interpolation can be used for values of the effective depth d different than those tabulated.

The lower limit of the punching resistance ($v_{\min} + k_1\sigma_{cp}$) has been introduced because in slabs with low reinforcement ratios, like prestressed slabs, $v_{Rd,c}$ resistance is very small if compared with experimental evidence.

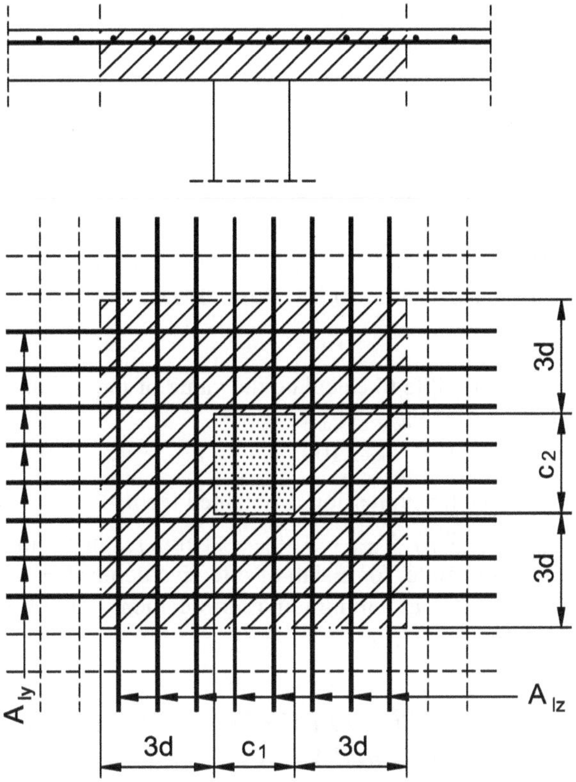

Fig. 9.42 Slab strips for the calculation of flexural reinforcement ratios in both orthogonal directions

Table 9.12 Values of v_{min} (N/mm^2) versus f_{ck} and effective depth d

f_{ck} (N/mm^2)	d (mm)						
	≤ 200	250	300	350	400	450	500
20	0.443	0.408	0.383	0.364	0.349	0.337	0.326
25	0.495	0.456	0.428	0.407	0.390	0.377	0.365
28	0.524	0.483	0.453	0.431	0.413	0.398	0.386
30	0.542	0.500	0.469	0.446	0.428	0.412	0.400
32	0.560	0.516	0.485	0.461	0.442	0.426	0.413
35	0.586	0.540	0.507	0.482	0.462	0.446	0.432
40	0.626	0.577	0.542	0.515	0.494	0.476	0.462
45	0.664	0.612	0.575	0.546	0.524	0.505	0.490
50	0.700	0.645	0.606	0.576	0.552	0.533	0.516
55	0.734	0.677	0.635	0.604	0.579	0.558	0.541
60	0.767	0.707	0.664	0.631	0.605	0.583	0.565
70	0.828	0.764	0.717	0.681	0.653	0.630	0.611

Table 9.13 Minimum values of ρ_l (%) that make $v_{Rd,c} \geq v_{min}$

d (mm)	f_{ck} (N/mm²)											
	20	25	28	30	32	35	40	45	50	55	60	70
≤ 200	0.314	0.351	0.371	0.384	0.397	0.415	0.444	0.471	0.496	0.521	0.544	0.587
250	0.289	0.324	0.342	0.354	0.366	0.383	0.409	0.434	0.458	0.480	0.501	0.541
300	0.272	0.304	0.321	0.333	0.344	0.359	0.384	0.408	0.430	0.451	0.471	0.508
350	0.258	0.289	0.306	0.316	0.327	0.342	0.365	0.387	0.408	0.428	0.447	0.483
400	0.248	0.277	0.293	0.303	0.313	0.327	0.350	0.371	0.391	0.410	0.429	0.463
500	0.231	0.259	0.274	0.284	0.293	0.306	0.327	0.347	0.366	0.384	0.401	0.433

It is possible to calculate the value of the minimum flexural reinforcement ratio to assure that $v_{Rd,c} \geq v_{min}$.

$v_{Rd,c} = 0.12 \cdot k \cdot (100 \cdot \rho_l \cdot f_{ck})^{1/3} \geq v_{min} = 0.035\ k^{3/2}\ f_{ck}^{1/2}$,

the following expression is obtained for ρ_l

$\rho_l \geq \left(\frac{0.035}{0.12}\right)^3 \frac{f_{ck}^{1/2} k^{3/2}}{100} \cong \frac{f_{ck}^{1/2} k^{3/2}}{4030}$, being f_{ck} in N/mm² (Table 9.13).

For example, for a slab of effective depth $d = 300$ mm, concrete class C25/30, Table 9.13 gives a minimum reinforcement ratio of 0.3% to assure that $v_{Rd,c} \geq v_{min}$.

9.10.2 *Verification Problem: Calculation of* $v_{Rd,c}$ *Using Tabulated Values*

Table 9.14 lists $v_{Rd,c}$ versus concrete class and effective depth, for three values of the flexural reinforcement ratio ($\rho_l = \sqrt{\rho_{ly}\,\rho_{lz}} = 0.5\% - 1\% - 2\%$).

For ρ_l values different than those tabulated, just multiply values corresponding to $\rho_l = 1\%$ by $(\rho_l)^{1/3}$. For example when $f_{ck} = 28$ N/mm² and $d = 300$ mm, $v_{Rd,c}$ value is 0.662 N/mm² if $\rho_l = 1\%$ and when $\rho_l = 1.5\%$, $v_{Rd,c} = 0.662 \cdot (1.5)^{1/3} = 0.758$ N/mm².

The Appendix to this chapter also reports some Tables for different concrete classes. Tables supply $v_{Rd,c}$ values depending on the distance and diameter of the tensile flexural reinforcement, supposed to be the same in both y and z directions.

9.10.2.1 Example No. 6—Punching Resistance of a Slab Without Transverse Reinforcement on a Square Column

Calculate the shear punching resistance of a 300 *m thick slab on a* 400 *mm square column. The flexural reinforcement of the slab is made of* 16 *mm bars at* 150 *mm spacing* (1ϕ16/150) *in both directions, the concrete class is C28/35 and the concrete cover is equal to* 30 mm.

Effective depth in x direction

Table 9.14 Values of $v_{Rd,c}$ (N/mm^2) for $\rho_l = 0.5\%$, 1% e 2% versus f_{ck} and effective depth of the slab

f_{ck} (N/mm^2)	ρ_l (%)	d (mm)						
		≤ 200	250	300	350	400	450	500
20	0.5	0.517	0.490	0.470	0.454	0.441	0.431	0.422
	1	0.651	0.617	0.592	0.572	0.556	0.543	0.532
	2	0.821	0.777	0.745	0.721	0.701	0.684	0.670
25	0.5	0.557	0.528	0.506	0.489	0.475	0.464	0.455
	1	0.702	0.665	0.637	0.616	0.599	0.585	0.573
	2	0.884	0.837	0.803	0.776	0.755	0.737	0.722
28	0.5	0.578	0.548	0.525	0.508	0.494	0.482	0.472
	1	0.729	0.690	0.662	0.640	0.622	0.607	0.595
	2	0.918	0.870	0.834	0.806	0.784	0.765	0.749
30	0.5	0.592	0.561	0.538	0.520	0.505	0.493	0.483
	1	0.746	0.706	0.677	0.655	0.637	0.621	0.609
	2	0.940	0.890	0.853	0.825	0.802	0.783	0.767
32	0.5	0.605	0.573	0.549	0.531	0.516	0.504	0.494
	1	0.762	0.722	0.692	0.669	0.650	0.635	0.622
	2	0.960	0.909	0.872	0.843	0.819	0.800	0.784
35	0.5	0.623	0.590	0.566	0.547	0.532	0.519	0.509
	1	0.785	0.744	0.713	0.689	0.670	0.654	0.641
	2	0.989	0.937	0.898	0.868	0.844	0.824	0.807
40	0.5	0.651	0.617	0.592	0.572	0.556	0.543	0.532
	1	0.821	0.777	0.745	0.721	0.701	0.684	0.670
	2	1.034	0.980	0.939	0.908	0.883	0.862	0.844
45	0.5	0.678	0.642	0.615	0.595	0.578	0.565	0.553
	1	0.854	0.809	0.775	0.749	0.729	0.711	0.697
	2	1.076	1.019	0.977	0.944	0.918	0.896	0.878
50	0.5	0.702	0.665	0.637	0.616	0.599	0.585	0.573
	1	0.884	0.837	0.803	0.776	0.755	0.737	0.722
	2	1.114	1.055	1.012	0.978	0.951	0.928	0.909
55	0.5	0.724 (0.734)[a]	0.686	0.658	0.636	0.618	0.604	0.591
	1	0.913	0.865	0.829	0.801	0.779	0.761	0.745
	2	1.150	1.089	1.044	1.010	0.982	0.958	0.939
60	0.5	0.746 (0.767)[a]	0.706 (0.707)[a]	0.677	0.655	0.637	0.621	0.609
	1	0.940	0.890	0.853	0.825	0.802	0.783	0.767
	2	1.184	1.121	1.075	1.039	1.010	0.986	0.966

(continued)

Table 9.14 (continued)

f_{ck} (N/mm^2)	ρ_1 (%)	d (mm)						
		≤ 200	250	300	350	400	450	500
70	0.5	0.785 (0.828)[a]	0.744 (0.764)[a]	0.713 (0.717)[a]	0.689	0.670	0.654	0.641
	1	0.989	0.937	0.898	0.868	0.844	0.824	0.807
	2	1.246	1.180	1.132	1.094	1.064	1.038	1.017

[a] When $v_{Rd,c} \leq v_{\min}$, the value of $v_{\min}$, that should be used in place of $v_{Rd,c}$ is also reported in round brackets

$$d_y = s - c - \phi/2 = 300 - 30 - 8 = 262\,\text{mm}$$

Effective depth in y direction

$$d_y = s - c - \phi/2 = 300 - 30 - 16 - 8 = 246\,\text{mm}$$

Slab effective depth

$$d = (d_y + d_z)/2 = 254\,\text{mm}$$

Reinforcement ratio

$$1\phi16/150\,\text{mm} \Rightarrow A_{1,ly} = A_{1,lz} = 201\,\text{mm}^2\; i_y = i_z = 150\,\text{mm}$$
$$\rho_1 = \rho_{ly} = \rho_{lz} = A_{1,ly}/(i_y, d) = 201/(150 \cdot 254) = 5.27‰ < 2\%$$

Analytical calculation

$$k = 1 + \sqrt{\frac{200}{d}} = 1.887$$

Shear-punching strength

$$\begin{aligned} v_{Rd,c} &= C_{Rd,c} k (100 \rho_l f_{ck})^{1/3} \\ &= 0.12 \cdot 1.887 \cdot (100 \cdot 0.00527 \cdot 28)^{1/3} = 0.555\,\text{N/mm}^2 \end{aligned}$$

Tabular calculation

The evaluation of $v_{Rd,c}$ is repeated using Table 9.14.

For $f_{ck} = 28$ N/mm^2, $d = 250$ mm (slightly lower than the actual value of 254 mm) and $\rho_l = 0.5\%$, the following value is obtained

$$v_{Rd,c} = 0.548\,\text{N/mm}^2$$

To consider the exact reinforcement ratio ($\rho_l = 0.527\%$), the value corresponding to $\rho_l = 0.5\%$ must be multiplied by the ratio $(0.527/0.5)^{1/3} = 1.018$.

$$v_{Rd,c} = 0.548 \cdot 1.018 = 0.558\,\text{N/mm}^2$$

This value is a little higher than the value calculated analytically because tabulated values are valid for an effective depth of 250 mm, which is a little lower than the actual depth of 254 mm.

9.10.2.2 Example No. 7—Punching Resistance of a Slab Without Transverse Reinforcement on a Rectangular Column

Calculate the shear-punching resistance of a 250 mm *thick slab on a* 300 × 500 mm *rectangular column. The flexural reinforcement of the slab is made of* 16 *mm bars at* 200 mm *spacing* (1ϕ16/200) *in both directions, the concrete class is C28/30 and the concrete cover is equal to* 25 mm.

Effective depth in x direction

$$d_y = s - c - \phi/2 = 250 - 25 - 8 = 217\,\text{mm}$$

Effective depth in y direction

$$d_z = s - c - \phi - \phi/2 = 250 - 25 - 16 - 8 = 201\,\text{mm}$$

Slab effective depth

$$d = (d_y + d_z)/2 = 209\,\text{mm}$$

Reinforcement ratio

$$1\phi 16/200\,\text{mm} \Rightarrow A_{1,ly} = A_{1,lz} = 201\,\text{mm}^2\; i_y = i_z = 200\,\text{mm}$$
$$\rho_1 = \rho_{ly} = \rho_{lz} = A_{1,ly}/(i_y, d) = 201/(200 \cdot 209) = 4.81‰ < 2\%$$

Analytical calculation

$$k = 1 + \sqrt{\frac{200}{d}} = 1.978$$
$$v_{Rd,c} = C_{Rd,c}\,k\,(100\,\rho_l\,f_{ck})^{1/3}$$
$$= 0.12 \cdot 1.978 \cdot (100 \cdot 0.00481 \cdot 25)^{1/3} = 0.54\,\text{N/mm}^2$$

Tabular calculation

For $fck = 25$ N/mm$_2$ and $\rho_l = 0.5\%$, Table 9.14 gives

$$v_{rd,c} = 0.557\,\text{N/mm}^2\,\text{per}\,d = 200\,\text{mm}$$
$$v_{rd,c} = 0.528\,\text{N/mm}^2\,\text{per}\,d = 250\,\text{mm}$$

Linearly interpolating these values, for d = 209 mm the following value is obtained

$$v_{Rd,c} = 0.552\,\text{N/mm}^2$$

As in Example no. 6, by multiplying the result by the ratio $(0.481/0.5)^{1/3}$ to consider the actual value of the flexural reinforcement ratio ($\rho_l = 0.481\%$), the following punching resistance is obtained

$$v_{Rd,c} = 0.552 \cdot (0.481/0.5)^{1/3} = 0.545\,\text{N/mm}^2$$

9.11 Comparison of the Shear Forces $V_{Rd,\max}$ and $V_{Rd,c}$

The punching shear verification on the control perimeter u_1 is satisfied if

$$V_{\text{Ed}} \leq v_{Rd,c} \cdot u_1 \cdot d \leq \max(v_{Rd,c}) \cdot u_1 \cdot d$$

where $\max(v_{Rd,c})$ is the value of the shear-punching resistance obtained with $\rho_l = 2\%$, that is the maximum value of the reinforcement ratio that can be used in the calculation of $v_{Rd,c}$ according to EC2.

On the perimeter u_0 of the column or loaded area, the punching verification is expressed as

$$V_{Ed} \leq v_{Rd,\max} \cdot u_0 \cdot d$$

The comparison of above expressions shows that the verification is more onerous on u_1 than on u_0, if the following condition is satisfied

$$\max(v_{Rd,c}) \cdot u_1 \cdot d \leq v_{Rd,\max} \cdot u_0 \cdot d$$

the expression can be rewritten in the following form

$$\max(v_{Rd,c}) \cdot u_1/(v_{Rd,\max} \cdot u_0) \leq 1$$

The last expression is useful to calculate the values of effective depth d for which $V_{Rd,\max}$ is equal to $V_{Rd,c}$. For this purpose, it is sufficient to fix the characteristic compressive strength of the concrete, the shape and dimensions of the column cross-section and then calculate $V_{Rd,c}$ and $V_{Rd,\max}$ for different values of the slab effective depth d, up to identify the value of d for which the two punching shear resistant forces have about the same magnitude. Particularly for d values in the range 150 mm $\div$ 1300 mm, with 10 mm increments, values of Table 9.15 are obtained.

Tabulated values represent the maximum effective depth $d_{\max}$ where $V_{Rd,c} \leq V_{Rd,\max}$, that is values below which the punching shear failure occurs on the perimeter u_1. For effective depths greater than tabulated values, it results that $V_{Rd,c} > V_{Rd,\max}$, and the punching failure occurs on the perimeter u_0 due to concrete crushing. The analysis of tabulated values shows that, once the concrete compressive strength and dimensions of the column cross-section have been assigned, the verification on u_1 is more onerous for small effective depths, while the verification on u_0 is more onerous for high depths.

Please note that the table has been built assuming $\rho_l = 2\%$, so that for $\rho_l < 2\%$, tabulated values of $d_{\max}$ are underestimated.

9.11.1 *Example No. 8—Maximum Value of the Effective Depth to Have* $V_{Rd,c} \leq V_{Rd,max}$

Identify the most severe punching verification (along the perimeter u_0 or the control perimeter u_1) for a slab without transverse reinforcement with effective depth d = 250 mm *supported by a* 300 × 400 mm *rectangular column or a* 350 mm *diameter circular column. The concrete strength class is* C25/30 *and the ratio of flexural reinforcement is* 0.02.

***Rectangular column with dimensions* 300 x 400 mm**

$$u_0 = 2(300 + 400) = 1400\,\text{mm}$$

$$v_{Rd,\max} = 0.2 f_{cd} = 0.2 \cdot 0.85 \cdot 25/1.5 = 2.833\,\text{N/mm}^2$$

$$u_1 = u_0 + 2\pi(2d) = 1400 + 2\pi(2 \cdot 250) = 4542\,\text{mm}$$

$$V^*_{Rd,\max^3} = v_{Rd,\max} \text{u}_0 = 2.833 \cdot 1400 = 3966\,\text{N/mm}$$

$$k = 1 + (200/250)^{0.5} = 1.894$$

$$v_{Rd,c} = C_{Rd,c} k (100 \rho_l f_{ck})^{1/3} = 0.12 \cdot 1.894 \cdot (100 \cdot 0.02 \cdot 25)^{1/3}$$
$$= 0.837\,\text{N/mm}^2$$

$$V^*_{Rd,c} = v_{Rd,c} \cdot u_1 = 0.837 \cdot 4542 = 3800\,\text{N/mm}$$

Being $V^*_{Rd,c} < V^*_{Rd,\max}$, and $V_{Rd,c} < V_{Rd,\max}$, the failure occurs on the control perimeter u_1.

Table 9.15 Maximum values d_{max} (mm) of the effective depth below which $V_{Rd,c} \leq V_{Rd,max}$ for $\rho_l = 0.02$ (values are given with an accuracy of ± 10 mm)

Rectangular column

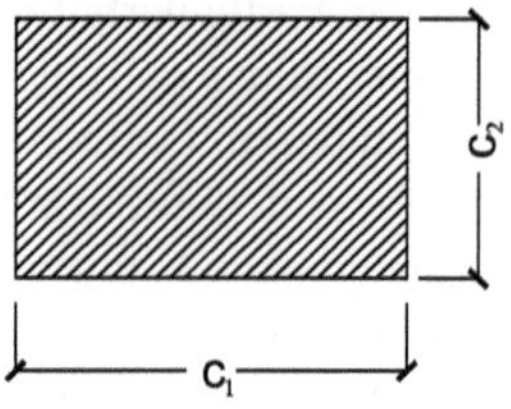

$c_1 + c_2$ (mm)	f_{ck} (N/mm^2)								
	20	25	28	30	32	35	40	45	50
500	140	170	190	210	230	260	300	350	390
550	150	190	220	240	260	290	340	390	440
600	160	210	250	270	300	330	390	440	490
650	180	240	280	310	330	370	430	490	550
700	190	270	310	340	370	410	480	540	600
800	240	330	380	410	440	490	570	640	720
900	280	390	450	480	520	580	660	750	830
1000	330	450	510	560	600	660	760	850	950

Circular column

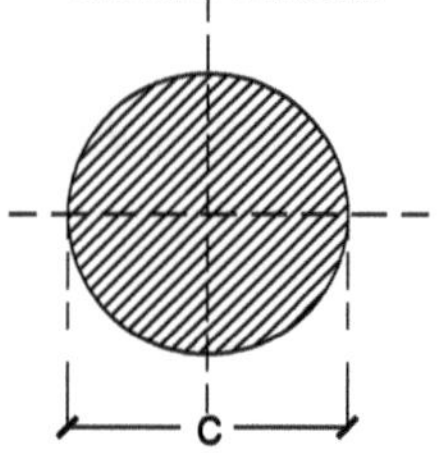

Diameter c (mm)	f_{ck} (N/mm^2)								
	20	25	28	30	32	35	40	45	50
250	110	130	150	160	170	180	210	250	280
300	130	160	180	190	210	230	280	320	360
350	150	190	220	240	260	290	340	390	440
400	170	230	270	290	320	350	410	470	530
450	190	270	320	350	380	420	480	550	610

(continued)

Table 9.15 (continued)

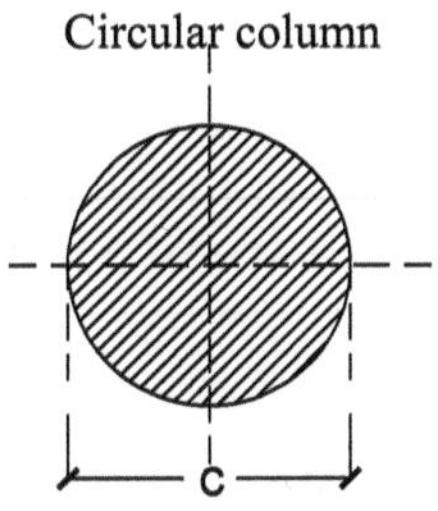

Diameter c (mm)	f_{ck} (N/mm^2)								
	20	25	28	30	32	35	40	45	50
500	230	320	370	400	430	480	560	630	700
550	270	370	420	460	490	550	630	710	790
600	300	410	480	520	550	610	700	790	880

Table 9.16 Values of $f_{wd,ef}$ (N/mm^2) steel versus slab effective depth

d (mm)	200[a]	250	300	350	400	450	500	600	700	800
$f_{wd,ef}$ (N/mm^2)	300	312.5	325	337.5	350	362.5	375	391 (400)[b]	391 (425)[b]	391 (434.8)[b]

[a] In the table and every table concerning punching shear reinforcement, only values of d higher than 200 mm have been considered because this is the minimum depth required by EC2 for reinforced slabs to guarantee the reinforcement anchorage [9.3.2(1)]
[b] Values in round brackets are valid for B500C steel; for $d \leq 565$ mm, $f_{wd,ef}$ values are the same for B450C and B500C

Verification using Table 9.15

For $f_{ck} = 25$ N/mm^2 and $c_1 + c_2 = 700$ mm, Table 9.15 gives $d_{\max} = 270$ mm, therefore being $d = 250$ mm < $d_{\max} = 270$ mm, it follows that $V_{Rd,c} < V_{Rd,\max}$ and failure occurs on the perimeter u_1.

Circular column with diameter d = 350 mm

$$u_0 = \pi c = 1100\,\text{mm}$$
$$u_1 = \pi(c + 4d) = \pi(350 + 4 \cdot 250) = 4241\,\text{mm}$$
$$V^*_{Rd,\max} = v_{Rd,\max} u_0 = 2.833 \cdot 1100 = 3116\,\text{N/mm}$$
$$V^*_{Rd,c} = v_{Rd,c} \cdot u_1 = 0.837 \cdot 4241 = 3550\,\text{N/mm}$$

it results that

$V^*_{Rd,c} > V^*_{Rd,\max}$, then $V_{Rd,c} > V_{Rd,\max}$, and failure occurs along u_0.

Verification with Table 9.15.

For $f_{ck} = 25$ N/mm^2 and $c = 350$ mm, it results that $d_{\max} = 190$ mm, therefore being $d = 250$ mm $> d_{\max} = 190$ mm, it follows that $V_{Rd,c} > V_{Rd,\max}$ and failure occurs with brittle cracks of the concrete on the column periphery.

9.12 Punching Shear Resistance of Slabs with Punching Shear Reinforcement

The punching shear resistance $v_{Rd,cs}$ of slabs with transverse reinforcement is given by the sum of two terms

$$v_{Rd,cs} = v'_{Rd,cs} + v''_{Rd,cs}$$

being $v'_{Rd,cs}$ and $v''_{Rd,cs}$ the concrete and steel contributions respectively.

In general, the question arises about how to combine steel and concrete contributions. As the width of inclined cracks which separate the slab from the column increases, the total tension force decreases. Then, when steel yields, the concrete resistance is less than the initial one.

In literature, two different approaches are proposed to consider this aspect (CEB-FIP, 2001). Some researchers propose to reduce the contribution of transverse reinforcement through the introduction of efficiency factors between 0.8 and 0.25; other researchers propose to reduce the concrete contribution of an amount between 20 and 40% compared to slabs without transverse reinforcement (efficiency factors between 0.6 and 0.8).

EC2 has adopted the second approach, with an efficiency factor for concrete strength equal to 0.75.

A further aspect to be considered is that not all the transverse reinforcement strength can be considered because it is difficult to well anchor the reinforcement on both sides of a punching shear crack. Therefore, the effective strength of the transverse reinforcement is considered to vary linearly with the effective depth d, up to the steel yield strength.

According to EC2, the punching resistance of a slab with transverse reinforcement is expressed as

$$v_{Rd,cs} = 0.75\, v_{Rd,c} + 1,5\,(d/s_r)\, A_{sw}\, f_{ywd,ef} \frac{1}{u_1\, d} \sin\alpha \qquad [(6.52)]$$

where it is assumed that the concrete contribution is like the slab without reinforcement reduced of 25% (0.75 $v_{Rd,c}$) and the parameters which appear in the second term (steel contribution) have the following meaning:

A_{sw} is the area of one perimeter of shear reinforcement around the column

s_r is the radial spacing of perimeters of shear reinforcement

$f_{ywd,ef}$ $= 250 + 0.25\,d \leq f_{ywd}$ (N/mm^2) with d in mm is the effective design strength of the punching shear reinforcement (Table 9.16); the design value of the steel stress varies with the slab effective depth, as the anchorage efficiency increases with the effective depth d (for steel B450C, $f_{ywd,ef}$ is equal to f_{ywd} if $d \geq 565$ mm, while for steel B500C, often used for studs, $f_{ywd,ef} = f_{ywd}$ if $d \geq 739$ mm)

d is the mean of the effective depths in the orthogonal directions (mm)

α is the angle between the punching shear reinforcement and the plane of the slab.

Studs are often prefabricated and made of steel B500C; the effective stress for B500C steel is the same as steel B450C if $d \leq 565$ mm.

In Model Code 1990 $f_{ywd,\mathrm{ef}}$ is equal to 300 N/mm^2 for every value of d; EC2 gives the same value of Model Code 1990 for $d = 200$ mm, lower values for $d < 200$ mm and higher values for $d > 200$ mm.

Anchorage conditions are not equal for every type of reinforcement, but EC2 does not take into account this aspect; in other codes, as e.g. in DIN 1045-1, better bond conditions are considered for bent-up bars instead of studs.

If the reinforcement is given by a unique perimeter of bent-up bars, the ratio d/s_r takes the value of 0.67.

Values of $v'_{Rd,cs}$ can be obtained quickly, by multiplying $v_{Rd,c}$ values of Table 9.14 by 0.75.

In the following two tables, values of the shear resistance with transverse reinforcement $v''_{Rd,cs} = (v''_{Rd,cs} d\, u_1)$ are given for 45° bent-up bars (Table 9.17) and studs (Table 9.18)

If in the y direction there are n_y bent-up bars (with 2 n_y inclined legs) and in the z direction there are n_z bent-up bars (with 2 n_z inclined legs), it is sufficient to multiply values given in the table by $(n_y + n_z)$

For different values of the α inclination of bent-up bars with the horizontal, it is sufficient to multiply values of Table 9.17 by the ratio (sen α/sen 45°); e.g. for $\alpha \geq 30°$ the ratio is equal to $1/\sqrt{2} \cong 0.707$ and for $\alpha = 60°$ it is equal to 1.225

Table 9.18 collects $v''_{Rd,cs}$ values for studs placed on two or more perimeters at radial spacing 0.75 d, which represents the maximum spacing allowed in the radial direction for studs.

It is worth noting that the number of stud perimeters (always more than two according to rules given in [9.4.3(1)]) does not have any influence on the calculation of $v''_{Rd,cs}$, which only depends on the radial spacing s_r between perimeters and reinforcement area of each perimeter. This is due to the requirement that each perimeter should be capable to assure the strength required for the equilibrium as if the other perimeters do not exist and under the hypothesis that the punching shear surface crosses just one perimeter. It is but necessary that studs follow all the geometrical rules collected in para. 9.13.1, so the number of perimeters is chosen to cover the entire region around the column or loaded area where punching shear failure could occur

Table 9.17 Values (kN) of $v''_{Rd,cs} = 1.5\,(d/s_r)\,A_{sw}\,f_{ywd,ef}\,\sin\alpha$ for 45° bent-up bars on one perimeter. The table is valid for just one bent-up bar (2 inclined legs) in only one direction (*y* or *z*, being *yz* the plane of the slab)

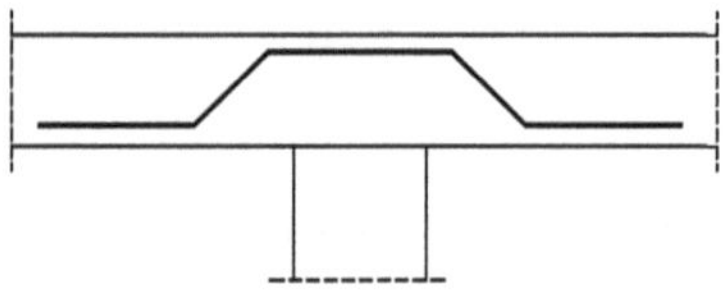

Diameter of bent-up bars (mm)	d (mm)						
	200	250	300	350	400	450	500
ϕ12	48.2	50.2	52.2	54.3	56.3	58.3	60.3
ϕ14	65.6	68.4	71.1	73.8	76.6	79.3	82.0
ϕ16	85.7	89.3	92.9	96.4	100.0	103.6	107.2
ϕ18	108.5	113.0	117.5	122.1	126.6	131.1	135.6
ϕ20	134.0	139.5	145.1	150.7	156.3	161.9	167.4
ϕ22	162.1	168.8	175.6	182.3	189.1	195.9	202.6
ϕ24	192.9	200.9	209.0	217.0	225.0	233.1	241.1
ϕ26	226.4	235.8	245.2	254.7	264.1	273.5	283.0

[a] Values are valid both for steel B450C and B500C, because thickness values not higher than 500 mm are considered and, in those cases, the effective stress is the same for both types of steel

To calculate the shear resistant force $v''_{Rd,cs}$ given by studs, it is sufficient to multiply tabulated values by the number of studs within each perimeter (e.g., for links like in the sketch of Table 9.18, the number is eight). For values of the radial spacing s_r less than $0.75d$, tabulated values should be multiplied by $(0.75d/s_r)$; e.g., if studs are at distance s_r equal to $0.5d$, values of the table have to be amplified by 1.5.

9.13 Arrangement of the Punching Shear Reinforcement

The arrangement of the punching shear reinforcement follows rules given in [9.4.3]:

1. the reinforcement must be placed between the column—or loaded area—and the perimeter at distance $1.5d$ from the perimeter u_{out}, where u_{out} is the perimeter at which punching shear reinforcement is no more required (Fig. 9.43); the limit of $1.5d$ is useful to avoid the development of a failure surface between the perimeter u_{out} and the last reinforcement perimeter, without crossing any reinforcement perimeter (it is assumed that the failure surface is inclined of 26.6° (Fig. 9.44) and then it extends of $2d$ in the horizontal radial direction)

Table 9.18 $v''_{Rd,cs} = [1.5\,(d/s_r)\,A_{sw}\,f_{ywd,ef}]$ (kN) for studs placed on one or more perimeters with radial spacing $s_r = 0.75$d ($v''_{Rd,cs} = 2\,A_{sw}\,f_{ywd,ef}$). Table collects values obtained considering the area of only one stud (black circles in the sketch) for each perimeter

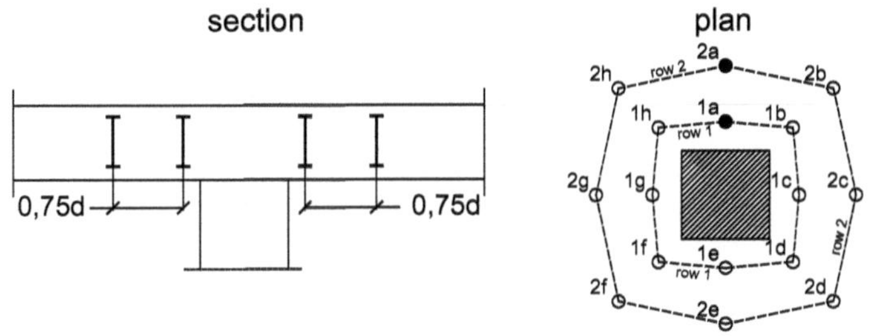

Diameter of studs (mm)	d (mm)						
	200	250	300	350	400	450	500
ϕ10	47.12	49.09	51.05	53.01	54.98	56.94	58.90
ϕ12	67.86	70.69	73.51	76.34	79.17	82.00	84.82
ϕ14	92.36	96.21	100.06	103.91	107.76	111.61	115.45
ϕ16	120.64	125.66	130.69	135.72	140.74	145.77	150.80
ϕ18	152.68	159.04	165.40	171.77	178.13	184.49	190.85
ϕ20	188.50	196.35	204.20	212.06	219.91	227.77	235.62
ϕ22	228.08	237.58	247.09	256.59	266.09	275.60	285.10
ϕ24	271.43	282.74	294.05	305.36	316.67	327.98	339.29
ϕ26	318.56	331.83	345.10	358.38	371.65	384.92	398.20

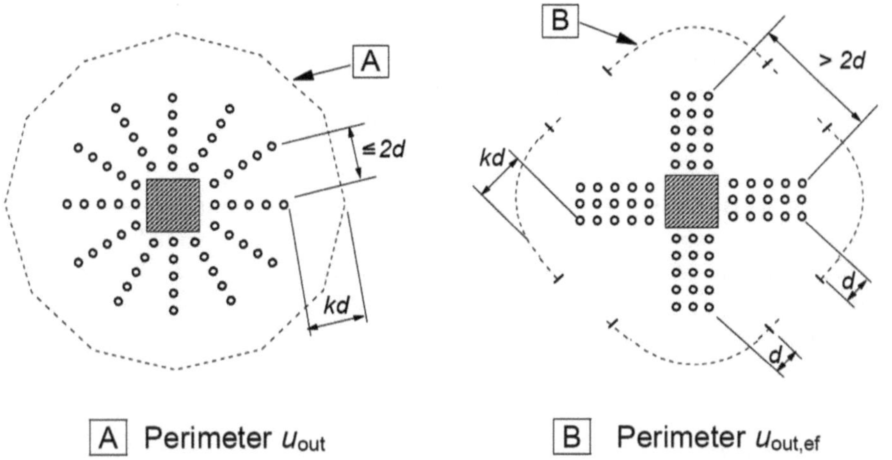

a) b)

Fig. 9.43 **a** Perimeters u_{out} and $u_{out,ef}$ at which the punching shear verification is satisfied without transverse reinforcement. The configuration **b** creates the least interference with the flexural reinforcement, which is normally parallel with the column sides [Figure 6.22]

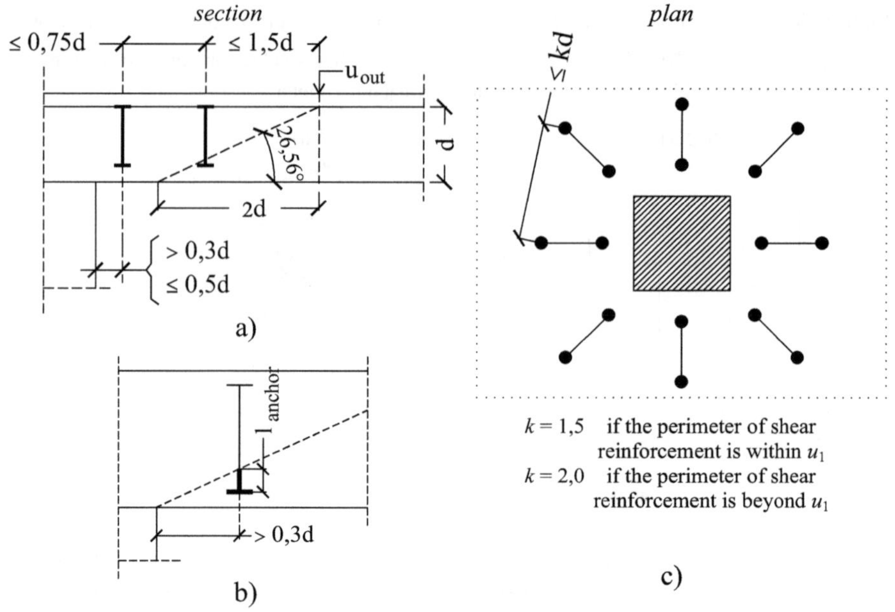

Fig. 9.44 Geometric limitations for studs

2. studs should be placed on at least two perimeters at radial spacing not higher than 0.75 d; this limit avoids that the failure surface arises between two adjacent perimeters without crossing any reinforcement perimeter
3. bent-up bars can be placed only on one perimeter; if bent-up bars are placed on more perimeters, the radial spacing should not exceed 0.75 d like studs;
4. the minimum area of a single stud or bent-up bar is given by

$$A_{sw1,\min} = 0.08\frac{\sqrt{f_{ck}}}{f_{yk}}\frac{s_r s_t}{1.5\,\text{sen}\alpha + \cos\alpha} \qquad [(9.11)]$$

where s_r and s_t are radial and tangential spacings.

In [9.4.3(2)] EC2 uses the symbol $A_{sw,\min}$ to indicate the minimum area of a single stud or bent-up bar, while in [6.4.5] EC2 uses the symbol A_{sw} to indicate the area of one perimeter of shear reinforcement. As the symbol $A_{sw,\min}$ could suggest the minimum area of one perimeter and not a single bar, $A_{sw1,\min}$ is preferred to $A_{sw,\min}$ to avoid misunderstandings.

9.13.1 Studs

Punching shear reinforcement made of studs should follow geometrical rules shown in Fig. 9.44 and Table 9.19.

It is worth noting that the first reinforcement perimeter should be far from the column perimeter more than 0.3d to guarantee that it is crossed by the failure surface

Table 9.19 Geometric limitations for studs

Perimeter u_{out} at which punching shear reinforcement is no more required	$u_{\text{out}} = \beta\, V_{Ed}/(v_{Rd,c}\, d)$	[(6.54)]
Distance (k'_d) of the perimeter u_{out} from the column perimeter	$k'd = [u_{\text{out}} - 2(c_1 + c_2)]/(2\,\pi)$	
Minimum number of reinforcement perimeters	2	[9.4.3 (1)]
Maximum radial spacing between two adjacent reinforcement perimeters	$s_r \leq 0.75\, d$	[9.4.3 (1)]
Distance a_1 of the first reinforcement perimeter from the column perimeter	$0.30\, d \leq a_1 \leq 0.50\, d$	[Figure (9.10a)] [9.4.3 (4)]
Distance b_u of the last reinforcement perimeter from the perimeter u_{out} at which punching shear reinforcement is no more required	$b_u \leq 1.5\, d$	[6.4.5 (4)]
Distance s_t between two adjacent studs within the same perimeter (corresponding to the stud spacing in the circumferential direction)	$s_t \leq 1.5\, d$ (within the control perimeter u_1); $s_t \leq 2.0\, d$ (beyond the control perimeter u_1, only in those parts of the perimeter considered in the calculation of punching shear strength)	[9.4.3 (1)]
Minimum area of each stud	$A_{sw1,\min} \cdot \frac{(1.5 \cdot \sin\alpha + \cos\alpha)}{s_r \cdot s_t} \geq \frac{0.08 \cdot \sqrt{f_{ck}}}{f_{yk}}$	[(9.11)]

and the bottom part of the studs is well anchored. Moreover, the punching shear surface crosses studs at a minimum distance from the bottom of the slab (Fig. 9.44b).

Concerning the minimum diameter of studs, if $s_r = 0.75\ d$ and $s_t = 1.5\ d$, it follows that

$$A_{sw1,\min} = 0.08\frac{\sqrt{f_{ck}}}{f_{yk}}0.75d^2 = 0.06d^2\frac{\sqrt{f_{ck}}}{f_{yk}}$$

Table 9.20 collects values of the minimum diameter of a single stud made of steel B450C versus effective depth d and concrete strength, for $s_r = 0.75\ d$ and $s_t = 1.5\ d$.

9.13.2 Bent-Up Bars

Instead of studs, punching shear reinforcement can be made of bent-up bars (Fig. 9.45, Table 9.21).

Table 9.20 Minimum diameter of a stud (made of B450C steel) versus effective depth of the slab and characteristic compressive strength of concrete ($s_r = 0.75\ d$, $s_t = 1.5\ d$)[a]

d (mm)	f_{ck} (N/mm^2)								
	20	25	28	30	32	35	40	45	50
200	6	6	6	8	8	8	8	8	8
250	8	8	8	8	8	8	10	10	10
300	10	10	10	10	10	10	10	12	12
350	10	12	12	12	12	12	12	12	14
400	12	12	12	14	14	14	14	14	14
450	14	14	14	14	14	16	16	16	16
500	14	16	16	16	16	16	18	18	18
550	16	18	18	18	18	18	20	20	20
600	18	18	18	20	20	20	20	22	22

[a] The table has been built by considering all the diameters from 6 to 22 mm; studs are usually available in the following diameters: 10, 12, 14, 16, 18, 20 and 25 mm

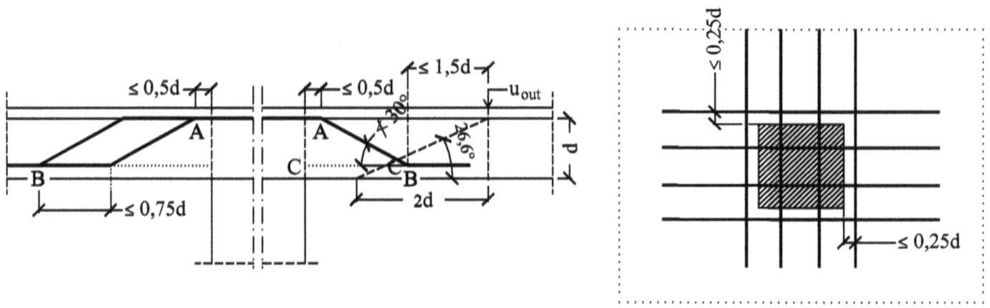

Fig. 9.45 Spacing and geometric limitations for bet-up bars (A = outer control perimeter requiring shear reinforcement, B = first control perimeter not requiring shear reinforcement u_{out})

Table 9.21 Geometric limitations for bent-up bars

Minimum number of reinforcement perimeters	1	[9.4.3 (1)]
Maximum radial spacing between two adjacent reinforcement perimeters	$s_r \leq 0.75\ d$	[9.4.3 (1)]
Distance measured at the level of the tensile reinforcement of the first perimeter of bent-up bars from the column or loaded area	$a_1 \leq 0.50\ d$	[Figure (9.10b)] [9.4.3 (4)]
Horizontal distance from the column face or loaded area	$b_1 \leq 0.25\ d$	[Figure (9.10b)] [9.4.3 (3)]
Minimum area of a single bent-up bar	$A_{sw1,min} \cdot \frac{(1.5 \cdot \sin\alpha + \cos\alpha)}{s_r \cdot s_t} \geq \frac{0.08 \cdot \sqrt{f_{ck}}}{f_{yk}}$	[(9.11)]

Table 9.22 Minimum diameter of a bent-up bar with a slope of 45° (steel grade B450C) arranged on one perimeter versus the effective slab depth and concrete characteristic compressive strength ($s_r = s_t = 1.5\ d$)

d (mm)	f_{ck} (N/mm²)								
	20	25	28	30	32	35	40	45	50
200	8	8	8	8	10	10	10	10	10
250	10	10	10	10	12	12	12	12	12
300	12	12	12	12	14	14	14	14	14
350	14	14	14	14	16	16	16	16	16
400	16	16	16	16	18	18	18	18	20
450	18	18	18	18	20	20	20	20	22
500	18	20	20	20	22	22	22	22	24
550	20	22	22	22	24	24	24	26	26
600	22	24	24	24	26	26	27	27	27

When bars are placed on only one perimeter, as shown in Fig. 9.45, their slope with the horizontal can be reduced up to 30°. If one perimeter only it is not possible to cover the whole region between the perimeter at distance $0.5d$ from the column or loaded area and the perimeter at distance $1.5d$ from u_{out} (i.e. the distance of point B in Fig. 9.45 from u_{out} is more than $1.5d$), it will be necessary to use more perimeters of bent-up bars. The first perimeter will have point A at a distance less than $0.5d$ from the column, the last perimeter will have point B at a distance not more than $1.5d$ from u_{out} and the intermediate perimeters shall be at spacing less than $0.75d$.

For bent-up bars on one perimeter, d/s_r is equal to 0.67 [6.4.5(1)], i.e. $s_r = 1.5\ d$; if the slope of bars is 45° and $s_t = 1.5\ d$, the minimum area for a single bar should be $A_{sw1,\min} = 0.08\frac{\sqrt{f_{ck}}}{f_{yk}}1.27d^2 = 0.102d^2\frac{\sqrt{f_{ck}}}{f_{yk}}$.

Table 9.22 collects values of the minimum diameter of a bent-up bar made of steel grade B450C with a slope of 45° with varying the effective depth d and the concrete compressive strength.

9.14 Maximum Area of Transverse Reinforcement

In a slab or foundation with punching shear reinforcement, the punching verification is satisfied if the following three conditions are met:

- $V_{Ed} \leq V_{Rd,cs}$ on perimeter u_1
- $V_{Ed} \leq V_{Rd,\max}$ on perimeter u_0 of the column or loaded area
- $V_{Ed} \leq V_{Rd,c,out}$ on perimeter u_{out} at which transverse reinforcement is no more required.

The three conditions can be synthesized in the following expression

$$V_{Ed} \leq \min(V_{Rd,cs};\ V_{Rd,\max};\ V_{Rd,c,\mathrm{out}}).$$

On one side, if the area A_{sw} of transverse reinforcement increases, the punching shear strength $V_{Rd,cs}$ increases. On the other side there is a maximum value $A_{sw,\max}$ beyond which also if A_{sw} is increased, the punching shear strength does not increase anymore, because one of the two following conditions holds: $V_{Rd,\max} \leq V_{Rd,\mathrm{cs}}$ or $V_{Rd,c,\mathrm{out}} \leq V_{Rd,cs}$.

In the first case ($V_{Rd,\max} \leq V_{Rd,\mathrm{cs}}$) the punching failure is governed by concrete compressive strength on the column perimeter. In the second case ($V_{Rd,c,\mathrm{out}} \leq V_{Rd,cs}$) the punching failure is governed by the failure beyond the region equipped with transverse reinforcement.

If the arrangement of the transverse reinforcement is given (e.g. the number of perimeters is assigned), it is possible to evaluate the maximum area of one perimeter of shear reinforcement through the following condition

$$V_{Rd,cs} \leq \min(V_{Rd,\max};\ V_{Rd,c,\mathrm{out}})$$

which, considering the concrete contribution, becomes

$$V_{Rd,cs} = v_{Rd,cs} \cdot u_1 d = 0.75\, v_{Rd,c} u_1 d + 1.5\,(d/s_r)\, A_{sw} f_{ywd,ef} \sin\alpha$$

The maximum area A_{sw} of one perimeter to assure that $V_{Rd,cs} \leq V_{Rd,\max}$ can be obtained by setting $V_{Rd,cs} = V_{Rd,\max}$

$$0.75\, v_{Rd,c} u_1 d + 1.5\,(d/s_r)\, A_{sw} f_{ywd,ef} \sin\alpha = 0.2\, f_{cd} u_0\, d$$

from which

$$A_{sw} = \frac{0.2\, f_{cd} u_0\, d - 0.75\, v_{Rd,c} u_1 d}{1.5\,(d/s_r)\, f_{ywd,ef} \sin\alpha}$$

The maximum amount of transverse reinforcement A_{sw} for each perimeter to obtain $V_{Rd,cs} \leq V_{Rd,c,\mathrm{out}}$ can be achieved by setting $V_{Rd,cs} = V_{Rd,c,\mathrm{out}}$

$$0.75\, v_{Rd,c} u_1 d + 1.5\,(d/s_r)\, A_{sw} f_{ywd,ef} \sin\alpha = v_{Rd,c} u_{out} d$$

from which

$$A_{sw} = \frac{v_{Rd,c}(u_{out} d - 0.75\, u_1 d)}{1.5\,(d/s_r)\, f_{ywd,ef} \sin\alpha}$$

Then, the maximum area of one perimeter of shear reinforcement matching both conditions will be the minimum of calculated values

$$A_{sw,\max} = \min\left[\frac{0.2\ f_{cd}u_0\,d - 0.75\,v_{Rd,c}u_1 d}{1.5\,(d/s_r)\ f_{ywd,ef}\ \sin\alpha};\ \frac{v_{Rd,c}(u_{out}d - 0.75\,u_1 d)}{1.5\,(d/s_r)\ f_{ywd,ef}\ \sin\alpha}\right]$$

9.14.1 Case A: Studs on Two Perimeters

If the reinforcement is made of studs in two perimeters only at maximum spacing of EC2 (Fig. 9.44), it follows that

- the perimeter u_{out} is placed at the following distance from the column perimeter

$$0.5d + 0.75d + 1.5d = 2.75d$$

- $\sin\alpha = 1$; $d/s_r = 1/0.75 = 4/3$; $1.5\ d/s_r = 2$

and the expression of $A_{sw,\max}$ becomes

$$A_{sw,\max} = \min\left[\frac{\left(0.2\ f_{cd}u_0 - 0.75\,v_{Rd,c}u_1\right)d}{2\ f_{ywd,ef}};\ \frac{v_{Rd,c}(u_{\text{out}} - 0.75\,u_1)\,d}{2\ f_{ywd,ef}}\right]$$

Table 9.23 collects values of $A_{sw,\max}$ of each perimeter of shear reinforcement versus slab effective depth *d,* and the sum of dimensions of the rectangular column, for concrete class C25/30 and flexural reinforcement ratio of 0.02.

Grey boxes indicate when the reinforcement area satisfying the condition $V_{Rd,cs} = V_{Rd,\max}$ is lower than the area satisfying the condition $V_{Rd,cs} = V_{Rd,c,\text{out}}$. Then, if $A_{sw} > A_{sw,\max}$, the punching failure occurs at the perimeter u_0 for grey boxes and the perimeter u_{out} for white boxes. In the same table, empty boxes indicate when the failure arises at u_0 independently from the value of A_{sw}, because the resistant shear force $V_{Rd,c}$ without traversal reinforcement for a flexural reinforcement ratio of $\rho_l = 2\%$ is higher than $V_{Rd,\max}$.

For example, for a 350 mm square column ($c_1 + c_2 = 700$ mm) and a slab effective depth *d* of 250 mm, from Table 9.23 the maximum area of reinforcement $A_{sw,\max}$ on each of the two perimeters is equal to 446 mm^2 (≅ nine 8 mm studs on each perimeter). The grey box also indicates that, once this limit value of A_{sw} is exceeded, the failure occurs at u_0.

For a rectangular column with dimensions 400 × 550 mm ($c_1 + c_2 = 950$ mm) and for effective depth of the slab of 250 mm, it results that $A_{sw,\max} = 817$ mm^2 (≅ eight 12 mm studs on each of the two perimeters); the white box indicates that for $A_{sw} > 817$ mm^2 the failure occurs at u_{out}.

Table 9.23 Values of $A_{sw,max}$ (mm^2) on each perimeter for studs placed on two perimeters and maximum spacing of EC2 for rectangular columns

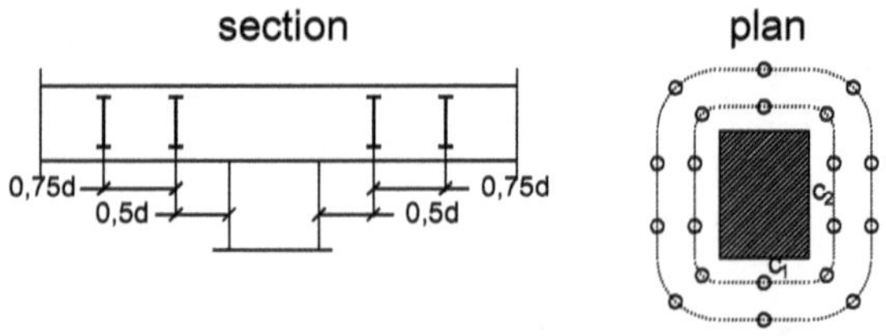

c_1+c_2 (mm)	d (mm)						
	200	250	300	350	400	450	500
500	–[a]	–	–	–	–	–	–
550	–	–	–	–	–	–	–
600	313	–	–	–	–	–	–
650	385	–	–	–	–	–	–
700	457	446	–	–	–	–	–
750	530	534	497	–	–	–	–
800	581	622	600	–	–	–	–
850	588	710	703	657	–	–	–
900	596	798	806	773	–	–	–
950	603	817	909	890	836	–	–
1000	610	825	1011	1007	965	892	–
1050	618	834	1068	1123	1095	1033	–
1100	625	842	1077	1240	1225	1175	1094
1150	632	850	1086	1338	1354	1316	1247
1200	640	859	1096	1348	1484	1458	1400
1250	647	867	1105	1358	1613	1599	1553
1300	655	876	1114	1368	1635	1741	1706
1350	662	884	1123	1378	1646	1883	1858
1400	669	892	1133	1388	1657	1936	2011
1450	677	901	1142	1398	1667	1948	2164
1500	684	909	1151	1408	1678	1959	2250

Values have been calculated for a slab made of concrete C25/30 and flexural reinforcement ratio ρ_l = 2% (maximum value allowed by EC2 in the calculation of $v_{Rd,c}$)

9.14.2 Case B: Bent-Up Bars on One Perimeter

If the shear reinforcement is made of bent-up bars inclined at 45° and arranged in one perimeter at the maximum spacing of EC2, it follows that:

- the perimeter u_{out} is placed at a distance ($0.5\ d + 1.0\ d + 1.5\ d = 3d$) from the column perimeter (where $0.5\ d$ is the distance between the bar and the column; $1.0\ d$ is the horizontal projection of the inclined segment of the bar)
- $\sin \alpha = \sqrt{2}/2$
- $d/s_r = 0.67$

- $1.5\ d/s_r = 1$

and the expression for $A_{sw,\max}$ becomes

$$A_{sw,\max} = \min\left[\frac{\sqrt{2}\,\left(0.2\ f_{cd}u_0 - 0.75\,v_{Rd,c}u_1\right)d}{f_{ywd,ef}};\ \frac{\sqrt{2}\,v_{Rd,c}}{f_{ywd,ef}}(u_{\text{out}}d - 0.75\,u_1 d)\right]$$

It is worth to precise that $A_{sw,\max}$ is given by the sum of areas of both legs of all the bent-up bars arranged on one perimeter around the column.

Table 9.24 lists $A_{sw,\max}$ values versus slab effective depth and column dimensions.

Grey boxes again refer to values when the reinforcement area which satisfies the condition $V_{Rd,cs} = V_{Rd,\max}$ is lower than the area satisfying the condition $V_{Rd,cs} = V_{Rd,c,\text{out}}$. Then, when $A_{sw} > A_{sw,\max}$ the punching failure occurs at u_0 for grey boxes and u_{out} for white boxes.

In the same table, empty boxes refer to situations where the failure occurs at u_0 independently from the value of A_{sw}, because $V_{Rd,c}$ for a flexural reinforcement ratio $\rho_l = 2\%$ is higher than $V_{Rd,\max}$.

For example, in the case of a 450 mm square column ($c_1 + c_2 = 900$ mm) and slab effective depth d equal to 350 mm, from Table 9.24 the maximum reinforcement area $A_{sw,\max}$ of one perimeter is equal to 2187 mm^2 ($\cong$ six 16 mm bars, which with their two legs give an area of $6 \times 2 \times 201 = 2412$ mm^2); the grey box indicates that once this limit value is exceeded, the failure occurs at u_0.

In case of a 450 × 500 mm rectangular column ($c_1 + c_2 = 950$ mm) and slab effective depth equal to 250 mm, the maximum reinforcement area $A_{sw,\max}$ of one perimeter is equal to 2311 mm^2 ($\cong$ eight 14 mm bars, which with their two legs give an area of $8 \times 2 \times 154 = 2464$ mm^2) and the white box indicates that for $A_{sw} > 2311$ mm^2, the failure occurs at u_{out}.

9.14.2.1 Example No. 9—Punching Shear Reinforcement (Bent-Up Bars)

Slab made of concrete C28/35 *with effective depth* d = 250 mm *supported by a* 300 × 500 mm *rectangular column. The flexural reinforcement of the slab is given by* 16 *mm bars at* 200 *mm spacing* (1ϕ16/200) *in both directions and the reinforcement is at* $d' = 40$ mm *from the edge.*

Design the bent-up bars for punching shear reinforcement for: (a) $V_{Ed} = 800$ kN, (b) $V_{Ed} = 920$ kN.

Case (a) $\boldsymbol{V_{Ed} = 800}$ **kN**

Verification on the column perimeter

Maximum punching shear strength

$$v_{Rd,\max} = 0.2 f_{cd} = 0.2 \cdot 0.85 \cdot 28/1.5 = 3.17\ \text{N/mm}^2$$

Table 9.24 $A_{sw,\max}$ (mm^2) for bent-up bars with a slope of 45° arranged on only one perimeter at the maximum distances of EC2 for rectangular columns

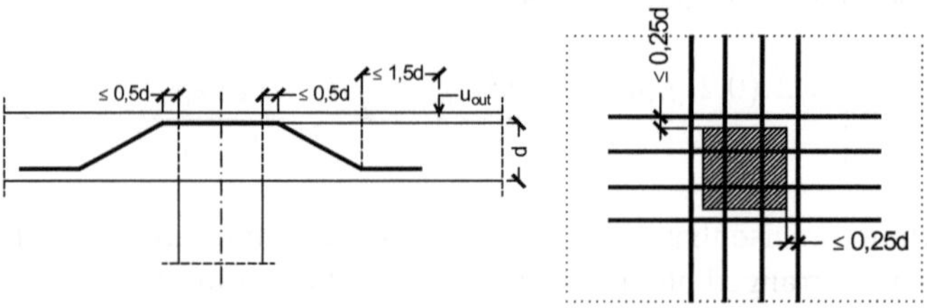

c_1+c_2 (mm)	d (mm)						
	200	250	300	350	400	450	500
500	–[a]	–	–	–	–	–	–
550	–	–	–	–	–	–	–
600	884	–	–	–	–	–	–
650	1089	–	–	–	–	–	–
700	1293	1260	–	–	–	–	–
750	1498	1510	1405	–	–	–	–
800	1643	1759	1696	–	–	–	–
850	1664	2009	1987	1857	–	–	–
900	1685	2258	2278	2187	–	–	–
950	1705	2311	2570	2517	2364	–	–
1000	1726	2334	2861	2848	2731	2522	–
1050	1747	2358	3020	3178	3097	2922	–
1100	1768	2382	3047	3508	3464	3323	3096
1150	1789	2405	3073	3784	3830	3723	3528
1200	1810	2429	3099	3813	4197	4124	3960
1250	1830	2453	3125	3841	4563	4524	4392
1300	1851	2476	3151	3870	4625	4924	4824
1350	1872	2500	3178	3898	4655	5325	5257
1400	1893	2524	3204	3926	4686	5477	5689
1450	1914	2547	3230	3955	4716	5509	6121
1500	1935	2571	3256	3983	4747	5542	6365

Values have been calculated for concrete C25/30 and flexural reinforcement ratio $\rho_l = 2\%$ (maximum value allowed by EC2 in the calculation of $v_{Rd,c}$)

Punching shear strength on the column perimeter

$$v_{Ed,0} = V_{Ed}/u_0 d = 800{,}000/[2 \cdot (300 + 500) \cdot 250] = 2\ \text{N/mm}^2 < v_{Rd,\max}$$

Being $v_{Ed,0} < v_{Rd,\max}$, the verification is satisfied.

Verification on the control perimeter u_1

Control perimeter

$$u_1 = 2(c_1 + c_2) + 2\pi(2d) = 2(300 + 500) + 2\pi 500 = 4742\,\text{mm}$$

Punching shear strength on the control perimeter

$$v_{Ed,1} = v_{Ed}/u_1 d = 800{,}000/(4742 \cdot 250) = 0.67\,\text{N/mm}^2$$

Punching shear strength without transverse reinforcement

For $d = 250$ mm and 1ϕ16/200 in both directions, Table 9.29 of the Appendix to this chapter gives the following value of $v_{Rd,c}$

$$v_{Rd,c} = 0.510\,\text{N/mm}^2$$

Since $v_{Ed,1} > v_{Rd,c}$, the verification is not satisfied and bent-up bars are adopted as punching shear reinforcement.

Perimeter u_{out} and its distance k'_d from the column perimeter[3]

$$u_{out} = \beta V_{Ed}/(v_{Rd,c} d) = 1.0 \cdot 800{,}000/(0.510 \cdot 250) = 6275\,\text{mm}$$
$$k'_d = [u_{out} - 2(c_1 + c_2)]/(2\pi) = [6275 - 2(300 + 500)]/(2\pi) = 744\,\text{mm}$$
$$k' = 744/250 = 2.98$$

Number of reinforcement perimeters

The outermost perimeter of bent-up bars shall have a distance from u_{out} not greater than $1.5d$, therefore bent-up bars shall cover a segment BC (Fig. 9.45) of length $(k' - 1.5)\,d = 1.48d = 370$ mm.

One perimeter of bent-up bars placed at distance $0.5d$ from the column perimeter covers a distance BC with length $0.5d + (d - d') = 335$ mm $< 1.48d$ if bars are bent at 45° and length $0.5d + (d - d')/\tan 30° \cong 489$ mm $> 1.48d$ if bars are bent at 30° (minimum slope allowed by EC2 for bent-up bars on one perimeter).

Therefore, one perimeter of 30° bent-up bars is enough as punching shear reinforcement.

Design of 30° bent-up bars arranged on one perimeter

Punching shear reinforcement shall absorb the difference of shear stress $v_{Ed,1} - 0.75\, v_{Rd,c}$

[3] Bent-up bars are arranged in two orthogonal directions, then the perimeter to consider in the calculation is $u_{out,ef}$ (Fig. 9.43b); nevertheless, for the sake of simplicity, the value of u_{out} has been used, because for one or two perimeters of bent-up bars it results that $u_{out,ef} \cong u_{out}$, although for two perimeters the approximation is less good.

$$\Delta v_{Ed} = v_{Ed,1} - 0.75\, v_{Rd,c} = 0.67 - 0.75 \cdot 0.510 = 0.29\,\text{N/mm}^2$$

i.e., the resistant shear force offered by the reinforcement shall be equal at least to

$$V''_{Rd,cs} = \Delta v_{Ed} \cdot u_1 \cdot d = 0.29 \cdot 4742 \cdot 250 = 343{,}795\,\text{N} \cong 344\,\text{kN}$$

then 16 mm bent bars are adopted.

For $d = 250$ mm and one 16 mm bar with two legs at 45°, Table 9.17 gives

$$V''_{Rd,cs(1)} = [1.5(d/S_r)A_{sw}\,f_{ywd,ef}\,\sin\alpha] = 89.3\,\text{kN}$$

while, for a slope of 30°,

$$V''_{Rd,cs(1)} = 89.3 \cdot (\sin 30°/\sin 45°) = 63.1\,\text{kN}$$

then the required number of perimeters is

$$n = V''_{Rd,cs}/V''_{Rd,cs(1)} = 344/63.1 = 5.5 \cong 6$$

as shown in Fig. 9.46, three bars are arranged in the direction of c_1 and three bars are arranged in the direction of c_2.

Case (b) $\boldsymbol{V_{Ed}}$ **= 920 kN**

Verification on the column perimeter

$v_{Ed,0} = V_{Ed}/u_0\, d = 920{,}000/[2 \cdot (300 + 500) \cdot 250] = 2.3\ \text{N/mm}^2 < v_{Rd,\max}$.

being $v_{Ed,0} < v_{Rd,\max}$, the verification is satisfied.

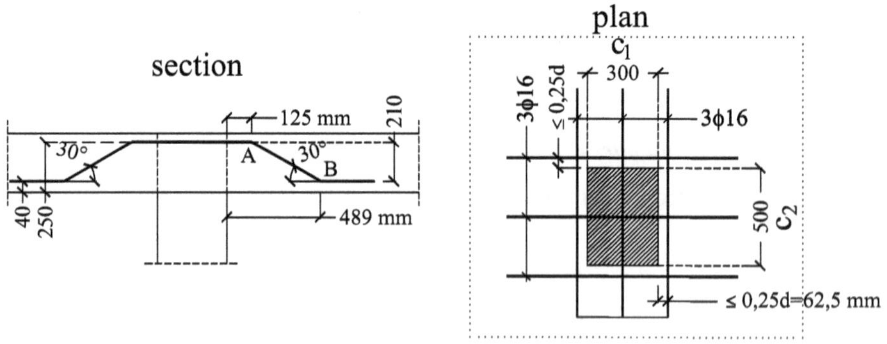

Fig. 9.46 Punching reinforcement of Example 9 formed by one row of bent-up bars with a slope of 30°

Verification on the control perimeter u_1

Control perimeter

$$u_1 = 4742\,\text{mm}$$

Punching shear strength on the control perimeter

$$v_{Ed,1} = V_{Ed}/u1d = 920{,}000/(4742 \cdot 250) = 0.78\,\text{N/mm}^2$$

it results that

$$v_{Ed,1} > v_{Rd,c}$$

then punching shear reinforcement is required.

Perimeter u_{out} and its distance k'd from the column perimeter [4]

$$u_{out} = \beta V_{Ed}/(v_{Rd,c}d) = 1.0 \cdot 920{,}000/0.510 \cdot 250 = 7216\,\text{mm}$$
$$k'_d = [u_{out} - 2(c_1 + c_2)]/(2\pi) = [7216 - 2(300 + 500)]/(2\pi) = 894\,\text{mm}$$
$$k' = 894/250 = 3.58$$

Number of perimeters

Bent-up bars shall cover a segment BC with length $(k' - 1.5)\,d = 2.08d \cong 520$ mm, therefore one perimeter of 30° bent-up bars is not enough.

Two perimeters of 45° bent-up bars are adopted: the first perimeter at distance $0.5d$ from the column and the second perimeter at distance $1.25d$ from the column to cover a segment BC with length of $(0.5d + 0.75d + d - d') = 2.25d - d' =$ 522.5 mm > 520 mm.

Design of 45° bent-up bars arranged on two perimeters

Punching shear reinforcement shall absorb the difference of shear stress $v_{Ed,1} - 0.75\, v_{Rd,c}$

$$\Delta v_{Ed} = v_{Ed,1} - 0.75\, v_{Rd,c} = 0.78 - 0.75 \cdot 0.510 = 0.40\,\text{N/mm}^2$$

i.e. the resistant shear force offered by reinforcement shall at least be equal to

$$V''_{Rd,cs} = \Delta v_{Ed} \cdot u_1 \cdot d = 0.40 \cdot 4742 \cdot 250 = 474{,}200\,\text{N} \cong 474\,\text{kN}$$

[4] Bent-up bars are arranged in two orthogonal directions, then the perimeter to consider in the calculation is $u_{out,ef}$ (Fig. 9.43b); nevertheless, for the sake of simplicity, the value of u_{out} has been used, because for one or two perimeters of bent-up bars it results that $u_{out,ef} \cong u_{out}$, although for two perimeters the approximation is less good.

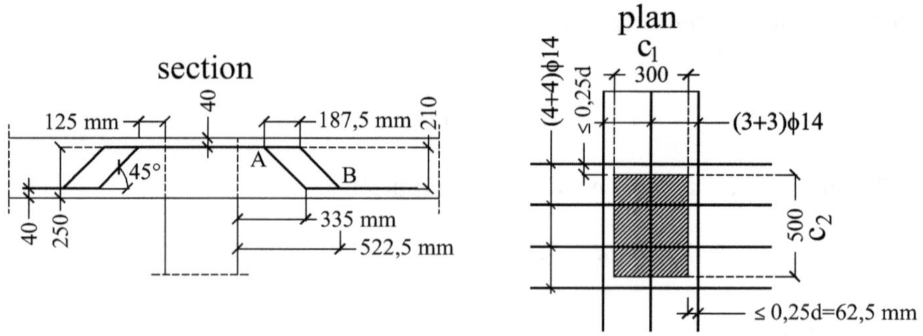

Fig. 9.47 Punching reinforcement of Example 9 formed by 45° bent-up bars on two lines

45° bent-up bars with diameter of 14 mm are adopted; for $d = 250$ mm and one 14 mm bar with two legs at 45°, Table 9.17 gives

$$V''_{Rd,cs(1)} = [1.5(d/S_r)A_{sw}f_{ywd,ef}\sin\alpha] = 68.4\,\text{kN}$$

then the required number of bars is equal to

$$n = V''_{Rd,cs}/V''_{Rd,cs(1)} = 474/68.4 = 6.9 \cong 7$$

as shown in Fig. 9.47, each perimeter has four bars in the direction of c_1 and three bars in the direction of c_2.

9.14.2.2 Example No. 10—Punching Shear Reinforcement (Studs)

For the slab of Example no. 9, design studs to support a design punching load of 1200 kN.

Perimeter u_{out} at which punching shear reinforcement is no more required

$$u_{\text{out}} = \frac{\beta V_{Ed}}{v_{Rd,c}\,d} = \frac{1.0 \cdot 1{,}200{,}000}{0.510 \cdot 250} = 9412\,\text{mm}$$

(where $v_{Rd,c} = 0.510$ N/mm^2 is obtained from Table 9.29 of Appendix 1 to this chapter for $d = 250$ mm and flexural reinforcement made of 16 mm bars at 200 mm spacing (1ϕ16/200) in both directions).

Distance k'_d of the perimeter u_{out} from the column perimeter

$$k'_d = [u_{\text{out}} - 2(c_1 + c_2)]/(2\pi) = [9412 - 2(300 + 500)]/(2\pi) = 1243\,\text{mm}$$

$k' = 1243/250 = 4972$

Maximum radial spacing $s_{r,\max}$

$$S_{r,\max} = 0.75d = 0.75 \cdot 250 = 187.5\,\text{mm}$$

Distance a_1 ***of the first perimeter of studs from the column perimeter***

$$0.30d = 75\,\text{mm} \leq a_1 \leq 0.50d = 125\,\text{mm}$$

Distance b_u ***of the last perimeter of studs from the perimeter*** u_{out}

$$b_u \leq 1.5d = 1.5 \cdot 250 = 375\,\text{mm}$$

Distance a_u ***of the last perimeter of studs from the column perimeter***

$$a_u = k'_d - b_u = 1243 - 375 = 868\,\text{mm}(\cong 3.47d)$$

Minimum distance from first and last perimeter of studs

$$a_u - a_1 = 868 - 125 = 743\,\text{mm}(\cong 2.97\,\text{d})$$

Number and spacing of stud perimeters

The total number of spacings between reinforcement perimeters is given by

$$n = (a_u - a_1)/S_{r,\max} = 743/187.5 = 3.96 \cong 4$$

so, the number of perimeters is

$$n + 1 = 5$$

therefore, between the fist and the last reinforcement perimeter other three perimeters are required.

The radial spacing is given by

$$S_r = (a_u - a_1)/n = 743/4 \cong 186\,\text{mm}$$

Table 9.25 collects the distances from the column perimeter of all perimeters of shear reinforcement.

Table 9.25 Distances of bar perimeters from the periphery of the column

Perimeter	1	2	3	4	5
Distance (mm)	125 (0.5 *d*)	311 (1.24 *d*)	497 (1.99 *d*)	683 (2.73 *d*)	868 (3.47 *d*)

Tangential spacing s_t between studs of each perimeter of shear reinforcement

The first three reinforcement perimeters are placed within the control perimeter u_1 (at distance 2*d* from the column perimeter, see Table 9.25), while the fourth and the fifth perimeter have a distance greater than 2*d*; according to [9.4.3(1)] s_t shall satisfy following rules

$$S_t \leq 1.5d = 375\,\text{mm for the first three perimeters}$$
$$S_t \leq 2.0d = 500\,\text{mm for the other two perimeters.}$$

Choice of the stud diameter

The area of each stud shall satisfy [(9.11)]

$$A_{sw,\min} \cdot \frac{(1,5 \cdot \sin\alpha + \cos\alpha)}{s_r \cdot s_t} \geq \frac{0.08 \cdot \sqrt{f_{ck}}}{f_{yk}} (f_{ck} \text{ e } f_{yk} \text{ in N/mm}^2)$$

where

sen $\alpha = 1$, cos $\alpha = 0$
$s_r = 186$ mm
$s_t = 375$ mm for the first, the second and the third perimeter
$s_t = 500$ mm for the fourth and the fifth perimeter
therefore, the area of each stud should satisfy:

$$A_{sw,\min} \geq \frac{0.08 \cdot \sqrt{f_{ck}}}{f_{yk}} \cdot \frac{s_r \cdot s_t}{1.5 \cdot \sin\alpha + \cos\alpha}$$
$$= \frac{0.08 \cdot \sqrt{28}}{450} \cdot \frac{186 \cdot 375}{1.5 \cdot 1 + 0} = 43.7\,\text{mm}^2$$

within perimeters no 1, 2 and 3, and

$$A_{sw,\min} \geq \frac{0.08 \cdot \sqrt{f_{ck}}}{f_{yk}} \cdot \frac{s_r \cdot s_t}{1,5 \cdot \sin\alpha + \cos\alpha}$$
$$= \frac{0.08 \cdot \sqrt{28}}{450} \cdot \frac{186 \cdot 500}{1.5 \cdot 1 + 0} = 58.3\,\text{mm}^2$$

within perimeters no 4 and 5.

A diameter not smaller than 8 mm ($A_s = 50$ mm$^2 >$ 43.7 mm^2) is required for the first three perimeters and 10 mm ($A_s = 78.5$ mm$^2 >$ 58.3 mm^2) for the further two perimeters.

12 mm studs can be used for all perimeters.

Area of each perimeter of shear reinforcement

From Example no. 9 the punching-shear capacity provided by the reinforcement shall be at least equal to

$$V''_{Rd,cs} = 747\,\text{kN}$$

For $d = 250$ mm and one 12 mm stud, Table 9.18 gives

$$V''_{Rd,cs(1)} = 1.5(d/S_r)A_{sw}f_{ywd,ef} = 70.69\,\text{kN}$$

The number of studs for each perimeter is then given by

$$n = (747/70.69) = 10.6 \cong 11$$

If the five perimeters are drawn at the calculated distances, studs must be arranged as shown in Fig. 9.48, with 18 studs for each perimeter (seven more than those calculated), otherwise the maximum tangential spacing is not satisfied. Then, the bar diameter can be reduced from 12 to 10 mm; for $\phi = 10$ mm, $V''_{Rd,cs(1)} = 49.09$ kN (Table 9.18) and the minimum number of studs for each perimeter is now given by

$$n = (747/49.09) = 15.2 \cong 16 < 18$$

Alternatively to the configuration of Fig. 9.48, studs could be arranged as shown in Fig. 9.43b, where the perimeter B ($u_{\text{out,ef}}$) has the same length but it is more distant from the column than the perimeter A (u_{out}). In this way, the region with transverse reinforcement would extend for a longer segment from the column perimeter.

9.15 Foundations

In [6.4.4(2)] EC2 gives the expression of the punching shear resistance of column bases. For slender slabs, the control perimeter is placed at the distance $2d$. For column bases, the control perimeter is not known a priori, and it is placed at the distance a from the column perimeter not greater than twice the effective depth of the foundation; its position shall be found by trial and error (Fig. 9.2b).

The different position and length of the control perimeter for foundations is due first to their lower slenderness than slabs.

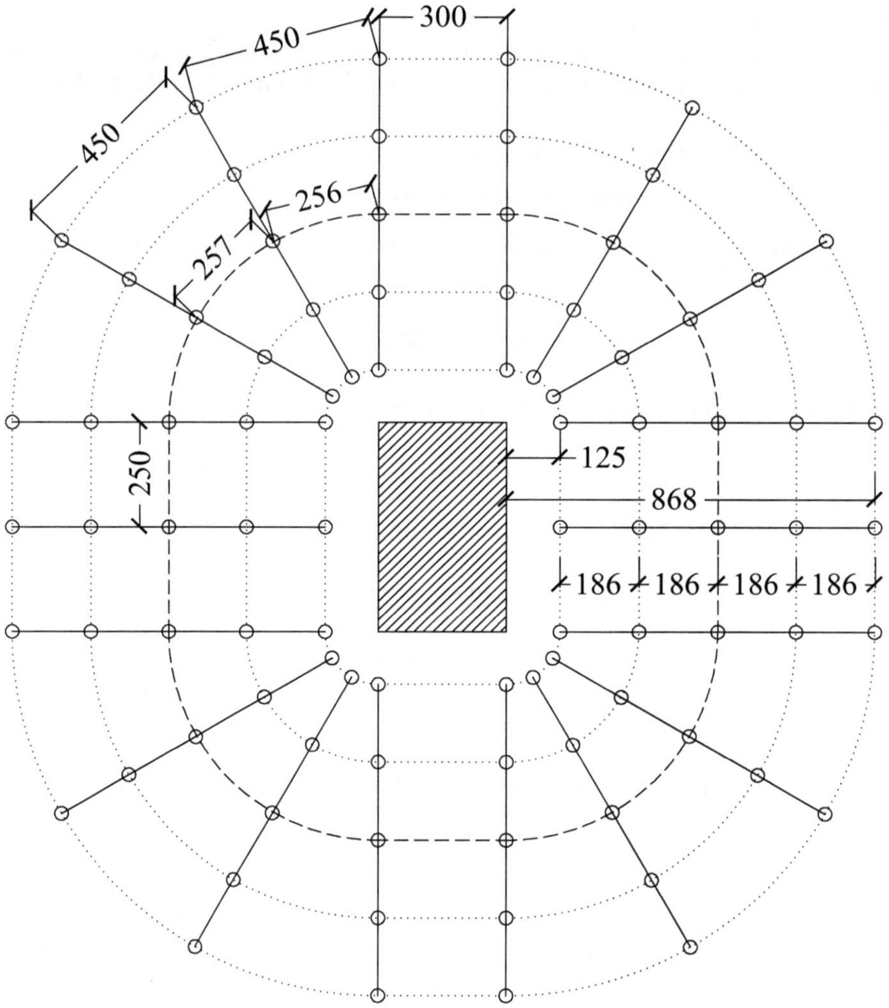

Fig. 9.48 Punching-shear reinforcement of Example 10 formed by studs (dashed line = control perimeter u_1; within u_1 the spacing of reinforcement perimeters is not higher than $1.5d = 375$ mm, while the spacing of perimeters beyond u_1 is not higher than $2d = 500$ mm)

Mechanical models to calculate punching shear resistance are based on results of experimental tests on relatively slender slabs with ratio l_V/d between length and effective depth greater than $3 \div 4$.

Foundations are usually characterized by a lower ratio l_V/d and the soil bearing pressure modifies the slope of the punching surface. Tests have shown that in foundations with reduced values of l_V/d ratio, the slope of the punching surface is steeper than slabs.

EC2 adopts for foundations the solution proposed in Model Code 1990. The distance a of the control perimeter from the column perimeter, never greater than $2d$, is investigated by trial and error with the following procedure. As a varies in the range $0 < a \leq 2d$, the slope of the punching surface changes, and the design shear stress and punching shear resistance change, too. Several trials are performed until the control perimeter u is identified as the perimeter for which the ratio between the design punching shear stress and the punching shear strength is maximum.

For centred load, the shear force is

$$V_{Ed,red} = V_{Ed} - \Delta V_{Ed} \qquad [(6.48)]$$

where

V_{Ed} is the design shear force (equal to the compression load transmitted by the column);

ΔV_{Ed} is the net upward force within the considered control perimeter; this force is equal to the resultant force of the soil bearing pressure minus the foundation self-weight.

The design punching shear stress is given by

$$v_{Ed} = \frac{V_{Ed,red}}{ud} \qquad [(6.49)]$$

while the punching shear resistance is expressed as

$$v_{Rd} = C_{Rd,c}\, k\, (100\, \rho\, f_{ck})^{1/3} \frac{2d}{a} = v_{Rd,c} \frac{2d}{a} \geq v_{\min} \frac{2d}{a} \qquad [(6.50)]$$

where

a is the distance from the column perimeter to the considered control perimeter

$C_{Rd,c}$, k, $v_{\min}$ are defined in 9.10.1.

For foundations with variable thickness, the effective depth is assumed as the depth at the column or loaded area perimeter [6.2.4 (6)] (Fig. 9.49).

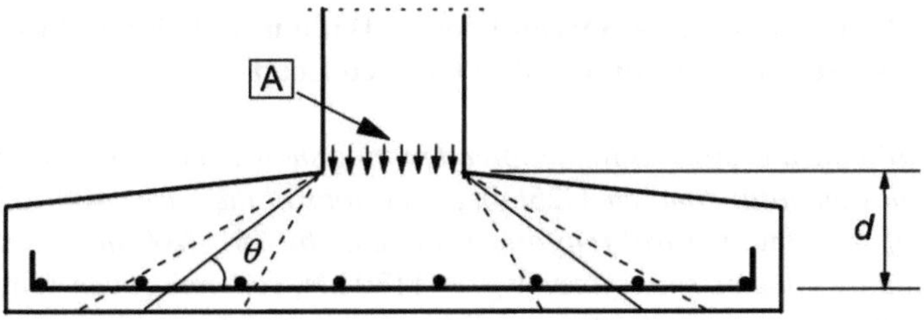

Fig. 9.49 Effective depth of a plinth with variable thickness

For eccentric load, the design punching shear resistance is obtained like slabs

$v_{Ed} = \frac{V_{Ed,red}}{u \cdot d}\left[1 + k\frac{M_{Ed}\,u}{V_{Ed,red}\,W}\right]$ [(6.51)]

where W is similar to W_1 but for perimeter u and k coefficient has already been defined in 9.7.1.

Where shear reinforcement is required, it should be calculated with the same expression of slabs

$$v_{Rd,cs} = 0.75\,v_{Rd} + 1.5\,(d/s_r)\,A_{sw}\,f_{ywd,ef}\frac{1}{u\,d}\sin\alpha$$

where $v_{Rd,c}$ is replaced by $v_{Rd} = v_{Rd,c}\,2d/a$, and u_1 is replaced by the control perimeter u placed at a distance $a \le 2d$ from the column perimeter.

As EC2 does not indicate the control perimeter to consider for the design of punching shear reinforcement of foundations, the following procedure can be adopted. The following equality is imposed: $v_{Rd,cs} = v_{Ed}$, then the formula is inverted to obtain the following lower limit for the area A_{sw} of one perimeter of shear reinforcement

$$A_{sw} \ge \frac{(v_{Ed} - 0.75\,v_{Rd})\,u\,d}{1.5\,(d/s_r)\,f_{ywd,ef}\,\sin\alpha}$$

which shall be satisfied for any perimeter $u \le u_1$.

Finally, after some iterations, the perimeter u maximizing A_{sw} is identified. To this aim, in the previous expression the distance a is varied between 0 and $2d$, considering that for each value of a, a different value of both the reduced shear resistance $V_{Ed,red}$ and the punching shear resistance $v_{Rd} = v_{Rd,c}\,2d/a$ is obtained.

The perimeter u maximizing A_{sw} is generally different from the control perimeter maximizing the ratio v_{Ed}/v_{Rd} which is used in the punching verification without transverse reinforcement.

9.15.1 Examples

9.15.1.1 Example No. 11—Spread Footing Without Punching Shear Reinforcement Subjected to Centred Load

Verify shear punching stresses in the spread footing shown in Fig. 9.50, *assuming the following design data: concrete* C25/30, *dimensions of the base* 2400 × 2400 mm, *thickness of* 400 mm, *flexural reinforcement made by* 1ϕ16/200 *in both directions,* 300 mm *square column, shear force* $V_{Ed} = 1170$ kN, *concrete cover* 40 mm.

The punching shear verification shall be made considering the control section at distance $a_{crit} \le 2d$, which a priori is unknown; therefore some tentative control

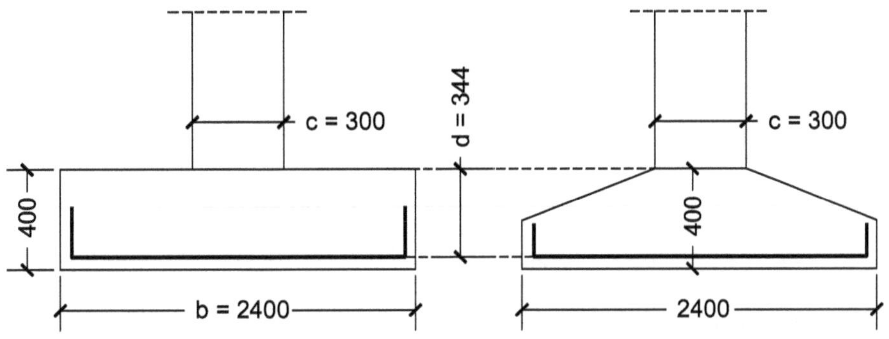

Fig. 9.50 Geometry of the spread footing (Example no. 11 is valid for both the rectangular and the symmetrical trapezoidal geometry)

perimeters are considered until the control perimeter maximizing the ratio $v_{Ed}/v_{Rd,c}$ is identified.

Effective depth, reinforcement ratio, soil bearing pressure

Effective depth in y direction: $d_y = 400 - 40 - 8 = 352$ mm.
Effective depth in z direction: $d_z = 400 - 40 - 16 - 8 = 336$ mm.
Slab effective depth: $d = (352 + 336)/2 = 344$ mm.
Reinforcement ratio

$$1\,\phi\,16/200\,\text{mm} \Rightarrow \text{A}_{1,\text{ly}} = \text{A}_{1,\text{lz}} = 200\,\text{mm}^2\ \text{i}_\text{y} = \text{i}_\text{z} = 200\,\text{mm}$$

$$\rho_l = \rho_{ly} = \rho_{lz} = \frac{A_{1,ly}}{i_y d} = \frac{200}{200 \cdot 344} = 0.291 \cdot 10^{-2} \leq 0.02$$

The soil bearing pressure without considering the foundation self-weight is

$$p_{\text{net}} = \frac{V_{Ed}}{2400 \cdot 2400} = \frac{1{,}170{,}000}{2400 \cdot 2400} \cong 0.203\,\text{N/mm}^2$$

while the total soil pressure, including the foundation self-weight ($2.5 \cdot 10^{-5}$ N/mm^3), is

$$p_{\text{tot}} = p_{\text{net}} + 400 \cdot 2.5 \cdot 10^{-5} = 0.213\,\text{N/mm}^2$$

Punching shear stress at the column perimeter (u_0)

Column perimeter (c is the value common to two sides of the column cross-section, $c = c_1 = c_2$)

$$u_0 = 2(c_1 + c_2) = 2(300 + 300) = 1200\,\text{mm}$$

Base area within the perimeter u_0[5]

$$A_0 = c^2 = 90000\ \text{mm}^2$$

Resultant force of the soil bearing pressure within the control perimeter u_0

$$\Delta V_{Ed,0} = p_{\text{netta}} A_0 = 0.203 \cdot 90{,}000 = 18{,}270\,\text{N}$$
$$V_{Ed,red,0} = V_{Ed} - \Delta V_{Ed,0} = 1{,}170{,}000 - 18{,}270 = 1{,}151{,}730\,\text{N}$$

Design punching shear stress [(6.49)]

$$v_{Ed,0} = \frac{\beta V_{Ed,red,0}}{u_0 d} = \frac{1{,}151{,}730}{1200 \cdot 344} = 2.79\,\text{N/mm}^2 (\beta = 1 \text{ for axial load})$$

Maximum value of the punching shear resistance

$$v_{Rd,\max} = 0.2\, f_{cd} = 0.2 \cdot \frac{0.85 \cdot 25}{1.5} \cong 2.83\,\text{N/mm}^2$$

as $v_{Ed,0} \leq v_{Rd,\max}$, the verification at the column perimeter is satisfied.

The verification is now performed at the control perimeter at distance $a \leq 2d$ from the column perimeter.

First attempt control perimeter (a = 2d)

First attempt control perimeter is assumed at distance $2d$

$$a = 2d$$
$$u_1 = 4\,c + 2\,\pi\,(2\,d) = 4 \cdot 300 + 2 \cdot \pi \cdot (2 \cdot 344) = 5523\,\text{mm}$$

Base area within the control perimeter

$$A_1 = c^2 + 4\,c\,(2\,d) + \pi\,(2\,d)^2 = 300^2 + 4 \cdot 300 \cdot (2 \cdot 344) + \pi \cdot (2 \cdot 344)^2 = 2{,}402{,}654\,\text{mm}^2$$

Resultant force of the soil bearing pressure within the control perimeter

$$\Delta V_{Ed,1} = p_{\text{netta}} A_1 = 0.203 \cdot 2{,}402{,}654 = 487{,}739\,\text{N}$$
$$V_{Ed,red,1} = V_{Ed} - \Delta V_{Ed,1} = 1{,}170{,}000 - 487{,}739 = 682{,}261\,\text{N}$$

Design punching shear stress

---

[5] Forces and stresses evaluated at the perimeter u_0 are labelled with subscript "0", those at the perimeter u_1 with subscript "1" and so on.

$$v_{Ed,1} = \frac{V_{Ed,red,1}}{u_1 d} = \frac{682{,}261}{5523 \cdot 344} = 0.36\,\text{N/mm}^2$$

Punching shear resistance without reinforcement

$$\begin{aligned} v_{Rd,1} &= C_{Rd,c}\, k\, (100\, \rho\, f_{ck})^{1/3} \frac{2\,d}{a} \\ &= \frac{0.18}{1.5}\left(1 + \sqrt{\frac{200}{344}}\right)\left(100 \cdot 0.291 \times 10^{-2} \cdot 25\right)^{1/3} \cdot 1 = 0.41\,\text{N/mm}^2 \\ &\geq v_{\min} = 0.035 \cdot k^{3/2} \cdot f_{ck}^{1/2} \cong 0.41\,\text{N/mm}^2 \end{aligned}$$

as $v_{Ed,1} = 0.36\,\text{N/mm}^2 \leq v_{Rd,1} = 0.41\,\text{N/mm}^2$ and $v_{Ed,1}/v_{Rd,1} = 0.88$, at perimeter u_1, the verification is satisfied.

Second attempt control perimeter (a = 1.5d)

Second attempt control perimeter is assumed at distance 1.5 d

$$\begin{aligned} a &= 1.5d \\ u_2 &= 4\,c + 2\,\pi\,(1.5\,d) \\ &= 4 \cdot 300 + 2 \cdot \pi \cdot (1.5 \cdot 344) = 4442\,\text{mm} \end{aligned}$$

Base area within the control perimeter is

$$\begin{aligned} A_2 &= c^2 + 4\,c\,(1.5\,d) + \pi\,(1.5\,d)^2 \\ &= 300^2 + 4 \cdot 300 \cdot (1.5 \cdot 344) + \pi\,(1.5 \cdot 344)^2 \\ &= 1{,}545{,}668\,\text{mm}^2 \end{aligned}$$

Resultant force of the soil bearing pressure within the control perimeter

$$\begin{aligned} \Delta V_{Ed,2} &= p_{netta} A_2 = 0.203 \cdot 1{,}545{,}668 = 313{,}771\,\text{N} \\ V_{Ed,red,2} &= V_{Ed} - \Delta V_{Ed,2} = 1{,}170{,}000 - 313{,}771 = 856{,}229\,\text{N} \end{aligned}$$

Design punching shear stress

$$v_{Ed,2} = \frac{V_{Ed,red,2}}{u_2 d} = \frac{856{,}229}{4442 \cdot 344} = 0.56\,\text{N/mm}^2$$

Punching shear resistance without reinforcement

$$v_{Rd,2} = C_{Rd,c}\, k\, (100\, \rho\, f_{ck})^{1/3} \frac{2\,d}{a}$$

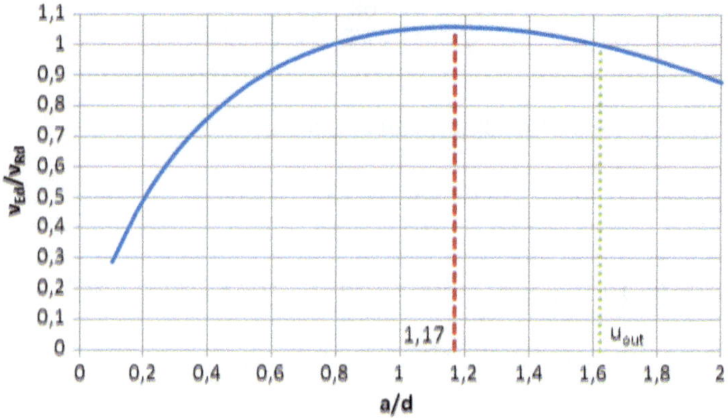

Fig. 9.51 Ratio v_{Ed}/v_{Rd} versuss *a/d* for the 2400 × 2400 × 400 mm plinth of Example no. 11

$$= \frac{0.18}{1.5}\left(1+\sqrt{\frac{200}{344}}\right)\left(100\cdot 0.291\times 10^{-2}\cdot 25\right)^{1/3}\frac{2}{1.5}$$

$$= 0.55\,\text{N/mm}^2 \geq v_{\min}\cdot\frac{2}{1,5}$$

$$= 0.55\,\text{N/mm}^2$$

as $v_{Ed,2} = 0.56$ N/mm^2 > $v_{Rd,2} = 0.55$ N/mm^2, at perimeter u_2 the verification is not satisfied, and it is necessary to adopt punching shear reinforcement.

In Example no. 14 the design of punching shear reinforcement is discussed.

If the calculation is repeated for other values of *a*, the curve v_{Ed}/v_{Rd} can be built at varying the ratio *a/d* (Fig. 9.51).

The ratio v_{Ed}/v_{Rd} takes its maximum value for $a = 1.17\ d$, which represents the distance of the control perimeter from the column perimeter. The control perimeter u_{out} at which shear reinforcement is not required is placed at 1.62 *d* from the column perimeter.

If dimensions of the base foundation are increased and the verification is repeated, the control perimeter is placed at a greater distance. In general, the distance of the control perimeter from the column perimeter is varying with ratios *c*/*d* and *b*/*d*, where *c* is the dimension of the column cross-section and *b* is the dimension of the foundation base once values of f_{ck} and ρ_l are assigned.

Tables 9.26 and 9.27 collect values of ratio a_{crit}/d with varying *c*/*d* and *b*/*d* for a square column and a circular column (c = diameter); in both cases, the foundation has a square base. These values depend neither on the concrete class, neither on the flexural reinforcement ratio.

Table 9.26 Ratios a_{crit}/d with varying c/d and b/d for a square column on a square foundation

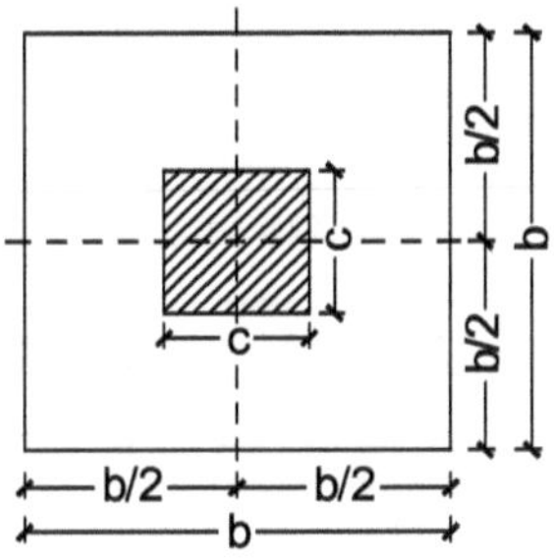

c/d	b/d										
	5	5.5	6	6.5	7	7.5	8.0	8.5	9.0	9.5	10.0
0.5	0.82	0.89	0.96	1.03	1.09	1.16	1.22	1.28	1.34	1.40	1.45
0.6	0.84	0.91	0.98	1.06	1.12	1.19	1.26	1.32	1.39	1.45	1.51
0.7	0.85	0.92	1.00	1.08	1.15	1.22	1.29	1.36	1.42	1.49	1.55
0.8	0.85	0.93	1.01	1.09	1.17	1.24	1.31	1.38	1.45	1.52	1.59
0.9	0.85	0.94	1.02	1.10	1.18	1.25	1.33	1.40	1.48	1.55	1.62
1.0	0.85	0.93	1.02	1.10	1.19	1.27	1.34	1.42	1.49	1.57	1.64
1.1	0.84	0.93	1.02	1.11	1.19	1.27	1.35	1.43	1.51	1.58	1.66
1.2	0.83	0.92	1.02	1.10	1.19	1.27	1.36	1.44	1.52	1.60	1.67
1.3	0.82	0.92	1.01	1.10	1.19	1.28	1.36	1.44	1.52	1.61	1.68
1.4	0.81	0.91	1.00	1.09	1.18	1.27	1.36	1.45	1.53	1.61	1.69
1.5	0.79	0.89	0.99	1.09	1.18	1.27	1.36	1.45	1.53	1.61	1.70

9.15.1.2 Example No. 12—Evaluation of Control Perimeter for a Foundation

Using Table 9.26 *evaluate the distance of the control perimeter for the foundation of Example no. 11.*

Geometrical data: $d = 344$ mm, $c = 300$ mm, $b = 2400$ mm

$$c/d = 300/344 = 0.872 \cong 0.9$$
$$b/d = 2400/344 = 6.97 \cong 7$$

For $c/d = 0.9$ and $b/d = 7$, Table 9.26 gives

$$a_{crit}/d = 1.18.$$

Table 9.27 Ratios a_{crit}/d with varying c/d and b/d for a circular column on a square foundation

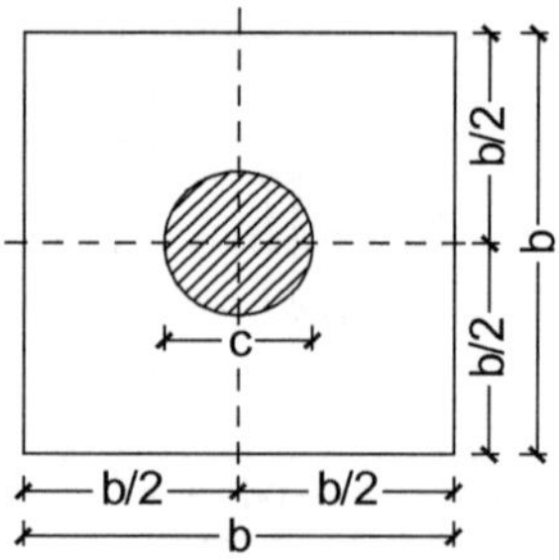

c/d	b/d										
	5	5.5	6	6.5	7	7.5	8.0	8.5	9.0	9.5	10.0
0.5	0.79	0.86	0.92	0.98	1.04	1.10	1.16	1.21	1.27	1.32	1.38
0.6	0.81	0.88	0.95	1.02	1.08	1.14	1.20	1.26	1.32	1.38	1.44
0.7	0.83	0.90	0.97	1.04	1.11	1.17	1.24	1.30	1.36	1.42	1.48
0.8	0.84	0.91	0.99	1.06	1.13	1.20	1.27	1.33	1.40	1.46	1.52
0.9	0.84	0.92	1.00	1.08	1.15	1.22	1.29	1.36	1.43	1.49	1.56
1.0	0.85	0.93	1.01	1.09	1.16	1.24	1.31	1.38	1.45	1.52	1.58
1.1	0.85	0.93	1.01	1.09	1.17	1.25	1.32	1.40	1.47	1.54	1.61
1.2	0.84	0.93	1.02	1.10	1.18	1.26	1.33	1.41	1.48	1.56	1.63
1.3	0.84	0.93	1.02	1.10	1.18	1.26	1.34	1.42	1.50	1.57	1.64
1.4	0.83	0.93	1.01	1.10	1.19	1.27	1.35	1.43	1.51	1.58	1.66
1.5	0.83	0.92	1.01	1.10	1.19	1.27	1.35	1.43	1.51	1.59	1.67

For a more accurate evaluation of a/d, it is possible to linearly interpolate values of a_{crit}/d ratio between $c/d = 0.8$ and $c/d = 0.9$: for $c/d = 0.872$ it results that $a_{crit}/d = 1.172$, which is very close to the value calculated in Example no. 11.

9.15.1.3 Example No. 13—Foundation Subjected to Eccentric Load

Execute the punching shar verification of the foundation of Example no. 11 subjected to $V_{Ed} = 500$ kN *and* $M_{Ed} = 100$ kNm.

From Example no. 11, $d = 344$ mm, $\rho_{ly} = \rho_{lz} = 0.291 \times 10^{-2}$; moreover, in the present case,

$$e = M_{Ed}/V_{Ed} = 100/500 = 0.2\,\text{m} = 200\ \text{min} \leq b/6 = 400\,\text{mm}$$

Net mean soil bearing pressure: $p_{net} = 500{,}000/(2400 \cdot 2400) = 0.087\ \text{N/mm}^2$.

Net maximum bearing pressure: $p_{\max} = 0.087 + M_{Ed}/(b^3/6) = 0.13$ N/mm^2.

Design punching shear stress at the column perimeter (u_0)

Column perimeter: $u_0 = 1200$ mm.

Base area within the perimeter u_0: $A_0 = 90{,}000$ mm^2.

Resultant force of the bearing pressure within the perimeter u_0.

$$\Delta V_{Ed,0} = p_{\text{netta}} A_0 = 0.087 \cdot 90{,}000 = 7830\,\text{N}$$
$$V_{Ed,red,0} = V_{Ed} - \Delta V_{Ed,0} = 500{,}000 - 7830 = 492{,}170\,\text{N}$$

$\beta = 1.37$ (for the calculation of β see the following verification on the perimeter u_1)

Design punching shear stress

$$v_{Ed,0} = \frac{\beta V_{Ed,red,0}}{u_0 d} = \frac{1.37 \cdot 492{,}170}{1200 \cdot 344} = 1.63\,\text{N/mm}^2$$

Maximum value of punching shear resistance

$$v_{Rd,\max} = 0.2\, f_{cd} = 0.2 \cdot \frac{0.85 \cdot 25}{1.5} \cong 2.83\,\text{N/mm}^2$$

Being $v_{Ed,0} \le v_{Rd,\max}$ the verification at the column perimeter is satisfied; then, the verification is performed at the control perimeter placed at a distance $a \le 2d$ from the column perimeter.

For the evaluation of the control perimeter, various attempts are made, by setting the first attempt control perimeter at distance $2d$.

First attempt control perimeter (a = 2d)

$$(c = c_1 = c_2 = 300\,\text{mm})$$
$$u_1 = 4c + 2\pi\,(2d) = 4 \cdot 300 + 2 \cdot \pi \cdot (2 \cdot 344)$$
$$= 5523\,\text{mm}$$

Base area within the control perimeter

$$A_1 = c^2 + 4c\,(2d) + \pi\,(2d)^2 = 300^2 + 4 \cdot 300 \cdot (2 \cdot 344)$$
$$+ \pi \cdot (2 \cdot 344)^2 = 2{,}402{,}654\,\text{mm}^2$$

Resultant force of the soil bearing pressure within the control perimeter

$$\Delta V_{Ed,1} = p_{\text{netta}} A_1 = 0.087 \cdot 2{,}402{,}654 = 209{,}031\,\text{N}$$
$$V_{Ed,red,1} = V_{Ed} - \Delta V_{Ed,1} = 500{,}000 - 209{,}031 = 290{,}969\,\text{N}$$

$$W_1 = \frac{3}{2}c_1^2 + (4 + 2\pi)\, c_1\, d + 16\, d^2$$
$$= \frac{3}{2}300^2 + (4 + 2\pi) \cdot 300 \cdot 344 + 16 \cdot 344^2$$
$$= 3{,}089{,}601\ \text{mm}^2$$

$$\beta = 1 + k\frac{M_{Ed}}{V_{Ed,red,1}}\frac{u_1}{W_1} = 1 + 0.6\frac{1 \cdot 10^8}{290{,}969} \cdot \frac{5523}{3{,}089{,}601} = 1.37$$

Design punching shear stress

$$v_{Ed,1} = \frac{\beta V_{Ed,red,1}}{u_1 d} = \frac{1,37 \cdot 290{,}969}{5523 \cdot 344} = 0.21\ \text{N/mm}^2$$

Punching shear resistance without reinforcement

$$v_{Rd,1} = C_{Rd,c}\, k\, (100\, \rho\, f_{ck})^{1/3} \frac{2\, d}{a}$$
$$= 0.12 \left(1 + \sqrt{\frac{200}{344}}\right) \left(100 \cdot 0.291 \times 10^{-2} \cdot 25\right)^{1/3}$$
$$= 0.41\ \text{N/mm}^2 \geq v_{\min} = 0.41\ \text{N/mm}^2$$

Being $v_{Ed,1} = 0.21\ \text{N/mm}^2 \leq v_{Rd,1} = 0.41\ \text{N/mm}^2$, the verification is satisfied at the perimeter u_1.

Second attempt control perimeter (a = d)

$$u_2 = 4\, c + 2\, \pi\, d = 4 \cdot 300 + 2 \cdot \pi \cdot 344 = 3361\ \text{mm}$$

Base area within the control perimeter

$$A_2 = c^2 + 4\, cd + \pi\, d^2 = 300^2 + 4 \cdot 300 \cdot 344 + \pi \cdot 344^2 = 874{,}564\ \text{mm}^2$$

Resultant force of the soil bearing pressure within the control perimeter

$$\Delta V_{Ed,2} = p_{netta} A_2 = 0.087 \cdot 874{,}564 = 76{,}087\ \text{N}$$
$$V_{Ed,red,2} = V_{Ed} - \Delta V_{Ed,2} = 500{,}000 - 76{,}087 = 423{,}913\ \text{N}$$
$$W_2 = \frac{3}{2}c_1^2 + (2 + \pi)\, c_1\, d + 4d^2 = \frac{3}{2}300^2 + (2 + \pi) \cdot 300 \cdot 344 + 4 \cdot 344^2$$
$$= 1{,}138{,}956\ \text{mm}^2$$

Design punching shear stress

$$v_{Ed,2} = \frac{\beta V_{Ed,red,2}}{u_2 d} = \frac{1.37 \cdot 423{,}913}{3361 \cdot 344} = 0.50\ \text{N/mm}^2$$

where the value of β calculated at the perimeter u_1 has been used

Punching shear resistance without reinforcement

$$v_{Rd,2} = C_{Rd,c}\, k\,(100\,\rho\, f_{ck})^{1/3}\frac{2\,d}{a}$$
$$= 0.12\left(1+\sqrt{\frac{200}{344}}\right)\left(100 \cdot 0.291 \times 10^{-2} \cdot 25\right)^{1/3} \cdot 2$$
$$= 0.82\,\text{N/mm}^2 \geq v_{\min} \cdot 2$$
$$= 0.82\,\text{N/mm}^2$$

Being $v_{Ed,2} = 0.50\,\text{N/mm}^2 \leq v_{Rd,2} = 0.82\,\text{N/mm}^2$, the verification is satisfied at the perimeter u_2, too.

If the calculations are repeated for any other distance $a \leq 2d$ from the control perimeter, it is always obtained that $v_{Ed} \leq v_{Rd}$, therefore the verification is satisfied.

9.15.1.4 Example No. 14—Punching *R*einforcement *D*esign with *E*ccentric *L*oad

Design the punching shear reinforcement for the foundation of Example no. 11, considering a 400 mm *square column subjected to* $V_{Ed} = 1000$ kN *and* $M_{Ed} = 200$ kNm.

$$e = M_{Ed}/V_{Ed} = 200/1000 = 0.2\,\text{m}$$
$$= 200\,\text{mm} \leq b/6 = 400\,\text{mm}$$

Mean soil bearing pressure: $p_{net} = 1{,}000{,}000/(2400 \cdot 2400) = 0.174\ \text{N/mm}^2$.
Maximum soil bearing pressure: $p_{max} = 0.174 + M_{Ed}/(b^3/6) = 0.26\ \text{N/mm}^2$.

Punching shear stress at the column perimeter (u_0)

Column perimeter: $u_0 = 1600$ mm.
Base area within the perimeter u_0: $A_0 = 160{,}000\ \text{mm}^2$.
Resultant force of the soil bearing pressure within the perimeter u_0

$$\Delta V_{Ed,0} = p_{netta} A_0 = 0.174 \cdot 160{,}000 = 27{,}840\,\text{N}$$
$$V_{Ed,red,0} = V_{Ed} - \Delta V_{Ed,0} = 1{,}000{,}000 - 27{,}840 = 972{,}160\,\text{N}$$

$\beta = 1.38$ (for the calculation of β see the following verification on the perimeter u_1)

Design punching stress

$$v_{Ed,0} = \frac{\beta V_{Ed,red,0}}{u_0 d} = \frac{1.38 \cdot 972{,}160}{1600 \cdot 344} = 2.44\,\text{N/mm}^2$$

Maximum punching shear stress

$$v_{Rd,\max} = 0.2\, f_{cd} = 0.2 \cdot \frac{0.85 \cdot 25}{1.5} \cong 2.83\,\text{N/mm}^2$$

Being $v_{Ed,0} \leq v_{Rd,\max}$, the verification at the column perimeter is satisfied; then, the verification on the control perimeter at distance $a \leq 2d$ from the column perimeter is performed.

To search for the control perimeter, some attempts are made, by choosing the perimeter at distance $2d$ as the first attempt control perimeter.

First attempt control perimeter (a = 2d)

$$(c = c_1 = c_2 = 400\,\text{mm})$$
$$u_1 = 4\,c + 2\,\pi\,(2\,d) = 4 \cdot 400 + 2 \cdot \pi \cdot (2 \cdot 344) = 5923\,\text{mm}$$

Base area within the control perimeter

$$A_1 = c^2 + 4\,c\,(2\,d) + \pi\,(2\,d)^2 = 400^2 + 4 \cdot 400 \cdot (2 \cdot 344) + \pi \cdot (2 \cdot 344)^2 = 2{,}747{,}854\,\text{mm}^2$$

Resultant force of the soil bearing pressure within the control perimeter

$$\Delta V_{Ed,1} = p_{\text{netta}} A_1 = 0.174 \cdot 2{,}747{,}854 = 478{,}127\,\text{N}$$
$$V_{Ed,red,1} = V_{Ed} - \Delta V_{Ed,1} = 1{,}000{,}000 - 478{,}127 = 521{,}873\ \text{N}$$
$$W_1 = \frac{3}{2} c_1^2 + (4 + 2\,\pi)\,c_1\,d + 16\,d^2 = \frac{3}{2} 400^2 + (4 + 2\,\pi) \cdot 400 \cdot 344 + 16 \cdot 344^2 = 3{,}548{,}342\,\text{mm}^2$$
$$\beta = 1 + k \frac{M_{Ed}}{V_{Ed,red,1}} \frac{u_1}{W_1} = 1 + 0.6 \frac{2 \cdot 10^8}{521{,}873} \frac{5923}{3{,}548{,}342} = 1.38$$

Design punching shear stress

$$v_{Ed,1} = \frac{\beta V_{Ed,red,1}}{u_1 d} = \frac{1.38 \cdot 521{,}873}{5923 \cdot 344} = 0.35\,\text{N/mm}^2$$

Punching shear resistance without reinforcement

$$v_{Rd,1} = C_{Rd,c}\, k\,(100\,\rho\, f_{ck})^{1/3} \frac{2\,d}{a} = 0.12 \left(1 + \sqrt{\frac{200}{344}}\right) \left(100 \cdot 0.291 \times 10^{-2} \cdot 25\right)^{1/3}$$

$$= 0.41\,\text{N/mm}^2 \geq v_{\min}$$
$$= 0.41\,\text{N/mm}^2$$

Being $v_{Ed,1} = 0.35\,\text{N/mm}^2 \leq v_{Rd,1} = 0.41\,\text{N/mm}^2$, at the perimeter u_1, the verification is satisfied.

Second attempt control perimeter (a = d)

$$\mathrm{u_2 = 4\,c + 2\,\pi\,d = 4 \cdot 400 + 2 \cdot \pi \cdot 344 = 3761\ mm}$$

Base area within the control perimeter

$$A_2 = c^2 + 4\,cd + \pi\,d^2 = 400^2 + 4 \cdot 400 \cdot 344 + \pi \cdot 344^2 = 1{,}082{,}164\,\text{mm}^2$$

Resultant force of the soil bearing pressure within the control perimeter

$$\Delta V_{Ed,2} = p_{\text{netta}} A_2 = 0.174 \cdot 1{,}082{,}164 = 188{,}297\,\text{N}$$
$$V_{Ed,red,2} = V_{Ed} - \Delta V_{Ed,2} = 1{,}000{,}000 - 188{,}297 = 811{,}703\,\text{N}$$
$$W_2 = \frac{3}{2} c_1^2 + (2 + \pi)\, c_1\, d + 4d^2 = \frac{3}{2} 400^2$$
$$+ (2 + \pi) \cdot 400 \cdot 344 + 4 \cdot 344^2$$
$$= 1{,}420{,}827\,\text{mm}^2$$

Design punching shear stress

$$v_{Ed,2} = \frac{\beta V_{Ed,red,2}}{u_2 d} = \frac{1.38 \cdot 811{,}703}{3761 \cdot 344} = 0.87\,\text{N/mm}^2$$

where it has been used the value of β calculated at the perimeter u_1.

Punching shear resistance without reinforcement

$$v_{Rd,2} = C_{Rd,c}\, k (100\, \rho\, f_{ck})^{1/3} \frac{2\,d}{a}$$
$$= 0.12 \left(1 + \sqrt{\frac{200}{344}}\right) \left(100 \cdot 0.291 \times 10^{-2} \cdot 25\right)^{1/3} \cdot 2$$
$$= 0.82\,\text{N/mm}^2 \geq v_{\min} \cdot 2$$
$$= 0.82\,\text{N/mm}^2$$

Being $v_{Ed,2} = 0.87\,\text{N/mm}^2 > v_{Rd,2} = 0.82\,\text{N/mm}^2$, at the perimeter u_2, the verification is not satisfied, therefore punching shear reinforcement is required.

To this aim, the perimeter u_{out}, beyond which shear reinforcement is no more required, is searched for.

Through some attempts, the perimeter u_{out} is found at a distance of $1.65d$ from the column perimeter (by comparison, in Example no. 11 the perimeter u_{out} is placed at distance $1.62d$—Fig. 9.51). The punching shear reinforcement shall then be extended from the column perimeter to $1.65d - 1.5d = 0.15d$, which means that the shear reinforcement could be provided by just one perimeter of bent-up bars or two perimeters of studs (minimum number prescribed by EC2) at spacing not greater than $0.75d$.

The area of each perimeter of shear reinforcement is calculated through the procedure described at the end of 9.15. With varying the distance a between 0 and $2d$, the perimeter u which maximizes A_{sw} is given by the following expression

$$A_{sw} \geq \frac{(v_{Ed} - 0.75\, v_{Rd})\; u\, d}{1.5\,(d/s_r)\; f_{ywd,ef}\; \sin\alpha}$$

(it is worth noting that if u varies also v_{Ed} and v_{Rd} vary, as already mentioned in 9.15).

Repeating the calculation of A_{sw} for different values of a using a worksheet, it is found that the searched perimeter u is placed at distance 0.843 d ($\cong$ 290 mm) from the column perimeter and it is given by

$$\mathrm{u(0.843d)} = 4c + 2\pi(0.843d) = 3422\,\mathrm{mm}$$

Base area within the control perimeter u

$$\begin{aligned} A_{(0.843d)} &= c^2 + 4\,c \cdot 0.843d + \pi\,(0.843d)^2 \\ &= 400^2 + 4 \cdot 400 \cdot 290 + \pi \cdot 290^2 \\ &= 888{,}181\,\mathrm{mm}^2 \end{aligned}$$

Resultant force of the soil bearing pressure within the control perimeter

$$\begin{aligned} \Delta V_{Ed(0.843d)} &= p_{\mathrm{netta}} A_{(0.843d)} = 0.174 \cdot 888{,}181 = 154{,}543\,\mathrm{N} \\ V_{Ed,red(0.843d)} &= V_{Ed} - \Delta V_{Ed(0.843d)} = 1{,}000{,}000 - 154{,}543 = 845{,}457\,\mathrm{N} \end{aligned}$$

Punching stress on the perimeter $u_{(0.843\mathrm{d})}$

$$v_{Ed(0.843d)} = \frac{\beta V_{Ed,red(0.843d)}}{u_{(0.843d)}\, d} = \frac{1.38 \cdot 845{,}457}{3422 \cdot 344} = 0.99\,\mathrm{N/mm}^2$$

Shear resistance without reinforcement

$$\begin{aligned} v_{Rd(0.843d)} &= C_{Rd,c}\, k\,(100\,\rho\, f_{ck})^{1/3} \frac{2\,d}{a} \\ &= 0.12\left(1 + \sqrt{\frac{200}{344}}\right)(100 \cdot 0.291 \times 10^{-2} \cdot 25)^{1/3} \cdot \frac{2}{0.843} \end{aligned}$$

$$= 0.972\,\text{N/mm}^2 \geq v_{\min} \cdot \frac{2}{0.843}$$

$$= 0.972\,N/mm^2 < v_{Ed(0.843)} \text{ then shear reinforcement is required.}$$

Design of 45° bent-up bars

The required reinforcement area for 45° bent-up bars is

$$A_{sw} \geq \frac{\left(v_{Rd(0.843d)} - 0.75 v_{Rd(0.843d)}\right) \cdot u_{(0.843d)} \cdot d}{f_{ywd,ef} \cdot sen\alpha}$$
$$= \frac{(0.99 - 0.75 \cdot 0.972) \cdot 3422 \cdot 344}{336 \cdot \sqrt{2}/2} = 1293\text{ mm}^2$$

where

$$f_{ywd,ef} = 250 + 0.25 \cdot 344 = 336\,\text{N/mm}^2 < f_{ywd} = 391\,\text{N/mm}^2$$

14 mm bars are adopted (that diameter corresponds to the minimum diameter taken from Table 9.22 for $d = 350$ mm and $f_{ck} = 25$ N/mm^2), then the number of inclined legs shall not be less than: 1293/154 = 8.4 ≅ 9.

Five bent-up bars are then adopted (10 inclined legs), three bars in one direction and two bars in the other direction.

Design of studs on two perimeters

Alternatively, if studs on two perimeters are used, the following result is obtained

$$A_{sw,cs} \geq \frac{\left(v_{Ed(0.843d)} - 0.75 v_{Rd(0.843d)}\right) \cdot u_{(0.843d)} \cdot d}{1.5 \cdot d/s_r \cdot f_{ywd,ef}}.$$
$$= \frac{(0.99 - 0.75 \cdot 0.972) \cdot 3422 \cdot 344}{1.5 \cdot (1/0.75) \cdot 336} = 457\text{ mm}^2$$

Minimum diameter: 12 mm (from Table 9.20 for $d = 350$ mm and $f_{ck} = 25$ N/mm^2).

Number of studs per each perimeter: $n = 457/113 = 4.04 \cong 5$.

Verification of the maximum tangential spacing within each perimeter

The first perimeter of studs will be placed at a distance 0.5 d from the column perimeter (maximum distance indicated in [9.4.3(4)]), while the second perimeter is placed at a distance (0.5 + 0.75) d = 1.25 d.

Then, at least nine studs per perimeter are required, so the tangential spacing within the second perimeter satisfy the upper limit of 1.5d (= 516 mm) required by EC2 in [9.4.3(1)]:

$$S_t = [u_0 + 2\pi(1.25d)]/9 = 478\,\text{mm} \leq 516\,\text{mm}.$$

Appendix 1: Tables for Rapid Calculation of $v_{Rd,c}$ (N/mm^2) with Varying Diameter and Spacing of Flexural Reinforcement

Concrete Class: C25/30

See Table 9.28.

For $f_{ck} > 25$ N/mm^2, $v_{Rd,c}$ can be obtained by multiplying tabulated values by $(f_{ck}/25)^{1/3}$ and by comparing the result with $v_{\min}$; if $v_{Rd,C} < v_{\min}$, $v_{Rd,C} = v_{\min}$ is adopted.

For the sake of simplicity, the following tables collect values of $v_{Rd,c}$ for the following concrete classes: C28/35, C32/40, C35/45, C40/50, C45/55.

Concrete Class: C28/35

See Table 9.29.

Concrete Class: C32/40

See Table 9.30.

Concrete Class: C35/45

See Table 9.31.

Concrete Class: C40/50

See Table 9.32.

Concrete Class: C45/55

See Table 9.33.

Table 9.28 Values of $v_{Rd,c}$ (N/mm^2) for concrete class C25/30 with varying flexural reinforcement (same diameter and spacing in both directions y and z) and slab effective depth

Spacing 100 mm	*d* (mm) 150	200	250	300	350	400	450	500
ϕ12/100	0.639	0.580	0.510	0.460	0.423	0.393	0.369	0.349
ϕ14/100	0.708	0.643	0.566	0.510	0.469	0.436	0.409	0.387
ϕ16/100	0.774	0.703	0.618	0.558	0.512	0.476	0.447	0.423
ϕ18/100	0.837	0.760	0.669	0.603	0.554	0.515	0.484	0.457
ϕ20/100	**0.884**[a]	0.816	0.717	0.647	0.594	0.553	0.519	0.491
ϕ22/100	**0.884**	0.869	0.764	0.690	0.633	0.589	0.553	0.523
ϕ24/100	**0.884**	**0.884**	0.810	0.731	0.671	0.624	0.586	0.554
ϕ26/100	**0.884**	**0.884**	0.854	0.771	0.708	0.658	0.618	0.584

[a]For $\rho_l > 0.02$, the calculation has been performed by assuming $\rho_l = 0.02$ (maximum value allowed by EC2 for the calculation of $v_{Rd,c}$). Those values are highlighted in bold.

Spacing 150 mm	*d* (mm) 150	200	250	300	350	400	450	500
ϕ12/150	0.558	0.507	0.446	0.402	0.369	0.343	0.322	0.305
ϕ14/150	0.618	0.562	0.494	0.446	0.409	0.381	0.357	0.338
ϕ16/150	0.676	0.614	0.540	0.487	0.447	0.416	0.391	0.369
ϕ18/150	0.731	0.664	0.584	0.527	0.484	0.450	0.422	0.400
ϕ20/150	0.784	0.713	0.627	0.565	0.519	0.483	0.453	0.429
ϕ22/150	0.836	0.759	0.668	0.603	0.553	0.514	0.483	0.457
ϕ24/150	**0.884**	0.805	0.708	0.639	0.586	0.545	0.512	0.484
ϕ26/150	**0.884**	0.849	0.746	0.673	0.618	0.575	0.540	0.510

Spacing 200 mm	*d* (mm) 150	200	250	300	350	400	450	500
ϕ12/200	0.507	0.461	0.405	0.365	0.336	0.312	0.293	0.277
ϕ14/200	0.562	0.510	0.449	0.405	0.372	0.346	0.325	0.307
ϕ16/200	0.614	0.558	0.491	0.443	0.407	0.378	0.355	0.336
ϕ18/200	0.664	0.604	0.531	0.479	0.440	0.409	0.384	0.363
ϕ20/200	0.713	0.647	0.569	0.514	0.472	0.439	0.412	0.389
ϕ22/200	0.759	0.690	0.607	0.547	0.503	0.467	0.439	0.415
ϕ24/200	0.805	0.731	0.643	0.580	0.533	0.495	0.465	0.440
ϕ26/200	0.849	0.771	0.678	0.612	0.562	0.522	0.490	0.464

Spacing 250 mm	*d* (mm) 150	200	250	300	350	400	450	500
ϕ12/250	0.471	0.428	0.376	0.339	0.312	0.290	0.272	0.257
ϕ14/250	0.522	0.474	0.417	0.376	0.345	0.321	0.301	0.285
ϕ16/250	0.570	0.518	0.455	0.411	0.377	0.351	0.329	0.312
ϕ18/250	0.617	0.560	0.493	0.445	0.408	0.380	0.356	0.337
ϕ20/250	0.662	0.601	0.529	0.477	0.438	0.407	0.382	0.361
ϕ22/250	0.705	0.640	0.563	0.508	0.467	0.434	0.407	0.385
ϕ24/250	0.747	0.679	0.597	0.539	0.495	0.460	0.432	0.408
ϕ26/250	0.788	0.716	0.630	0.568	0.522	0.485	0.455	0.431

(continued)

Table 9.28 (continued)

Spacing 300 mm	d (mm)							
	150	200	250	300	350	400	450	500
ϕ12/300	0.443	0.402	0.354	0.319	0.293	0.273	0.256	0.242
ϕ14/300	0.491	0.446	0.392	0.354	0.325	0.302	0.284	0.268
ϕ16/300	0.536	0.487	0.429	0.387	0.355	0.330	0.310	0.293
ϕ18/300	0.580	0.527	0.464	0.418	0.384	0.357	0.335	0.317
ϕ20/300	0.623	0.566	0.497	0.449	0.412	0.383	0.360	0.340
ϕ22/300	0.663	0.603	0.530	0.478	0.439	0.408	0.383	0.362
ϕ24/300	0.703	0.639	0.562	0.507	0.465	0.433	0.406	0.384
ϕ26/300	0.742	0.674	0.592	0.535	0.491	0.456	0.428	0.405
	d (mm)							
Minimum resistance	150	200	250	300	350	400	450	500
v_{min}	0.495	0.495	0.456	0.428	0.407	0.390	0.377	0.365

Values in grey boxes are less than v_{min} and shall be replaced by v_{min} (v_{min} values are collected in the last row)

Table 9.29 Values of $v_{Rd,c}$ (N/mm^2) for concrete class C28/35 with varying flexural reinforcement (same diameter and spacing in both directions *y* and *z*) and slab effective depth

Spacing 100 mm	*d* (mm) 150	200	250	300	350	400	450	500
ϕ12/100	0.663	0.603	0.530	0.478	0.439	0.408	0.383	0.362
ϕ14/100	0.735	0.668	0.587	0.530	0.487	0.452	0.425	0.402
ϕ16/100	0.804	0.730	0.642	0.579	0.532	0.495	0.464	0.439
ϕ18/100	0.869	0.790	0.694	0.627	0.575	0.535	0.502	0.475
ϕ20/100	**0.918**[a]	0.847	0.745	0.672	0.617	0.574	0.539	0.509
ϕ22/100	**0.918**	0.903	0.794	0.716	0.658	0.612	0.574	0.543
ϕ24/100	**0.918**	**0.918**	0.841	0.759	0.697	0.648	0.608	0.575
ϕ26/100	**0.918**	**0.918**	0.887	0.801	0.735	0.684	0.642	0.607

[a] For $\rho_l > 0.02$. the calculation has been performed by assuming $\rho_l = 0.02$ (maximum value allowed by EC2 for the calculation of $v_{Rd,c}$). Those values are highlighted in bold.

Spacing 150 mm	*d* (mm) 150	200	250	300	350	400	450	500
ϕ12/150	0.579	0.526	0.463	0.418	0.384	0.357	0.335	0.317
ϕ14/150	0.642	0.583	0.513	0.463	0.425	0.395	0.371	0.351
ϕ16/150	0.702	0.638	0.561	0.506	0.465	0.432	0.406	0.384
ϕ18/150	0.759	0.690	0.607	0.547	0.503	0.467	0.439	0.415
ϕ20/150	0.815	0.740	0.651	0.587	0.539	0.501	0.471	0.445
ϕ22/150	0.868	0.789	0.693	0.626	0.575	0.534	0.502	0.474
ϕ24/150	**0.918**	0.836	0.735	0.663	0.609	0.566	0.531	0.503
ϕ26/150	**0.918**	0.882	0.775	0.699	0.642	0.597	0.561	0.530

Spacing 200 mm	*d* (mm) 150	200	250	300	350	400	450	500
ϕ12/200	0.526	0.478	0.421	0.380	0.348	0.324	0.304	0.288
ϕ14/200	0.583	0.530	0.466	0.421	0.386	0.359	0.337	0.319
ϕ16/200	0.638	0.579	0.510	0.460	0.422	0.393	0.369	0.348
ϕ18/200	0.690	0.627	0.551	0.497	0.457	0.425	0.399	0.377
ϕ20/200	0.740	0.672	0.591	0.534	0.490	0.456	0.428	0.404
ϕ22/200	0.789	0.717	0.630	0.568	0.522	0.485	0.456	0.431
ϕ24/200	0.836	0.759	0.668	0.602	0.553	0.514	0.483	0.457
ϕ26/200	0.882	0.801	0.704	0.635	0.584	0.543	0.509	0.482

Spacing 250 mm	*d* (mm) 150	200	250	300	350	400	450	500
ϕ12/250	0.489	0.444	0.390	0.352	0.324	0.301	0.282	0.267
ϕ14/250	0.542	0.492	0.433	0.390	0.359	0.333	0.313	0.296
ϕ16/250	0.592	0.538	0.473	0.427	0.392	0.364	0.342	0.324
ϕ18/250	0.640	0.582	0.512	0.462	0.424	0.394	0.370	0.350
ϕ20/250	0.687	0.624	0.549	0.495	0.455	0.423	0.397	0.375
ϕ22/250	0.732	0.665	0.585	0.528	0.485	0.451	0.423	0.400
ϕ24/250	0.776	0.705	0.620	0.559	0.514	0.478	0.448	0.424
ϕ26/250	0.818	0.744	0.654	0.590	0.542	0.504	0.473	0.447

(continued)

Table 9.29 (continued)

Spacing 300 mm	d (mm)							
	150	200	250	300	350	400	450	500
ϕ12/300	0.460	0.418	0.367	0.332	0.304	0.283	0.266	0.251
ϕ14/300	0.510	0.463	0.407	0.367	0.337	0.314	0.295	0.279
ϕ16/300	0.557	0.506	0.445	0.402	0.369	0.343	0.322	0.304
ϕ18/300	0.603	0.548	0.481	0.434	0.399	0.371	0.348	0.329
ϕ20/300	0.647	0.587	0.517	0.466	0.428	0.398	0.374	0.353
ϕ22/300	0.689	0.626	0.550	0.497	0.456	0.424	0.398	0.376
ϕ24/300	0.730	0.663	0.583	0.526	0.483	0.449	0.422	0.399
ϕ26/300	0.770	0.700	0.615	0.555	0.510	0.474	0.445	0.421
Minimum resistance	**d (mm)**							
	150	200	250	300	350	400	450	500
v_{min}	0.524	0.524	0.483	0.453	0.431	0.413	0.398	0.386

Values in grey boxes are less than v_{min} and shall be replaced by v_{min} (v_{min} values are collected in the last row)

Table 9.30 Values of $v_{Rd,c}$ (N/mm^2) for concrete class C32/40 with varying flexural reinforcement (same diameter and spacing in both directions y and z) and slab effective depth

Spacing 100 mm	*d* (mm) 150	200	250	300	350	400	450	500
ϕ12/100	0.694	0.630	0.554	0.500	0.459	0.427	0.401	0.379
ϕ14/100	0.769	0.698	0.614	0.554	0.509	0.473	0.444	0.420
ϕ16/100	0.840	0.763	0.671	0.606	0.556	0.517	0.485	0.459
ϕ18/100	0.909	0.826	0.726	0.655	0.602	0.559	0.525	0.497
ϕ20/100	**0.960**[a]	0.886	0.779	0.703	0.645	0.600	0.563	0.533
ϕ22/100	**0.960**	0.944	0.830	0.749	0.688	0.639	0.600	0.568
ϕ24/100	**0.960**	**0.960**	0.879	0.794	0.729	0.678	0.636	0.602
ϕ26/100	**0.960**	**0.960**	0.928	0.837	0.769	0.715	0.671	0.634

[a]For $\rho_l > 0.02$. the calculation has been performed by assuming $\rho_l = 0.02$ (maximum value allowed by EC2 for the calculation of $v_{Rd,c}$). Those values are highlighted in bold.

Spacing 150 mm	*d* (mm) 150	200	250	300	350	400	450	500
ϕ12/150	0.606	0.550	0.484	0.437	0.401	0.373	0.350	0.331
ϕ14/150	0.671	0.610	0.536	0.484	0.444	0.413	0.388	0.367
ϕ16/150	0.734	0.667	0.586	0.529	0.486	0.452	0.424	0.401
ϕ18/150	0.794	0.721	0.634	0.572	0.525	0.489	0.459	0.434
ϕ20/150	0.852	0.774	0.680	0.614	0.564	0.524	0.492	0.465
ϕ22/150	0.907	0.825	0.725	0.654	0.601	0.559	0.524	0.496
ϕ24/150	**0.960**	0.874	0.768	0.693	0.637	0.592	0.556	0.525
ϕ26/150	**0.960**	0.922	0.810	0.731	0.671	0.624	0.586	0.554
Spacing 200 mm	*d* (mm) 150	200	250	300	350	400	450	500
ϕ12/200	0.550	0.500	0.440	0.397	0.364	0.339	0.318	0.301
ϕ14/200	0.610	0.554	0.487	0.440	0.404	0.375	0.352	0.333
ϕ16/200	0.667	0.606	0.533	0.481	0.441	0.410	0.385	0.364
ϕ18/200	0.721	0.655	0.576	0.520	0.477	0.444	0.417	0.394
ϕ20/200	0.774	0.703	0.618	0.558	0.512	0.476	0.447	0.423
ϕ22/200	0.825	0.749	0.659	0.594	0.546	0.508	0.476	0.451
ϕ24/200	0.874	0.794	0.698	0.630	0.578	0.538	0.505	0.477
ϕ26/200	0.922	0.837	0.736	0.664	0.610	0.567	0.533	0.504
Spacing 250 mm	*d* (mm) 150	200	250	300	350	400	450	500
ϕ12/250	0.511	0.464	0.408	0.368	0.338	0.315	0.295	0.279
ϕ14/250	0.566	0.515	0.452	0.408	0.375	0.349	0.327	0.309
ϕ16/250	0.619	0.562	0.495	0.446	0.410	0.381	0.358	0.338
ϕ18/250	0.670	0.608	0.535	0.483	0.443	0.412	0.387	0.366
ϕ20/250	0.718	0.653	0.574	0.518	0.475	0.442	0.415	0.392
ϕ22/250	0.765	0.695	0.611	0.552	0.507	0.471	0.442	0.418
ϕ24/250	0.811	0.737	0.648	0.585	0.537	0.499	0.469	0.443
ϕ26/250	0.856	0.777	0.684	0.617	0.566	0.527	0.494	0.467

(continued)

Table 9.30 (continued)

Spacing 300 mm	d (mm)							
	150	200	250	300	350	400	450	500
ϕ12/300	0.481	0.437	0.384	0.347	0.318	0.296	0.278	0.263
ϕ14/300	0.533	0.484	0.426	0.384	0.353	0.328	0.308	0.291
ϕ16/300	0.583	0.529	0.465	0.420	0.386	0.359	0.337	0.318
ϕ18/300	0.630	0.572	0.503	0.454	0.417	0.388	0.364	0.344
ϕ20/300	0.676	0.614	0.540	0.487	0.447	0.416	0.391	0.369
ϕ22/300	0.720	0.654	0.575	0.519	0.477	0.443	0.416	0.394
ϕ24/300	0.763	0.694	0.610	0.550	0.505	0.470	0.441	0.417
ϕ26/300	0.805	0.732	0.643	0.580	0.533	0.496	0.465	0.440
	d (mm)							
Minimun resistance	150	200	250	300	350	400	450	500
v_{min}	0.560	0.560	0.516	0.485	0.461	0.442	0.426	0.413

Table 9.31 Values of $v_{Rd,c}$ (N/mm^2) for concrete class C35/45 with varying flexural reinforcement (same diameter and spacing in both directions y and z) and slab effective depth

Spacing 100 mm	d (mm)							
	150	200	250	300	350	400	450	500
ϕ12/100	0.715	0.649	0.571	0.515	0.473	0.440	0.413	0.390
ϕ14/100	0.792	0.719	0.633	0.571	0.524	0.487	0.458	0.433
ϕ16/100	0.866	0.786	0.692	0.624	0.573	0.533	0.500	0.473
ϕ18/100	0.936	0.851	0.748	0.675	0.620	0.576	0.541	0.512
ϕ20/100	**0.989**[a]	0.913	0.802	0.724	0.665	0.618	0.580	0.549
ϕ22/100	**0.989**	0.972	0.855	0.772	0.708	0.659	0.618	0.585
ϕ24/100	**0.989**	**0.989**	0.906	0.818	0.751	0.698	0.655	0.620
ϕ26/100	**0.989**	**0.989**	0.956	0.862	0.792	0.736	0.691	0.654

[a]For $\rho_l > 0.02$. the calculation has been performed by assuming $\rho_l = 0.02$ (maximum value allowed by EC2 for the calculation of $v_{Rd,c}$). Those values are highlighted in bold.

Spacing 150 mm	d (mm)							
	150	200	250	300	350	400	450	500
ϕ12/150	0.624	0.567	0.499	0.450	0.413	0.384	0.361	0.341
ϕ14/150	0.692	0.629	0.553	0.499	0.458	0.426	0.400	0.378
ϕ16/150	0.756	0.687	0.604	0.545	0.501	0.465	0.437	0.413
ϕ18/150	0.818	0.743	0.653	0.590	0.541	0.503	0.473	0.447
ϕ20/150	0.877	0.797	0.701	0.633	0.581	0.540	0.507	0.479
ϕ22/150	0.935	0.850	0.747	0.674	0.619	0.576	0.540	0.511
ϕ24/150	**0.989**	0.900	0.792	0.714	0.656	0.610	0.573	0.541
ϕ26/150	**0.989**	0.950	0.835	0.753	0.692	0.643	0.604	0.571

(continued)

Table 9.31 (continued)

Spacing 200 mm	d (mm)							
	150	200	250	300	350	400	450	500
ϕ12/200	0.567	0.515	0.453	0.409	0.375	0.349	0.328	0.310
ϕ14/200	0.629	0.571	0.502	0.453	0.416	0.387	0.363	0.343
ϕ16/200	0.687	0.624	0.549	0.495	0.455	0.423	0.397	0.375
ϕ18/200	0.743	0.675	0.594	0.536	0.492	0.457	0.429	0.406
ϕ20/200	0.797	0.724	0.637	0.575	0.528	0.491	0.461	0.436
ϕ22/200	0.850	0.772	0.679	0.612	0.562	0.523	0.491	0.464
ϕ24/200	0.900	0.818	0.719	0.649	0.596	0.554	0.520	0.492
ϕ26/200	0.950	0.863	0.759	0.685	0.629	0.584	0.549	0.519
Spacing 250 mm	d (mm)							
	150	200	250	300	350	400	450	500
ϕ12/250	0.526	0.478	0.421	0.380	0.348	0.324	0.304	0.288
ϕ14/250	0.583	0.530	0.466	0.421	0.386	0.359	0.337	0.319
ϕ16/250	0.638	0.579	0.510	0.460	0.422	0.393	0.369	0.348
ϕ18/250	0.690	0.627	0.551	0.497	0.457	0.425	0.399	0.377
ϕ20/250	0.740	0.672	0.591	0.534	0.490	0.456	0.428	0.404
ϕ22/250	0.789	0.717	0.630	0.568	0.522	0.485	0.456	0.431
ϕ24/250	0.836	0.759	0.668	0.602	0.553	0.514	0.483	0.457
ϕ26/250	0.882	0.801	0.704	0.635	0.584	0.543	0.509	0.482
Spacing 300 mm	d (mm)							
	150	200	250	300	350	400	450	500
ϕ12/300	0.495	0.450	0.396	0.357	0.328	0.305	0.286	0.271
ϕ14/300	0.549	0.499	0.439	0.396	0.363	0.338	0.317	0.300
ϕ16/300	0.600	0.545	0.479	0.433	0.397	0.369	0.347	0.328
ϕ18/300	0.649	0.590	0.519	0.468	0.430	0.400	0.375	0.355
ϕ20/300	0.696	0.633	0.556	0.502	0.461	0.429	0.402	0.381
ϕ22/300	0.742	0.674	0.593	0.535	0.491	0.457	0.429	0.406
ϕ24/300	0.786	0.715	0.628	0.567	0.521	0.484	0.454	0.430
ϕ26/300	0.830	0.754	0.663	0.598	0.549	0.511	0.479	0.453
Minimum resistance	d (mm)							
	150	200	250	300	350	400	450	500
v_{min}	0.586	0.586	0.540	0.507	0.482	0.462	0.446	0.432

Values in grey boxes are less than v_{min} and shall be replaced by v_{min} (v_{min} values are collected in the last row)

Table 9.32 Values of $v_{Rd,c}$ (N/mm^2) for concrete class C40/50 with varying flexural reinforcement (same diameter and spacing in both directions y and z) and slab effective depth

Spacing 100 mm	*d* (mm) 150	200	250	300	350	400	450	500
ϕ12/100	0.747	0.679	0.597	0.539	0.495	0.460	0.432	0.408
ϕ14/100	0.828	0.752	0.661	0.597	0.548	0.510	0.478	0.452
ϕ16/100	0.905	0.822	0.723	0.652	0.599	0.557	0.523	0.494
ϕ18/100	0.979	0.889	0.782	0.706	0.648	0.603	0.566	0.535
ϕ20/100	**1.034**[a]	0.954	0.839	0.757	0.695	0.646	0.607	0.574
ϕ22/100	**1.034**	1.017	0.894	0.807	0.741	0.689	0.647	0.611
ϕ24/100	**1.034**	**1.034**	0.947	0.855	0.785	0.730	0.685	0.648
ϕ26/100	**1.034**	**1.034**	0.999	0.902	0.828	0.770	0.723	0.683

[a] For $\rho_l > 0.02$. the calculation has been performed by assuming $\rho_l = 0.02$ (maximum value allowed by EC2 for the calculation of $v_{Rd,c}$). Those values are highlighted in bold.

Spacing 150 mm	*d* (mm) 150	200	250	300	350	400	450	500
ϕ12/150	0.653	0.593	0.521	0.470	0.432	0.402	0.377	0.357
ϕ14/150	0.723	0.657	0.578	0.521	0.479	0.445	0.418	0.395
ϕ16/150	0.791	0.718	0.632	0.570	0.523	0.487	0.457	0.432
ϕ18/150	0.855	0.777	0.683	0.616	0.566	0.526	0.494	0.467
ϕ20/150	0.917	0.834	0.733	0.661	0.607	0.565	0.530	0.501
ϕ22/150	0.978	0.888	0.781	0.705	0.647	0.602	0.565	0.534
ϕ24/150	**1.034**	0.941	0.828	0.747	0.686	0.638	0.599	0.566
ϕ26/150	**1.034**	0.993	0.873	0.788	0.723	0.673	0.631	0.597

Spacing 200 mm	*d* (mm) 150	200	250	300	350	400	450	500
ϕ12/200	0.593	0.539	0.474	0.427	0.392	0.365	0.343	0.324
ϕ14/200	0.657	0.597	0.525	0.474	0.435	0.404	0.380	0.359
ϕ16/200	0.718	0.653	0.574	0.518	0.475	0.442	0.415	0.392
ϕ18/200	0.777	0.706	0.621	0.560	0.514	0.478	0.449	0.425
ϕ20/200	0.834	0.757	0.666	0.601	0.552	0.513	0.482	0.455
ϕ22/200	0.888	0.807	0.710	0.640	0.588	0.547	0.513	0.485
ϕ24/200	0.941	0.855	0.752	0.679	0.623	0.579	0.544	0.514
ϕ26/200	0.993	0.902	0.793	0.716	0.657	0.611	0.574	0.542

Spacing 100 mm	*d* (mm) 150	200	250	300	350	400	450	500
ϕ12/250	0.550	0.500	0.440	0.397	0.364	0.339	0.318	0.301
ϕ14/250	0.610	0.554	0.487	0.440	0.404	0.375	0.352	0.333
ϕ16/250	0.667	0.606	0.533	0.481	0.441	0.410	0.385	0.364
ϕ18/250	0.721	0.655	0.576	0.520	0.477	0.444	0.417	0.394
ϕ20/250	0.774	0.703	0.618	0.558	0.512	0.476	0.447	0.423
ϕ22/250	0.825	0.749	0.659	0.594	0.546	0.508	0.476	0.451
ϕ24/250	0.874	0.794	0.698	0.630	0.578	0.538	0.505	0.477
ϕ26/250	0.922	0.837	0.736	0.664	0.610	0.567	0.533	0.504

(continued)

Table 9.32 (continued)

Spacing 300 mm	d (mm)							
	150	200	250	300	350	400	450	500
ϕ12/300	0.518	0.471	0.414	0.373	0.343	0.319	0.299	0.283
ϕ14/300	0.574	0.522	0.459	0.414	0.380	0.353	0.332	0.314
ϕ16/300	0.627	0.570	0.501	0.452	0.415	0.386	0.363	0.343
ϕ18/300	0.679	0.617	0.542	0.489	0.449	0.418	0.392	0.371
ϕ20/300	0.728	0.662	0.582	0.525	0.482	0.448	0.421	0.398
ϕ22/300	0.776	0.705	0.620	0.559	0.514	0.478	0.448	0.424
ϕ24/300	0.822	0.747	0.657	0.593	0.544	0.506	0.475	0.449
ϕ26/300	0.867	0.788	0.693	0.625	0.574	0.534	0.501	0.474

Minimum resistance	d (mm)							
	150	200	250	300	350	400	450	500
v_{min}	0.626	0.626	0.577	0.542	0.515	0.494	0.476	0.462

Values in grey boxes are less than v_{min} and shall be replaced by v_{min} (v_{min} values are collected in the last row)

Table 9.33 Values of $v_{Rd,c}$ (N/mm²) for concrete class C45/55 with varying flexural reinforcement (same diameter and spacing in both directions y and z) and slab effective depth

Spacing 100 mm	*d* (mm) 150	200	250	300	350	400	450	500
ϕ12/100	0.777	0.706	0.621	0.560	0.514	0.478	0.449	0.425
ϕ14/100	0.861	0.782	0.688	0.621	0.570	0.530	0.498	0.470
ϕ16/100	0.941	0.855	0.752	0.679	0.623	0.579	0.544	0.514
ϕ18/100	1.018	0.925	0.813	0.734	0.674	0.627	0.588	0.556
ϕ20/100	**1.076**[a]	0.992	0.873	0.787	0.723	0.672	0.631	0.597
ϕ22/100	**1.076**	1.057	0.930	0.839	0.770	0.716	0.672	0.636
ϕ24/100	**1.076**	**1.076**	0.985	0.889	0.816	0.759	0.713	0.674
ϕ26/100	**1.076**	**1.076**	1.039	0.938	0.861	0.801	0.752	0.711

[a]For $\rho_l > 0.02$, the calculation has been performed by assuming $\rho_l = 0.02$ (maximum value allowed by EC2 for the calculation of $v_{Rd,c}$). Those values are highlighted in bold.

Spacing 150 mm	*d* (mm) 150	200	250	300	350	400	450	500
ϕ12/150	0.679	0.617	0.542	0.489	0.449	0.418	0.392	0.371
ϕ14/150	0.752	0.683	0.601	0.542	0.498	0.463	0.435	0.411
ϕ16/150	0.822	0.747	0.657	0.593	0.544	0.506	0.475	0.449
ϕ18/150	0.889	0.808	0.711	0.641	0.589	0.547	0.514	0.486
ϕ20/150	0.954	0.867	0.762	0.688	0.632	0.587	0.551	0.521
ϕ22/150	1.017	0.924	0.812	0.733	0.673	0.626	0.587	0.556
ϕ24/150	**1.076**	0.979	0.861	0.777	0.713	0.663	0.623	0.589
ϕ26/150	**1.076**	1.033	0.908	0.819	0.752	0.700	0.657	0.621
Spacing 200 mm	*d* (mm) 150	200	250	300	350	400	450	500
ϕ12/200	0.617	0.560	0.493	0.445	0.408	0.380	0.356	0.337
ϕ14/200	0.683	0.621	0.546	0.493	0.452	0.421	0.395	0.373
ϕ16/200	0.747	0.679	0.597	0.539	0.495	0.460	0.432	0.408
ϕ18/200	0.808	0.734	0.646	0.583	0.535	0.497	0.467	0.442
ϕ20/200	0.867	0.788	0.693	0.625	0.574	0.534	0.501	0.474
ϕ22/200	0.924	0.839	0.738	0.666	0.611	0.569	0.534	0.505
ϕ24/200	0.979	0.889	0.782	0.706	0.648	0.603	0.566	0.535
ϕ26/200	1.033	0.938	0.825	0.744	0.684	0.636	0.597	0.564
Spacing 100 mm	*d* (mm) 150	200	250	300	350	400	450	500
ϕ12/250	0.572	0.520	0.457	0.413	0.379	0.352	0.331	0.313
ϕ14/250	0.634	0.576	0.507	0.457	0.420	0.391	0.367	0.347
ϕ16/250	0.694	0.630	0.554	0.500	0.459	0.427	0.401	0.379
ϕ18/250	0.750	0.682	0.599	0.541	0.497	0.462	0.433	0.410
ϕ20/250	0.805	0.731	0.643	0.580	0.533	0.495	0.465	0.440
ϕ22/250	0.858	0.779	0.685	0.618	0.568	0.528	0.495	0.469
ϕ24/250	0.909	0.826	0.726	0.655	0.602	0.559	0.525	0.497
ϕ26/250	0.959	0.871	0.766	0.691	0.635	0.590	0.554	0.524

(continued)

Table 9.33 (continued)

Spacing 300 mm	d (mm)							
	150	200	250	300	350	400	450	500
ϕ12/300	0.539	0.489	0.430	0.388	0.357	0.332	0.311	0.294
ϕ14/300	0.597	0.542	0.477	0.430	0.395	0.367	0.345	0.326
ϕ16/300	0.653	0.593	0.521	0.470	0.432	0.402	0.377	0.357
ϕ18/300	0.706	0.641	0.564	0.509	0.467	0.435	0.408	0.386
ϕ20/300	0.757	0.688	0.605	0.546	0.501	0.466	0.438	0.414
ϕ22/300	0.807	0.733	0.645	0.582	0.534	0.497	0.466	0.441
ϕ24/300	0.855	0.777	0.683	0.616	0.566	0.526	0.494	0.467
ϕ26/300	0.902	0.820	0.721	0.650	0.597	0.555	0.521	0.493

Minimum resistance	d (mm)							
	150	200	250	300	350	400	450	500
v_{min}	0.664	0.664	0.612	0.575	0.546	0.524	0.505	0.490

Values in grey boxes are less than v_{min} and shall be replaced by v_{min} (v_{min} values are collected in the last row)

Appendix 2: Tables with the Maximum Area of Studs Within Each Perimeter for Slabs on Rectangular Columns Equipped with Two Reinforcement Perimeters

(For values of the stud area higher than those listed in following tables, the punching failure occurs along the perimeter u_0 of the column or loaded area—box with grey background—or the perimeter u_{out} out for the reinforced region—boxes with white background).

Concrete Class: C28/35

See Table 9.34.

Concrete Class: C32/40

See Table 9.35.

Concrete Class: C35/45

See Table 9.36.

Concrete Class: C40/50

See Table 9.37.

Concrete Class: C45/55

See Table 9.38.

Table 9.34 Values of $A_{sw,\max}$ (mm^2) within each perimeter for studs distributed on two perimeters at the maximum spacing allowed by EN1992-1-1; concrete class C28/35, reinforcement ratio $\rho_l = \sqrt{\rho_{ly}\,\rho_{lz}} = 2\%$ (maximum value of ρ allowed by EC2 in the calculation of $v_{Rd,c}$)

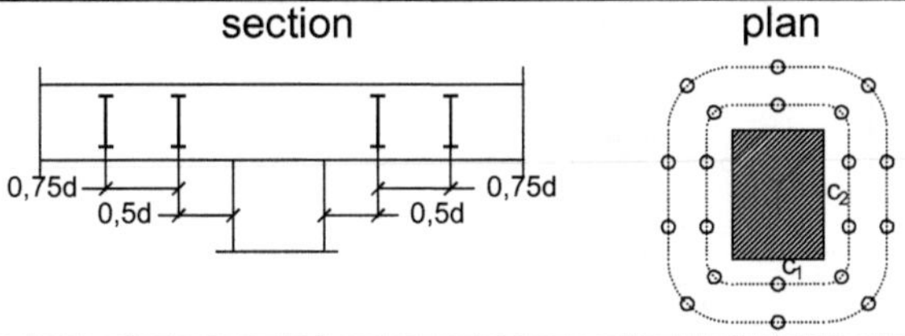

c_1+c_2 (mm)	d (mm) 200	250	300	350	400	450	500
500	–[a]	–	–	–	–	–	–
550	334	–	–	–	–	–	–
600	417	390	–	–	–	–	–
650	500	491	–	–	–	–	–
700	583	592	558	–	–	–	–
750	596	693	676	619	–	–	–
800	603	794	793	752	–	–	–
850	611	831	911	885	823	–	–
900	619	840	1028	1019	971	890	–
950	626	848	1090	1152	1119	1051	–
1000	634	857	1099	1285	1267	1213	1127
1050	641	866	1109	1369	1414	1374	1301
1100	649	874	1119	1379	1562	1535	1475
1150	657	883	1128	1389	1664	1697	1649
1200	664	892	1138	1400	1676	1858	1823
1250	672	901	1147	1410	1687	1975	1998
1300	680	909	1157	1421	1698	1987	2172
1350	687	918	1167	1431	1709	1999	2299
1400	695	927	1176	1442	1720	2011	2312
1450	703	935	1186	1452	1732	2023	2324
1500	710	944	1196	1463	1743	2035	2337

Table 9.35 Values of $A_{sw,max}$ (mm^2) within each perimeter for studs distributed on two perimeters at the maximum spacing allowed by EN1992-1-1; concrete class C32/40, reinforcement ratio $\rho_l = \sqrt{\rho_{ly}\,\rho_{lz}} = 2\%$ (maximum value of ρ allowed by EC2 in the calculation of $v_{Rd,c}$)

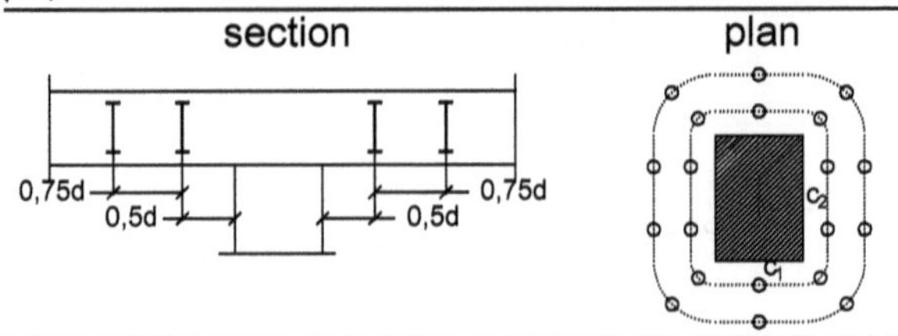

c_1+c_2 (mm)	d (mm) 200	250	300	350	400	450	500
500	366	–[a]	–	–	–	–	–
550	463	439	–	–	–	–	–
600	559	556	509	–	–	–	–
650	607	674	646	–	–	–	–
700	615	792	783	732	–	–	–
750	623	851	920	887	817	–	–
800	631	860	1057	1043	989	–	–
850	639	869	1119	1198	1161	1088	–
900	647	878	1129	1353	1333	1276	1185
950	655	887	1139	1409	1505	1463	1388
1000	663	896	1149	1420	1677	1651	1590
1050	671	905	1159	1431	1717	1839	1793
1100	679	914	1170	1442	1729	2027	1995
1150	687	923	1180	1453	1740	2040	2198
1200	695	932	1190	1464	1752	2053	2365
1250	703	942	1200	1474	1764	2065	2378
1300	711	951	1210	1485	1775	2078	2391
1350	719	960	1220	1496	1787	2090	2404
1400	727	969	1230	1507	1799	2103	2417
1450	735	978	1240	1518	1810	2115	2430
1500	743	987	1250	1529	1822	2127	2443

Table 9.36 Values of $A_{sw,\max}$ (mm^2) within each perimeter for studs distributed on two perimeters at the maximum spacing allowed by EN1992-1-1; concrete class C35/45, reinforcement ratio $\rho_l = \sqrt{\rho_{ly}\,\rho_{lz}} = 2\%$ (maximum value of ρ allowed by EC2 in the calculation of $v_{Rd,c}$)

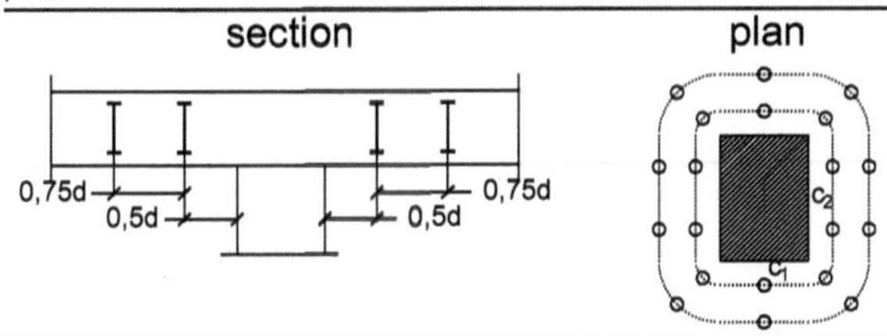

c_1+c_2 (mm)	d (mm)						
	200	250	300	350	400	450	500
500	453	423	–[a]	–	–	–	–
550	561	553	–	–	–	–	–
600	617	684	651	–	–	–	–
650	625	814	803	749	–	–	–
700	633	867	955	921	848	–	–
750	642	876	1107	1093	1039	948	–
800	650	886	1143	1265	1229	1156	–
850	658	895	1153	1429	1420	1363	1273
900	666	904	1164	1440	1610	1571	1497
950	675	914	1174	1452	1745	1779	1721
1000	683	923	1184	1463	1757	1987	1945
1050	691	933	1195	1474	1769	2077	2169
1100	699	942	1205	1485	1781	2090	2393
1150	707	951	1215	1497	1793	2102	2423
1200	716	961	1226	1508	1805	2115	2437
1250	724	970	1236	1519	1817	2128	2450
1300	732	979	1246	1530	1829	2141	2463
1350	740	989	1257	1542	1841	2154	2477
1400	749	998	1267	1553	1853	2166	2490
1450	757	1008	1278	1564	1865	2179	2504
1500	765	1017	1288	1575	1877	2192	2517

Table 9.37 Values of $A_{sw,\max}$ (mm^2) within each perimeter for studs distributed on two perimeters at the maximum spacing allowed by EN1992-1-1; concrete class C40/50, reinforcement ratio $\rho_l = \sqrt{\rho_{ly}\,\rho_{lz}} = 2\%$ (maximum value of ρ allowed by EC2 in the calculation of $v_{Rd,c}$)

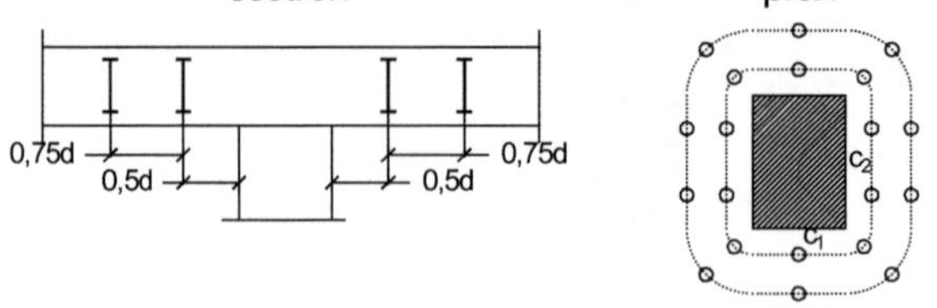

c_1+c_2 (mm)	d (mm)						
	200	250	300	350	400	450	500
500	603	596	541	—[a]	—	—	—
550	636	748	718	—	—	—	—
600	645	887	895	844	—	—	—
650	653	897	1072	1044	974	—	—
700	662	906	1173	1244	1196	1109	—
750	671	916	1184	1443	1417	1350	1249
800	679	926	1195	1482	1638	1592	1509
850	688	936	1206	1494	1799	1833	1769
900	697	946	1216	1506	1812	2074	2029
950	705	955	1227	1518	1824	2145	2289
1000	714	965	1238	1530	1837	2158	2491
1050	722	975	1249	1541	1849	2171	2505
1100	731	985	1260	1553	1862	2185	2519
1150	740	995	1271	1565	1875	2198	2533
1200	748	1004	1282	1577	1887	2211	2547
1250	757	1014	1292	1588	1900	2225	2562
1300	766	1024	1303	1600	1912	2238	2576
1350	774	1034	1314	1612	1925	2252	2590
1400	783	1044	1325	1624	1938	2265	2604
1450	791	1053	1336	1635	1950	2278	2618
1500	800	1063	1347	1647	1963	2292	2632

Table 9.38 Values of $A_{sw,\max}$ (mm^2) within each perimeter for studs distributed on two perimeters at the maximum spacing allowed by EN1992-1-1; concrete class C45/55, reinforcement ratio $\rho_l = \sqrt{\rho_{ly}\,\rho_{lz}} = 2\%$ (maximum value of ρ allowed by EC2 in the calculation of $v_{Rd,c}$)

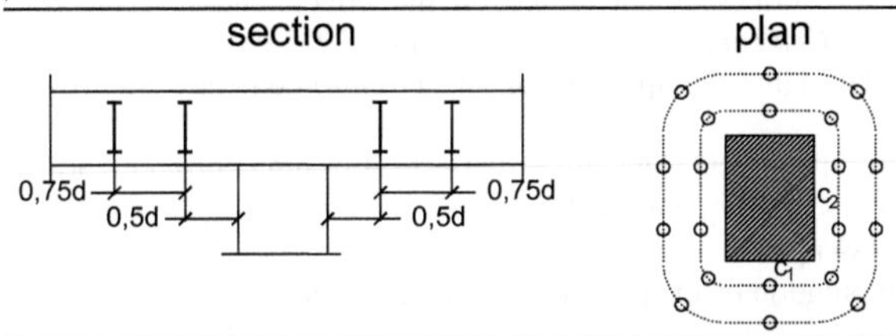

c_1+c_2 (mm)	d (mm) 200	250	300	350	400	450	500
500	653	774	741	662	–[a]	–	–
550	662	912	943	890	–	–	–
600	671	922	1144	1118	1047	–	–
650	680	933	1209	1345	1299	1213	–
700	689	943	1220	1517	1552	1488	1388
750	698	953	1231	1530	1804	1763	1684
800	707	963	1243	1542	1858	2038	1980
850	716	973	1254	1554	1871	2203	2276
900	724	984	1265	1566	1884	2217	2562
950	733	994	1276	1579	1897	2230	2576
1000	742	1004	1288	1591	1910	2244	2591
1050	751	1014	1299	1603	1923	2258	2606
1100	760	1024	1310	1615	1937	2272	2620
1150	769	1034	1322	1627	1950	2286	2635
1200	778	1045	1333	1640	1963	2300	2649
1250	787	1055	1344	1652	1976	2314	2664
1300	796	1065	1355	1664	1989	2328	2679
1350	805	1075	1367	1676	2002	2342	2693
1400	814	1085	1378	1689	2015	2356	2708
1450	823	1096	1389	1701	2028	2369	2723
1500	832	1106	1400	1713	2041	2383	2737

References

Beutel, R., & Hegger, J. (2002). The effect of anchorage on the effectiveness of the shear reinforcement in the punching zone. *Cement & Concrete Composites, 24*, 539–549.

Broms, C.E. (2007). Ductility of flat slabs: Comparison of shear reinforcement systems. *ACI Structural Journal, V, 104*(6).

CEB-FIP (2001). Punching of structural concrete slabs, fib Bulletin 12, Losanna, Svizzera, 307 pp.

Guandalini, S. (2005), Poinçonnement symétrique des dalles en béton armé, Tesi di Dottorato, No. 3380, EPFL, Losanna, Svizzera, 289 pp.

Menetréy, P. (1996). Analytical computation of the punching strength of reinforced concrete. *ACI Structural Journal, 93*(5), 503–511.

Chapter 10
Strut-And-Tie Models

Abstract The chapter deals with Strut-and-Tie (S&T) models for the design of those RC structural elements which are not classifiable as Saint–Venant beams, like rigid spread footings, deep beams, corbels, or those regions of slender beams subjected to nonlinear strain distribution (e.g., at concentrated loads or reactions, at abrupt cross-sectional changes). The mechanical model assumed in S&T models is built by identifying a triangulated truss made of struts and ties, corresponding to compressive and tensile stress fields, respectively. The idea of using S&T models has been first introduced at the end of the nineteenth century and the beginning of the twentieth century by Ritter and Mörsch for the shear design of reinforced concrete beams. Later the use of S&T models has been extended to the design of all reinforced concrete structures by Jörg Schlaich at the University of Stuttgart. According to EC2, S&T models can be used for ULS design of both "continuity" and "discontinuity" regions. Continuity or B-regions are those where the Euler–Bernoulli hypothesis is satisfied ("plane sections remain plane"), while discontinuity or D-regions are characterized by static or geometric discontinuities. S&T models can also be used for the verification of the serviceability limit states (SLS). The chapter includes many case studies about the identification of the S&T geometry and the design of the reinforcement layout for common RC structural members.

10.1 Introduction

Strut-and-tie models (S&T) are used for the design of non-standard RC structural elements which cannot be assumed as slender elements or Saint–Venant beams, like rigid spread footings, deep beams, corbels or those regions of slender beams subjected to a nonlinear strain distribution (e.g. at concentrated loads or reactions, at abrupt cross-sectional changes).

The mechanical model assumed in S&T design is built by identifying a triangulated truss made of struts and ties, corresponding to compressive and tensile stress fields, respectively. Figure 10.1 shows the S&T model for a deep beam.

This chapter was authored by Franco Angotti and Maurizio Orlando.

© The Author(s), under exclusive license to Springer Nature Switzerland AG 2022
F. Angotti et al., *Reinforced Concrete with Worked Examples*,
https://doi.org/10.1007/978-3-030-92839-1_10

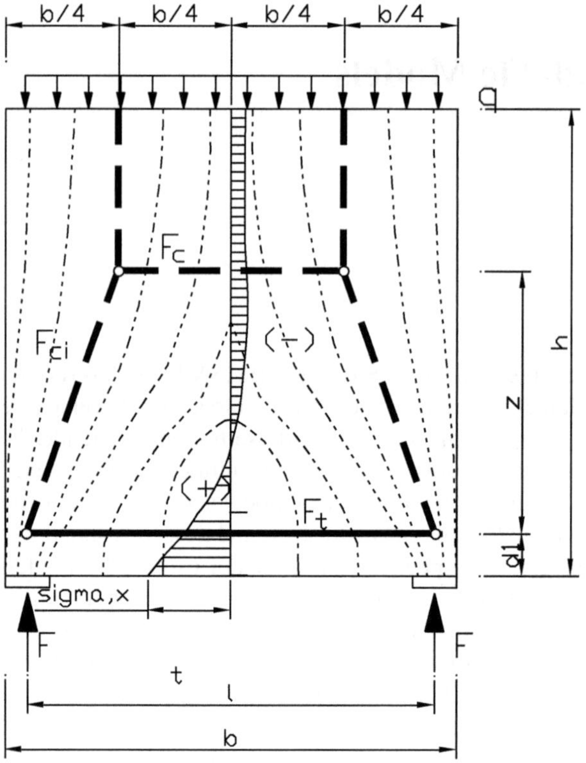

Fig. 10.1 S&T model of a deep beam

The idea of using S&T models has been first introduced at the end of the nineteenth century and the beginning of the twentieth century by Ritter and Mörsch, who introduced the truss analogy method (Fig. 10.2) for the shear design of reinforced concrete beams. Such a method has been successively developed by Jörg Schlaich and his team at the University of Stuttgart, extending its application to the design of all reinforced concrete structural elements.

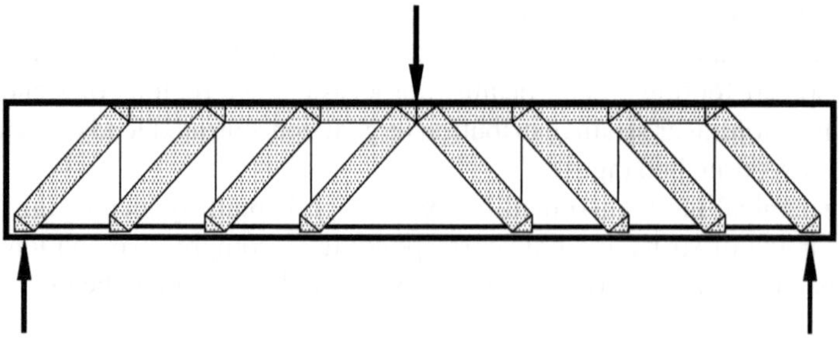

Fig. 10.2 Ritter-Mörsch truss model

Compressive stress fields are represented by concrete struts, while tensile stress fields (ties) correspond to reinforcing bars.

According to EC2, S&T models can be used for ULS design of both "continuity" and "discontinuity" regions of reinforced concrete elements.

"Continuity" regions are called "B" regions (from "Bernoulli" o from the word "beam") and correspond to those regions where hypotheses of Saint–Venant problem are fulfilled. In the cracked state those regions are modelled with the truss analogy by Mörsch, with struts representing compressive stress fields and ties representing the reinforcement.

Discontinuity regions are characterized by the presence of static or geometrical discontinuities ("D" regions, from the word "discontinuity"), where the Euler–Bernoulli hypothesis is not satisfied.

Static discontinuities comprehend concentrated loads, support reactions, anchorage zones of prestressing cables, etc., while geometric discontinuities correspond to abrupt changes of cross-section or axis direction, holes, thick elements (corbels, deep beams, Gerber hinges, etc.).

S&T models can also be used for the verification of the serviceability limit states (SLS). To this purpose, EC2 suggests aligning struts and ties with principal stress trajectories in the uncracked state; actually, it is important to follow this rule also for the verification of the ultimate limit states (ULS) (see 10.1.1).

10.1.1 Strut and Tie Method as an Application of the Lower Bound (Static) Theorem of Limit Analysis

Once distributed loads are substituted with equivalent concentrated loads, the design of a reinforced concrete structural element through the S&T method consists in modelling stress fields with a truss model made of linear elements (struts and ties) in equilibrium with external loads.

Curvatures of isostatic lines are concentrated in special points called nodes, which are given by the intersections of truss axes with applied loads or restraint reactions.

Once the truss geometry is defined, the forces in the elements of the strut-and-tie model are calculated, reinforcements are designed, and the strength of struts and nodes is checked.

S&T modelling is a method of plastic analysis of reinforced concrete structures, and it can be seen as an application of the first theorem of the limit analysis (static theorem or theorem on the lower edge of the limit loads).

This theorem can be stated as follows: if there is any stress distribution satisfying all equilibrium conditions (internal and external) and not exceeding the material strength (*plasticity condition*), then the corresponding load is less or equal to the failure load.

Limit analysis can be employed for reinforced concrete structures only if they are sufficiently ductile.

For example, for a frame structure, cross-sections must have a sufficient rotation capacity to allow the development of all plastic hinges predicted by the limit analysis, and bending failure occurs with the whole reinforcement in the plastic range. On the contrary, limit analysis is not suitable for frame structures where cross-sections are subjected to shear failure or flexural compression failure with crushing of the concrete and reinforcing bars in the elastic range or slightly above the elastic limit.

Analogously, the geometry of an S&T model shall be selected so that all the elements of the truss reach their design resistance and the early failure of a strut or node does not occur. For this purpose, it is sufficient to follow some practical rules (see 10.5), which guarantee that concrete withstands inelastic strains associated with the S&T model.

It is worth noting that EC2 does not give any rule on the control of concrete strain capacity and, therefore, any rule for the geometrical definition of the truss.

EC2 only indicates:

- in [6.5.1(1)P] the possibility of using S&T models in regions where a nonlinear stress distribution exists (e.g., at supports, in the proximity of loaded areas or in plane stress elements);
- in [5.6.4(5)], among the possible tools for the development of suitable S&T models, the evaluation of isostatic lines and stress fields obtained through the elasticity theory or the Load Path Method (LPM); it is worth noting that all S&T models may be optimized through energy criteria.

EC2 does not pay attention to the identification of the truss geometry to reproduce in the best way the stress field obtained through the elastic analysis in the uncracked state for a reinforced concrete structural element; the method is but presented as one of the "possible tools" for the development of S&T models.

It is preferable to define the truss geometry starting from stress fields in the elastic uncracked state, otherwise, the truss ductility can be not sufficient to activate the design strength of each strut and each tie. In other words, the discrete field of tension and compression forces in an S&T model could be not activated if a premature failure occurs in some elements (struts or nodes).

If different methods are used to define the truss geometry, the structural element ductility shall then be verified, e.g. through finite-element nonlinear analysis.

10.2 Identification of the Geometry of the S&T Model

The definition of the S&T model within a reinforced concrete structure requires first the identification of discontinuity regions and the definition of their extension. Further on, the structure is divided into "B" continuity regions and "D" discontinuity regions; therefore, steps suggested in Table 10.1 are followed.

Table 10.1 Procedure for the application of the S&T method

1	Identification of discontinuity regions ("D") corresponding to concentrated loads and/or geometrical discontinuities
2	Definition of the extension of discontinuity regions with the application of Saint–Venant's principle and subdivision of the structure into continuity regions ("B") and discontinuity regions ("D")
3	Evaluation of the stress field and design of the reinforcement in "B" regions
4	Calculation of forces applied at the boundary of "D" regions
5	Definition of the geometry of the S&T model for each "D" region
6	Calculation of forces in struts and ties of each "D" region
7	Design of the reinforcement (amount and layout), eventual refinement of the model (fi. to simplify the reinforcement layout), new calculation of internal forces in struts and ties, and new reinforcement design
8	Check of struts and nodes, eventual new design of struts and nodes based on geometrical considerations, refinement of the model, new calculation of internal forces in struts and ties, and new verification of struts and nodes
9	Design of reinforcement anchorage and distributed reinforcement to avoid cracking

10.2.1 Position and Extension of "D" Regions

"D" regions are placed at static discontinuities (concentrated loads) and/or geometrical discontinuities (e.g., a sharp variation of the axis line). Their extension can be determined by applying Saint–Venant's principle, which states that the stress filed at a sufficient distance from the loaded area does not depend on the load distribution, but only on the resultant force and moment. The distance at which this condition can be considered fulfilled is almost equal to the largest dimension of the loaded area (Figs. 10.3, 10.4, 10.5).

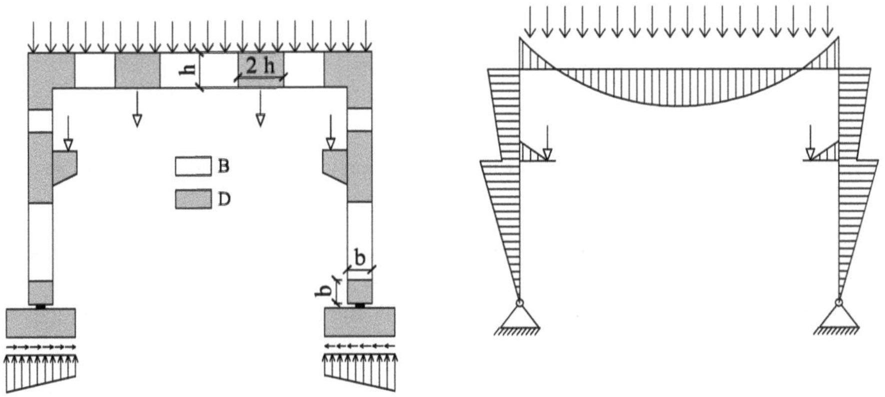

Fig. 10.3 Subdivision of a frame into "D" and "B" regions (each "D" region extends on both sides of the discontinuity for a length equal to the cross-section height)

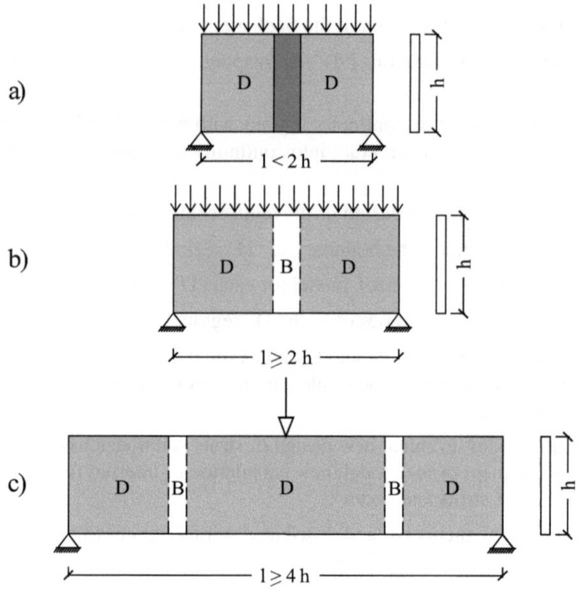

Fig. 10.4 Subdivision into "D" and "B" regions for simply supported beams with different slenderness (each "D" zone is extended on both sides of the discontinuity—represented by the concentrated load or constraint reaction—for a length equal to the height *h* of the structural element)

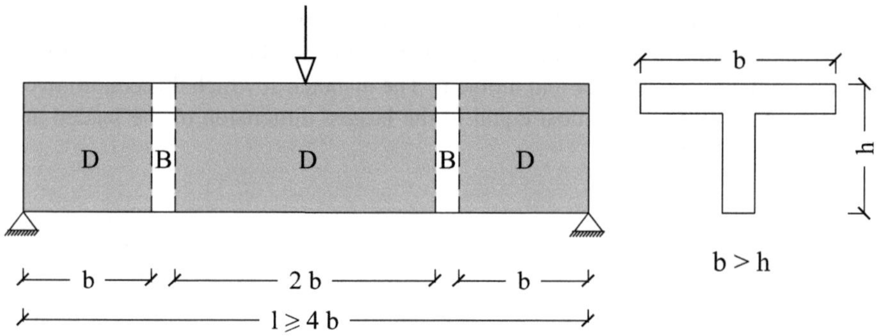

Fig. 10.5 Subdivision of a T beam with width *b* larger than the height *h* into "D" and "B" regions

Saint–Venant's principle is used as a qualitative tool to identify "D" regions and to develop the S&T model. If the element is not slender (e.g., a deep beam with a span not longer than twice its height, Fig. 10.4a), the discontinuity region reduces to the element itself (Fig. 10.6).

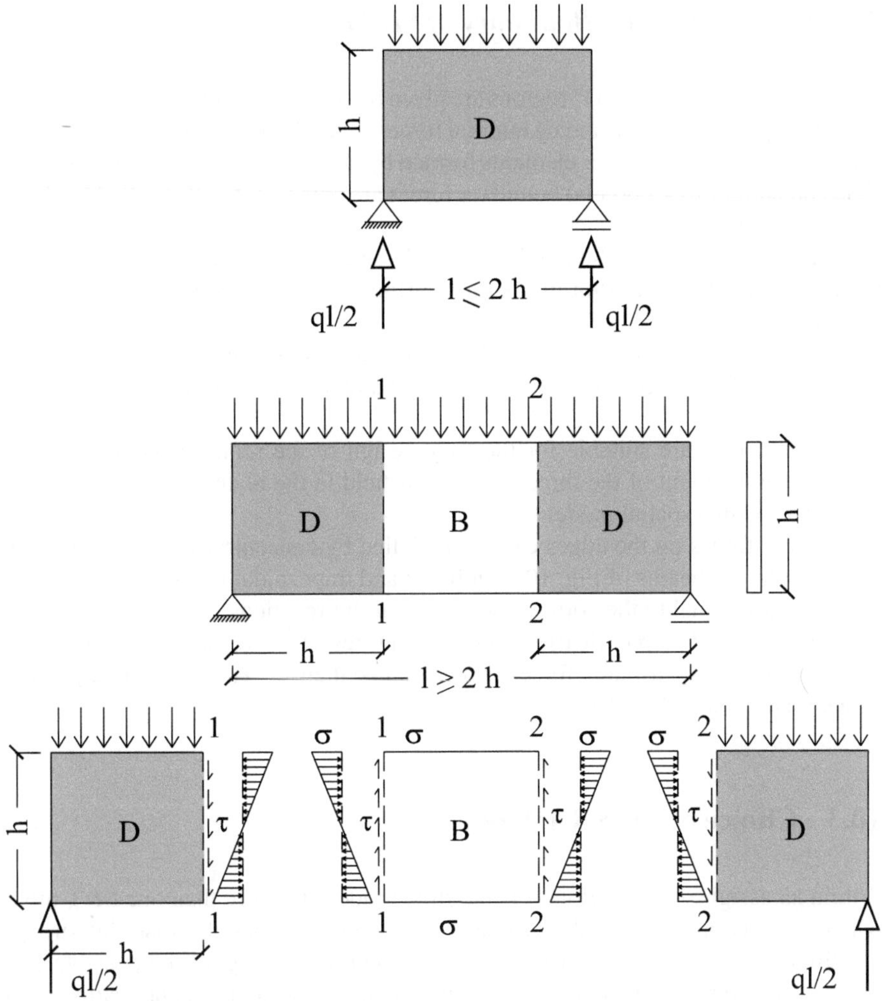

Fig. 10.6 Calculation of forces at the boundary of a "D" region. For the first beam ($l \leq 2\ h$), boundary forces are coincident with external loads and constraint reactions. For the second beam ($l > 2\ h$), forces at the interface of "D" regions with "B" regions are given by resultant moment and resultant shear force in cross-sections 1–1 and 2–2 of the "B" region

10.2.2 Evaluation of the Stress Field and Design of Reinforcement in "B" Regions

Once the structure is subdivided into "B" and "D" regions, "B" regions can be analysed with models which are suitable for slender beams. It is then possible to evaluate the internal forces of each section, particularly in those sections corresponding to the boundaries with "D" regions and proceed to the design of reinforcement.

10.2.3 Forces at the Boundary of "D" Regions

Forces at the boundary of "D" regions are given both by loads and constraint reactions directly applied over them and by internal forces applied at the interface with adjacent "B" regions (see 10.2.2). For elements formed by "D" regions only (e.g., deep beams, rigid spread footings, corbels) boundary forces are coincident with applied loads and restraint reactions.

Once all the forces applied on a given "D" region are identified, it is necessary to verify the rigid body equilibrium before starting with the identification of the S&T model.

A 3D structure can be analysed using plane models, which are obtained from the subdivision of the three-dimensional model and identifying an S&T model on each plane.

These models are suitable for the easy design of the reinforcement but do not allow the assessment of the three-axial stress field in the nodes, which can only be studied through a spatial model.

Possible forces on the edges can be modelled by concentrated equivalent loads: for example, for beams of Fig. 10.7, uniform and trapezoidal loads are both divided into two parts equal to the corresponding constraint reactions.

Finally, possible forces distributed within "D" regions, like the self-weight, can be replaced by edge distributed forces, which can be themselves reduced to equivalent concentrated loads (Fig. 10.8).

10.3 Choice of the S&T Model

Within an assigned "D" region, it is possible to identify more than one S&T model which is in equilibrium with boundary forces and satisfies the material strength condition (Fig. 10.9). Then, it is necessary to choose among all the possible S&T models. As already shown, the S&T design method represents an application of the lower bound (static) theorem, which is suitable for rigid-plastic materials. However, concrete is not an indefinitely plastic material, but only allows small plastic deformations. The static theorem can, therefore, be applied to reinforced concrete structural elements only if the compatibility of plastic deformations is also verified. Briefly, the S&T model has to be chosen so that the material strain capability shall not be exceeded in concrete struts before all the tension forces in the ties reach the design value.

In the most stressed regions, this requirement is fulfilled if struts and ties are oriented along isostatic lines obtained from the elastic analysis in the uncracked phase. On the contrary, in less stressed regions, struts and ties can deviate significantly from isostatic lines of the uncracked phase, without the strain capability of the concrete is exceeded.

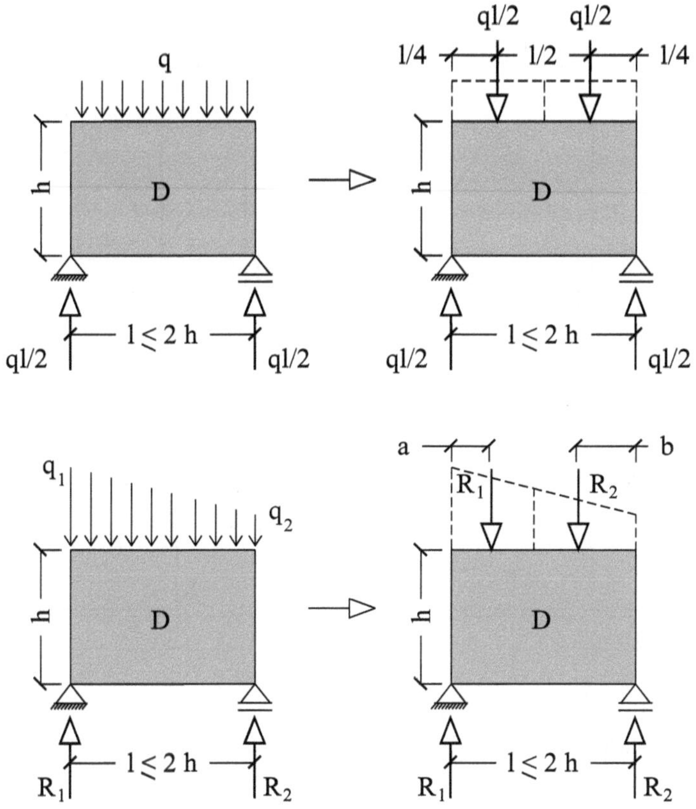

Fig. 10.7 Equivalence of edge load distributions (on the left) with edge concentrated forces (on the right)

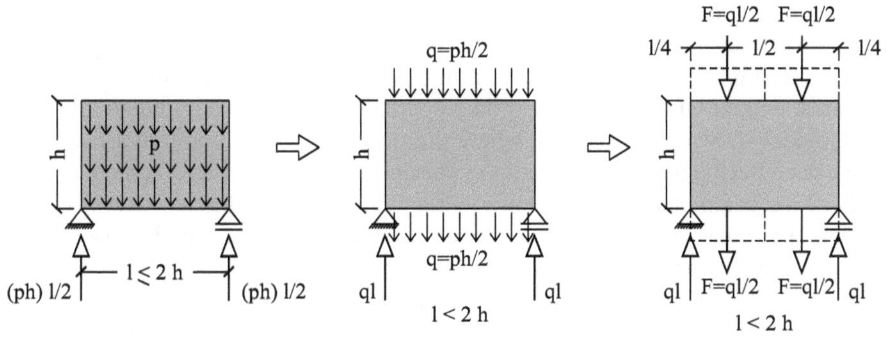

Fig. 10.8 Model of a deep beam with distributed loads for unit surface within the beam (*p* load on the left), with loads distributed on the edges (*q* loads in the middle) and with equivalent concentrated forces (*F* loads on the right)

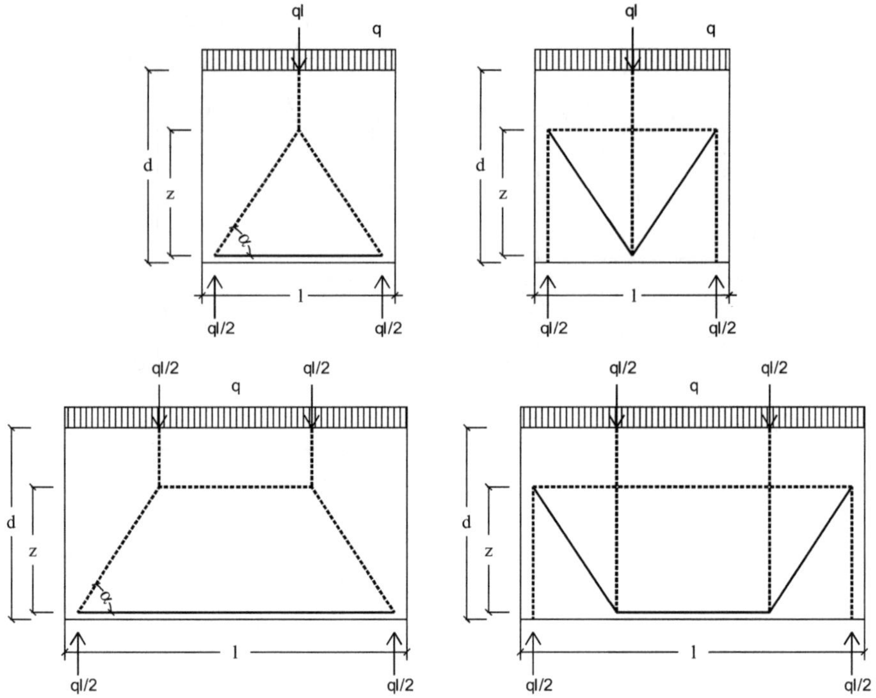

Fig. 10.9 Possible S&T models for a deep beam (dashed lines = struts, solid lines = ties)

The choice of the S&T model based on the elasticity theory leads to neglect portion of the ultimate resistance of the structure, which could be considered by applying the plasticity theory. If the geometry of an S&T model drawn from isostatic lines of the uncracked phase is adapted to the stress distribution of the ultimate limit state, it is possible to evaluate the ultimate load capacity very accurately. For example, for the deep beam shown in Fig. 10.10, in the cracked state the horizontal strut moves upwards, and, at ULS, it is close to the upper edge. It is then possible to adopt the S&T model shown in Fig. 10.10, where the tie F_{t2} placed at about half the height of the deep beam takes into account the presence of diffused reinforcement on both sides of the beam. This model allows getting an estimate of the ultimate load very close to the actual load-bearing capacity of the structure, while a model based on the elastic behaviour gives an ultimate strength lower than half the actual one.

On the other hand, an S&T model identified in the uncracked phase has the advantage to be used for the analysis at both ULS and SLS, while a model based on the ULS stress field does not guarantee the fulfilment of SLS requirements.

In general, "D" regions designed using S&T models identified in the cracked phase show excessive cracking and deformations at SLS. Moreover, the identification of the S&T model in the elastic phase has the advantage of not requiring *a priori* knowledge

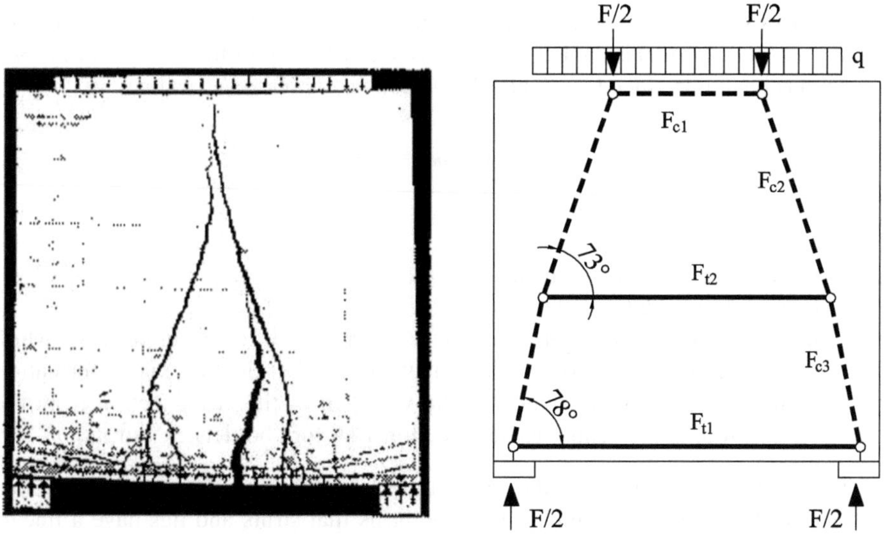

Fig. 10.10 Model of the deep beam of Fig. 10.1 adapted to the stress distribution in the ULS (struts = dashed lines, ties = solid lines)

of the reinforcement distribution, which, on the contrary, is required to identify the model in the cracked state.

As suggested by Schlaich, struts and ties shall be positioned and oriented along isostatic lines from the elastic uncracked state to avoid strain incompatibility and to adapt the S&T model to the actual behaviour of the structural element.

As the load increases, and before failure, the structural element can adapt its stress field to the stress field considered in the design of the S&T model, only if it can withstand large plastic deformation. Reinforcing bars can assure high plastic deformation without loss of strength, while concrete ductility is very low. If the S&T model has been oriented along isostatic lines of the elastic uncracked state, the plastic deformation required for stress redistribution is easily tolerated by materials, otherwise, the available ductility should be checked through a nonlinear finite-element analysis.

If the chosen S&T model does not match the stress field of the uncracked phase, the stress redistribution capacity can be extremely low or even null and failure can occur prematurely, at a load lower than the design load. Moreover, the structural element exhibits a very pronounced cracking at SLS.

However, structural elements can experience a premature failure also if the geometry of the S&T model is oriented along isostatic lines from elastic analysis, but it is not sufficiently accurate. For example, if the model does not contain ties suitable to resist tensile stresses due to the transverse spread of strut compression forces, reinforcement design can be inadequate, and the anticipated failure of struts or nodes can occur.

While selecting the S&T model for a "D" region, it can be considered that loads follow the path corresponding to the minimum stresses and strains. Being ties much more deformable than concrete struts, the model with the lower total length of ties (fewer ties and shorter) is the best one. This easy criterion for the choice of the S&T model has been introduced by Schlaich and can be expressed as follows:

$$\sum F_i l_i \varepsilon_{mi} = \min$$

where:

F_i force in the *i-th* truss element,
l_i length of the *i-th* truss element,
ε_{mi} mean strain of the *i-th* truss element ($\varepsilon_{mi} = \varepsilon_i =$ constant for ties, while for struts, except for those with prismatic cross-section, the axial deformation varies along the axis because of the variation of the cross-section from the edges towards the centre, as described in 10.9 for "bottleneck" struts).

This equation comes from the minimum elastic deformation energy principle applied to the S&T model, under the hypothesis that struts and ties have a linear-elastic behaviour:

$$1/2\Sigma(F_i \Delta l_i) = 1/2 \sum (F_i l_i \varepsilon_{mi}) = \min$$

where Δl_i is the length variation of the *i-th* truss element.

The strain in concrete struts can generally be neglected because it is much smaller than steel ties.[1]

If the tie design is performed by adopting steel areas close to the calculated value ($A_{sij} = F_i / f_{yd}$ for the *i-th* element), the strain is almost the same in all ties and equal to the steel yield strain (ε_{yd}) and the previous expression becomes:

$$\Sigma F_i l_i \varepsilon_{mi} = \Sigma F_i l_i \varepsilon_{yd} = \varepsilon_{yd} \Sigma F_i l_i = \text{ minimum } \rightarrow \Sigma F_i l_i = \text{ minimum }.$$

Finally, it is desirable to orient struts and ties the model along isostatic lines of the uncracked state. For this purpose, one of the following methods can be employed.

[1] According to Schlaich e al. (1987), the strain energy of struts can be neglected in the calculation of the strain energy of a S&T model. Nevertheless, some researchers (Brown and Bayrak, 2008) have verified that the Schlaich hypothesis is not always fulfilled. They calculated the strain energy for a series of 596 samples of beams with shear length not higher than twice the height of the cross-section; for each specimen they considered both the S&T model shown in Fig. 10.18b with only two inclined struts connecting the applied load to the supports and the S&T model shown in Fig. 10.18c with more struts and two vertical ties. For the first model, which has a very simple geometry, the strain energy of struts exceeds the strain energy of ties in 39% of cases; on the contrary, for the second model, which has a more complex geometry, only in 5% of cases the strain energy of struts overcomes that of ties. It can be deduced that, if the complexity of the S&T model increases, the contribution of concrete struts to the strain energy decreases with respect to the strain energy of ties and it can be neglected; for simple models with a few struts, on the contrary, the strain energy of struts cannot in general be neglected (Schlaich's hypothesis is not fulfilled).

Method based on the use of a known S&T model

This is the quickest method to define an S&T model and it consists of using one of the known S&T models for the element under investigation (10.6).

Load path method

This method developed by Schlaich is based on curvilinear load paths connecting forces applied on the boundary of the "D" region: each load applied on one side of the region finds its counterpart on the opposite side. Some paths can start and end on the same side showing a "U" shape, like for example the path B-B of the deep beam in Fig. 10.11. To draw the load paths, it is necessary to avoid intersections and it is appropriate to follow the shorter route. Curvilinear load paths are then substituted by polygonal load paths, which are finally integrated with other elements to guarantee the equilibrium of nodes.

Isostatic line method

The isostatic line method places struts and ties of the S&T model along isostatic lines from an elastic analysis of the "D" region in the uncracked phase.

For this purpose, a finite-element linear elastic analysis is performed and, where the principal stresses have almost the same intensity, sign and direction, the nodes of the finite element model are grouped. Each group of nodes is then subdivided into sub-groups formed by adjacent nodes and each of them is finally substituted by a

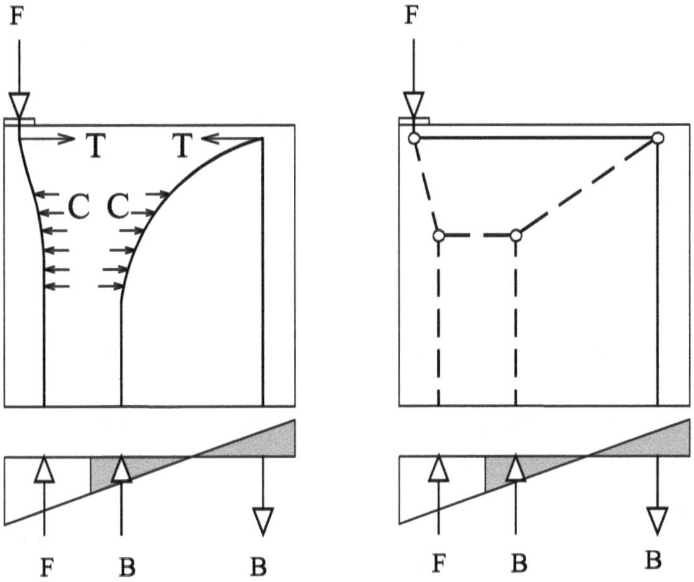

Fig. 10.11 Load paths for a deep beam (struts = dashed lines, ties = solid lines)

straight line passing through the centroid of the principal stresses of the sub-group and having direction equal to the mean direction of the group stresses.[2]

Elastic modulus reduction (iterative method)

A finite-element linear elastic analysis of the "D" region is performed, at the end of which the less stressed FE elements are identified, i.e. those where the strain energy is less than a small target value. The FE model is then modified by assigning to those elements a very small value of the elastic modulus, that is equivalent to remove them from the model.

The elastic analysis of the modified model is repeated; at the end of the second analysis, again the elastic modulus of the less stressed elements is reduced, and a new elastic analysis is performed. Further elastic analyses are made, and the elastic modulus is reduced in further elements, until only the most stressed elements, which identify the geometry of the S&T model, are kept. To accurately identify the S&T model, a dense FE mesh should be chosen.[3]

10.4 Kinematically Unstable S&T Models

S&T model can be kinematically unstable, being equilibrium possible only for the considered load condition. For example, the S&T model shown in Fig. 10.12b is suitable for a deep beam under two symmetrical concentrated loads, but not for the analysis of the same beam under two non-symmetrical concentrated loads, under which the S&T model is not in equilibrium. It is then necessary to define a new S&T model, as shown in Fig. 10.13.

Alternatively, the S&T model for a deep beam under two non-symmetrical concentrated loads can be obtained as the mirror of the bending moment diagram of a slender beam with the same span and subjected to the same loads (Fig. 10.14).

It is worth noting that also the S&T model obtained by this way is kinematically unstable, but statically determined for the considered load condition.

10.5 Practical Rules for the Identification of the S&T Model

For the identification of the S&T model, Model Code 1990 (CEB/FIP, 1991) suggests some practical rules, discussed below.

---

[2] Harisis and Fardis (1991), Computer-Aided Automatic Construction of Strut-and-Tie Models. Structural Concrete, IABSE Colloquium, Stuttgart, International Association for Bridge and Structural Engineering, Zürich, p. 533–538, Mar.

[3] Angotti and Spinelli (2001), The method of the elastic modulus reduction for the identification of S&T models in RC structures: some considerations and experiences (in Italian), Proceedings of the Workshop S&T-2001, Florence, March 16, 2001, Ed. Centro Stampa 2P—Florence, pp. 3–14.

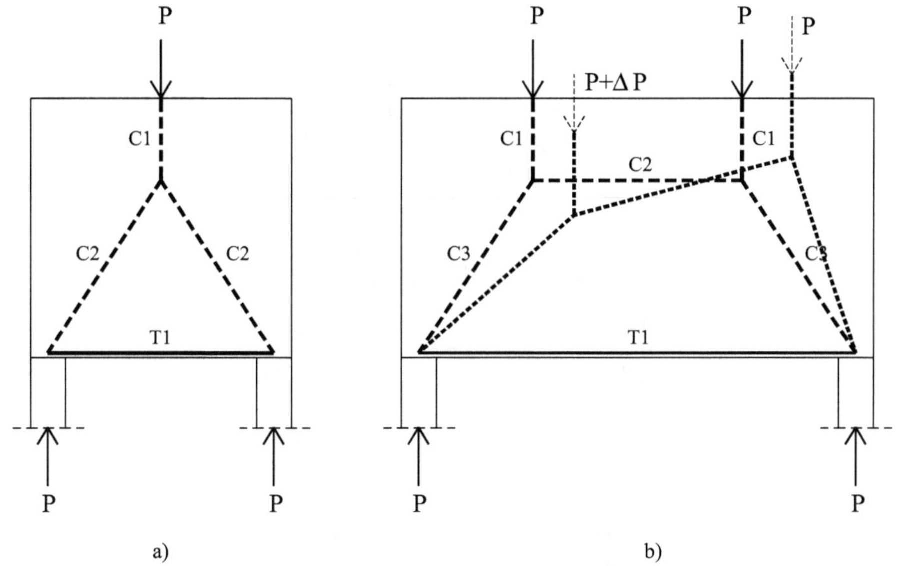

Fig. 10.12 Example of an isostatic S&T model (on the left) and of a kinematically unstable but statically determined S&T model (on the right) (*the Figure on the right also shows the mechanism due to the application of a load increment* ΔP *only to the left load*)

Simple S&T models, with a few elements, can be adopted in the beginning and improved later (Fig. 10.15). Struts and ties shall be oriented along stress trajectories of the elastic uncracked state to well reproduce the mean stress fields in the elastic range. However, to simplify the reinforcement layout it is convenient to adopt parallel or orthogonal ties at the edges of the structural element (Fig. 10.15c), where stresses are low. In less stressed regions, the structure is capable to adapt itself to the geometry of the selected model, even if this is a bit far away from the stress field of the uncracked phase.

Angles between struts and ties shall be at least 45°, except for nodes where a strut intercepts two orthogonal ties when EC2 provides a reduction of the concrete design strength of 25% compared to nodes subjected to compression loads only (see 10.10.5); angles less than 30° shall be avoided (Fig. 10.16).

The limitation of the angle between struts and ties at nodes is useful to limit cracking and avoiding that the shortening of struts and the elongation of ties occur almost in the same direction. The same limitation on the angle can be also found in the variable inclination truss model for the shear design of slender beams, where the angle θ between struts and longitudinal reinforcing bars cannot take values lower than 21.8° ($\cot\theta \leq 2.5$).

The angle of spread of concentrated loads is about 32.5°, as can be deduced from the elasticity theory in an elastic half-space (Fig. 10.17). Nevertheless, the angle of spread of concentrated loads varies as a function of shape and boundary conditions of the "D" region, consequently, the S&T model should be adapted. With reference

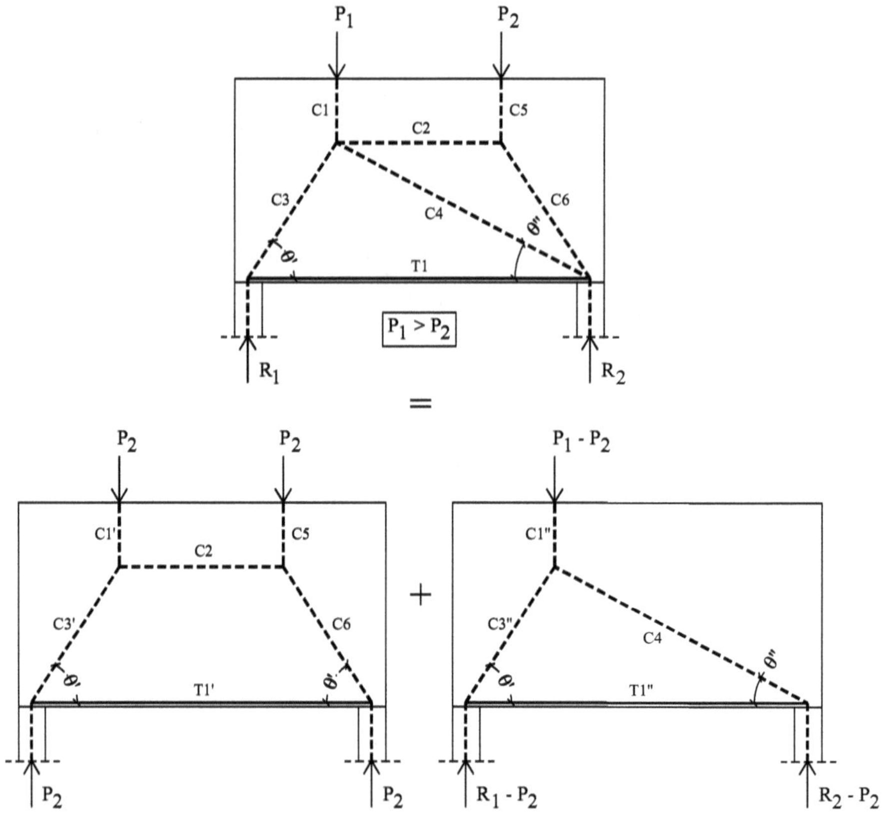

Fig. 10.13 Example of an S&T model statically determined for a deep beam subjected to two non-symmetrical concentrated loads; truss can be imagined as the superposition of a kinematically unstable truss (on the left, at the bottom) and an isostatic truss (on the right, at the bottom)

to the deep beam shown in Fig. 10.18, at increasing the ratio L/h, the angle δ of spread increases and the angle γ between inclined strut and horizontal tie decreases; for $L = h$ (Fig. 10.18a), $\delta \cong 26.5°$ and $\gamma \cong 63.5° > 45°$, while for $L = 3h$ (Fig. 10.18b), $\delta \cong 56.5° e\ \gamma \cong 33.5° < 45°$.

In case (b) of Fig. 10.18, it is convenient to modify the S&T model to obtain an angle γ not smaller than 45°, by inserting on each side a vertical tie placed in the middle between the support and the applied load. Then the truss shown in Fig. 10.18c is drawn, which can also be combined with the model (b) to obtain a statically indeterminate truss. For the evaluation of strut and tie forces in the last S&T model, the stiffness of struts and ties should be considered.

The axes of struts shall be placed at a minimum distance from the edges of the structural element to take into account the transverse size of struts themselves (Fig. 10.19); analogously, the same holds for ties formed by reinforcing bars arranged on more layers and for nodes.

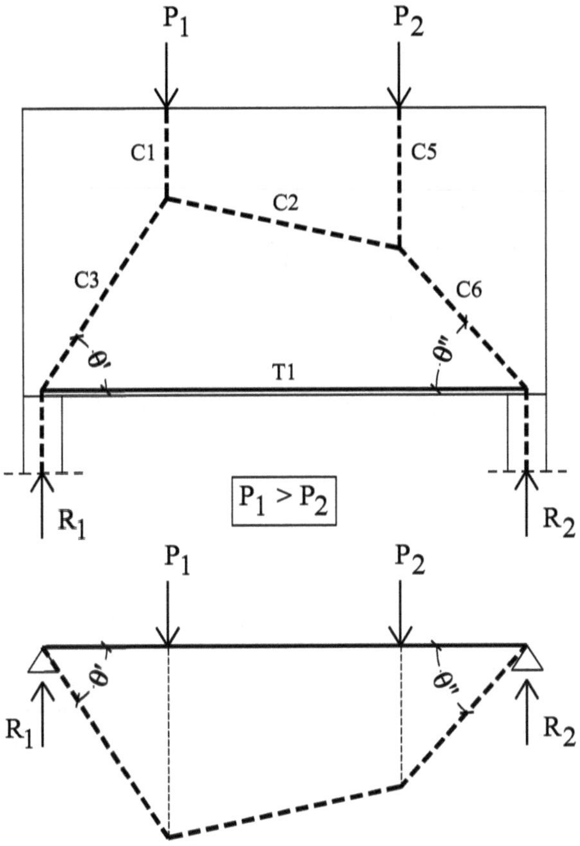

Fig. 10.14 Deep beam under two non-symmetrical concentrated loads: identification of the S&T model starting from the bending moment diagram for a slender beam with identical span and loads

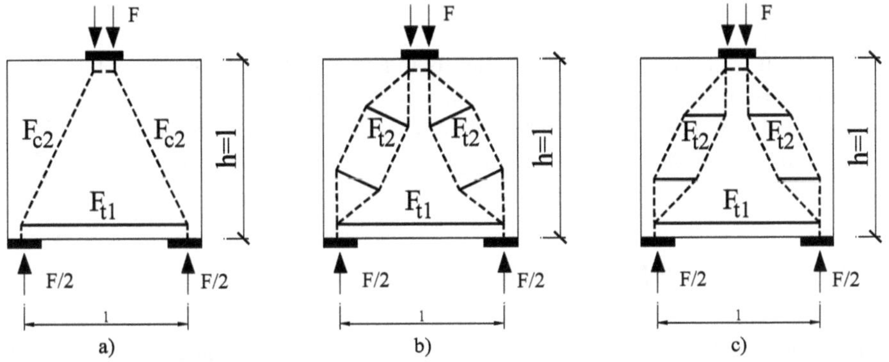

Fig. 10.15 Deep beam under a concentrated load: **a** S&T model, **b** refined model with inclined ties, **c** refined model with horizontal ties (struts = dashed lines, ties = solid lines)

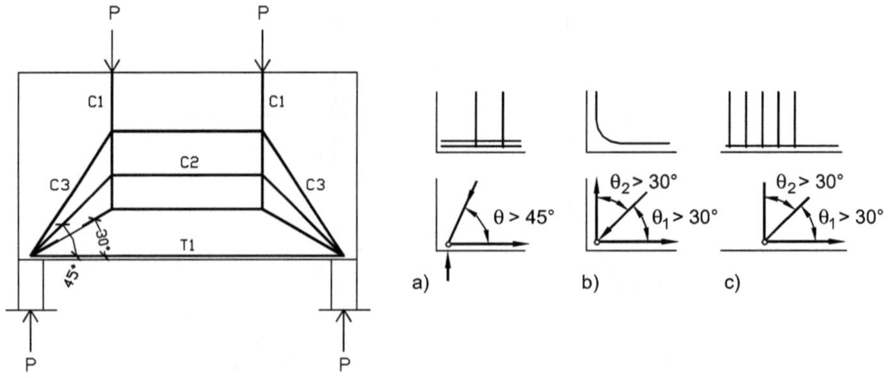

Fig. 10.16 S&T models for a deep beam with different angles between inclined struts and horizontal ties, variable from 30° (minimum value) to more than 45°: the intersection of a strut with one **a** or two orthogonal ties **b** and **c**

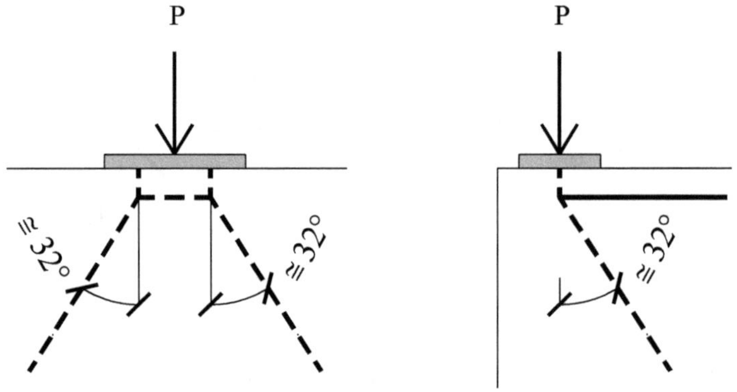

Fig. 10.17 Angle of spread of a concentrated load (from elasticity theory)

Finally, it is possible to adopt a statically indeterminate truss, given by the superposition of two different S&T models, each of which is subjected to a portion of the applied load. Examples are given by the S&T model of the deep beam in Fig. 10.18d and by the S&T model used for the design of corbels (see Examples no. 7, 8 and 9).

10.6 Common S&T Models

It is often possible to identify the same S&T model in structures that seem very different from each other. In the following, three common S&T models are provided, regarding Table 10.2.

- the spread of a centred concentrated load (D1);

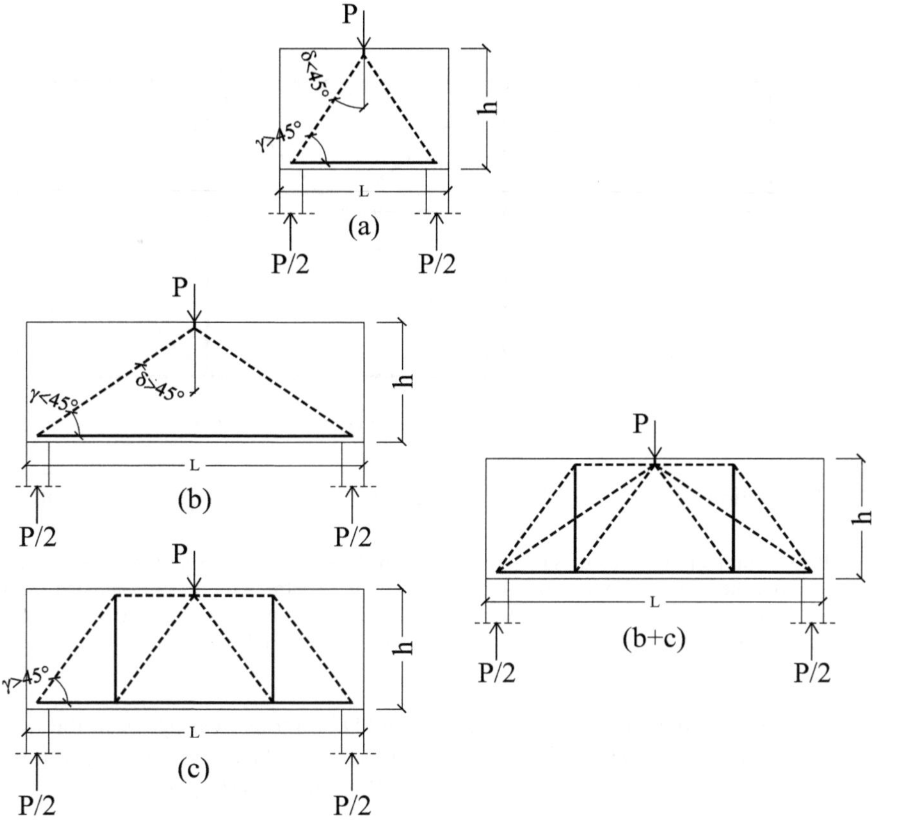

Fig. 10.18 S&T models of deep beams with different span-to-depth ratios

– the spread of an eccentric concentrated load (D2);
– a single-span deep beam uniformly loaded on the top edge (D3).

10.6.1 Spread of a Concentrated Load Within a Strut (D1)

A strut with width b and height H is assumed to be subjected to two opposite forces F at both ends. Within the strut, it is possible to have "B" and "D" regions or only "D" regions, depending on the ratio H/b.

Each force F spreads within a "D" region, which extends for a length b from the loaded end according to Saint Venant's principle. If the strut height H is greater than twice its width b (*partial discontinuity*) (Fig. 10.20a), the regions "D" at the two ends do not occupy the whole strut volume, but a "B" region with height $(H - 2b)$exists in the middle, where isostatic lines are parallel to the strut axis.

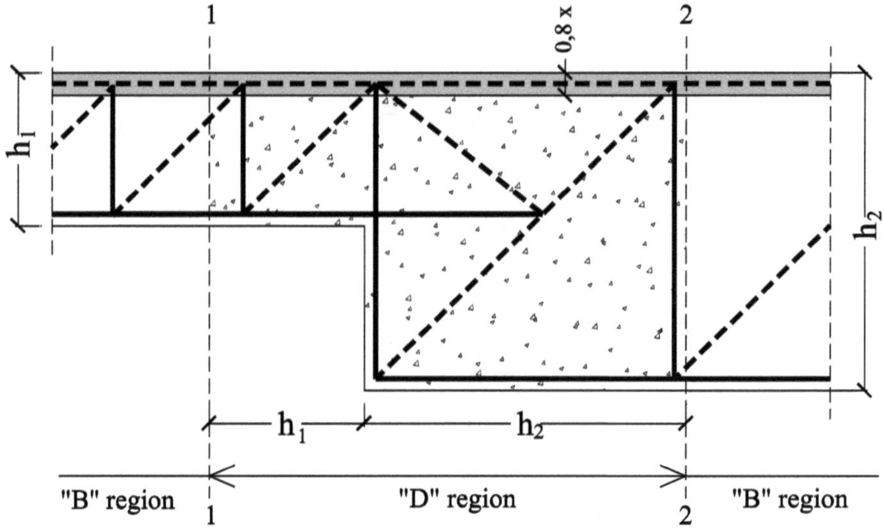

Fig. 10.19 Minimum distance of struts and ties from the edges: the top horizontal strut has a distance from the edge not less than 0.4 x, where x is the neutral axis depth calculated in the adjacent "B" region (it can be assumed $x = \max(x_1; x_2)$ where x_1 and x_2 are the neutral axis depths in section 1–1 on the left of the "D" region and section 2–2 on the right)

Conversely, in the case of $H \leq 2b$, the two "D" regions overlap (or at the limit they are contiguous for $H = 2b$) and the two loads spread over a reduced width b_{ef}, while the central "B" region disappears (*total discontinuity*) (Fig. 10.20b).

10.6.1.1 Partial Discontinuity Regions

If the strut has a width b less than half of the height H, the tensile force T orthogonal to the strut axis can be evaluated by adopting an S&T model which reflects the scheme for the spread of a concentrated load as already adopted by Mörsch (Fig. 10.21a).

The rotational equilibrium of the inclined strut 1–2 gives (Fig. 10.22c):

$$T\frac{b}{2} = \frac{F}{2}\left(\frac{b}{4} - \frac{a}{4}\right) \Rightarrow T = \frac{F}{4}\frac{b-a}{b} \Rightarrow \frac{T}{F} = 0.25\left(1 - \frac{a}{b}\right) \quad [(6.58)]$$

The slope of struts is measured from the angle θ with the horizontal line; the tangent of θ is a function of the ratio *a*/*b* between the slab width and cross-section width:

$$\tan\theta = \frac{b/2}{b/4 - a/4} = \frac{2}{1 - a/b}$$

Table 10.2 Common S&T models

Tag	Description	Geometry	Where it can be used
D1	Spread of concentrated loads within a strut (10.6.1)	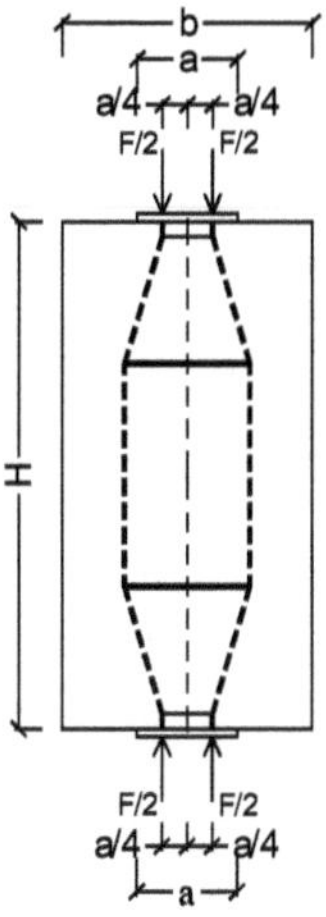	Struts of all S&T models (except prismatic or fan-shaped compression fields)
D2	Spread of an eccentric load (10.6.2)	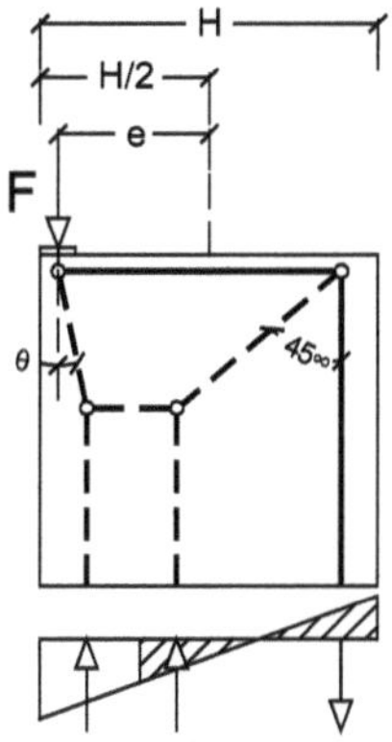	Eccentric load or prestress force
D3	Single-span deep beam uniformly loaded on the top edge (10.6.3)	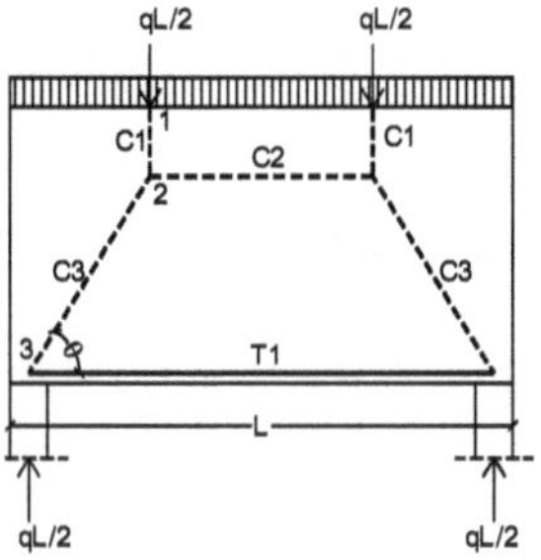	Deep beam Concentric prestress force with two symmetric tendons close to the cross-section edges (*imagine replacing columns with two cable heads in the figure*)

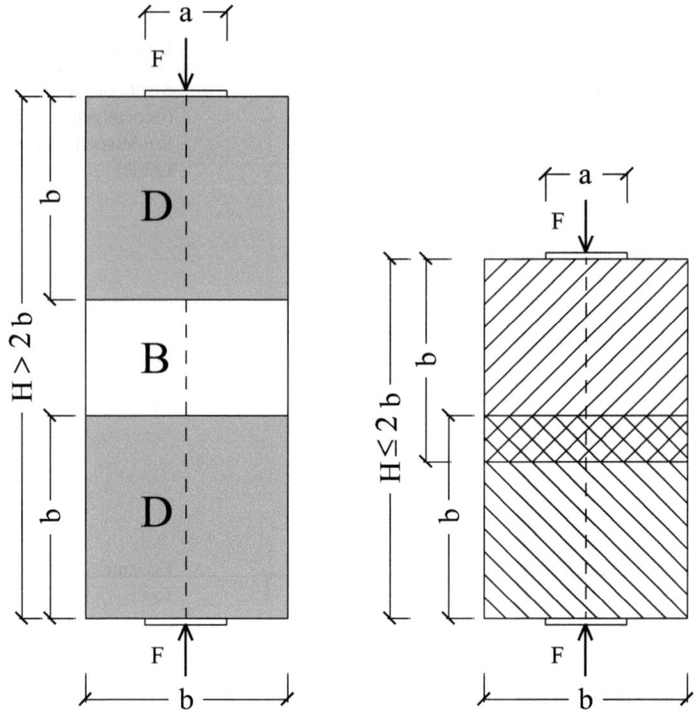

Fig. 10.20 Strut with partial ($H > 2b$) or total discontinuity ($H \leq 2b$))

Table 10.3 collects the values of the ratio T/F, tanθ and θ angle as the ratio a/b varies; for $a/b = 0$ following values hold: $T = 0.25F$, $\tan\theta = 2$ and $\theta = 63.43°$.

The diagram of Fig. 10.23 shows the theoretical elastic curves and the approximated curves obtained with the S&T model for the ratio T/F and the angle θ as the ratio *a*/*b* varies.

The anti-burst reinforcement is arranged as shown in Fig. 10.24.

10.6.1.2 Total Discontinuity Regions

When $H \leq 2b$, the spread of the force F on the width b_{ef} is assumed to be given by (Fig. 10.25b):

$$b_{ef} = 0.5H + 0.65a$$

As for the previous case, the lever arm z of internal forces is assumed to be half of the height of the spreading region: $z = h/2 = H/4$.

The rotation equilibrium of the inclined strut gives:

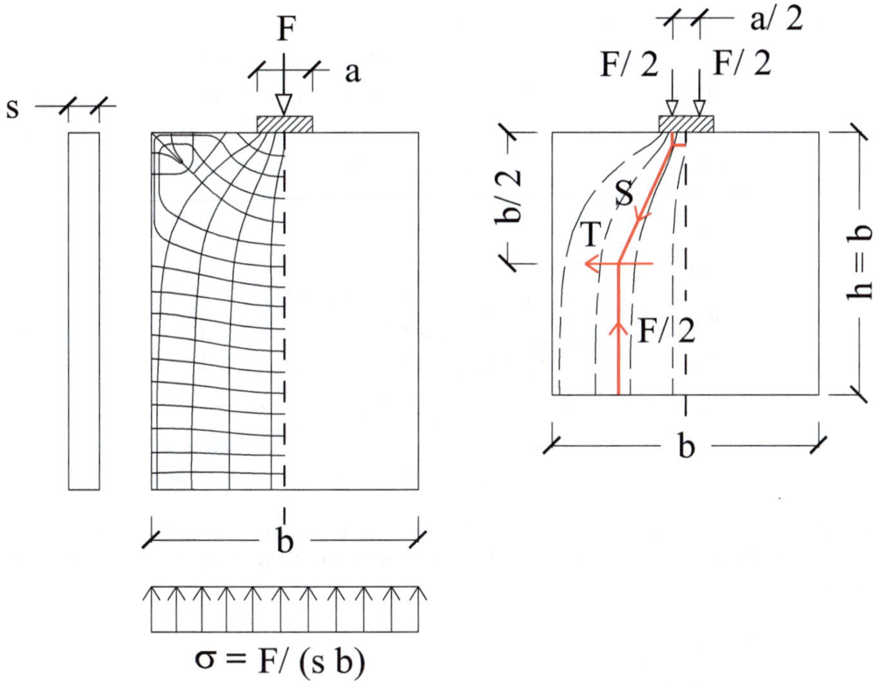

Fig. 10.21 Spread of a concentrated load according to Mörsch

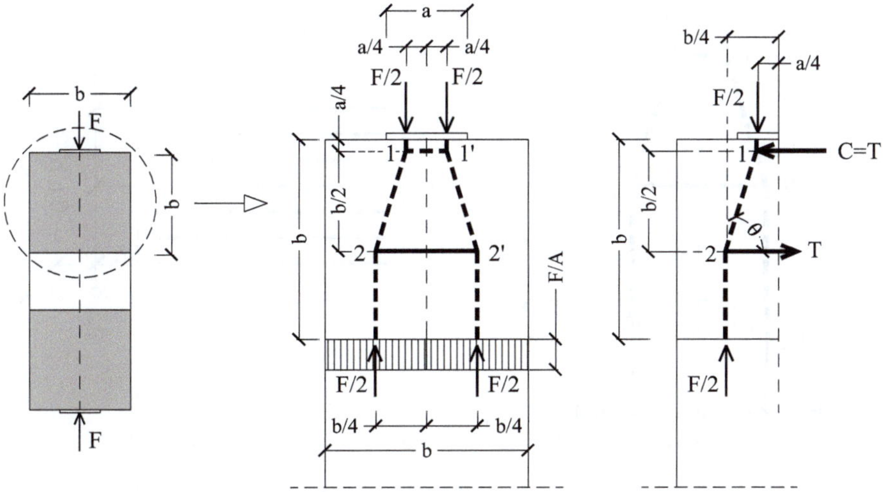

Fig. 10.22 S&T model of a strut with partial discontinuity (struts = dashed lines, ties = solid lines) [Fig. 6.25]

Table 10.3 T/F and θ angle of the slope of struts for partial discontinuity versus a/b ratio

a/b	0	0.1	0.2	0.3	0.4	0.5	0.6	0.7	0.8	0.9
T/F	0.25	0.23	0.20	0.18	0.15	0.13	0.10	0.08	0.05	0.03
$\tan\theta$	2.00	2.22	2.50	2.86	3.33	4.00	5.00	6.67	10.00	20.00
$\theta(°)$	63.43	65.77	68.20	70.71	73.30	75.96	78.69	81.47	84.29	87.14

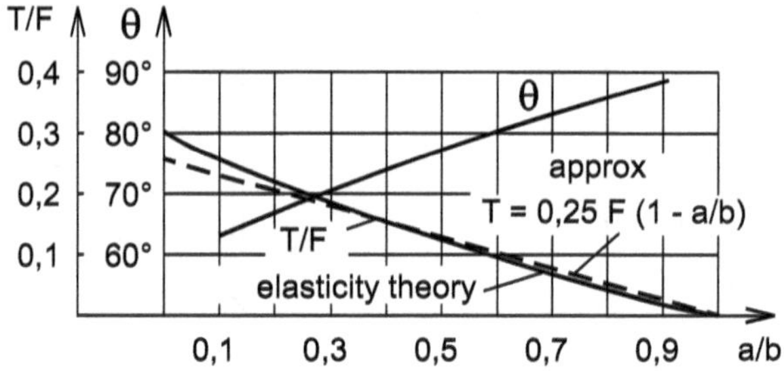

Fig. 10.23 Curves of transverse force and slope of concrete struts versus *a/b* ratio (continuous curves are obtained with the S&T model of Fig. 10.22, while the dashed curve is obtained by elastic analysis)

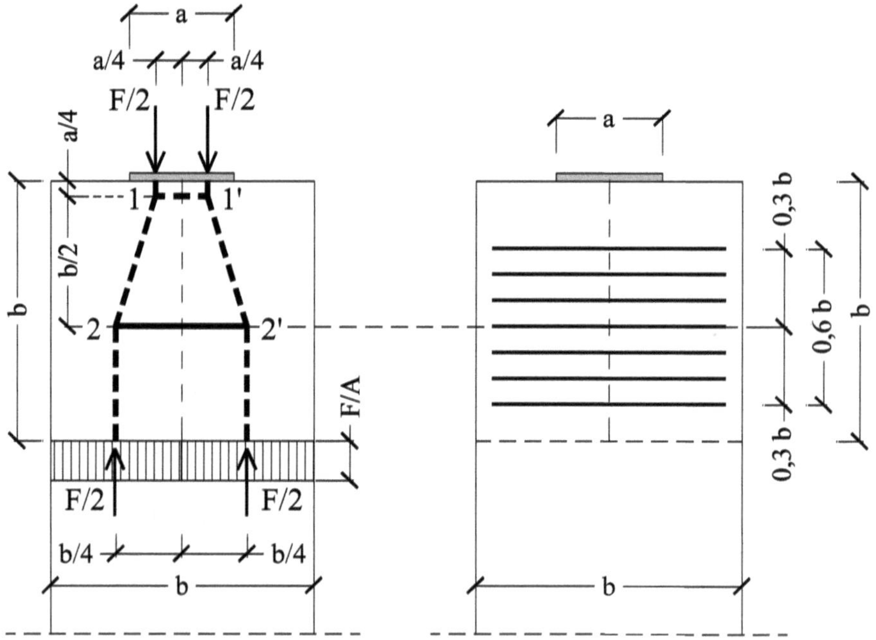

Fig. 10.24 Layout of anti-burst reinforcement (reinforcing bars are centred on the horizontal strut 2–2' of the S&T model and are smeared over a length equal to 0.6 *b*)

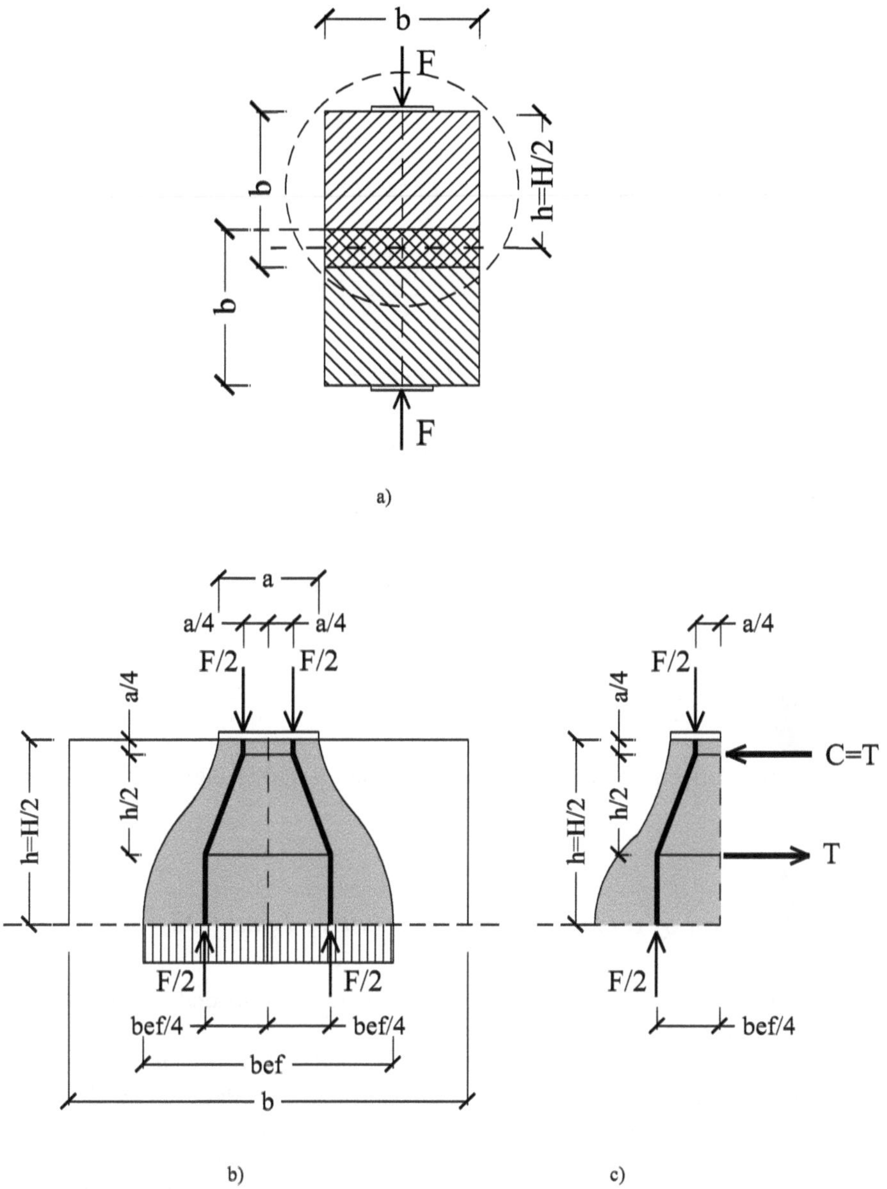

Fig. 10.25 S&T model of a strut with total discontinuity [Fig. 6.25]: **a** superposition of "D" regions in the strut central zone; **b** S&T model; **c** rotation equilibrium of half model

Table 10.4 T/F, tan θ and θ angle of the slope of struts in case of total discontinuity versus ratio *a*/*H*

a/*H*	0	0.05	0.1	0.15	0.2	0.25	0.3	0.35	0.4	0.45	0.5
T/F	0.25	0.24	0.23	0.22	0.22	0.21	0.20	0.19	0.18	0.17	0.16
tan θ	2.00	2.07	2.15	2.23	2.33	2.42	2.53	2.65	2.78	2.92	3.08
θ(°)	63.43	64.24	65.06	65.89	66.73	67.58	68.45	69.32	70.20	71.09	72.00

$$T\,\frac{h}{2} = \frac{F}{2}\left(\frac{b_{ef}}{4} - \frac{a}{4}\right)$$

which becomes, by substituting $h/2 = H/4$ and $b_{ef} = 0.5H + 0.65a$:

$$T\frac{H}{4} = \frac{F}{2}\frac{0.5H - 0.35a}{4} \Rightarrow T = \frac{F}{2}\left(0.5 - 0.35\frac{a}{H}\right) = \frac{F}{4}\left(1 - 0.7\frac{a}{H}\right)$$

$$\Rightarrow \frac{T}{F} = 0.25\left(1 - 0.7\frac{a}{H}\right) \quad [(6.59)]$$

The tangent of the angle θ measures the slope of struts 1–2 and 1'-2' and is given by the following expression:

$$\tan\,\theta = \frac{H/4}{b_{ef}/4 - a/4} = \frac{H}{(0.5H + 0.65a) - a} = \frac{1}{0.5 - 0.35a/H}.$$

Table 10.4 collects the values of ratio T/F, tan θ and θ angle as the ratio *a*/*H* varies.

10.6.2 Spread of a Concentrated Eccentric Load (D2)

A "D" region, bounded by a "B" region at the bottom and loaded on the top by an eccentric load close to one of the corners, is considered.

The S&T model can be identified by using the load path method by Schlaich. For this purpose, the diagram of normal stresses on the bottom edge is divided into two parts (Fig. 10.26): the left part with a resultant force equal to F and the right part with a resultant force equal to zero (two equal and opposite forces B).

The load path associated with the two equal and opposite forces B enters in the "D" region in point B_1 and exits it in point B_2, following a curvilinear U-shaped path (Fig. 10.26a), associated with a distribution of transverse loads C which deviate the compression forces. By approximating the curvilinear load path with a polygon made of compression and tension members, the S&T model to be used for the design of the "D" region is drawn (Fig. 10.26b).

In drawing the geometry of the S&T model, angular deviations of compressed struts shall not be greater than 60° (Fig. 10.27), as the angle between struts and ties at

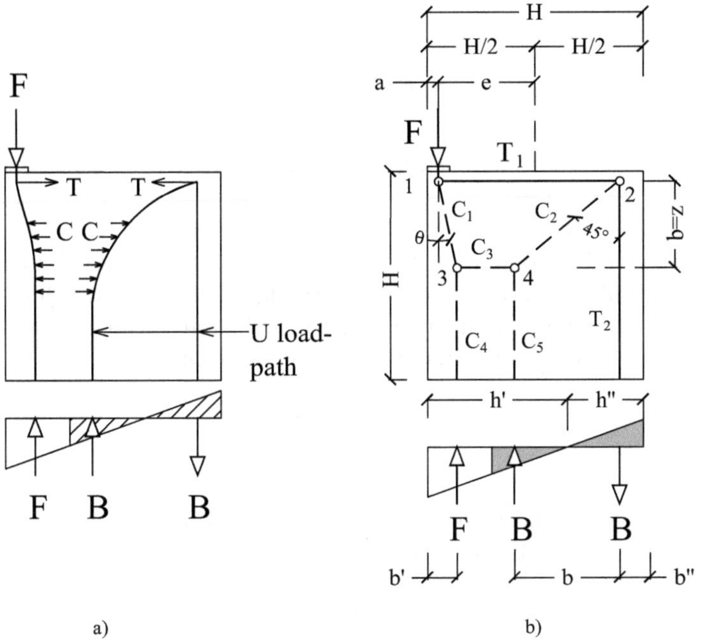

Fig. 10.26 "D2" model: **a** load paths; **b** S&T model with the identification of struts, ties and nodes

Fig. 10.27 Maximum deviation (< 60°) of the strut axis

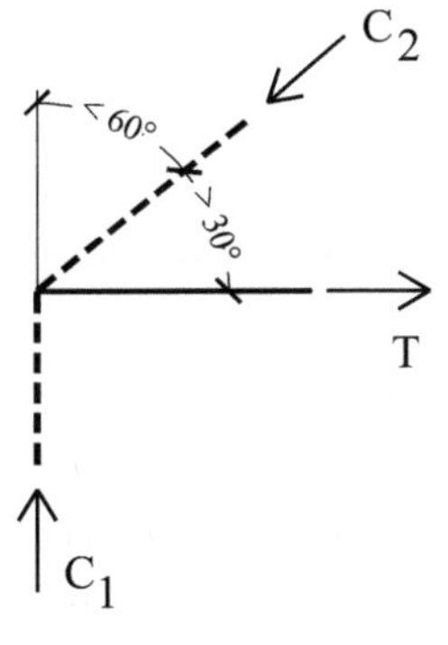

the node where the deviation occurs shall be at least equal to 30° (this is the minimum value to assure the strain compatibility, as already underlined in para. 10.5 devoted to the practical rules for the identification of the S&T geometry).

The lever arm z of internal forces can be assumed equal to the distance between the two forces B, which is equivalent to assume that the angle between strut C_2 in Fig. 10.26b and the horizontal is 45°.

In the following, the calculation of forces in the truss members is described.

In Fig. 10.26b, lengths h' and h'' identify the point where the normal stress is zero on the cross-section at distance H from the top, where H is the width of the D region.

They are given by:

$$h' = (H/2)\,[1 + H/(6e)]$$

$$h'' = (H/2)\,[1 - H/(6e)]$$

with the limitation $H/6 < e < H/2$, where the lower bound $(H/6)$ guarantees the normal stress distribution shown in Fig. 10.26b (otherwise the normal stress diagram would be triangular for $e = H/6$ and trapezoidal for $e < H/6$), while the upper bound $(H/2)$ corresponds to the limit case where the force is applied on the corner.

The distance of the force F from the left edge is given by:

$$b' = \left(\frac{H}{e}\right)^2 \frac{18e - H}{216},$$

while the distance b of the couple of forces B and the distance b'' from the right edge of the force B applied in B_2 are given by:

$$b'' = h''/3 = (H/6)[1 - H/(6e)],$$

$$b = 4h''/3 = (2H/3)[1 - H/(6e)].$$

Once $z = b$ is assigned, the angle θ is finally given by:

$$\theta = \arctan\left[\left(b' - a\right)/z\right].$$

Table 10.5 collects the values b'/H, b''/H, a/H and b/H as the ratio e/H varies.

Table 10.5 $b'/H, b''/H, a/H, b/H$ and θ versus e/H

e/H	b'/H	b''/H	a/H	$b/H(=z/H)$	θ (°)
0.1666	0.333	0.000	0.333	0.000	0.000
0.2	0.301	0.028	0.300	0.111	0.477
0.25	0.259	0.056	0.250	0.222	2.386
0.3	0.226	0.074	0.200	0.296	5.080
0.35	0.200	0.087	0.150	0.349	8.197
0.4	0.179	0.097	0.100	0.389	11.539
0.45	0.162	0.105	0.050	0.420	14.981
0.5[a]	0.148	0.111	0.000	0.444	18.435

[a]The value $e/H = 0.5$ is theoretical, because the maximum eccentricity of the force F applied on the region D is strictly less than 0.5; nevertheless, the values obtained for $e/H = 0.5$ can be useful for design purposes

Once the geometry of the model is known, it is possible to calculate all the forces of the truss members; for this purpose, it is sufficient to write down the equilibrium equations to vertical and horizontal translation for every node.

***Node* 1**

Vertical translational equilibrium:

$$C_1 \cos\theta = F \; C_1 = F/\cos\theta (< 1.054F),$$

where 1.054 F is the upper bound obtained for the limit case $e/H = 0.5$ (from Table 10.5, $\theta = 18.435°$, $\cos\theta = 0.9487$, $1/\cos\theta = 1.054$).

Horizontal translational equilibrium:

$$T_1 = C_1 \sin\theta \rightarrow T_1 = (F/\cos\theta)\sin\theta = F\tan\theta$$

***Node* 2**

Horizontal translational equilibrium:

$$T_1 = C_2 \sin 45° \rightarrow C_2 = \sqrt{2}T_1 = \sqrt{2}F\tan\theta$$

Vertical translational equilibrium:

$$T_2 = C_2 \cos 45° \rightarrow T_2 = (\sqrt{2}F\tan\theta)/\sqrt{2} = F\tan\theta = B$$

***Node* 3**

Vertical translational equilibrium:

$$C_4 = C_1 \cos\theta = F$$

Horizontal translational equilibrium:

$$C_3 = C_1 \sin\theta = (F/\cos\theta)\sin\theta = F\tan\theta$$

***Node* 4**

Vertical translational equilibrium:

$$C_5 = C_2 \cos 45° = \sqrt{2}F\tan\theta/\sqrt{2} = F\tan\theta$$

Figure 10.28 shows the diagram of the ratio B/F, as a function of the ratio e/H; Table 10.6 lists values of forces of the truss members expressed as functions of F and B.

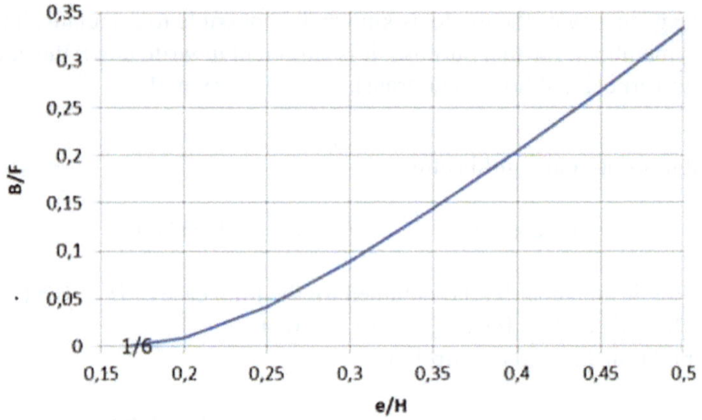

Fig. 10.28 Ratio *B/F* versus ratio *e/H* (normal forces in ties T_1 and T_2 and struts C_3 and C_5 have the same intensity of force *B*)

Table 10.6 Forces of the truss members expressed as functions of *F* and *B*

C_1	$F/\cos\theta$	$B/\sin\theta$
C_2	$\sqrt{2}F\tan\theta$ θ	$\sqrt{2}B$
C_3	$F\tan\theta$	B
C_4	F	$B/\tan\theta$
C_5	$F\tan\theta$	B
T_1	$F\tan\theta$	B
T_2	$F\tan\theta$	B

10.6.3 Single-Span Deep Beam Uniformly Loaded on the Top Edge (D3)

A single-span deep beam uniformly loaded on the top edge is considered. The S&T model is shown in Fig. 10.29.

Once the distributed load is discretized into two concentrated forces (10.2.3) and the position of the bottom tie T_1 is fixed, the truss geometry is known only after fixing the position of the horizontal strut C_2. For this purpose, according to paragraph 6.8.2.1 of Model Code 1990, the lever arm of internal forces can be adopted with a length equal to $0.6 \div 0.7$ times the distance L' between the supports, but not greater than the lever arm of a slender beam with the same span: $(0.6 \div 0.7)\ L' \leq 0.67H$. Forces in the truss members are collected in Table 10.7.

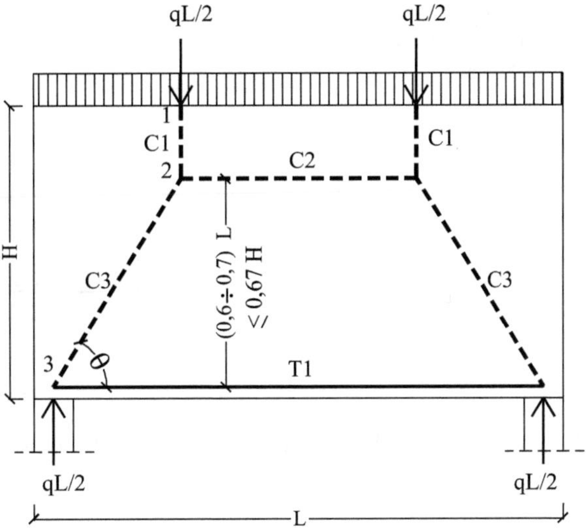

Fig. 10.29 Single-span deep beam uniformly loaded on the top edge

Table 10.7 Forces in the truss members of the deep beam (D3)

C_1	$qL/2$
C_2	$qL/(2\tan\theta)$
C_3	$qL/(2\sin\theta)$
T_1	$qL/(2\tan\theta)$

10.7 Verification of Members and Nodes of the S&T Model

Failure of an S&T model can be due to:

- yielding of one or more ties,
- crushing of a concrete strut,
- crushing of a node,
- pulling out of one or more ties in a node.

If a reinforced concrete structural element has been correctly designed by using an S&T model, failure occurs due to the yielding of one or more ties.

The limitation of compression forces for concrete struts (see 10.9 and 10.10), the choice of suitable angles (not too small) between struts and ties, and all the considerations already discussed in 10.5 allow extending the plasticity theory to reinforced concrete structures, even if this material does not completely fulfil the hypotheses of this theory.

Concrete stresses are limited to avoid concrete localized crushing or longitudinal cracking ("splitting") of struts and nodes; maximum allowable concrete stresses are chosen based on the confinement level of the concrete. In this way, in nodes with three or more struts, it is possible to accept higher stress levels, thanks to the

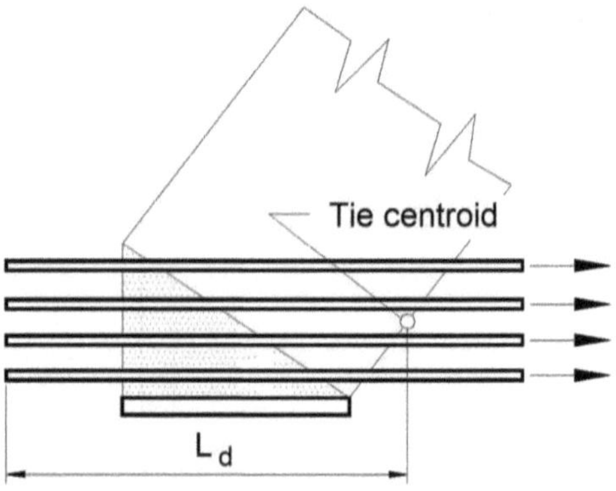

Fig. 10.30 Straight anchorage of a reinforcing bar in a node

high confinement level of the material, while in nodes where one or more ties are convergent, the stress level has to be smaller.

10.8 Reinforcement Design

Reinforcing bars are used both for ties of S&T models (main reinforcement) and for elements capable to support tension forces produced by the spread of the load, which occurs in the direction normal to compression fields (anti-burst reinforcement, 10.6.1). The cross-sectional area A of each tie is obtained by dividing the design normal force N_{Ed} by the steel design strength: $A \geq N_{\mathrm{Ed}}/f_{xd}$.

Main reinforcement shall be smeared over the node height where it is anchored; ties formed by more than one bar shall be arranged on more layers to avoid the "congestion" of reinforcing bars and to enhance the shape of the node. In each node, reinforcing bars can be anchored by steel anchor plates, loops or U-bar loops; these last ones shall have a minimum curvature radius satisfying [8.3]. For straight anchorages, the anchorage length is chosen according to rules given in [8.4], as discussed in Chap. 12.

It is assumed that the anchorage length starts at the cross-section where the centre-line of the reinforcing bar or bars intercepts the internal edge of the strut (Fig. 10.30). Reinforcing bars shall be extended at least to the opposite face of the node, namely they shall pass through it completely.

10.9 Verification of Struts

Compression fields, i.e. struts, are basically of three types (Fig. 10.31): parallel or prismatic, "bottle-neck" shaped and "fan-shaped".

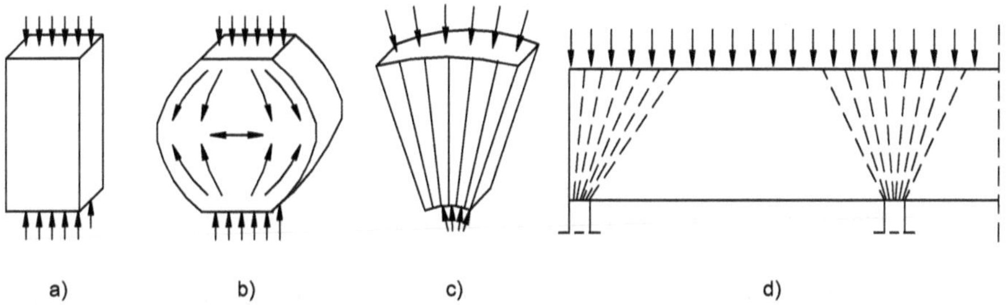

Fig. 10.31 Compression fields: **a** "prismatic", **b** "bottle-neck", **c, d** "fan-shaped"

Prismatic fields (a) are typical of "B" regions and can be treated as columns subjected to axial compression load or as the compression chord of the truss model for the shear design of slender beams.

On the contrary, stress fields (b), (c) and (d) reproduce the transverse spread of strut compression forces: "bottle-neck" shaped struts represent bi-axial or three-axial stress fields in regions where concentrated loads are applied, while "fan-shaped" struts represent mainly straight (negligible curvature) compression fields and therefore with negligible transverse forces. For example, the last case well represents compression fields in a deep beam under a uniformly distributed load, where "fan-shaped" struts develop from the loaded surface at the top to the supports (Fig. 10.31d).

The verification of struts is performed by checking that the maximum concrete compression stress is smaller than the design strength. Moreover, for "bottle-neck" compression fields it is necessary to design a reinforcement suitable to support transverse tension forces calculated using models of 10.6.1.

In the absence of transverse tension forces or presence of transverse compression forces, the design strength of a concrete strut is equal to the concrete design strength:

$$\sigma_{\mathrm{Rd,max}} = f_{cd} \quad [(6.55)]$$

while, in the presence of transverse tensile stresses, the design strength is smaller than the concrete strength and it is given by:

$$\sigma_{\mathrm{Rd,max}} = 0.60\nu' \mathrm{f_{cd}} \quad [(6.56)]$$

where:

$$\nu' = 1 - f_{ck}/250 = (5/3)\nu \quad [(6.57\mathrm{N})]$$

with

$$\nu = 0.6\,(1 - f_{ck}/250)\,\left(f_{ck} \text{ in N/mm}^2\right) \quad [(6.6\mathrm{N})].$$

Table 10.8 Design strength of concrete struts

Transverse stresses	Design strength of struts
Zero or compressive	f_{cd}
Tensile	$0.60\nu' f_{cd}$

The reduced value of the strut strength in presence of transverse tensile stresses takes into account that cracking induced by transverse forces, by interrupting the strut solidity, reduces the concrete strength, even when cracks are parallel to the strut. The strength reduction is therefore emphasized for cracks inclined to the strut direction. Table 10.8 collects expressions to be used for the design strength of concrete struts.

Finally, Fig. 10.32 shows the diagram of the design strength of concrete struts at varying the angle θ_s between the strut axis and the direction of transverse strains (i.e., associated to a tie crossing the strut); the diagram is obtained by applying the expression proposed by AASHTO (2007):

$$\sigma_{\mathrm{Rd,max}} = \frac{f_{cd}}{0.8 + 170\,\varepsilon_l} \leq 0.85\, f_{cd}$$

with

$$\varepsilon_l = \varepsilon_5 + (\varepsilon_5 + 0.002)\mathrm{cotg}^2\theta_s$$

where ε_1 main tensile strain in the direction orthogonal to the strut axis,
ε_s strain in reinforcing bars inclined at an angle θ_s to the strut axis.

The expression of ε_1 is suitable for principal compressive strain $\varepsilon_2 = 0.002$ in the strut direction; as θ_s decreases, strain ε_1 increases, and the strength $\sigma_{\mathrm{Rd,max}}$ decreases.

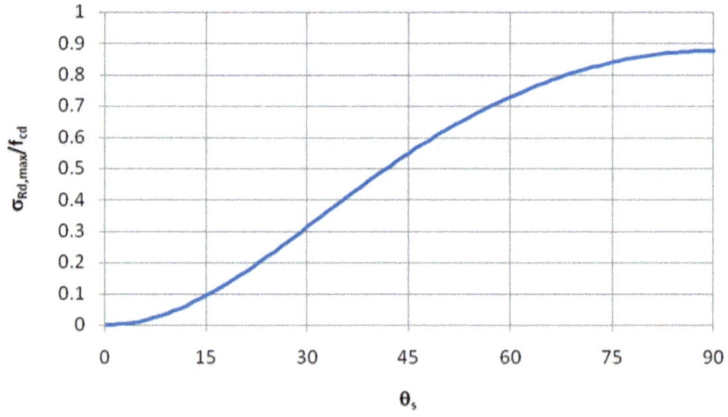

Fig. 10.32 The strength of concrete struts in presence of a transverse strain $\varepsilon_s = 2\%$ versus the angle θ_s between the strut axis and the direction of the transverse strain (AASHTO, 2007)

10.9.1 Transverse Reinforcing Bars

In general, it is not necessary to perform the verification of concrete struts if the verification of nodes is satisfied and a suitable transverse reinforcement is arranged. The total transverse force between the end and the mid of a concrete strut can be assumed not higher than 25% of the strut compression force, as per the S&T model "D1" for the spread of a concentrated load (see 10.6.1).

10.10 Verification of Nodes

In an S&T model, a node is defined as the concrete volume within the intersection of strut compression fields among them and with reinforcing bars and/or external forces.

Nodes are "critical zones" because they experience abrupt changes in the direction of forces and are then subjected to a concentration of internal forces. According to their geometry and extension, nodes are classified as "concentrated" and "diffused": in "concentrated" nodes, internal forces are deviated in a very restricted area compared to the length of converging elements (Fig. 10.33a, b), while in "diffused" nodes the intersection area is more extended (Fig. 10.33c, d). Normally, "diffused" nodes are less critical, and they do not require the verification of concrete compression forces; in "concentrated" nodes, on the contrary, both the verification of the maximum concrete compression force and the anchorage of reinforcements are required.

Examples of concentrated nodes are the following: application points of concentrated loads, supports, anchorage zones with a concentration of ordinary or prestressing steel, reinforcement hooks and bends, frame corners.

Forces in a node must be in equilibrium; moreover, for three-dimensional nodes, it is necessary to consider the spread of forces in two orthogonal planes (10.10.4.1).

The node strength is strictly connected to the dimensions and the layout of reinforcing bars, especially to the anchorage type.

10.10.1 Types of Nodes

It is possible to identify four different types of nodes, depending on the members merging on them (Fig. 10.34):

- CCC: three struts,
- CCT: two struts and one tie,
- CTT: one strut and two ties,
- TTT: three ties.

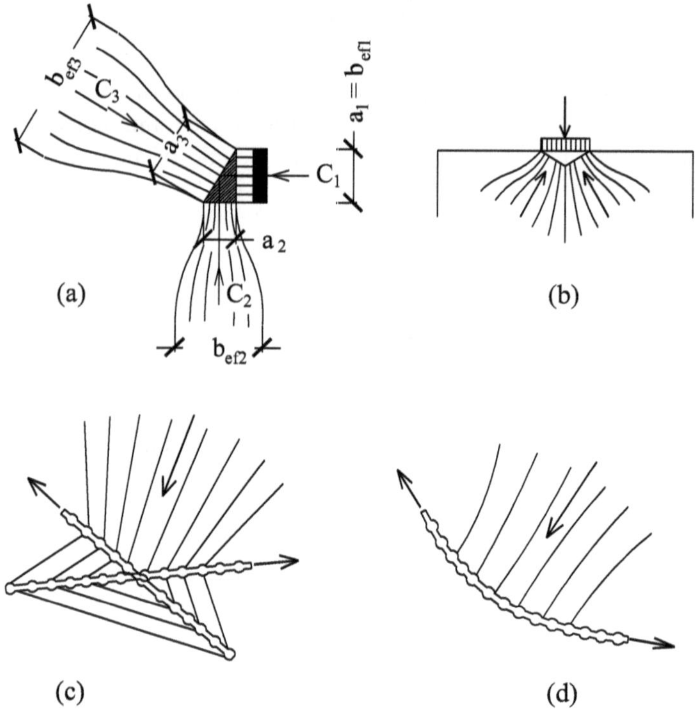

Fig. 10.33 Concentrated nodes **a, b** and diffused nodes **c, d**

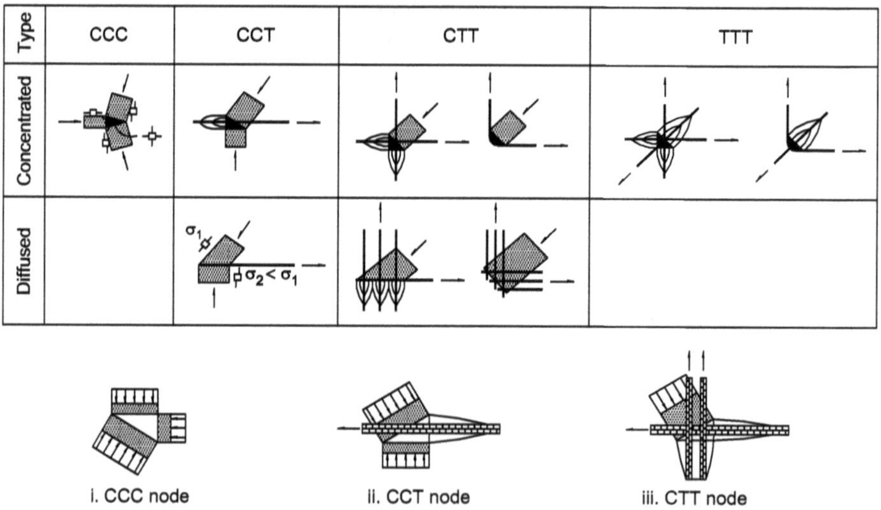

Fig. 10.34 Types of nodes

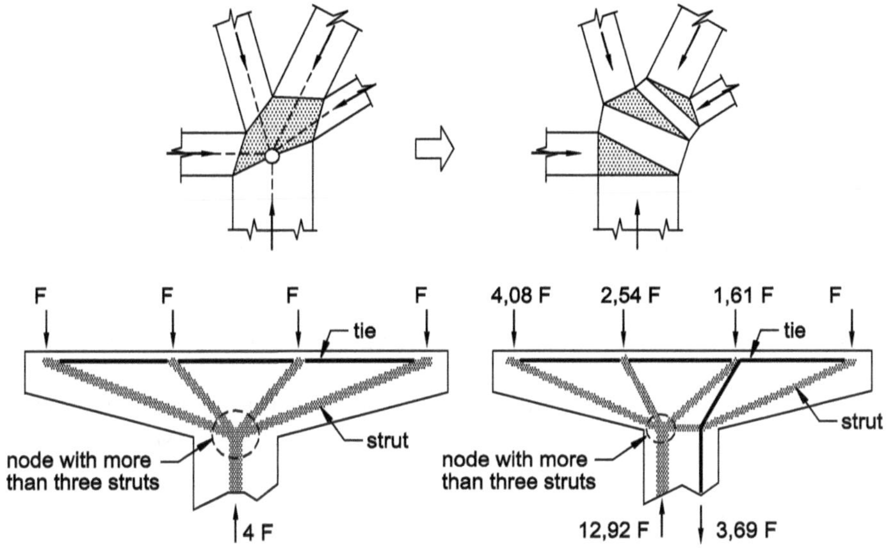

Fig. 10.35 Examples of nodes where more than three members are merging (at the top the subdivision of a polygonal node into three triangular nodes is shown)

It is also possible that more than three members are merging in one node, as shown in Fig. 10.35.

10.10.2 Strength of Nodes

The verification of concrete crushing is required once the geometry and dimensions of the node have been defined. The design strength is a portion ν of the concrete compressive strength, where the coefficient ν is called the *efficiency factor*.

Efficiency factors take different values for nodes CCC, CCT and CTT: nodes CCT and CTT have efficiency factors smaller than CCC nodes because tension forces transmitted by reinforcing bars decrease their strength.

Three different strength values are therefore obtained: $\sigma_{1\text{Rd}}$, max for nodes CCC, $\sigma_{2\text{Rd,max}}$ for nodes CCT and $\sigma_{3\text{Rd,max}}$ for nodes CTT (Table 10.9). Table 10.10 collects the values of $\sigma_{1\text{Rd,max}}$, $\sigma_{2\text{Rd,max}}$, $\sigma_{3\text{Rd,max}}$ for different concrete classes.

In the following, the four types of nodes are described.

Table 10.9 Design strength of nodes

Type of node	Efficiency factor	Design strength	
CCC	$k_1 = 1.0$	$\sigma_{1\text{Rd,max}}$	$1.0(\nu' f_{cd})$
CCT	$k_2 = 0.85$	$\sigma_{2\text{Rd,max}}$	$0.85(\nu' f_{cd})$
CTT	$k_3 = 0.75$	$\sigma_{3\text{Rd,max}}$	$0.75(\nu' f_{cd})$
TTT	–	–	$\min\left(f_{y\,d}, f_{y,\text{ader}}\right)$
Three-axial compression (3D)		$\sigma_{\text{Rd,max}}$	$3.00(\nu' f_{cd})$

Table 10.10 Values of stress $\sigma_{1\text{Rd,max}}$, $\sigma_{2\text{Rd,max}}$, $\sigma_{3\text{Rd,max}}$ in N/mm^2 ($\sigma_{1\text{Rd,max}} > \sigma_{2\text{Rd,max}} > \sigma_{3\text{Rd,max}}$)

f_{ck}(N/mm^2)	f_{cd}(N/mm^2)	$\sigma_{1\text{Rd,max}}$ (N/mm^2)	$\sigma_{2\text{Rd,max}}$ (N/mm^2)	σ_3Rd, max(N/mm^2)
12	6.80	6.47	5.50	4.86
16	9.07	8.49	7.21	6.36
20	11.33	10.43	8.86	7.82
25	14.17	12.75	10.84	9.56
28	15.87	14.09	11.98	10.57
30	17.00	14.96	12.72	11.22
35	19.83	17.06	14.50	12.79
40	22.67	19.04	16.18	14.28
45	25.50	20.91	17.77	15.68
50	28.33	22.67	19.27	17.00
55	31.17	24.31	20.66	18.23
60	34.00	25.84	21.96	19.38
70	39.67	28.56	24.28	21.42

10.10.3 Compression Nodes (CCC)

Compression nodes are placed at concentrated loads and on intermediate supports of slender beams and deep beams, as well as in frame corners under negative bending moment, corbels, and at openings. These nodes are characterized by bi-axial compression fields; their edges are schematized with straight lines, which define triangular or polygonal regions.

The maximum compression stress which can be applied to the edges of compression nodes is given by:

$$\sigma_{1\text{Rd,max}} = k_1 \nu' f_{cd} \quad [(6.60)]$$

Figure 10.36 shows two different types of compression nodes with three struts on the same plane. The first node refers to the intermediate support of a deep beam while the second is typical of a frame corner under a negative bending moment.

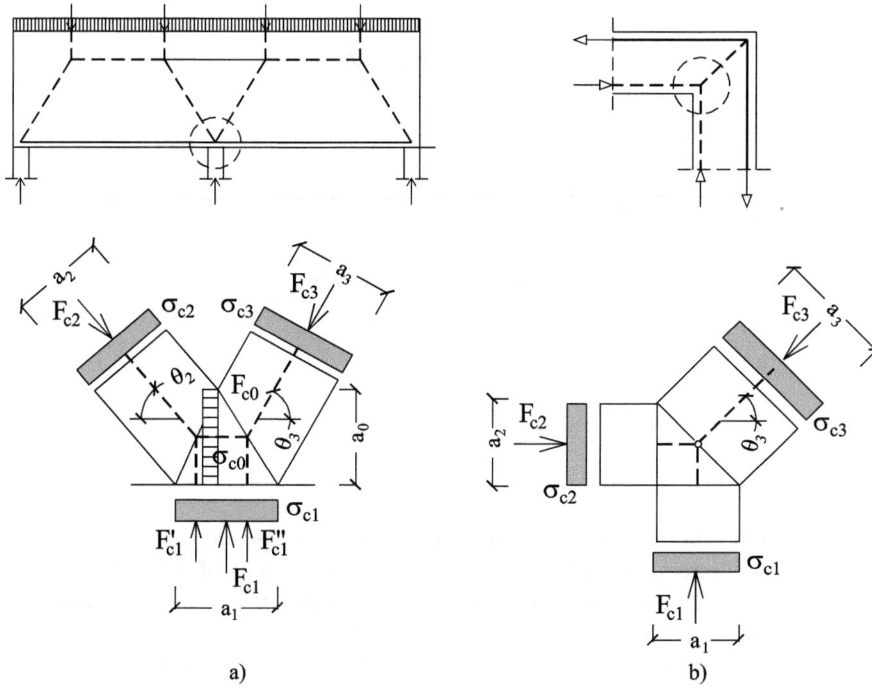

Fig. 10.36 Compression nodes with three struts on the same plane: **a** node on the continuity support for a deep beam, **b** node on a frame corner under a negative bending moment (which tends to close the node)

10.10.3.1 Hydrostatic Compression Nodes

If the faces of a compression node are orthogonal to struts and the dimensions of the strut cross-sections are proportional to compression forces, normal stresses on all the nodal faces are equal. In this case, the node stress field is hydrostatic, i.e. normal stresses are the same on each plane; this kind of node is called "hydrostatic node" (Fig. 10.37).

Strictly speaking, the word hydrostatic implies that all the three principal stresses are identical; nevertheless, in S&T models, the word "hydrostatic" is used to describe a stress field where only the two principal stresses in the plane of the S&T model are equal.

The transverse spread of the force through the thickness of the structural element is considered using a different S&T model (see 10.10.4.1).

Within a hydrostatic node, shear stresses are not present, however, the concrete of the nodal region is capable to support them, so non-hydrostatic nodes are acceptable, too. For non-hydrostatic nodes, Schlaich (1987) suggests that the ratio between the maximum and minimum compressive stresses shall not be larger than two.

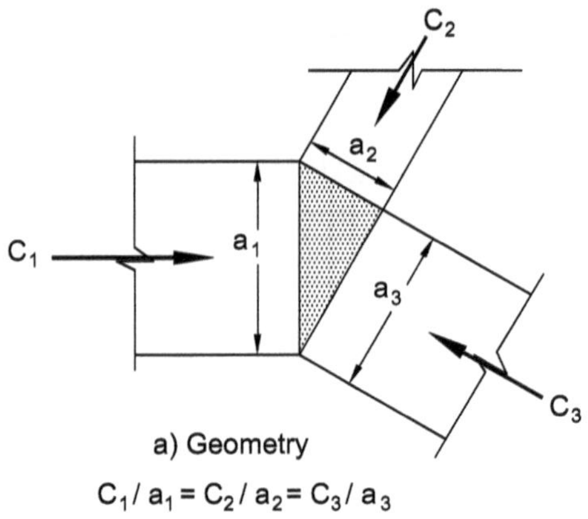

Fig. 10.37 Hydrostatic node

However, it is not easy to have hydrostatic nodes, because the geometry of a node is defined by the reinforcement layout and/or the dimensions of support plates, the dimensions of the column cross-section and the geometry of the S&T model. For example, for node C_1-C_2-C_3 of the deep beam with thickness b shown in Fig. 10.38, the width a_1 of the vertical strut C_1 and the width a_2 of the horizontal strut C_2 can be chosen so that the normal stresses on the nodal faces 1–2 and 1–3 are both equal to $\sigma_{1\text{Rd,max}}$ (i.e. the concrete strength in a CCC node):

$$a_1 = C_1/\left(\sigma_{1\text{Rd,max}} \text{b}\right)$$

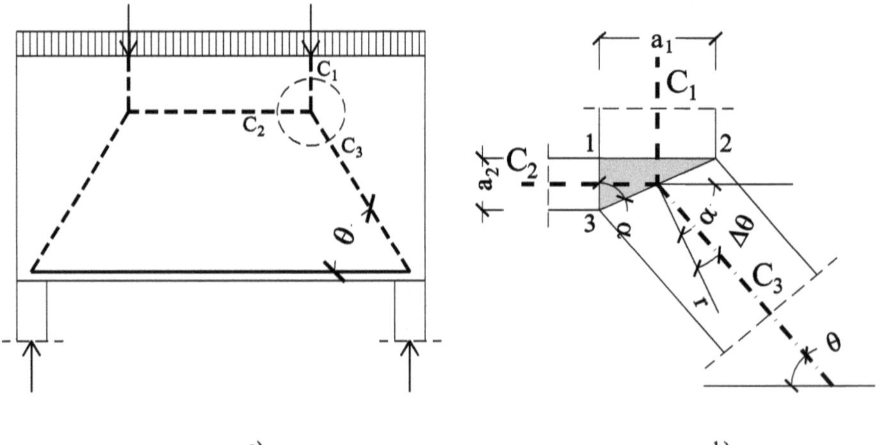

Fig. 10.38 Example of a non-hydrostatic node

$$a_2 = C_2/(\sigma_{1\text{Rd,max}} b)$$

The width a_3 of the third nodal face (face 2–3) is equal to

$$a_3 = \sqrt{a_1^2 + a_2^2}$$

and it has the slope $\alpha = \arctan(a_1/a_2)$ to the horizontal line.

Therefore, the node is hydrostatic only if the third strut is directed along the r direction, orthogonal to the face 2–3 (Fig. 10.38b), i.e. its slope θ to the horizontal is equal to α. Nevertheless, the strut direction is already defined by the truss geometry and, in general, it is inclined by the angle $\Delta\theta$ to r direction.

10.10.3.2 Compression Non-Hydrostatic Nodes

For non-hydrostatic nodes, the nodal faces are not orthogonal to strut axes and compression forces within the struts have different values. The verification of a non-hydrostatic node can be reduced to the verification of a hydrostatic one with the same base, following the procedure described below (Fig. 10.39).

A node with three struts is considered, representing, for example, the continuity node of a two-span deep beam. If the two faces AB and AC of the inclined struts are not normal to the directions of corresponding struts, only on face BC the pressure $\sigma_c 1$ is orthogonal to the nodal face and then is equal to one of the principal stresses in the nodal region.

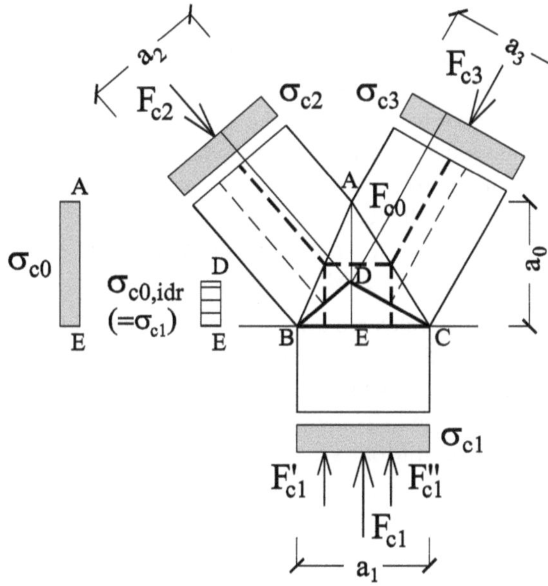

Fig. 10.39 Non-hydrostatic node (ABC) and hydrostatic node (DBC): the two nodes have the same base, but different heights (the height of the hydrostatic node is smaller)

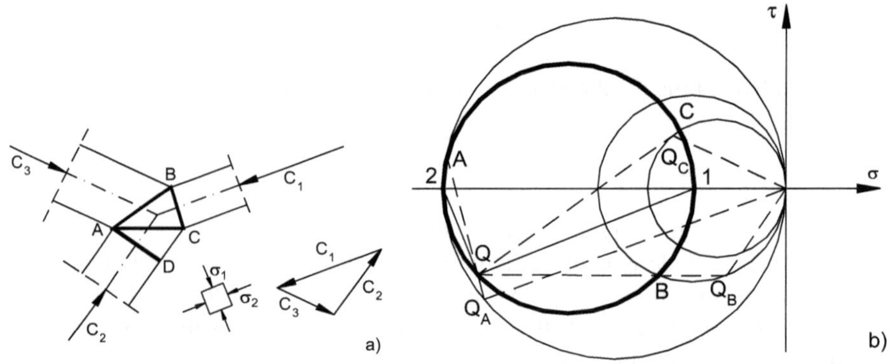

Fig. 10.40 Mohr graphical construction for the evaluation of the stress field in a CCC node (Marti, 1985)

It can be demonstrated that the value of the other principal stress σ_{c0}, applied on the face a_0 and normal to the base face BC, satisfies the following condition:

$$\sigma_{c0} \leq \sigma_{c1} \text{ if } a_0 \geq a_{0,\,hydr}$$

$$\sigma_{c0} > \sigma_{c1} \text{ if } a_0 < a_{0,\,hydr}$$

where a_0 is the height of the node and $a_{0,hydr}$ is the height of the hydrostatic node with the same base a_1.

The proof is immediate in Fig. 10.39, which shows a non-hydrostatic node ABC and a hydrostatic node DBC with the same base.

For the hydrostatic node DBC, the stress on the face $\text{DE}(a_{0,\text{idr}})$ is equal to the hydrostatic stress σ_{c1}, while for the non-hydrostatic node ABC, the stress σ_{c0} on the face AE (a_0) is certainly smaller than σ_{c1}, being $\text{AE} \geq \text{DE}(a_0 \geq a_{0,\text{idr}})$. In this case, it is sufficient to verify only the value of the stress σ_{c1}.

Analogously, it can be proved that, if the height of the node is smaller than the hydrostatic node with the same base a_1, then the principal stress σ_{c0} on the face AE is greater than the face BC ($\sigma_{c0} > \sigma_{c1}$) and it is necessary to verify the stress σ_{c0}.

More in general, forces in a non-hydrostatic node can be found according to the following procedure (Marti, 1985) (Fig. 10.40):

1. Mohr's circles are drawn for the stress field of each strut (for example Mohr's circle of the strut C_2 passes through the origin of the coordinate system and the point with abscissa $\sigma_{c2} = C_2/(\text{AD S})$, where σ_{c2} is the compressive stress in the strut and s is the thickness of the node),
2. the pole of Mohr's circle for each strut (Q_A, Q_B e Q_C)[4] is identified,

[4] The pole P of Mohr's circle is a special point located on the circumference of the circle having the following property: the point of intersection of Mohr's circle and the line drawn through the pole parallel to a given plane, gives the stresses on that plane.

3. lines parallel to sides BC, CA and AB of the node are drawn through poles Q_A, Q_B and Q_C of Mohr's circles; intersection points A, B and C of these lines with the corresponding Mohr's circles define the Mohr's circle of the bi-axial stress field within the node ABC; the centre of this circle lies on the normal stress axis and lines Q_AA, Q_BB and Q_CC intercept in the same pole Q.

10.10.3.3 Strength of a CCC Node

For practical purposes, the calculation of the strength of a node can be performed according to what is suggested by ACI (2019). Except for hydrostatic nodes, where the strength of the node can be calculated on any plane, whatever its inclination, for non-hydrostatic nodes the strength can be obtained by multiplying the concrete strength (which is a function of the type of node—CCC, CCT, CTT) by the smallest of following dimensions:

- the area of the nodal face on which the compression force F_c is applied, projected orthogonally to the direction of F_c; e.g. as the two inclined struts in Fig. 10.41a are not orthogonal to the corresponding nodal faces, for each of them the orthogonal projection to the strut axis is considered (a_2 for the strut on the left and a_3 for the strut on the right), or
- the area of a section, drawn normally to one of the faces, dividing the node into two portions; in Fig. 10.41b–d the sections passing through the three vertexes of the node and orthogonal to the nodal faces are drawn; for example, for section AH_1, the resultant force is coincident with the horizontal projection of the force in the strut 3 (or equivalently, with the horizontal projection of the force in the strut 2).

The strength σ_R of the node shown in Fig. 10.41 is given by:

$$\sigma_R = \min\left(\frac{F_{c1}}{a_1 \cdot b}; \frac{F_{c2}}{a_2 \cdot b}; \frac{F_{c3}}{a_3 \cdot b}; \frac{F_{AH}1}{AH_1 \cdot b}; \frac{F_{CH2}}{CH_2 \cdot b}; \frac{F_{BH3}}{BH_3 \cdot b}\right)$$

where b is the width of the structural element in the direction normal to the plane of the node.

10.10.3.4 Subdivision of the Nodal regions

In many cases, for the sake of simplicity, it is useful to divide a node into several parts; each resulting portion transfers a part of forces applied to the node. For example, for the node shown in Fig. 10.41b, the reaction F_{c1} is subdivided into a component F_{c1}, which is in equilibrium with the vertical component of F_{c2}, and a component F'_{c1}, which is in equilibrium with the vertical component of F_{c3}.

In many cases, the nodal region has a polygonal form, and the analysis of the stress field requires a subdivision into triangular regions connected by intermediate struts,

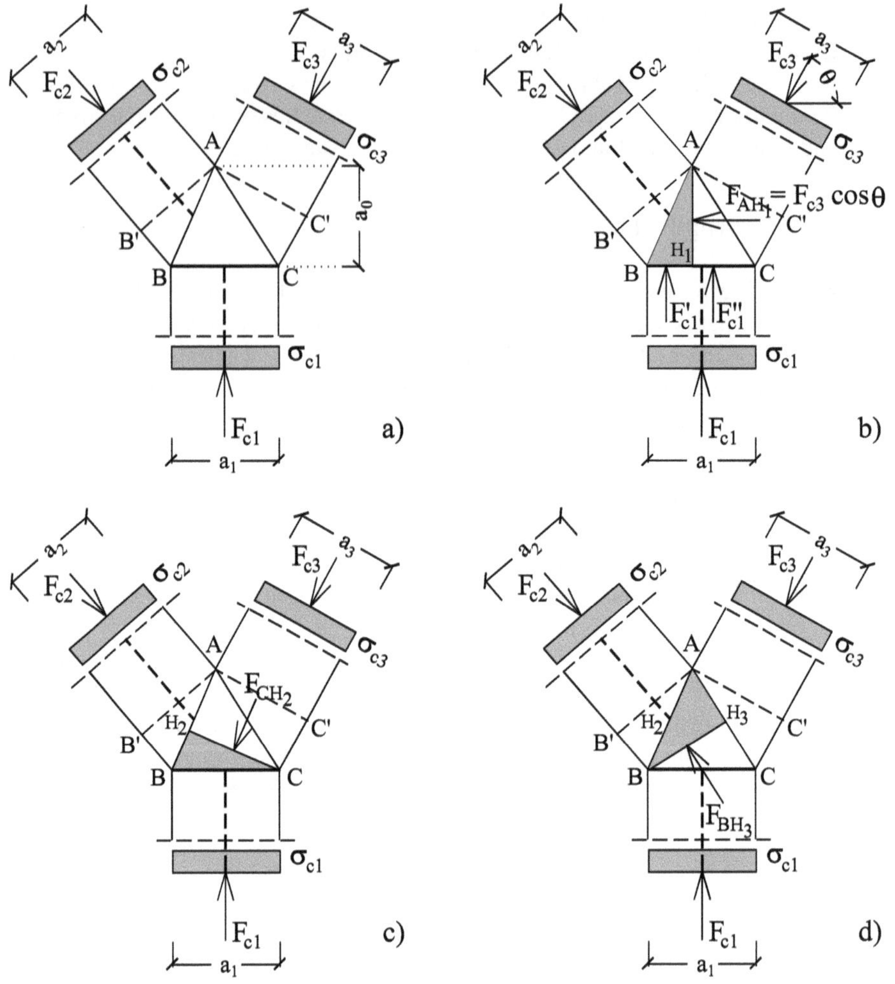

Fig. 10.41 Node CCC: **a** verification on the faces orthogonal to the strut axis (AB', AC', BC); **b**, **c**, **d** verification on the sections orthogonal to the nodal faces

which are added to the original ones (Fig. 10.42). This operation requires verifying that the new geometry is suitable with the geometry of the structural element because the amount of space occupied by the node after the subdivision is larger than the initial one (dashed areas in Fig. 10.42c). In the node shown in Fig. 10.42, the resultant force F_c of forces F_{c1} and F_{c2} is coincident with the compression force within the strut BCEF inside the node; the force F_c is moreover in equilibrium with forces F_{c3} and F_{c4}.

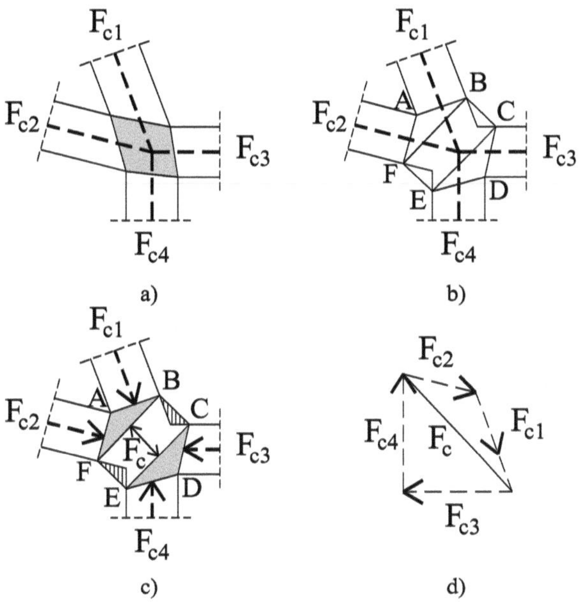

Fig. 10.42 Quadrangular node: **a** geometry, **b** subdivision of the node, **c** strut BCEF within the node, with dashed portions not included in the original geometry, **d** polygon of forces

10.10.4 Compression-Tension Nodes with Anchored Ties Provided in One Direction (CCT)

In compression-tension nodes, reinforcement is anchored or deviated; there are nodes where reinforcing bars are anchored in one direction (CCT) (Fig. 10.43) and nodes with ties arranged in two directions (CTT). CCT nodes are typical of beam external supports, but they can be also found under load plates of corbels, in frame corners and Gerber hinges (see Examples no. 5 and 7). The analysis of a CCT node can be performed like a CCC node if the tie is anchored using a plate placed on the opposite side of the tie itself (Fig. 10.43b).

The design compressive stress of a CCT node is given by the following expression:

$$\sigma_{2\mathrm{Rd,wax}} = k_2 \nu' f_{cd} = 0.85\nu' f_{cd} \quad [(6.61)]$$

where $\sigma_{2\mathrm{Rd,max}} = \max\left(\sigma_{\mathrm{Rd},1};\ \sigma_{\mathrm{Rd},2}\right)$ (Fig. 10.43a).

Differently from a CCC node, the strength is 15% smaller because of the cracking induced within the node by the spread of the tie tension force.

By analysing the node in its middle plane, the reinforcement can be imagined as uniformly distributed across the total depth b of the structural element ($b =$ dimension in the plane orthogonal to the plane of S&T model) and with an effective height u, over which the deviation of the compression field occurs. Expressions to be used for the effective height u are defined according to Model Code 1990 and they are listed in Table 10.11.

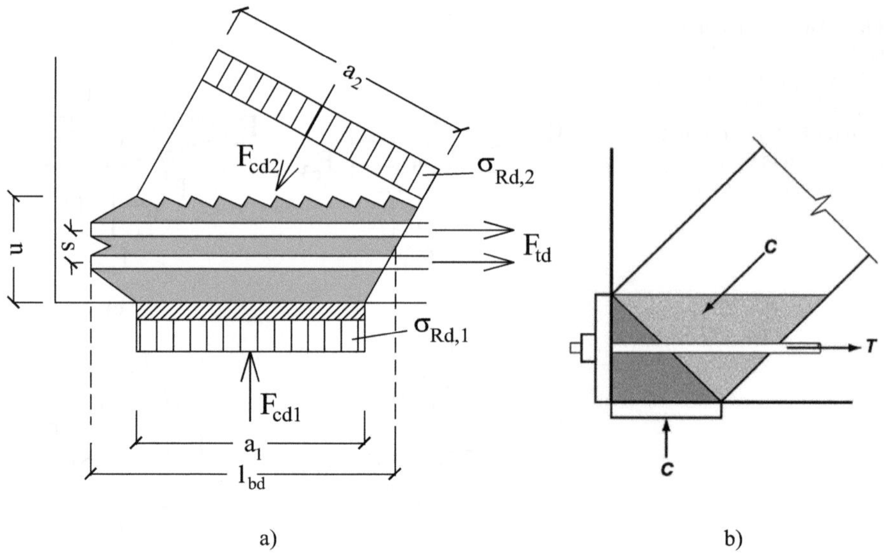

Fig. 10.43 CCT node for end supports: **a** tie anchored by bond, **b** anchorage through an external steel plate

It is necessary to consider that the node strength can be increased by 10% (see 10.10.6), when reinforcing bars are arranged on more than one layer.

The support width a_1, the strut angle θ, the effective height u and the diagonal width a_2 of the compression field are related as follows (Fig. 10.44)

$$a_2 = a_1 \sin\theta + u \cos\theta.$$

For the verification of the node, the calculation of the contact pressure at the base of the node is required:

$$\sigma_{c1} = \frac{F_{c1}}{a_1 b}$$

together with the compressive stress in the diagonal strut:

$$\sigma_{c2} = \frac{F_{c2}}{a_2 b} = \frac{F_{c1}/\text{sen}\theta}{a_2 b}$$

and the reinforcement anchorage length.

The expressions of σ_{c1} and σ_{c2} have been calculated under the hypothesis that the plate has a width b equal to the depth of the structural element; this hypothesis excludes the spread of the compression field in the plane orthogonal to the plane of the S&T model.

Table 10.11 Effective height u of a CCT node

Type of node		Effective height
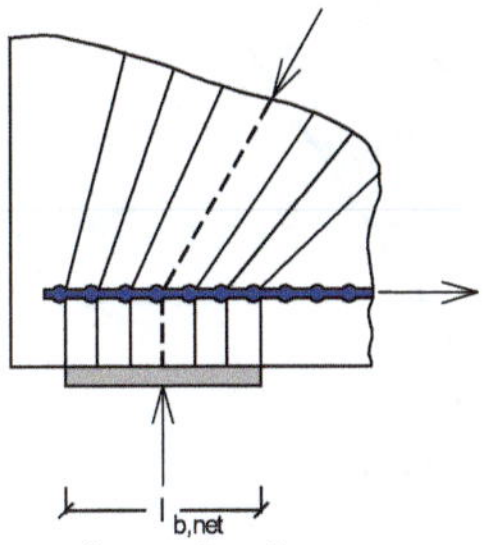	One reinforcement layer ending within the node	$u = 0$
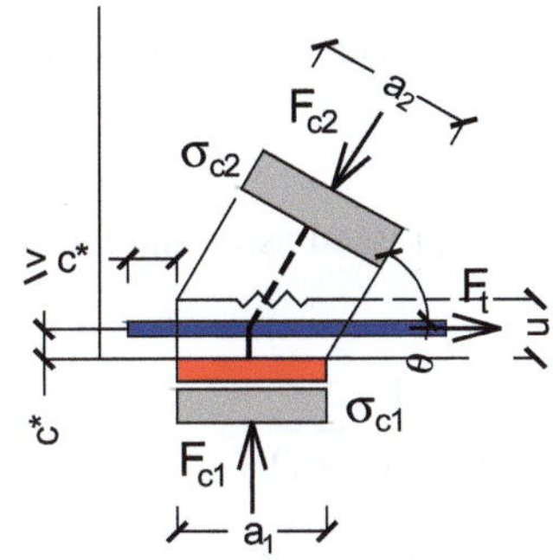	One reinforcement layer extended beyond the support by at least c^*, where c^* is the distance of the reinforcement axis from the bottom edge	$u = 2c^*$ the extension of the bar for the length c^* beyond the deviated compression field allows to distribute the tie force from the bar end section over a length not less than 2 c^*, under the hypothesis that the force spreads at an angle of 45° on both sides of the bar
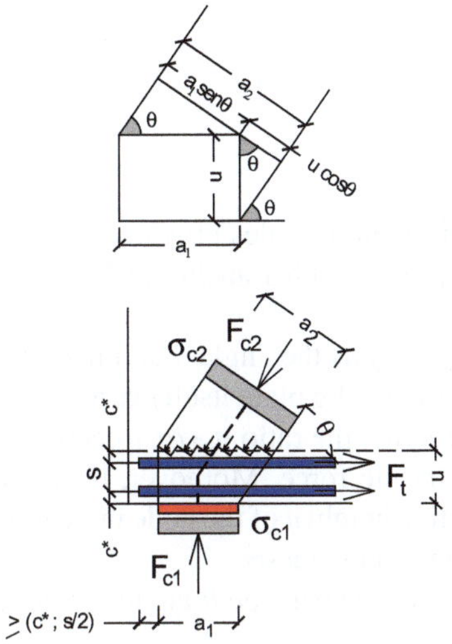	n reinforcement layers arranged with spacing s, extended beyond the deviated compression field for a length equal to max (c^*, $s/2$)	$u = 2c^* + (n - 1)s$ like the previous case, considering that the force spreads at an angle of 45° from the end section of bars

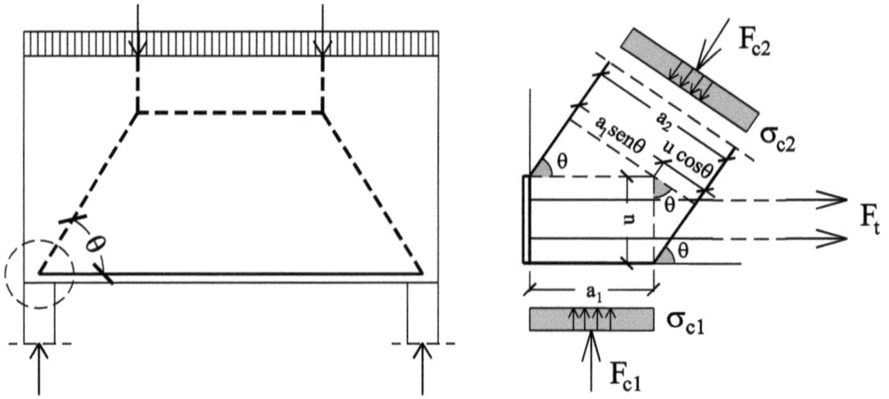

Fig. 10.44 Scheme for the calculation of the width a_2 of the diagonal strut for a CCT node at the end support of a deep beam

The ratio of compressive stresses σ_{c2} and σ_{c1} can be expressed as a function of the angle θ and of the ratio u/a_1, through the formula

$$\frac{\sigma_{c2}}{\sigma_{c1}} = \frac{(F_{c1}/\text{sen}\theta)}{a_2 b} \cdot \frac{a_1 b}{F_{c1}} = \frac{a_1}{\text{sen}\theta \cdot a_2} = \frac{1}{\text{sen}\theta \cdot (a_2/a_1)}$$

and being $\frac{a_2}{a_1} = \text{sen}\theta + \frac{u}{a_1}\cos\theta$, the following expression is obtained: $\frac{\sigma_{c2}}{\sigma_{c1}} = \frac{a_1}{\text{sen}\theta \cdot a_2} = \frac{1}{\text{sen}\theta} \cdot \frac{1}{\text{sen}\theta + \frac{u}{a_1}\cos\theta}$.

The verification of the node is satisfied if

$$\max(\sigma_{c1}, \sigma_{c2}) \leq \sigma_{2\text{Rd,max}}$$

For the anchorage length of reinforcing bars in the node, rules provided in [8.4] "Longitudinal reinforcement anchorage" and [8.6] "Welded bar anchorage" of EC2 can be followed.

Table 10.12 collects the values of the ratio $(\sigma_{c2}/\sigma_{c1})$ as the angle θ and the ratio (u/a_1) (between the height of the node and the width of the plate itself) vary.

It is worth noting that, for an assigned value of u/a_1, the ratio σ_{c2}/σ_{c1} decreases with the increase of the θ angle, as it occurs for the tie force. Moreover, if the tie reinforcement is arranged on more layers, the effective height u of the node increases, and the width a_2 of the diagonal strut and the stress σ_{c2} decrease.

Figure 10.45 shows the diagram of the ratio σ_{c2}/σ_{c1} versus the θ angle for three different values of the ratio u/a_1.

Table 10.12 Ratio σ_{c2}/σ_{c1} versus θ angle and ratio (u/a_1) (grey boxes indicate values less than one)

θ(°)	u/a_1									
	0.1	0.2	0.3	0.4	0.5	0.6	0.7	0.8	0.9	1.0
30	3.41	2.97	2.63	2.36	2.14	1.96	1.81	1.68	1.56	1.46
32.5	2.99	2.64	2.35	2.13	1.94	1.78	1.65	1.54	1.44	1.35
35	2.66	2.36	2.13	1.93	1.77	1.64	1.52	1.42	1.33	1.25
37.5	2.39	2.14	1.94	1.77	1.63	1.51	1.41	1.32	1.24	1.17
40	2.16	1.95	1.78	1.64	1.52	1.41	1.32	1.24	1.17	1.10
42.5	1.98	1.80	1.65	1.53	1.42	1.32	1.24	1.17	1.11	1.05
45	1.82	1.67	1.54	1.43	1.33	1.25	1.18	1.11	1.05	1.00
47.5	1.69	1.55	1.44	1.35	1.26	1.19	1.12	1.06	1.01	0.96
50	1.57	1.46	1.36	1.28	1.20	1.13	1.07	1.02	0.97	0.93
52.5	1.48	1.38	1.29	1.22	1.15	1.09	1.03	0.98	0.94	0.90
55	1.39	1.31	1.23	1.16	1.10	1.05	1.00	0.96	0.91	0.88
57.5	1.32	1.25	1.18	1.12	1.07	1.02	0.97	0.93	0.89	0.86
60	1.26	1.20	1.14	1.08	1.03	0.99	0.95	0.91	0.88	0.85
62.5	1.21	1.15	1.10	1.05	1.01	0.97	0.93	0.90	0.87	0.84
65	1.16	1.11	1.07	1.03	0.99	0.95	0.92	0.89	0.86	0.83
67.5	1.12	1.08	1.04	1.01	0.97	0.94	0.91	0.88	0.85	0.83
70	1.09	1.06	1.02	0.99	0.96	0.93	0.90	0.88	0.85	0.83
72.5	1.07	1.03	1.00	0.98	0.95	0.92	0.90	0.88	0.86	0.84
75	1.04	1.02	0.99	0.97	0.95	0.92	0.90	0.88	0.86	0.85

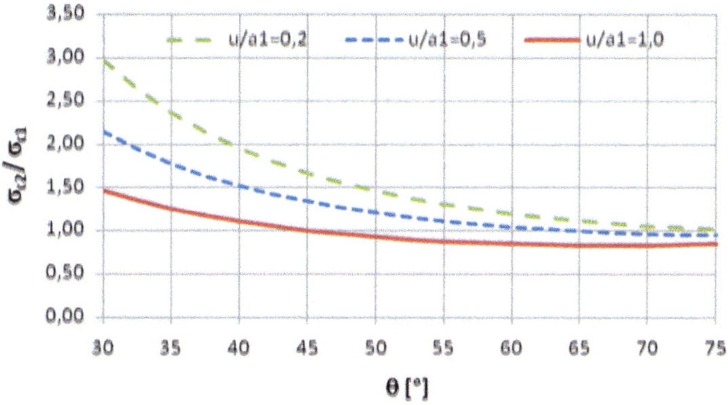

Fig. 10.45 Evolution of internal forces σ_{c1} and σ_{c2} within the two struts of a node CCT on the support at the end section versus θ angle

10.10.4.1 CCT Nodes with Anchored Ties Provided in One Direction and Spread of the Load in the Plane Orthogonal to the Node Plane

A special case of the CCT node is represented by end supports provided with a load plate whose width b_1 is smaller than the depth b of the structural element. This situation cannot only occur at the end support of a beam but also under the load plate of a corbel.

It is necessary to consider the transverse spread of the force through the thickness of the structural element (Fig. 10.46), by adopting the same S&T model already described in 10.6.1.1.

Therefore, the transverse tension force F_t can be calculated according to the S&T model shown in Fig. 10.46a, where the lever arm of internal forces between the tie and the horizontal strut is assumed to be half of the thickness:

$$F_t = \frac{b - b_1}{4b} F_{c1}$$

Reinforcing bars of the tie F_t shall be distributed over a height equal to the depth b of the structural element.

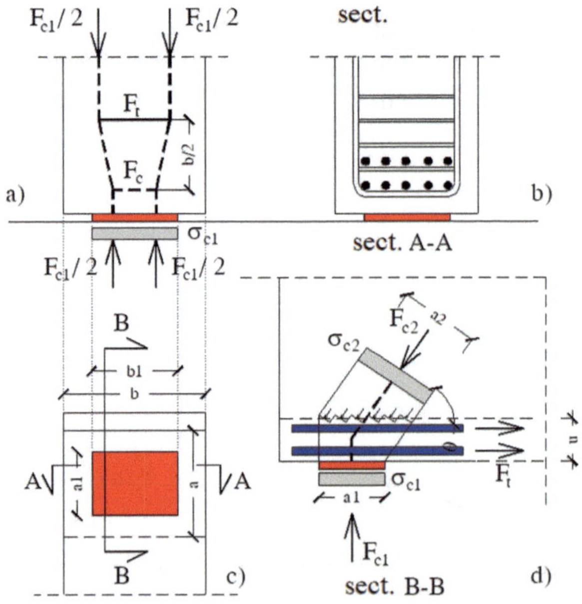

Fig. 10.46 CCT node with the load spread both on the plane of the structural element (at the bottom, on the right) and in the plane orthogonal to the thickness (at the top, on the left)

10.10.5 Compression-Tension Nodes with Ties Arranged in Two Directions (CTT)

In compression-tension nodes with anchored ties provided in two directions, the concrete compressive stress shall not exceed the following value

$$\sigma_{3\text{Rd,max}} = k_3 \nu' f_{cd} = 0.75 \cdot \nu' f_{cd} \quad [(6.62)]$$

A typical example of a CTT node is the corner joint of a frame that tends to be closed by applied moments (for example, node A in Fig. 10.50f). In a CTT node with a bent bar (Fig. 10.47) the strut has the following length (FIB, 1999)

$$a = d_m \sin\theta$$

where

d_m is the minimum mandrel diameter
θ is given by min(θ_1; θ_2), being θ_1 and θ_2 the angles of the reinforcement with the strut

and the strut stress σ_c is:

$$\sigma_c = \frac{F_c}{ab}$$

where b is the depth of the node.

For nodes with orthogonal ties ($\theta_1 + \theta_2 = 90°$) (Fig. 10.48), the concrete compressive stress is given by

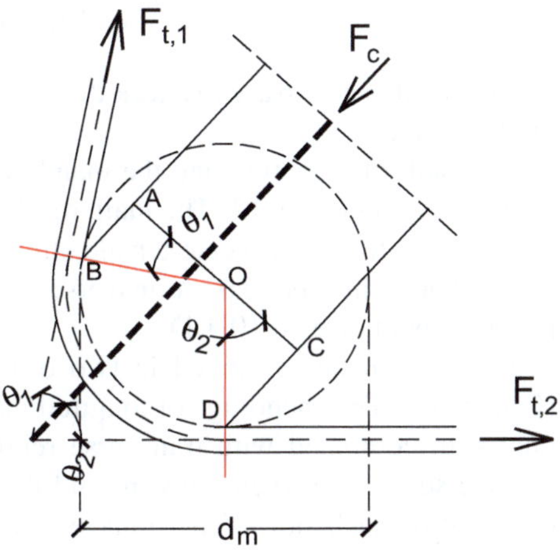

Fig. 10.47 CTT node with bent bar

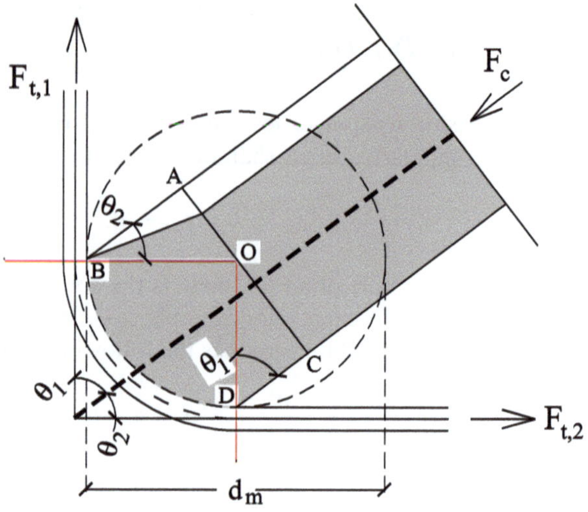

Fig. 10.48 CTT node with orthogonal ties ($\theta_1 + \theta_2 = 90°$)

$$\sigma_c = \frac{\max\ F_t}{b\, d_m \operatorname{sen}\theta_1\ \cos\theta_1} = \frac{\max\ F_t}{b\, d_m \operatorname{sen}\theta_2\ \cos\theta_2}$$

where $\max\ F_t = \max(F_{t1}; F_{t2})$ and $\sin\theta_1 \cos\theta_1 = \sin\theta_2 \cos\theta_2$, being in this case θ_1 and θ_2 complementary angles.

The last expression can be immediately obtained by the general one, by the following procedure. If $\theta_1 < \theta_2$, $F_{t1} = \max F_t$ and $F_c = F_{t1}/\cos\theta_1 = \max\ F_t/\cos\theta_1$; by substituting the last expression and the expression for a ($a = d_m \sin\theta$) in the expression of σ_c ($\sigma_c = F_c/(ab)$), the following expression is obtained

$$\sigma_c = \frac{\max\ F_t}{b\,(d_m \operatorname{sen}\theta_1)\cos\theta_1}$$

Finally, it is necessary to check that $\sigma_c \leq \sigma_{3\mathrm{Rd}_2\,\max}$, where σ_c is the stress induced by design loads.

The radii of curvature and the mandrel diameters shall follow the rules of [8.3], as illustrated in Chap. 12. The main reinforcement shall be uniformly smeared over the depth to limit transverse tensile stresses due to the spread of the load through the thickness; in every case, transverse tensile stresses may be resisted by a suitable reinforcement (see 10.10.4.1).

Nodes with ties arranged in two orthogonal directions are typical for beams (Fig. 10.49). The anchorage of longitudinal bars is guaranteed by the concrete diagonal strut combined with transverse reinforcement (stirrups). The length $l_{b,\mathrm{net}}$ of the intersection between the strut and the flexural reinforcement is assumed to be extended on each side of the transverse reinforcement for not more than six times the diameter of longitudinal bars (Fig. 10.49) (AASHTO, 2007).

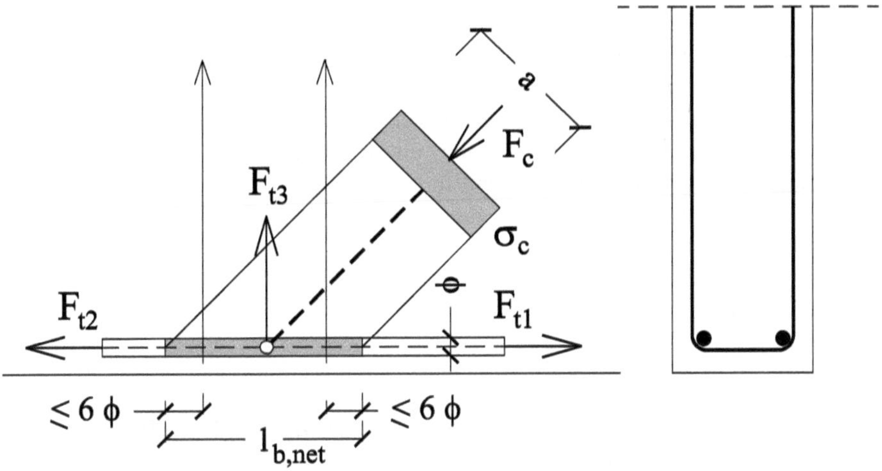

Fig. 10.49 Nodes with ties placed in two orthogonal directions

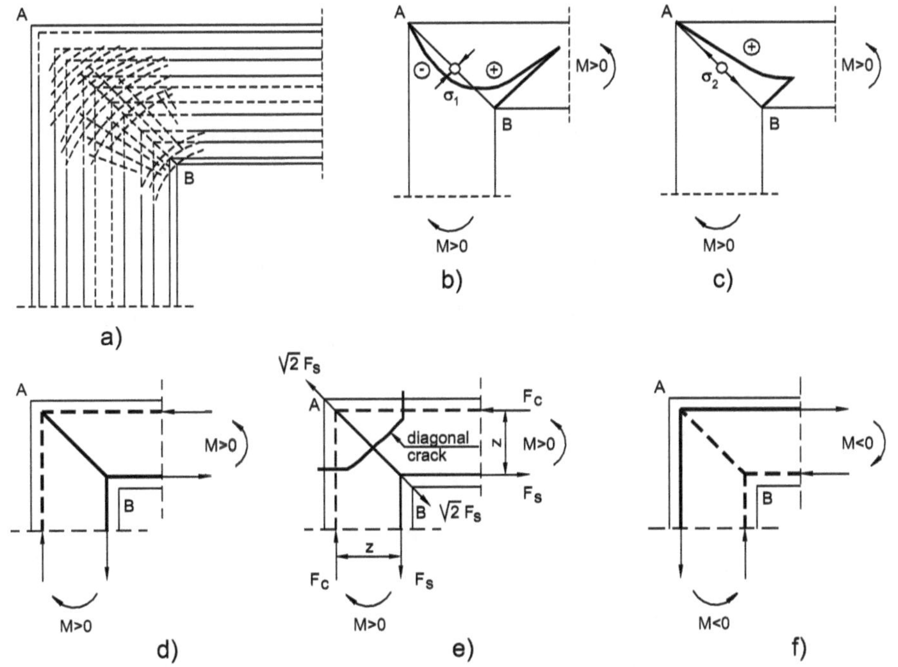

Fig. 10.50 Frame corner under a *positive bending moment (opening moment)*: **a** isostatic lines, **b** normal tensile stresses parallel to the diagonal, **c** normal compressive stresses in the diagonal section, **d** S&T model, **e** diagonal crack. **f** Frame corner under a *negative bending moment (closing moment)*: S&T model

If transverse bars (i.e., vertical stirrups that form the tie F_{t3} in Fig. 10.49) are arranged over an adequate length of the longitudinal reinforcement, the node is of the "diffused" type and it is not necessary to proceed with its verification. Otherwise, if stirrups are very close, the anchorage of high tension forces on a very short length produces high compressive stresses in the concrete diagonal strut.

The anchorage of the reinforcement in compression-tension nodes, like in end supports, starts at the internal face of the node, Fig. 10.30. The anchorage length is extended over the entire node length. In some cases, the reinforcement may also be anchored behind the node. For anchorage and bending of reinforcement, see [8.4] and [8.6], discussed in Chap. 12.

10.10.6 Conditions for Increasing the Strength of Nodes and Confined Nodes

The design strength of nodes may be increased up to 10% if at least one of the following conditions applies [6.5.4(5)]:

1. triaxial compression is assured,
2. all angles between struts and ties are not lower than 55°,
3. stresses at supports or concentrated loads are uniform, and the node is confined by stirrups,
4. the reinforcement is arranged in multiple layers,
5. the node is reliably confined through adequate bearing devices or friction (EC2 does not give any information about the required constraint).

It is worth noting that the strength increase of nodes in the above-mentioned cases does not provide any advantage if the strength of the S&T model is driven by the strength of struts.

For the verification of triaxially compressed nodes, the confined concrete strength can be assumed as design strength:

$$f_{cd,c} = \frac{f_{ck,c}}{\gamma_C}$$

where

$$f_{ck,c} = f_{ck}\left(1.00 + 5\frac{\sigma_2}{f_{ck}}\right) \quad \text{for } \sigma_2 \leq 0.05 \cdot f_{ck} \quad [(3.24)]$$

$$f_{ck,c} = f_{ck}\left(1.125 + 2.50\frac{\sigma_2}{f_{ck}}\right) \quad \text{for } \sigma_2 > 0.05 \cdot f_{ck} \quad [(3.25)]$$

being σ_2 the lateral compressive stress at ULS.

Table 10.13 Ratio between the characteristic strength of confined and of unconfined concrete versus σ_2/f_{ck}

σ_2/f_{ck}	0.05	0.1	0.2	0.3	0.4	0.5	0.55
$f_{ck,c}$ / f_{ck}	1.25	1.375	1.625	1.875	2.125	2.375	2.5

Table 10.13 gives the ratio between the characteristic strength for confined ($f_{ck,c}$) and unconfined concrete (f_{ck}) as σ_2/f_{ck} varies. It is worth noting that a lateral compression equal to just 5% of f_{ck} provides a strength increase up to 25% of the monoaxial compression strength.

In every case, the node design strength shall not exceed the following maximum value

$$\sigma_{2\text{Rd,max}} = k_4 v' f_{cd} = 3v' f_{cd} \quad [(6.5.4(6))]$$

10.11 Frame Corners

Frame corners are characterized by the presence of a geometric discontinuity due to an abrupt change of the axis direction, which passes from the beam horizontal direction to the column vertical direction.

For approximately equal depths of the column and beam cross-sections and in the theoretical case of pure bending, the diagonal cross-section of the node AB is subjected to compressive stresses σ_1 on the external fibres and tensile stresses on the internal fibres (Fig. 10.50b). Orthogonally to AB, σ_2 compressive stresses arise for positive bending moment, which tends to open the node (Fig. 10.50c), and tensile stresses for negative bending moment, which tends to close the node. The sign of σ_2 is clear if the stress field is analysed through the simple S&T model shown in Fig. 10.50d–f: from the equilibrium of nodes A and B along the direction AB, for $M > 0$ the diagonal element AB is under tension ($\sigma_2 < 0$), while for $M < 0$ it is under compression ($\sigma_2 < 0$).

For $M > 0$, failure may occur through the formation of a crack orthogonal to AB (Fig. 10.50e), therefore it is necessary to arrange suitable stirrups parallel to AB. However, this reinforcement is not able to hold also the concrete cover on the external edge, which can crush due to high compression forces, i.e. for high values of the bending moment (Fig. 10.51). For this reason, in the design of a frame corner under a positive bending moment, it is recommended to neglect the concrete cover.

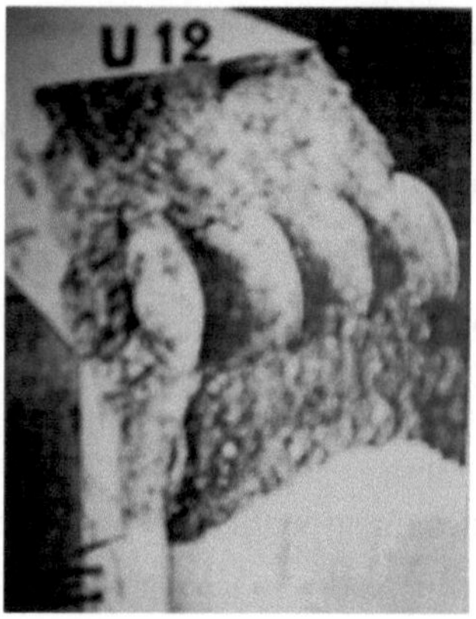

Fig. 10.51 Failure of the concrete cover for a frame corner under a positive bending moment (opening moment) (Nilsson, 1973)

10.11.1 Frame Corner with Closing Moments

In the case of closing moments and for approximately equal depths of column and beam cross-sections ($2/3 < h_1/h_2 < 3/2$), the node stress field can be analysed through the S&T model of Fig. 10.52. If all the reinforcement of the beam is bent around the corner, no verification of the link reinforcement or anchorage lengths within the beam-column joint is required. This model represents a refinement of the model already shown in Fig. 10.50f: the diagonal strut AB is subdivided into two struts to better follow the curvilinear reinforcement.

Writing down the equilibrium equations, it is necessary to consider the shear of the beam and the normal force and the shear of the column. The shear beam and the column axial force enter the vertical translational equilibrium equation together with forces F_{t2} and F_{c2}, while the column shear is considered together with forces F_{t1} and F_{c1} in the horizontal translational equilibrium equation.

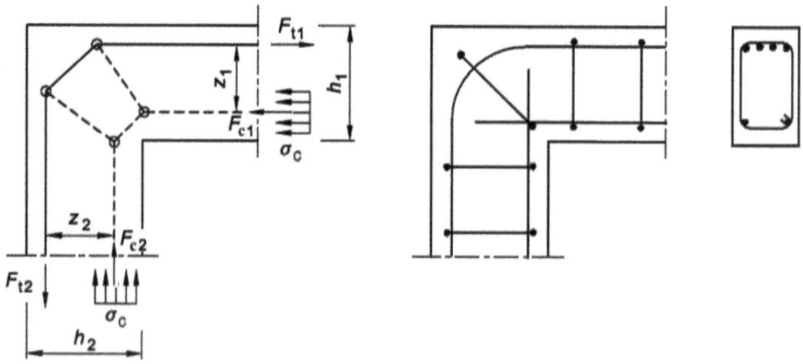

Fig. 10.52 S&T model and reinforcement layout for a beam-column joint under closing moments with $2/3 < h_1/h_2 < 3/2$ [Figure J.2a]

For very different depths of beam and column cross-sections, the S&T model shown in Fig. 10.50 is not suitable, because the angle between the diagonal strut and the vertical tie would be too small. The model shall be modified as shown in Figs. 10.53b and 10.54a, with θ angle not larger than 45°.

In appendix J [J.2.2(3)], EC2 recommends values of tan θ between 0.4 and 1, i.e. θ angles between 21.8° and 45°.

It is worth noting that the S&T model for the design of frame corners with closing moments is the same used for the design of corbels because the node can be seen as a vertical corbel jutting from the column and loaded at the top free end by a concentrated force due to the beam top reinforcement (Fig. 10.55).

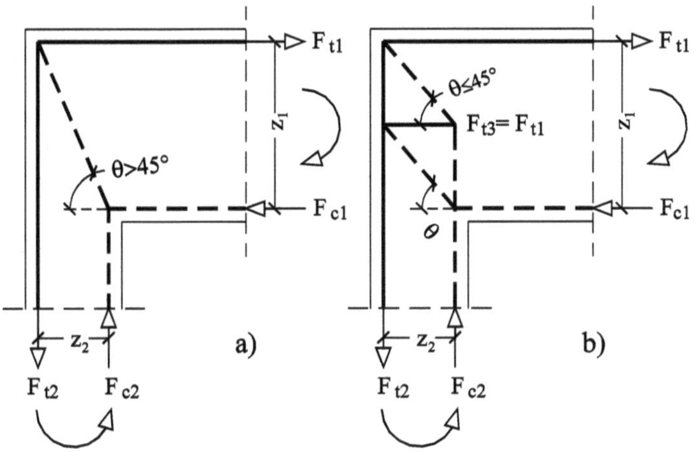

Fig. 10.53 S&T model for a node with closing bending moment with $h_{beam} \gg h_{col}$: **a** unacceptable model for the excessive amplitude of θ angle and small amplitude of the angle between the diagonal strut and the column reinforcement; **b** acceptable model

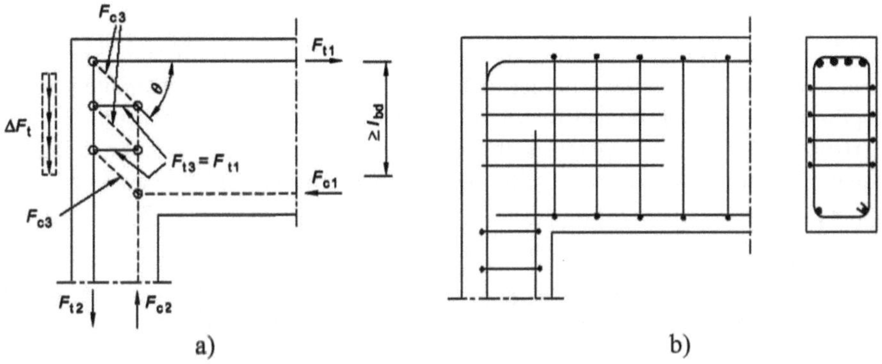

Fig. 10.54 Node with $h_{\mathrm{beam}} \gg h_{\mathrm{col}}$ under closing bending moment: S&T model **a** and reinforcement layout **b** [Figure J.2b]

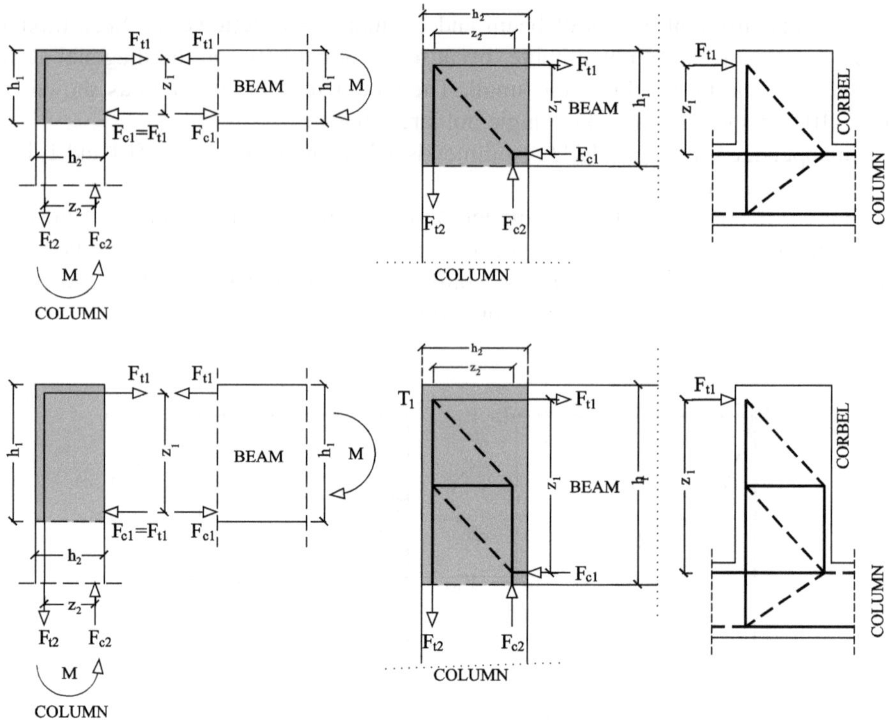

Fig. 10.55 S&T model for frame corners with closing bending moments: analogy with S&T models of corbels

10.11.2 Frame Corners with Opening Moments

10.11.2.1 Frame Corners with Moderate Opening Moments ($A_s/Bh \leq 2\%$)

In the case of moderate opening moments, the stress field in a frame corner can be analysed using the S&T model of Fig. 10.56, where, for approximately equal depths of column and beam cross-sections, links are inclined at an angle of 45° to the horizontal.[5] Under this hypothesis, the total tension force in the two inclined links is equal to the force F_t within the reinforcement of the beam (which is assumed to be equal to the force in the column reinforcement) multiplied by $\sqrt{2}\left(\sqrt{2}F_t \cong 1.4F_t\right)$, so the force in each link is about 0.7 F_t.

The reinforcement can be arranged as shown in Fig. 10.56b and c, e.g. as a loop in the corner region or as two overlapping U bars in combination with inclined links.

[5] Even if the column and the beam cross-sections have approximately the same depth, the lever arm of internal forces is not always the same, depending on the depth of the compressed concrete area.

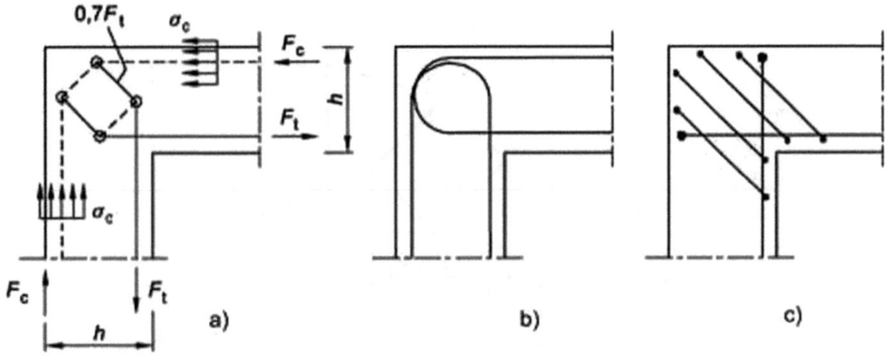

Fig. 10.56 Frame corner with moderate opening moment: **a** S&T model, **b** e **c** detailing of reinforcement [Figure J.3]

10.11.2.2 Frame Corners with Large Opening Moments ($A_s/Bh \geq 2\%$)

For large opening moments, the S&T model shown in Fig. 10.57 can be considered. A diagonal bar passes through the beam and the column stirrups until it reaches the compression chords of the beam and the column. Stirrups gradually deviate compression forces of the two chords before the nodal region, therefore it is necessary to strengthen the regions of the beam and the column adjacent to the node.

For approximately equal depths of column and beam cross-sections, to draw the S&T model, first, the position of the horizontal tie 1–2 is fixed, and the diagonal tie 2–4, which resists tension forces on the internal side of the corner, is drawn at an angle of 45°. Then the bisecting line of the angle $1\hat{2}4$ is drawn from node 2 until the intersection with the beam horizontal tie in node 5; finally, from node 5, a 45° inclined tie is drawn until node 3 where it reaches the strut 1–3, which is inclined by α (α is assumed equal to 16.5° in Fig. 10.57). It is then possible to draw the other struts and ties, accounting for the symmetry of the truss about the bisector plane of the node.

In case of pure bending moment, $F_c = F_t$ and from the translational equilibrium of node 1, forces in the tie 1–2 and the strut 1–3 are calculated:

$$F_{t1} = F_c \tan 16.5° = F_t \tan 16.5° \cong 0.3 F_t$$

$$C_{13} = F_c / \cos 16.5° = F_t / \cos 16.5° \cong 1.043 F_t$$

Writing down the translational equilibrium of node 2 in the direction 2–5 of the bisecting line of angle $1\hat{2}4$, the force in the diagonal tie 2–4 is calculated:

$$F_{t3} = F_{t1} = 0.3 F_t$$

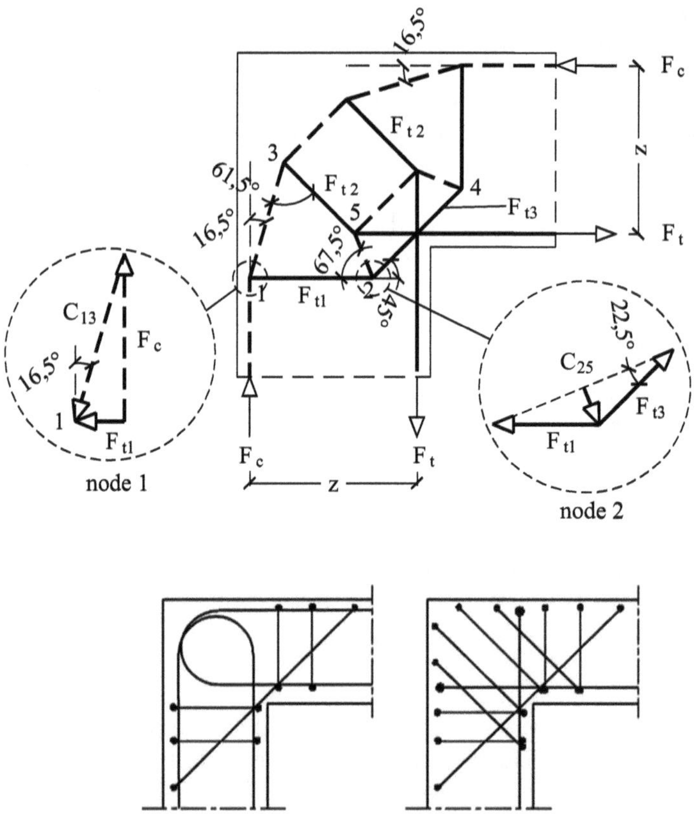

Fig. 10.57 Frame corner with large opening moment: S&T model (above) and detailing of reinforcement (below) [Figure J.4]

Finally, to obtain the value of the force in the tie 3–5, it is sufficient to write down the equilibrium of node 3 along the direction 3–5:

$$F_{t2} = C_{13} \cos 61.5° = F_t \cos 61.5°/\cos 16.5° \cong 0.5 F_t$$

10.11.2.3 Efficiency of Nodes with Opening Moments

The efficiency of a node is defined as the ratio between the resistant moment of the node and the resistant moment of the converging elements—beam and column. Experimentally, nodes with the reinforcement layout shown in Fig. 10.58a have a resistant moment equal to the beam resistant moment and deformations are limited. Diagonal bars limit the width of the crack on the internal side of the corner and decelerate its propagation within the node; the beam and the column reinforcement are

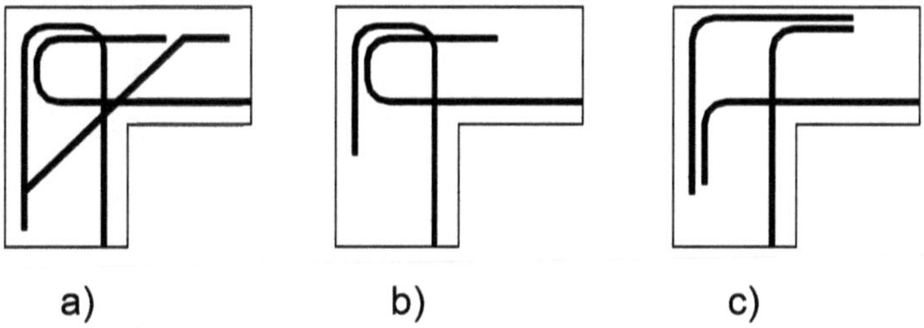

Fig. 10.58 Efficiency of a frame corner with an opening moment as a function of the reinforcement type: **a** 100%, **b** e **c**: 25 ÷ 35%

provided by loops confining the node. Vice versa, nodes provided with the reinforcement layouts of Fig. 10.58b, c have a low efficiency: for reinforcement geometrical ratios around 1%, they are capable to transfer only a percentage of 25 ÷ 35% of the resistant moment (Nilsson & Losberg, 1976).

10.12 Examples

10.12.1 Example No. 1—Simply Supported Deep Beam Under a Uniformly Distributed Load of 280 kN/m

Design the deep beam with dimensions 8000 × 5500 × 300 mm, *under a uniformly distributed load of* 280 kN/m *at ULS (included the self-weight of the beam). The deep beam is supported by two* 300 × 500mm *columns.*

Materials: concrete grade C25/30 $f_{ck} = 25\ \mathrm{N/mm^2}$, steel grade B450C $f_{yk} = 450\ \mathrm{N/mm^2}$.

Concrete design strength

$$f_{cd} = \frac{0.85\, f_{ck}}{1.5} = \frac{0.85 \cdot 25}{1.5} = 14.17\ \mathrm{N/mm^2}$$

Steel design strength

$$f_{yd} = \frac{f_{yk}}{1.15} = \frac{450}{1.15} = 391.3\ \mathrm{N/mm^2}$$

Node compression strength

From Table 10.10, the following values of the node compressive strengths are obtained:

nodes CCC: $\sigma_{1\mathrm{Rd,max}} = 12.75\mathrm{N/mm^2}.$

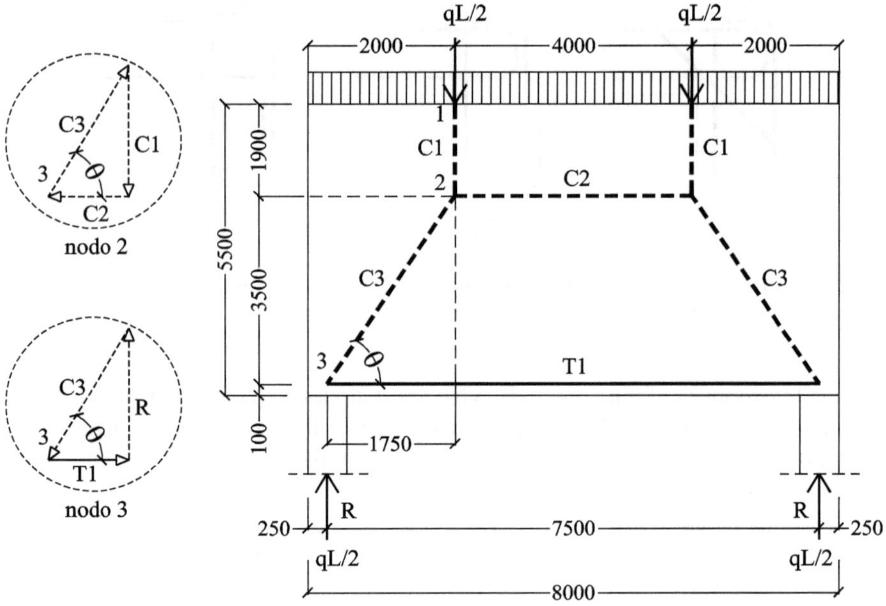

Fig. 10.59 S&T model of the deep beam

nodes CCT: $\sigma_{2Rd,max} = 10.84 \text{N/mm}^2$.
nodes CTT: $\sigma_{3Rd,max} = 9.56 \text{N/mm}^2$.

Geometry of the S&T model

As already mentioned in 10.6.3, Model Code 1990 (6.8.2.1) suggests adopting an internal lever arm equal to $0.6 \div 0.7$ times the span of the beam, but not higher than the internal lever arm of a slender beam with the same span: $(0.6 \div 0.7)\ L \leq 0.67H$; in the present case (Fig. 10.59):

$L = 8000$ mm (total span),
$L' = 7500$ mm (distance between axes of columns),
$H = 5500$ mm,
$(0.6 \div 0.7)L' = (4500 \div 5250)$ mm,
$0.67H = 3685$ mm.

It is assumed that $z = 3500$ mm ($\cong 0.64$H), so the distance of the strut C_2 from the bottom of the beam is equal to $100 + 3500 = 3600$ mm and angle θ is equal to:

$$\theta = \arctan(3500/1750) = 63.43°$$

Column reactions

$$R = qL/2 = 280 \cdot 8_200/2 = 1120 \text{ kN}$$

Calculation of internal forces in the elements of the S&T model (see Table 10.7).

Equilibrium at node 1

$$C_1 = \frac{q\,L}{2} = 1120 \text{ kN}$$

Equilibrium at node 3.

$C_3 = \frac{R}{\text{sen}\theta} = 1252$ kN where $\theta = \text{arctg}\frac{3500}{1750} = 63.43° \geq 45°$,

$$T_1 = C_3 \cos\theta = \frac{R}{\tan\theta} = 560 \text{ kN}$$

Equilibrium at node 2

$$C_2 = C_3 \cos\theta = T_1 = 560 \text{ kN}$$

Design of the tie

The steel reinforcement of the lower tie is arranged over a depth of 200 mm; it is necessary to adopt a steel area not less than:

$$A_{s1} \geq \frac{T_1}{f_{yd}} = \frac{560000}{391.3} = 1431 \text{ mm}^2$$

then six 18 mm bars $\left(6\phi 18 = 1524 \text{ mm}^2\right)$, arranged on three layers, are used.

Verification of node 3 over the left column

The geometry of the node is univocally defined by the width of the column, the depth of the deep beam (300 mm), the depth of the bottom reinforcement layers and the slope of the strut C_3.

Node 3 is a tension–compression node with anchored reinforcing bars provided in one direction (CCT node), therefore it is necessary to verify that the maximum compressive stress is not higher than

$$\sigma_{2\text{Rd,max}} = 10.84 \text{ N/mm}^2$$

It results that stresses σ_{cl} and σ_{cl} (Fig. 10.60) are both lower than $\sigma_2\text{Rd}, max$:

$$\sigma_{c1} = \frac{1,120,000}{300 \cdot 500} = 7.47 \text{ N/mm}^2 \leq \sigma_{2\text{Rd,max}}$$

$$\sigma_{c2} = \frac{1,252,000}{300 \cdot (200 \cdot \cos\theta + 500 \cdot sen\theta)} = 7.78 \text{ N/mm}^2 \leq \sigma_{2\text{Rd,max}}$$

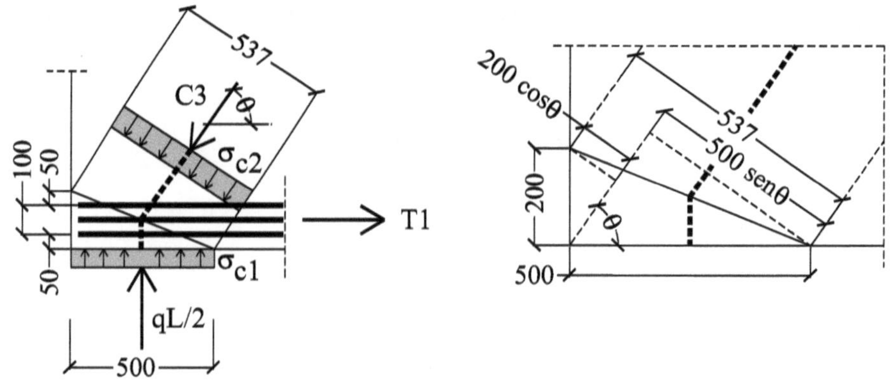

Fig. 10.60 Node 3

where 200 mm is the height u of the node, accounting for the vertical depth of the bottom reinforcement layers.

Note. Once the stress σ_{c1} is evaluated, the stress σ_{c2} may be approximately evaluated by using Table 10.12; for $u/a_1 = 200/500 = 0.4$ and for $\theta = 62.5°$

$$\sigma_{c2}/\sigma_{c1} = 1.05 \quad \rightarrow \quad \sigma_{c2} = 1.05 \cdot 7.47 = 7.84 \text{ N/mm}^2$$

This value is slightly greater than the calculated value, because it is corresponding to a slightly smaller an angle θ than the actual one (62.5° instead of 63.43°).

Minimum reinforcement on the two faces of the deep beam

EC2 recommends (section [9.7]) to provide an orthogonal reinforcement mesh on each face of the deep beam, with a minimum cross-sectional area $A_{s\,\text{dbmin}} = \max\left(0.001A_c;\ 150\text{mm}^2/\text{m}\right)$ along each direction.

The distance s between two adjacent bars of the mesh shall respect the following limit:

$$s \leq \min(2t;\ 300 \text{ mm})$$

where t is the thickness of the deep beam.

In the present case

$$A_{s_d,\text{bmin}} = 0.001A_c = 0.001 \cdot 300 \cdot 1000 = 300 \text{ mm}^2/\text{m} > 150 \text{ mm}^2/\text{m}$$

A 150 mm by 150 mm 8 mm wire mesh $\left(= 333\text{mm}^2/\text{m}\right)$ is adopted on each face.

Moreover, the wire meshes on the two faces are connected by cross ties, whose diameter and spacing satisfy the rules given in EC2 for walls (see 12.2.5.3).

To guarantee the equilibrium of the node, the bottom bars shall be fully anchored beyond the node, by using hooks or U-shaped bars or anchorage devices, except when

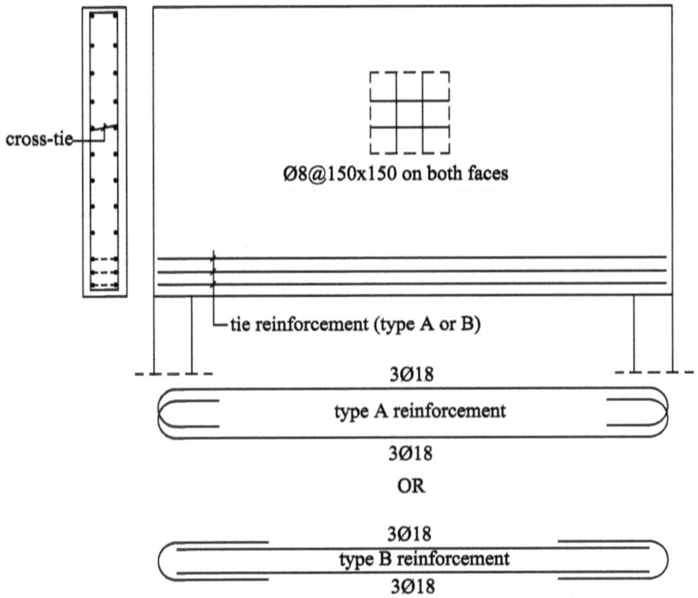

Fig. 10.61 Reinforcement layout for the deep beam

the length between the node and the lateral side of the beam is so high to allow a straight anchorage. In the present case, a straight anchorage cannot be used, because there is not enough space. As a consequence, it is necessary to adopt shaped bars, for example like those in Fig. 10.61, i.e. bars with U-loops at both ends or straight bars which overlap U-bars.

10.12.2 Example No. 2—Simply Supported Deep Beam Under a Uniformly Distributed Load of 420 kN/m

Repeat the design of the deep beam of Example no. 1 *assuming a ULS design load equal to* 420 kN/m.

All the results are increased by the ratio $420/280 \cong 1.5$; the tie force is equal to

$$T_1 = 560 \cdot 1_25 \cong 840 \text{ kN}$$

Therefore, the amount of required steel is given by

$$A_{s1} \geq \frac{840000}{391.3} = 2147 \text{ mm}^2$$

six 18 mm bars and two 20 mm bars $\left(6\phi18 + 2\phi20 = 2152\ \text{mm}^2\right)$ are adopted on four layers.

Concerning the verification of node 3, the following stress values are obtained:

$$\sigma_{c1} = \frac{1120000 \cdot 1.5}{300 \cdot 500} = 11.2\ \text{N/mm}^2 > \sigma_{2\text{Rd,max}}$$

$$\sigma_{c2} = \frac{1252000 \cdot 1.5}{300 \cdot (250 \cdot \cos\vartheta + 500 \cdot \text{sen}\vartheta)} = 11.2\ \text{N/mm}^2 > \sigma_{2\text{Rd,max}}$$

where the calculation of σ_{c2} is performed by assuming a height u of the node equal to 250 mm and not to 200 mm, accounting for the additional reinforcement layer compared with the previous example. Stresses σ_{c1} and σ_{c2} are higher than the design strength $\sigma_{2\text{Rd,max}}$.

To make $\sigma_{c2} \le \sigma_{2\text{Rd,max}}$, it would be possible to increase the spacing between reinforcement layers, while to reduce σ_{c1}, the column cross-section could be enlarged. Nevertheless, being all the angles between struts and ties greater than 55° ($\theta = 63.43°$), the node strength can be increased by 10% [6.5.4(5)] (see 10.11.2.3) and the verification of node 3 is satisfied without any change to its geometry.

10.12.3 Example No. 3—Rigid Spread Footing

Design the reinforcement of the rigid spread footing (1500 × 1500 × 650 mm) *of a column with a square cross-section of* 300 × 300 mm; *the column is subjected to a normal load of* 400 kN *at ULS. Two cases are considered, the one with null eccentricity and the one with eccentricity* 0.25 m *or* 0.20 *m. Materials are the following: concrete grade* C25/30 *and steel grade* B450C Fig. 10.62.

Materials:

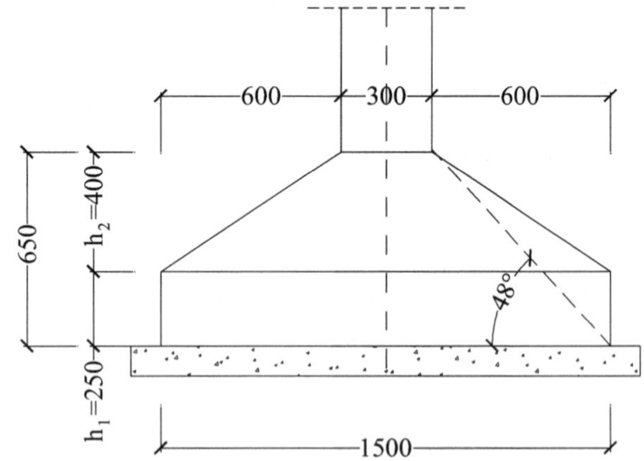

Fig. 10.62 Rigid spread footing

$$\begin{aligned}&\text{concrete } f_{ck} = 25\ \text{N/mm}^2\\&\text{steel} \quad\ \ f_{yk} = 450\ \text{N/mm}^2.\end{aligned}$$

Case a: null eccentricity (centred axial load)

$$N = F_d = 544\ \text{kN}$$

The bearing pressure to be used for the calculation of internal forces is equal to

$$p = N/A_{\text{base}} = 544/2.25 = 242\ \text{kN/m}^2 = 0.242\ \text{N/mm}^2.$$

The axis of the bottom horizontal reinforcement is assumed to be at 50 mm[6] from the bottom face of the footing.

Geometry of the S&T model

Figure 10.63 shows the three-dimensional S&T model of the footing under the centred vertical load. If a is the size of the column square cross-section, the top ends of the four inclined struts are placed at the horizontal distance $a/4$ from column faces and at the vertical distance $a/4$ from the top face of the footing. The bottom ends of the struts are at the same level as the bottom reinforcement and are at the horizontal distance $b/4$ from the corners of the footing base (b is the size of the footing base).

The spatial S&T model may be subdivided into two planar trusses, arranged along diagonals of the footing base (Fig. 10.63b), and each of them subjected to half of the vertical load. The four diagonal struts are subjected to the same compression force:

$$C_{1,2,3,4} = (N/4)/\sin\theta_{\text{diag}}$$

where θ_{diag} is the slope of the four struts to the horizontal.

The tension force in the bottom reinforcement is calculated by projecting the compression forces of the inclined struts on the horizontal plane and by decomposing the projected forces along the twoss directions x and y. For example, denoting by C' the projection of the compression force C_1 on the horizontal plane (Fig. 10.63c), the following expression is obtained for $T'_x (= T'_y)$:

$$T'_x = T'_y = C'_1 \ \cos 45° = \left(C_1 \cos\theta_{\text{diag}}\right)\ \cos 45° = \frac{N/4}{\tan\theta_{\text{diag}}}\cos 45°$$

and, being $\tan\theta_{\text{diag}} = \frac{z}{\frac{(b-a)\sqrt{2}}{4}} = \frac{4z}{(b-a)\sqrt{2}}$,

[6] Minimum concrete cover for a footing is equal to 40 mm (see Chap. 3); by adopting a diameter not larger than 20 mm for the bottom reinforcement, the minimum distance between the reinforcement centroid and the bottom face of the footing is not lower than $10 + 40 = 50$ mm.

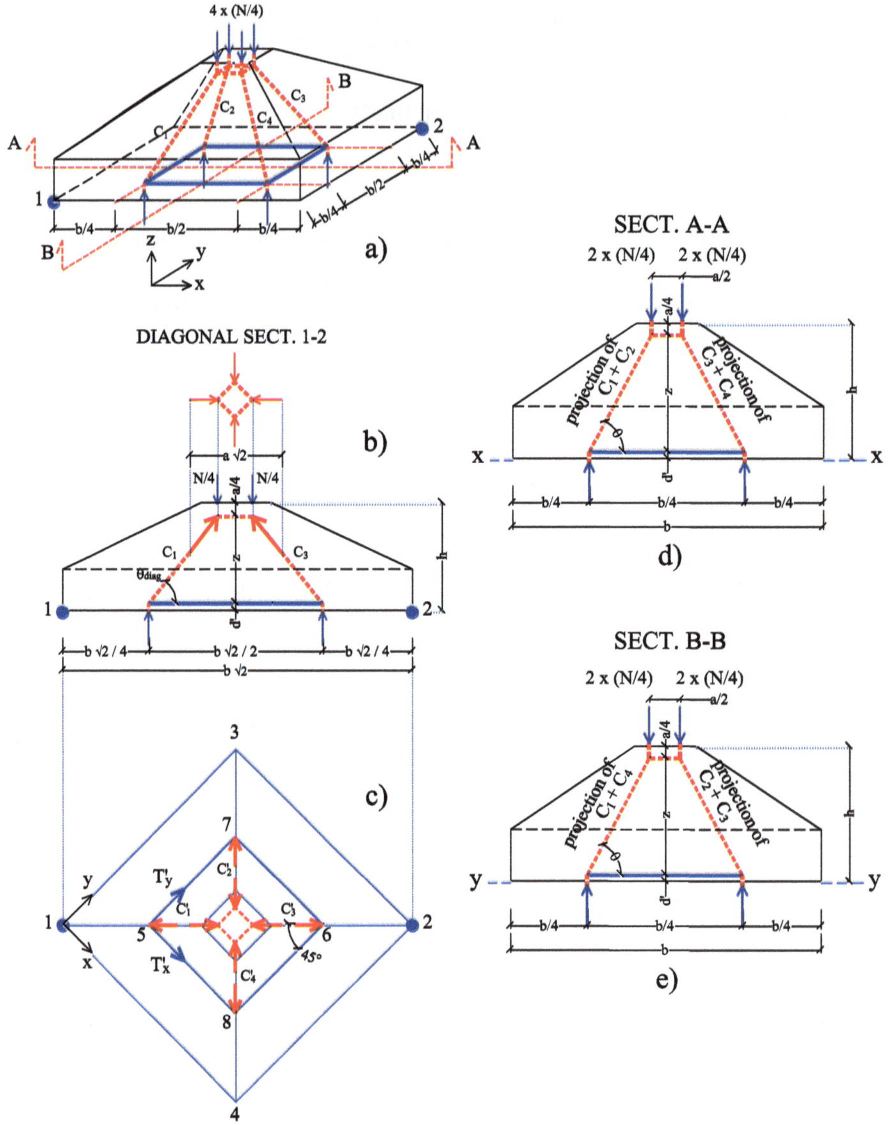

Fig. 10.63 S&T model for a rigid spread footing under centred vertical load

$$T'_x = T'_y = \frac{N}{4}\frac{(b-a)\sqrt{2}}{4z}\frac{\sqrt{2}}{2} = \frac{N(b-a)}{16z}$$

The total force T_x in the reinforcing bars arranged along the x-direction (ties 5–8 and 6–7 in Fig. 10.63c) is obtained by multiplying $T'x$ by two:

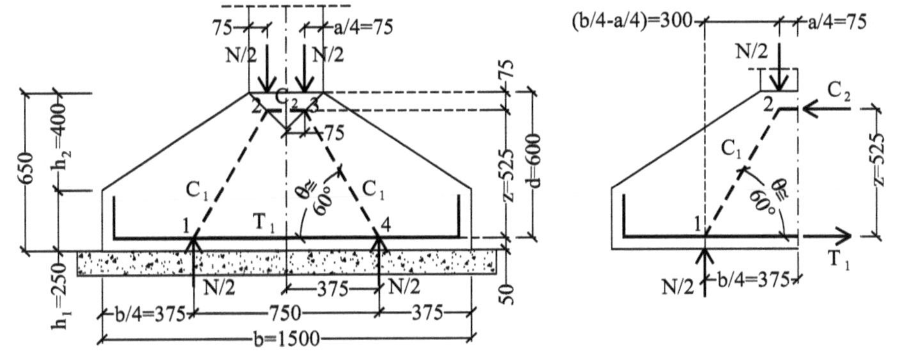

Fig. 10.64 S&T model obtained by bisecting the footing indifferently along A-A or B-B; tie T_1 is corresponding to force T_x if section A-A is considered and to force T_y if section B-B is considered

$$T_x = 2T'_x = \frac{N(b-a)}{8z}$$

and the same holds for the total force T_y of ties 5–7 and 6–8 along the y-direction.

Tie forces are more easily obtained considering the two fictitious trusses of Fig. 10.63d and e, which represent the projections of the 3D truss on sections A-A and B-B parallel to the footing sides. These trusses allow a rapid calculation of the tie forces; indeed, the rotational equilibrium gives (Fig. 10.64):

$$T_1 = T_x = T_y = \frac{N}{2} \cdot \frac{(b-a)}{4z}$$

The design of the footing reinforcement is performed considering the plane S&T model of Fig. 10.64.

Planar truss geometry

The force N is divided into two equal parts, applied symmetrically at a distance $a/4$ from the column axis, where a is the size of the column cross-section. Nodes 2 and 3 are placed on the action lines of the two forces N/2 and at the vertical distance $a/4$ from the top face of the footing (Fig. 10.64).

The internal lever arm and the slope θ of the strut C_1 with respect to the horizontal are

$$z = d - a/4 = 600 - 75 = 525 \text{ mm}$$

$$\vartheta = \text{atan}\frac{\text{z}}{(\text{b}-\text{a})/4} = \frac{525}{(1500-300)/4} = 60.25°$$

Calculation of forces in struts and ties

Equilibrium at node 1

$$C_1 = \frac{N/2}{\sin\theta} \cong 313 \text{ kN}$$

$$T_1 = \frac{N/2}{\tan\theta} = 155 \text{ kN}$$

Equilibrium at node 2

$$C_2 = C_1 \cos\theta = T_1 = 155 \text{ kN}$$

Tie design

The steel amount shall not be smaller than

$$A_{s1} \geq \frac{T_1}{f_{yd}} = \frac{155000}{391.3} = 396 \text{ mm}^2$$

therefore, four 12 mm bars $\left(4\phi 12 = 452 \text{ mm}^2\right)$ are adopted.

Being $T_1 = (N/2)/\tan\theta$ and $\tan\theta = z/[(b-a)/4]$, the tie area may be expressed also as follows

$$A_{s1} \geq \frac{T_1}{f_{yd}} = \frac{N/2}{\tan\theta \cdot f_{yd}} = 0.125\frac{N \cdot (b-a)}{z \cdot f_{yd}}$$

Case b: eccentricity *e* = 250 mm

$$N = F_d = 544 \text{ kN}$$

$$M = Ne = 544 \cdot 0.25 = 136 \text{ kNm}$$

Being $e = b/6$, the bearing pressure has a triangular shape with maximum value equal to

$$\begin{aligned} p_{\max} &= N/A_{\text{base}} + M/W_{\text{base}} \\ &= 544/2.25 + 136/\left(1.50^3/6\right) \\ &= 484 \text{ kN/m}^2 = 0.484 \text{ N/mm}^2 \end{aligned}$$

the distance of the centroid of the horizontal reinforcement from the bottom face of the footing is equal to 50 mm.

Geometry of the S&T model

The compression force N_1 in the base cross-section of the column acts at a distance of 0.4 × from the compression edge of the same cross-section, where x is the depth of the neutral axis at failure.

The tension force N_2 acts at the centroid of the column tensile bars: these bars are at distance $c + \phi_{st} + \phi_1/2$ from the tension edge of the cross-section, where c is the concrete cover, ϕ_{st} is the diameter of stirrups and ϕ_l is the diameter of longitudinal bars.

The direction of the eccentric load N divides the bearing pressure distribution into two parts with resultant forces R_1 and R_2.

Node 1 and node 3 are identified by the intersection of R_1 and R_2 action lines with the axis of the bottom tie T_1. Node 2 is placed on the action line of N_1 at depth $a/4$ from the base cross-section of the column. Finally, node 4 is placed at the intersection of the action line of N_2 with the bottom tie T_1.

Like the previous example with centred axial load, being node 2 placed at vertical distance $a/4$ from the top face of the footing, the lever-arm is equal to z = 525 mm.

Determination of the action line of N_2

The concrete cover of the column is assumed to be equal to 25 mm. The distance between the centroid of vertical bars and the face of the column is equal to the sum of the concrete cover, the stirrup diameter, and half the vertical bar diameter: $25 + 8 + 16/2 = 41$ mm. Therefore, the effective depth of the column cross-section is equal to 259 mm and the force N_2 acts at distance $d_{N2} = (150 - 41) = 109$ mm from the axis of the column.

Determination of the action line of N_2

The column is provided with symmetrical vertical reinforcement equal to 4ϕ16 (one bar for each corner) and 8 mm stirrups at 200 mm spacing.

The position of the neutral axis at failure is calculated by imposing the translational equilibrium along the axis of the column

$$(0.8x)af_{cd} + A'_s f_{yd} - A_s f_{yd} = N$$

where the concrete stress distribution is replaced by an equivalent rectangular distribution with uniform stress f_{cd} ("stress-block") and both steels are assumed to be yielded, then

$$x = N/(0.8af_{cd}) = 544000/(0.8 \cdot 300 \cdot 0.85 \cdot 25/1.5) = 160 \text{ mm}$$

$$x/d = 160/259 = 0.62 < 0.641$$

being $x/d < 0.641$, tension steel yielded[7] it is possible to easily verify that also the compression steel is yielded, as assumed in the equilibrium condition.

The resultant compression force in the concrete is given by[8]

$$N_{1,c} = 0.8 \text{ a } x f_{cd} = 0.8 \cdot 300 \cdot 160 \cdot 14.17 = 544128 \text{ N}$$

while the resultant tension force in the tension reinforcement is equal to

$$N_2 = A_s f_{yd} = 400 \cdot 391.3 = 156520 \text{ N}$$

which has the same intensity of the force $N_{1,s}$ in the compression steel.

Therefore, the total compression force is equal to

$$N_1 = N_{1,c} + N_{1,s} = 544128 + 156520 = 700648 \text{ N}$$

The flexural strength is equal to[9]

$$\begin{aligned} M_{\text{Rd}} &= (0.8x)af_{cd}(h/2 - 0.4x) + A'_s f_{yd}(h/2 - d') \\ &\quad + A_s f_{yd}(d - h/2) = 0.8 \cdot 160 \cdot 300 \cdot 14.17 \cdot (150 - 0.4 \cdot 160) \\ &\quad + 400 \cdot 391.3 \cdot (150 - 41) + 400 \cdot 391.3 \cdot (259 - 150) \\ &= 80916368 \text{ Nmm} \cong 80.92 \text{ kNm} \end{aligned}$$

The resultant compression force in the concrete $N_{1,c}$ acts at distance $0.4x = 64$ mm from the compression edge, while the resultant force in the compression steel $N_{1,\text{acc}}$ acts at the distance of 41 mm; therefore, the resultant of compression forces N_1 is placed at following distance from the compression edge

$$(544128 \cdot 64 + 156520 \cdot 41)/(544128 + 156520) \cong 59 \text{ mm}$$

and its distance from the column axis is equal to $d_{N1} = (150 - 59) = 91$ mm.

Determination of the action lines of R_1 and R_2

R_2 is the resultant force of the triangular distribution of the bearing pressure on the right with respect to the direction of the load N

[7] Ratio x/d is equal to 0.641 when the concrete reaches its ultimate strain of 3.5 ‰ and the steel attains its strain at the elastic limit ($\varepsilon_{\text{el}} = 1.96$ ‰, for $f_{yd} = 391.3$ N/mm^2 and $Es = 200000$ N/mm^2 – for the steel modulus of elasticity the value suggested by EN1992-1-1 at 3.2.7(4) has been adopted).

[8] $N_{1,c}$ shall be coincident with $N = 544$ kN (being $As = A'_s$ and both steels yielded); the small difference is due to calculation approximations.

[9] It is worth to note that, if an axial force is also present, the rotational equilibrium to evaluate the flexural strength shall be imposed with respect to the geometrical centroid of the concrete cross-section.

$$R_2 = [(2/3b) \cdot (2/3p_{\max}) \cdot b]/2 = [1000 \cdot (2/3 \cdot 0.484) \cdot 1500]/2$$
$$= 242000 \text{ N} = 242 \text{ kN}$$

The direction of R_2 acts at distance (4/9 b) from the right corner

$$y_2 = 4/9b = 667 \text{ mm}$$

and at distance x_2 from the N direction

$$x_2 = (2/3b) - y_2 = 1000 - 667 = 333 \text{ mm}$$

The resultant force R_1 is equal to

$$R_1 = N - R_2 = 544 - 242 = 302 \text{ kN}$$

and its direction can be found from the rotational equilibrium (Fig. 10.65)

$$R_1 \cdot x_1 = R_2 \cdot x_2$$

$$x_1 = (R_2/R_1) \cdot x_2 = (242/302) \cdot 333 = 267 \text{ mm}$$

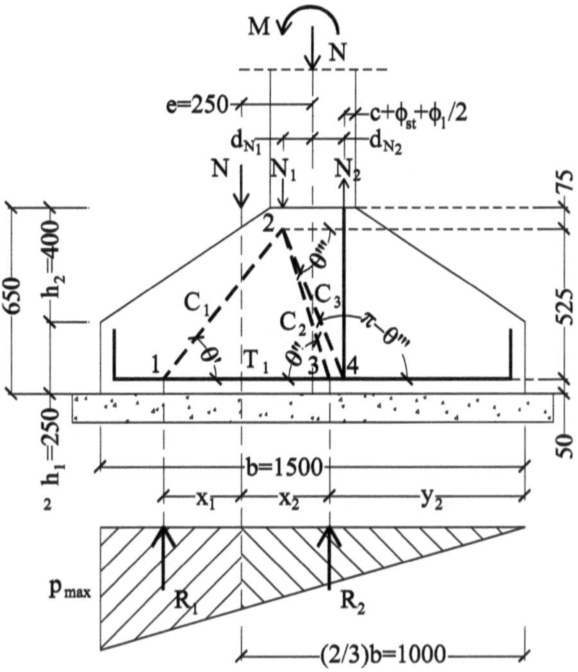

Fig. 10.65 S&T model for the isolated footing with eccentric load ($e = b/6$)

Distances of R_1 and R_2 from the column axis are, respectively

$$d_{R1} = (x_1 + e) = 267 + 250 = 517 \text{ mm}$$

$$d_{R2} = (x_2 - e) = 333 - 250 = 83 \text{ mm}$$

Calculation of the strut slope

The slope of the strut C_1 to the horizontal line is given by

$$\theta' = \arctan \frac{z}{d_{R1} - d_{N1}} = \arctan \frac{525}{517 - 91} = 50.94^\circ > 45^\circ$$

while for struts C_2 and C_3, the following values are obtained

$$\theta'' = \arctan \frac{z}{d_{R2} + d_{N1}} = \arctan \frac{525}{83 + 91} = 71.66^\circ > 45^\circ$$

$$\theta''' = \arctan \frac{z}{d_{N2} + d_{N1}} = \arctan \frac{525}{109 + 91} = 69.14^\circ > 45^\circ$$

Calculation of forces in struts and ties

Equilibrium at node 1

$$C_1 = \frac{R_1}{\sin \theta'} = 389 \text{ kN}$$
$$T_{1(1-2)} = \frac{R_1}{\tan \theta'} = 245 \text{ kN} \quad \text{(tension force in the tie } 1 - 2\text{)}$$

Equilibrium at node 4

$$\begin{aligned} C_3 &= \frac{N_2}{\sin \theta'''} = 167 \text{ kN} T_{1(3-4)} \\ &= \frac{N_2}{\tan \theta'''} \\ &= 60 \text{ kN} \quad \text{(tension force in the tie } 3 - 4\text{)} \end{aligned}$$

Design of the steel tie

The required steel area for the tie T_1 shall not be less than:
$A_{s1} \geq \frac{T_{1(1-2)}}{f_{yd}} = \frac{245000}{391.3} = 626 \text{ mm}^2$ then six 12 mm bars ($6\phi 12 = 678 \text{ mm}^2$) are adopted.

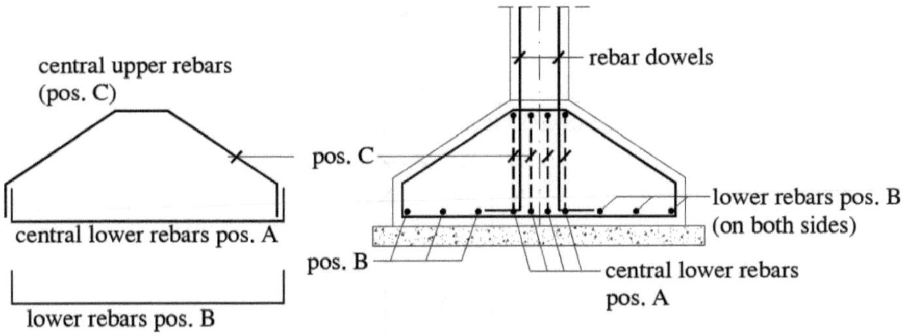

Fig. 10.66 Reinforcement layout in a rigid spread footing

For the design of the reinforcement in the plane perpendicular to the plane of the axial load eccentricity, a symmetrical S&T model can be identified as the case of null eccentricity.

Figure 10.66 shows the bar layout of the footing.

Case c: eccentricity *e* = 200 mm.

The only significant difference compared to the previous case is given by the shape of the bearing pressure distribution, which is not triangular but trapezoidal. The S&T model is drawn in Fig. 10.67; the calculation procedure is the same as the case $e = 250$ mm.

10.12.4 Example No. 4—Isolated Footing on Four Piles

Identify the S&T model suitable for the isolated footing on four piles shown in Fig. 10.68, under a vertical load N_{Ed} and a bending moment M_{Ed}. Calculate also forces in struts and ties.

It can be assumed that the spread of column loads occurs according to the following scheme: column loads are transmitted from the π_1 plane (Fig. 10.68) to the π_2 and π_3 orthogonal planes, which pass through the piles. The force transfer continues within each plane π_2 and π_3 until the piles; the strut-and-tie model shown in Fig. 10.69 is corresponding to the transfer in the π_1 plane, while the S&T model in Fig. 10.70 is corresponding to the transfer in π_2 and π_3 planes. In the S&T model lying on the π_1 plane, under the hypothesis that $M_{Ed} / L > N_{Ed} / 2$,[10] following forces are obtained:

compression: $A' = (M_{Ed}/L + N_{Ed}/2)$,

tension: $B' = (M_{Ed}/L - N_{Ed}/2)$,

so, it follows that:

on each compressed pile: $A = A'/2$;

[10] M_{Ed} and N_{Ed} denote the intensity (without the sign) of the bending moment and vertical load, respectively.

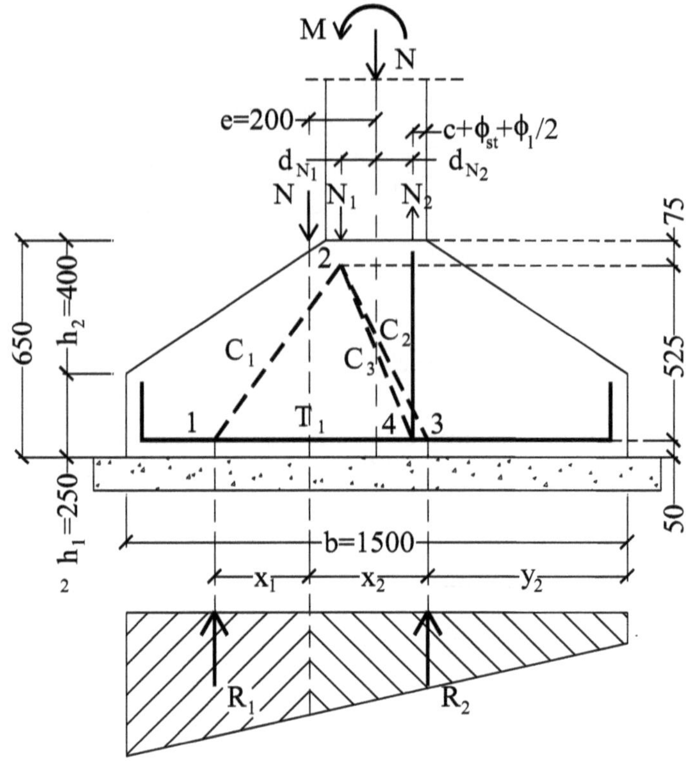

Fig. 10.67 S&T model for a rigid spread footing under low eccentric load ($e < b/6$)

on each tensile pile: $B = B'/2$.

In the following, expressions of angles shown in Fig. 10.69 and 10.70 are collected:

$$\theta_{11} = \arctan(h/e) \quad \text{on the } \pi_1 \text{ plane,}$$

$$\theta_{12} = \arctan(h/f) \quad \text{on the } \pi_2 \text{ plane,}$$

$$\theta_{13} = \arctan(h/l) \quad \text{on the } \pi_3 \text{ plane.}$$

Table 10.14 collects the tie forces, as a function of the angles θ_{11}, θ_{12} and θ_{13} and forces A, A', B and B' within the piles.

Figure 10.71 shows the reinforcement layout.

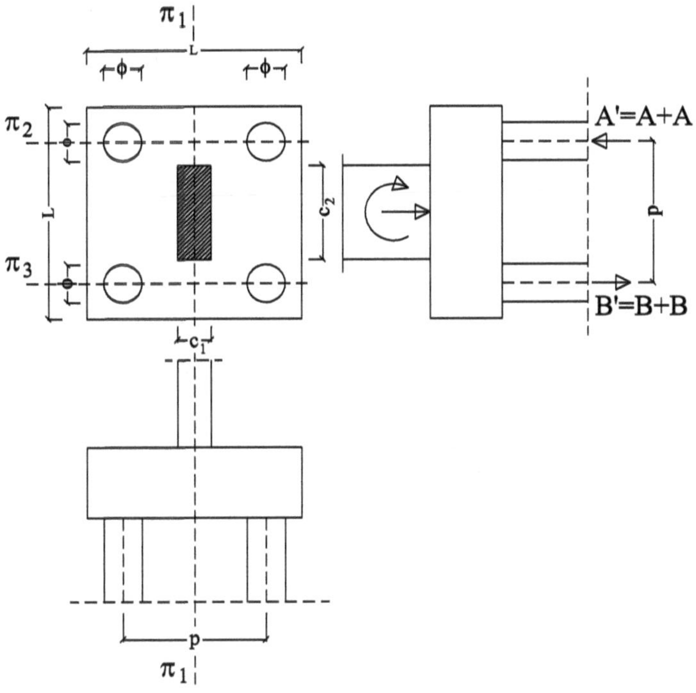

Fig. 10.68 Isolated footing on piles

10.12.5 Example No. 5—Gerber Hinge

Define the S&T model for a Gerber hinge and draw the reinforcement layout.

A Gerber hinge can be designed using two different S&T models, which, eventually, can be combined each other [10.9.4.6(1)]: a model (a) where the suspension reinforcement is vertical (Fig. 10.72a) and a model (b) where it is inclined (Fig. 10.72b). Even if EC2 leaves the possibility to use just one of the two trusses and to adopt one reinforcement layout only, the design based on one model only could not be satisfactory.

Indeed, if only the truss (a) is used, it is necessary to consider a top longitudinal reinforcement for the anchorage of the vertical tie 3–4 of Fig. 10.72a and to add the confining reinforcement for the inclined strut C_1. Vice versa, by using model (b), the bottom edge of the hinge remains fully free of reinforcement, so its behaviour is poor at SLS.

The reinforcement design can be more correctly performed by combining the two trusses, giving to each of them the 50% of the hinge reaction.

***Truss* (a)**

R_a denotes the rate of the hinge reaction assigned to the truss (a); it is worth observing that R_a may be coincident with the total hinge reaction R if only the truss (a) is

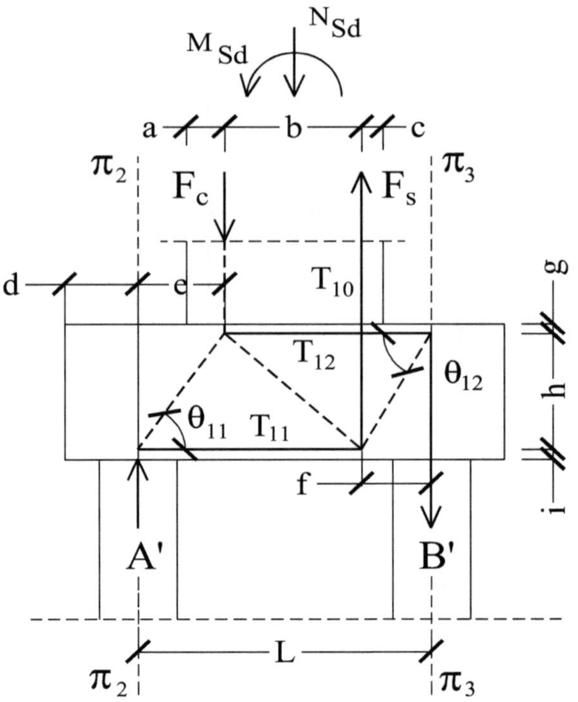

Fig. 10.69 S&T model on the π_1 plane (**a**: distance of the concrete compression resultant from the column edge, **b**: lever arm of the column cross-section, **c**: distance of column tensile rebars from the edge, **d**: distance of axes of compressed piles from the footing edge, **e**: lever arm between the column compression resultant and the compression resultant in piles, **f**: lever arm between the column tension resultant and the resultant in tension piles, **i**: distance of bottom rebars from footing bottom face, **h**: distance between the bottom and top rebars of the footing, **g**: distance of the top rebars from the footing top face)

considered, or with a rate of R (normally 50%) if a combination of truss (a) and truss (b) is considered.

Identification of the position of struts and ties

The position of the strut C_1 is known once the flexural strength of the beam cross-section is known; it is obtained by considering the centroid of the compression forces within the concrete and the top longitudinal bars. The other geometrical data can be obtained through elementary calculations.

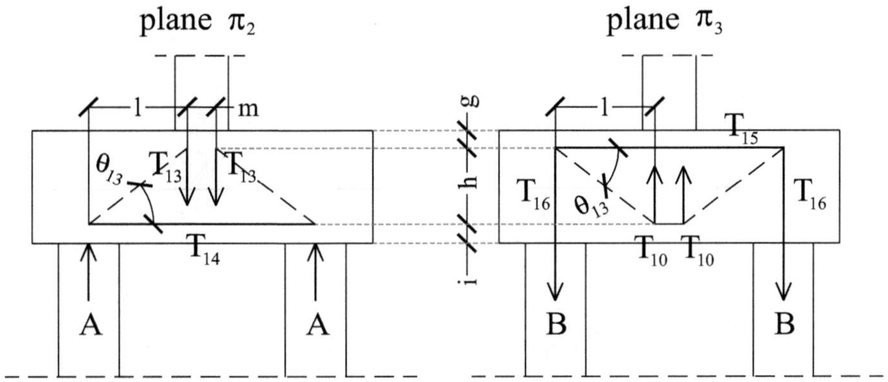

Fig. 10.70 S&T models on the π_2 and π_3 planes (*i*: distance of bottom rebars from footing bottom face, *h*: distance between the bottom and top rebars of the footing, *g*: distance of the top rebars from the footing top face, *l*: distance between pile axis and nearest column rebars, *m*: distance between column rebars)

Table 10.14 Tie forces of the S&T models for an isolated footing on four piles

Tie	Force
T_{10}	Fs[a]
T_{11}	$A'/\tan\theta_{11} = A'e/h$
T_{12}	$B'/\tan\theta_{12} = B'f/h$
T_{13}	A
T_{14}	$A/\tan\theta_{13} = Al/h$
T_{15}	$B/\tan\theta_{13} = Bl/h$
T_{16}	B

[a] F_s = tension force in the column reinforcement

Calculation of internal forces in the members of the truss **(a)**

Equilibrium at node 1

$$C_1 = R_a/\mathrm{sen}\,\theta_1 \quad T_1 = R_a\,/\tan\theta_1$$

Equilibrium at node 2

$$C_2\cos\theta_2 + C_3\cos 45^\circ = T_1$$

$$C_2\sin\theta_2 = C_3\sin 45^\circ$$

so, it follows that

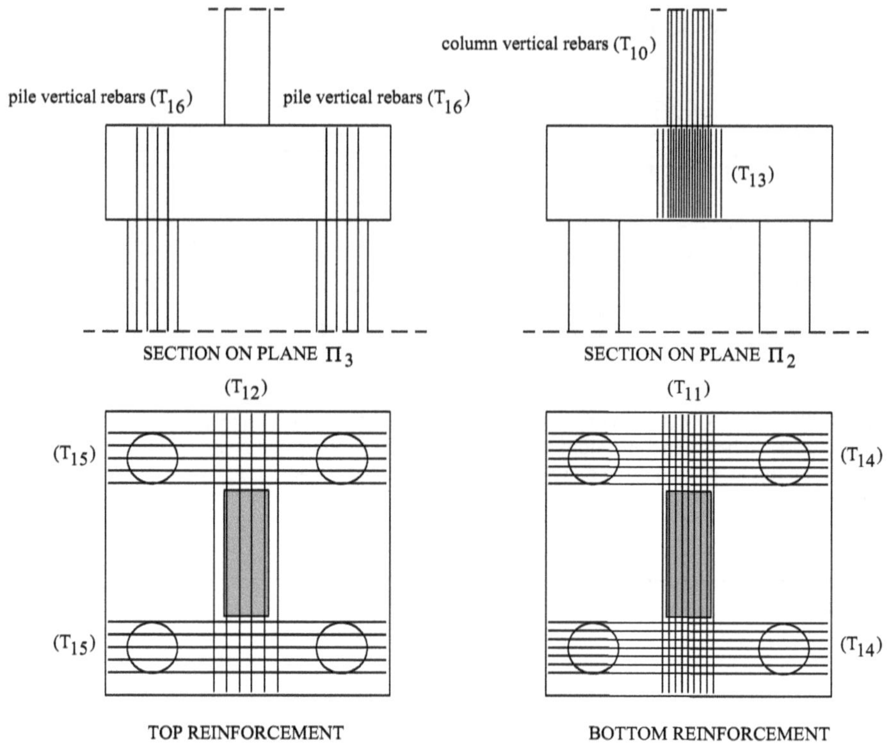

Fig. 10.71 Reinforcement layout for the isolated footing on four piles

$$C_2 = T_1/(\sin\theta_2 + \cos\theta_2) = \frac{R_a}{\tan\theta_1(\sin\theta_2 + \cos\theta_2)}$$

$$C_3 = C_2\sin\theta_2/\sin 45° = \frac{R_a\sin\theta_2/\sin 45°}{\tan\theta_1(\sin\theta_2 + \cos\theta_2)} = \frac{\sqrt{2}R_a}{\tan\theta_1(1 + \cot\theta_2)}$$

Equilibrium at node 3

$$T_2 = C_1\sin\theta_1 + C_2\sin\theta_2 = R_a + \frac{R_a\sin\theta_2}{\tan\theta_1(\sin\theta_2 + \cos\theta_2)}$$
$$= R_a + \frac{R_a}{\tan\theta_1(1 + \cot\theta_2)}$$

Equilibrium at node 4

$$C_4\sin 45° + C_3\sin 45° = T_2$$

it is obtained

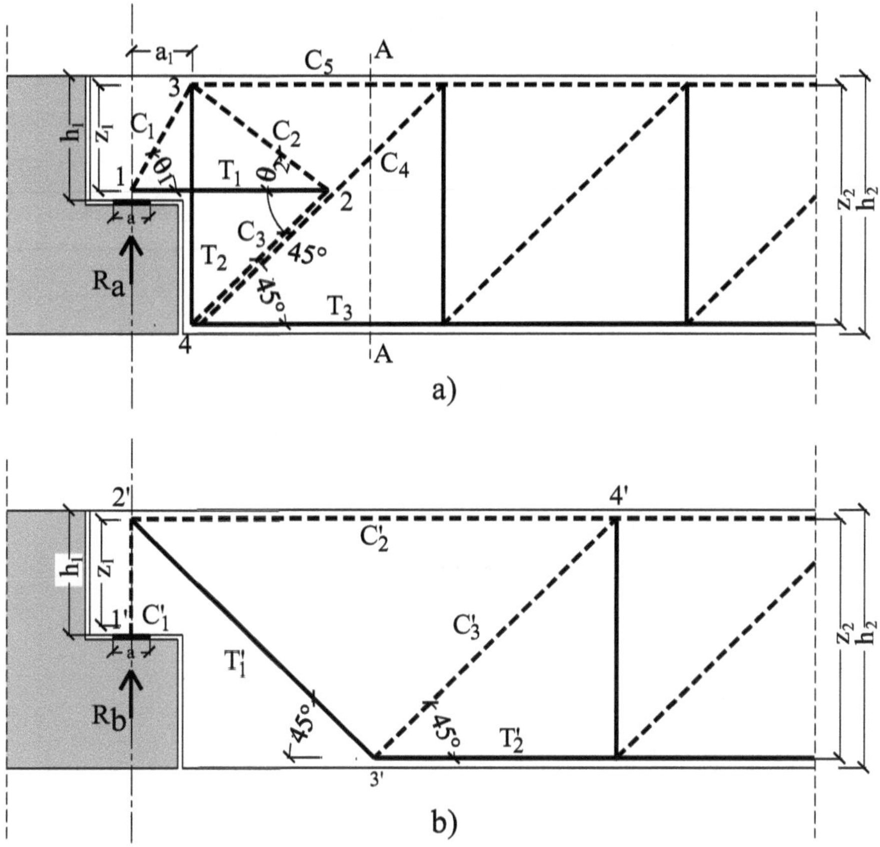

Fig. 10.72 S&T models for a Gerber hinge

$$C_4 = T_2/\sin 45^\circ - C_3$$

and, by substituting values of C_3 and T_2[11]

$$C_4 = \left[R_a + \frac{R_a}{\tan\theta_1(1+\cot\theta_2)}\right]\sqrt{2} - \frac{\sqrt{2}R_a}{\tan\theta_1(1+\cot\theta_2)} = \sqrt{2}R_a$$

finally,[12]

$$T_3 = C_3\cos 45^\circ + C_4\cos 45^\circ = \frac{\sqrt{2}R_a}{\tan\theta_1(1+\cot\theta_2)}\frac{\sqrt{2}}{2} + \sqrt{2}R_a\frac{\sqrt{2}}{2}$$

[11] The expression of C_4 can be obtained directly from the vertical equilibrium at section A-A: $C_4 \sin 45^\circ = R_a$, from it follows $C_4 = \sqrt{2}R_a$.

[12] $T_3 = T_2$ for the translational equilibrium in the direction of struts C_3 and C_4.

Table 10.15 Strut and tie forces of the truss (a) for the Gerber hinge

C_1	$R_a/\sin\theta_1$
C_2	$R_a/[\tan\theta_1(\sin\theta_2 + \cos\theta_2)]$
C_3	$\sqrt{2}R_a/[\tan\theta_1(1 + \cot\theta_2)]$
C_4	$\sqrt{2}R_a$
T_1	$R_a/\tan\theta_1$
T_2	$R_a + R_a/[\tan\theta_1 \cdot (1 + \cot\theta_2)]$
T_3	$R_a + R_a/[\tan\theta_1(1 + \cot\theta_2)]$

$$= \frac{R_a}{\tan\theta_1(1 + \cot\theta_2)} + R_a$$

Table 10.15 collects expressions of member forces of the truss (a).

Truss (b)

R_b denotes the reaction rate assigned to the truss (b).

***Calculation of forces in the members of the truss* (b)**

Equilibrium at node 1'

$$C_1' = R_b$$

Equilibrium at node 2'

$$T_1' \sin 45° = C_1'$$

therefore,

$$T_1' = \sqrt{2}\, C_1' = \sqrt{2}\, R_b$$

$$C_2' = T_1' \cos 45° = R_b$$

Equilibrium at node 3'

$$T_1'\text{sen}45° = C_3'\text{sen}45°,\, C_3' = T_1' = \sqrt{2}\, R_b$$

$$T_2' = T_1' \cos 45° + C_3' \cos 45° = 2\left(\sqrt{2}\, R_b \frac{\sqrt{2}}{2}\right) = 2\, R_b$$

Table 10.16 collects expressions of the member forces of the truss (b) Fig. 10.73.

Table 10.16 Strut and tie forces of the truss (b)

C'_1	R_b
C'_2	R_b
C'_3	$\sqrt{2}R_b$
T'_1	$\sqrt{2}R_b$
T'_2	$2\,R_b$

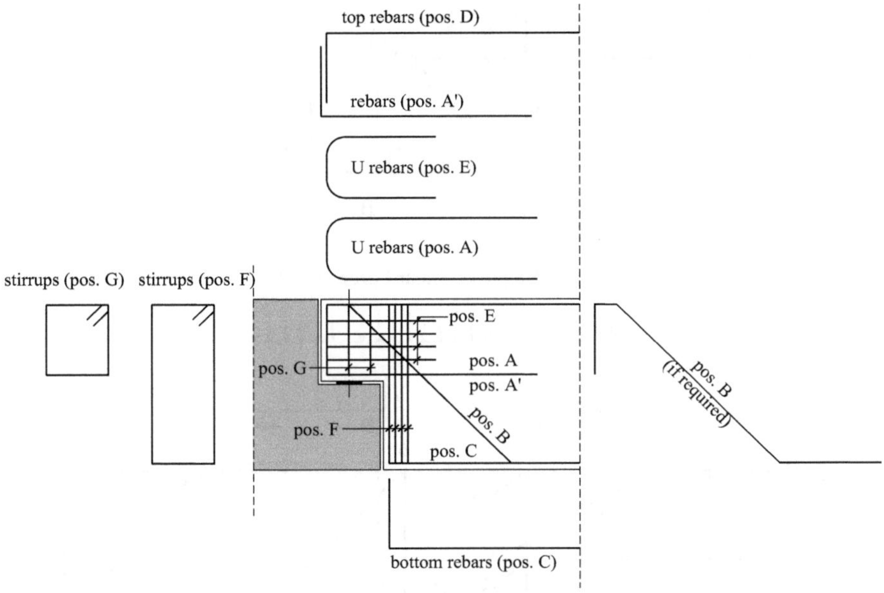

Fig. 10.73 Reinforcement layout in the Gerber hinge. If the design is performed only considering the truss **a**, the inclined reinforcement **b** is not required

10.12.6 Example No. 6—Abrupt Change of the Height of a Slender Beam

Identify the strut-and-tie model corresponding to an abrupt change of the height of a slender beam.

The S&T model for the analysis of the "D" region corresponding to an abrupt change of the height of a beam, subjected to positive bending moment and positive shear, is coincident with the model (a) used in the design of the Gerber hinge in Example no. 5. The only difference relies on the value of angle θ_1 in Fig. 10.74a, which is assumed equal to 45°, while in the Gerber hinge its value changes with the aspect ratio of the hinge itself.

Figure 10.74 shows both S&T models for positive or negative shear, under the assumption that the slope of struts is 45°. In the following, strut and tie forces for both trusses are evaluated.

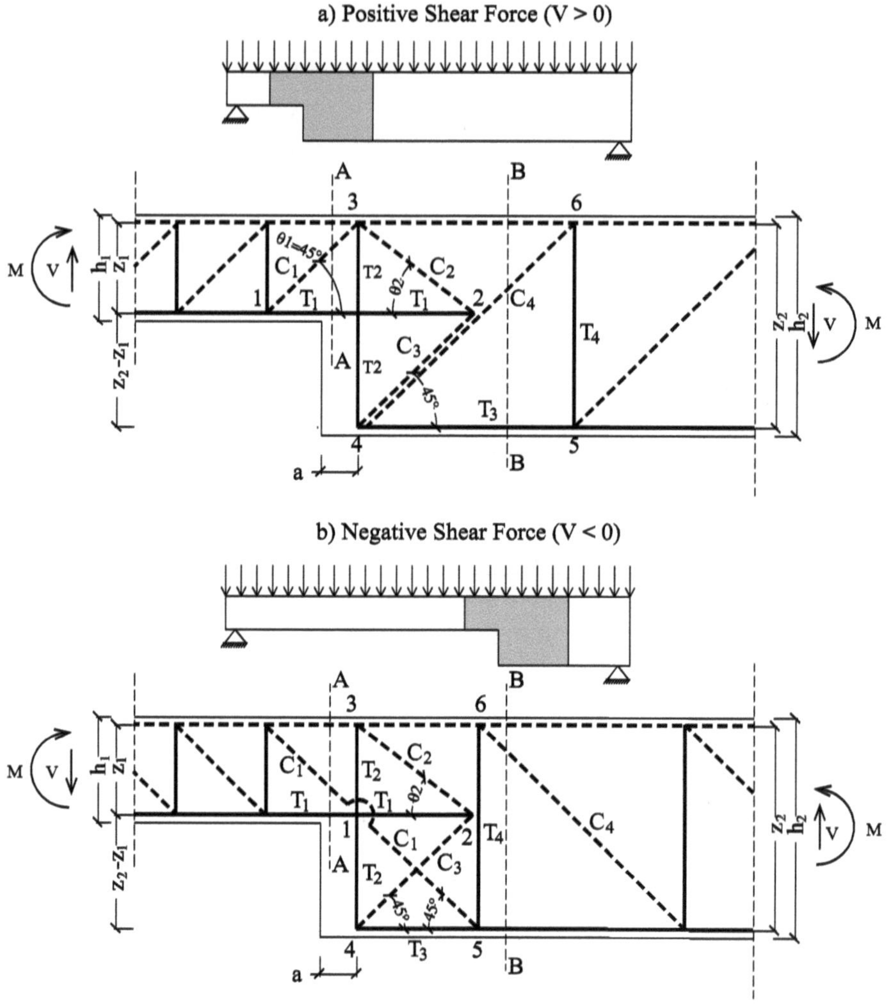

Fig. 10.74 S&T model for the abrupt change of the height of a slender beam with $h_2 > 2h_1$ $h_2 > 2h_1$: with positive **a** or negative **b** shear (the vertical tie T_2 between nodes 3 and 4 is at a distance a from the cross-section change because vertical stirrups forming the tie T_2 are distributed over the length $2a$, Fig. 10.75)

Calculation of forces in the truss members with positive shear (Fig. 10.74a)

Section (A-A)

$$C_1 = \sqrt{2}\, V$$
$$T_1 = M/z_1$$

Equilibrium at node 2

$$C_2 \cos\theta_2 + C_3 \cos 45^\circ = T_1$$
$$C_2 \sin\theta_2 = C_3 \sin 45^\circ$$
$$C_2 = T_1/(\sin\theta_2 + \cos\theta_2) = \frac{M}{z_1(\sin\theta_2 + \cos\theta_2)}$$

$$C_3 = C_2 \sin\theta_2 / \sin 45^\circ = \frac{M}{z_1} \frac{\sin\theta_2}{\sin 45^\circ(\sin\theta_2 + \cos\theta_2)} = \frac{\sqrt{2}M}{z_1(1 + \cot\theta_2)}$$

Equilibrium at node 3

$$T_2 = C_1 \sin 45^\circ + C_2 \sin\theta_2$$
$$= \sqrt{2}V\frac{\sqrt{2}}{2} + \frac{M \sin\theta_2}{z_1(\sin\theta_2 + \cos\theta_2)} = V + \frac{M}{z_1(1 + \cot\theta_2)}$$

Equilibrium at node 4

$$C_4 \sin 45^\circ + C_3 \sin 45^\circ = T_2$$

then

$$C_4 = T_2/\sin 45^\circ - C_3$$

and, by substituting expressions of C_3 and T_2,[13]

$$C_4 = \left[V + \frac{M}{z_1(1 + \cot\theta_2)}\right]\sqrt{2} - \frac{\sqrt{2}M}{z_1(1 + \cot\theta_2)} = \sqrt{2}V$$

finally

$$T_3 = C_3 \cos 45^\circ + C_4 \cos 45^\circ = \frac{\sqrt{2}M}{z_1(1 + \cot\theta_2)}\frac{\sqrt{2}}{2} + \sqrt{2}V\frac{\sqrt{2}}{2}$$
$$= \frac{M}{z_1(1 + \cot\theta_2)} + V$$

which is coincident with the expression of T_2.[14]

[13] The expression of C_4 can be obtained directly from the vertical equilibrium at the section B-B: $V = C_4 \sin 45^\circ$ from which: $C_4 = \sqrt{2}V$.

[14] The equation $T_3 = T_2$ can also be obtained from the translational equilibrium at node 4 in the direction of struts C_3 and C_4 (whose slope with respect to the horizontal line is 45°).

Equilibrium at node 6

$$T_4 = C_4 \sin 45° = \sqrt{2}\, V \frac{\sqrt{2}}{2} = V$$

Calculation of forces in the truss members with negative shear (Fig. 10.74b).
Section (A-A)

$$C_1 = \sqrt{2}\, V;\, T_1 = M/z_1$$

Equilibrium at node 2

$$C_2 \cos\theta_2 + C_3 \cos 45° = T_1$$
$$C_2 \sin\theta_2 = C_3 \sin 45°$$

and

$$C_2 = T_1/(\sin\theta_2 + \cos\theta_2) = \frac{M}{z_1} \frac{1}{(\sin\theta_2 + \cos\theta_2)}$$

$$C_3 = C_2 \sin\theta_2 / \sin 45° = \frac{M}{z_1} \frac{\sqrt{2}}{(1 + \cot\theta_2)}$$

Equilibrium at node 3[15]

$$T_2 = C_2 \sin\theta_2 = \frac{M}{z_1} \frac{1}{(1 + \cot\theta_2)}$$

Equilibrium at node 4

$$T_3 = C_3 \cos 45° = \frac{M}{z_1} \frac{\sqrt{2}}{(1 + \cot\theta_2)} \frac{\sqrt{2}}{2} = \frac{M}{z_1} \frac{1}{(1 + \cot\theta_2)} = T_2$$

Equilibrium at node 5

$$T_4 = C_1 \sin 45° = V$$

Equilibrium at node 6

$$C_4 \sin 45° = T_4$$

therefore,

[15] Alternatively, the force in the T_2 tie can be obtained by the vertical equilibrium at node 4: $T_2 = C_3 \mathrm{sen}45° = \frac{M}{z_1} \frac{\sqrt{2}}{(1+\cot g\theta_2)} \frac{\sqrt{2}}{2} = \frac{M}{z_1} \frac{1}{(1+\mathrm{cotg}\,\theta_2)}$

Table 10.17 Force in the members of the S&T model shown in Fig. 10.74

Member	Truss (a) with positive shear V	Truss (b) with negative shear V
C_1	$\sqrt{2}V$	$\sqrt{2}/V$
C_2	$(M/z_1)/(\sin\theta 2{+}\cos\theta 2)$	$(M/z_1)/(\sin\theta_2+\cos\theta_2)$
C_3	$\sqrt{2}(M/z_1)/(1+\text{cotg}\theta_2)$	$\sqrt{2}(M/z_1)/(1+\cot g\theta_2)$
C_4	$\sqrt{2}V$	$\sqrt{2}V$
T_1	M/z_1	M/z_1
T_2	$V+$ $(M/z_1)/(1+\text{cotg}\theta_2)$	$(M/z_1)/(1+\text{cotg}\theta_2)$
T_3	$V+$ $(M/z_1)/(1+\text{cotg}\theta_2)$	$(M/z_1)/(1+\text{cotg}\theta_2)$
T_4	V	V

$$C_4 = \sqrt{2}\,T_4 = \sqrt{2}\,V$$

Table 10.17 collects expressions of the member forces for both trusses (with positive or negative shear).

In the special case that $z_1 = 0.5\,z_2$ and $\theta_2 = 45°$, expressions of tie forces become.

for $V > 0$ (truss a)

$$T_1 = M/z_1 \quad T_2 = T_3 = V + 0.5M/z_1 = V + M/z_2; \quad T_4 = V$$

for $V < 0$ (truss b)

$$T_1 = M/z_1 \quad T_2 = T_3 = 0.5M/z_1 = M/z_2; \quad T_4 = V$$

It is worth noting that the tie T_2 of the truss (a) gives the contribution to the equilibrium both to the vertical component of the inclined struts C_2 and C_3 and to the tie T_1 anchorage, while the analogous tie in the truss (b) T_2 is in equilibrium only with this last one.

Figure 10.75 shows a possible arrangement of reinforcing bars at an abrupt change of cross-section.

10.12.7 Example No. 7—Corbel

Design the corbel shown in Fig. 10.76. *Materials: concrete* grade C35/45 – $f_{ck} =$ 35 N/mm^2 and *steel grade* B450C—$f_{yk} = 450$ N/mm^2. *Vertical load is* $V_{Ed} =$ 700000N = 700kN *and horizontal load is* $H_{Ed} = 70000$N = 70kN.

The resultant force of horizontal and vertical loads is equal to

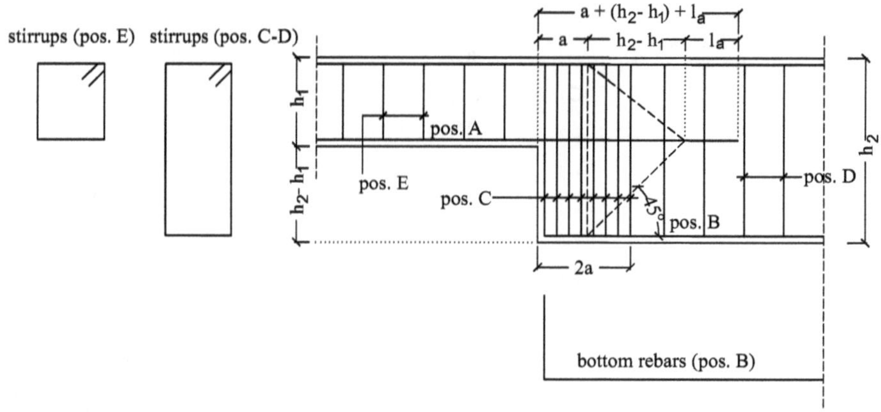

Fig. 10.75 Reinforcement layout at an abrupt change of the beam height with $h_2 > 2h_1$ (bars in pos. A shall be elongated of an amount equal to $(a + h_2 - h_1 + l_a)$, where a is half of the length over which stirrups C are distributed and l_a is the anchorage length)

$$F_{Ed} = \left(V_{Ed}^2 + H_{Ed}^2\right)^{0.5} = 703491 \text{ N}$$

and the slope α of the resultant force to the vertical is equal to

$$\alpha = \arctan(H_{Ed}/V_{Ed}) = 5.71°$$

The main reinforcement is arranged on two layers with a height u equal to 100 mm (Fig. 10.77); in this way, the centroid of the tension bars is placed at a distance $d' = 50$ mm from the external edge of the corbel. The effective depth is $d = h_c - 50 = 350$ mm, where $h_c = 400$ mm is the corbel depth.

Concrete design strength

$$f_{cd} = \frac{0.85\, f_{ck}}{1.5} = \frac{0.85 \cdot 35}{1.5} = 19.83 \text{ N/mm}^2$$

Steel design strength

$$f_{yd} = \frac{f_{yk}}{1.15} = \frac{450}{1.15} = 391.3 \text{ N/mm}^2$$

Node compressive strength

From Table 10.10 and for $f_{ck} = 35 \text{ N/mm}^2$, following values are obtained.
Compression nodes

$$\sigma_{1\text{Rd,max}} = 17.06 \text{ N/mm}^2$$

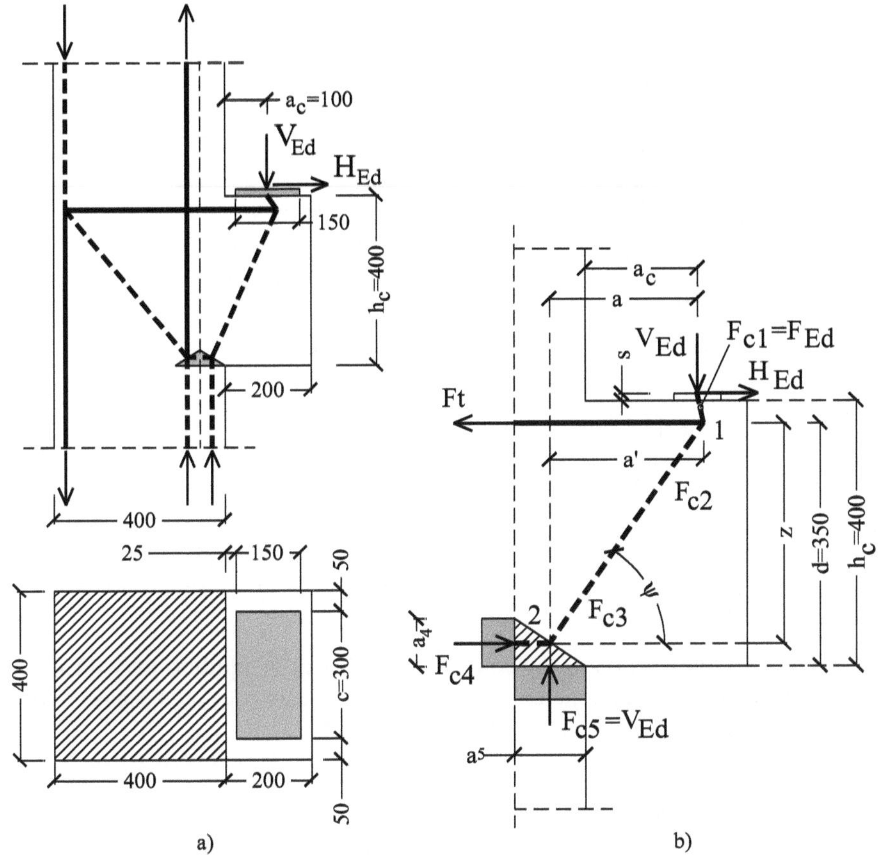

Fig. 10.76 200 × 400 mm corbel: **a** geometrical dimensions and S&T model for the corbel and the column (S&T model of the column assumes that bending moments at top and bottom cross-sections are counter clockwise), **b** corbel S&T model; for the sake of simplicity, the compression force in the inclined strut is denoted as F_{c2} at node 1 and F_{c3} ($= F_{c2}$) at node 3

Compression-tension nodes with anchored ties provided in one direction

$$\sigma_{2Rd,max} = 14.50 \text{ N/mm}^2$$

Compression-tension nodes with anchored ties provided in more than one direction

$$\sigma_{3Rd,max} = 12.79 \text{ N/mm}^2$$

(see also values collected in Table 10.10 for $f_{ck} = 35 \text{ N/mm}^2$)

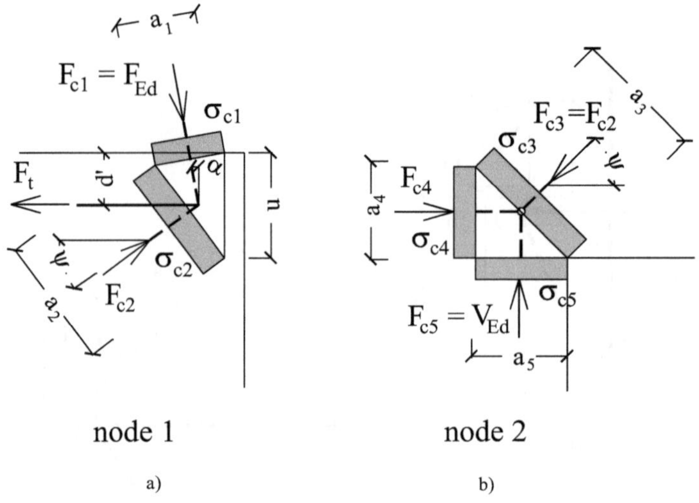

Fig. 10.77 Nodes 1 and 2 of the S&T model

Evaluation of forces F_t and F_{c4}

The width a_5 of the portion of the column vertical strut in equilibrium with V_{Ed} is calculated by imposing that the compressive stress equals $\sigma_{\text{1Rd,max}}$ because node 2 is a compression node

$$a_5 = \frac{V_{Ed}}{\sigma_{\text{1Rd,max}} b} = \frac{700000}{17.06 \cdot 400} = 102 \text{ mm}$$

Node 2 is placed at distance $a_5/2 \cong 51$ mm from the external face of the column, therefore the horizontal distance of the load application point from node 2 is equal to (Fig. 10.76b)

$$a = a_c + a_5/2 = 100 + 51 = 151 \text{ mm}$$

being the centroid of the top reinforcement at distance $d' = 50$ mm from the external edge of the corbel and under the hypothesis that the loading plate is 20 mm thick, the horizontal distance of node 1 from the load application point on the plate is equal to (Fig. 10.78b)

$$e = \left(d' + s\right) \tan \alpha = 70 \cdot (H_{Ed}/V_{Ed}) = 7 \text{ mm}$$

where α is the angle between the resultant of external loads F_{Ed} and the vertical direction, while the horizontal distance of node 2 from node 1 is equal to

$$a' = a + e = 151 + 7 = 158 \text{ mm}$$

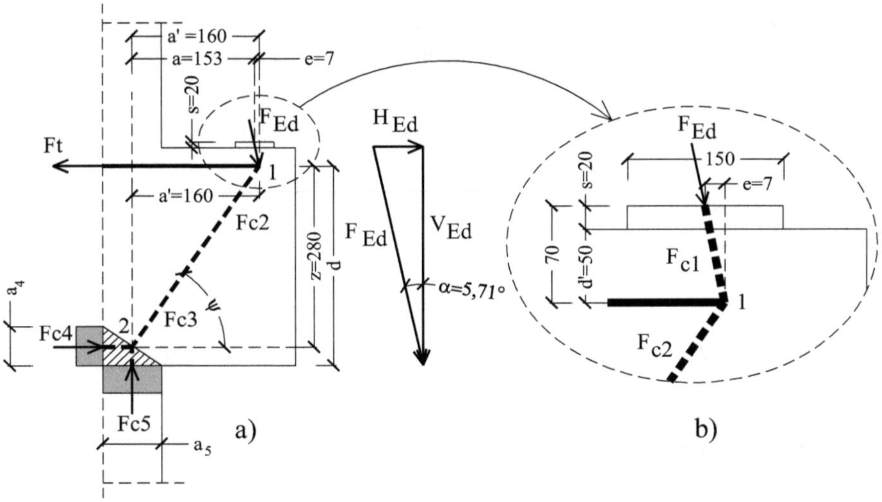

Fig. 10.78 Force polygon for the calculation of the tension force in the main reinforcement

If the internal lever arm z is equal to $0.8.d$ ($z = 0.8 \cdot 350 = 280$ mm) and the height of node 2 is indicated with a_4 (Fig. 10.77b), the distance of the same node 2 from the bottom edge of the corbel is equal to

$$a_4/2 = d - z = 0.2d = 0.2 \cdot 350 = 70 \text{ mm}$$

then $a_4 = 140$ mm.

Following expressions hold (Fig. 10.78):

– rotational equilibrium

$$V_{Ed}a' = (F_t - H_{Ed})\,z$$

$$700000 \cdot 158 = F_t \cdot 280 - 70000 \cdot 280$$

$$F_t = \frac{700000 \cdot 158 + 70000 \cdot 280}{280} = 465000\text{N} \cong 465 \text{ kN}$$

– translational equilibrium in the horizontal direction

$$F_{c4} = F_t - H_{Ed} = 465000 - 70000 = 395000\text{N} = 395 \text{ kN}$$

Evaluation of the force in the inclined strut ($F_{c2} = F_{c3}$)

The angle ψ between the strut and the horizontal direction is equal to:

$$\psi = \arctan\frac{z}{a'} = \arctan\frac{280}{158} = 60.56°$$

from the vertical equilibrium at node 1

$$F_{c2}\sin\psi = V_{Ed}$$

$$F_{c2} = \frac{V_{Ed}}{\sin\psi} = \frac{700000}{\sin 60.56°} = 803793\ \text{N} \cong 804\ \text{kN}$$

Verification of node 2

See Fig. 10.77b.

$$a_3 = \sqrt{a_4^2 + a_5^2} = \sqrt{140^2 + 106^2} \cong 176\ \text{mm}$$

$$\sigma_{c3} = \frac{F_{c3}}{a_3 b} = \frac{F_{c2}}{a_3 b} = \frac{806187}{176 \cdot 400} = 11.45\ \text{N/mm}^2 \le \sigma_{1\text{Rd,max}}$$

$$\sigma_{c4} = \frac{F_{c4}}{a_4 b} = \frac{400000}{140 \cdot 400} = 7.14\ \text{N/mm}^2 \le \sigma_{1\text{Rd,max}}$$

Design of the main reinforcement at the top

$$A_s = \frac{F_t}{f_{yd}} = \frac{470000}{391.3} = 1201\ \text{mm}^2$$

eight 14 mm bars $\left(8\phi14 = 1232\ \text{mm}^2\right)$ are adopted.Verification of node 1 (under the loading plate)

Dimensions a_1 and u are equal to
$a_1 = 150\cos\alpha = 149$ mm (Fig. 10.78)

$$u = 100\ \text{mm}$$

moreover (Fig. 10.80a)

$$a_1\cos\alpha = a_2\cos\beta$$

$$a_1\sin\alpha + a_2\sin\beta = u$$

and therefore

$$\beta = \arctan \frac{u - a_1 \text{sen}\alpha}{a_1 \cos \alpha} = 29.83°$$

$$a_2 = \frac{u - a_1 \sin \alpha}{\sin \beta} = 171 \text{ mm}$$

Stress verification of node 1 is performed considering the plate width $c = 300$ mm: Fig. 10.79

$$\sigma_{c1} = \frac{F_{Ed}}{a_1 c} = \frac{703491}{149 \cdot 300} = 15.74 \text{ N/mm}^2 > \sigma_{2\text{Rd,max}} = 14.50 \text{ N/mm}^2$$

The verification on face a_1 is not satisfied; therefore, the width of the plate is enlarged to 350 mm

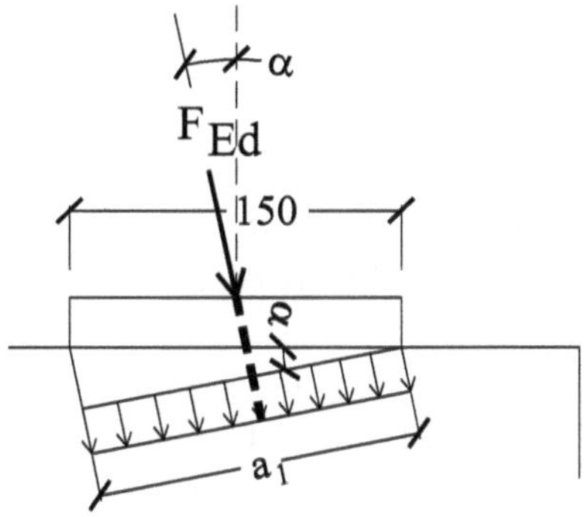

Fig. 10.79 Width a_1 of the strut F_{c1}

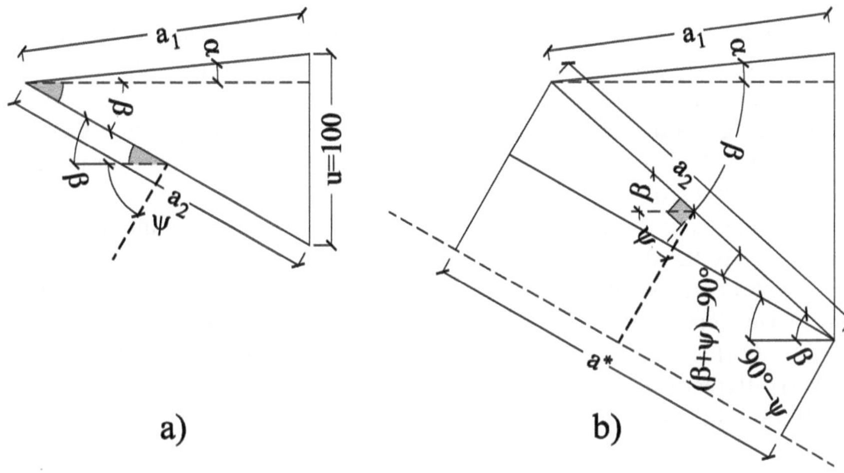

Fig. 10.80 Geometry of the node 1 (node under the loading plate): **a** u = 100 mm, **b** u > 100 mm

$$\sigma_{c1} = \frac{F_{Ed}}{a_1 c} = \frac{703491}{149 \cdot 350} = 13.49 \text{ N/mm}^2 \le \sigma_{2\text{Rd,max}}$$

The verification of the node is completed by considering the compressive stress on the face a_2.

With reference to Fig. 10.80a, $\beta + \psi = 29.83° + 60.56° = 90.39° \cong 90°$, therefore the strut axis can be considered orthogonal to face a_2 of node 1

$$\sigma_{c2} = \frac{F_{c2}}{a_2 c} = \frac{803793}{171 \cdot 350} = 13.43 \text{ N/mm}^2 \le \sigma_{2\text{Rd,max}}$$

If $(\beta + \psi) > 90°$, as for values of $u > 100$ mm, compressive stress on the nodal face shall be evaluated on the cross-section a^* of the inclined strut, i.e., orthogonal to the strut axis (Fig. 10.80b).

Reinforcement layout

The reinforcement layout is presented in Example no. 8.

10.12.8 Example No. 8—Design of the Corbel Secondary Reinforcement

Design the secondary reinforcement of the corbel of Example no. 7.

The S&T model of Example no. 7 has to be completed to consider the spread of the compression force within the inclined strut between nodes 1 and 2, assuming one of the S&T models shown in Fig. 10.81. These S&T models allow designing transverse, horizontal or vertical reinforcement, suitable to resist tensile stresses associated with the spread of the load. Without loss of generality, only the vertical load is considered, because all the considerations below do not change if the horizontal load is also present, except for the position of node 1 under the loading plate, which would be moved to the right or the left, as already seen in Example no. 7. The force in the main reinforcement will be therefore increased or decreased in the amount corresponding to the horizontal force.

The secondary reinforcement can be designed considering one of the two trusses shown in Fig. 10.82: the truss 2*a* will be adopted for corbels with $a_c \le h_c/2$ and the truss 2*b* for corbels with $a_c \le h_c/2$. The two trusses are different because of the different span–depth ratios a_c/h_c: EC2 suggests using horizontal or inclined stirrups for $a_c \le h_c/2$ (Fig. 10.85) and vertical closed stirrups for $a_c > h_c/2$ (Fig. 10.86). These two different layouts derive from the different slopes of the compressed strut (which transfers the load from the top face of the corbel to the column) in the two cases.

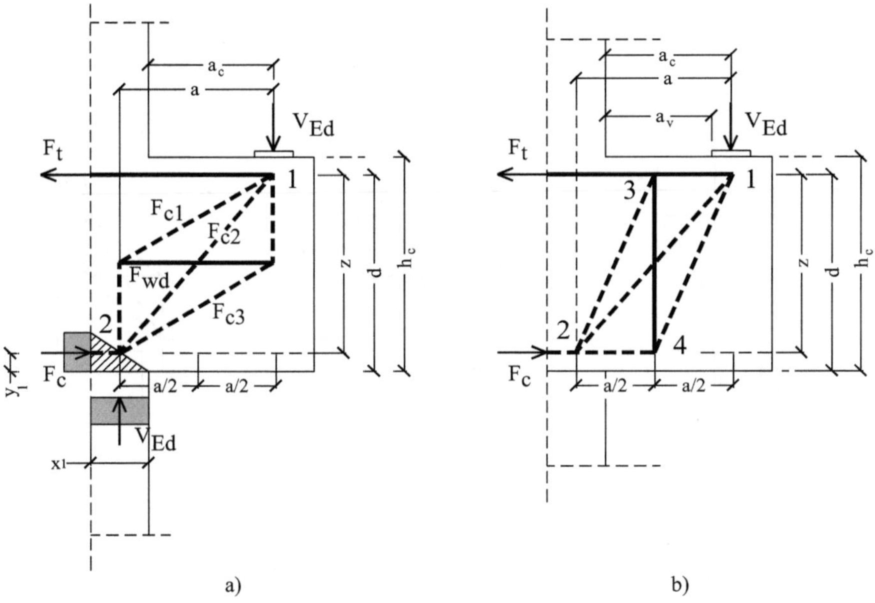

Fig. 10.81 Corbel: **a** $a_c \leq h_c/2$, **b** $a_c > h_c/2$

In the first case, the strut has a moderate slope to the vertical direction, therefore stresses induced by the spread of the compression force are almost horizontal; in the second case, the strut axis tends to approach the horizontal direction and therefore vertical stirrups are more suitable to resist tensile stresses due to the spread of the compression force in the strut.

Moreover, for $a_c > h_c/2$, the use of secondary reinforcement is suggested by EC2 only when the design shear exceeds the shear strength $V_{Rd,ct}$ without transverse reinforcement; however, for crack width limitation, it is better to adopt vertical stirrups.

The combination of truss 2 with truss 1, which has been already used in Example no. 7 for the design of the primary top reinforcement, gives rise to a statically indeterminate truss, where normal forces in struts and ties cannot be calculated through equilibrium conditions only.

For each of the two cases, it is necessary to identify a criterion to subdivide the load between the two trusses. In the following, the procedure for the load subdivision is described.

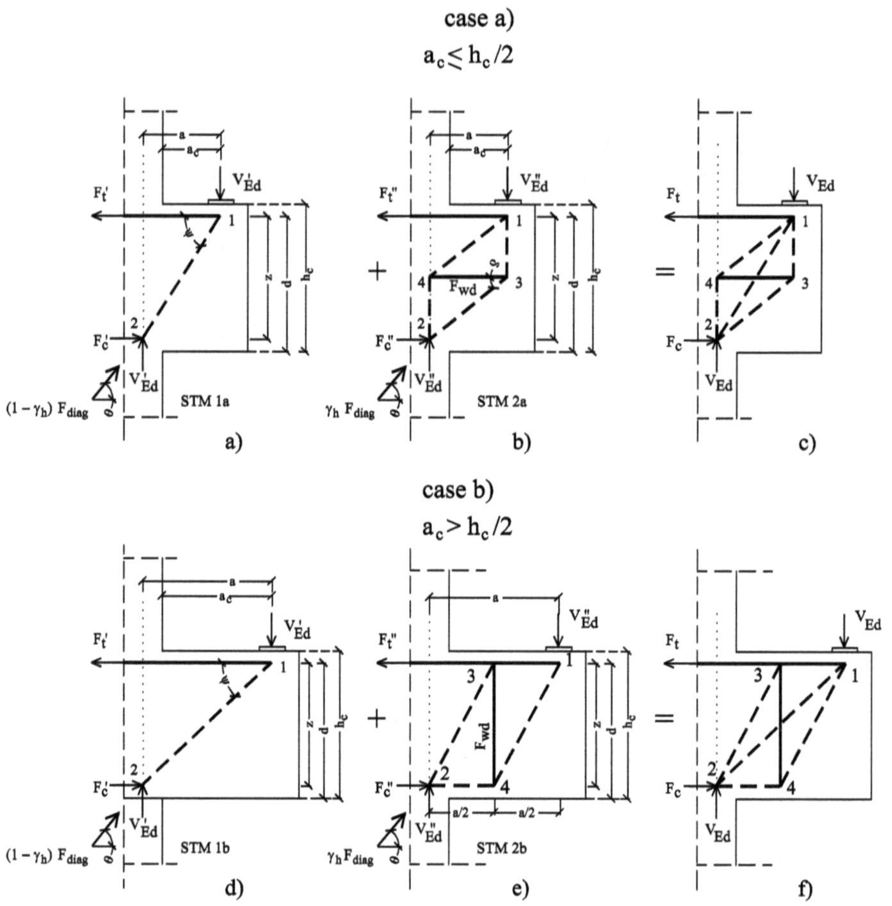

Fig. 10.82 S&T model for a corbel as the sum of two elementary trusses: case **a** (above) for corbel with $a_c \leq h_c/2$; case **b** (below) for corbel with $a_c > h_c/2$

Case a: $a_c \leq h_c/2$

By considering the principal compression field obtained from finite-element linear elastic analyses, the load V_{Ed} can be subdivided into the following two portions (Fig. 10.83):

$$V'_{Ed} = \frac{4 - \frac{z}{a}}{\frac{z}{a} + 3} \cdot V_{Ed} \quad \text{for truss } 1a$$

$$V''_{Ed} = \frac{2\frac{z}{a} - 1}{\frac{z}{a} + 3} \cdot V_{Ed} \quad \text{for truss } 2\,a$$

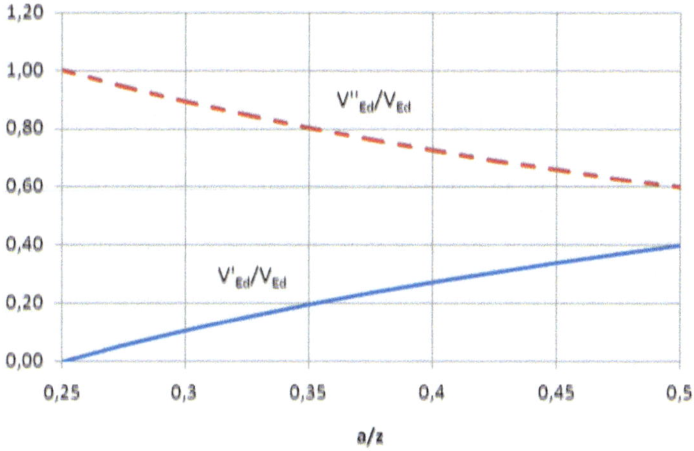

Fig. 10.83 Corbel with $a_c \leq h_c/2$: ratios V'_{Ed}/V_{Ed} (solid line) and $V''Ed/V_{Ed}$ (dashed line) with varying *a/z* (V'_{Ed} is the rate of the vertical load on truss 1a and V''_{Ed} is the rate on truss 2a shown in Fig. 10.82)

The force in the secondary reinforcement can be evaluated by considering truss 2a under the load V''_{Ed}, imposing the rotational equilibrium (Fig. 10.82b):

$$F''_t = V''_{Ed}\frac{a}{z}$$

and, from the translational equilibrium in the horizontal direction for nodes 1 and 4, the tension force in both the top tie and tie 3–4 is equal to the horizontal component of the compression force in strut 1–4:

$$F_{wd} = F''_t = V''_{Ed}\frac{a}{z}$$

by substituting the above expression in V''_{Ed}, the following expression is obtained

$$F_{wd} = \left(\frac{2\frac{z}{a} - 1}{3 + \frac{z}{a}} \cdot V_{Ed}\right) \cdot \frac{a}{z}$$

The expression given by Model Code 1990 for F_{wd} is the following

$$F_{wd} = \left(2\frac{z}{a} - 1\right)\frac{F_t}{3 + V_{Ed}/F_t}$$

where F_t is the total tie force; if F_t is explicitly written down as ($F_t = V_{Ed}a/z$), it can be easily proved that the expression of F_{wd} given by Model Code 1990 reduces itself to the same expression

$$F_{wd} = \left(2\frac{z}{a} - 1\right)\frac{V_{Ed}\frac{a}{z}}{3 + \frac{V_{Ed}}{V_{Ed}}\frac{z}{a}} = \frac{2\frac{z}{a} - 1}{3 + \frac{z}{a}} \cdot V_{Ed} \cdot \frac{a}{z}$$

Figure 10.84 shows the trend of the ratio F_{wd} / V_{Ed} at varying a /z: the maximum value can be found for a /z = 0.5 and it is equal to a 0.3.

EC2 suggests a minimum amount of secondary reinforcement not smaller than 25% of the main reinforcement [see J.3(2)].

If main reinforcement strength is indicated with F_t, secondary reinforcement strength shall meet the condition

$$F_{wd\,\mathrm{min}} \geq 0.25 F_t$$

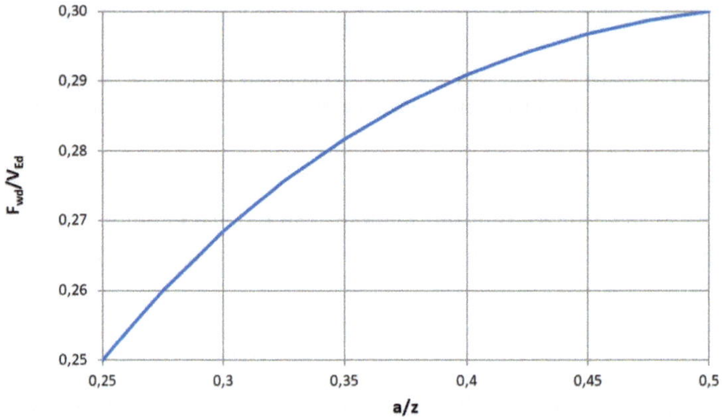

Fig. 10.84 Corbel with $a_c \leq h_c/2$: ratio F_{wd} / V_{Ed} *vs* a/z according to MC90 formula

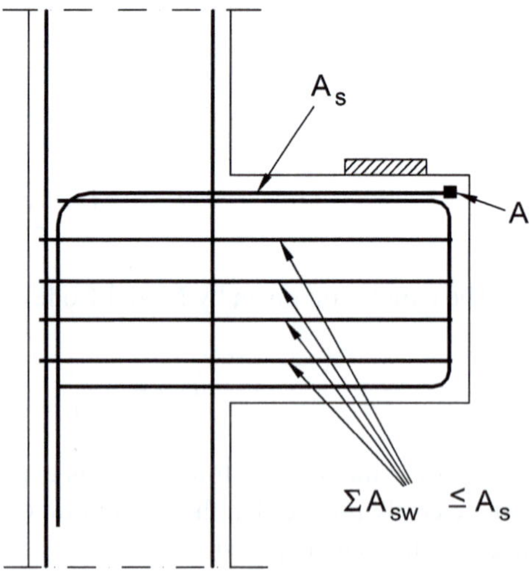

Fig. 10.85 Corbel with horizontal secondary reinforcement: A = anchorage device or U-bars (see Fig. 10.87), $A_{s\,\mathrm{main}} = mainreinforcement$, $\sum A_{s,\mathrm{lnk}}$ = horizontal secondary reinforcement

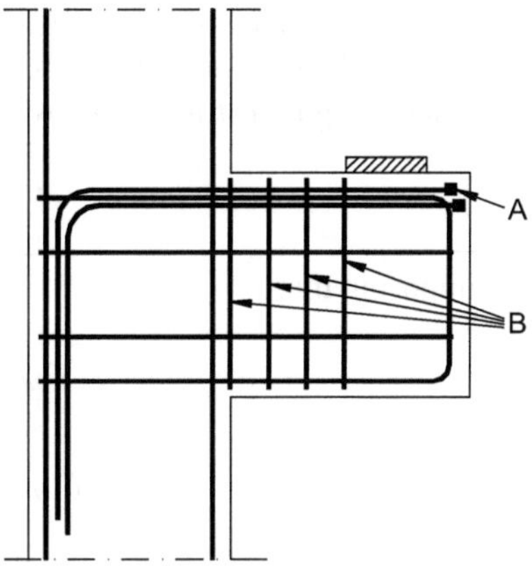

Fig. 10.86 Corbel with vertical secondary reinforcement: $A =$ anchorage device or U-bars (see Fig. 10.87), $B =$ secondary reinforcement made with closed vertical stirrups

Being $F_{\mathrm{t}} = V_{\mathrm{Ed}}a/z$ (from the rotational equilibrium, Fig. 10.82c), the inequality can be rewritten as

$$F_{wd\,\min}/V_{\mathrm{Ed}} \geq 0.25a/z$$

For a/z in the range [0.25; 0.5], the ratio $F_{wd\,\min}/V_{Ed}$ takes values between 0.0625 and 0.125, i.e. values lower than obtained using the MC90 design formula (MC90 results are collected in the diagram of Fig. 10.84). Therefore, the amount of secondary reinforcement given by the design formula is always greater than the minimum amount recommended by EC2.

When a significant horizontal force is present, the minimum reinforcement may be greater than the calculated amount. In this case, the main reinforcement force takes the following expression

$$F_t = V_{Ed}(a/z) + H_{Ed}$$

and then

$$F_{wd\,\min}/V_{Ed} \geq 0.25(a/z + H_{Ed}/V_{Ed})$$

***Case b*: $a_c > h_c/2$**

The force F_{wd} in the vertical tie is calculated under the hypothesis that it varies linearly with a and that

$$F_{wd} = 0 \text{ for } a = z/2$$

$$F_{wd} = V_{Ed} \text{ for } a = 2z$$

In other words, for $a \leq z/2$ (corbel), the resistant truss reduces to truss 1*b* (Fig. 10.82d) and for $a \geq 2z$ to truss 2*b* (Fig. 10.82e); the equation for F_{wd} is corresponding to a straight line with angular coefficient F_{w1} and ordinate F_{w2}

$$F_{wd} = F_{w1}a + F_{w2}$$

by imposing that

$$F_{wd} = 0 \; for \; a = z/2$$

$$F_{wd} = V_{Ed} \; for \; a = 2z$$

the following expressions are obtained

$$F_{w1} = \frac{2}{3}\frac{V_{Ed}}{z} \text{ and } F_{w2} = -\frac{V_{Ed}}{3}$$

finally, the expression of the straight line describing the variation of F_w with a becomes

$$F_{wd} = \frac{2}{3}\frac{V_{Ed}}{z}a - \frac{V_{Ed}}{3} = V_{Ed}\frac{2a/z - 1}{3}$$

The vertical translational equilibrium for nodes 1 and 4 of truss 2*b* gives

$$F_{\text{wd}} = V''_{Ed}$$

the ratio $[(2a/z - 1)/3]$ in the F_{wd} expression is corresponding to the portion of vertical load V''_{Ed} applied to truss 2*b*; in particular, for $a/z = 1$, this ratio takes the value of 1/3.

The minimum value of F_{wd} is given by the minimum amount of stirrups suggested by EC2 $\left(A_{s_{\text{lnk}}} \geq 0.5\ V_{Ed}/f_{yd}\right)$ [J.3(3)], which means a minimum force in the stirrups equal to $F_{wd,min} = 0.5\ V_{Ed}$.

The maximum value of F_{wd} is equal to $V_{Ed}/3$ for $a/z = 1$; this value is smaller than the minimum vertical secondary reinforcement recommended by EC2, therefore, for $a_c > h_c/2$, the secondary reinforcement is always coincident with the minimum one.

To design the secondary reinforcement, first, it is required to identify which is the case to be considered (a or b) for the corbel of Example no. 7; to this aim, a_c is compared with $h_c/2$:

$$a_c = 100 \text{ mm}, h_c/2 = 200 \text{ mm}$$

being $a_c = h_c/2$, the secondary reinforcement could be arranged in the horizontal direction considering the truss 2*a* or in the vertical direction using the truss 2*b*.

Rigorously, the choice between the two different reinforcement layouts shall be based on the ratio a/z (as indicated in Model Code 1990) and not on the ratio a_c/h_c (which is suggested by EC2 because of its easier evaluation). Indeed, the actual slope of the compressed strut is defined by the ratio a/z and it is depending also on the concrete cover of the main reinforcement and the size of the bottom node.

Considering geometrical data of Example no. 7, but substituting a with a' because the horizontal load moves outwards node 1, the following quantities are calculated:

$$a' = 158 \text{ mm} \quad z = 280 \text{ mm} \quad \tan\psi = 280/158 = 1.77 \leq 2.5$$

being $a' > z/2$[16] the secondary reinforcement is designed considering the truss 2*b* of Fig. 10.82. However, for the sake of completeness and to obtain an easy comparison of results, the calculation is performed with both trusses 2*a* and 2*b*.

Design of the horizontal secondary reinforcement (truss 2a)

The force in the horizontal secondary reinforcement can be calculated using the design formula which has been demonstrated above

$$F_{wd} = \frac{2\frac{z}{a} - 1}{3 + \frac{z}{a}} \cdot \frac{a}{z} \cdot V_{Ed}$$

or with the formula proposed in the Model Code 1990

$$F_{wd} = \left(2\frac{z}{a} - 1\right)\frac{F_t}{3 + \frac{V_{Ed}}{F_t}}$$

by using the difference $(F_t - H_{Ed})$ instead of F_t.

Moreover, because of the horizontal load, in both expressions it is useful to put $a' = 160$ mm (Fig. 10.78a), instead of using the distance $a = 153$ mm; therefore, the following expression of F_{wd} is obtained[17]

$$F_{wd} = \frac{2\frac{280}{158} - 1}{3 + \frac{280}{158}} \cdot \frac{158}{280} \cdot 700000 = 210597 \text{ N}$$

[16] Appendix J of EC2 at J.3(1) gives limits to the application field of the S&T model for the design of corbels with values of tan Ψ in the range 1 ÷ 2.5 (*N.B. in EC2 the symbol* θ *is used instead of* Ψ *to indicate the angle between the inclined strut and the horizontal*).

[17] By using the formula of Model Code 1990, with $(F_t - -H_{Ed})$ in place of F_t, the same result is obtained:

$$F_{wd} = \left(2\frac{z}{a} - 1\right)\frac{F_t - H_{Ed}}{3 + \frac{V_{Ed}}{F_t - H_{Ed}}} = \left(2\frac{280}{158} - 1\right)\frac{465000 - 70000}{3 + \frac{700000}{465000 - 70000}} = 210597 \text{ N}$$

The secondary reinforcement will have a total area not smaller than

$$A_{sw} = \frac{F_{wd}}{f_{yd}} = \frac{210597}{391.3} \cong 538 \text{ mm}^2 \geq k_1 \cdot A_s = 0.25 \cdot 1232 = 308 \text{ mm}^2$$

four 10 mm stirrups with two legs (4 ϕ 10 – A_{sw} = 628 mm^2) are adopted (Fig. 10.85).[18]

***Design of the vertical secondary reinforcement* (*truss* 2b)**

By applying the equation valid for the truss 2*b*, the force in the vertical secondary reinforcement is calculated as

$$F_{wd} = V_{Ed}\frac{2a/z - 1}{3}$$

As for the previous case, due to the presence of the horizontal load H_{Ed}, in the expression of F_{wd} it is required to use $a' = 160$ mm instead of $a = 153$ mm, obtaining

$$F_{wd} = 700000\frac{2 \cdot 158/280 - 1}{3} \cong 30000 \text{ N}$$

the vertical secondary reinforcement will have a total area equal to

$$A_{sw} = \frac{F_{wd}}{f_{yd}} = \frac{30000}{391.3} \cong 77 \text{ mm}^2$$

which is less than the minimum [see J.3(3)][19]

$$A_{sw,\min} = k_1 \cdot A_s = 0.25 \cdot 1232 = 308 \text{ mm}^2 \geq A_{sw}$$

therefore four 8 mm stirrups with two legs ($4\phi8 - A_{sw} = 400\text{mm}^2 \geq A_{\text{sw,min}}$).

In drawing the reinforcement layout, it is worth noting that the anchorage of the main reinforcement shall be suitably provided under the loading plate. The anchorage may be guaranteed by (Fig. 10.87):

(a) U-bars,
(b) straight bars welded to a transverse bar,
(c) bent bars in the vertical plane,
(d) mechanically anchored bars or headed bars.

[18] The lower limit $A_{sw} \geq 0.25\, A_s$ is indicated in Appendix J at [J.3(2)].

[19] For $a_c > h_c/2$, it is always $A_{sw} < A_{sw\,\min}$, as already mentioned above.

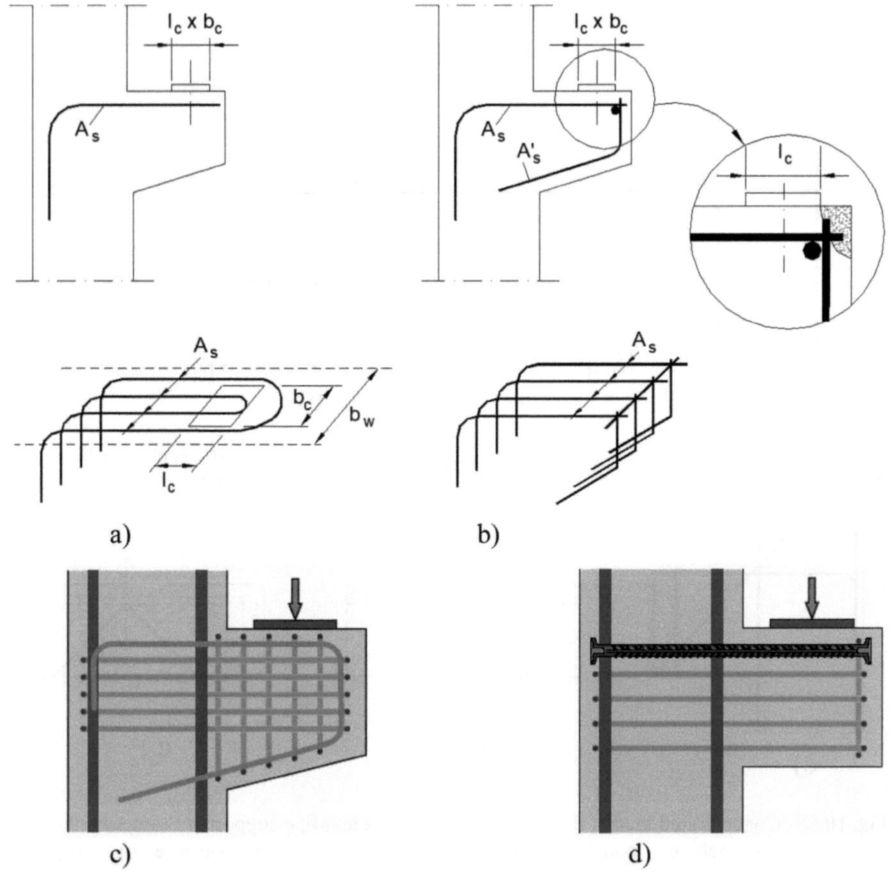

Fig. 10.87 Reinforcement layouts for a corbel: **a** U-bars, **b** welded transverse bar, **c** bent bars, **d** headed bars

10.13 Corbel Subjected to a Concentrated Load at the Bottom

If the load is applied at the bottom surface of the corbel, it can be transmitted to the top by a vertical tie, thus the same calculation scheme valid for the load at the top can be used (Fig. 10.88a), or by an inclined reinforcement (Fig. 10.88b). When the corbel supports a beam with the same height (Fig. 10.88c), a combination of the two models can be adopted, bearing in mind that the load transmission methods vary with the reinforcement layout in the supported beam: the load is transmitted to the bottom by inclined concrete struts in the beam (Fig. 10.88c_1), while bent bars can transmit the load to the top (Fig. 10.88c_2).

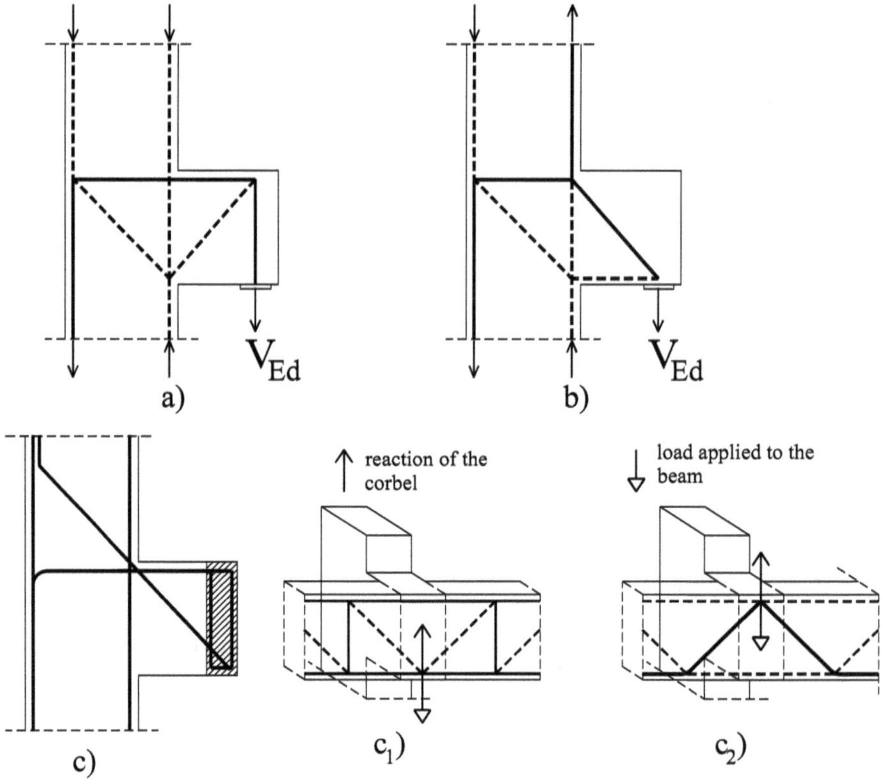

Fig. 10.88 Concentrated load at the bottom of a corbel **a** and **b**, **c** supported beam with the same height as the corbel: ($\mathbf{c}_1$) load transmitted at the bottom by inclined concrete struts, ($\mathbf{c}_2$) load transmitted at the top by bent-up bars

References

AASHTO LRFD. (2007) *Bridge design specifications*, (4th Edn). American Association of State Highway and Transportation Officials.

ACI 318-19, *Building code requirements for structural concrete and commentary.* ACI Committee 318

Angotti, D., & Spinelli, P. (2001). The method of reducing the elastic modulus for the identification of strut-and-tie models in reinforced concrete structures: Some considerations and experiences (in Italian). In *Proceedings of the Workshop S&T-2001*, School of Engineering, Florence (Italy), pp. 3–14

Brown, M. D., & Bayrak, O. (July–August 2008). Design of deep beams using strut-and-tie models-part I: Evaluating U.S. provisions. *ACI Structural Journal, 105*(4).

CEB/FIP. (1991). Model Code 1990—Final draft (Vol. 3).

FIB. (1999). *Structural concrete.* Textbook on behavior, design and performance updated knowledge of the CEB/FIP Model Code 1990

Harisis, A., & Fardis, M. N. (1991). Computer-aided automatic construction of strut-and-tie models. Structural Concrete, IABSE Colloquium, Stuttgart, International association for bridge and structural engineering (pp. 533–538), Zürich.

Marti, P. (January–February 1985). Basic tools of reinforced concrete beam design. *ACI Journal.*
Nilsson, I. (1973). Reinforced concrete corners and joints subjected to bending moment. *National Swedish Building Research, document D7*
Nilsson, I., & Losberg, A. (June 1976). Reinforced concrete corners and joints subjected to bending moment, proceedings ASCE. *Journal of the Structural Division, 102*(ST6), 1229–1254.
Schlaich, J., Schäfer, K., & Jennewein, M. (1987), Towards a Consistent Design of Structural Concrete, *PCI Journal*, V. 32, No. 3, Chicago, IL, pp. 74–150

Chapter 11
Serviceability Limit States (SLS)

Abstract The chapter deals with the behaviour of structures in service, which should fulfil the fundamental requirements of limited deflections and limited cracking. The above-mentioned purposes are pursued through the limitation of stresses on concrete and steel under service load combinations and through direct evaluation of deflections and width of cracks. EC2 also introduces simplified methods based on the L/d slenderness ratio. Several worked examples are developed and some calculation tools, such as tables and interaction diagrams, are provided.

11.1 General

The capability of structures to perform their function during service conditions is guaranteed through stress limitation, crack opening control and deflection control.

The values suggested by EC2 are recalled in the following.

- Stress limitation

  Concrete

- for structures with exposure class XD, XF, or XS (Table [4.1]), in the characteristic load combination, $\sigma_c \leq 0.60 f_{ck}$ is required [7.2(2) and note];
- in every case, in the quasi-permanent load combination, if $\sigma_c \leq 0.45\, f_{ck}$, the hypothesis of linear creep is valid. Otherwise, non-linearity should be assumed [7.2(3) and note].

Reinforcement in tension: for characteristic load combinations $\sigma_s \leq 0.80 f_{yk}$ is required by [7.2(4)P], [7.2(5) and note].

This chapter was authored by Piero Marro and Matteo Guiglia.

© The Author(s), under exclusive license to Springer Nature Switzerland AG 2022
F. Angotti et al., *Reinforced Concrete with Worked Examples*,
https://doi.org/10.1007/978-3-030-92839-1_11

- Crack control

For reinforced concrete structures, depending on the exposure class, for quasi-permanent load combinations, control of the crack opening (less than 0.4–0.3 mm) is required. For bonded prestressed structures, frequent combination, limits are 0.2 mm or decompression, depending on the exposure class (see Table [7.1N]). The main criteria to design the required minimum reinforcement in tension are collected in [7.3.2], always to limit crack opening.

- Deflection control

Limit values, taken from ISO 4356, are the following:

- for esthetical reasons, deflection for quasi-permanent load combinations shall not be greater than $\frac{1}{250}$ span;
- for structures supporting brittle elements like glass, thin walls, etc., the reference limit is $\frac{1}{500}$ span, under quasi-permanent load applied after the construction.

11.2 Stress Limitation

EC2 does not provide any specific procedure to calculate stresses at SLS. However, it is logical to assume the hypothesis of linear elasticity and, where tensile strength is reached, to use a conventional modular ratio between the modulus of elasticity of steel and concrete for defining a cracked transformed section. The temporary edition of EC2, ENV 1992-1-1, accounting for concrete classes until $f_{ck} = 50$ N/mm^2, uses to suggest modular ratio $\alpha_e = E_s/E_c = 15$, which is a value taking into account a creep amount. For concretes with higher strength, until 90 N/mm^2, the application of such a ratio seems not correct because, with the increase of the strength, elastic modulus E_c increases and creep decreases. Considering that EC2 does not provide any suggestion, it is possible to refer to French Code BEAL/BPEL (see References) which suggests:

- $\alpha_e = 15$ for $f_{ck} < 60$ N/mm^2
- $\alpha_e = 9$ for $f_{ck} \geq 60$ N/mm^2

such values being scientifically justified in (Lab.P.C. 1996).

A more detailed discussion and a comparison with the general method (non-linear analysis) can be found in (fib Bulletin no. 92 "Serviceability Limit State of Concrete Structures").

In the present text, therefore, the following values will be accounted for:

- $\alpha_e = 15$ for $f_{ck} \leq 50$ N/mm^2
- $\alpha_e = 9$ for $f_{ck} > 50$ N/mm^2

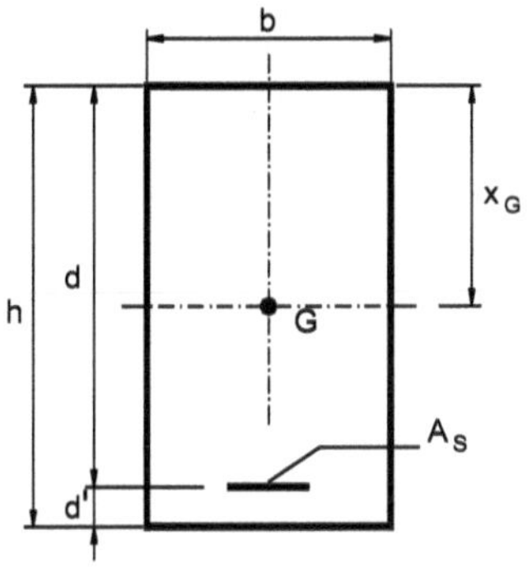

Figure. 11.1 Rectangular cross-section with reinforcement in tension

11.2.1 Bending–Solving Formulas

To speed up stress calculation, formulas for geometric and mechanic parameters are expressed through dimensionless quantities (neutral axis, second moment of area, applied loads, etc.). In the case of cracked cross-section (phase II), suitable Tables given in the Appendix provide the relevant numerical values.

11.2.1.1 Formulas for the Calculation of Neutral Axis and Second Moment of Area in Uncracked Phase I

The characteristics of uncracked cross-sections are used in prestressed structures and deflection calculation.

Various parameters are identified through subscript 1.

Rectangular Cross-Section (b, h) with Reinforcement in Tension A_s

Positions (Fig. 11.1):

$$\rho_1 = \frac{A_s}{bh};\ \xi_1 = \frac{x_G}{h};\ p = \frac{d}{h};\ i_1 = \frac{I_1}{bh^3};$$

$\alpha_e = E_s/E_c$ (ratio between steel and concrete moduli) $= 15$ for $f_{ck} \leq 50\ \text{N/mm}^2$; $= 9$ for $f_{ck} > 50\ \text{N/mm}^2$.

Relative depth of the neutral axis: $\xi_1 = \frac{0.5+\alpha_e\rho_1 p}{1+\alpha_e\rho_1}$.

Reduced second moment of area: $i_1 = \frac{1}{3}\left(1 - 3\xi_1 + 3\xi_1^2\right) + \alpha_e\rho_1\left(p^2 + \xi_1^2 - 2p\xi_1\right)$.

Section modulus on the reinforcement side: $W_1 = \frac{i_1}{(1-\xi_1)} \cdot bh^2$.

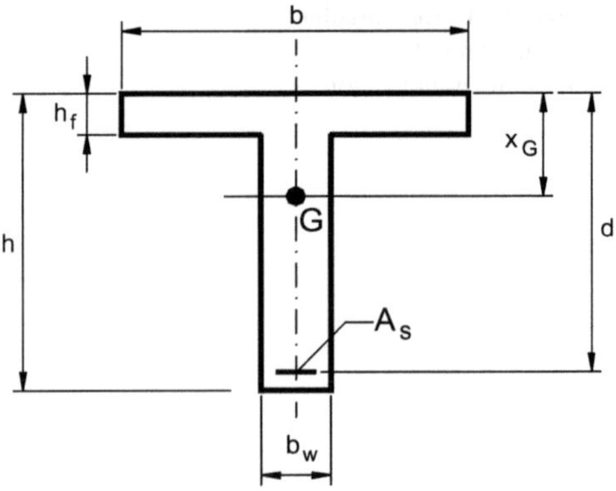

Fig. 11.2 T-shaped cross-section with reinforcement in tension

T-shaped Cross-Section with Reinforcement in Tension A_s

Positions (Fig. 11.2): $\rho_1 = \frac{A_s}{bh}$; $n = \frac{b}{b_w}$; $w_1 = \frac{h_f}{h}$; $p = \frac{d}{h}$; $\xi_1 = \frac{x_G}{h}$; $i_1 = \frac{I_1}{bh^3}$;
α_e(ratio between steel and concrete moduli) $= 15$ for $f_{ck} \leq 50$ N/mm^2; $= 9$ for $f_{ck} > 50$ N/mm^2.

Relative depth of the neutral axis: $\xi_1 = 0.5 \cdot \frac{1+(n-1)\cdot w_1^2+2n\alpha_e\rho_1 p}{1+(n-1)\cdot w_1+n\alpha_e\rho_1}$ $(\xi_1 > w_1)$

Reduced second moment of area:

$$i_1 = \frac{1}{3}\cdot\left[\xi_1^3 - (\xi_1 - w_1)^3\right] + \frac{1}{3n}\cdot\left[(\xi_1 - w_1)^3 + (1-\xi_1)^3\right] + \alpha_e\rho_1\cdot(p-\xi_1)^2$$

Section modulus on the reinforcement side: $W_1 = \frac{i_1}{(1-\xi_1)}\cdot bh^2$.

11.2.1.2 Formulas for the Calculation of the Neutral Axis, Second Moment of Area End Stresses for Cross-Sections Under the Action of Bending Moment in Phase II (Cracked Phase)

The characteristics of cracked cross-sections are used in stress verifications, calculation of crack openings and deflections.

Rectangular Cross-Section with Only Reinforcement in Tension

Notations: b width; h total height; d effective depth; $\rho = \frac{A_s}{bd}$;
$\alpha_e = \frac{E_s}{E_c}$, modular ratio between elasticity moduli of steel and concrete; $\xi = \frac{x}{d}$; $i = \frac{I}{b\cdot d^3}$.

Relative depth of the neutral axis: $\xi = \alpha_e\rho\left(-1 + \sqrt{1 + \frac{2}{\alpha_e\rho}}\right)$.

Reduced second moment of area: $i = \frac{I}{bd^3} = \frac{\xi^3}{3} + \alpha_e\rho(1-\xi)^2$.

Stresses:

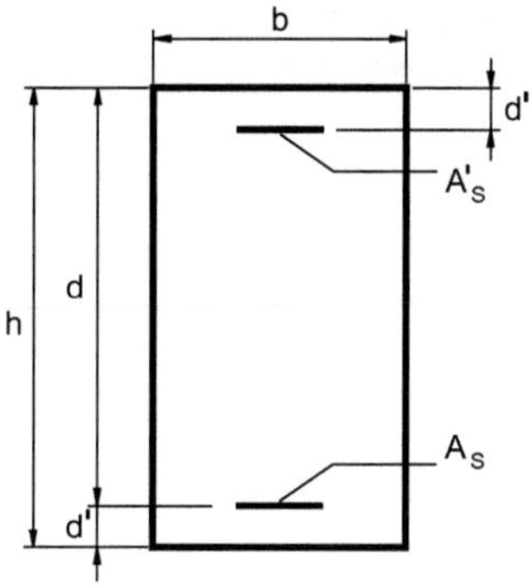

Fig. 11.3 Rectangular cross-section—notations

$$\sigma_c = \frac{M \cdot x}{I} = \frac{\mu \cdot b \cdot d^2 \cdot f_{cd}}{i \cdot b \cdot d^3} \cdot \xi \cdot d = \mu \cdot \frac{\xi}{i} \cdot f_{cd} \tag{11.1}$$

$$\sigma_s = \frac{M \cdot (d - x)}{I} \alpha_e = \frac{\mu \cdot b \cdot d^2 \cdot f_{cd}}{i \cdot b \cdot d^3} \cdot \alpha_e \cdot (1 - \xi) \cdot d = \mu \cdot \frac{1 - \xi}{i} \cdot \alpha_e \cdot f_{cd} \tag{11.2}$$

being μ the reduced moment in service conditions

$$\mu = \frac{M_{Ed,es}}{b \cdot d^2 \cdot f_{cd}} \tag{11.3}$$

Table E1 (in the Appendix) collects $\xi, i, \frac{\xi}{i}, \frac{1-\xi}{i}$ values as a function of ρ, for $\alpha_e = 15$ and $\alpha_e = 9$.

Rectangular Cross-Section with Reinforcement in Tension and Compression

Further notations (Fig. 11.3): $\delta' = \frac{d'}{d}$; $\rho' = \frac{A'_S}{b \cdot d}$.

Relative depth of the neutral axis: $\xi = \alpha_e \cdot (\rho + \rho') \cdot \left[-1 + \sqrt{1 + \frac{2(\rho + \rho'\delta')}{\alpha_e \cdot (\rho + \rho')^2}}\right]$.

Dimensionless second moment of area:

$$i = \frac{\xi^3}{3} + \xi^2(\rho' + \rho)\alpha_e - 2\alpha_e \xi(\rho + \rho'\delta') + \alpha_e \rho + \alpha_e \rho'(\delta')^2$$

Stresses: formulas (11.1), (11.2) and (11.3) are valid.

Tables E2, E3, E4 collect the values of $\xi, i, \frac{\xi}{i}, \frac{1-\xi}{i}$ as a function of ρ with (ρ'/ρ) equal to 0.1–0.2 and 0.3 for $\alpha_e = 15$ and $\alpha_e = 9$.

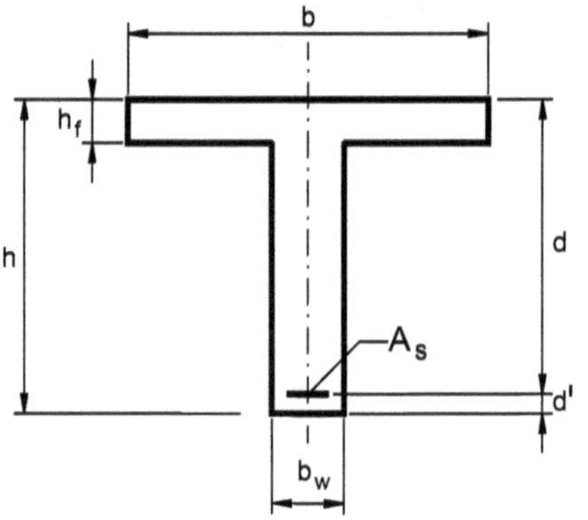

Figure. 11.4 T-shaped cross-section–notations

T-shaped Cross-Section with Only Reinforcement in Tension

Notations (Fig. 11.4):

$$\xi = \frac{x}{d};\, n = \frac{b}{b_w};\, w = \frac{h_f}{d};\, \rho = \frac{A_s}{bd};\, i = \frac{I}{b \cdot d^3}$$

$$A = n(w - \frac{w}{n} + \alpha_e \rho)$$

$$C = w^2(n-1) + 2n\alpha_e \rho$$

Relative depth of the neutral axis: $\xi = -A + \sqrt{A^2 + C}$.
Reduced second moment of area:
If $x_G > h_f$, i.e. $\xi > w$, which is verified if: $w^2 < 2\alpha_e \rho(1 - w)$,

$$i = \frac{1}{3}\left[\xi^3 - (\xi - w)^3\right] + \frac{1}{3n}(\xi - w)^3 + \alpha_e \rho \cdot (1 - \xi)^2$$

If $x_G < h_f$ relationships 11.2.1.2.1 are valuable.
Stresses:
Relationships (11.1), (11.2) and (11.3) are valid.
Tables E5, E6, E7 collect the values ξ, i, $\frac{\xi}{i}$, $\frac{1-\xi}{i}$ as a function of ρ, for $\alpha_e = 15$.

11.2.2 Axial Force Combined with Bending–Rectangular Cross-Section–Solving Formulas

A rectangular cross-section under axial load combined with bending moment with big eccentricity and with reinforcements A_s (and, it depends, A_s') designed at ULS is considered.

The verification of stresses under service loads, both for characteristic combination and quasi-permanent combination, requires the calculation of the neutral axis, which is the antipolar line of the pressure centre about to the central ellipse of inertia of the cracked cross-section. Practically, the distance between the centre and the axis is given by the ratio $\frac{I_x}{S_x}$ between the second and the first moment of area of the reacting cross-section calculated about the axis itself.

With reference to Fig. 11.5, where eccentricity is evaluated from the reinforcement in tension, the equation to calculate x becomes:

$$\frac{I_x}{S_x} = e - d + x \tag{11.4}$$

Expressions of I_x and S_x are the following:

$$I_x = b\frac{x^3}{3} + \alpha_e A_s'(x - d')^2 + \alpha_e A_s(d - x)^2 \tag{11.5}$$

$$S_x = b\frac{x^2}{2} + \alpha_e A_s'(x - d') - \alpha_e A_s(d - x) \tag{11.6}$$

Stresses are given by:
Concrete

$$\sigma_c = \frac{N \cdot x}{S_x} \tag{11.7}$$

$$\text{Reinforcement in compression} \quad \sigma_s' = \frac{\alpha_e N \cdot (x - d')}{S_x} \tag{11.8}$$

$$\text{Reinforcement in tension} \quad \sigma_s = \frac{\alpha_e N \cdot (d - x)}{S_x} \tag{11.9}$$

Figure. 11.5 Rectangular cross-section under bending moment combined with big eccentricity axial force

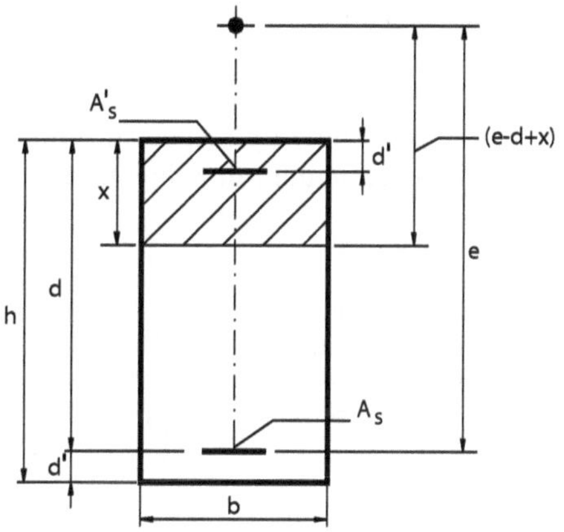

where S_x is the value calculated with x found by solving Eq. (11.4).

In this case, it is not possible to provide tables analogous to those of bending, because the solution is depending on a cubic equation containing various parameters. For this reason, dimensionless symbols are not adopted.

11.2.3 Service Interaction Diagrams $\nu - \mu$ for Rectangular Cross-Sections with Double Symmetrical Reinforcement

In this case, the tool to check stress is provided by interaction diagrams at SLS, built by considering the stress limitations $0.6 f_{ck}$ and $0.8 f_{yk}$. Diagrams are shown in Tables E8, E9, E10, E11, corresponding respectively to f_{ck} 20, 25, 30, 40 N/mm^2. For $f_{ck} > 40$ N/mm^2 such diagrams are unnecessary because of the reasons explained in 11.2.4.

It is worth noting that for diagrams in SLS (Tables from E8 to E11), dimensionless values of internal forces have at the denominator the characteristic value f_{ck} and not the design value f_{cd}, as requested for ULS diagrams.

11.2.4 Cases When SLS Stress Verifications Are Implicitly Satisfied by ULS Verifications

11.2.4.1 Bending

(a) Rectangular cross-section with only reinforcement in tension or compression $\rho' \leq 0.3\rho$

Stress limits $0.6 f_{ck}$ and $0.8 f_{yk}$ of characteristic combination are automatically satisfied for concretes $f_{ck} \geq 50$ N/mm^2 also in presence of redistributions at ULS with $\delta \geq 0.85$ if the ratio of the elastic moments at SLS and ULS is $\leq$ 0.70.

(b) In presence of concretes $f_{ck} \leq 30$ N/mm^2, stress limitations $0.6 f_{ck}$ and 0.8 f_{yk} are implicitly respected with limitations of ρ (reinforcement in tension) collected in the following Table:

Table 11.1 Limitations of the reinforcement ratio ρ in tension as a function of the redistribution coefficient δ

f_{ck}	$\delta = 1$	$\delta = 0.9$
20	$\rho \leq 0.90/100$	$\rho \leq 0.70/100$
25	$\rho \leq 1.30/100$	$\rho \leq 1.10/100$
30	$\rho \leq 1.60/100$	$\rho \leq 1.30/100$

For cases non included in Table 11.1 and for concretes f_{ck} 35–40–45 N/mm^2, it is necessary to perform verification calculations using the formulas recalled in 11.2.1.2.

11.2.4.2 Axial Force Combined with Bending—Rectangular Cross-Sections with Double Symmetrical Reinforcement

For concretes with strength f_{ck} $\geq$40 N/mm^2, stress limitations under service loads–0.6 f_{ck} and 0.8 f_{yk}—are implicitly satisfied if the verification at ULS is fulfilled with the corresponding interaction diagrams $\nu - \mu$. The two diagrams of Fig. 11.6 demonstrate this fact. Each of these diagrams describes the strength domain at ULS in the plane ν–μ (continuous curves) for two reinforcement percentages. On the same Figures, dashed lines represent, with a suitable scaling, SLS domains conditioned by 0.6f_{ck} and 0.8f_{yk} stress limits.

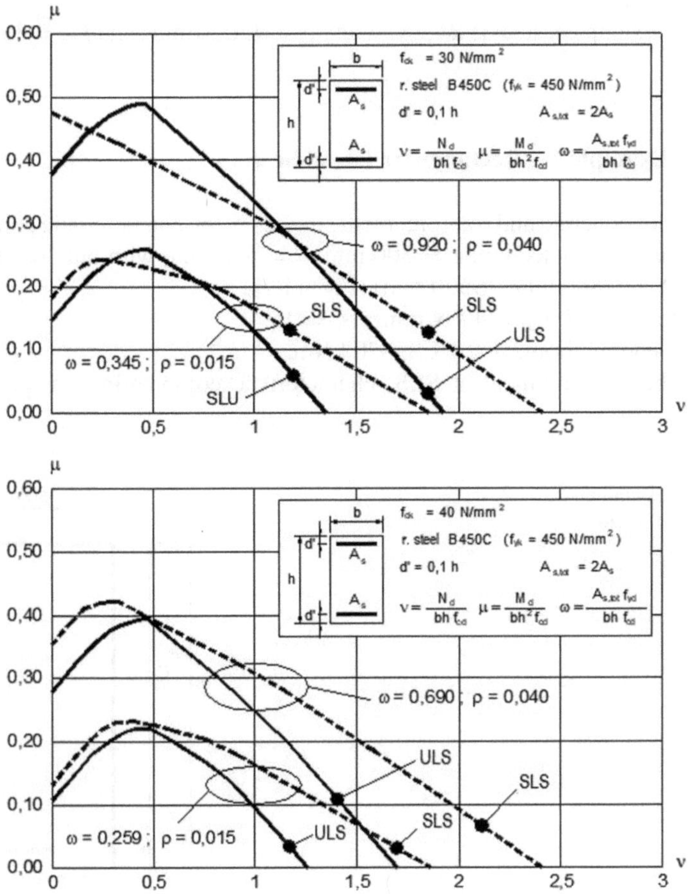

Fig. 11.6 Interaction diagrams ULS–SLS

Finally, for rectangular cross-sections with equal symmetrical reinforcements, stress verifications are automatically fulfilled ($f_{ck} \geq 40$ N/mm^2) or can be obtained using suitable interaction diagrams ($f_{ck} \leq 40$ N/mm^2 Tables E8, E9, E10, E11) without any calculation.

11.2.5 Application Examples

In the following examples, referred to cross-sections with bending moment or bending moment combined with an axial force whose reinforcement is calculated at ULS, the ratio between the service load at characteristic combination and ULS combination is assumed equal to 0.70. Such a value is well-established in literature (see CEB 127-1995). On the contrary, in the case of the quasi-permanent combination, the ratio with the ULS combination is not uniquely defined. The value can vary between 0.40 and 0.55.

In all the examples, steel is B450C with $f_{yk} = 450$ N/mm^2 and $f_{yd} = 391$ N/mm^2. Bending.

11.2.5.1 Example 1—Rectangular Cross-Section

Calculation of concrete and reinforcement stresses in the characteristic combination.

Rectangular cross-section: $b = 300$ mm, $d = 500$ mm, with only reinforcement in tension $A_s = 1886$ mm^2 ($6\phi20$ mm) (Fig. 11.7).

Concrete: C30/37; $f_{cd} = 17$ N/mm^2; $\alpha_e = 15$.

Design bending moment at ULS: 320 kNm.

Design bending moment at SLS, characteristic combination: $M = 3200.7 = 224$ kNm.

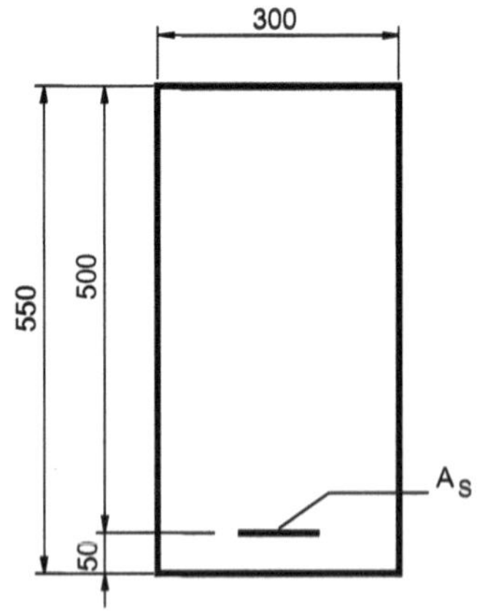

Figure. 11.7 Bending: example 1, rectangular cross-section

Reduced bending moment at SLS, through (11.3), is given by:

$$\mu_{SLE} = \frac{224 \cdot 10^6}{300 \cdot 500^2 \cdot 17} = 0.1757$$

From the given data, we get

$$\rho = \frac{1886}{300 \cdot 500} = \frac{1.26}{100}$$

From Table E1 corresponding to $\rho' = 0$ and $\alpha_e = 15$, by interpolation, we get:

$$\frac{\xi}{i} = 5.190; \frac{1-\xi}{i} = 6.244$$

from which we get, by applying (11.1) and (11.2):

$$\sigma_c = \mu \cdot \frac{\xi}{i} \cdot f_{cd} = 0.1757 \cdot 5.19 \cdot 17 = 15.50\ \text{N/mm}^2 < 18\text{N/mm}^2$$

$$\sigma_s = \mu \cdot \frac{1-\xi}{i} \cdot \alpha_e \cdot f_{cd} = 0.1757 \cdot 6.244 \cdot 17 \cdot 15 = 280\text{N/mm}^2 < 360\text{N/mm}^2$$

Stresses are therefore acceptable. Such a conclusion could be obtained directly from Table 11.1, being $\rho = \frac{1.26}{100} < \frac{1.60}{100}$.

11.2.5.2 Example 2—Rectangular Cross-Section

Example 3 of 7.6.3.3, discussed for ULS, is here recalled, calculating concrete and reinforcement stresses due to the service moment in the characteristic combination.

Rectangular cross-section: $b = 300$ mm; $h = 550$ mm; $d = 500$ mm (Fig. 11.8).

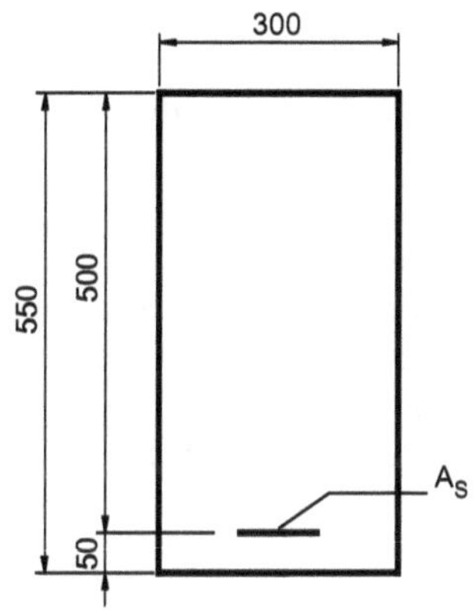

Figure. 11.8 Bending: example 2, rectangular cross-section

Concrete: C70/85; $f_{cd} = 39.6$ N/mm^2; $\alpha_e = 9$.
Design bending moment at ULS: $M_{Ed} = 500$ kNm.
Reinforcements required at ULS are

$$A_s = 2864\text{mm}^2;\ A_s' = 0$$

Service moment at characteristic combination is given by
$M = 500 \cdot 0.7 = 350$ kNm
Moment reduced by (11.3) is

$$\mu_{SLE} = \frac{350 \cdot 10^6}{300 \cdot 500^2 \cdot 39.6} = 0.1177$$

$$\rho = \frac{2864}{300 \cdot 500} = \frac{1.91}{100}$$

From Table E1 for $\rho' = 0$ and $\alpha_e = 9$, by interpolating, it is obtained

$$\frac{\xi}{i} = 5.30;\ \frac{1-\xi}{i} = 6.80$$

from which, by applying (11.1) and (11.2)

$$\sigma_c = \mu \cdot \frac{\xi}{i} \cdot f_{cd} = 0.1177 \cdot 5.30 \cdot 39.6 = 24.7\text{N/mm}^2 < 42\text{N/mm}^2$$

$$\begin{aligned}\sigma_s &= \mu \cdot \frac{1-\xi}{i} \cdot \alpha_e \cdot f_{cd}\\ &= 0.1177 \cdot 6.80 \cdot 9 \cdot 39.6 = 285.2\text{N/mm}^2 < 360\text{N/mm}^2\end{aligned}$$

Therefore, stresses are acceptable.

Considering 11.2.4.1 (a), stress verification is implicitly fulfilled by USL verification.

11.2.5.3 Example 3—Rectangular Cross-Section

Example 4 of 7.6.3.4, discussed for ULS, is here recalled, calculating concrete and reinforcement stresses due to the service moment in the characteristic combination.

Rectangular cross-section: $b = 300$ mm; $h = 550$ mm; $d = 500$ mm (Fig. 11.9).
Concrete: C25/30; exposure class XD; $f_{cd} = 14.16$ N/mm^2; $\alpha_e = 15$.
Internal force at ULS: $M_{Ed} = 450$ kNm.
Reinforcements are the following

$$A_s = 2814 + 261 = 3075\text{ mm}^2,\ A_s' = 261\text{ mm}^2$$

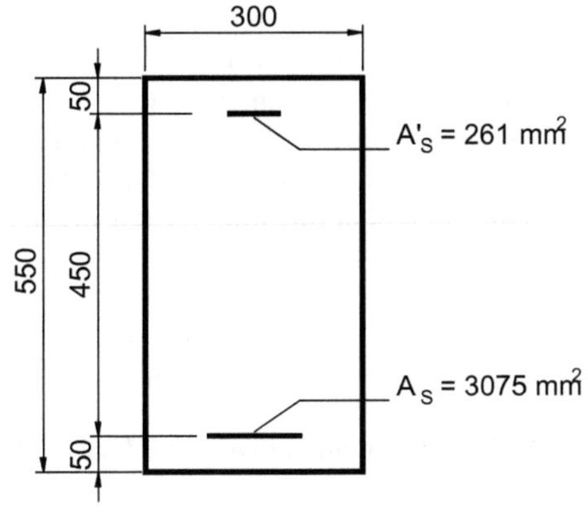

Fig. 11.9 Bending: example 3, rectangular cross-section

corresponding to: $\rho = \frac{3075}{500 \cdot 300} = \frac{2.05}{100}$ and $\frac{\rho'}{\rho} = 0.085$.
The reduced moment at ULS is given by $\mu_{SLU} = 0.424$.
from which service moment can be obtained $\mu_{SLE} = 0.7 \cdot 0.424 = 0.297$.
corresponding to $M_{SLE} = 0.297 \cdot 300 \cdot 500^2 \cdot 14.16 = 315$ kNm.
Being $\frac{\rho'}{\rho} = 0.085 \cong 0.1$, from Table E2 (Appendix) corresponding to $\frac{\rho'}{\rho} = 0.1$ and $\alpha_e = 15$ it is obtained, by interpolating for $\rho = \frac{2.05}{100}$:

$$\frac{\xi}{i} = 4.218; \quad \frac{1-\xi}{i} = 3.905$$

from which it results, applying (11.1) and (11.2):

$$\sigma_c = 0.297 \cdot 4.218 \cdot 14.16 = 17.73 \text{N/mm}^2 > 0.6 \cdot 25 = 15 \text{N/mm}^2$$

$$\sigma_s = 0.297 \cdot 3.905 \cdot 14.16 \cdot 15 = 246.3 \text{N/mm}^2 < 360 \text{N/mm}^2$$

EC2 requires that such a moment is reduced of the following amount if the structure is exposed in an environment classified as XD, XF, XS according to Table [4.1].

$\mu_{SLE} = 0.297 \cdot 15/17.73 = 0.251$ (reduction of 16%).
and therefore $M_{SLE} = 0.251 \cdot 300 \cdot 500^2 \cdot 14.16 = 267$ kNm $0.251 \cdot 300 \cdot 500^2 \cdot 14.16 = 267$ kNm.

11.2.5.4 Example 4—T-shaped Cross-Section

Example 1 of 7.7.2.1, discussed for ULS, is here recalled, calculating concrete and reinforcement stresses due to service moment in the characteristic combination.

Slab with ribs spaced by 0.50 m.

Considering two ribs we get: $h = 280$ mm; $d = 250$ mm; $h_f = 50$ mm; $b_w = 200$ mm; $b = 1000$ mm (Fig. 11.10).

Dimensionless parameters: $w = h_f/d = 0.20$; $n = b/b_w = 5$.

Concrete: C25/30; $f_{cd} = 14.16$ N/mm^2.

Design bending moment at ULS: $M_{Ed} = 60$ kNm.

Reinforcement fulfilling ULS is given by $A_s = 636$ mm^2/m, i.e. 318 mm^2 for each rib.

$$\rho = \frac{A_s}{b \cdot d} = \frac{636}{1000 \cdot 250} = \frac{0.25}{100}$$

Service moment is given by $M_{SLS} = 60 \cdot 0.7 = 42$ kNm, and therefore, according to (11.3)

$$\mu_{SLE} = \frac{42 \cdot 10^6}{1000 \cdot 250^2 \cdot 14.16} = 0.0475$$

From Table E6 (Appendix) valid for $w = 0.20$, $n = 5$ and $\alpha_e = 15$, we get:

$$\frac{\xi}{i} = 9.19; \frac{1-\xi}{i} = 28.89$$

from which following stresses result, by applying (11.1) and (11.2)

$$\sigma_c = 0.0475 \cdot 9.19 \cdot 14.16 = 6.18\text{N/mm}^2 < 0.6 \cdot 25 = 15\text{N/mm}^2$$

$$\sigma_s = 0.0475 \cdot 28.89 \cdot 14.16 \cdot 15 = 291\text{N/mm}^2 < 0.8 \cdot 450 = 360\text{N/mm}^2$$

These stresses are allowable.

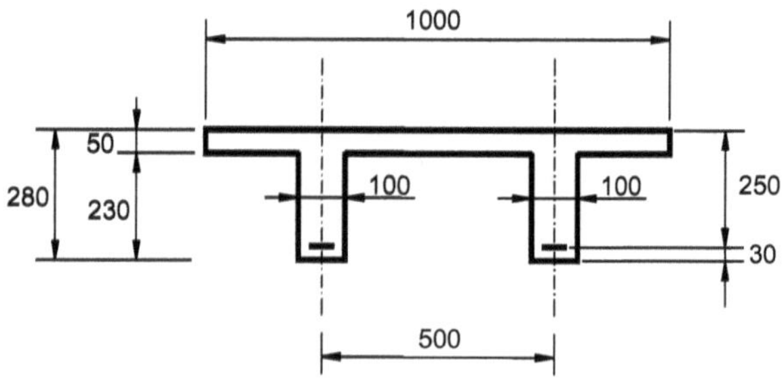

Fig. 11.10 Bending: example 1, slab section

11.2.5.5 Example 5—T-shaped Cross-Section

Example 2 of 7.7.2.2, discussed for ULS, is here considered to calculate concrete and reinforcement stresses at serviceability in the characteristic combination.

T-shaped cross-section: $b = 800$ mm; $b_w = 200$ mm; $h = 850$ mm; $d = 800$ mm; $h_f = 120$ mm (Fig. 11.11).

Dimensionless parameters (see Fig. 11.4): $n = 4$; $w = 0.15$.

Concrete: C25/30; $f_{cd} = 14.16$ N/mm^2.

Bending moment at ULS: $M_{Ed} = 1050$ kNm.

Reinforcement required at ULS: $A_s = 3645$ mm^2.

$$\rho = \frac{3645}{800 \cdot 800} = \frac{0.57}{100}$$

Service moment is given by

$$M_{SLE} = 0.7 \cdot 1050 = 735 \text{ kNm}$$

Service moment reduced according to (11.3) is given by

$$\mu_{SLE} = \frac{735 \cdot 10^6}{800 \cdot 800^2 \cdot 14.16} = 0.1014$$

From Table E5 (Appendix) corresponding to $w = h_f/d = 0.15$, $n = 4$ and $\alpha_e = 15$, for $\rho = \frac{0.57}{100}$ by interpolating it we obtain:

$$\frac{\xi}{i} = 7.94; \quad \frac{1-\xi}{i} = 12.86$$

from which we get, by applying (11.1) and (11.2)

$$\sigma_c = 0.1014 \cdot 7.94 \cdot 14.16 = 11.4\text{N/mm}^2 < 15\text{N/mm}^2$$

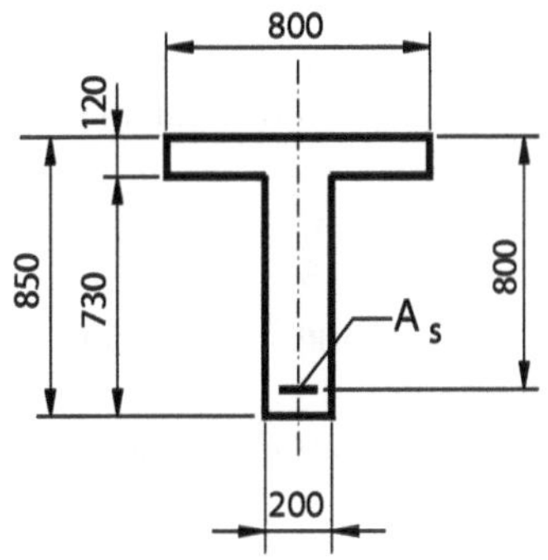

Fig. 11.11 Bending: example 2, T-shaped cross-section

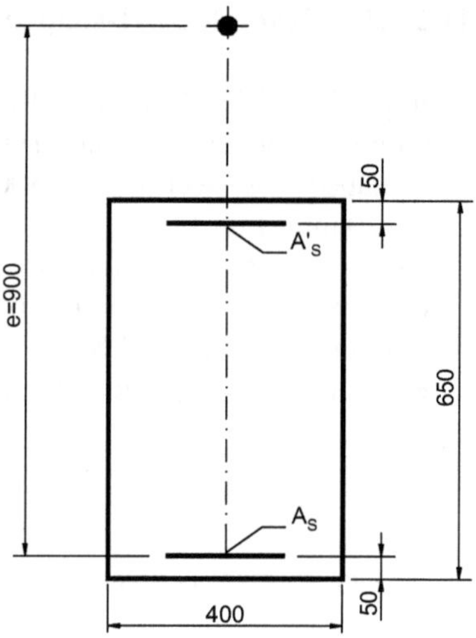

Fig. 11.12 Axial force combined with bending: example 1, rectangular cross-section

$$\sigma_s = 0.1014 \cdot 12.86 \cdot 14.16 \cdot 15 = 277\text{N/mm}^2 < 360\text{N/mm}^2$$

Therefore, stresses are allowable.

11.2.5.6 Axial Force Combined with Bending

Example 1—Rectangular Cross-Section

Rectangular cross-section: $b = 400$ mm, $h = 650$ mm; $d = 600$ mm (Fig. 11.12).
Concrete: C30/37; $f_{cd} = 17$ N/mm^2; $\alpha_e = 15$.
Load at ULS: $N = -300$ kN, with eccentricity: $e = 1500$ mm to the reinforcement in tension.
Reinforcement design requires: $A_s = 1377$ mm^2; $A'_s = 0$.

- Stress verification at SLS, characteristic combination

  In this case: $N = 0.7{\cdot}300 = 210$ kN; $e = 1500$ mm.
  The equation to calculate x is (11.4) which, once developed, gives:

$$\left[400 \cdot \frac{x^3}{3} + 15 \cdot 1377 \cdot (600 - x)^2\right] / \left[400 \cdot \frac{x^2}{2} - 15 \cdot 1377 \cdot (600 - x)\right]$$
$$= (1500 - 600 + x)$$

This equation is satisfied if $x = 239$ mm, from which we get the first moment of area:

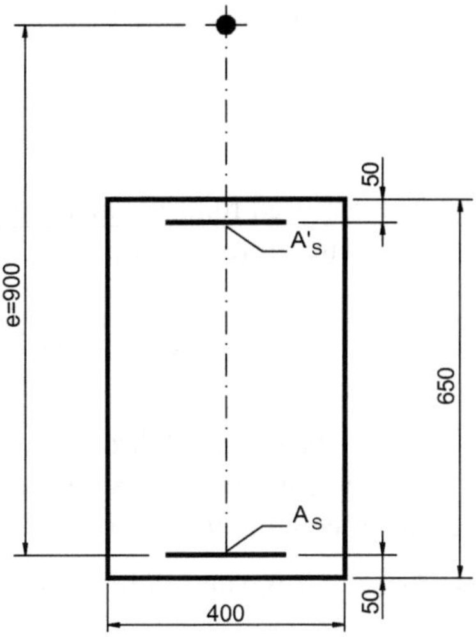

Fig. 11.13 Axial force combined with bending: example 2, rectangular cross-section

$$S_x = \frac{400}{2} \cdot 239^2 - 15 \cdot 1377 \cdot (600 - 239) = 3967745 \text{ mm}^3$$

Stresses are the following:

$$\sigma_c = \frac{N \cdot x}{S_x} = \frac{210000 \cdot 239}{3967745} = 12.6 \text{ N/mm}^2 < 18 \text{ N/mm}^2$$

$$\sigma_s = \frac{\alpha_e \cdot N \cdot (d - x)}{S_x} = \frac{15 \cdot 210000 \cdot (600 - 239)}{3967745}$$
$$= 286.6 \text{ N/mm}^2 < 360 \text{ N/mm}^2$$

and therefore, they are allowable.

Example 2—Rectangular Cross-Section

Rectangular cross-section: $b = 400$ mm, $h = 650$ mm, $d = 600$ mm, $d' = 50$ mm (Fig. 11.13).

Concrete: C60/75; $f_{cd} = 34$ N/mm^2.

Internal force at ULS: $N = -1900$ kN, with eccentricity $e = 900$ mm to the reinforcement in tension.

- Reinforcement design at ULS

$M = 1900 \cdot 0.9 = 1710$ kNm

$$\mu = \frac{1710 \cdot 10^6}{400 \cdot 600^2 \cdot 34} = 0.350$$

which requires (Table U2) $\omega = 0.4390$; $\omega' = 0.1\ \omega = 0.0439$

$$\nu = \frac{1900000}{400 \cdot 600 \cdot 34} = 0.2328$$

The required reinforcement in tension is given by $\omega - \nu = 0.4390 - 0.2328 = 0.2062$. Therefore, we get:

$$A_s = \frac{0.2062 \cdot 400 \cdot 600 \cdot 34}{391} = 4302\ \text{mm}^2$$

which can be provided through seven bars of 28 mm of diameter, equal to 4305 mm^2; reinforcement in compression (from ω') is equal to 916 mm^2, fulfilled by 3 bars 20 mm diameter equal to 942 mm^2.

- Stress verification in SLS, characteristic combination

  $N = 0.7 \cdot 1900 = 1330$ kN; $\alpha_e = 9$.
  With reference to 11.2.2, Eq. (11.4) giving x can be written:

$$\frac{400 \cdot \frac{x^3}{3} + 9 \cdot 942 \cdot (x - 50)^2 + 9 \cdot 4305 \cdot (600 - x)^2}{400 \cdot \frac{x^2}{2} + 9 \cdot 942 \cdot (x - 50) - 9 \cdot 4305 \cdot (600 - x)} = (900 - 600 + x)$$

  Solving the equation, we get $x = 326$ mm and,

$$S_x = \frac{400}{2} \cdot 326^2 + 9 \cdot 942 \cdot (326 - 50) - 9 \cdot 4305 \cdot (600 - 326)$$
$$= 12978998\ \text{mm}^3$$

  Stress is given by:

$$\sigma_c = \frac{N \cdot x}{S_x} = \frac{1330000 \cdot 326}{12978998} = 33.4\ \text{N/mm}^2 < 0.6 \cdot 60 = 36\ \text{N/mm}^2 \quad (11.10)$$

$$\sigma_s = \frac{\alpha_e \cdot N \cdot (d - x)}{S_x} = \frac{9 \cdot 1330000 \cdot (600 - 326)}{12978998} = 253 < 360\ \text{N/mm}^2 \quad (11.11)$$

  and therefore they are allowable.

Example 3—Rectangular Cross-Section

Rectangular cross-section: $h = 600$ mm; $b = 400$ mm; $d' = 60$ mm (Fig. 11.14).
Symmetrical reinforcement.
Concrete: C30/37; $f_{cd} = 17$ N/mm^2.

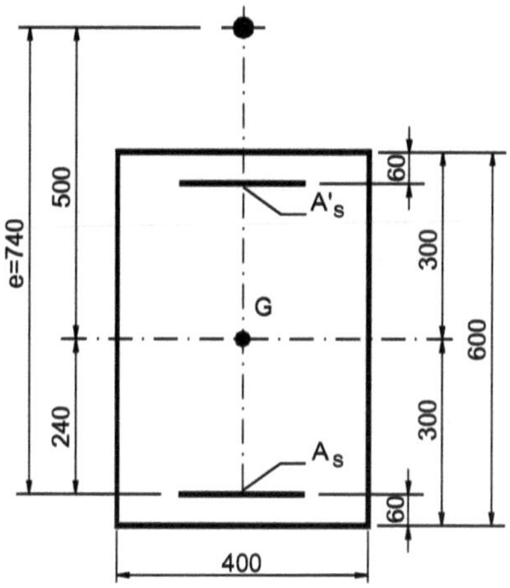

Fig. 11.14 Axial force combined with bending: Example 3, rectangular cross-section

Internal forces at ULS: $N_{Ed} = -1000$ kN; $M_{Ed} = 500$ kNm.

We get: $e* = 500/1000 = 0.50$ m, from the centroid of the concrete section.

For the design of the reinforcement at ULS, an interaction diagram with $d' = 0.1d$ is used, valid for $f_{ck} \leq 50\text{N/mm}^2$(Table U10).

Dimensionless internal forces are the following

$$\nu = \frac{1000000}{400 \cdot 600 \cdot 17} = 0.2451$$

$$\mu = \frac{500 \cdot 10^6}{400 \cdot 600^2 \cdot 17} = 0.2042$$

Based on the previous results, the Table gives $\omega = 0.30$.

Total reinforcement is $A_{s,\text{tot}} = 0.30 \cdot 400 \cdot 600 \cdot \frac{17}{391} = 3130\ \text{mm}^2$; on each side $A_s = 3130/2 = 1565\ \text{mm}^2$, fulfilled, for example, by 5 bars 20 mm diameter (1570 mm^2).

- Service verification

It is assumed that the characteristic service condition is expressed by M and N at ULS reduced by 0.7 coefficient. We get:

$N_{Ed} = -700$ kN

$M_{Ed} = 350$ kNm

The verification is performed by means of the interaction diagram E10 corresponding to rectangular cross-section with $d' = 0.1\ h$ and $f_{ck} = 30\ \text{N/mm}^2$.

Dimensionless internal forces are given by

$$\nu = \frac{N}{bhf_{ck}} = \frac{700000}{400 \cdot 600 \cdot 30} = 0.0972$$

$$\mu = \frac{M}{bh^2 f_{ck}} = \frac{350 \cdot 10^6}{400 \cdot 600^2 \cdot 30} = 0.0810$$

Positioning the point with coordinates ν, μ on the service interaction diagram, reinforcement ratio should be fixed $\rho = \frac{1.10}{100}$, i.e. $\omega = \frac{1.10}{100} \cdot \frac{391}{17} = 0.253$, smaller than the one required at ULS. Therefore, verification is fulfilled implicitly.

However, in the following the verification at SLS is developed with the reinforcement required at ULS.

Being $e* = \frac{500}{1000} = 0.50$ m from the centroid of the cross-section, the eccentricity to the reinforcement in tension is $e = 740$ mm (Fig. 11.17).

$$e - d = 740 - 540 = 200 \text{ mm}$$

Recalling Eq. (11.4)

$$I_x = S_x \cdot (200 + x)$$

by substituting, we get:

$$400 \cdot \frac{x^3}{3} + 15 \cdot 1570 \cdot (x - 60)^2 + 15 \cdot 1570 \cdot (540 - x)^2$$
$$= \left[400 \cdot \frac{x^2}{2} + 15 \cdot 1570 \cdot (x - 60) - 15 \cdot 1570 \cdot (540 - x)\right]$$
$$\times (200 + x)$$

Solving the equation we get: $x = 256$ mm. Therefore $S_x = 1103480 \text{ mm}^3$.
Stress are given by

$$\sigma_c = \frac{N \cdot x}{S_x} = \frac{700000 \cdot 256}{11034800} = -16.2 \text{N/mm}^2 < 18 \text{N/mm}^2 \quad (11.12)$$

$$\sigma_{s,\text{ top}} = \frac{\alpha_e \cdot N \cdot (x - d')}{S_x} = \frac{15 \cdot 700000 \cdot (256 - 60)}{11034800} = -186 \text{N/mm}^2 \quad (11.13)$$

$$\sigma_{s,\text{ bottom}} = \frac{\alpha_e \cdot N \cdot (d - x)}{S_x} = \frac{15 \cdot 700000 \cdot (540 - 256)}{11034800} = 270 < 360 \text{N/mm}^2 \quad (11.14)$$

Example 4—Rectangular Cross-Section

Rectangular cross-section: $h = 600$ mm; $d = 540$ mm; $d' = 0.1\ h = 60$ mm; $b = 400$ mm) (Fig. 11.15).

Symmetrical reinforcement.

Concrete C20/25; $f_{cd} = 11.33 \text{ N/mm}^2$.

Internal forces at ULS: $N_{Ed} = -1400$ kN; $M_{Ed} = 500$ kNm.

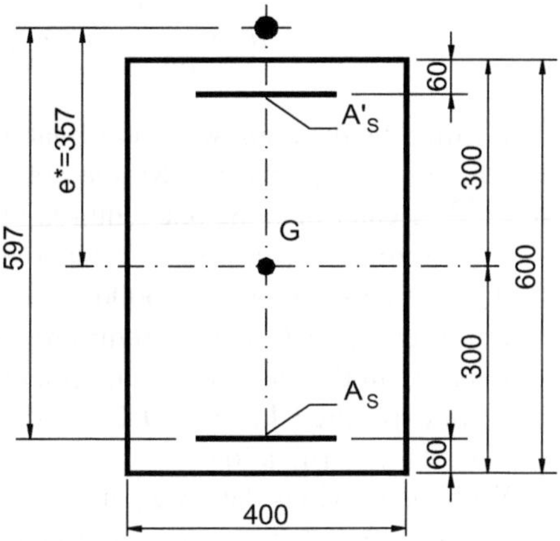

Fig. 11.15 Axial force combined with bending: example 4, rectangular cross-section

The design at ULS is performed through the interaction diagram shown in Table U10, with *d'* = 0.1 *h* valid for $f_{ck} \leq 50$ N/mm^2, which gives the required reinforcement.

Dimensionless internal forces are

$$\nu = \frac{1400000}{400 \cdot 600 \cdot 11.33} = 0.5149$$

$$\mu = \frac{500 \cdot 10^6}{400 \cdot 600^2 \cdot 11.33} = 0.3065$$

Interaction diagram requires $\omega = 0.50$, i.e.

$$\rho = \omega \frac{f_{cd}}{f_{yd}} = 0.5 \frac{11.33}{391} = \frac{1.45}{100}$$

Therefore, total reinforcement is given by
$\frac{1.45}{100} \cdot 600 \cdot 400 = 3480$ mm^2, equal to 1740 mm^2 on each side.

- Service verification

It is assumed the characteristic service condition expressed by *M* and *N* at ULS multiplied by 0.7 coefficient:

$$N_{Ed} = -980 \text{ kN}; \ M_{Ed} = 350 \text{ kNm}$$

In dimensionless terms

$$\nu = \frac{N}{bhf_{ck}} = \frac{980000}{400 \cdot 600 \cdot 20} = 0.204$$

$$\mu = \frac{M}{bh^2 f_{ck}} = \frac{350 \cdot 10^6}{400 \cdot 600^2 \cdot 20} = 0.1215$$

Positioning the point with coordinates ν, μ on the service interaction diagram $f_{ck} = 20$ N/mm^2 (Appendix, E8), the corresponding reinforcement ratio is obtained: $\rho = \frac{2.5}{100}$, greater than the one required by the ULS verification.

Therefore, it is necessary to arrange the reinforcement given by

$A_s = 0.025 \cdot 400 \cdot 600 = 6000$ mm^2, i.e. 3000 mm^2 on each side.

In this case, 6026 = 3181 mm^2 are arranged.

Remark: in this case, the requirement at SLS is prevailing on the requirement at ULS for exposure Classes XD, XS and XF.

Analytical verification

With the previous data, we get

$e^* \frac{500}{1400} = 0.357$ m, eccentricity to the centroid

$e = 357{+}240 = 597$ mm eccentricity to the reinforcements in tension

$$e - d = 597 - 540 = 57 \text{ mm}$$

The Eq. (11.4) gives the solution $x = 342$ mm; therefore we get,

$$S_x = 27400860 \text{ mm}^3$$

Stresses are given by

$$\sigma_c = \frac{N \cdot x}{S_x} = \frac{980000 \cdot 342}{27400860} = -12.2\text{N/mm}^2 \approx 0.6 f_{ck} = 12\text{N/mm}^2 \quad (11.15)$$

$$\sigma_{s,\,\text{sup}} = \frac{\alpha_e \cdot N \cdot (x - d')}{S_x} = \frac{15 \cdot 980000 \cdot (342 - 60)}{2740086} = -151\text{N/mm}^2 \quad (11.16)$$

$$\sigma_{s,\,\text{inf}} = \frac{\alpha_e \cdot N \cdot (d - x)}{S_x} = \frac{15 \cdot 980000 \cdot (540 - 342)}{2740086} = +106 < 360\text{N/mm}^2 \quad (11.17)$$

Remark: the greater amount required at SLS is necessary to limit concrete compression.

11.3 Crack Control

11.3.1 General Considerations

To limit crack width, EC2 provides four procedures:

(a) for reinforced or prestressed slabs under bending actions with depth not greater than 200 mm, special controls are not necessary if the detailing of [9.3] is fulfilled (concerning minimum main and secondary reinforcements and distance between bars).
(b) for members under bending actions, if reinforcements have been calculated at ULS and stress σ_s has been calculated at SLS with a cracked cross-section for the selected load combination (characteristic, quasi-permanent), Table [7.2N] and Table [7.3N] guarantee crack width control $w_k = 0.2 - 0.3 - 0.4$ mm, in the first case with the diameter limitation, in the second case with the limitation of the distance between bars;

the procedure seems easy, but Tables [7.2N] and [7.3N] have strong limitations because they are valid only for $f_{ck} = 30$ N/mm^2, concrete cover $c = 25$ mm, h_{cr}=0.5h, h_{cr} being the depth of the area in tension before cracking.
(c) Calculation of the crack width w_k: this topic will be further discussed in the following.
(d) Definition of a minimum reinforcement for crack width control.

11.3.2 Calculation of Crack Widths

According to EC2, crack width w_k may be calculated by the expression

$$w_k = s_{r,\max}(\varepsilon_{sm} - \varepsilon_{cm}) \quad [7.8]$$

where w_k is the characteristic crack value, 95% fractile, i.e. with the 5% probability to be exceeded. $s_{r,\max}$ is defined as "maximum crack final spacing"; $(\varepsilon_{sm} - \varepsilon_{cm})$ is the mean strain in the reinforcement, minus the strain of the concrete in tension between cracks, given by:

$$\varepsilon_{sm} - \varepsilon_{cm} = \frac{\sigma_s - k_t \frac{f_{ctm}}{\rho_{p,\text{eff}}}\left(1 + \alpha_e \rho_{p,\text{eff}}\right)}{E_s} \geq 0.6\frac{\sigma_s}{E_s} \quad [7.9]$$

with $\alpha_e = \frac{E_s}{E_{cm}}$.

$\rho_{p,\text{eff}} = A_s/A_{c,\text{eff}}$ (for ordinary reinforced concrete) where $A_{c,\text{eff}}$ is the area of the concrete tensile zone where the reinforcement controls cracking. Typical cases of $A_{c,\text{eff}}$ are collected in Fig. 7.1 of EC2. $k_t = 0.6$ for short-term loading, 0.4 for long-term loading.

Explanation of the formula [(7.11)] giving $s_{r,\max}$.

A concrete prism with a bonded coaxial steel bar coming out from the end sections is considered. A tensile force is applied on the prism through the bar coming out. With the increase of the force, a crack arises where concrete reaches its tensile strength f_{ct}. On the two sides of the crack tensile force spread out in the concrete and no cracks can arise at a distance smaller than transmission length s_o, where concrete stress takes the value f_{ct}. The equilibrium condition is given by:

$s_o \pi\, \phi \cdot f_b = A_c f_{ct}$, with f_b = bond stress and A_c = area of the prism section.

By introducing the reinforcement ratio $\rho = \frac{\pi\, \phi^2}{4A_c}$ and being tensile strength and bond stress proportional, we can set $k = \frac{f_{ct}}{f_b}$. By substituting, we get:

$$s_o = 0.25k \cdot \frac{\phi}{\rho}$$

By specializing and generalizing the formula, k can be articulated in the product of k_1, depending on the bond characteristic of the bar (for rebar $k_1 = 0.8$), and k_2 depending on the internal force of the member (0.5 for bending, 1.0 for tension). Moreover, the term ρ is specified by the symbol ρ_{eff}. We can define $\rho_{\text{eff}} = A_s/A_{c,eff}$, being $A_{c,\text{eff}}$ the portion of concrete cross-section where the reinforcement influences cracking—see Figure [7.1] of EC2.

The length s_o is not complete, because it is necessary to add the term $2c$ (twice the concrete cover for members in bending) which takes into account the experimental fact that, in the first segment starting from the crack, there is the complete detachment of the bar from the concrete, as a consequence of the sudden stress jump experienced by the bar when a crack occurs. Therefore, we get

$$s_o = 2c + 0.25k \cdot \frac{\phi}{\rho_{\text{eff}}}$$

The characteristic value $s_{r,k}$ (indicated as $s_{r,\max}$ by EC2) is given by the product of the mean value s_o for $(1 + 1.645\ c_v)$ where c_v is the coefficient of variation of value ≈ 0.4; therefore, we get $s_o(1 + 1.645 \cdot 0.4) = 1.66\ s_o$, rounded up as $1.7\ s_o$. Therefore, we get [7.11]

$$s_{r,k} = 1.7\ s_o = 3.4c + 0.425 k_1 k_2 \cdot \frac{\phi}{\rho_{\text{eff}}}$$

The mean strain of reinforcement $\varepsilon_{sm} - \varepsilon_{cm}$ is given by [7.9] and does not require special comments.

According to EC2, the values calculated should not exceed values of maximum crack width ($w_{\max}$) given in Table [7.1N], here partially recalled.

Exposure classes are defined as a function of environmental conditions in EN 206-1 and are collected in [4.2] of EC2 Table 11.2.

The application of the procedures described above is developed in Example 4 of 4.3.5.4 and Example of 4.3.7, both in Chap. 4.

11.3.3 Minimum Reinforcement Areas

A minimum amount of reinforcement is required for two main reasons: the first one is a safety reason, to avoid that a failure of the whole structure can occur at the

Table 11.2 Synthesis of table [7.1N]

Exposure class	Reinforced members and prestressed members with unbonded tendons	Prestressed members with bonded tendons
	Quasi-permanent load combination	Frequent load combination
X, XC1	0.4 mm	0.2 mm
XC2, XC3, XC4	0.3 mm	0.2 mm
XD1, XD2, XS1, XS2, XS3	0.3 mm	Decompression

first crack (non-brittleness criterion). The second reason is to guarantee that, for a spontaneous cracking due to constrained and not allowed deformations, or also in case of cracking due to loads, reinforcement stresses are not exceeding a suitable value (functionality criterion).

In the first case, EC2 provides, for bending, formula [9.1 N]

$$A_{s,\min} = 0.26 \cdot \frac{f_{ctm}}{f_{yk}} \cdot b_t \cdot d \ (\geq 0.0013 \cdot b_t \cdot d)$$

where b_t is the width of the area in tension.

For the second case, EC2 gives formula [(7.1)]

$$A_{s,\min}\sigma_s = k_c k f_{ct,\text{eff}} A_{ct}$$

where A_{ct} is the area in tension or a portion of the area in tension of the cross-section, σ_S is the value of the maximum allowable stress in the reinforcement immediately after the opening of the crack. This value can be assumed as f_{yk} or as a smaller value to limit the opening of the crack, by using Tables [7.2] and [7.3].

11.3.3.1 Cracking of a Wall Due to Shrinkage—Reinforcement Design

The wall of Fig. 11.16 is 10 m in length between two dilatation joints, 3 m in height, and 25 cm in depth. It is restrained at the foundation and is subjected to shortening due to shrinkage.

A complete study would require taking into account geometric and mechanic foundation characteristics, geotechnics soil parameters, and the adoption of complex procedures.

The Example is instead developed as an easy application of [7.3.2(1)P] which says: if crack control is required, it is necessary to arrange a minimum amount of bonded reinforcement to control cracking in areas where tension is expected. The reinforcement amount can be estimated by equating the tensile force in the concrete before cracking and the tensile force that the reinforcement can provide if it works at its elastic limit or at lower stress to control the crack width (see Formula [(7.1)]

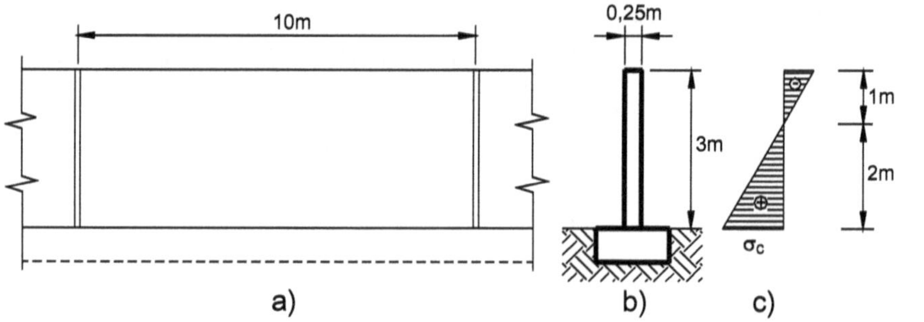

Fig. 11.16 Wall geometry

Fig. 11.17 Cross-section geometry

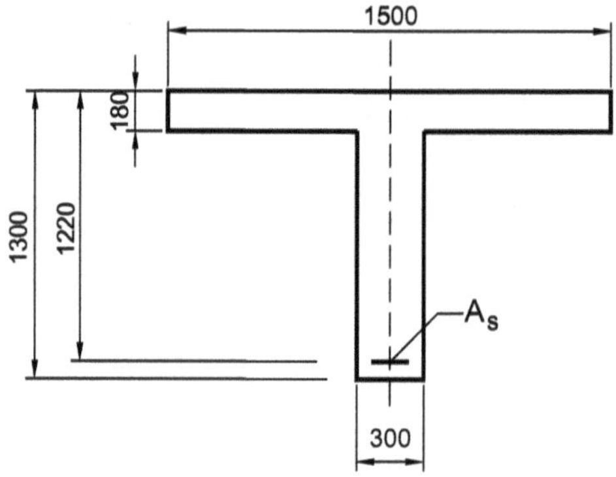

which is here applied assuming $k = 1$ and $k_c = 1$ for the sake of simplicity and to be on the safe side).

$$\text{Materials: concrete C25/30 with } f_{ctm} = 2.6\ \text{N/mm}^2;$$
$$\text{steel B450C with } f_{yk} = 450\ \text{N/mm}^2.$$

If the wall is released from the foundation through a cut at the joint level, under the hypothesis of null friction, the wall would shrink freely. As the shrinkage is not free, a horizontal force H should be put at the two end sections, at the wall base, to reestablish the compatibility of horizontal displacements with the foundation. Equal and opposite forces should apply to the foundation. Force H, whose value is here not necessary, applied with eccentricity $h/2$ originates in the wall cross-sections a bending action combined with a tensile force with neutral axis placed at $h/3$ from the top, being the 1/3 upper in compression and the 2/3 lower in tension.

The stress diagram is shown in Fig. 11.16c. Bending combined with tensile force produces curvature of the wall with centre up. The compatibility requires vertical stresses at the joint level, which are not considered in this simple example.

When horizontal tensile stress reaches the value f_{ctm}, the wall cracks. The resultant tensile force is given by: $\frac{1}{2} \cdot \left(\frac{2}{3} \cdot h \cdot s\right) \cdot f_{ctm}$. This force is in equilibrium with the reinforcement, whose total area is given by

$$A_s\sigma_s = \frac{1}{2} \cdot \left(\frac{2}{3} \cdot 3000 \cdot 250\right) \cdot 2.6$$

Cautiously, it is assumed $\sigma_S = 280$ N/mm^2. We get

$$A_s = \frac{650000}{280} = 2321 \text{ mm}^2$$

corresponding to 10 couples of bars 12 mm diameter arranged on the height of 2 m, i.e. 2 bars every 200 mm.

The remark at [7.3.3(2)] says that, with a good probability, for cracking due to impressed deformations, as in the examined case, with bars diameter 12 mm working at 280 N/mm^2, crack width is below 0.3 mm.

11.3.4 Surface Reinforcements in High Beams

When the height of the beam is greater than 1 m, according to [7.3.3(3)], it is suitable to arrange a supplementary surface reinforcement to limit crack width along the depth of the rib. If only the reinforcement calculated for bending is arranged, at service conditions, distribution of small close cracks in the proximity of the side in tension of the beam could arise as a consequence of the high amount of reinforcement locally present, while on the rest of the height, due to lack of reinforcement, strongly separated big cracks could arise.

11.3.4.1 Calculus Example

T-shaped beam: $h = 1300$ mm; $d = 1220$ mm; $b = 1500$ mm; $b_w = 300$ mm; $h_f = 180$ mm. Concrete: $f_{ck} = 30$ N/mm^2, $f_{cd} = 17$ N/mm^2 (Fig. 11.17).

Load at ULS: $M_{Ed} = 2400$ kNm

The verification at ULS is developed according to the approximated method described in 7.7.1.1.2.

With the above-recalled data, we get:

$$z = 1220 - \frac{180}{2} = 1130 \text{ mm}; \quad A_s = \frac{2400 \cdot 10^6}{1130 \cdot 391} = 5431 \text{ mm}^2$$

We arrange 8 bars 30 mm diameter in two superimposed layers. Geometric reinforcement ratio is therefore given by

$$\rho = \frac{A_s}{b \cdot d} = \frac{5640}{1500 \cdot 1220} = 0.0031$$

Verification of the flange

$$\sigma_c = \frac{2400 \cdot 10^6}{1130} \cdot \frac{1}{1500 \cdot 180} = 7.9 < 17 \text{ N/mm}^2$$

At SLS, characteristic combination, the moment is given by

$$M_{Ed} = 0.7 \cdot 2400 = 1680 \text{ kNm}$$

Corresponding dimensionless value is given by:

$$\mu = \frac{M_{Ed}}{b \cdot d^2 \cdot f_{cd}} = \frac{1680 \cdot 10^6}{1500 \cdot 1220^2 \cdot 17} = 0.0443$$

Being $h_f/d = 0.15$ and $b/b_w = 5$, Table E5 of Appendix can be used, obtaining, for $\rho = 0.0031$, $\xi = 0.28$ from which $x = 336$ mm; $\frac{1-\xi}{i} = 24$.

We get

$$\sigma_s = \mu \cdot \frac{1-\xi}{i} \cdot \alpha_e \cdot f_{cd} = 0.0443 \cdot 24 \cdot 15 \cdot 17 = 270 \text{ N/mm}^2$$

which is allowable being smaller than 360 N/mm^2.

Surface reinforcements are designed according to 7.3.3(3) of EC2 which requires the arrangement of horizontal bars within the stirrups distributed between the neutral axis and the main reinforcements. The area of such reinforcements is the one required to sustain, at stress f_{yk}, the tensile force of the web beam between the considered levels, which is calculated according to (7.1)

$$A_s = k_c \cdot k \cdot f_{ct} \cdot A_{ct}$$

assuming $k = 0.5$, $k_c = 1$, $f_{ct} = f_{ctm} = 2.9$ N/mm^2 (corresponding to $f_{ck} = 30$ N/mm^2).

The height of the web between the neutral axis and the main reinforcement of the beam is given by: 1300 − 336 − 120 = 844 mm.

The area of the surface reinforcement required to receive the stress triangular diagram is given by:

$$A_s = \frac{300 \cdot 844 \cdot 0.5 \cdot 2.9 \cdot 1}{450} = 815 \text{ mm}^2$$

(5 + 5) bars 10 mm diameter are required with distance 170 mm (Fig. 11.18).

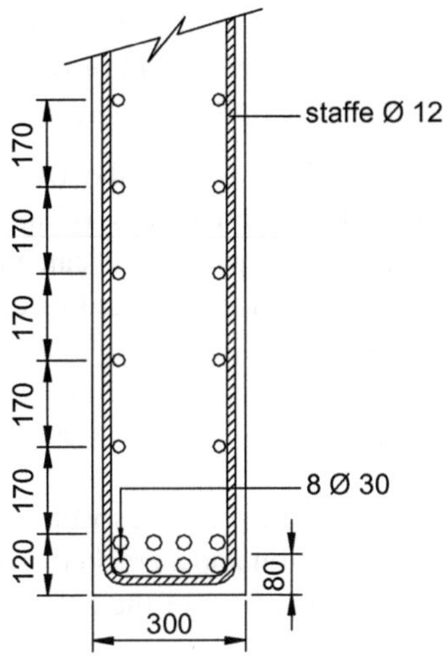

Fig. 11.18 Arrangement of the surface reinforcements

11.4 Deflection Control

To verify structures at SLS, the designer should verify structural deflection to:

- guarantee their proper functioning;
- avoid damages to the brittle elements supported by the structure;
- give eventual upward deflections during constructing phase.

The deflection verification is required because a structure that satisfies the ULS, could show an excessive deflection under service loads. If for a slab used for a storehouse no specific requirements are necessary, this does not occur for a slab for a residential building neither for a structure supporting a scientific laboratory with special needs of rigidity. The requirements for common structures, object of distinguished codes, are recalled in 11.1.

The excessive deformation can be due, in general, to the following causes:

- high slenderness, measured by the ratio span / effective depth $\frac{\ell}{d}$;
- presence of cracks, anyway unavoidable, reducing stiffness.

The verification procedure is, in a general way, founded on the calculation of the deflection through a double integration of the load-induced and shrinkage curvatures, modified by creep. The calculation is very cumbersome, and results are often uncertain, because of the uncertainties on the following aspects:

- concrete tensile strength, which defines the transition from the phase I, uncracked, to phase II, cracked;
- shrinkage values, depending on materials, hardening and environmental conditions of the structure;
- concrete elasticity modulus and creep estimation;

- extension of the transition between phase I and phase II: this is difficult to schematize, and, in most cases, service conditions lie in such transition,
- the effectivity of restraints in statically indeterminate structures.

For all the aspects above recalled, deflections calculated can be also very different from the effective ones, of an amount of 20–30%

A lighter procedure can be used to calculate the deflection, assuming all the cross-sections uncracked or cracked, and combining results through the following formula:

$$\alpha = \zeta\alpha_{II} + (1-\zeta)\alpha_I \quad [(7.18)]$$

where α_I e α_{II} are, respectively, the deflection values calculated for an uncracked cross-section and a cracked cross-section and ζ is the coefficient of tension stiffening (stiffening effect of the concrete in tension between cracks) given by the formula

$$\zeta = 1 - \beta\left(\frac{M_{cr}}{M}\right)^2 \quad [(7.19)]$$

where β is a coefficient that takes the value 1 for a short application and 0.5 for a long duration; M_{cr} is the cracking moment, M is the moment produced by loads.

Moreover, EC2 describes a synthetic verification procedure, founded on the calculation of the slenderness $\frac{\ell}{d}$, which allows avoiding deflection calculation. Practically, once a beam is given with assigned geometric and mechanic characteristics and the end-sections restraints, slenderness is calculated and compared with the limit values guaranteeing the respect of the deflections allowable for the quasi-permanent service conditions ($\frac{\ell}{250}$ total, $\frac{\ell}{500}$ for the loads occurring after the construction). The procedure provides two formulas to calculated slenderness $\frac{\ell}{d}$, the first one for beams with a low reinforcement ratio, the second one for beams with a high reinforcement ratio:

$$\frac{\ell}{d} = K\left[11 + 1.5\sqrt{f_{ck}}\frac{\rho_0}{\rho} + 3.2\sqrt{f_{ck}}\left(\frac{\rho_0}{\rho} - 1\right)^{1.5}\right] \quad \text{if } \rho \le \rho_0 \quad [(7.16a)]$$

$$\frac{\ell}{d} = K\left[11 + 1.5\sqrt{f_{ck}}\frac{\rho_0}{\rho - \rho'} + \frac{1}{12}\sqrt{f_{ck}}\sqrt{\frac{\rho'}{\rho_0}}\right] \quad \text{if } \rho > \rho_0 \quad [(7.16b)]$$

being $\rho_0 = 10^{-3}\sqrt{f_{ck}}$ the reinforcement ratio and K a factor depending on the beam supports, as shown in Table 11.3.

Moreover, the following corrective coefficients are provided:

- $\left(\frac{500}{f_{yk}} \cdot \frac{A_{s,prov}}{A_{s,req}}\right)$ where f_{yk} is the effective yielding value while 500 N/mm^2 is the value considered in EC2, $A_{s,\text{prov}}$ and $A_{s,\text{req}}$ are the reinforcement area respectively arranged and required for calculation. The correction for yielding is given by: $\frac{500}{450}$ = 1.11, under the assumption of steel $f_{yk} = 450$ N/mm^2.

Table 11.3 Excerpt of table [7.4N] – $\frac{\ell}{d}$ values

Structural system	K	Highly stressed concrete $\rho = 1.5\%$	Lightly stresses concrete $\rho = 0.5\%$
Simply supported beam	1.0	14	20
End span of continuous beam	1.3	18	26
Interior span of continuous beam	1.5	20	30
Cantilever beam	0.4	6	8

- (0.8) if, for a T-shaped beam, the width of the web is greater than three times the one of the rib;
- ($7/\ell$) if the beam supporting brittle structures has a span greater than 7 m.

An example of the procedure is shown in Table 11.4 collecting values $\frac{\ell}{d}$ associated with two different values of reinforcement ratios ρ for various types of beams, one under the hypothesis that concrete is "highly stressed", with a $\rho = 1.5\%$, another one under the hypothesis of a "lightly stressed" concrete, associated to a value of $\rho = 0.5\%$, with the possibility of interpolation. The Table also collects the values of coefficient K which characterize support conditions and are valuable also for the expressions [(7.16)]. Corrective coefficients indicated above are applied to $\frac{\ell}{d}$ values.

To analyse the capacity of formulas [(7.16a)] and [(7.16b)], three different cases have been developed for a simply supported beam:

- $\rho = \rho_o$, the boundary of the domain of both formulas, being $\rho_o = 10^{-3}\sqrt{f_{ck}}$, with the employ of [(7.16a)]
- $\rho = \rho_o/2$ with the employ of [(7.16a)]
- $\rho = 3\rho_o$, $\rho' = 0.1\ \rho$ with the employ of [(7.16b)].

From Tables 11.4, 11.5 and 11.6 containing calculation results (ℓ/d values and corresponding concrete stresses under the hypothesis of bending moment in service conditions equal to half of the ultimate resistant bending moment), we get that:

- configurations $\rho = \rho_o$ are reasonable and corresponding to situations with concrete lightly stressed, with values ℓ/d according to Table [7.4N];
- configurations $\rho = \rho_o/2$ gives very high values ℓ/d, not suitable for reinforced concrete beams. Geometric reinforcement ratios around $\rho = \rho_o/2$ or similar are suitable for T-shaped cross-sections but with smaller ℓ/d values. The high ℓ/d values given by the formula signify that there is no problem with excessive deflections and it is possible to avoid further verifications;
- finally, cases as the one developed with $\rho = 3\rho_o$, $\rho' = 0.1\ \rho$, are characterized by highly stressed concrete, at least for not too high values of f_{ck}. With the increase of f_{ck} relevant stresses are decreasing and slenderness ℓ/d increase moderately. It is possible to assume that [(7.16b)] is reliable for the slenderness limitation.

Table 11.4 $\rho = \rho_0$

f_{ck}	16	20	25	30	40	50	60	70	80	90
ρ	0.0040	0.0045	0.0050	0.0055	0.0063	0.0070	0.0077	0.0083	0.0089	0.0095
ℓ/d	18.9	19.6	20.5	21.3	22.8	24.0	25.1	26.1	27.1	28.0
σ_c/f_{ck}	0.34	0.30	0.25	0.23	0.19	0.16	0.17	0.15	0.14	0.13

Table 11.5 $\rho = \rho_0/2$

f_{ck}	16	20	25	30	40	50	60	70	80	90
ρ	0.0020	0.0022	0.0025	0.0028	0.0031	0.0035	0.0038	0.0042	0.0044	0.0048
ℓ/d	39.7	43.0	46.6	49.9	55.7	60.9	65.5	69.8	73.7	77.5

Table 11.6 $\rho = 3\rho_0$

f_{ck}	16	20	25	30	40	50	60	70	80	90
ρ	0.0120	0.0135	0.0150	0.0165	0.0189	0.0210	0.0231	0.0249	0.0267	0.0285
ℓ/d	14.9	15.2	15.5	15.9	16.4	16.9	17.4	17.8	18.2	18.5
σ_c/f_{ck}	0.57	0.50	0.45	0.40	0.34	0.30	0.31	0.28	0.26	0.24

11.4.1 Application Examples

11.4.1.1 Design of a Beam Sustaining Brittle Superstructures Which Require Small Deflections

A simply supported beam 10 m span with T-shaped cross-section (Fig. 11.19) is considered:

– $h = 860$ mm, $d = 800$ mm, $h_f = 200$ mm, $b = 400$ mm, $b_w = 200$ mm.

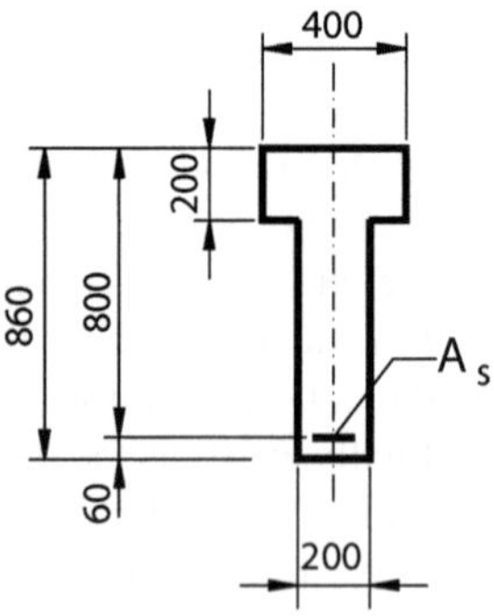

Fig. 11.19 Cross-section geometry

- Materials: concrete C25/30 ($f_{ck} = 25$ N/mm^2) corresponding, according to Table [3.1], to: $E_{cm} = 31.000$ N/mm^2 and $f_{ctm} = 2.6$ N/mm^2. Steel: $f_{yk} = 450$ N/mm^2; $f_{yd} = 391$ N/mm^2.
- Structural self-weight: $g_1 = 20$ kN/m
- Superstructure weight: $g_2 = 5$ kN/m
- Variable load: $q = 12$ kN/m
- Load for design at ULS: $20{\cdot}1.3 + 5{\cdot}1.5 + 12{\cdot}1.5 = 26.0 + 7.5 + 18.0 = 51.5$ kN/m
- Load for characteristic combination at SLS: $20.0 + 5.0 + 12.0 = 37.0$ kN/m
- Load for quasi-permanent combination at SLS: coefficient $\psi_2 = 0.6$ indicated by EN 1990 for commercial building is assumed, therefore variable load is $0.6 \cdot 12 = 7.2$ kN/m. Totally: $25 + 7.2 = 32.2$ kN/m.
- Bending moment in the mid-span section for the above-defined loads:

ULS	$M_{Ed} = \frac{10^2}{8} \cdot 51.5 = 644.0$ kNm.
SLS, characteristic combination	$M_{Ed} = 462.5$ kNm.
SLS, quasi-permanent combination	$M_{Ed} = 402.5$ kNm.
With only permanent load	$M_{Ed} = 312.5$ kNm.

Design at ULS.
From Table U8 (T-shaped cross-sections at ULS)

$$w = \frac{h_f}{d} = \frac{200}{800} = 0.25; \; \frac{b}{b_w} = \frac{400}{200} = 2$$

$$\mu = \frac{644000000}{400 \cdot 800^2 \cdot 14.16} = 0.1777$$

Being $\mu < \mu_1 = 0.2188$, the neutral axis cuts the flange. In this case, the lever arm of the internal couple can be evaluated in 0.70 m (from the centre of the flange to the centre of the reinforcement in tension). The required tensile reinforcement is therefore

$$A_s = \frac{644000000}{700 \cdot 391} = 2352\text{mm}^2$$

This is satisfied using six bars of 20 mm diameter and two 18 mm bars, with $A_s = 2392$ mm^2. The reinforcement ratio is

$$\rho = \frac{A_s}{b \cdot d} = \frac{2392}{400 \cdot 800} = \frac{7.5}{1000}$$

For the verifications at SLS, [7.4.2] identifies the situations when the calculation of deflections can be omitted. To check if this is the case, the ratio $\frac{\ell}{d}$ is evaluated using [(7.16b)], being $\rho > \rho_o = 10^{-3}\sqrt{f_{ck}} = 10^{-3}\sqrt{25} = 0.005$. Because $k = 1$ for a simply supported beam (coefficient taken from Table [7.4N]), we get:

$$\frac{\ell}{d} = 1\left[11 + 1.5 \cdot \sqrt{25} \cdot \frac{0.0050}{0.0075}\right] = 16$$

Being, in this case, $f_{yk} = 450$ N/mm^2, [(7.17)] allows the following adjustment:

$$\frac{500}{f_{yk}} \cdot \frac{A_{s,\mathrm{req}}}{A_{s,\mathrm{prov}}} = \frac{500}{450} \cdot \frac{2392}{2352} = 1.13;\ 16 \cdot 1.13 = 18.0$$

This ratio, according to [7.4.2(2)], shall be multiplied by $7/l_{\mathrm{eff}} = 0.7$ being the span of the beam longer than 7 m, employed for supporting brittle superstructures.

Therefore, it must be $\frac{\ell}{d} < 18.0 \cdot 0.7 = 12.60$.

In this case study, slenderness is given by $\frac{\ell}{d} = \frac{10}{0.80} = 12.50 < 12.60$.

According to EC2, deflection control would not be necessary. However, as an example, this is developed in the following three cases:

(a) quasi-permanent load combination at an infinite time;
(b) effect of a load after beam construction, quasi-permanent load combination at an infinite time;
(c) short-duration loading test with variable load application corresponding to the characteristic combination.

In the three cases we apply the formula:

$$f = \zeta \cdot f_2 + (1 - \zeta) \cdot f_1 \quad [7.18]$$

where f_1 is the displacement in the mid-span cross-section evaluated with uncracked cross-sections and f_2 is evaluated with all cracked sections. The coefficient ζ is evaluated according to [7.19] $\zeta = 1 - \beta \cdot \left(\frac{M_{cr}}{M}\right)^2$, where M_{cr} is the cracking moment and β is a coefficient taking into account the duration of load application.

In cases (a) and (b), to account for long-duration loads, the ratio between steel and concrete moduli $\alpha_e = \frac{E_s}{E_{c,\mathrm{eff}}}$ is used for the calculation of f_1 and f_2, being $E_{c,\mathrm{eff}}$ expressed by [7.20] $E_{c,\mathrm{eff}} = \frac{E_{cm}}{1+\varphi(\infty,t_0)}$, and in [7.19] we introduce $\beta = 0.5$.

In case (c), for a short-duration load, the calculation of f_1 and f_2 is developed with a modular ratio between steel and concrete moduli $\alpha_e = \frac{E_s}{E_{cm}} = 6.45$ and with $\beta = 1$ in [7.19].

Because cracking occurs generally in the short period under the action of the first load which causes the reaching of the concrete tensile strength, cracking moment M_{cr} can be evaluated with $\alpha_e = \frac{E_s}{E_{cm}} = \frac{200000}{31000} = 6.45$.

- *Calculation of the cracking moment M_{cr}*

By applying dimensionless formulas corresponding to the uncracked cross-section (see 11.2.1.1) we get:

$$\rho_1 = \frac{A_s}{bh} = \frac{2392}{400 \cdot 860} = 0.00695$$

$$w_1 = \frac{200}{860} = 0.2326;\ n = \frac{400}{200} = 2;\ p = \frac{800}{860} = 0.93;\ \alpha_e = \frac{200000}{31000} = 6.45$$

and then

$$\xi_1 = 0.462;\ i = 0.0667$$

Dimensional values are:
depth of the neutral axis $x_1 = 0.462 \cdot 860 = 397$ mm
second moment of area $I_1 = ibh^3 = 0.0667 \cdot 0.40 \cdot 0.860^3 = 0.01696\ \text{m}^4$
Section modulus on the tensile reinforcement side: $W = \frac{i}{(1-\xi_1)} \cdot bh^2 = 3.66 \cdot 10^7\ \text{mm}^3$
Being the cracking moment $M_{cr} = W \cdot f_{ctm}$, with $f_{ctm} = 2.6\ \text{N/mm}^2$, we get:

$$M_{cr} = 3.66 \cdot 10^7 \cdot 2.6 \cdot 10^{-6} = 9.52 \cdot 10^7 \text{Nmm} = 95.2\ \text{kNm}$$

- Cases (a) and (b): calculation of geometrical characteristics and ζ coefficient (long duration, quasi-permanent combination)

Uncracked cross-section (phase I):

$$\rho_1 = 0.00695$$

$$w_1 = 0.2326;\ n = 2;\ p = 0.93;\ \alpha_e = \frac{E_s}{E_{c,\text{eff}}}$$

For the evaluation at infinite time, we introduce the elastic modulus corrected according to [7.20]. We assume $\varphi(\infty, t_0) = 2$, therefore $E_{c,eff} = \frac{E_{cm}}{3} = 10333\ \text{N/mm}^2$ and $\alpha_e = 19.35$.
In dimensionless terms, we get
$\xi_1 = 0.5177;\ i = 0.0840.$
Dimensional values are the following:
depth of the neutral axis $x_1 = 0.5177 \cdot 860 = 445$ mm
second moment of area

$$I_1 = ibh^3 = 0.0840 \cdot 0.40 \cdot 0.860^3 = 2.1369 \cdot 10^{10} \text{mm}^4 = 0.02137 \text{m}^4$$

Cracked cross-section (phase II)

$$\rho = \frac{A_s}{bd} = \frac{2392}{400 \cdot 800} = 0.00748;\ w = \frac{200}{800} = 0.25;\ n = \frac{400}{200} = 2;\ \alpha_e = 19.35$$

By applying dimensionless formulas for cracked cross-sections, (see 11.2.1.1) we get:

$$\xi = 0.4261; \quad i = 0.0725$$

Therefore, we get

$$x = 341 \text{ mm}; I = 0.0725 \cdot 400 \cdot 800^3 = 1.4854 \cdot 10^{10}\text{mm}^4 = 0.01485 \text{ m}^4$$

The coefficient ζ is calculated according to [7.19] under the assumption that $\beta = 0.5$, $M_{cr} = 95.2$ kNm and $M = 402.5$ kNm. We obtain:

$$\zeta = 1 - 0.5 \cdot \left(\frac{95.2}{402.5}\right)^2 = 0.972$$

- *Case (c): calculation of the geometrical characteristics and ζ coefficient (short duration, characteristic combination)*

Uncracked cross-section (phase I)
As already obtained for the evaluation of cracking moment, we get:
depth of the neutral axis $\xi_1 = 0.462$; $x_1 = 397$ mm.
second moment of area $i = 0.0667$; $I_1 = 1.696 \cdot 10^{10}\text{mm}^4$.
Cracked cross-section (phase II)

$$\rho = \frac{A_s}{bd} = \frac{2392}{400 \cdot 800} = 0.00748; w = \frac{200}{800} = 0.25; n = \frac{400}{200} = 2; \alpha_e = 6.45$$

By applying dimensionless formulas for cracked cross-section (see 11.2.1.1), we get:

$$\xi = 0.2663; \quad i = 0.0322$$

Therefore, we get: $x = 213$ mm; $I = 0.0322 \cdot 400 \cdot 800^3 = 6.5945 \cdot 10^9\text{mm}^4$.
Coefficient ζ is calculated by assuming
$\beta = 1$, $M_{cr} = 95.2$ kNm and $M = 462.5$ kNm
obtaining

$$\zeta = 1 - 1 \cdot \left(\frac{95.2}{462.5}\right)^2 = 0.958$$

Following Tables 11.7 and 11.8 collect dimensionless and dimensional values of the geometrical characteristics calculated previously.

- *Calculation of deflections: case (a)*

Design load referred to quasi-permanent load combination is given by: $q = 32.2$ kN/m. Therefore, we get:

Table 11.7 Excerpt of dimensionless values

Cases	α_e	ξ_1	i_1	ξ	i
M_{cr}	6.45	0.4620	0.0667	–	–
a), b)	19.35	0.5177	0.0840	0.4261	0.0725
c)	6.45	0.4620	0.0667	0.2663	0.0322

Table 11.8 Excerpt of dimensional values

Cases	α_e	x_1 (m)	I_1 (m^4)	x (m)	I (m^4)
M_{cr}	6.45	0.397	0.01,696	–	–
a), b)	19.35	0.445	0.02,137	0.341	0.01,485
c)	6.45	0.397	0.01,696	0.213	0.006,594

$$f_1 = \frac{5}{384} \cdot \frac{32.2 \cdot 10^{16}}{2.1369 \cdot 10^{10} \cdot 10333} = 19.0 \text{ mm}$$

$$f_2 = \frac{5}{384} \cdot \frac{32.2 \cdot 10^{16}}{1.4854 \cdot 10^{10} \cdot 10333} = 27.3 \text{ mm}$$

$$f* = 19.0 \cdot (1 - 0.972) + 27.3 \cdot 0.972 = 27.1 \text{ mm}$$

f * is smaller than $\frac{\ell}{250} = \frac{10000}{250} = 40$ mm.

If the load effect is added to the shrinkage effect, as it will be calculated in 11.4.1.2, we get the deflection:

$$f = 27.1 + 5.5 = 32.6 \text{ mm}$$

smaller than $\frac{\ell}{250} = 40$ mm.

- *Deflection calculation: case (b)*

With reference to the quasi-permanent combination for loads present after construction, we get $q = 32.2 - 20.0 = 12.2$ kN/m.

$$f_1 = \frac{5}{384} \cdot \frac{12.2 \cdot 10^{16}}{2.1369 \cdot 10^{10} \cdot 10333} = 7.2 \text{ mm}$$

$$f_2 = \frac{5}{384} \cdot \frac{12.2 \cdot 10^{16}}{1.4854 \cdot 10^{10} \cdot 10333} = 10.3 \text{ mm}$$

$$f* = 7.2 \cdot (1 - 0.972) + 10.3 \cdot 0.972 = 10.2 \text{ mm}$$

This value is smaller than $\frac{\ell}{500} = \frac{10000}{500} = 20$ mm.

To this value we should add a deflection amount due to shrinkage developing after the building construction and which is, therefore, a fraction of the total value calculated in 11.4.12.

- *Deflection calculation: case (c)*

For the short duration load $q = 12.0$ kN/m, the following values are obtained

$$f_1 = \frac{5}{384} \cdot \frac{12.0 \cdot 10^{16}}{1.696 \cdot 10^{10} \cdot 31000} = 2.97\,\text{mm}$$

$$f_2 = \frac{5}{384} \cdot \frac{12.0 \cdot 10^{16}}{6.5945 \cdot 10^{9} \cdot 31000} = 7.64\,\text{mm}$$

$$f* = 2.97 \cdot (1 - 0.958) + 7.64 \cdot 0.958 = 7.44\,\text{mm}$$

11.4.2 Deflection Calculation Due to Shrinkage

Here we calculate the deflection due to shrinkage of the T-shaped beam with a 10 m span of 11.4.1.1. We apply formula (7.21) of EC2 with the calculation of curvature in phase I and phase II and their combination with the formula [7.18]; then we integrate along the span. For the application of [7.18], it is assumed that shrinkage develops in the presence of a quasi-permanent load which gives at the mid-span the moment $M = 402.5$ kNm.

It is assumed that the relative humidity of the environment is 70%, exposure time is $t_s = 10$ days, the time for the valuation of the shrinkage effect is $t = 1000$ days.

The curvature induced by shrinkage is calculated using the formula [7.4.3(6)]

$$\frac{1}{r_{cs}} = \varepsilon_{cs}\alpha_e \frac{S}{I} \quad [(7.21)]$$

where ε_{cs} is the shrinkage strain, α_e is the ratio $\frac{E_s}{E_{c,eff}}$, S is the first moment of area of the cross-sectional area of the reinforcement about the barycentric axis of the transformed cross-section, I is the second moment of area of the transformed cross-section.

Explanation of the formula

By definition, the curvature is given by $\frac{1}{r_{cs}} = \frac{\int \varepsilon_{cs} y dA_c}{I}$ and, because ε_{cs} is not depending on the cross-sectional area A_c, it can be written that $\frac{1}{r_{cs}} = \frac{\varepsilon_{cs} \int y dA_c}{I}$. The last integral represents the first moment of area of the concrete section (which undergoes shrinkage) about the neutral axis (passing through the centroid) of the transformed cross-section. Numerically, it is equal to the first moment of area of the transformed reinforcement ($\alpha_e A_s$), as it is evident by comparison with the formula [(7.21)].

The amount of the shrinkage is indicated as follows in 3.1.4(6) of EC2: sum of the shrinkage due to drying and autogenous shrinkage

$$\varepsilon_{cs} = \varepsilon_{cd} + \varepsilon_{ca} \quad [(3.8)]$$

where ε_{cd} is the shrinkage due to drying and ε_{ca} is the autogenous shrinkage

$$\varepsilon_{cd}(t) = \beta_{ds}(t, t_s) \cdot k_h \cdot \varepsilon_{cd,0}[(3.9)]$$

where k_h is a coefficient depending on h_0; k_h values are collected in Table [3.3]

$$\beta_{ds}(t, t_s) = \frac{(t - t_s)}{(t - t_s) + 0.4\sqrt{h_0^3}} \quad [(3.10)]$$

being t the age expressed in days at the considered time, t_s is the age (in days) at the beginning of the shrinkage due to drying, practically coincident with the end of the "curing" time

$$\varepsilon_{ca}(t) = \beta_{as}(t) \cdot \varepsilon_{ca}(\infty) \quad (3.11)$$

where

$$\varepsilon_{ca}(\infty) = 2.5 \cdot (f_{ck} - 10) \cdot 10^{-6} \quad (3.12)$$

$$\beta_{as}(t) = 1 - \exp\left(-0.2t^{0.5}\right) \quad [(3.13)]$$

In [7.4.3(6)] it is specified that S and I should be calculated in the two hypotheses of uncracked cross-section $\left(\frac{1}{r_1}\right)$ and completely cracked cross-section, i.e. partially cracked because of bending $\left(\frac{1}{r_2}\right)$, being the final curvature evaluated through

$$\frac{1}{r_{cs}} = \frac{1}{r_1} \cdot (1 - \zeta) + \frac{1}{r_2} \cdot \zeta \quad [(7.18)]$$

where

$$\zeta = 1 - \beta \cdot \left(\frac{M_{cr}}{M}\right)^2 \quad [(7.19)]$$

- *Development of calculations*

Evaluation of shrinkage.

From Table [3.2], given data ($f_{ck} = 25$ N/mm^2, relative humidity 70%), allow to deduce

$$\varepsilon_{cd,0} = \frac{0.4}{1000}$$

The area of the transverse cross-section A_c is given by (Fig. 11.19).

$$200 \cdot (400 + 660) = 212000 \text{ mm}^2$$

The exposed perimeter is corresponding to the bottom (slab and under-the-flange) and to the sides

$$u = (400 + 2 \cdot 660) = 1720 \text{ mm}$$

We get

$$h_0 = \frac{2A_c}{u} = \frac{2 \cdot 212000}{1720} = 246 \, \text{mm}$$

From Table [3.3] with $h_0 = 246$ mm we obtain the coefficient: $k_h = 0.80$.
Shrinkage at 1000 days with $t_s = 10$ days

$$\beta_{ds} = \frac{(1000 - 10)}{(1000 - 10) + 0.04\sqrt{246^3}} = 0.86$$

$$\varepsilon_{cd} = 0.86 \cdot 0.80 \cdot \frac{0.4}{1000} = \frac{0.275}{1000}$$

$$\varepsilon_{ca}(\infty) = 2.5 \cdot (25 - 10) \cdot 10^{-6} = 37.5 \cdot 10^{-6}$$

$$\beta_{as}(1000) = 1 - e^{-0.2 \cdot 1000^{0.5}} = 0.998 \approx 1$$

$$\varepsilon_{cs} = (0.275 + 0.037) \cdot 10^{-3} = 0.31 \cdot 10^{-3} \quad [(3.8)]$$

For the modular ratio between steel and concrete moduli and geometric parameters of the cross-sections, we adopt those of the Example treated in 11.4.1.1, cases (a) and (b).

$E_{c,\text{eff}} = 10333 \, \text{N/mm}^2$ from which $\alpha_e = \frac{200000}{10333} = 19.35$.
Uncracked cross-section: $x_1 = 445$ mm; $I_1 = 21.37 \cdot 10^9 \text{ mm}^4 \text{ mm}^4$
First moment of area $S_1 = 2392 \cdot (860 - 445) = 992680 \text{ mm}^3$
Cracked cross-section: $x_2 = 341$ mm; $I_2 = 14.85 \cdot 10^9 \text{ mm}^4$
First moment of area $S_2 = 2392 \cdot (800 - 341) = 1098000 \text{ mm}^3$
Curvatures are the following:

$$\frac{1}{r_1} = 0.31 \cdot 10^{-3} \cdot 19.35 \cdot \frac{849160}{21.37 \cdot 10^9} = 0.278 \cdot 10^{-6} \text{mm}^{-1}$$

$$\frac{1}{r_2} = 0.31 \cdot 10^{-3} \cdot 19.35 \cdot \frac{1098000}{14.85 \cdot 10^9} = 0.443 \cdot 10^{-6} \text{mm}^{-1}$$

By combination with the formula [(7.18)], we get:

$$\frac{1}{r^*} = [0.278 \cdot (1 - 0.97) + 0.443 \cdot 0.97] \cdot 10^{-6} = 0.438 \cdot 10^{-6} \text{mm}^{-1}$$

under the assumption that $\zeta = 0.97$ as in the Example of 11.5.1.1 (case b).

- *Deflection calculation*

We apply the corollary of Mohr theorem: the deflection is measured by the bending moment at the mid-span of a beam loaded by the fictitious load given by curvatures.

We get:

$$f = \left(\frac{1}{r^*}\right) \cdot \frac{\ell^2}{8} = 0.438 \cdot 10^{-6} \cdot \frac{10000^2}{8} = 5.47 \text{mm}$$

11.5 Further Verifications at SLS

Further verifications at SLS can be found in Chap. 4, paragraphs 4.3.5.4.4 and 4.3.7.8.

References

(Lab.P.C. 1996) Extension du domaine d'application des règlements de calcul BAEL/BPEL aux bétons à 80 MPa—Bulletin des laboratoires des Ponts et Chaussées—Numéro spécial XIX—mai 1996 Paris.

(CEB 127-1995) CEB Bulletin n. 127 Analysis of beams and frames, August 1995.

(CEB-*fib* 92-2019) (September 2019). *fib* Bulletin n. 92. *Serviceability limit state of concrete structures*. ISBN 978-2-88394-136-6.

Chapter 12
Detailing of Reinforcement and Structural Members for Buildings

Abstract The chapter collects the rules of EC2 for detailing of ordinary and prestressing steel and structural members for buildings. The first part presents rules for detailing of ordinary and prestressing steel, like spacing of bars, permissible mandrel diameters for bent bars, minimum anchorage lengths for longitudinal reinforcement, minimum anchorage lengths for stirrups and shear reinforcement, minimum superposition length for bars, rules for large diameter bars, rules for bundled bars. Detailing rules for reinforcement steel and prestressing steel apply to structural members of normal buildings and bridges subject to predominantly static loads. The same rules are not appropriate for members subjected to seismic actions, vibrating machines, or impacts, and members including special painted bars or bars covered by epoxy resin or zinc, where bond conditions are less favourable. The second part discusses the minimum amount of reinforcement to avoid brittle failures and excessive cracks in structural members of buildings: beams, columns, slabs, and walls. The minimum reinforcement is also required to absorb internal forces associated with restrained strains, like those due to drying shrinkage, thermal variations, and settlements. Some case studies show the application of rules for detailing of steel and structural members for buildings.

This chapter collects the rules of EC2 for detailing of ordinary and prestressing steel and structural members for buildings.

12.1 Detailing of Reinforcement

Following rules for detailing of ordinary and prestressing steel are dealt with:

1. spacing of bars,
2. permissible mandrel diameters for bent bars,

This chapter was authored by Franco Angotti and Maurizio Orlando.

© The Author(s), under exclusive license to Springer Nature Switzerland AG 2022

F. Angotti et al., *Reinforced Concrete with Worked Examples*,
https://doi.org/10.1007/978-3-030-92839-1_12

3. minimum anchorage lengths for longitudinal reinforcement,
4. minimum anchorage lengths for stirrups and shear reinforcement,
5. minimum superposition length for bars,
6. rules for large diameter bars,
7. rules for bundled bars, are described.

Rules concerning detailing for reinforcement steel (*only for deformed and weldable bars, including mesh sheets*) and prestressing steel apply to structural members of normal buildings and bridges subject to predominantly static loads.

The same rules cannot be appropriate in the following cases:

- members subjected to seismic actions (EN1998 rules should be considered for buildings in seismic prone areas), vibrating machines or impacts,
- members including special painted bars or bars covered by epoxy resin or zinc, where bond conditions are less favourable.

EC2 gives additional rules for large diameter bars, i.e., for bars with a diameter ϕ larger than 32 mm. It is worth noting that in lightweight aggregate concrete, the diameter of the bars shall be limited to 32 mm and bundles of bars shall not be made of more than two bars and their equivalent diameter (defined in 12.1.9) shall not exceed 45 mm.

12.1.1 Spacing of Bars

The spacing of bars shall be such that the concrete can be placed and compacted satisfactorily to assure an adequate bond between concrete and bars.

The clear distance (horizontal a_h and vertical a_v) between individual parallel bars or horizontal layers of parallel bars should satisfy the following limitation

$$a_h, a_v \geq \max\,(\phi;\, d_g + 5\text{ mm};\, 20\text{ mm})$$

where ϕ is the maximum bar diameter and d_g is the maximum aggregate size.

If bars are positioned in separate horizontal layers, the bars in each layer should be located vertically above each other. There should be enough space between the resulting vertical alignments of bars to allow for access of vibrators and good compaction of the concrete (Fig. 12.1).

Lapped bars may be allowed to touch one another within the lap length.

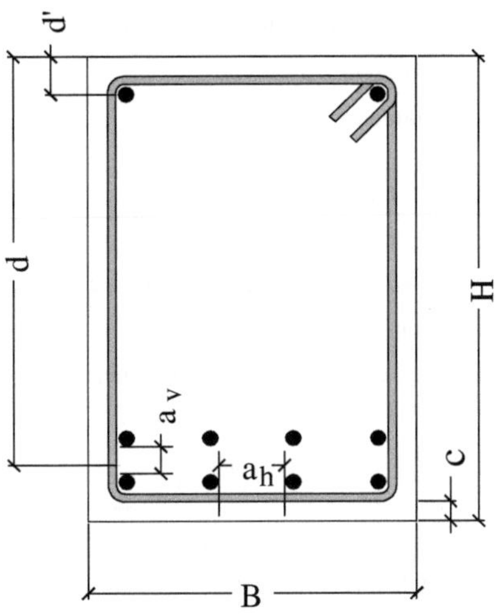

Fig. 12.1 Clear horizontal and vertical spacing between bars. If bars are positioned in separate horizontal layers, the bars in each layer should be located vertically above each other

12.1.2 Permissible Mandrel Diameters for Bent Bars

The minimum diameter to which a bar is bent shall be such as to avoid bending cracks in the bar, and to avoid failure of the concrete inside the bend of the bar.

If the mandrel diameter is too small, the stress concentration in the bent of the bar could lead to cracks. The risk of cracks increases as the temperature decreases because the steel toughness decreases with the temperature, so EN13670 requires a minimum temperature to perform bending tests. Bending is discouraged if the temperature is lower than –5 °C, unless special cautions are taken, as the reduction of the bending speed and/or the increase of the mandrel diameter.

Table 12.1 collects the values of the minimum bending diameter $\phi_{\mathrm{m,min}}$ to avoid bar damaging.

The values of $\phi_{\mathrm{m,min}}$ for welded bent reinforcement and mesh bent after welding are different depending on the position of the transverse bar relative to the longitudinal bar segment affected by the welding process, which extends for a length equal to 3ϕ on each side of the welding point.

The verification of concrete inside the bend of the bar can be considered fulfilled if at least one of the following conditions is satisfied:

- the anchorage of the bar does not require a length more than 5ϕ past the end of the bend, so bar stresses are small and radial pressures on the concrete inside the bend are small, too (Fig. 12.2),
- the bar is not positioned at the edge, i.e., the plane of the bend is not close to the concrete face, and there is a transverse bar with a diameter $\geq \phi$ inside the bend; this condition is normally fulfilled inside the bend of beams and columns.

If these conditions are not fulfilled, the minimum mandrel diameter $\phi_{m,\mathrm{min}}$ shall satisfy the following condition:

Table 12.1 Minimum mandrel diameter $\phi_{m,min}$ to avoid damage to reinforcement

For bars and wires	
Bar diameter	Minimum mandrel diameter $\phi_{m,min}$ for bends, hooks and loops
$\phi \leq 16$ mm	4ϕ
$\phi > 16$ mm	7ϕ
For welded bent reinforcement and mesh bent after welding	
Position of welded bar	Minimum mandrel diameter $\phi_{m,min}$
On the same side of the bent zone, out or within the curved zone 	5ϕ
On the opposite side of the bent zone, out or within the curved zone 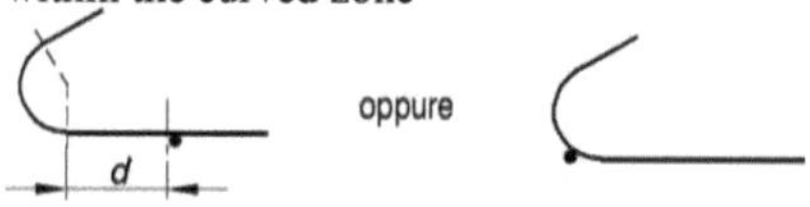	5ϕ, if $d \geq 3\phi$ 20ϕ, if $d < 3\phi$ or the welding is on the curved zone

Note The mandrel size for welding within the curved zone may be reduced to 5ϕ where the welding is carried out according to Annex B of EN ISO 17660

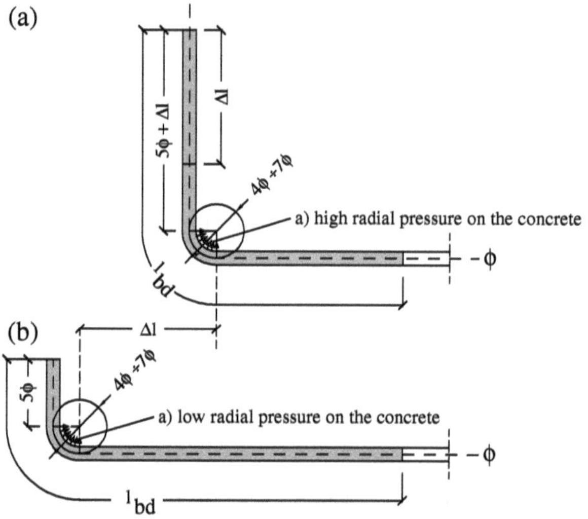

Fig. 12.2 **a** Anchorage of a bar requiring a length more than 5ϕ past the end of the bend, so inside the bend the bar is still subjected to high stresses and radial pressures on the concrete are high, too; **b** anchorage of a bar not requiring a length more than 5ϕ past the end of the bend, so in the bend stresses in the bar are small as well as radial pressures on the concrete.

$$\phi_{m,\min} \geq F_{bt}\frac{1/a_b + 1/2\phi}{f_{cd}} \quad [(8.1)]$$

where:

F_{bt} is the tensile force at ULS in a bar or group of bars in contact at the start of a bend

a_b for a given bar (or group of bars in contact) is half of the centre-to-centre distance between bars (or groups of bars) perpendicular to the plane of the bend (Fig. 12.3); for a given bar (or group of bars) adjacent to the face of the member, a_b should be taken as the cover (referred to the considered bar) plus $\phi/2$.

For concrete classes exceeding C55/67, $\phi_{m,min}$ is calculated by using the value of f_{cd} corresponding to concrete class C55/67.

Radial compressive pressure σ inside the bend, which has radius R equal to half of the mandrel diameter ($R = \phi_{m,min}/2$), is equal to

$$\sigma = p_{rad}/\phi = F_{bt}/(\phi \cdot R) = F_{bt}/(\phi \cdot \phi_{m,\min}/2)$$

where $p_{rad} = F_{bt}/R$ is the well-known Mariotte's law; therefore, the following expression for $\phi_{m,min}$ holds

$$\phi_{m,\min} = 2F_{bt}/(\phi \cdot \sigma)$$

and by comparing this expression with the limit [(8.1)] given by EC2,

$$\phi_{m,\min} = \frac{2\ F_{bt}}{\phi \cdot \sigma} \geq F_{bt}\frac{\frac{1}{a_b} + \frac{1}{2\phi}}{f_{cd}}$$

from which

$$\phi \cdot \sigma \leq \frac{2\ f_{cd}}{\frac{1}{a_b} + \frac{1}{2\phi}}$$

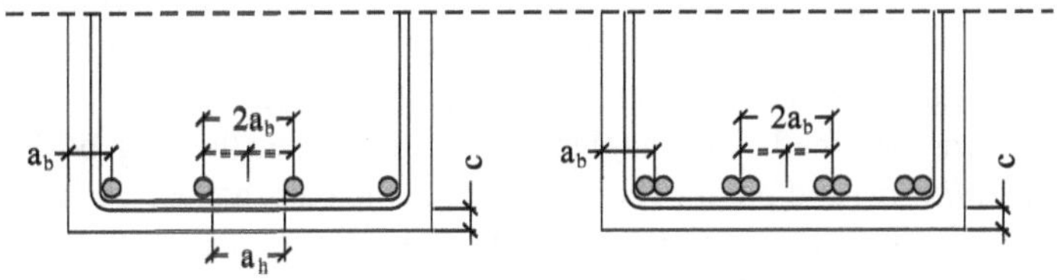

Fig. 12.3 Spacing a_b for single bars (on the left) and groups of bars (on the right)

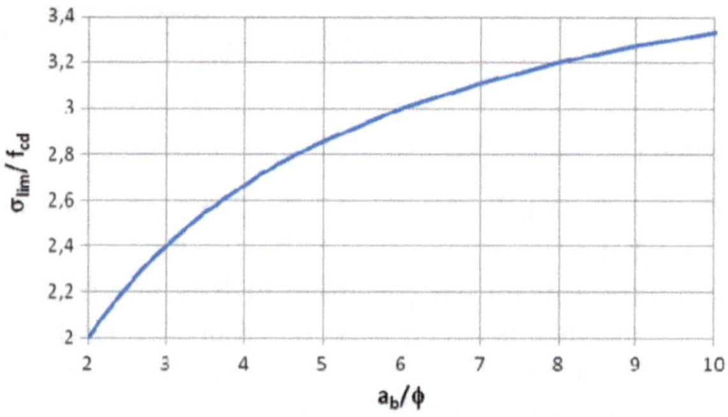

Fig. 12.4 Ratio between the limit radial pressure σ_{lim} in the concrete inside the bend and the design concrete compressive strength f_{cd} at varying the ratio a_b/ϕ; $\sigma_{\text{lim}}/f_{cd}$ tends asymptotically to 4 as a_b/ϕ increases.

and

$$\sigma \leq \frac{2\ f_{cd}}{\frac{\phi}{a_b}+\frac{1}{2}}=\frac{4\ f_{cd}}{\frac{2\phi}{a_b}+1}$$

The last expression shows that the limit given on the mandrel diameter to avoid failure of the concrete inside the bend of the bar is equivalent to limit the concrete radial pressure to $s_{\text{lim}} = 4f_{cd}/\ (2\text{f}/a_b + 1)$. Figure 12.4 shows the curve of the ratio $\sigma_{\text{lim}}/f_{cd}$ at varying a_b/ϕ in the interval 2 ÷ 10: if a_b/ϕ increases, the ratio tends asymptotically to 4.

The expression [(8.1)] leads to high values of the mandrel diameter: for example, for concrete class C25/30 and maximum aggregate size of 20 mm, it takes values between 23 and 30 times the diameter ϕ of the bar as ϕ varies between 10 and 20 mm.

Special attention should be paid to those cases where the bar has a large diameter that requires a large curvature radius so that the corner of the structural member is not equipped with any reinforcement. In those cases, it is required to add a suitable reinforcement, made of small diameter bars, to avoid the separation of the corner.

12.1.3 Anchorage of Longitudinal Reinforcement

To provide the transfer of bond forces to concrete and to avoid longitudinal cracking or spalling, reinforcing bars, wires or welded mesh fabrics shall be suitably anchored.

The transfer of bond forces is mainly assured by the mechanical interlock between the ribs on the surface of the bar and the surrounding concrete.

Bond forces are inclined by the angle θ to the bar axis and they can be decomposed into normal (*splitting*) and parallel components to the bar; the last are in equilibrium with the axial force within the bar. Normal components act on the concrete and are directed towards the outside of the bar; they are in equilibrium with tensile circumferential forces within the surrounding concrete (Fig. 12.5).

The problem can be treated as that of a thick cylinder subjected to internal outward radial pressure. The internal radius of the cylinder is coincident with the radius of

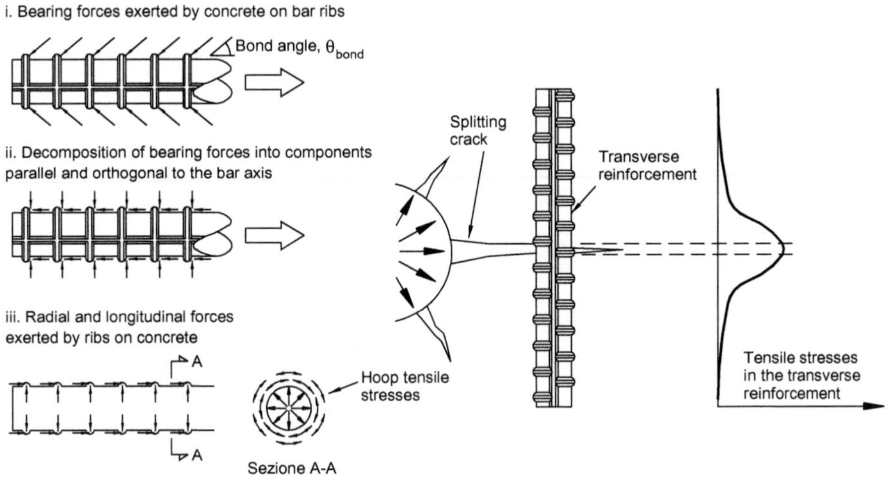

Fig. 12.5 Transfer of bond internal forces

the bar, while the external radius is equal to the radius of the bar increased by the minimum length between the concrete cover and half the clear distance between two adjacent bars. Therefore, denoted by a_h the clear distance between two adjacent bars, the bond failure between steel and concrete can occur with one of the following mechanisms (Fig. 12.6):

1. if $c \geq a_h/2$, a horizontal crack arises at the level of the bar centre of gravity
2. if $c < a_h/2$, the failure is characterized by vertical cracks within the concrete cover followed by the horizontal crack on the plane containing the centre of gravity of bars

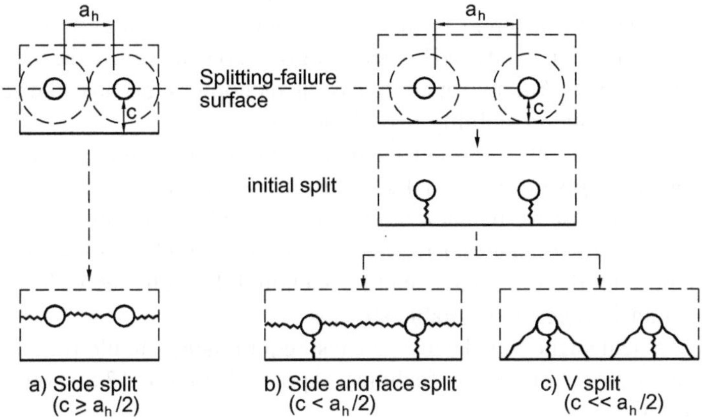

Fig. 12.6 Shape of the bond failure surfaces for single bars

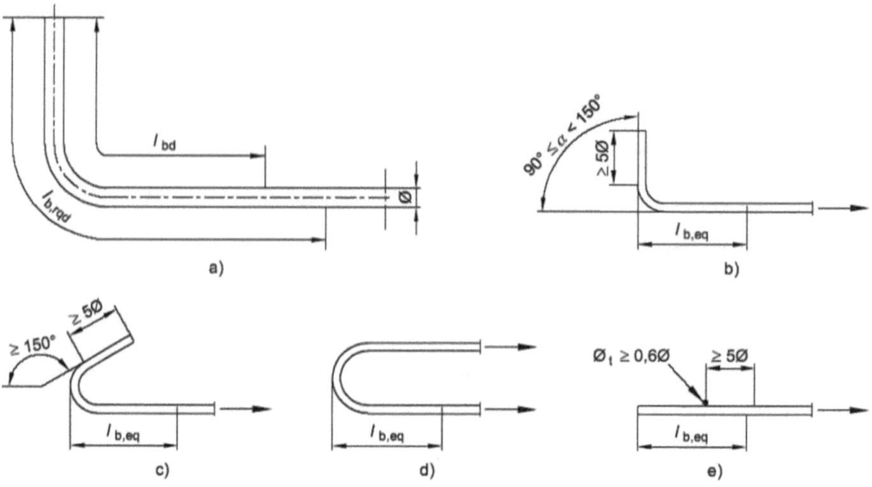

a) Basic tension anchorage length, lb, for any shape measured along the centreline (see § 12.6)
b) Equivalent anchorage length for standard bend
c) Equivalent anchorage length for standard hook
d) Equivalent anchorage length for standard loop
e) Equivalent anchorage length for welded transverse bar

Fig. 12.7 Methods of anchorage other than by a straight bar according to EC2 [Fig. 8.1]

3. if $c << a_h/2$, the failure arises with V-shaped cracks, i.e. with vertical cracks in the concrete cover followed by V-inclined cracks.

Then the transfer mechanism of bond stresses is strongly affected by the value of concrete cover, but it is also depending on other factors like the anchorage shape (straight bars, bent bars) and transverse reinforcement (welded or not to the bar to be anchored); transversal pressures like those transmitted by direct supports of beams. Figure 12.7 shows the different methods of anchorage according to EC2.

It is worth noting that bends and hooks only give a contribution to the bar anchorage for tensile anchorages, but not for compression anchorages, because the eccentricity at the bend can produce a stability loss of the bar.

Moreover, EC2 allows considering neither the contribution to the transfer of compression force given by normal stresses at the end section of the bar, neither the favourable increase of the bar diameter due to the Poisson effect.

In the case of large diameter bars ($\phi > 32$ mm), *splitting* forces and dowel actions are very important, therefore bars shall be anchored through mechanical devices or with straight anchorages and suitable stirrups.

For mechanical anchorage devices, the test requirements should fulfil the relevant product standard or a European Technical Approval. Figure 12.8 shows some on-the-market devices.

For the transfer of the prestress force to concrete, see Chap. 6.

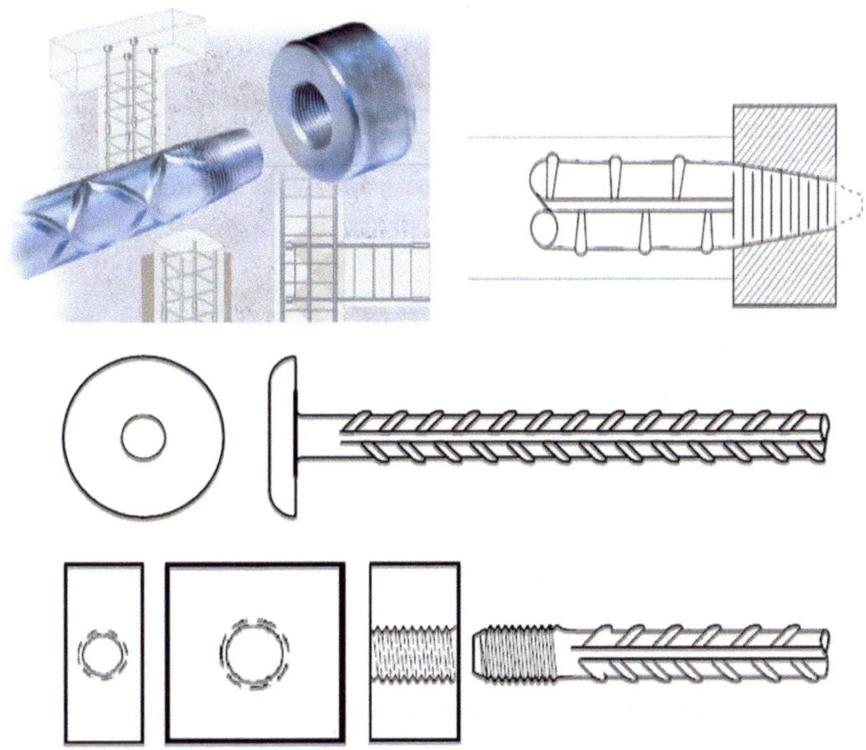

Fig. 12.8 Examples of mechanical anchorage devices (LENTON system by ERICO—http://www.erico.com/ and system DAYTONSUPERIOR—http://www.daytonsuperior.com/)

12.1.4 Ultimate Bond Stress

Anchorage length for steel bars is calculated by adopting a design value of the ultimate bond stress f_{bd}, assumed as uniform along the anchorage length:

$$f_{bd} = 2.25\eta_1\eta_2 f_{ctd} \quad [(8.2)]$$

where:

f_{ctd} is the design value of concrete tensile strength ($f_{ctd} = \alpha_{ct} f_{ctk;0.05}/\gamma_C$); due to the increasing brittleness of higher strength concretes, $f_{ctk;0.05}$ should be limited here to the value for class C60/75 ($f_{ctk;0.05\text{-}C60/75} = 3.1$ N/mm^2), unless it can be verified that the average bond strength increases above this limit;

η_1 is a coefficient related to the quality of the bond condition and the position of the bar during concreting (Fig. 12.9):

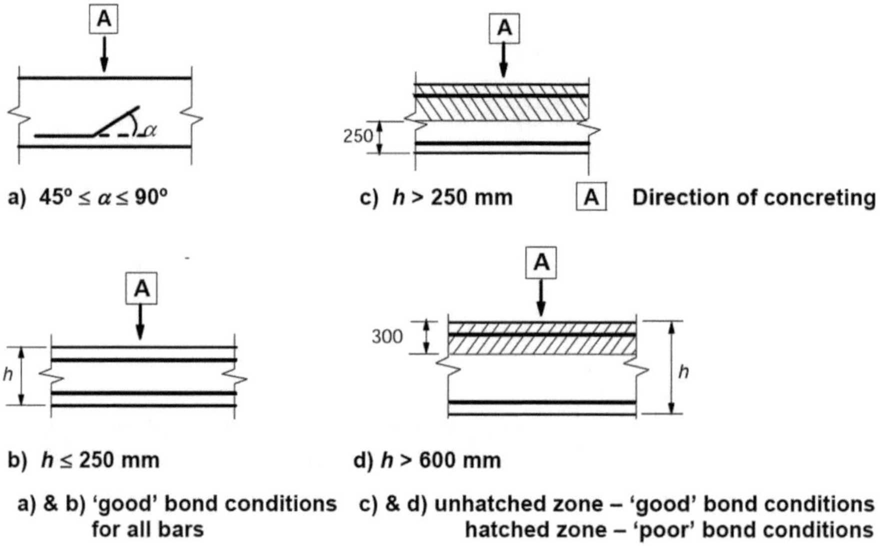

Fig. 12.9 Description of bond conditions: **a** and **b** good bond conditions for all the bars; **c** and **d** poor bond conditions for the bars *in the hatched zone*; good bond conditions for the bars out of the hatched zone [Figure 8.2]

$\eta_1 = 1.0$ when "good" conditions are obtained (for example horizontal bars of cases *a* and *b* of Fig. 12.9, bottom bars of cases *c* and *d* in the same figure and vertical bars),

$\eta_1 = 0.7$ for all other cases and for bars in structural elements built with slip-forms,[1] unless it can be shown that "good" bond conditions exist;

η_2 is related to the bar diameter:

$\eta_2 = 1.0$ for $\phi \leq 32$ mm,

$\eta_2 = (132 - \phi)/100$ for $\phi > 32$ mm.

Good bond conditions are obtained in the following cases:

- vertical bars (for example longitudinal bars of columns and walls),
- horizontal bars with a slope between 45° and 90° to the horizontal (Fig. 12.9a),
- bars with a slope less than 45° to the horizontal line but in the following conditions:
  - in elements with thickness not exceeding 250 mm (Fig. 12.9b),
  - at a distance not exceeding 250 mm from the bottom face of members with a thickness between 250 and 600 mm (Fig. 12.9c),

[1] Structural elements made by extrusion with horizontal sliding formworks, as for example prestressed alveolar slabs, require a consistency almost dried (S1 – S2) and therefore a strong compaction through vibro-compression, compared to concretes with plastic or plastic-fluid consistency (S3 – S4). If compacting operations are not sufficiently accurate and/or suitable additives are not employed, cast can be porous and barely bonded to reinforcement bars.

Table 12.2 Ultimate bond stress for bars with a diameter $\phi \leq 32$ mm[a]

f_{ck} (N/mm^2)	16	20	25	28	30	32	35	40	45	50	55	60
Good bond conditions ($\eta_1 = 1.0$)												
f_{bd} (N/mm^2)	1.95	2.25	2.70	2.90	3.00	3.18	3.30	3.75	4.05	4.35	4.50	4.65
Poor bond conditions ($\eta_1 = 0.7$)												
f_{bd} (N/mm^2)	1.365	1.575	1.89	2.03	2.10	2.22	2.31	2.625	2.835	3.045	3.15	3.255

[a] For bars with diameter $\phi > 32$ mm, the collected values shall be multiplied by $(132 - \phi)/100$

- at more than 300 mm from the top face of members with a thickness exceeding 600 mm (Fig. 12.9d).

Bond conditions for bars in the hatched zones of Fig. 12.9c, d are poor because the concrete in the upper part of the member is less compacted than in the bottom part and because of the phenomenon of *bleeding*. Bleeding is a form of segregation where some of the water in the concrete tends to rise to the surface of the freshly placed material. The phenomenon is due to the inability of the solid components of the concrete to hold all of the mixing water when they settle downwards. In the process of going towards the top, the water can get accumulated below the aggregates and the horizontal reinforcement bars, so once it evaporated some voids remain below the bars and reduce the bond surface between bars and concrete. The combined effect of segregation and water rising reduces the bond strength for the horizontal bars placed in the upper part of the element.

The design value of the ultimate bond stress for bars within lightweight aggregate concrete can be evaluated with the analogous expression already proposed for normal density concretes, by using f_{lctd} instead of f_{ctd}, where

$$f_{lctd} = f_{lctk;0.05}/\gamma_C$$

$f_{lctk,0.05}$ values are collected in [Table 11.3.1] provided by EC2 (Table 12.2).

12.1.5 Anchorage Length

EC2 defines the following anchorage lengths:

- basic required anchorage length ($l_{b,rqd}$),
- design anchorage length (l_{bd}),
- minimum anchorage length ($l_{b,\min}$).

The basic required anchorage length ($l_{b,rqd}$) is the required length for anchoring the force $A_s\sigma_{Ed}$ (A_s: cross-sectional area of the bar, σ_{Ed}: design normal stress of the bar) in a straight bar assuming constant bond stress equal to f_{bd}, being the concrete cover equal to the bar diameter (Fig. 12.10).

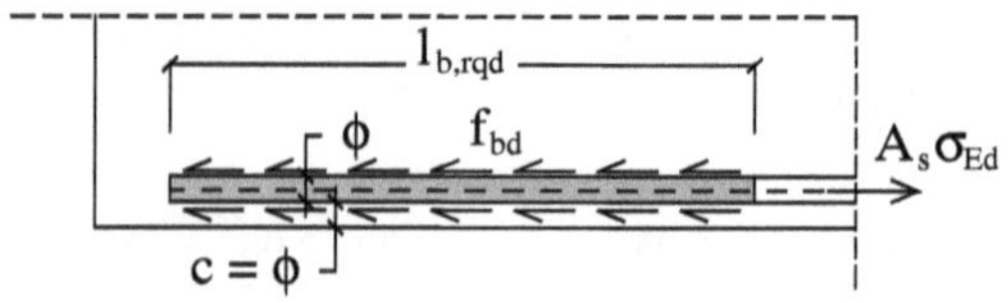

Fig. 12.10 Basic required anchorage length

The design anchorage length (l_{bd}) is the required length for anchoring the design force $A_s\sigma_{Ed}$ considering the favourable effect of:

- bars whose profile is not straight,
- concrete cover higher than the bar diameter,
- presence of confining transverse reinforcement (welded or not to the bar to be anchored),
- presence of transverse pressures (as in the case of direct supports of beams).

The design anchorage length is always less or at least equal to the basic required anchorage length ($l_{bd} \leq l_{b,rqd}$).

The minimum anchorage length $l_{b,\min}$ is the lower limit of the design anchorage length l_{bd}, introduced to guarantee the bar minimum anchorage and for practical/constructive reasons. $L_{b,\min}$ takes into account constructive tolerances and possible translations of bending moments due to horizontal actions not considered during design or support subsidence, etc. EC2 provides the following limits for the minimum anchorage length for anchorages under tensile forces

$$l_{b,\min} \geq \max\,(0.3 l_{b,rqd};\ 10\phi;\ 100\ \text{mm})$$

while for anchorages under compression forces

$$l_{b,\min} \geq \max\,(0.6 l_{b,rqd};\ 10\phi;\ 100\ \text{mm})$$

12.1.5.1 Basic Required Anchorage Length $l_{b,rqd}$

The basic required anchorage length $l_{b,rqd}$ depends on the steel type and bond properties of the bars. It is given by.

$$l_{b,rqd} = \frac{\phi}{4} \cdot \frac{\sigma_{sd}}{f_{bd}} \quad [(8.3)],$$

being σ_{sd} the design normal stress in the cross-section starting from which the anchorage is measured. For example, for a bottom bar with area A_s, anchored at the end section of a slender simply supported beam, the design stress σ_{sd} is given by $\sigma_{sd} = V_{Ed}/A_s$ if the member is without shear reinforcement and $\sigma_{sd} = 0.5\, V_{Ed}$ (cot

Table 12.3 Ratios $l_{b,rqd}/\phi$, 0.6 $l_{b,rqd}/\phi$ and 0.3 $l_{b,rqd}/\phi$ for steel grade B450C (f_{yd} = 450/1.15 = 391 N/mm^2) and bars with diameter $\phi \leq 32$ mm[a]

f_{ck} (N/mm^2)	16	20	25	28	30	32	35	40	45	50	55	60
Good bond conditions (η_1 = 1.0)												
$l_{b,rqd}/\phi$	50.2	43.5	36.3	33.7	32.6	30.8	29.7	26.1	24.2	22.5	21.8	21.1
0.6 $l_{b,rqd}/\phi$	30.1	26.1	21.8	20.2	19.6	18.5	17.8	15.7	14.5	13.5	13.1	12.7
0.3 $l_{b,rqd}/\phi$	15.1	13.1	10.9	10.1	9.8	9.3	8.9	7.9	7.3	6.8	6.6	6.4
Other bond conditions (η_1 = 0.7)												
$l_{b,rqd}/\phi$	71.7	62.1	51.8	48.1	46.6	44	42.4	37.3	34.5	32.2	31.1	30.1
0.6 $l_{b,rqd}/\phi$	15.1	13.1	10.9	28.9	9.8	26.4	8.9	7.9	7.3	6.8	6.6	6.4
0.3 $l_{b,rqd}/\phi$	21.5	18.7	15.6	14.5	14	13.2	12.7	11.2	10.4	9.7	9.4	9.1

[a] For bars with diameter $\phi > 32$ mm, values of $l_{b,rqd}/\phi$ in the table shall be divided by $(132 - \phi)/10$

$\theta - \cot\alpha)/A_s$ if the member is shear-reinforced (see Chap. 8 for the meaning of the symbols; V_{Ed} is the shear force at ULS).

The basic anchorage length $l_{b,rqd}$ and the design length l_{bd} for bent bars shall be measured along the axis of the bar (Fig. 12.7a). Table 12.3 collects the values of the ratio $l_{b,rqd}/\phi$ as a function of the concrete strength under the hypothesis that the axial force in the bar equals its design strength. Moreover, Table 12.3 collects the values 0.6 $l_{b,rqd}$ and 0.3 $l_{b,rqd}$ for the calculation of the minimum anchorage length (see the expression of $l_{b,min}$ at the end of 12.1.5).

For welded double wire[2] mesh, the diameter ϕ to be used in the calculation of $l_{b,rqd}$ shall be substituted by the equivalent diameter $\phi_n = \phi\sqrt{2}$.

12.1.5.2 Design Anchorage Length l_{bd}

The design anchorage length l_{bd} is obtained by modifying the basic required anchorage length $l_{b,rqd}$ to take into account the positive effects of different factors like the concrete cover exceeding the minimum cover, the presence of transverse reinforcement welded or not to the bar to be anchored, the presence of transverse pressures. The design anchorage length l_{bd} is then expressed as:

$$l_{bd} = \alpha_1\alpha_2\alpha_3\alpha_4\alpha_5 l_{b,rqd} \geq l_{b,\min}$$

[2] The word "double wire" indicates two wires made of the same steel and with identical dimensions placed in contact with each other, employed in couple in a welded mesh. A welded mesh made with single bars of identical diameter placed at 90° cannot be advantageously used in all those cases where the required reinforcement in one direction is higher than in the orthogonal direction. Then, it is better to use a mesh with coupled bars ("double wire") in one direction and single bars in the other direction, which should be preferred to a mesh with single bars in both directions with different diameters. A larger diameter of the bars in one direction gives to the mesh a reduced flexibility when mounted on curved surfaces.

where coefficients α_1, α_2, α_3, α_4 and α_5 are here listed together with the effects they account for

α_1 anchorage shape (under the hypothesis of suitable concrete cover),
α_2 concrete cover,
α_3 confinement due to transverse reinforcement,
α_4 influence of one or more welded transverse bars ($\phi_t > 0.6\ \phi$) along the design anchorage length l_{bd} (see also 8.6)
α_5 pressure transverse to the plane of splitting along the design anchorage length l_{bd}.

Coefficients α_i take the following values (Table 12.4):

- $\alpha_1 = 0.7$ or $\alpha_1 = 1.0$
- $0.7 \leq \alpha_2 \leq 1.0$
- $0.7 \leq \alpha_3 \leq 1.0$
- $\alpha_4 = 0.7$ or $\alpha_4 = 1.0$
- $0.7 \leq \alpha_5 \leq 1.0$.

moreover, formula [(8.5)] of EC2 indicates that the product ($\alpha_2\ \alpha_3\ \alpha_5$) cannot be assumed smaller than 0.7.

If some coefficients α_i are equal to one, the design anchorage length l_{bd} can be expressed only considering those coefficients α_i different from one, as indicated in Table 12.5. For example, in the absence of confining transverse reinforcement, welded transverse bars and transverse pressures, as it occurs in many cases, l_{bd} can be expressed as

$$l_{bd} = \alpha_1 \alpha_2 l_{b,rdq}$$

which provides values of l_{bd} between 0.49 l and $l_{b,rqd}$.

A possible simplified design approach consists of adopting a design anchorage length l_{bd} equal to the basic required anchorage length $l_{b,rqd}$, i.e. by assuming all the coefficients α_i equal to 1, if this hypothesis does not lead to excessive values of l_{bd}.

12.1.5.3 Alternative Simplified Expression for the Calculation of the Design Anchorage Length l_{bd} Under Tensile Force

For the sake of simplicity, instead of using the expression $l_{bd} = \alpha_1\ \alpha_2\ \alpha_3\ \alpha_4\ \alpha_5\ l_{b,rqd} \geq l_{b,\min}$, the anchorage under tensile force for some type of anchorages can be calculated as an equivalent anchorage length $l_{b,eq}$ (Fig. 12.7):

$l_{b,eq} = \alpha_1\ l_{b,rqd}$ for bent or looped bars (Fig. 12.7b, c, d),
$l_{b,eq} = \alpha_4\ l_{b,rqd}$ for bars anchored by welded transverse bars (Fig. 12.7e).

Table 12.4 Coefficients $\alpha_1, \alpha_2, \alpha_3, \alpha_4$ and α_5

Influence factors	Anchorage type	Reinforcement bar	
		Under tensile force	Under compression force
Bar shape	Straight	$\alpha_1 = 1, 0$	$\alpha_1 = 1.0$
	Not straight (Fig. 12.7b, c, d)	$\alpha_1 = 0, 7$ se $c_d > 3\phi$ otherwise $\alpha_1 = 1, 0$ (see Fig. 12.11 for values of c_d)	$\alpha_1 = 1.0$
Concrete cover	Straight	$\alpha_2 = 1 - 0.15\ (c_d - \phi)/\ \phi$ $(0.7 \leq \alpha_2 \leq 1.0)$	$\alpha_2 = 1.0$
	Not straight (Fig. 12.7b, c, d)	$\alpha_2 = 1 - 0.15\ (c_d - 3\ \phi)/\ \phi$ $(0.7 \leq \alpha_2 \leq 1.0)$ (see Fig. 12.11 for values of c_d)	$\alpha_2 = 1.0$
Confinement through transverse reinforcement not welded to the main one	All the types	$\alpha_3 = 1 - K\lambda$ $(0.7 \leq \alpha_3 \leq 1.0)$	$\alpha_3 = 1.0$
Confinement through welded transverse reinforcement	All the types, positions and dimensions as specified in Fig. 12.7e	$\alpha_4 = 0.7$	$\alpha_4 = 0.7$
Confinement through transverse pressure	All the types	$\alpha_5 = 1 - 0.04p$ $(0.7 \leq \alpha_5 \leq 1.0)$	–

where:
$\lambda = (\Sigma A_{st} - \Sigma A_{st,min})/A_s$
ΣA_{st} = cross-sectional area of the transverse reinforcement along the anchorage design length l_{bd}
$\Sigma A_{st,min}$ = cross-sectional area of the minimum transverse reinforcement ($\Sigma A_{st,min} = 0.25\ A_s$ for beams and 0 for slabs)
A_s = area of the anchored bar with the maximum diameter
K = values collected in Fig. 12.13
p = transverse pressure (N/mm^2) at ULS along l_{bd}
N.B. See also 12.1.7: for direct supports l_{bd} can also be assumed smaller than $l_{b,min}$ if there is at least one transverse wire welded within the support zone. This wire should be at least 15 mm from the face of the support

12.1.5.4 Variation Intervals for Coefficients α_i for Anchorages in Tension

If c_1 is the horizontal concrete cover, c is the vertical concrete cover and a is the spacing between bars, c_d indicates the following quantity:

$c_d = \min\ (a/2, c_1, c)$ for straight bars (Fig. 12.11a),
$c_d = \min\ (a/2, c_1)$ for bent bars in the vertical plane (Fig. 12.11b),
$c_d = c$ for looped bars in the horizontal plane (Fig. 12.11c).

Table 12.5 Expressions to be employed for the calculation of l_{bd} and maximum and minimum values for l_{bd} in case of straight anchorages ($\alpha_1 = 1$) with or without transverse reinforcement and/or transverse pressures

Transverse reinforcement		Transverse pressures (α_5)	l_{bd} values	
Not welded to the anchored bar (α_3)	Welded to the anchored bar (α_4)		Anchorage under tensile force	Anchorage under compression force
No	No	No	$0.7\, l_{b,rqd} \leq l_{bd} = \alpha_1\, \alpha_2\, l_{b,rdq} \leq l_{b,rqd}$	$l_{b,rqd}$
Yes	No	No	$0.49\, l_{b,rqd} \leq l_{bd} = \alpha_1\, \alpha_2\, \alpha_3\, l_{b,rdq} \leq l_{b,rqd}$	$l_{b,rqd}$
No	Yes	No	$0.49\, l_{b,rqd} \leq l_{bd} = \alpha_1\, \alpha_2\, \alpha_4\, l_{b,rdq} \leq 0.7\, l_{b,rqd}$	$0.7\, l_{b,rqd}$
No	No	Yes	$0.49\, l_{b,rqd} \leq l_{bd} = \alpha_1\, \alpha_2\, \alpha_5\, l_{b,rdq} \leq l_{b,rqd}$	$l_{b,rqd}$
Yes	No	Yes	$0.7\, l_{b,rqd}$ [a] $\leq l_{bd} = \alpha_1\, \alpha_2\, \alpha_3\, \alpha_5\, l_{b,rdq} \leq l_{b,rqd}$	$l_{b,rqd}$
No	Yes	Yes	$0.343\, l_{b,rqd} \leq l_{bd} = \alpha_1\, \alpha_2\, \alpha_4\, \alpha_5\, l_{b,rdq} \leq 0.7\, l_{b,rqd}$	$0.7\, l_{b,rqd}$

[a] l_{bd} cannot be smaller than 0.49 $l_{b,rqd}$, because the product $\alpha_2\, \alpha_3\, \alpha_5$ cannot be smaller than 0.7 (formula [(8.5)])

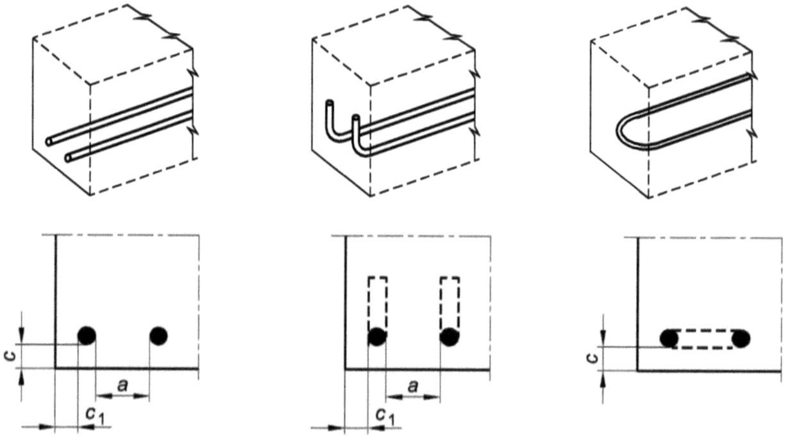

a) $c_d = \min(a/2, c_1, c)$

b) $c_d = \min(a/2, c_1)$

c) $c_d = c$

Fig. 12.11 c_d for beams and columns [Figure 8.3]

The minimum concrete cover for bond is equal to the diameter ϕ of the bar for $d_{max} \leq 32$ mm and to ($\phi + 5$ mm) for $d_{max} > 32$ mm (see Table 3.1), therefore it always turns out that $c_d \geq \phi$.

It is worth noting that, in the evaluation of c_d, the cover measured in the direction orthogonal to the bending plane should be used (Fig. 12.11).

Coefficient α_1 for bars in tension (anchorage shape)

- for straight bars $\alpha_1 = 1$,
- for bent and looped bars $\alpha_1 = 1$ if $c_d \leq 3\ \phi$; $\alpha_1 = 0.7$ if $c_d > 3\ \phi$.

Coefficient α_2 for bars in tension (concrete cover)

- for straight bars α_2 varies linearly between 1 and 0.7 if $\phi \leq c_d \leq 3\ \phi$, while it is constant and equal to 0.7 if $c_d > 3\ \phi$ (Fig. 12.12a)
- for bent and looped bars $\alpha_2 = 1$ if $c_d < 3\ \phi$, α_2 varies linearly between 1 and 0.7 for $3\ \phi \leq c_d \leq 5\ \phi$ and it is equal to 0.7 for $c_d > 5\ \phi$ (Fig. 12.12b).

Coefficient α_3 for bars in tension (transverse reinforcement)

The cross-section and the distance between transverse bars placed within the anchorage length significantly influence the anchorage strength. The anchorage strength increases at increasing the amount of transverse reinforcement, but there is an upper limit beyond which the further increase of transverse reinforcement does not provide any further increase in strength. The coefficient α_3 considers the presence of transverse reinforcement and it takes on the following values:

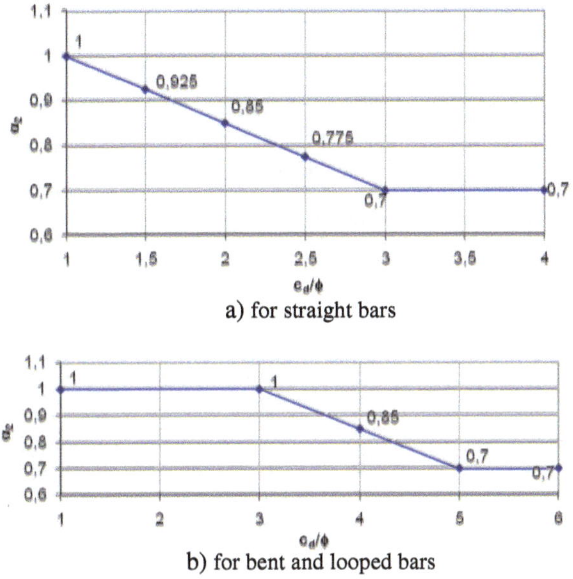

Fig. 12.12 α_2 versus c_d: **a** for straight bars, **b** for bent and looped bars

$\alpha_3 = 1$ if the transverse reinforcement is out of the anchored bar, i.e. on the side of the concrete cover,

$\alpha_3 = 1 - K\lambda$ if the transverse reinforcement is on the internal side, where $\lambda = \frac{\sum A_{st} - \sum A_{st,\min}}{A_s}$ with ΣA_{st} the cross-sectional area of transverse reinforcement within the design anchorage length,

$\Sigma A_{st,\min}$ cross-sectional area of the minimum transverse reinforcement, equal to $0.25A_s$ for beams and zero for slabs,

A_s cross-sectional area of the anchored bar with maximum diameter,

K coefficient accounting for the position of the transverse reinforcement relative to the reinforcement A_s to be anchored; K values are given in Fig. 12.13.

Figure 12.13 shows the relative position of the transverse reinforcement A_{st} and reinforcement A_s to be anchored on the plane of the transverse reinforcement, i.e. on the same plane used in Fig. 12.6 to represent the cracking patterns due to bond failure.

Figure 12.14 shows the variation of α_3 with λ for the first and the second case of Fig. 12.13, both corresponding to a transverse reinforcement A_{st} placed out of the anchored bar. The case of transverse reinforcement placed on the internal side is characterized by a null value of the coefficient K because the transverse bar is placed above the horizontal cracking plane passing through the bar centre of gravity (Fig. 12.15a) and the vertical and inclined cracking planes inside the concrete cover (Fig. 12.15b).

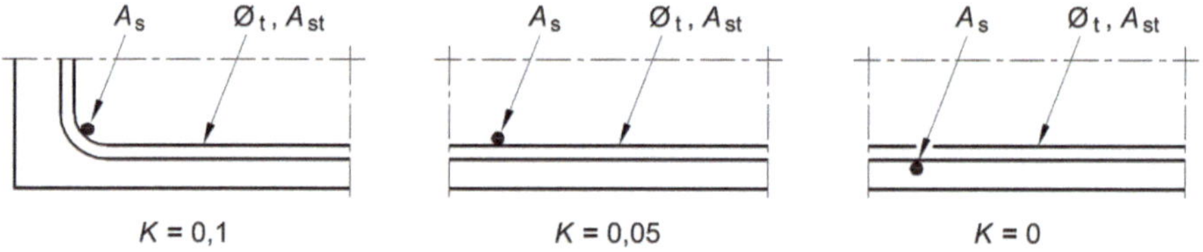

Fig. 12.13 K for beams and slabs [Figure 8.4]

Fig. 12.14 α3 at varying λ

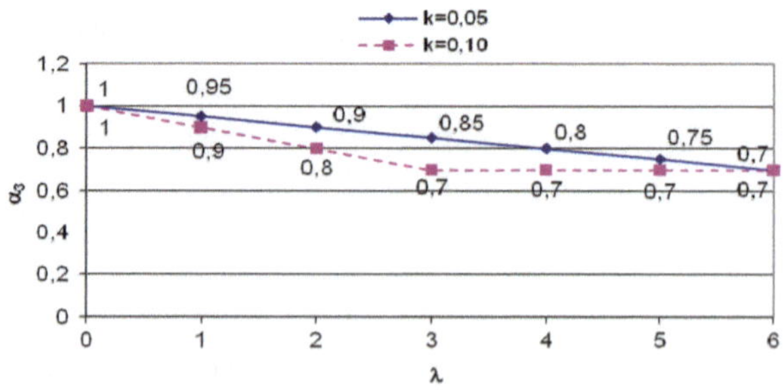

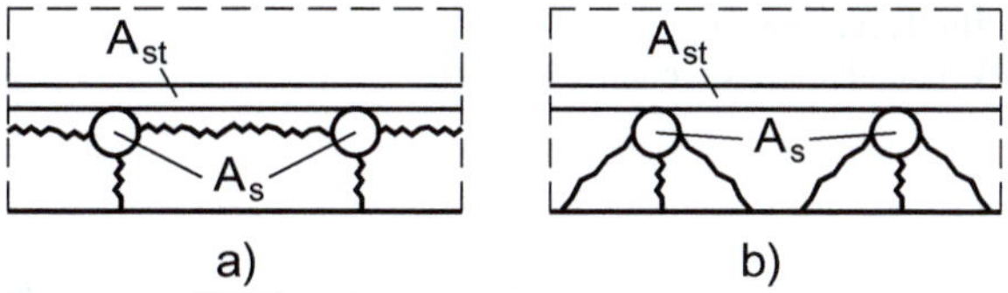

Fig. 12.15 Transverse bar placed within the bars to be anchored

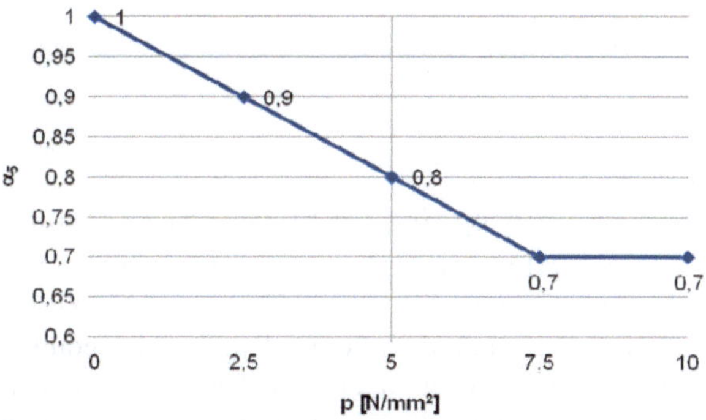

Fig. 12.16 α_5 diagram with varying transverse pressure p

The value of λ is zero for the minimum area of transverse reinforcement, equal to 0.25 A_s for beams and zero for slabs.

Coefficient α_4 for bars in tension (welded transverse reinforcement)

This coefficient accounts for the presence of welded transverse reinforcement: $\alpha_4 = 0.7$ for all the anchorage types, position and dimensions.

Coefficient α_5 for bars in tension (transverse pressure)

This coefficient accounts for a transverse pressure along l_{bd}, linearly varying between 1.0 and 0.7 being the transverse pressure p at ULS within the interval 0 ÷ 7.5 N/mm^2; for $p > 7.5$ N/mm^2 the coefficient α_5 is equal to 0.7 (Fig. 12.16).

12.1.5.5 Example No. 2. Calculation of the Anchorage Length

Calculate the anchorage length in tension for longitudinal bars of the beam shown in Fig. 12.17 *under the hypothesis that all the bars are working at the steel design strength and the concrete cover is* 33 *mm* (*concrete cover for stirrups* $\phi 8$ *is* 25 *mm*). *The steel grade is* B450C *and the concrete class is* C25/30.

Calculation of the basic required anchorage length

$l_{b,rqd}$ can be obtained directly from Table 12.3 (valid for $\phi \leq 32$ mm), under the hypothesis that bottom bars are in good bond conditions ($\eta_1 = 1.0$) and top bars are in poor bond conditions ($\eta_1 = 0.7$), as they are at a distance from the bottom face exceeding 250 mm (case *c* of Fig. 12.9):

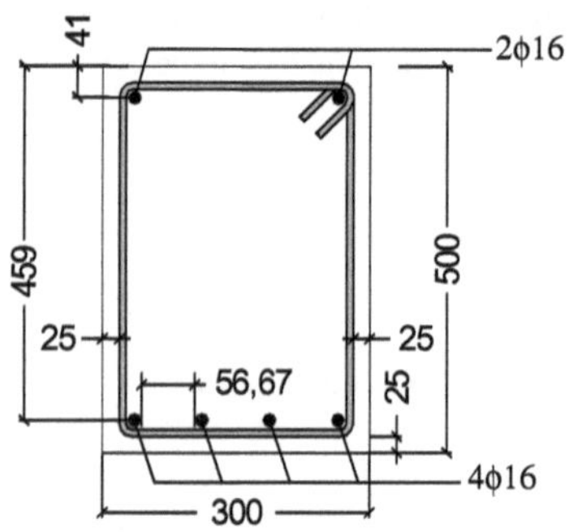

Fig. 12.17 Beam rectangular cross-section

$$\phi16 \text{ top} \quad l_{b,rqd} = 51.8\phi = 51.8 \cdot 16 = 828.8 \cong 829 \text{ mm},$$
$$\phi16 \text{ bottom} \quad l_{b,rqd} = 36.3\phi = 36.3 \cdot 16 = 580.8 \cong 581 \text{ mm}.$$

The basic required anchorage length is the required length to anchor a straight bar with a concrete cover equal to the bar diameter, i.e. equal to the minimum concrete cover to guarantee an adequate bond for $d_{max} \leq 32$ mm.

Calculation of the design anchorage length

As confining transverse reinforcement and transverse pressures are not present, the expression of $l_{\rm bd}$ reduces to:

$$l_{bd} = \alpha_1 \alpha_2 l_{b,rqd}$$

In the following, the anchorage length is calculated for different anchorage types.

Case (a): straight bar

$$c_d = \min(a/2, c_1, c) = \min(56.67/2; 33; 33) = 28.33 \text{ mm}$$
$$\alpha_1 = 1.0 \text{ (for straight anchorage)}$$
$$\alpha_2 = 1 - 0.15(c_d - \phi)/\phi = 1 - 0.15(28.33 - 16)/16 = 0.88$$

the design anchorage length is therefore equal to

$$\phi16 \text{ top} \quad l_{bd} = \alpha_1 \alpha_2 l_{b,rqd} = 1.0 \cdot 0.88 \cdot 829 = 730 \text{ mm}$$
$$\phi16 \text{ bottom} \quad l_{bd} = \alpha_1 \alpha_2 l_{b,rqd} = 1.0 \cdot 0.88 \cdot 581 = 511 \text{ mm}$$

Case (b): 90° bent bar

$$c_d = \min(a/2, c_1) = \min(56.67/2; 33) = 28.33 \text{ mm}$$
$$\alpha_1 = 1.0 \text{ (for not straight anchorage with } c_d = 28.33 \text{ mm } \leq 3\phi = 48 \text{ mm)}$$

$\alpha_2 = 1 - 0.15(c_d - 3\phi)/\phi = 1 - 0.15(28.33 - 3 \cdot 16)/16 = 1.18 > 1$
then $\alpha_2 = 1.0$

the design anchorage length is therefore equal to

$$\phi 16 \text{ top} \quad l_{bd} = \alpha_1 \alpha_2 l_{b,rqd} = 1.0 \cdot 1.0 \cdot 829 = 829 \text{ mm},$$
$$\phi 16 \text{ bottom} \quad l_{bd} = \alpha_1 \alpha_2 l_{b,rqd} = 1.0 \cdot 1.0 \cdot 581 = 581 \text{ mm}.$$

Case (c): 150° bent bar

As for case (b) $c_d = 28.33$ mm; $\alpha_1 = 1.0$; $\alpha_2 = 1.0$; the design anchorage length is equal to case (b).

Case (d): hooked bar

From Fig. 12.11c.
$c_d = c = 33$ mm.
From Table 12.4

$\alpha_1 = 1.0$ (as the anchorage is not straight and $c_d = 33 \text{ mm} \leq 3\phi = 48$ mm)
$\alpha_2 = 1 - 0.15(c_d - 3\phi)/\phi = 1 - 0.15(33 - 3 \cdot 16)/16 = 1.14 > 1$,
then $a_2 = 1.0$

The design anchorage length is therefore equal to

$$\phi 16 \text{ top} \quad l_{bd} = \alpha_1 \alpha_2 l_{b,rqd} = 1.0 \cdot 1.0 \cdot 829 = 829 \text{ mm},$$
$$\phi 16 \text{ bottom} \quad l_{bd} = \alpha_1 \alpha_2 l_{b,rqd} = 1.0 \cdot 1.0 \cdot 581 = 581 \text{ mm}.$$

Case (e): welded transverse bar

In bar welded to the anchored bar, but in absence of transverse pressures, the expression for the calculation of l_{bd} simplifies as follows:

$$l_{bd} = a_1 a_2 a_4 l_{b,rqd}$$

the transverse reinforcement diameter should satisfy the following limitation:

$$\phi_t \geq 0.6\phi = 9.6 \text{ mm}$$

therefore $\phi_t = 10$ mm is adopted.

$\alpha_1 = 1.0$ (for straight anchorage)
$c_d = \min(a/2, c_1) = \min(56.67/2;\ 33) = 28.33$ mm

For $c_d = 28.33$ mm.

$$\alpha_2 = 1 - 0.15\,(c_d - \phi)\,/\phi = 1 - 0.15\,(28.33 - 16)\,/\,16 = 0.88$$
$$\alpha_4 = 0.7$$

The design anchorage length is therefore equal to

$$\phi 16 \text{ top} \quad l_{bd} = \alpha_1 \alpha_2 \alpha_4 l_{b,rqd} = 1.0 \cdot 0.88 \cdot 0.7 \cdot 829 = 511 \text{ mm},$$
$$\phi 16 \text{ bottom} \quad l_{bd} = \alpha_1 \alpha_2 \alpha_4 l_{b,rqd} = 1.0 \cdot 0.88 \cdot 0.7 \cdot 581 = 358 \text{ mm}.$$

12.1.5.6 Example No. 3. Calculation of the Anchorage Length in Tension Using the Equivalent Anchorage Length

Repeat the calculation of Example no. 1 using the equivalent anchorage length.

$l_{b,eq} = \alpha_1\, l_{b,rqd}$ for bent bars or hooked bars (Fig. 12.7b, c, d),
$l_{b,eq} = \alpha_4\, l_{b,rqd}$ for bars anchored by welded transverse bars (Fig. 12.7e),

Case (a) (b) (c) (d)

$$\phi 16 \text{ top} \quad l_{b,eq} = \alpha_1 l_{b,rqd} = 1.0 \cdot 829 = 829 \text{ mm},$$
$$\phi 16 \text{ bottom} \quad l_{b,eq} = \alpha_1 l_{b,rqd} = 1.0 \cdot 581 = 581 \text{ mm}.$$

Case (e)

$$\phi 16 \text{ top} \quad l_{b,eq} = \alpha_4 l_{b,rqd} = 0.7 \cdot 829 = 580 \text{ mm}$$
$$\phi 16 \text{ bottom} \quad l_{b,eq} = \alpha_4 l_{b,rqd} = 0.7 \cdot 581 = 407 \text{ mm}.$$

12.1.6 Anchorage of Links and Shear Reinforcement

Links and shear reinforcement should normally be anchored with bends and hooks, by inserting a longitudinal bar within the bend or the hook (Fig. 12.18a, b), or through welded transverse reinforcement (Fig. 12.18c, d). The angles should be equal to 90° for bends and 135° for hooks.

Table 12.6 collects the values of the anchorage lengths for links with diameters equal to 6, 8, 10, 12, 14 and 16 mm for all four cases shown in Fig. 12.18.

Table 12.6 Minimum anchorage lengths (mm) for links

Anchorage type (Fig. 12.18)		Link diameter (mm)					
		6	8	10	12	14	16
(a)	Hook minimum length	50	50	50	60	70	80
(b)	Hook minimum length	70	80	100	120	140	160
(c)	Welded bars minimum diameter	5	6	7	9	10	12
	minimum spacing between welded bars	20	20	20	24	28	32
(d)	Welded bar minimum diameter	9	12	14	17	20	23
	Minimum concrete cover	50	50	50	50	50	50

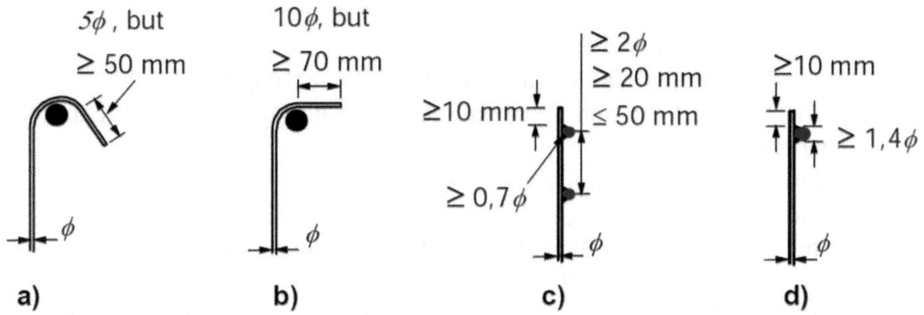

Note: For c) and d) the cover should not be less than either 3ϕ or 50 mm.

Fig. 12.18 Anchorage of links [Figure 8.5]

12.1.7 Anchorage by Welded Transverse Bars

The anchorage of a longitudinal bar can also be obtained using welded transverse bars (Fig. 12.19) bearing on the concrete, by verifying that the quality of welded joints is adequate. The diameter of the transverse bar should be at least 60% of the diameter of the bar to be anchored (see Fig. 12.7e). This type of anchorage has been studied in several experimental campaigns; the expression proposed by EC2 comes from the results of the research performed in Finland.

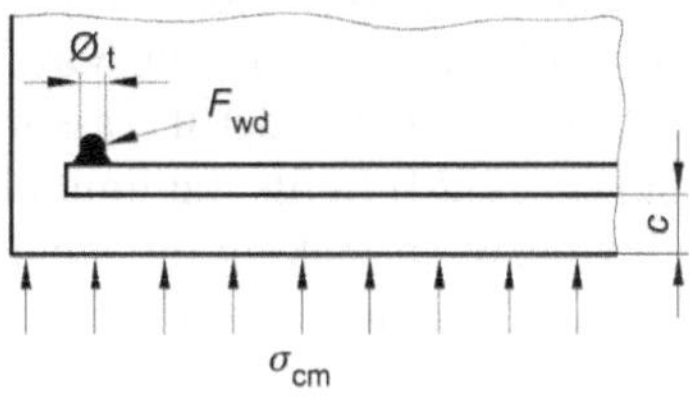

Fig. 12.19 Anchorage through welded transverse bar [Figure 8.6]

12.1.7.1 Welded Transverse Bars with a Diameter Between 14 and 32 mm

F_{btd} is the maximum force that can be transferred through a welded transverse bar, with a diameter between 14 and 32 mm, being the welded joint located on the internal side of the main bar, i.e. on the opposite side of the concrete cover. The force F_{btd} is given by the following expression.

$$F_{btd} = l_{td}\phi_t\sigma_{td} \leq F_{wd} = 0.5A_s f_{yd} \quad [(8.8\text{N})]$$

where

- F_{wd} is the design shear strength of the weld, equal to 50% of the yield strength of the anchored bar
- A_s is the cross-sectional area of the anchored bar
- f_{yd} is the steel design yield strength
- l_{td} is the design length of the transverse bar: $l_{td} = 1.16\ \phi_t\ (f_{yd}/\sigma_{td})^{0.5} \leq l_t$
- l_t is the length of the transverse bar, not exceeding the spacing of bars to be anchored
- ϕ_t is the diameter of the transverse bar
- σ_{td} is the concrete stress: $\sigma_{td} = (f_{ctd} + \sigma_{cm})/y \leq 3 f_{cd}$
- σ_{cm} is the compression in the concrete perpendicular to both bars (mean value, positive for compression)
- y $= 0.015 + 0.14\ e^{(-0.18x)}$, with $x = 2\ (c/\phi_t) + 1$ is a function accounting for the geometry
- c is the concrete cover perpendicular to both bars.

The basic required anchorage length $l_{b,rqd}$ is calculated by adopting the stress value σ_{sd} corresponding to F_{btd}

$$\sigma_{sd} = F_{btd}/A_s$$

where A_s is the cross-sectional area of the anchored longitudinal bar, therefore the following expression holds:

$$l_{b,rqd} = \frac{\phi}{4} \cdot \frac{\sigma_{sd}}{f_{bd}} = \frac{\phi}{4} \cdot \frac{F_{btd}}{A_s \cdot f_{bd}}$$

If two transverse bars with the same diameter are welded on the two opposite sides of the bar to be anchored (Fig. 12.20), the maximum anchorage force F_{btd} can be doubled under the assumption that the cover of the external bar (bar no. 2 in Fig. 12.20) fulfils the limitations on the minimum concrete cover (see Chap. 3). Obviously, in making the welds on the two sides, attention should be paid not to damage the bar in tension.

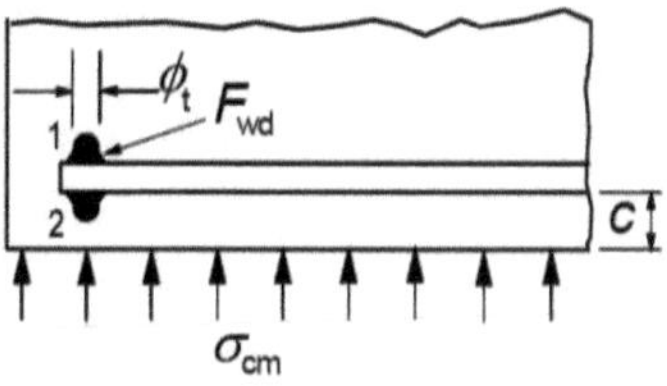

Fig. 12.20 Anchorage through two welded transverse bars

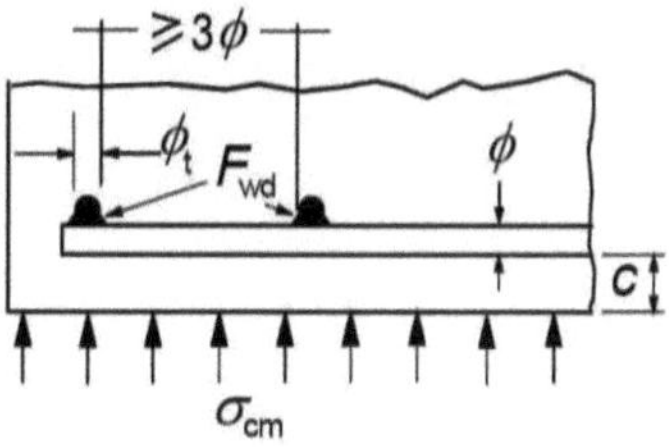

Fig. 12.21 Anchorage with two transverse bars welded on the internal side at a clear distance not smaller than 3ϕ

If two transverse bars are welded on the same side with a minimum spacing of 3ϕ, the force F_{btd} can be multiplied by 1.41 (Fig. 12.21).

12.1.7.2 Nominal bar Diameter of 12 mm and Less

For a nominal bar diameter of 12 mm and less of both the bar to be anchored and the welded transverse bar, the anchorage capacity F_{btd} of a welded transverse bar is equal to the design capacity F_{wd} of the welded joint.

$$F_{btd} = F_{wd} \leq 16 A_s f_{cd} \phi_t / \phi \quad [(8.9)]$$

where

F_{wd} design shear strength of the weld ($F_{wd} = 0.5\, A_s\, f_{yd}$),
ϕ_t nominal diameter of the transverse bar: $\phi_t \leq 12$ mm,
ϕ nominal diameter of the bar to be anchored: $\phi \leq 12$ mm.

If two welded transverse bars with a minimum spacing of ϕ_t are used, the anchorage capacity given by Expression [(8.9)] can be multiplied by 1.41.

12.1.7.3 Example No. 3. Calculation of the Anchorage Capacity for a Welded Transverse Bar

Calculate the anchorage capacity F_{btd} of three 24 mm bars with a 20 mm transverse bar welded on the internal side. The concrete cover is 25 mm*, spacing* s *between bars is* 105 mm *and transverse pressure is null (σ_{cm} = 0). The transverse bar extends of* *s/2 beyond the axis of each lateral bar. The steel grade is B450C, concrete class is C28/35.*

Concrete mechanical characteristics *C28/35*

$$f_{cd} = 0.5 \cdot 28/1.5 = 15.87 \text{ N/mm}^2$$
$$f_{ctm} = 0.30 \cdot f_{ck}^{2/3} = 2.77\,\text{N/mm}^2$$
$$f_{ctd} = 0.7 f_{ctm}/1.5 = 1.29 \text{ N/mm}^2$$

Design steel strength

$$f_{yd} = 450\,/1.15 = 391 \text{ N/mm}^2$$

Being

$$x = 2(c/\phi_t) + 1 = 3.5$$
$$y = 0.015 + 0.14\,\mathrm{e}^{(-0.18x)} = 0.0896$$
$$\sigma_{td} = (f_{ctd} + \sigma_{cm})\,/y = (1.29 + 0)\,/\,0.0896 = 14.4 \le 3 f_{cd} = 47.61 \text{ N/mm}^2$$
$$l_{td} = \min\,[1.16\phi_t\left(f_{yd}/\sigma_{td}\right)^{0,5};\, s] = \min\,[1.16 \times 20 \times (391/14.4)^{0.5};\, 105 \text{ mm}]$$
$$= \min\,[121 \text{ mm};\, 105 \text{ mm}] = 105 \text{ mm}$$

by substituting in the expression of F_{btd}, the following value is obtained for F_{btd}

$$F_{btd} = l_{td}\phi_t\sigma_{td} = 105 \cdot 20 \cdot 14.4 = 30{,}240 \text{ N}.$$

The design shear strength of the weld is given by

$$F_{wd} = 0.5 A_s f_{yd} = 88{,}366 N > F_{btd}.$$

12.1.8 Laps and Mechanical Couplers

Forces can be transmitted from one bar to another by:

- the lapping of bars, with or without bends or hooks,
- welding,
- mechanical devices.

12.1.8.1 Laps

The detailing of laps between bars shall be such that

- the transmission of forces from one bar to the next is assured
- spalling of the concrete in the neighbourhood of the joints does not occur
- large cracks which affect the performance of the structure do not occur.

Laps follow analogous rules to anchorages: the co-axial concrete cylinders of the two bars cross each other and originate a cylinder with an oval cross-section, but failure mechanisms remain similar to those indicated in Fig. 12.6 for the anchorage of a single bar.

The transfer of the tensile force F_s between two overlapping bars can be described through a strut-and-tie mechanism with struts inclined by 45° (Fig. 12.22). The equilibrium is fulfilled only if the concrete can absorb tensile forces orthogonal to the bars, with a resultant force equal to F_s, otherwise, a suitable transverse reinforcement should be provided.

Due to uncertainties on the effective inclination of the concrete struts between the two bars and the non-uniform distribution of bond stresses, rules given by EC2 on the design of laps come from experimental observations. Tests show that, by staggering the lap of two adjacent bars of a minimum length, the lap length is the same as the anchorage length, under the assumption of equal bar diameter, concrete cover, and concrete strength class.

Otherwise, if the lap has not a minimum staggering, or, in an equivalent manner, when in the same cross-section there is a high percentage of lapping bars, a

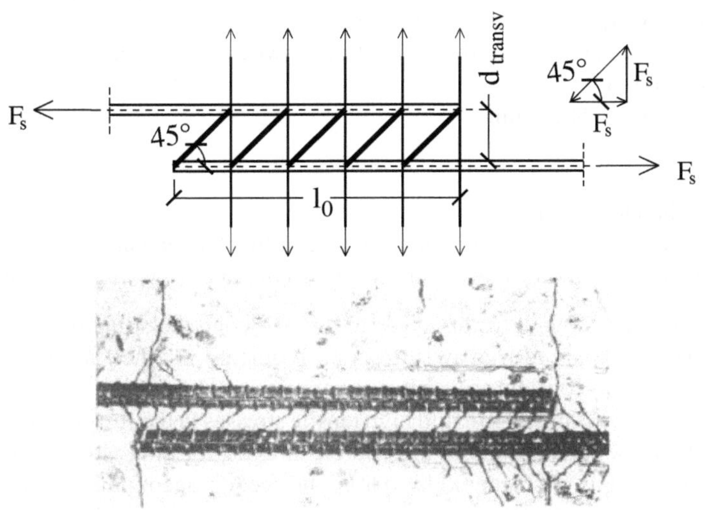

Fig. 12.22 Scheme of the transfer of the tensile force between two overlapping bars (above) and corresponding crack pattern (below) [Model Code 1990]

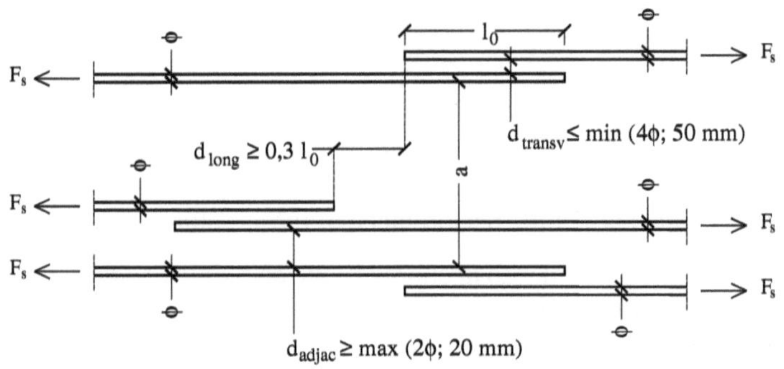

Fig. 12.23 Minimum distances between two lapping bars (d_{trasv}) and between adjacent bars of two different laps (d_{adjac}) (it is worth noting that distances a and d_{adjac} have a different meaning because a is the distance between two laps arranged in the same cross-section, while d_{adjac} is the distance between two adjacent staggered laps) [Figure 8.7]

greater lapping length should be adopted and, in some cases, a suitable transverse reinforcement could be required.

Under these assumptions, EC2 suggests staggering the laps and not placing them in areas under heavy loads, for example in areas where plastic hinges are possible. Moreover, if possible, laps shall be arranged symmetrically.

However, bars in compression and secondary (distribution) reinforcement may be lapped in one section.

In general, the lapping of bars in tension shall fulfil the following rules (Fig. 12.23):

- $d_{transv} \leq \min(4\ \phi;\ 50\ \text{mm})$ or, if $d_{transv} > \min(4\ \phi;\ 50\ \text{mm})$, lap length l_0 shall be increased by $d_{transv} - \min(4\ \phi;\ 50\ \text{mm})$,
- distance d_{long} in the longitudinal direction between two adjacent laps shall at least be equal to $0.3\ l_0$,
- $d_{adjac} \geq \max(2\ \phi;\ 20\ \text{mm})$,

where:

l_0	is the lap length,
d_{trasv}	is the transversal clear distance between two lapped bars,
d_{adjac}	is the clear distance between two adjacent bars of two adjacent laps.

The first rule accounts for an angle of 45° for concrete struts between the two lapped bars: if bars are moved away from each other, it is necessary to increase the lap length, keeping constant the strut inclination.

The second rule is useful to avoid high tensile stresses in the surrounding concrete; Fig. 12.24 shows the distribution of concrete stresses for three different couples of laps: the first couple is staggered by $0.5\ l_0$, the second one by l_0, and the third one by at least $1.3\ l_0$.

The minimum staggering of $1.3\ l_0$ proposed by EC2 allows avoiding a high concentration of stresses in the concrete, differently from the other two cases.

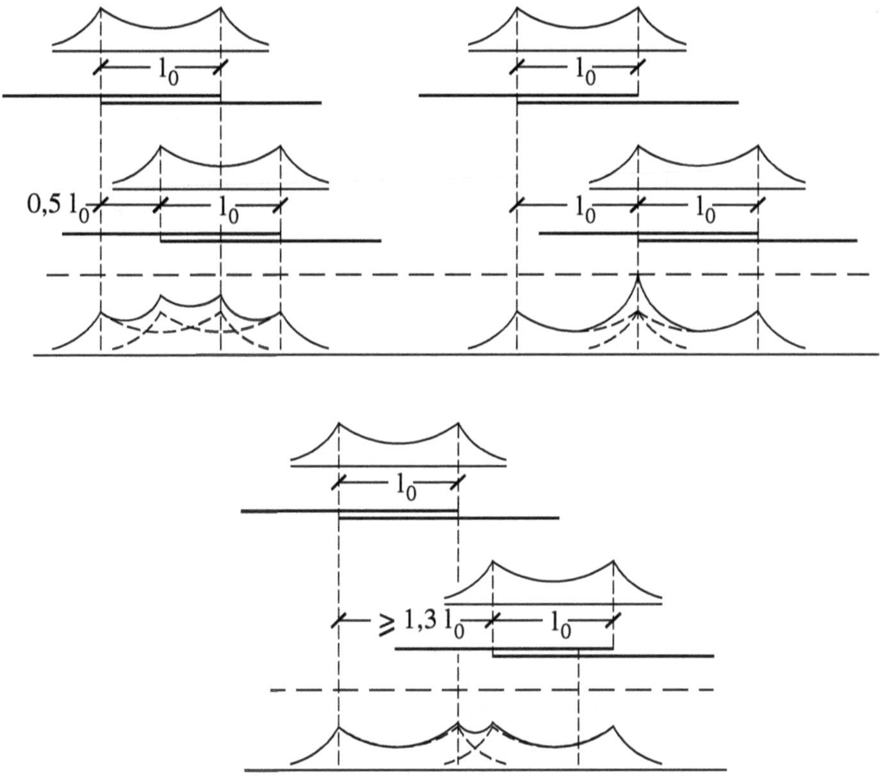

Fig. 12.24 Distribution of concrete tensile stresses at varying the distance of adjacent laps

If geometrical rules shown in Fig. 12.25 are followed, in the same cross-section it is possible to overlap the following percentages of bars in tension:

- 100% for bars arranged in one layer,
- 50% for bars arranged in more than one layer.

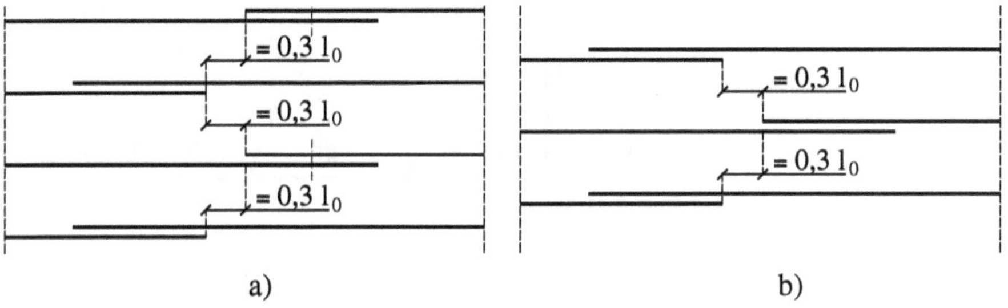

Fig. 12.25 Examples of laps fulfilling the rule on the minimum longitudinal distance d_{long} between adjacent laps: **a** four bars; **b** three bars with a symmetrical arrangement

EC2 does not give any indication for bars arranged in more than one layer; nevertheless, for laps of adjacent layers, the minimum distance of laps in the same layer could be used, i.e. not less than 0.3 l_0.

The limit of 50% of the spliced bars in the same cross-section shall not be applied to the two layers (top and bottom) of a slender beam, where only one of the two layers is in tension. It should but be applied to the bars in tension for a beam or a slab, if arranged on more than one layer, or to the top and bottom layer if they are both in tension due to a normal tensile force.

12.1.8.2 Calculation of the Lap Length

The lap length may be expressed as a function of the basic required anchorage length, through the same coefficients introduced in 12.1.5.2 for the calculation of l_{bd} (excluding the coefficient α_4) and adding a further coefficient (α_6), accounting for the ratio of lapping bars in the same cross-section:

$$l_0 = (\alpha_1 \alpha_2 \alpha_3 \alpha_5)\alpha_6 l_{b,rqd} \geq l_{0,\min}$$

Coefficients α_1, α_2, α_3 and α_5 are collected in Table 12.4 with the following remarks:

- for the calculation of α_3, it is assumed $\Sigma A_{st,\min} = A_s\ (\sigma_{sd}/f_{yd})$, where A_s is the area of the single lapped bar, because, as shown in Fig. 12.26, the same horizontal leg of the stirrup is enough to absorb tensile forces induced by more lapped bars arranged in the same horizontal layer; σ_{sd} is the normal stress in the bar;
- $\alpha_6 = (\rho_1/25)^{0.5}$ with $1.0 \leq \alpha_6 \leq 1.5$, being ρ_1 the geometrical ratio of lapped reinforcement arranged within the area A with length 1.30 l_0 (Fig. 12.27) and centred on the axis of the considered lap; α_6 values are collected in Table 12.7.

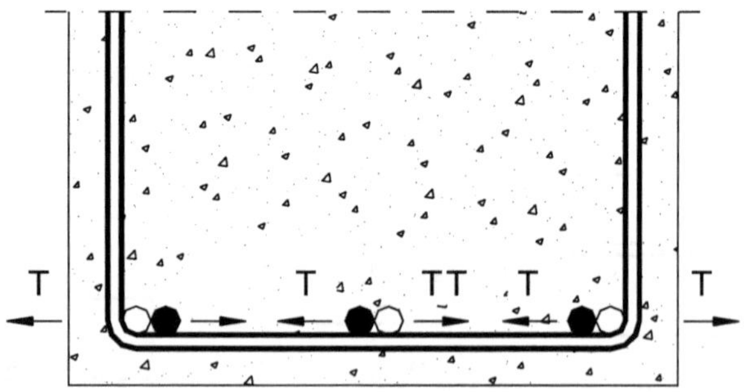

Fig. 12.26 The bottom horizontal leg of the stirrup carries on all the transverse forces generated by lapped bars arranged in the same layer

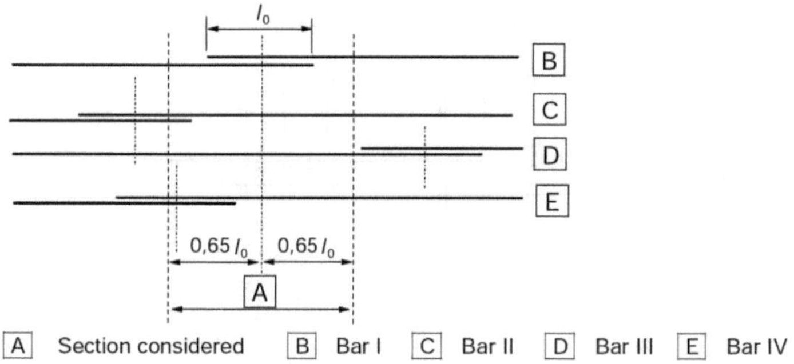

Fig. 12.27 Percentage of spliced bars in one lapped section (Example: Bars C and D are outside the section being considered: % = 50 and $\alpha_6 = 1.4$)

Table 12.7 Values of the coefficient α_6

Percentage of lapped bars relative to the total cross-section area[a]	<25%	33%	50%	>50%
α_6	1.0	1.15	1.4	1.5

[a] Intermediate values may be determined by interpolation

12.1.8.3 Transverse Reinforcement in the Lap Zone for Bars in Tension

Where the diameter ϕ of the lapped bars is less than 20 mm, or the percentage of lapped bars in any section is less than 25%, then any transverse reinforcement or links necessary for other reasons may be assumed sufficient for the transverse tensile forces without further justification.

Where the diameter ϕ of lapped bars is exceeding or is equal to 20 mm, the transverse reinforcement should have a total area ΣA_{st} (sum of all legs parallel to the layer of the spliced reinforcement) of not less than the area A_s of one lapped bar ($\Sigma A_{st} \geq 1.0\, A_s$). The transverse bars should be placed perpendicular to the direction of the lapped reinforcement, as normally it occurs for horizontal stirrup legs of beams.

If more than 50% of the reinforcement is lapped at one point and the distance a between adjacent laps is $\leq 10\phi$ (see Figure 8.7), transverse reinforcement should be formed by links or U bars anchored inside the body of the section.

EC2 does not give specific rules if each longitudinal bar shall be linked or not by a stirrup or U bar or if it is sufficient that it is arranged at a distance not exceeding 150 mm from a bar linked by a stirrup, as required for transverse reinforcement of columns. During the design phase, the problem can be solved by limiting to 50% the percentage of lapped bars in the same cross-section.

When required, the transverse reinforcement should be arranged at the outer sections of the lap as shown in Fig. 12.28. This arrangement can be used for beams, through stirrups, but not for laps of bars in the external layers of slabs and walls. Here the problem can be solved by staggering bars (Fig. 12.29) in a way that in the

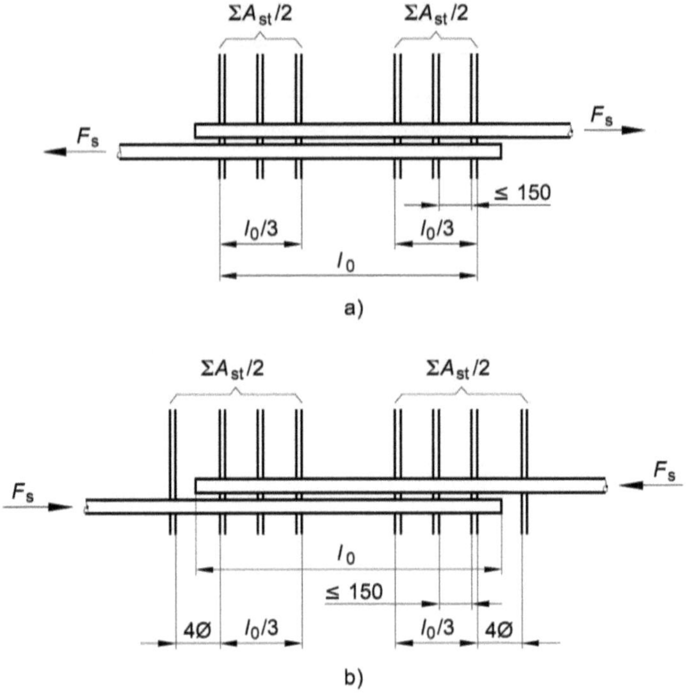

Fig. 12.28 Transverse reinforcement for lapped splices: **a** bars in tension, **b** bars in compression [Figure 8.9]

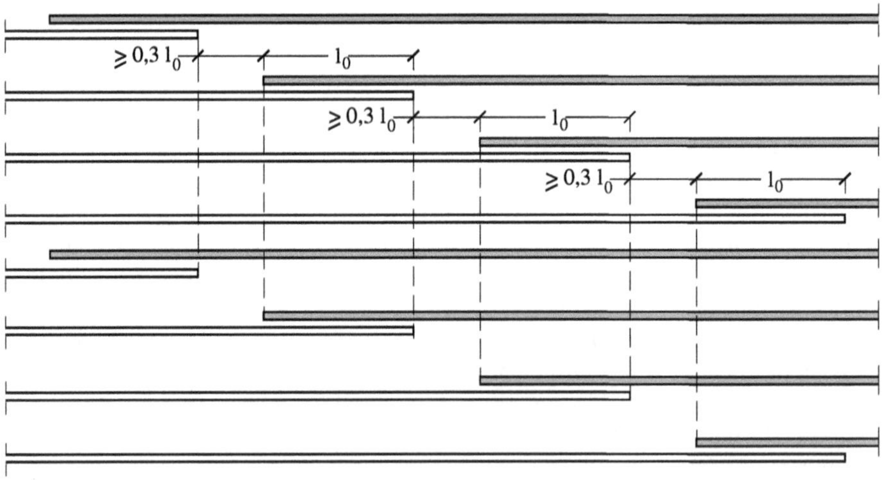

Fig. 12.29 Staggering of bars for a slab to limit to 25% the percentage of lapped bars in the same cross-section

same cross-section the percentage of lapped bars is not exceeding 25% and therefore transverse reinforcement is not required.

12.1.8.4 Transverse Reinforcement in the Lap Zone for Bars Permanently in Compression

The same rules for laps of bars in tension apply to laps of bars in compression. Besides, one bar of the transverse reinforcement should be placed outside each end of the lap length and within 4ϕ of the ends of the lap length (Fig. 12.28b).

12.1.8.5 Additional Rules for Large Diameter Bars

Crack control

For bars with $\phi > 32$ mm, the crack control may be achieved either by using surface reinforcement or by calculation of the crack width. Sometimes, as for deep foundations, the first solution is not possible, therefore the problem shall be solved by calculation.

Surface reinforcement can be designed by adopting the rules of Annex J (which is informative). In every case, its area should not be less than 0.01 $A_{ct,\mathrm{ext}}$ in the direction orthogonal to the large diameter bars and less than 0.02 $A_{ct,\mathrm{ext}}$ in the direction parallel to the same bars, where $A_{ct,\mathrm{ext}}$ is the concrete area in tension at ULS, arranged out of the links (dashed area in Fig. 12.30a).

Surface reinforcement can be made of welded mesh or small diameter bars and placed out of the links with longitudinal and vertical spacings not exceeding 150 mm.

Anchorage

Due to the high *splitting* forces, it is recommended to anchor large diameter bars by mechanical couplers (Fig. 12.8); alternatively, they can be anchored as straight bars if a suitable number of stirrups is added to those required for shear (Fig. 12.28).

For straight anchorage lengths (Fig. 12.31) following rules apply:

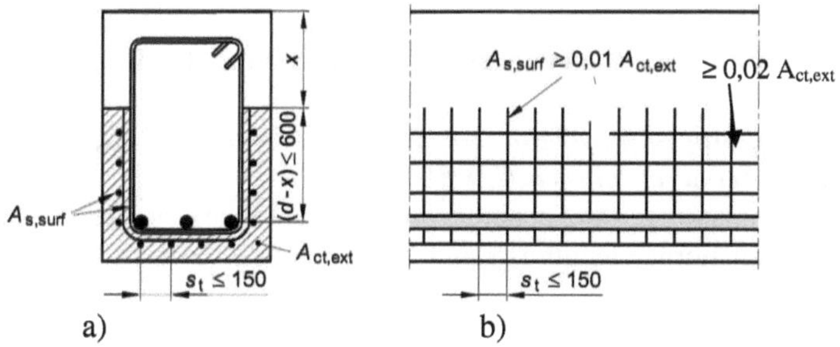

Fig. 12.30 Example of surface reinforcement [Figure J.1]

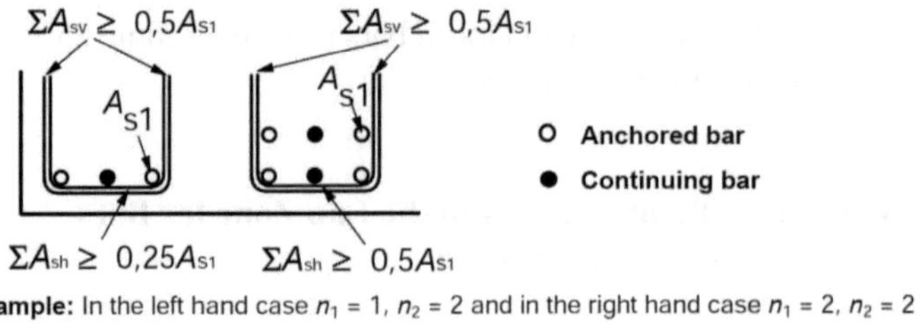

Fig. 12.31 Additional reinforcement in the anchorage zone of large diameter longitudinal bars without transverse compression [Figure 8.11]

- in the direction parallel to the tension face (horizontal legs in Fig. 12.31)

$$A_{sh} \geq 0.25 A_s n_1 \quad [(8.12)]$$

- in the direction perpendicular to the tension face (vertical legs in Fig. 12.31)

$$A_{sv} \geq 0.25 A_s n_2 \quad [(8.13)]$$

where

A_{sh} is the area of the additional reinforcement in the horizontal direction,
A_{sv} is the area of the transverse additional reinforcement in the vertical direction,
A_s is the cross-sectional area of an anchored bar,
n_1 is the number of layers with bars anchored in the same cross-section of the member,
n_2 is the number of bars anchored in each layer.

The additional transverse reinforcement should be uniformly distributed in the anchorage zone and the spacing of bars should not exceed 5 times the diameter of the longitudinal reinforcement.

However, no rules are given on the required number of legs as a function of the number of longitudinal bars; to this purpose, it is possible to refer to Model Code 1990 (Sect. 9.1.4), which, for each reinforcement layer, indicates a maximum of three bars between two vertical legs.

12.1.8.6 Example No. 4. Anchorage of a Large Diameter Bar

Design the anchorage of two external 36 mm bars made of steel grade B450C (Fig. 12.32)*; the concrete class is* C28/35.

For straight anchorages and concrete cover equal to the diameter of the anchored bar, the design anchorage length is equal to the required basic anchorage length

$$l_{bd} = l_{b,rqd} = \frac{\phi}{4} \cdot \frac{f_{yd}}{f_{bd}} = \frac{36}{4} \cdot \frac{391.3}{2.90} = 1214 \text{ mm}$$

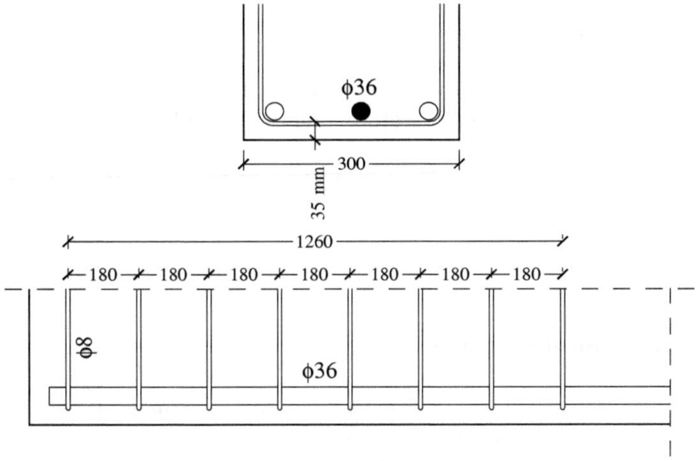

Fig. 12.32 Anchorage of 36 mm bars (Example no. 4)

where $f_{\text{yd}} = 450/1.15 = 391$ N/mm^2 for steel grade B450C and $f_{\text{bd}} = 2.9$ N/mm^2 for concrete class C28/35 (Table 12.2),

The transverse reinforcement shall fulfil the following limitations

- in the direction parallel to the tension face, i.e. in the horizontal direction

$$A_{sh} = 0.25 A_s n_1 = 0.25 \cdot 1018 \cdot 1 = 254.5 \text{ mm}^2$$

- in the direction orthogonal to the tension face, i.e. in the vertical direction

$$A_{sv} = 0.25 A_s n_2 = 0.25 \cdot 1018 \cdot 2 = 509 \text{ mm}^2$$

being

A_s = 1018 mm^2 the area of each 36 mm bar,
n_1 = 1 the number of layers with bars anchored in the same cross-section,
n_2 = 2 the number of bars anchored in each layer.

If closed links are adopted as transverse reinforcement, their cross-sectional area A'_{st} shall satisfy both the following limitations:

$A'_{st} \geq A_{sh}$ (because in the horizontal direction only the bottom leg is interested)
$2A'_{st} \geq A_{sv}$ (because in the vertical direction both the vertical legs are interested)

then $A'_{st} \geq 254.5$ mm^2.

Additional links shall be uniformly distributed in the anchorage zone; if 8 mm links ($A'_{st} = 50$ mm^2) are used, the number of links should not be less than $n =$

254/50 = 5.09 is required, which leads to the following maximum spacing: $p \leq l_{bd}/n$ = 238.5 mm.

Moreover, spacing shall not exceed 5 times the diameter of the longitudinal reinforcement ($p \leq 5\phi = 180$ mm), therefore, eight links with a diameter of 8 mm and spacing of 180 mm are adopted.

In general, large diameter bars shall not be lapped. Laps are allowed only in cross-sections with a minimum dimension of 1.0 m. In all the other cases, mechanical couplers are required.

12.1.8.7 Mechanical Couplers

Joint mechanical devices of steel bars use mainly two joint methods, which guarantee a strength not lower than the strength of bars to couple.

The first method requires bars with threaded ends, directly coupled with each other through a male/female system (Fig. 12.33a), or it needs a cylindrical or truncated-conical threaded sleeve (Fig. 12.33b). In the first case, a mechanical device must be provided inside the sleeve to avoid the screwing of the first bar for a too-long segment, leaving to the second bar a connection length smaller than the design one. The risk is avoided if the truncated-conical geometry is adopted, which has also the advantage of facilitating the alignment operations of the bars with the sleeve. However, the second solution requires careful control of the bar tightening, because the truncated-conical shape allows to easily insert the bars inside the sleeve for a significant length, also without screwing them. Therefore, if the screwing is not performed immediately after the insertion of bars in the sleeve, the workmen could forget to perform it later and its omission could not be identified in further control.

The second method creates the joint through a sleeve closed on the bars by screw tightening (Fig. 12.33c) or with a pressure system requiring the use of a hydraulic press. The tightened sleeve with screws has the advantage of not requiring any preparation of the end-sections of the bar and therefore it can be used also for the connection of already installed bars (like in existing structures). Existing on the market systems use screws whose head is sheared off by the electric shear wrench at reaching the correct axial force in the screw.

12.1.8.8 Typical Values of Anchorage Lengths and Lap Lengths for Slabs, Beams, and Columns

In Tables 12.8, 12.9 and 12.10, typical values of anchorage lengths and lap lengths for slabs, beams and columns are collected. The tables are built under the following assumptions:

- steel grade B450C ($f_{yd} = 391$ N/mm^2)
- force to be anchored equal to $A_s f_{yd}$, where A_s is the cross-sectional area of the bar

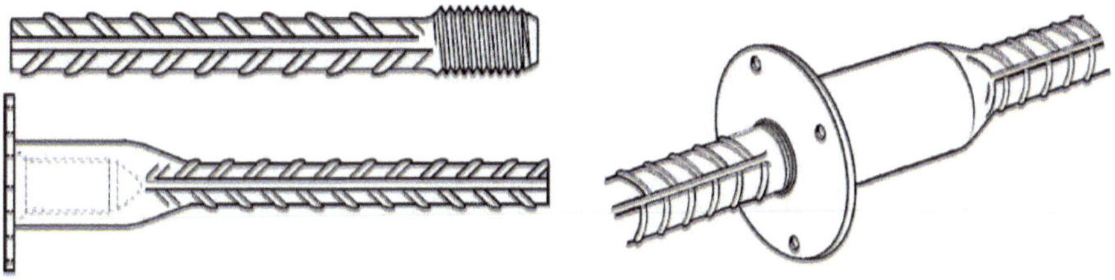

a) male/female threaded joint

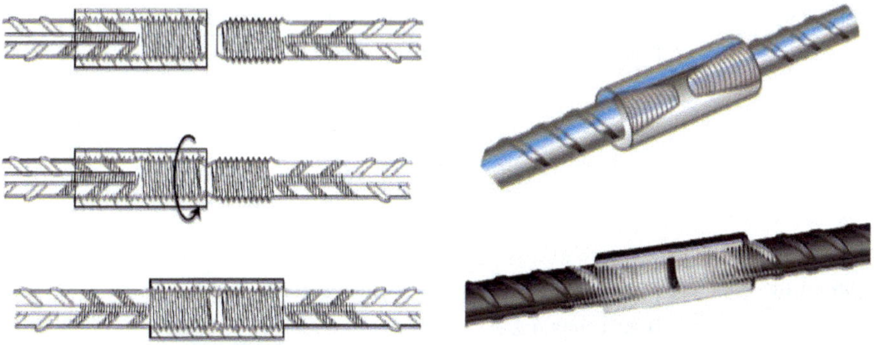

b) threaded joint with cylindrical sleeve (on the left) or truncated-conical (on the right)

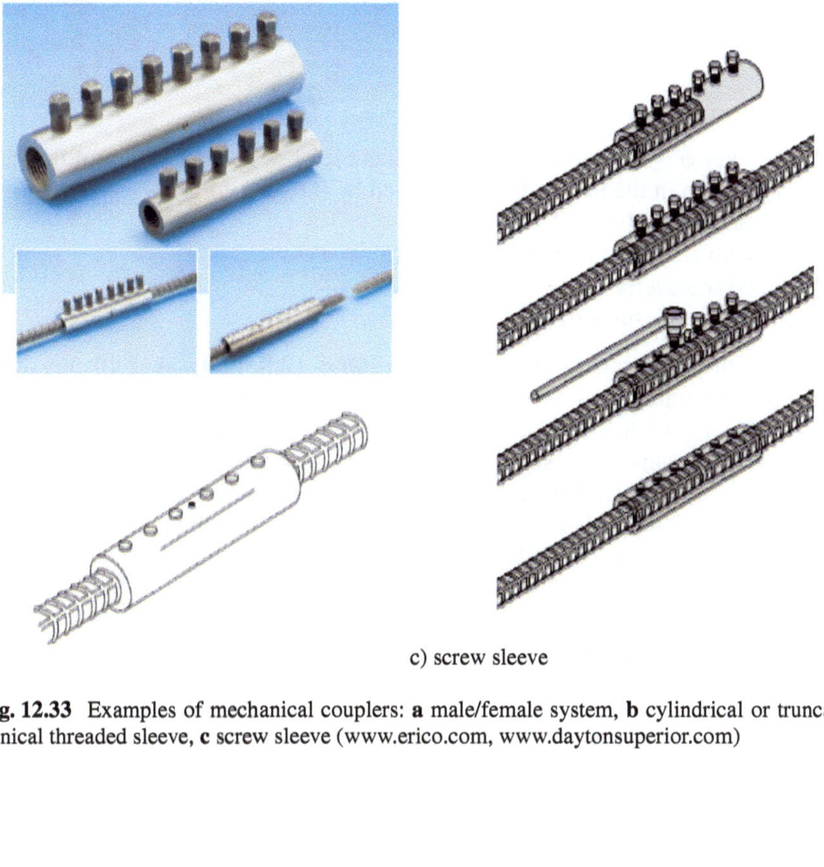

c) screw sleeve

Fig. 12.33 Examples of mechanical couplers: **a** male/female system, **b** cylindrical or truncated-conical threaded sleeve, **c** screw sleeve (www.erico.com, www.daytonsuperior.com)

Table 12.8 Slabs: anchorage length l_{bd} and lap length l_0 for bars with diameter ≤ 32 mm

	Bond conditions	C25/30	C28/35	C30/37	C32/40
l_{bd}/ϕ	Good	36	34	32	31
	Poor	52	48	46	44
$l_0/\phi = 1.15\ l_{bd}/\phi$	Good	42	39	37	35
	Poor	60	55	53	51

Table 12.9 Beams: anchorage length l_{bd} and lap length l_0 for bars with diameter ≤ 32 mm

	Bond conditions	C25/30	C28/35	C30/37	C32/40
l_{bd}/ϕ	Good	33	30	29	28
	Poor	47	43	41	40
$l_0/\phi = 1.15\ l_{bd}/\phi$	Good	38	35	33	32
	Poor	54	50	48	46

Table 12.10 Columns: anchorage length l_{bd} and lap length l_0 for bars with diameter ≤ 32 mm

	Bond conditions	C25/30	C28/35	C30/37	C32/40
l_{bd}/ϕ	Good	33	30	29	28
	Poor	47	43	41	40
$l_0/\phi = 1.15\ l_{bd}/\phi$	Good	49	45	43	42
	Poor	70	65	62	59

- bar diameter $\phi \leq 32$ mm (for bars with diameter $\phi > 32$ mm, values of the ratio ($l_{b,rqd}/\phi$) given in the tables shall be divided by [(132 – ϕ)/100]
- straight anchorage ($\alpha_1 = 1$)
- concrete cover equal to the bar diameter ($\alpha_2 = 1$)
- no welded transverse bars ($\alpha_4 = 1$)
- no transverse pressures ($\alpha_5 = 1$).

Values of the anchorage length collected in Tables 12.8, 12.9 and 12.10 only differ because of the eventual presence of transverse reinforcement, which is accounted for by α_3 coefficient, while the lap lengths are different depending on the percentage of lapped bars in the same cross-section, which is accounted for through α_6 coefficient.

Lap lengths shall be at least equal to the minimum lap length $l_{0,\min} \geq \max$ (0.3 $\alpha_6\ l_{b,rqd}$; 15 ϕ; 200 mm), therefore listed values shall be compared with $l_{0,\min}$ and, if lower, it is necessary to assume $l_{0,\min}$ as anchorage length.

Slabs

- no transverse reinforcement ($\alpha_3 = 1.0$)
- percentage of lapped bars in the same cross-section not exceeding 33% ($\alpha_6 =$ 1.15).

Beams

- confining action of stirrups ($\alpha_3 = 0.9$)
- percentage of lapped bars in the same cross-section not exceeding 33% ($\alpha_6 = 1.15$).

Columns

- confining action of stirrups ($\alpha_3 = 0.9$)
- percentage of lapped bars in the same cross-section exceeding 50% ($\alpha_6 = 1.5$).

12.1.9 Bundles of Bars

Reinforcement bars can also be arranged as bundles if they have the same characteristics (type and grade). In general, bundles will be formed by bars with the same diameter, but it is also possible to bundle bars of different sizes, provided that the ratio of diameters does not exceed 1.7 (see Table 12.11).

In design, the bundle is replaced by a notional bar having the same sectional area and the same centre of gravity as the bundle. The equivalent diameter ϕ_n of the notional bar made of n_b bars with the same diameter is given by

$$\phi_n = \phi\sqrt{n_b} \leq 55\,\text{mm} \quad [(8.14)]$$

where

n_b is the number of bars in the bundle, with the following limits:

$n_b \leq 4$ for vertical bars in compression (as in columns) and bars in a lapped joint,

$n_b \leq 3$ for all other cases.

To verify the accuracy of the equivalent diameter also for bond calculations, it is sufficient to compare failure bond surfaces of bundled bars evaluated on the effective bundle geometry (Fig. 12.34) with the surfaces calculated using the equivalent diameter.

Table 12.11 Minimum and maximum diameter of a bundle of bars

Minimum diameter (mm)	Maximum diameter (mm)
12	20
14	22
16	26
18	30
20	34
22	36
24	40

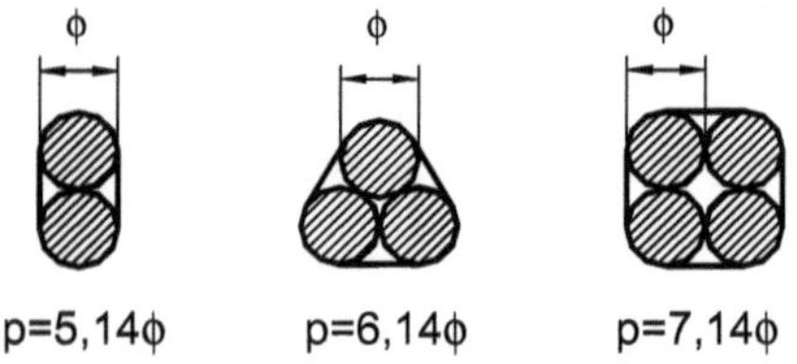

Fig. 12.34 Failure perimeter *p* for the bond failure of bundled bars

For bundles of two, three or four bars with the same diameter ϕ, the equivalent diameter is equal, respectively, to 1.41ϕ, 1.73ϕ and 2ϕ.

Effective perimeters of the bond failure surface are, respectively, 5.14ϕ, 6.14ϕ and 7.14ϕ (Fig. 12.34), while perimeters calculated using the equivalent diameter ϕ_n take the values 4.43ϕ, 5.4.36ϕ and 6.28ϕ. By comparing the values of the two series, it is evident that the equivalent diameter underestimates the circumscribed perimeter and therefore it can also be used for bond calculations and the evaluation of all the associated quantities, like the minimum concrete cover and the clear distance between two adjacent bars.

Under these assumptions, the clear distance between bundled bars will follow the same rules stated for single bars. For the calculation of a_h and a_v, the equivalent diameter of the group ϕ_n will be used, but the clear distance between bundled bars will be measured from the actual external perimeter of the bundle. Analogously, the concrete cover will be measured from the actual external perimeter of the bundle and will not be less than ϕ_n.

Finally, it is worth noting that, if two bars are arranged one over the other and bond conditions are good, they shall not necessarily be considered as a bundle.

12.1.9.1 Example No. 5. Clear Distance Between Bundled Bars

Evaluate the minimum clear distance between bundled bars formed each by 3 bars with a diameter of 24 mm, being the maximum aggregate size equal to 32 mm.

If a_h and a_v indicate the minimum horizontal and vertical clear distances between bundled bars, the following limitation holds:

$$a_h, a_v \geq \max\left(\phi_n; d_g + 5 \text{ mm}; 20 \text{ mm}\right)$$

being $\phi_n = \phi\sqrt{n_b} = 24\sqrt{3} \cong 42$ mm ($\leq$55 mm), a_h and a_v shall be not lower than:

$$a_h, a_v \geq \max\left(42; 37; 20\right) = 42 \text{ mm}$$

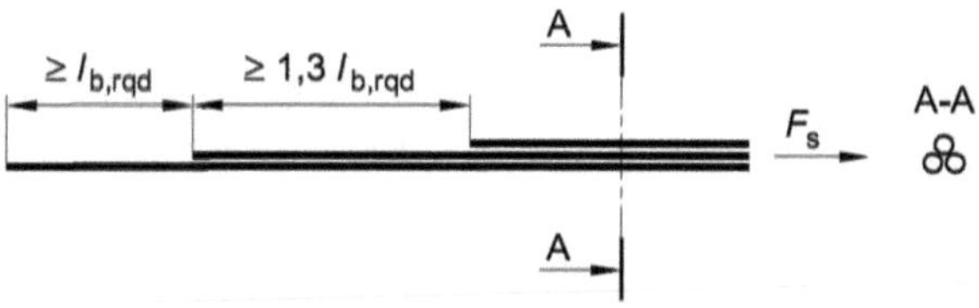

Fig. 12.35 Anchorage of three staggered bars in a bundle; staggering is obligatory for bundles with an equivalent diameter $\phi_n \geq 32$ mm and optional for bundles with $\phi_n < 32$ mm; $l_{b,rqd}$ is the basic required anchorage length for a single bar [Figure 8.12]

12.1.9.2 Anchorage of Bundled Bars

Anchorage in tension

Bundled bars in tension can be interrupted and anchored both at the end or intermediate supports. The staggering of bars within the bundle is not required if the equivalent diameter does not exceed 32 mm, otherwise bundled bars should be staggered as shown in Fig. 12.35.

If the configuration of Fig. 12.25 is adopted, the design anchorage length l_{bd} of the bundle can be obtained by using the diameter ϕ of the single bar in the calculation of coefficients α_i, otherwise it is necessary to adopt the equivalent diameter of the bundle ϕ_n.

Anchorage in compression

For anchorages in compression, it is not required to stagger the bars within the bundle; at the end-sections of the bundles with an equivalent diameter $\phi_n \geq 32$ mm, it is necessary to arrange at least four stirrups with a diameter not less than 12 mm and a further stirrup immediately after the end of the interrupted bundle.

12.1.9.3 Laps of Bundled Bars

The lap length for a bundle of bars is calculated considering a single bar with the equivalent diameter ϕ_n. For bundles of two bars with an equivalent diameter $\phi_n <$ 32 mm, bundles can be overlapped without staggering the bars within the group. In this case, the calculation of the lap length l_0 shall be performed considering the equivalent diameter of the bundle.

For bundles of two bars with an equivalent diameter $\phi_n \geq 32$ mm or bundles of three bars, bars within the bundle shall be staggered in the longitudinal direction of an amount at least equal to 1.3 l_0, as shown in Fig. 12.36, where l_0 represents the lap length of the single bar and the fourth bar is used for the lap. It is necessary to guarantee that there are not more than four bars in each transverse lapped cross-section, therefore, due to the need for the insertion of the fourth lap bar, it is not possible to lap bundles with more than three bars.

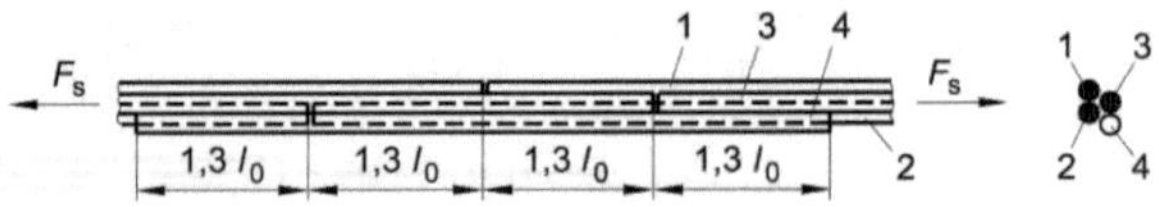

Fig. 12.36 Lapped joint in tension with the insertion of a fourth bar [Figure 8.13].

12.1.10 Rules for Prestressing Reinforcement

For anchorage rules of prestressing reinforcement, see Chap. 6, dedicated to prestressed members and structures.

12.2 Detailing of Beams, Columns, Slabs and Walls

All the members shall be provided with a minimum amount of reinforcement to avoid brittle failures and excessive cracks. Minimum reinforcement is also necessary to absorb internal forces due to restrained strains, like those due to drying shrinkage, thermal variations and settlements.

In the following, the rules for the minimum reinforcement and the reinforcement arrangement for typical members of buildings (beams, columns, slabs and walls) are described.

12.2.1 Beams

Following rules for minimum reinforcement of beams are described in subsequent paragraphs:

- longitudinal reinforcement,
- shear reinforcement,
- torsion reinforcement,
- surface reinforcement,
- reinforcement on indirect supports.

12.2.1.1 Longitudinal Reinforcement

Minimum area of longitudinal reinforcement in tension

The minimum area of longitudinal reinforcement in tension is given by

$$A_{s,\min} = 0.26\frac{f_{ctm}}{f_{yk}}b_t\,d \geq 0.0013\,b_t\,d \quad [(9.1\text{N})]$$

Table 12.12 Values (in ‰) of the ratio $A_{s,min}/(b_t\ d)$ for $f_{yk} = 450\ \text{N/mm}^2$ at varying the concrete strength class

f_{ck} (N/mm²)	16	20	25	28	32	35	40	45	50	55	60	70
f_{ctm} (N/mm²)	1.9	2.2	2.6	2.8	3.0	3.2	3.5	3.8	4.1	4.2	4.4	4.6
$A_{s,min}/(b_t\ d)$ (‰)	1.3	1.3	1.5	1.62	1.73	1.85	2.02	2.20	2.37	2.43	2.54	2.66

where

b_t is the mean width of the zone in tension: for a T beam with the flange in compression, b_t is coincident with the width b of the web, while in a T beam with the flange in tension, b_t is given by

$$b_t = B - \left(\frac{y_G - t}{y_G}\right)^2 \cdot (B - b)$$

where y_G is the distance of the geometrical centre of gravity[3] from the face in tension, t is the thickness of the flange, B is the width of the flange and b is the width of the web

f_{ctm} is the mean concrete strength
f_{yk} is the characteristic yield strength of the reinforcement steel

Table 12.12 collects the values of $A_{s,min}$ for $f_{yk} = 450\ \text{N/mm}^2$ at varying the concrete strength class.

The expression of the minimum longitudinal reinforcement is obtained by imposing that, at cracking, the design strength of reinforcement bars ($f_{yd} \cdot A_s$) equals the resultant tensile force in the cracked concrete. It is assumed that the transition from the uncracked state to the cracked state happens when the most stressed fibres take on the mean value of the concrete tensile strength.

With reference to Fig. 12.37, the bending moment immediately before cracking is equal to

$$M_{\text{before}} = T \cdot (2/3)h = [(f_{ctm} b_t h/2)/2] \cdot (2/3h) = \left(f_{ctm} b_t h^2\right)/6$$

while immediately after cracking it holds (Fig. 12.38)

$$M_{\text{after}} = T(0.9d) = \left(A_s f_{yd}\right)(0.9d)$$

[3] Strictly speaking, it should be considered the centre of gravity of the uncracked transformed section, which includes the contribution of steel bars with their transformed areas; however, the error due to the use of the geometrical section is usually negligible.

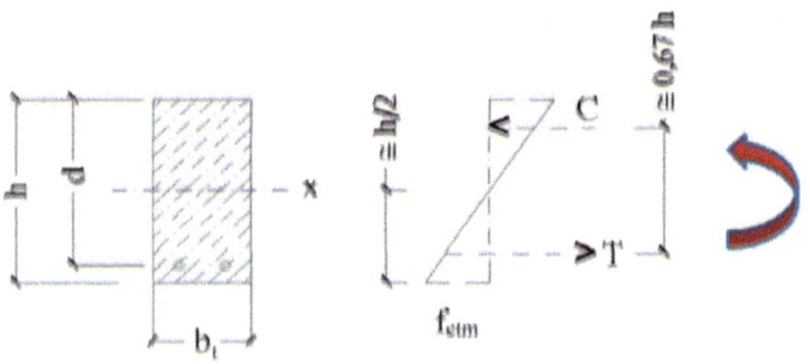

Fig. 12.37 Lever arm of an uncracked rectangular cross-section (neglecting the contribution of steel bars)

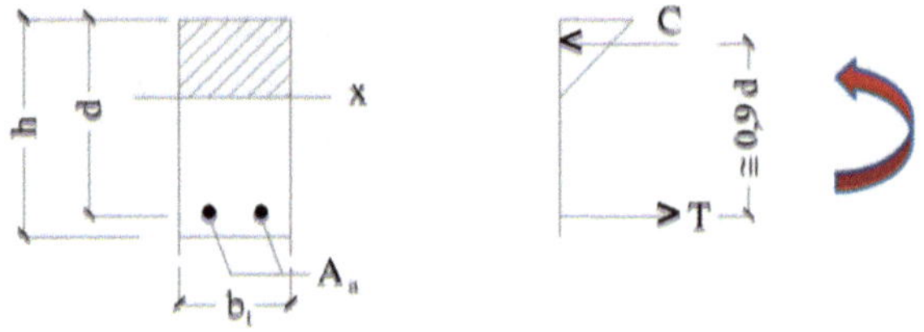

Fig. 12.38 Lever arm of a cracked rectangular cross-section

By imposing the non-brittleness condition, i.e. $M_{\text{after}} \geq M_{\text{before}}$, the following expression holds

$$0.9A_s f_{yd} d \geq \left(f_{ctm} b_t h^2\right)/6$$

if $d = 0.9\,h$, that is $h = d/0.9$, the expression simplifies as follows
$A_s \geq 0.2286\ (f_{ctm}/f_{yd})\ b_t\ d$.[4]

Finally, if f_{yd} is expressed as a function of the characteristic yield strength of steel ($f_{yd} = f_{yk}/1.15$), the following limitation holds: $A_s \geq 0.26\ (f_{ctm}/f_{yk})\ b_t\ d$.

For secondary members (like parapets), $A_{s,\min}$ can be taken as 1.2 times the area required for the verification at ULS.

Formula [(9.1 N)] for the minimum tensile longitudinal reinforcement does not apply to prestressed members with pretensioned bonded steel, like hollow-core slabs. Elements provided with a reinforcement amount less than $A_{s,\min}$ shall be considered like plain or lightly reinforced concrete members, so EC2 rules for those elements apply [Sect. 12].

For the calculation of the minimum reinforcement required to limit steel stresses under an assigned value when spontaneous cracking occurs due to restrained strains, refer to Chap. 11, dedicated to SLS.

[4] An analogous formula can be found in CEB Bulletin No. 228, High Performance Concrete, Recommended Extensions to the Model Code 90 - Research Needs, but with $f_{ctk,\max}$ (fractile 95%) instead of f_{ctm} and f_{yk} instead of f_{yd}.

(The same expression of $A_{s,min}$ is obtained if M_{after} is calculated using $z = 0.8d$ and the characteristic yield strength f_{yk} of steel).

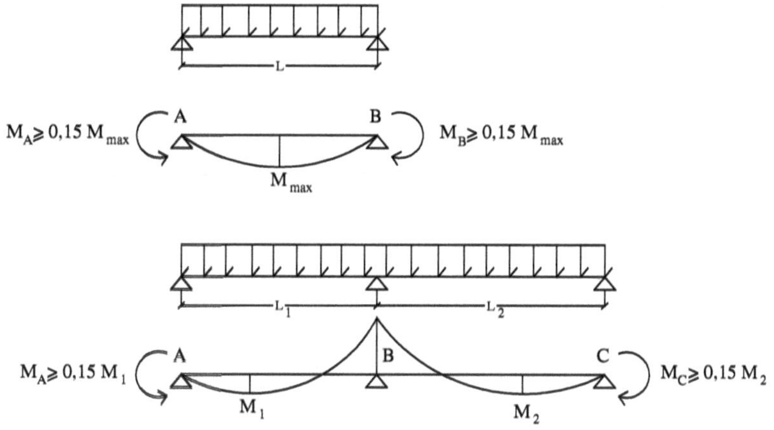

Fig. 12.39 Minimum bending moments for the design of bending reinforcement at end supports

Maximum reinforcement area

Outside lap zones, the areas of both reinforcement in tension and compression shall not exceed the ratio of 4% of the concrete area:

$$A_{s,\max} = 0.04A_c.$$

Other rules on the arrangement of longitudinal bars

In simply supported beams and continuous beams with simple end supports, the cross-sections at the end supports shall be designed considering a bending moment not less than 15% of the midspan bending moment (Fig. 12.39). The corresponding reinforcement shall be at least equal to the minimum area $A_{s,\min}$.

The total area of the tensile reinforcement A_s at intermediate supports of a continuous T-beam shall be spread over the effective width of the flange b_{eff}, which is defined in Fig. 12.40. Part of this reinforcement can be placed over the web width (Fig. 12.41).

Any longitudinal compressed bar with a diameter ϕ, considered in the resistance calculations, should be held by transverse reinforcement with spacing not greater than 15 ϕ to prevent instability.

Cut-off of longitudinal reinforcement in tension

The cut-off of longitudinal reinforcement shall be designed considering the effects of cracking. To this purpose, the "shift rule" is applied, that is the bending moment diagram is shifted by a segment with length a_l (Fig. 12.42), where:

- for members provided with transverse shear reinforcement

$$a_l = z(\cot\theta - \cot\alpha)/2$$

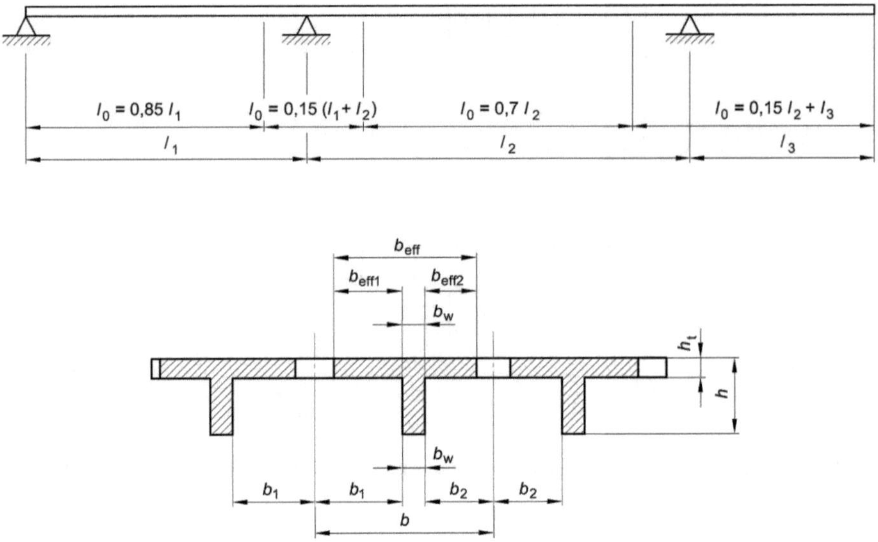

$b_{eff} = \Sigma\, b_{eff,i} + b_w \leq b$, where $b_{eff,i} = 0{,}2\ b_i + 0{,}1\ l_0 \leq \min(0{,}2\ l_0;\ b_i)$

Fig. 12.40 Effective flange width of a T cross-section [Figure 5.2 e 5.3]

Fig. 12.41 Arrangement of tension reinforcement in T cross-section at intermediate supports [Figure 9.1]

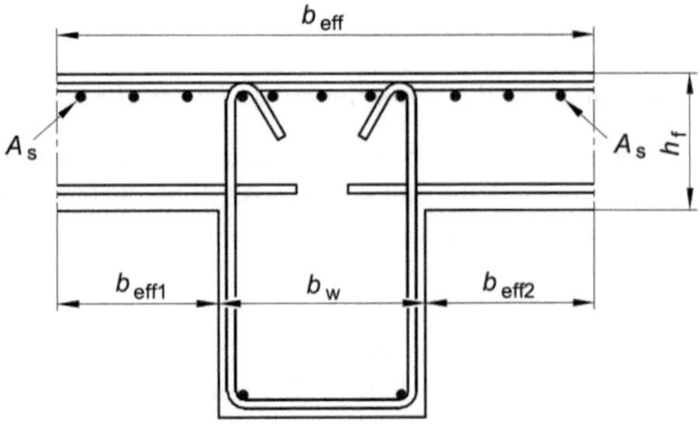

- for members without shear reinforcement (i.e. slabs, plates, walls)

$$a_1 = d$$

being,

θ the angle of concrete struts with the horizontal plane,
α the angle of the shear reinforcement.

The design length of longitudinal reinforcement is given by the length required to cover the envelope of maximum bending moments, once the envelope has been shifted by the amount a_1 on each side and adding the anchorage length at both ends.

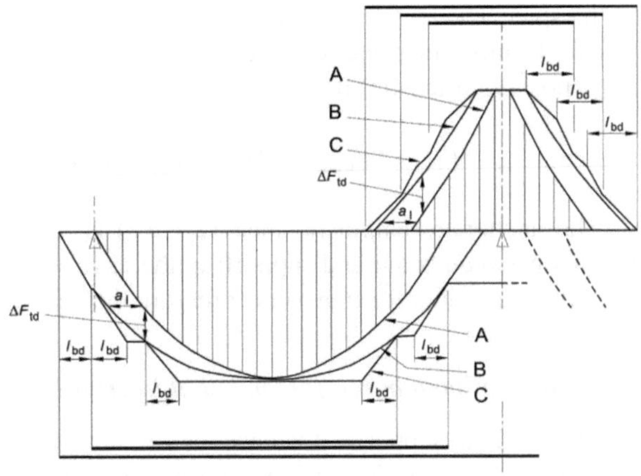

Fig. 12.42 Illustration of the cut-off of longitudinal reinforcement, taking into account the effect of inclined cracks and the resistance of reinforcement within anchorage lengths ("shift rule"): A = envelope of M_{Ed}/z, B = acting tensile force F_s, C = resisting tensile force F_{Rs} [Figure 9.2]

Anchorage of bent bars

The anchorage length of bent bars shall be at least equal to 1.3 l_{bd} in the tension zone and 0.7 l_{bd} in the compression zone. The anchorage length shall be measured starting from the intersection point of the axis of the bent bar with longitudinal reinforcement (Fig. 12.43a, b).

Anchorage of bottom reinforcement at end supports

At an end support with little or no end fixity assumed in design, the area of the bottom reinforcement shall be at least equal to 25% of the reinforcement provided

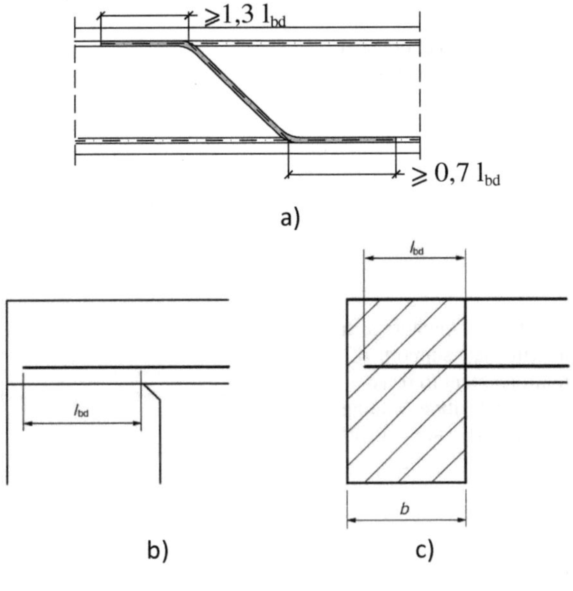

Fig. 12.43 Anchorage of bent bars: **a** tension zone at the top and compression zone at the bottom. Anchorage of bottom reinforcement at end supports, **b** direct support—beam supported by wall or column, **c** indirect support—beam intersecting another supporting beam [Figure 9.3]

in the span. The tensile force to be anchored can be evaluated by applying the shift rule for beams with shear reinforcement, including the contribution of the axial force N_{Ed} (positive if tensile) if any:

$$F_E = |V_{Ed}| \cdot a_1/z + N_{Ed}, \text{ with } a_1 = z(\cot\theta - \cot\alpha)$$

The anchorage length l_{bd} (see 12.1.5.2) is measured starting from the line of contact between beam and support; in the case of direct support (Fig. 12.43c), it is possible to account for transverse pressure through the coefficient $\alpha_5 < 1$ (see Table 12.4).

Anchorage of bottom reinforcement at intermediate supports

At intermediate supports, it is necessary to arrange a bottom reinforcement with an area at least equal to 25% of the reinforcement in the span; this rule is the same as end supports. Moreover, the anchorage length l shall fulfil the following limitations (Fig. 12.44):

- for straight bars: $1 \geq 10\,\phi$
- for hook and bent bars: $1 \geq d_m$ if $\phi \geq 16$ mm (d_m = mandrel diameter), $l \geq 2\,d_m$ if $\phi < 16$ mm.

The continuity of the bottom reinforcement at intermediate supports to resist possible positive bending moments, due for example to the settlement of the support, can be guaranteed by lapped bars. Laps can be partially outside and partially inside the support (Fig. 12.45a) or fully outside the support (Fig. 12.45b). In the first case, bars coming from the span shall be anchored inside the support for a length not smaller than 10ϕ.

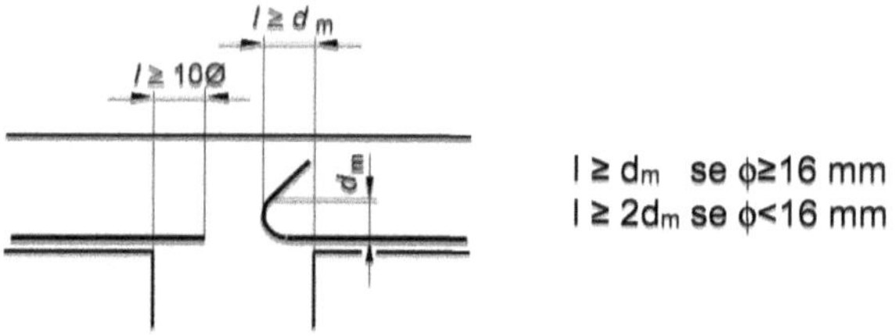

Fig. 12.44 Anchorage at intermediate supports [Figure 9.4a]

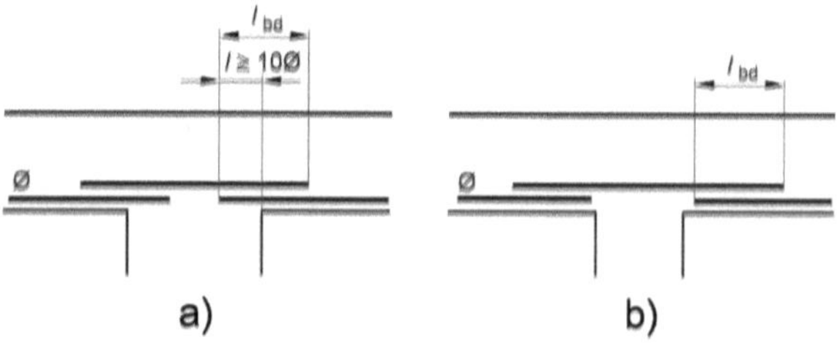

Fig. 12.45 Anchorage at intermediate supports of reinforcement to resist possible positive bending moments [Figure 9.4b, c]

12.2.1.2 Shear Reinforcement

Shear reinforcement shall form an angle α between 45° and 90° with the longitudinal axis of the structural member, according to [9.2.2(1)]. Such reinforcement can be made through a combination of (Fig. 12.46):

- links enclosing the longitudinal tension reinforcement and the compression zone
- bent-up bars
- cages, ladders, etc. which are cast in without enclosing the longitudinal reinforcement but are properly anchored in the compression and tension zones.

Links can be anchored as follows (Fig. 12.18):

- by hooks,
- by 90°-bent bars,
- by welded bars.

At least 50% of the required shear reinforcement shall be provided by links. The ratio of shear reinforcement is given by

$$\rho_w = A_{sw}/\left(s \cdot b_w \cdot \sin\alpha\right) \geq \rho_{w,\min}$$

where

A_{sw} is the area of shear reinforcement within length s,
s is the spacing of the shear reinforcement measured along the longitudinal axis of the member
b_w is the breadth of the web of the member
α is the angle between shear reinforcement and the longitudinal axis ($45° \leq \alpha \leq 90°$)

$$\rho_{w,\min} = \frac{0.08 \cdot \sqrt{f_{ck}}}{f_{yk}}$$

.

Table 12.13 collects the values of $\rho_{w,\min}$ versus the characteristic compressive strength f_{ck} of concrete.

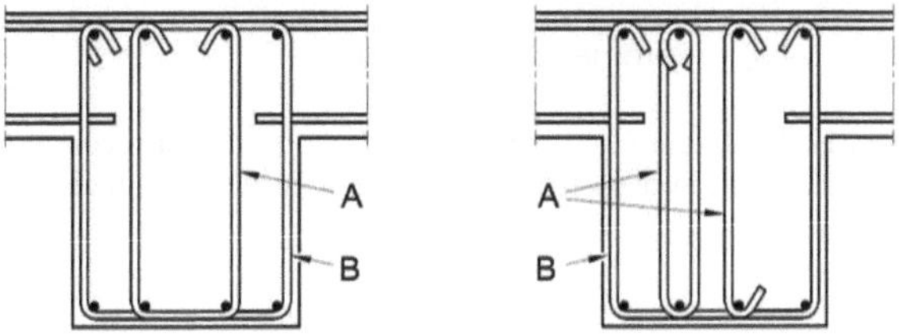

Fig. 12.46 Examples of shear reinforcement: A = alternatives for internal links, B = closing links [Figure 9.5]

Table 12.13 Minimum shear reinforcement $\rho_{w,\min}$ versus f_{ck} (steel grade B450C)

f_{ck} (N/mm^2)	12	16	20	25	28	30	32	35	40	45	50	70
$\rho_{w,\min}$ (‰)	0.62	0.71	0.80	0.89	0.94	0.97	1.01	1.05	1.12	1.19	1.26	1.48

Table 12.14 Maximum longitudinal spacing $s_{l,\max}$ for single legged stirrups and bent-up bars

	Single legged stirrups		Bent-up bars	
Angle α	$\alpha = 90°$	$\alpha \neq 90°$	$\alpha = 45°$	$\alpha = 60°$
$s_{l,\max}$	0.75 d	0.75d (1 + cot α)	1.2 d	0.95 d

If α is the slope of shear reinforcement to the longitudinal beam axis, the maximum longitudinal spacing should not exceed (Table 12.14).

- for shear assemblies $s_{l,\max} = 0.75d\,(1 + \cot\alpha)$
- for bent-up bars $s_{l,\max} = 0.6\,d\,(1 + \cot\alpha)$.

The transverse spacing of legs for a series of shear links shall not exceed

$$s_{t,\max} = 0.75d \leq 600 \text{ mm}$$

12.2.1.3 Torsion Reinforcement

Torsion links shall be closed and arranged perpendicularly to the axis of the structural member. Torsion links can be anchored through hooks at the end sections (Fig. 12.47a1 and a3) or by a lap along the top leg for the full width of the cross-section (Fig. 12.47a2). It is not allowed to employ stirrups with 90°-bends at the end sections, closed on the same corner (Fig. 12.47b).

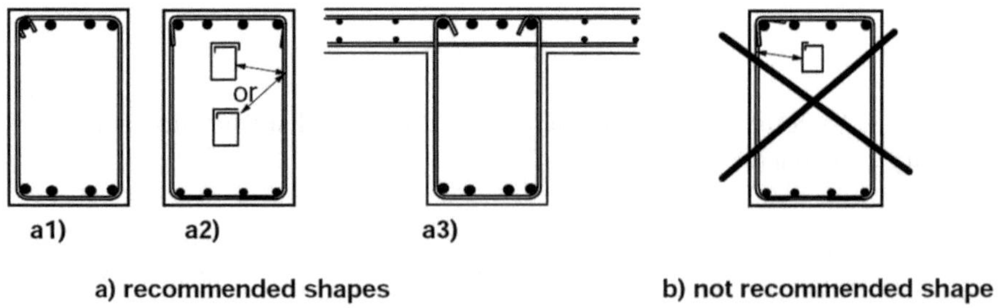

Fig. 12.47 Examples of shapes for torsion links: **a** recommended shapes (the second alternative for a2 in the lower sketch should have a full lap length along the top), **b** not recommended shape [Figure 9.6]

Fig. 12.48 Torsion resistant cross-section: A = centreline, B = outer edge of effective cross-section with circumference u, C = concrete cover [Figure 6.11]

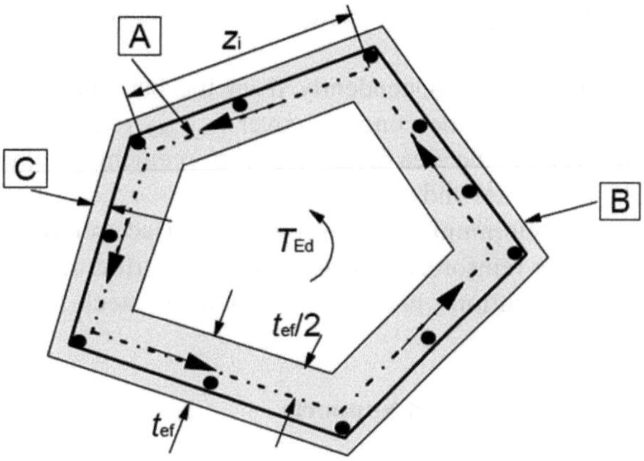

The provisions of minimum percentage of shear reinforcement (see Table 12.13) and maximum longitudinal spacing for vertical links ($s_{l,\max} = 0.75\ d$ if $\alpha = 90°$) are generally sufficient to provide the minimum torsion links required.

If torsion is present, longitudinal spacing of links shall fulfil the following requirement (Fig. 12.48):

$$s_{l,\max} \leq \max\ (u/8;\ 0.75d;\ b_{\min})$$

where

u is the outer edge of the effective cross-section (B in Fig. 12.48),
$b_{\min}$ is the smallest dimension of the cross-section.

Concerning the arrangement of longitudinal bars, at least one bar for each edge should be provided, while the other bars shall be uniformly distributed along the internal perimeter of links, with a spacing not exceeding 350 mm.

12.2.1.4 Surface Reinforcement

If large diameter bars ($\phi > 32$ mm) or bundles of bars with an equivalent diameter $\phi_n > 32$ mm are used, it is necessary to arrange surface reinforcement, both for cracking control and to guarantee a suitable strength against the spalling of the concrete over.

Surface reinforcement will be placed outside the links, as indicated in Fig. 12.30, and it will be made of wire mesh or small diameter bars. In both directions, parallel and orthogonal to the longitudinal reinforcement of the member, the area of surface reinforcement should satisfy the following limit:

$$A_{s,\text{surf}} \geq A_{s,\text{surf,min}} = 0.01 A_{ct,\text{ext}}$$

where $A_{ct,ext}$ is the concrete area in tension outside the links (dashed area in Fig. 12.30a).

Moreover, independently from the bar diameter, the surface reinforcement is needed when the concrete cover $c_{min,dur}$ required for durability exceeds 70 mm; in this case, the area of surface reinforcement should be equal to 0.005 $A_{c,ext}$ both in the horizontal and vertical directions.

The minimum concrete cover for surface reinforcement shall be evaluated as for the other reinforcement (see Chap. 3); furthermore, it may be appropriate to adopt special solutions, like the use of stainless steel.

12.2.1.5 Indirect Supports

Where a beam is supported by another beam instead of a wall or column, reinforcement should be provided and designed to resist the mutual reaction. This reinforcement should be added to that required for other reasons. This rule also applies to a slab supported at the bottom of a beam.

The supporting reinforcement between two beams should consist of links surrounding the primary reinforcement of the supporting member. Some of these links may be distributed outside the volume of the concrete, which is common to the two beams (Fig. 12.49).

12.2.2 Solid Slabs

EC2 defines as solid slabs those mono or bi-directional members where the width b and the effective span l_{eff} are not less than five times the height. The effective span is defined as [(5.8)]:

$$l_{eff} = l_n + a_1 + a_2$$

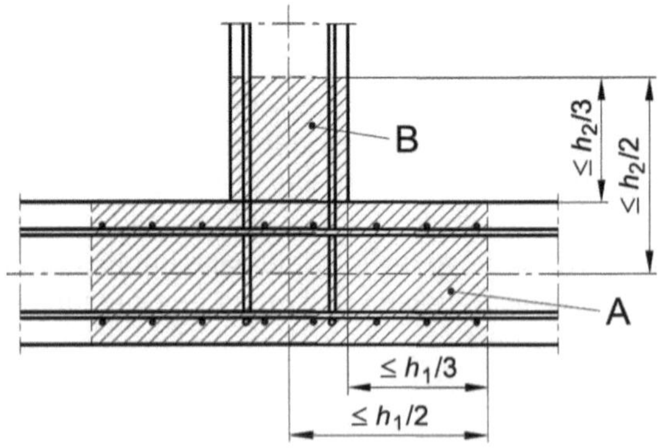

Fig. 12.49 Placing of supporting reinforcement in the intersection zone of two beams (plan view): A = supporting beam with height h_1, B = supported beam with height h_2 ($h_1 \geq h_2$) [Figure 9.7]

where the values of a_1 and a_2 at the two end-sections of the span can be obtained through the relevant values a_i indicated in Fig. 12.50.

Flexural reinforcement

For solid slabs, the same minimum and maximum steel percentages valid for beams apply. Moreover, for one-way slabs, a secondary transverse reinforcement of not less

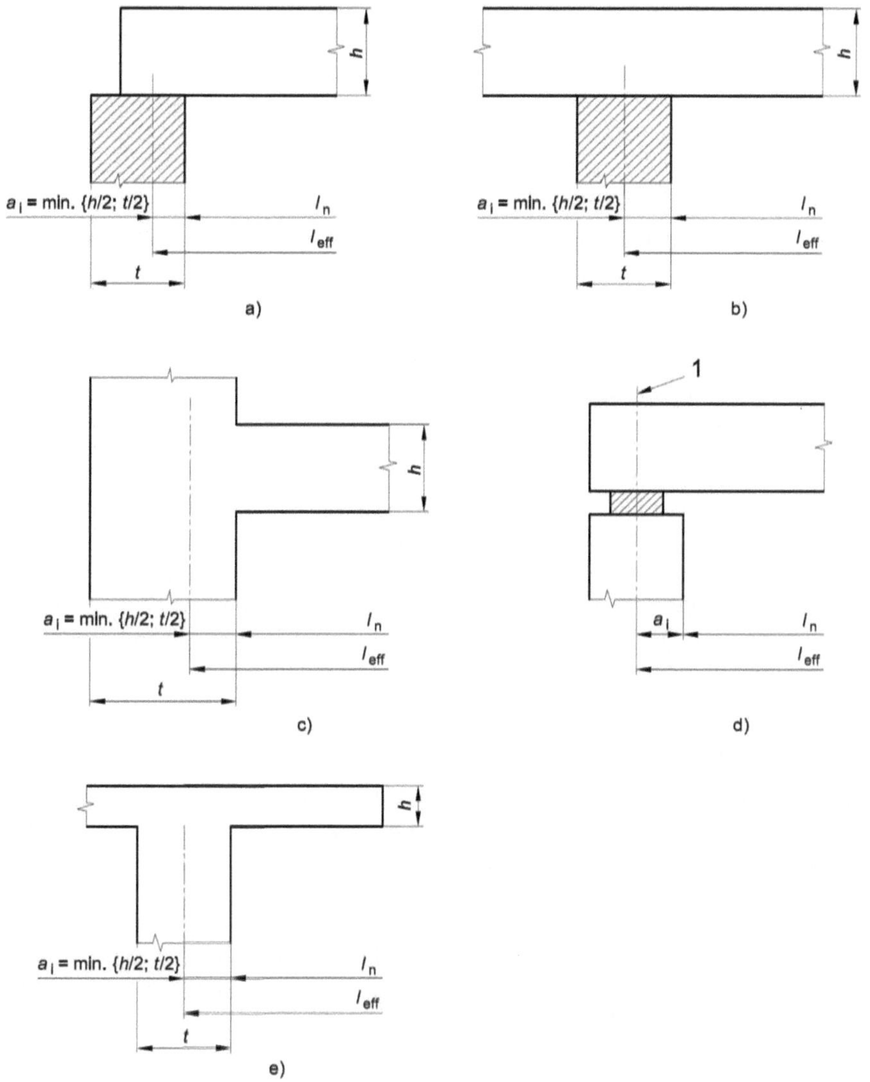

Fig. 12.50 Effective span (l_{eff}) for a slab with different support conditions (t = support width) [Figure 5.4]

than 20% of the primary reinforcement should be provided. In areas near supports, transverse reinforcement is not required if there is no transverse bending moment.

If h is the total depth of the slab, the following maximum spacings for primary and secondary reinforcement will be used:

$$s_{\max,\text{slabs}} = \min\,(2h;\,350\ \text{mm}) \quad \text{for primary reinforcement,}$$
$$s_{\max,\text{slabs}} = \min\,(3h;\,400\ \text{mm}) \quad \text{for secondary reinforcement.}$$

In areas subjected to point loads or maximum bending moments, spacing values reduce to:

$$s_{\max,\text{slabs}} = \min\,(2h;\,250\ \text{mm})$$

The same rules of beams apply for:

- the cut-off of longitudinal bars in tension (shift rule is applied under the assumption that $a_1 = d$),
- the anchorage of bottom reinforcement at end supports,
- the anchorage of bottom reinforcement at intermediate supports.

Reinforcement in slabs near supports

In simply supported slabs, half of the calculated span reinforcement should continue up to the support and be anchored therein according to 8.4.4.

Where partial fixity occurs along the edge of a slab but is not taken into account in the analysis, the top reinforcement should be capable of resisting at least 25% of the maximum moment in the adjacent span. This reinforcement should extend at least 0.2 times the length of the adjacent span, measured from the face of the support. It should be continuous across intermediate supports and anchored at end supports. At the end support, the moment to be resisted may be reduced to 15% of the maximum moment in the adjacent span.

Corner reinforcement

If the detailing arrangements at support are such that the lifting of the slab at a corner is restrained, suitable reinforcement should be provided, arranged along diagonals, and integrated by transverse distribution reinforcement.

Reinforcement at free edges

Along a free (unsupported) edge, a slab should normally contain longitudinal and transverse reinforcement, generally arranged as shown in Fig. 12.51. The normal reinforcement provided for a slab may act as edge reinforcement.

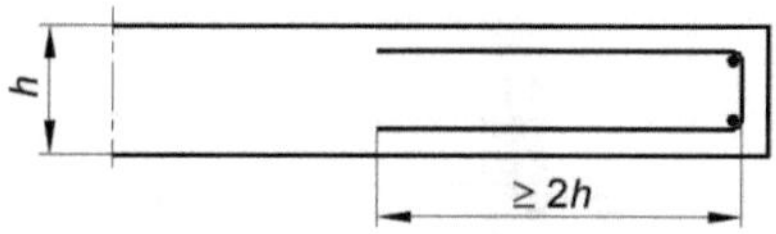

Fig. 12.51 Edge reinforcement for a slab [Figure 9.8]

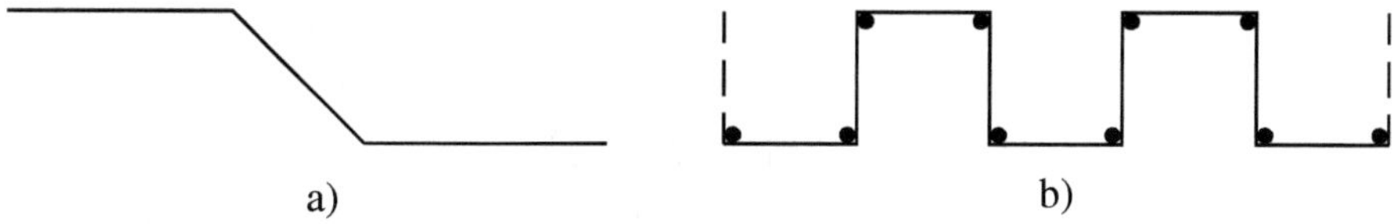

Fig. 12.52 Shear reinforcement for a slab: **a** bent-up bars, **b** shear assembly

12.2.2.1 Shear Reinforcement

A slab in which shear reinforcement is provided should have a depth of at least 200 mm to guarantee a sufficient anchorage of the punching shear reinforcement. In detailing the shear reinforcement, the minimum value and definition of reinforcement ratio for beams apply. For a slab with width b, the following limitation holds

$$\rho_w = A_{sw}/\,(s \cdot b \cdot \sin\alpha) \geq \rho_{w,\min}$$

where

A_{sw} is the area of shear reinforcement over the length s,
s is the spacing of shear reinforcement,
b is the width of the slab,
α is the angle between the shear reinforcement and the longitudinal axis ($45° \leq \alpha \leq 90°$ according to [9.2.2(1)],

$\rho_{w,\min} = \left(0.08 \cdot \sqrt{f_{ck}}\right) / f_{yk}$ (see Table 12.13).

If $|V_{Ed}| \leq 1/3\ V_{Rd,\max}$[5] the shear reinforcement may consist entirely of bent-up bars or shear reinforcement assemblies (Fig. 12.52).

The maximum longitudinal spacing of successive series of links is given by:

$$s_{\max} = 0.75d(1 + \cot\alpha)$$

where α is the inclination of the shear reinforcement ($\alpha = 90°$ in Fig. 12.52b), while the maximum longitudinal spacing of bent-up bars is given by

$$s_{\max} = d.$$

[5] Refer to Chap. 8 for the expression of $V_{Rd,\max}$.

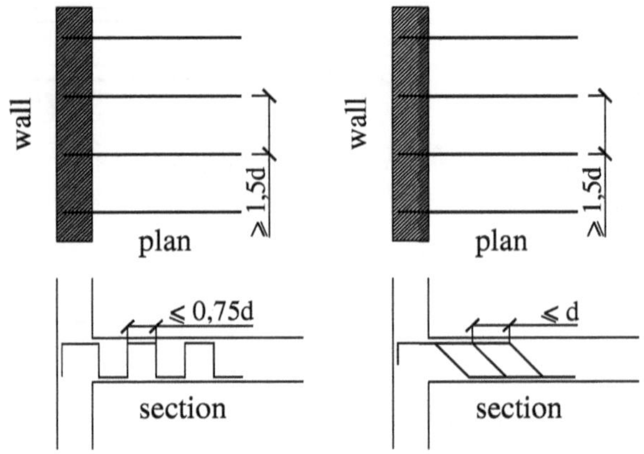

Fig. 12.53 Maximum longitudinal and transverse spacing for a slab shear reinforcement

The maximum transverse spacing of shear reinforcement should not exceed 1.5*d* (Fig. 12.53).

12.2.3 *Flat Slabs*

12.2.3.1 Slab at Internal Columns

The arrangement of flexural reinforcement in flat slabs should reflect their behaviour under working conditions. In general, this will result in a concentration of reinforcement over the columns.

At internal columns, unless rigorous serviceability calculations are carried out, top reinforcement of area 0.5 A_t should be placed in a width equal to the sum of 0.125 times the panel width on either side of the column. A_t represents the area of reinforcement required to resist the full negative moment applied to the sum of the two half panels [0.5 ($l_{y1} + l_{y2}$)], taken on each side of the column (Fig. 12.54).

In both directions, at least two bottom bars should be provided at internal columns and should pass through the column. These bars provide the minimum flexural strength to the slab against punching shear failure. In other words, loads that cannot be transferred to the bottom column, once a punching shear failure has occurred, can be transferred to adjacent columns thanks to the residual flexural strength of the slab.

12.2.3.2 Slab at the Edge and Corner Columns

The reinforcement perpendicular to a free edge required to transmit bending moments from the slab to an edge or corner column should be placed within the effective width b_e shown in Fig. 12.55.

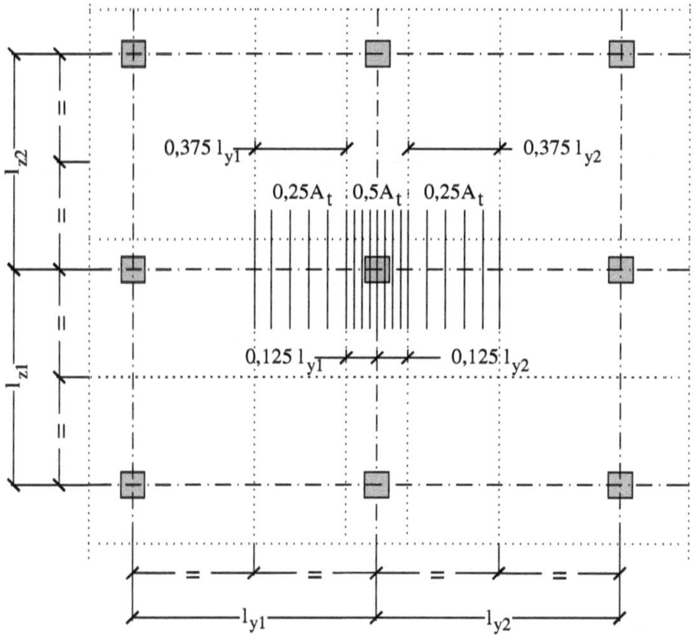

Fig. 12.54 Layout of the top reinforcement for a flat slab on an internal column (for the sake of clarity, only reinforcement in the z-direction is represented, being y–z the slab middle plane)

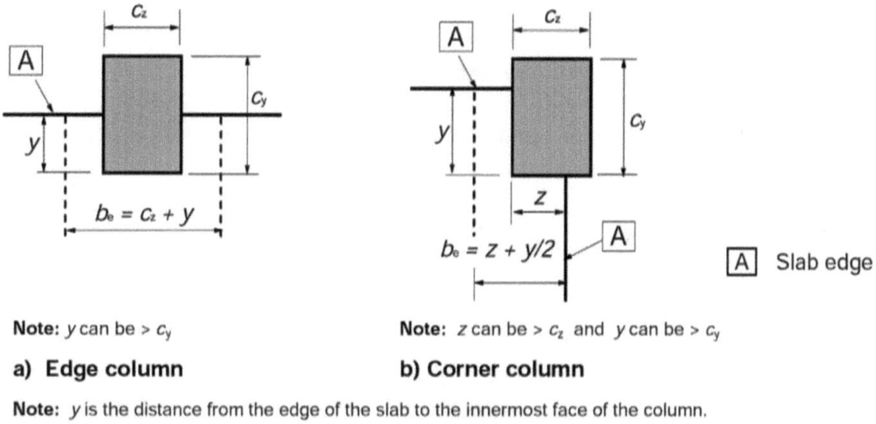

Fig. 12.55 Effective width, b_e, of a flat slab [Figure 9.9]

12.2.3.3 Punching-Shear Reinforcement

Refer to Chap. 10 which deals with the punching shear in slabs and foundations.

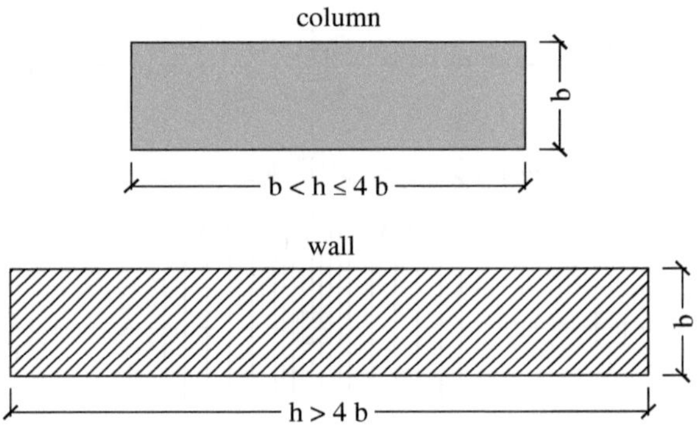

Fig. 12.56 Dimensions of the transverse cross-section for a column and a wall

Table 12.15 $A_{s,\min}/A_c$ (expressed in %) of columns for $N_{Ed} = f_{cd} \cdot A_c$ and steel grade B450C

f_{ck} (N/mm²)	12	16	20	25	30	35	40	45	50	55	60
$A_{s,\min}/A_c$ (%)	0.3	0.3	0.341	0.426	0.511	0.596	0.681	0.767	0.852	0.937	1.022

12.2.4 Columns

According to EC2, a column is a vertical member with a ratio *h*/*b* between the greater cross-sectional dimension (*h*) and the smaller one (*b*) not exceeding 4 (Fig. 12.56).

12.2.4.1 Longitudinal Reinforcement

The longitudinal reinforcement should have a diameter not smaller than 8 mm and its total area should not be smaller than

$$A_{s,\min} = \max\left(\frac{0.10N_{Ed}}{f_{yd}};\ 0.003A_c\right)$$

where

f_{yd} design yield strength of the steel reinforcement,
N_{Ed} design value of the compression force.[6]

For $N_{Ed} = f_{cd} \cdot A_c$ and steel grade B450C, $A_{s,\min}$ takes the values collected in Table 12.15.

The area of longitudinal reinforcement outside the lapping zone shall not exceed $A_{s,\max} = 0.04\,A_c$. A greater value can be taken only if it is possible to demonstrate that

[6] Even if EC2 does not give any precise indication, the design axial force shall be considered as due to the load combination at ULS.

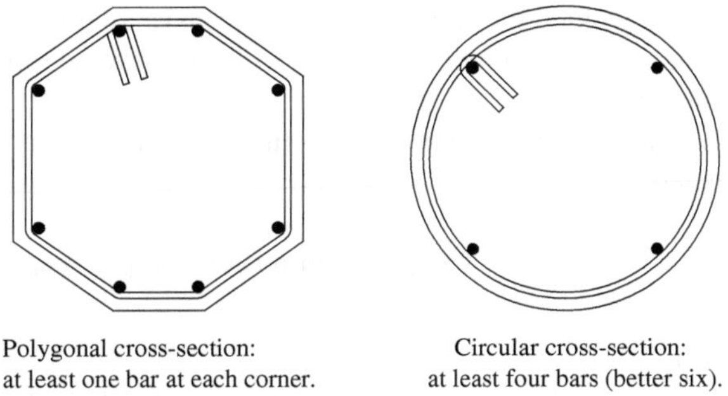

Fig. 12.57 Minimum number of longitudinal bars for columns

the greater amount of reinforcement does not affect the concrete integrity and that the design strength is reached at ULS. However, EC2 does not provide indications about this verification; therefore, it is suggested not to pass the maximum value given above.

At laps, the limit is increased to 0.08 *Ac*.

For columns with polygonal cross-section, at least one longitudinal bar should be provided at each corner, while in circular columns at least four bars are required (Fig. 12.57), nevertheless, it is better not to adopt less than six bars.

12.2.4.2 Transverse Reinforcement

The diameter ϕ_{transv} of transverse reinforcement (links, mandrel or helicoidal reinforcement) shall fulfil the following requirement

$$\phi_{transv} = \max\left(6\text{ mm}; \frac{\phi_{long,max}}{4}\right)$$

where $\phi_{long,max}$ is the maximum diameter of longitudinal bars (Fig. 12.58).

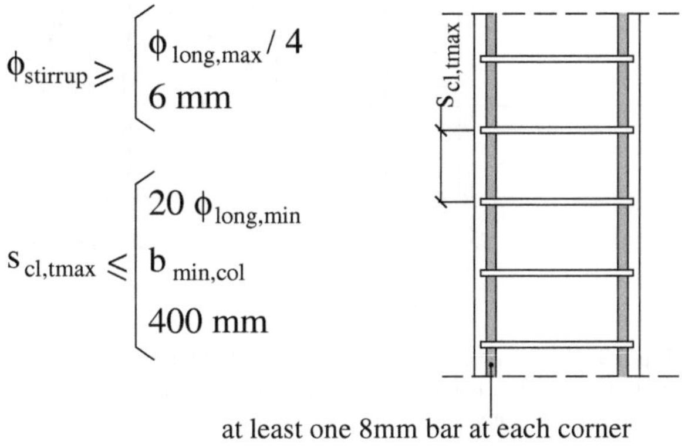

Fig. 12.58 Minimum diameter and maximum spacing of links for columns

If a welded mesh is employed as transverse reinforcement, mesh wires shall have a diameter not smaller than 5 mm. The transverse reinforcement shall be adequately anchored.

The spacing of transverse reinforcement for a column shall not exceed $s_{cl,tmax}$, where

$$s_{cl,t\,max} = \min\,(20\phi_{long,min};\ b_{min,col};\ 400\ \mathrm{mm})$$

with

$\phi_{long,min}$ minimum diameter of longitudinal reinforcement,
$b_{min,col}$ minimum dimension of column cross-section.

The maximum spacing $s_{cl,tmax}$ of transverse reinforcement shall be reduced by 40% in the following cases:

- for cross-sections immediately above or below a beam or a slab for a length equal to the greater dimension of the column cross-section,
- where bars are lapped, if the maximum diameter of bars exceeds 14 mm; nevertheless, at least three transverse bars should be arranged within the lapping length at constant spacing.

Where longitudinal bars change direction, for example, due to changes in the cross-section, the spacing of the transverse reinforcement shall be calculated by considering the transverse forces involved. These effects can be neglected if the change direction is lower or equal to 1/12 (Fig. 12.59).

Each longitudinal bar or bundle of longitudinal bars placed at a corner should be held by transverse reinforcement. No bar within a compression zone should be more than 150 mm away from a restrained bar (Fig. 12.60).

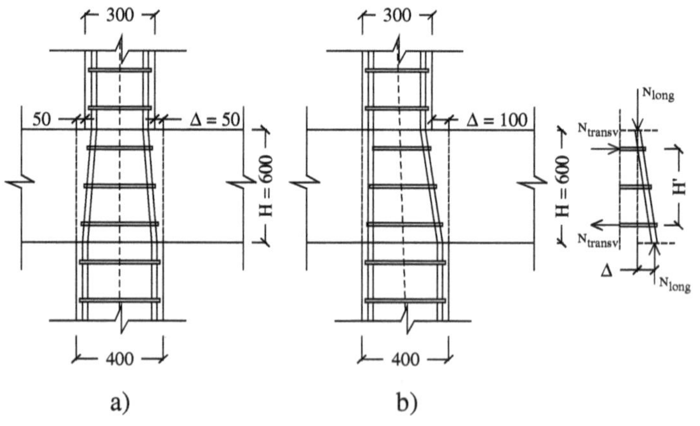

Fig. 12.59 Transverse reinforcement for columns where the cross-section changes: **a** $\Delta/H = 50/600 = 1/12$, it is not required any calculation of transverse reinforcement; **b** $\Delta/H = 100/600 = 1/6 > 1/12$, it is necessary to verify transverse reinforcement considering transverse forces

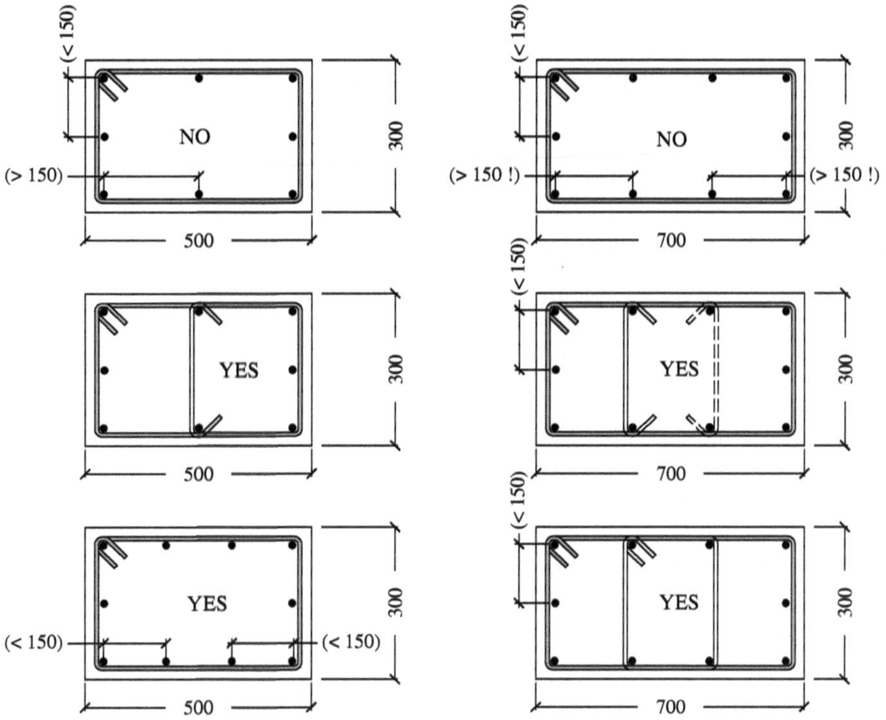

Fig. 12.60 Maximum distance of bars from a restrained bar

Fig. 12.61 Arrangement of reinforcement in a wall according to EC2

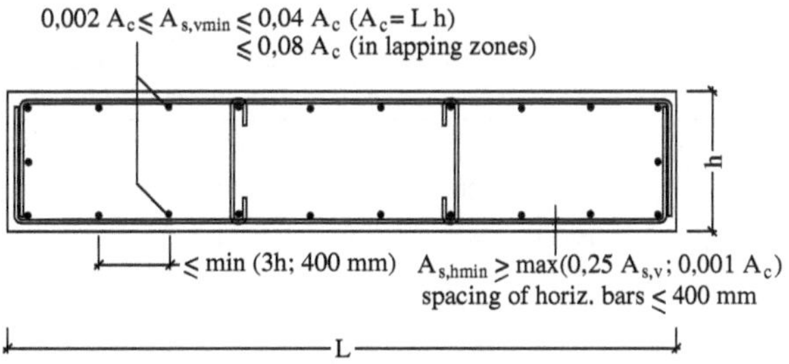

12.2.5 Walls

EC2 defines as walls those vertical elements with a length to thickness ratio of 4 or more (Fig. 12.61) and in which the reinforcement is taken into account in the strength analysis.[7]

---

[7] This remark on reinforcement is useful to differentiate walls from plain members, like those employed as gravity retaining walls and treated in Chap. 12 of EC2.

For the design of reinforcement in walls loaded in their plane, strut-and-tie models could be used (see Chap. 10), while for walls mainly subjected to out-of-plane loads (like retaining walls, tank walls, etc.), the same rules as for slabs apply (12.2.2).

12.2.5.1 Vertical Reinforcement

The area of the vertical reinforcement of a wall shall be between $A_{s,v\min}$ and $A_{s,v\max}$, where:

$A_{s,v\min} = 0.002\ A_c$ (Table 12.16).
$A_{s,v\max} = 0.04\ A_c$ outside the lapping zones (Table 12.16).
$0.08A_c$ in the lapping zones.

As for columns, it is possible to adopt a value of $A_{s,v\max}$ exceeding 0.04 A_c if the greater amount of reinforcement does not affect the concrete integrity and the design strength is reached. The maximum recommended value by EC2 should not be exceeded.

If the minimum reinforcement $A_{s,v\min}$ is sufficient for strength verifications, half of $A_{s,v\min}$ will be arranged on each face.

The spacing s_{long} of vertical bars shall fulfil the following requirement

$$s_{\mathrm{long}} \leq \min\ (3h;\ 400\ \mathrm{mm})$$

where h is the wall thickness.

Table 12.16 Minimum vertical reinforcement ($\geq 0.002\ A_c$) and maximum ($\leq 0.04\ A_c$) (between round brackets) at varying the wall thickness and vertical bar spacing. The reinforcement is referred to each face of the wall (*total reinforcement of the wall = tabulated value × 2*).

Wall thickness h (mm)	$A_{s,v}$	Vertical reinforcement spacing (mm)						
		100	150	200	250	300	350	400
150	min	$1\phi6$ [(a)]	$1\phi6$	$1\phi8$	$1\phi8$	$1\phi8$	$1\phi10$	$1\phi10$
	max	$(1\phi18)$	$(1\phi22)$	$(1\phi26)$	$(1\phi30)$	$(1\phi32)$		
200	min	$1\phi6$	$1\phi8$	$1\phi8$	$1\phi8$	$1\phi10$	$1\phi10$	$1\phi12$
	max	$(1\phi22)$	$(1\phi26)$	$(1\phi30)$				
250	min	$1\phi6$	$1\phi8$	$1\phi8$	$1\phi10$	$1\phi10$	$1\phi12$	$1\phi12$
	max	$(1\phi24)$	$(1\phi30)$					
300	min	$1\phi8$	$1\phi8$	$1\phi10$	$1\phi10$	$1\phi12$	$1\phi12$	$1\phi14$
	max	$(1\phi26)$	$(1\phi32)$					
350	min	$1\phi8$	$1\phi10$	$1\phi10$	$1\phi12$	$1\phi12$	$1\phi14$	$1\phi14$
	max	$(1\phi28)$						
400	min	$1\phi8$	$1\phi10$	$1\phi12$	$1\phi12$	$1\phi14$	$1\phi14$	$1\phi16$
	max	$(1\phi30)$						
450	min	$1\phi8$	$1\phi10$	$1\phi12$	$1\phi12$	$1\phi14$	$1\phi16$	$1\phi16$
	max	$(1\phi32)$						
500	min	$1\phi8$	$1\phi10$	$1\phi12$	$1\phi14$	$1\phi14$	$1\phi16$	$1\phi16$
	max							

[a] The diameter corresponding to 0.002 A_c is smaller than 6 mm

N. B. Grey boxes are corresponding to values of the reinforcement spacing exceeding the wall thickness. Maximum values have been indicated only when achievable with bars of diameter not exceeding 32 mm

It can be easily verified that, for walls with thickness h smaller than 133 mm, $s_{\mathrm{long}} = 3$ h, while for walls with thickness exceeding 134 mm, $s_{\mathrm{long}} = 400$ mm. In practical cases, the wall thickness is at least equal to 150 mm, therefore $s_{\mathrm{long}} \leq 400$ mm.

12.2.5.2 Horizontal Reinforcement

Horizontal reinforcement is required at each face, parallel to the wall faces and the free edges. EC2 suggests adopting the following minimum horizontal reinforcement

$$A_{s,hmin} \geq \max\left(0.25A_{s,v};\ 0.001\ \mathrm{A_c}\right)$$

where $A_{s,v}$ is the vertical reinforcement and A_c is the area of the transverse cross-section.

It is worth noting that the minimum vertical reinforcement is equal to 0.002 A_c and therefore

$$0.25A_{s,v} = 0.25 \cdot 0.002A_c = 0.0005A_c < 0.001A_c$$

so the expression of the minimum horizontal reinforcement simplifies as follows:

$$A_{s,hmin} \geq 0.001\ \mathrm{A_c}(= 0.1\%\ \mathrm{A_c}).$$

Finally, EC2 indicates that the spacing between two adjacent horizontal bars shall not exceed 400 mm.

12.2.5.3 Transverse Reinforcement

If the total amount of vertical reinforcement on the two wall faces exceeds 0.02 A_c, it is necessary to arrange transverse bars in the form of links, with rules analogous to those of columns (12.2.4.2). Where the main reinforcement is placed nearest to the wall faces, transverse reinforcement should also be provided in the form of links, at least 4 for each square meter of wall area. Figure 12.61 indicates the minimum reinforcement to adopt for a wall, according to EC2, for vertical, horizontal, and transverse directions.

12.2.5.4 Deep Beams

In [5.3.1 (3)], EC2 defines as deep beams those beams with a span/length ratio smaller than 3 (Fig. 12.62).

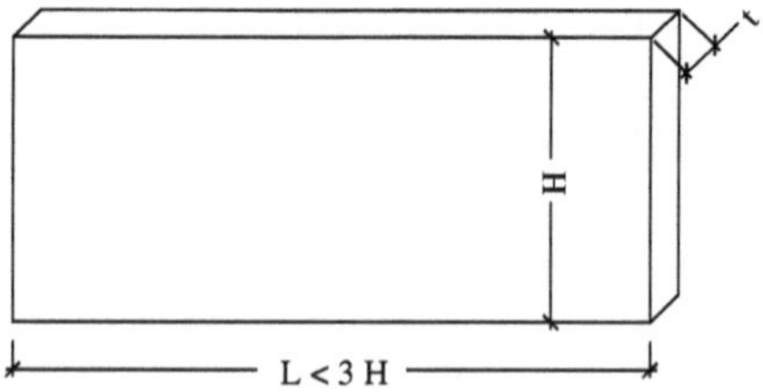

Fig. 12.62 Geometry of a deep beam

For deep beams, an orthogonal reinforcement mesh near each face should be provided, with a minimum cross-sectional area $A_{s,db\min} = \max\ (0.001\ A_c;\ 150\ \text{mm}^2/\text{m})$. If the thickness t of the deep beam is not smaller than 150 mm, it results that $0.001\ A_c \geq 150\ \text{mm}^2/\text{m}$.

Spacing between two adjacent wires of the mesh shall satisfy the following limitation:

$$s \leq \min(2t;\ 300\ \text{mm})$$

Table 12.17 collects the possible choices of the minimum orthogonal reinforcement mesh for different values of the thickness of the deep beam, considering the limitation on the maximum spacing.

Steel bars corresponding to ties considered in strut-and-tie models shall be anchored beyond the nodes [6.5.4], by raising the bars and by using U-bars or anchorage devices, if an adequate length l_{bd} between the node and the end section of the beam is not available. To this purpose, see Chap. 10, dedicated to the design of reinforced concrete members using strut-and-tie models.

Table 12.17 Minimum orthogonal reinforcement mesh $A_{s,db\min} = \max\ (0.001\ A_c;\ 150\ \text{mm}^2/\text{m})$ on each face for a deep beam at varying the thickness t

Thickness of the deep beam t (mm)	Orthogonal reinforcement mesh (mm)				
	100 × 100	150 × 150	200 × 200	250 × 250	300 × 300
150	1ϕ6	1ϕ6	1ϕ8	1ϕ8	1ϕ8
200	1ϕ6	1ϕ8	1ϕ8	1ϕ8	1ϕ10
250	1ϕ6	1ϕ8	1ϕ8	1ϕ10	1ϕ10
300	1ϕ8	1ϕ8	1ϕ10	1ϕ10	1ϕ12
350	1ϕ8	1ϕ10	1ϕ10	1ϕ12	1ϕ12
400	1ϕ8	1ϕ10	1ϕ12	1ϕ12	1ϕ14
450	1ϕ8	1ϕ10	1ϕ12	1ϕ12	1ϕ14
500	1ϕ8	1ϕ10	1ϕ12	1ϕ14	1ϕ14

Appendix
Tables and Diagrams

Abstract The Appendix collects tables and diagrams for the dimensionless verification of rectangular, circular and T sections subjected to bending or bending combined with axial force for both ultimate (U) and serviceability (E) limit state. For the ultimate limit state, tables contain the correlated values μ-ω-ξ at varying the ratio ω'/ω (being μ the dimensionless bending moment, ω and ω' the mechanical ratios of tensile and compressed reinforcement, ξ the relative depth of the neutral axis). Interaction diagrams ν-μ (being ν the dimensionless axial force) are given for axial force and monoaxial bending moment (rectangular and circular sections) as a function of ω and for axial force and biaxial bending moments (rectangular cross-section with reinforcement in the four corners). For the serviceability limit state, the geometric characteristics in dimensionless form, for rectangular and T sections in bending in state II (cracked), are tabulated. For rectangular sections with symmetrical reinforcement, interaction diagrams ν-μ representing the achievement of the stress limits $\sigma_c = 0.6 f_{ck}$ and/or $\sigma_s = 0.8 f_{yk}$ are also given. The diagrams are drawn for $f_{ck} = 20, 25, 30$ and 40 N/mm^2 since, for higher concrete strengths, the ULS verifications prevail over the SLS verifications, which become superfluous.

Tables and diagrams are distinguished with the abbreviations **U** (ultimate limit state) and **E** (serviceability limit state).[1]

[1] This appendix was authored by Piero Marro and Matteo Guiglia.

© The Editor(s) (if applicable) and The Author(s), under exclusive license to Springer Nature Switzerland AG 2022

F. Angotti et al., *Reinforced Concrete with Worked Examples*,
https://doi.org/10.1007/978-3-030-92839-1

Ultimate Limit State

Tables μ–ω–ξ for Rectangular Cross-Sections $d'/d = 0.10$

The tables, calculated referring to the generalized parabola-rectangle law for concrete [3.1.7 (1)] and the elastic-perfectly plastic law for steel [3.2.7], contain the correlated values $\mu - \omega - \xi$ (with $\omega'/\omega = 0.0$–0.1–0.2–0.3) for rectangular sections in bending, up to

$$\xi_{\lim} = \frac{\varepsilon_{cu2}}{\varepsilon_{cu2} + \varepsilon_{syd}}$$

relative depth of the neutral axis for which the deformation of the tensile reinforcement is equal to that corresponding to the design elastic limit. Furthermore, the values are highlighted which, while respecting the above limit, exceed the geometric ratio of reinforced reinforcement $A_s/A_c > 0.04$ which is the limit value specified in [9.2.1.1 (3)].

The dimensionless bending moment is:

$$\mu = \frac{M_d}{b \cdot d^2 \cdot f_{cd}}.$$

The mechanical ratios of tensile and compressed reinforcement are, respectively:

$$\omega = \frac{A_s \cdot f_{yd}}{b \cdot d \cdot f_{cd}}, \quad \omega' = \frac{A'_s \cdot f_{yd}}{b \cdot d \cdot f_{cd}}.$$

The relative depth of the neutral axis is:

$$\xi = \frac{x}{d}.$$

The tables presented below are:
U1—$f_{ck} \leq 50$ N/mm^2
U2—$f_{ck} = 60$ N/mm^2
U3—$f_{ck} = 70$ N/mm^2
U4—$f_{ck} = 80$ N/mm^2
U5—$f_{ck} = 90$ N/mm^2

- The values highlighted with gray shading correspond to the tensile reinforcements that exceed the maximum allowed design amount $A_S = 0.04\, A_c$ [9.2.1.1(3)].
- The values $\mu - \omega - \xi$ of the last box of each table correspond to (Tables U.1, U.2, U.3, U.4 and U.5):

$$\varepsilon_{syd} = \frac{f_{yd}}{E_S} = \frac{391}{200{,}000} = 0.00196$$

Table U.1 Rectangular section; $f_{ck} \leq 50\,\mathrm{N/mm^2}$; $d'/d = 0.10$

μ	$\omega'/\omega = 0.0$		$\omega'/\omega = 0.1$		$\omega'/\omega = 0.2$		$\omega'/\omega = 0.3$	
	ω	ξ	ω	ξ	ω	ξ	ω	ξ
0.0300	0.0305	0.0376	0.0301	0.0452	0.0299	0.0501	0.0304	0.0543
0.0400	0.0409	0.0505	0.0405	0.0569	0.0403	0.0612	0.0398	0.0640
0.0500	0.0514	0.0634	0.0511	0.0684	0.0510	0.0718	0.0509	0.0744
0.0600	0.0620	0.0766	0.0618	0.0798	0.0618	0.0822	0.0617	0.0840
0.0700	0.0727	0.0898	0.0727	0.0913	0.0726	0.0924	0.0726	0.0932
0.0800	0.0836	0.1033	0.0836	0.1028	0.0836	0.1024	0.0836	0.1021
0.0900	0.0946	0.1169	0.0946	0.1143	0.0946	0.1123	0.0946	0.1108
0.1000	0.1057	0.1306	0.1050	0.1251	0.1050	0.1215	0.1050	0.1188
0.1100	0.1170	0.1446	0.1166	0.1370	0.1166	0.1316	0.1166	0.1274
0.1200	0.1285	0.1587	0.1282	0.1490	0.1281	0.1416	0.1280	0.1358
0.1300	0.1401	0.1730	0.1396	0.1608	0.1394	0.1513	0.1392	0.1439
0.1400	0.1518	0.1876	0.1511	0.1726	0.1507	0.1610	0.1505	0.1518
0.1500	0.1638	0.2023	0.1628	0.1846	0.1622	0.1707	0.1618	0.1598
0.1600	0.1759	0.2173	0.1745	0.1966	0.1737	0.1804	0.1732	0.1676
0.1700	0.1882	0.2325	0.1864	0.2088	0.1853	0.1901	0.1846	0.1754
0.1800	0.2007	0.2479	0.1984	0.2210	0.1969	0.1998	0.1961	0.1832
0.1900	0.2134	0.2636	0.2105	0.2340	0.2087	0.2095	0.2076	0.1909
0.2000	0.2263	0.2796	0.2228	0.2478	0.2205	0.2193	0.2191	0.1986
0.2100	0.2395	0.2958	0.2354	0.2617	0.2323	0.2296	0.2307	0.2062
0.2200	0.2529	0.3123	0.2480	0.2758	0.2445	0.2416	0.2423	0.2138
0.2300	0.2665	0.3292	0.2609	0.2900	0.2568	0.2538	0.2540	0.2214
0.2400	0.2804	0.3464	0.2739	0.3045	0.2692	0.2660	0.2657	0.2297
0.2500	0.2946	0.3639	0.2871	0.3192	0.2817	0.2784	0.2777	0.2402
0.2600	0.3091	0.3818	0.3005	0.3341	0.2943	0.2909	0.2898	0.2506
0.2700	0.3239	0.4001	0.3142	0.3493	0.3071	0.3035	0.3020	0.2611
0.2800	0.3391	0.4189	0.3280	0.3647	0.3200	0.3163	0.3143	0.2717
0.2900	0.3546	0.4381	0.3421	0.3803	0.3331	0.3292	0.3266	0.2824
0.3000	0.3706	0.4578	0.3564	0.3962	0.3463	0.3423	0.3391	0.2932
0.3100	0.3869	0.4780	0.3709	0.4124	0.3597	0.3555	0.3517	0.3041
0.3200	0.4038	0.4988	0.3857	0.4288	0.3732	0.3688	0.3643	0.3150
0.3300	0.4211	0.5202	0.4008	0.4456	0.3869	0.3824	0.3771	0.3261
0.3400	0.4391	0.5424	0.4162	0.4627	0.4008	0.3961	0.3900	0.3372
0.3500	0.4576	0.5653	0.4319	0.4801	0.4148	0.4100	0.4029	0.3484
0.3600	0.4768	0.5890	0.4479	0.4980	0.4291	0.4240	0.4160	0.3597
0.3700	0.4968	0.6138	0.4643	0.5162	0.4435	0.4383	0.4292	0.3712
0.3800	0.5177	0.6396	0.4810	0.5348	0.4581	0.4528	0.4426	0.3827
0.3900	–	–	0.4982	0.5539	0.4730	0.4674	0.4560	0.3943
0.4000	–	–	0.5158	0.5735	0.4880	0.4823	0.4696	0.4060
0.4100	–	–	0.5339	0.5935	0.5033	0.4974	0.4833	0.4179
0.4200	–	–	0.5525	0.6142	0.5189	0.5128	0.4971	0.4298
0.4300	–	–	–	–	0.5346	0.5284	0.5111	0.4419
0.4400	–	–	–	–	0.5507	0.5442	0.5252	0.4541
0.4500	–	–	–	–	0.5670	0.5603	0.5394	0.4664
0.4600	–	–	–	–	0.5836	0.5768	0.5538	0.4789

(continued)

Table U.1 (continued)

μ	$\omega'/\omega = 0.0$		$\omega'/\omega = 0.1$		$\omega'/\omega = 0.2$		$\omega'/\omega = 0.3$	
	ω	ξ	ω	ξ	ω	ξ	ω	ξ
0.4700	–	–	–	–	0.6006	0.5935	0.5683	0.4915
0.4800	–	–	–	–	0.6178	0.6106	0.5831	0.5042
0.4900	–	–	–	–	0.6354	0.6280	0.5979	0.5170
0.5000	–	–	–	–	–	–	0.6130	0.5300
0.5100	–	–	–	–	–	–	0.6282	0.5432
0.5200	–	–	–	–	–	–	0.6435	0.5565
0.5300	–	–	–	–	–	–	0.6591	0.5700
0.5400	–	–	–	–	–	–	0.6749	0.5836
0.5500	–	–	–	–	–	–	0.6909	0.5974
0.5600	–	–	–	–	–	–	0.7070	0.6114
0.5700	–	–	–	–	–	–	0.7234	0.6256
0.5800	–	–	–	–	–	–	0.7401	0.6399
0.3807	0.5193	0.6414	–	–	–	–	–	–
0.4326	–	–	0.5766	0.6410	–	–	–	–
0.4975	–	–	–	–	0.6490	0.6413	–	–
0.5810	–	–	–	–	–	–	0.7418	0.6414

T-Section Tables

The tables, calculated referring to a rectangular diagram of the concrete stresses [3.1.8] and a elastic-perfectly plastic diagram of the steel [3.2.7], show the correlated values $\mu - \xi - \omega$ for sections in bending, starting from the values of ξ for which the flange is fully compressed. For lower ξ values, the compression block affects only a part of the flange and the sections behave as if they were rectangular. In these cases, Tables U.1, U.2, U.3, U.4 and U.5 apply. The last value ξ of the tables corresponds to the configuration

$$\xi_{\text{lim}} = \frac{\varepsilon_{cu2}}{\varepsilon_{cu2} + \varepsilon_{syd}}$$

The dimensionless bending moment is:

$$\mu = \frac{M_d}{b \cdot d^2 \cdot f_{cd}}$$

The mechanical ratios of tensile and compressed reinforcement are, respectively:

$$\omega = \frac{A_s \cdot f_{yd}}{b \cdot d \cdot f_{cd}}, \omega' = \frac{A'_s \cdot f_{yd}}{b \cdot d \cdot f_{cd}}$$

The relative depth of the neutral axis is:

$$\xi = \frac{x}{d}.$$

Table U.2 Rectangular section; $f_{ck} = 60\,\mathrm{N/mm^2}$; $d'/d = 0.10$

μ	$\omega'/\omega = 0.0$		$\omega'/\omega = 0.1$		$\omega'/\omega = 0.2$		$\omega'/\omega = 0.3$	
	ω	ξ	ω	ξ	ω	ξ	ω	ξ
0.0300	0.0305	0.0439	0.0310	0.0510	0.0308	0.0551	0.0308	0.0583
0.0400	0.0409	0.0589	0.0406	0.0634	0.0404	0.0667	0.0403	0.0694
0.0500	0.0514	0.0740	0.0513	0.0771	0.0512	0.0794	0.0512	0.0812
0.0600	0.0621	0.0893	0.0620	0.0906	0.0620	0.0916	0.0620	0.0925
0.0700	0.0729	0.1049	0.0729	0.1042	0.0729	0.1038	0.0729	0.1034
0.0800	0.0838	0.1206	0.0838	0.1179	0.0838	0.1158	0.0839	0.1141
0.0900	0.0949	0.1365	0.0949	0.1317	0.0949	0.1277	0.0949	0.1246
0.1000	0.1061	0.1527	0.1053	0.1446	0.1053	0.1390	0.1053	0.1343
0.1100	0.1175	0.1691	0.1169	0.1590	0.1169	0.1513	0.1169	0.1450
0.1200	0.1290	0.1857	0.1286	0.1735	0.1285	0.1635	0.1283	0.1554
0.1300	0.1408	0.2025	0.1402	0.1877	0.1398	0.1755	0.1396	0.1655
0.1400	0.1526	0.2197	0.1518	0.2021	0.1513	0.1876	0.1510	0.1756
0.1500	0.1647	0.2370	0.1636	0.2166	0.1628	0.1996	0.1623	0.1857
0.1600	0.1770	0.2547	0.1754	0.2312	0.1745	0.2118	0.1738	0.1957
0.1700	0.1895	0.2727	0.1875	0.2460	0.1862	0.2239	0.1853	0.2057
0.1800	0.2022	0.2909	0.1996	0.2610	0.1980	0.2362	0.1969	0.2157
0.1900	0.2151	0.3095	0.2119	0.2761	0.2099	0.2485	0.2085	0.2257
0.2000	0.2283	0.3285	0.2244	0.2915	0.2219	0.2609	0.2202	0.2357
0.2100	0.2417	0.3478	0.2370	0.3070	0.2340	0.2734	0.2319	0.2457
0.2200	0.2554	0.3675	0.2499	0.3236	0.2461	0.2859	0.2437	0.2557
0.2300	0.2694	0.3876	0.2630	0.3406	0.2584	0.2985	0.2556	0.2658
0.2400	0.2837	0.4082	0.2763	0.3578	0.2708	0.3118	0.2675	0.2758
0.2500	0.2983	0.4292	0.2897	0.3752	0.2836	0.3264	0.2795	0.2859
0.2600	0.3133	0.4508	0.3035	0.3930	0.2964	0.3412	0.2915	0.2960
0.2700	0.3286	0.4729	0.3174	0.4110	0.3094	0.3562	0.3036	0.3061
0.2800	0.3444	0.4955	0.3316	0.4294	0.3225	0.3713	0.3159	0.3182
0.2900	0.3606	0.5188	0.3460	0.4481	0.3359	0.3866	0.3285	0.3309
0.3000	0.3773	0.5428	0.3608	0.4672	0.3493	0.4021	0.3412	0.3436
0.3100	0.3945	0.5676	0.3758	0.4866	0.3630	0.4178	0.3539	0.3565
0.3200	0.4123	0.5932	0.3911	0.5065	0.3768	0.4338	0.3667	0.3694
0.3300	–	–	0.4067	0.5267	0.3908	0.4499	0.3797	0.3825
0.3400	–	–	0.4227	0.5474	0.4050	0.4663	0.3928	0.3956
0.3500	–	–	0.4391	0.5686	0.4195	0.4829	0.4060	0.4089
0.3600	–	–	0.4558	0.5903	0.4341	0.4997	0.4193	0.4224
0.3700	–	–	0.4604	0.5962	0.4489	0.5168	0.4328	0.4359
0.3800	–	–	–	–	0.4640	0.5342	0.4463	0.4496
0.3900	–	–	–	–	0.4794	0.5518	0.4601	0.4634
0.4000	–	–	–	–	0.4949	0.5697	0.4739	0.4774
0.4100	–	–	–	–	0.5108	0.5880	0.4879	0.4915
0.4200	–	–	–	–	–	–	0.5021	0.5057
0.4300	–	–	–	–	–	–	0.5164	0.5201
0.4400	–	–	–	–	–	–	0.5308	0.5347
0.4500	–	–	–	–	–	–	0.5455	0.5494
0.4600	–	–	–	–	–	–	0.5603	0.5643
0.4700	–	–	–	–	–	–	0.5752	0.5794
0.4800	–	–	–	–	–	–	0.5904	0.5947
0.3215	0.4150	0.5971			–	–	–	–
0.3630	–	–	0.4604	0.5962	–	–	–	–
0.4149	–	–	–	–	0.5187	0.5971	–	–
0.4816	–	–	–	–	–	–	0.5928	0.5971

Table U.3 Rectangular section; $f_{ck} = 70\,\text{N/mm}^2$; $d'/d = 0.10$

μ	$\omega'/\omega = 0.0$		$\omega'/\omega = 0.1$		$\omega'/\omega = 0.2$		$\omega'/\omega = 0.3$	
	ω	ξ	ω	ξ	ω	ξ	ω	ξ
0.0300	0.0305	0.0479	0.0308	0.0541	0.0307	0.0578	0.0306	0.0608
0.0400	0.0410	0.0643	0.0404	0.0676	0.0402	0.0704	0.0401	0.0727
0.0500	0.0515	0.0808	0.0514	0.0830	0.0514	0.0846	0.0513	0.0860
0.0600	0.0622	0.0976	0.0622	0.0979	0.0622	0.0981	0.0622	0.0983
0.0700	0.0730	0.1146	0.0730	0.1128	0.0731	0.1114	0.0731	0.1103
0.0800	0.0840	0.1318	0.0840	0.1279	0.0840	0.1247	0.0841	0.1221
0.0900	0.0951	0.1493	0.0951	0.1431	0.0951	0.1379	0.0951	0.1337
0.1000	0.1064	0.1670	0.1056	0.1573	0.1056	0.1504	0.1056	0.1445
0.1100	0.1179	0.1850	0.1172	0.1732	0.1172	0.1641	0.1172	0.1564
0.1200	0.1295	0.2033	0.1291	0.1894	0.1288	0.1777	0.1286	0.1680
0.1300	0.1414	0.2218	0.1406	0.2051	0.1402	0.1911	0.1400	0.1794
0.1400	0.1534	0.2407	0.1524	0.2211	0.1518	0.2046	0.1513	0.1907
0.1500	0.1656	0.2599	0.1642	0.2372	0.1634	0.2181	0.1628	0.2021
0.1600	0.1780	0.2794	0.1762	0.2535	0.1751	0.2317	0.1743	0.2134
0.1700	0.1907	0.2992	0.1884	0.2700	0.1869	0.2454	0.1859	0.2247
0.1800	0.2035	0.3194	0.2007	0.2867	0.1988	0.2591	0.1975	0.2360
0.1900	0.2167	0.3400	0.2132	0.3036	0.2108	0.2730	0.2093	0.2474
0.2000	0.2301	0.3611	0.2258	0.3207	0.2230	0.2870	0.2210	0.2588
0.2100	0.2438	0.3826	0.2386	0.3381	0.2352	0.3011	0.2329	0.2702
0.2200	0.2577	0.4045	0.2516	0.3557	0.2476	0.3153	0.2448	0.2816
0.2300	0.2720	0.4269	0.2648	0.3740	0.2600	0.3296	0.2568	0.2931
0.2400	0.2867	0.4499	0.2784	0.3932	0.2726	0.3441	0.2689	0.3046
0.2500	0.3017	0.4735	0.2921	0.4126	0.2853	0.3586	0.2810	0.3162
0.2600	0.3171	0.4977	0.3061	0.4324	0.2981	0.3743	0.2932	0.3278
0.2700	0.3330	0.5226	0.3204	0.4525	0.3114	0.3910	0.3055	0.3394
0.2800	0.3493	0.5482	0.3349	0.4730	0.3248	0.4078	0.3178	0.3511
0.2900	0.3662	0.5747	0.3497	0.4939	0.3383	0.4248	0.3303	0.3629
0.3000	–	–	0.3648	0.5153	0.3521	0.4420	0.3430	0.3768
0.3100	–	–	0.3803	0.5371	0.3660	0.4595	0.3559	0.3910
0.3200	–	–	0.3961	0.5594	0.3801	0.4772	0.3689	0.4053
0.3300	–	–	–	–	0.3944	0.4952	0.3821	0.4197
0.3400	–	–	–	–	0.4089	0.5134	0.3953	0.4343
0.3500	–	–	–	–	0.4237	0.5319	0.4087	0.4490
0.3600	–	–	–	–	0.4387	0.5508	0.4223	0.4639
0.3700	–	–	–	–	0.4539	0.5699	0.4360	0.4789
0.3800	–	–	–	–	–	–	0.4498	0.4941
0.3900	–	–	–	–	–	–	0.4637	0.5095
0.4000	–	–	–	–	–	–	0.4779	0.5250
0.4100	–	–	–	–	–	–	0.4922	0.5407
0.4200	–	–	–	–	–	–	0.5066	0.5565
0.4300	–	–	–	–	–	–	0.5212	0.5726
0.2919	0.3695	0.5798	–	–	–	–	–	–
0.3289	–	–	0.4104	0.5797	–	–	–	–
0.3750	–	–	–	–	0.4618	0.5798	–	–
0.4344	–	–	–	–	–	–	0.5264	0.5783

Table U.4 Rectangular section; $f_{ck} = 80\,\mathrm{N/mm^2}$; $d'/d = 0.10$

μ	$\omega'/\omega = 0.0$		$\omega'/\omega = 0.1$		$\omega'/\omega = 0.2$		$\omega'/\omega = 0.3$	
	ω	ξ	ω	ξ	ω	ξ	ω	ξ
0.0300	0.0306	0.0510	0.0308	0.0566	0.0307	0.0601	0.0306	0.0630
0.0400	0.0410	0.0684	0.0403	0.0710	0.0402	0.0734	0.0400	0.0755
0.0500	0.0516	0.0860	0.0515	0.0875	0.0515	0.0888	0.0514	0.0897
0.0600	0.0623	0.1039	0.0623	0.1034	0.0623	0.1031	0.0623	0.1028
0.0700	0.0732	0.1221	0.0732	0.1194	0.0732	0.1173	0.0732	0.1156
0.0800	0.0842	0.1405	0.0842	0.1356	0.0842	0.1315	0.0842	0.1282
0.0900	0.0954	0.1591	0.0953	0.1518	0.0953	0.1457	0.0953	0.1407
0.1000	0.1067	0.1781	0.1058	0.1671	0.1058	0.1591	0.1058	0.1523
0.1100	0.1183	0.1973	0.1175	0.1841	0.1174	0.1738	0.1174	0.1651
0.1200	0.1300	0.2169	0.1294	0.2015	0.1291	0.1886	0.1289	0.1776
0.1300	0.1419	0.2368	0.1411	0.2185	0.1406	0.2030	0.1403	0.1899
0.1400	0.1540	0.2570	0.1529	0.2356	0.1522	0.2175	0.1517	0.2021
0.1500	0.1664	0.2776	0.1649	0.2530	0.1639	0.2321	0.1632	0.2144
0.1600	0.1790	0.2986	0.1770	0.2706	0.1757	0.2468	0.1748	0.2267
0.1700	0.1918	0.3200	0.1893	0.2884	0.1876	0.2616	0.1864	0.2389
0.1800	0.2048	0.3418	0.2017	0.3064	0.1996	0.2765	0.1982	0.2513
0.1900	0.2182	0.3640	0.2143	0.3247	0.2117	0.2916	0.2099	0.2636
0.2000	0.2318	0.3868	0.2271	0.3433	0.2240	0.3068	0.2218	0.2760
0.2100	0.2458	0.4100	0.2401	0.3621	0.2364	0.3221	0.2338	0.2884
0.2200	0.2600	0.4338	0.2533	0.3812	0.2488	0.3375	0.2458	0.3009
0.2300	0.2747	0.4583	0.2667	0.4006	0.2615	0.3531	0.2579	0.3134
0.2400	0.2897	0.4833	0.2804	0.4210	0.2742	0.3689	0.2700	0.3260
0.2500	0.3051	0.5091	0.2944	0.4421	0.2871	0.3848	0.2823	0.3387
0.2600	0.3210	0.5356	0.3087	0.4636	0.3001	0.4008	0.2946	0.3514
0.2700	0.3374	0.5629	0.3233	0.4854	0.3133	0.4182	0.3071	0.3642
0.2800	–	–	0.3381	0.5077	0.3270	0.4365	0.3196	0.3770
0.2900	–	–	0.3533	0.5305	0.3408	0.4548	0.3322	0.3899
0.3000	–	–	0.3688	0.5538	0.3547	0.4734	0.3448	0.4029
0.3100	–	–	–	–	0.3689	0.4924	0.3578	0.4179
0.3200	–	–	–	–	0.3833	0.5116	0.3710	0.4333
0.3300	–	–	–	–	0.3979	0.5311	0.3843	0.4489
0.3400	–	–	–	–	0.4127	0.5509	0.3978	0.4646
0.3500	–	–	–	–	–	–	0.4114	0.4805
0.3600	–	–	–	–	–	–	0.4251	0.4965
0.3700	–	–	–	–	–	–	0.4390	0.5128
0.3800	–	–	–	–	–	–	0.4531	0.5292
0.3900	–	–	–	–	–	–	0.4673	0.5458
0.4000	–	–	–	–	–	–	0.4817	0.5626
0.2728	0.3420	0.5706	–	–	–	–	–	–
0.3070	–	–	0.3799	0.5705	–	–	–	–
0.3497	–	–	–	–	0.4274	0.5705	–	–
0.4046	–	–	–	–	–	–	0.4885	0.5705

Table U.5 Rectangular section; $f_{ck} = 90\,\text{N/mm}^2$; $d'/d = 0.10$

μ	$\omega'/\omega = 0.0$		$\omega'/\omega = 0.1$		$\omega'/\omega = 0.2$		$\omega'/\omega = 0.3$	
	ω	ξ	ω	ξ	ω	ξ	ω	ξ
0.0300	0.0306	0.0524	0.0308	0.0579	0.0307	0.0614	0.0306	0.0641
0.0400	0.0410	0.0703	0.0403	0.0726	0.0402	0.0750	0.0400	0.0769
0.0500	0.0516	0.0885	0.0515	0.0897	0.0515	0.0907	0.0515	0.0915
0.0600	0.0624	0.1069	0.0623	0.1061	0.0624	0.1054	0.0624	0.1049
0.0700	0.0732	0.1256	0.0732	0.1225	0.0733	0.1200	0.0733	0.1180
0.0800	0.0843	0.1445	0.0843	0.1391	0.0843	0.1346	0.0843	0.1309
0.0900	0.0955	0.1638	0.0954	0.1558	0.0954	0.1492	0.0954	0.1437
0.1000	0.1069	0.1833	0.1059	0.1716	0.1059	0.1629	0.1059	0.1557
0.1100	0.1185	0.2031	0.1176	0.1890	0.1176	0.1781	0.1175	0.1688
0.1200	0.1303	0.2233	0.1297	0.2070	0.1293	0.1932	0.1291	0.1816
0.1300	0.1422	0.2438	0.1414	0.2245	0.1408	0.2081	0.1405	0.1942
0.1400	0.1544	0.2647	0.1532	0.2422	0.1524	0.2230	0.1519	0.2068
0.1500	0.1668	0.2860	0.1652	0.2601	0.1642	0.2380	0.1634	0.2194
0.1600	0.1795	0.3077	0.1774	0.2782	0.1760	0.2532	0.1750	0.2320
0.1700	0.1924	0.3298	0.1897	0.2966	0.1879	0.2684	0.1867	0.2446
0.1800	0.2056	0.3524	0.2022	0.3152	0.2000	0.2838	0.1984	0.2573
0.1900	0.2190	0.3755	0.2149	0.3341	0.2122	0.2993	0.2103	0.2700
0.2000	0.2328	0.3991	0.2278	0.3533	0.2245	0.3150	0.2222	0.2827
0.2100	0.2469	0.4232	0.2409	0.3728	0.2369	0.3308	0.2341	0.2955
0.2200	0.2613	0.4480	0.2542	0.3926	0.2494	0.3467	0.2462	0.3083
0.2300	0.2761	0.4734	0.2677	0.4130	0.2621	0.3628	0.2583	0.3212
0.2400	0.2914	0.4995	0.2816	0.4345	0.2749	0.3791	0.2705	0.3342
0.2500	0.3070	0.5264	0.2957	0.4563	0.2879	0.3955	0.2828	0.3472
0.2600	0.3232	0.5541	0.3102	0.4785	0.3010	0.4128	0.2952	0.3603
0.2700	–	–	0.3249	0.5013	0.3145	0.4314	0.3077	0.3735
0.2800	–	–	0.3399	0.5245	0.3282	0.4501	0.3202	0.3867
0.2900	–	–	0.3553	0.5483	0.3421	0.4692	0.3329	0.4000
0.3000	–	–	–	–	0.3562	0.4885	0.3456	0.4147
0.3100	–	–	–	–	0.3705	0.5081	0.3589	0.4307
0.3200	–	–	–	–	0.3851	0.5281	0.3722	0.4466
0.3300	–	–	–	–	0.3998	0.5484	0.3856	0.4627
0.3400	–	–	–	–	0.4149	0.5690	0.3992	0.4790
0.3500	–	–	–	–	–	–	0.4129	0.4955
0.3600	–	–	–	–	–	–	0.4268	0.5121
0.3700	–	–	–	–	–	–	0.4408	0.5289
0.3800	–	–	–	–	–	–	0.4550	0.5460
0.3900	–	–	–	–	–	–	0.4693	0.5632
0.2658	0.3329	0.5706	–	–	–	–	–	–
0.2991	–	–	0.3697	0.5704	–	–	–	–
0.3407	–	–	–	–	0.4149	0.5690	–	–
0.3942	–	–	–	–	–	–	0.4740	0.5689

The tables, relating to concrete $f_{ck} \leq 50$ N/mm^2, are classified according to $w = h_f/d$ (h_f being the thickness of the flange):
U6—$h_f/d = 0.15; f_{ck} \leq 50$ N/mm^2
U7—$h_f/d = 0.20; f_{ck} \leq 50$ N/mm^2
U8—$h_f/d = 0.25; f_{ck} \leq 50$ N/mm^2
U9—$h_f/d = 0.30; f_{ck} \leq 50$ N/mm^2

- The values highlighted with gray shading correspond to the tensile reinforcements that exceed the maximum allowed design amount $A_S = 0.04\,A_c$ [9.2.1.1(3)] with $f_{ck} = 50\,\text{N/mm}^2$ (Tables U.6, U.7, U.8 and U.9).

U.10–U.22—Interaction Diagrams $\nu - \mu$ for Rectangular and Circular Cross-Sections

Interaction diagrams $\nu - \mu$ for rectangular and circular sections, as a function of ω. They are calculated referring to the generalized parabola—rectangle law of concrete [3.1.7 (1)] and the elastic-perfectly plastic diagram of steel [3.2.7]. The diagram also represents the deformed configurations corresponding to:

- $\xi_{\lim}$, for which the maximum resistant moment is achieved with tensile reinforcement at the design elastic limit,
- tensile reinforcement at zero stress,
- top fiber of concrete with zero stress,
- uniformly compressed section.

ν, μ, ω are expressed as a function of f_{cd}.
The diagrams presented below are:
Rectangular section with symmetrical reinforcement
U10—$d'/h = 0.10; f_{ck} \leq 50$ N/mm^2
U11—$d'/h = 0.10; f_{ck} = 60$ N/mm^2
U12—$d'/h = 0.10; f_{ck} = 70$ N/mm^2
U13—$d'/h = 0.10; f_{ck} = 80$ N/mm^2
U14—$d'/h = 0.10; f_{ck} = 90$ N/mm^2
U15—$d'/h = 0.05; f_{ck} \leq 50$ N/mm^2
Rectangular section with reinforcement distributed on 4 sides
U16—$d'/h = 0.10; f_{ck} \leq 50$ N/mm^2
U17—$d'/h = 0.05; f_{ck} \leq 50$ N/mm^2
Circular section
U18—$d'/h = 0.10; f_{ck} \leq 50$ N/mm^2
U19—$d'/h = 0.10; f_{ck} = 60$ N/mm^2
U20—$d'/h = 0.10; f_{ck} = 70$ N/mm^2
U21—$d'/h = 0.10; f_{ck} = 80$ N/mm^2
U22—$d'/h = 0.10; f_{ck} = 90$ N/mm^2.

Table U.6 T-Section; $f_{ck} \leq 50\,\mathrm{N/mm^2}$; $w = h_f/d = 0.15$

$n = b/b_w = 2$				
μ	ξ	ω	$1000\ \varepsilon_c$	$1000\ \varepsilon_s$
0.1386	0.1871	0.1498	−3.5	15.21
0.1450	0.2060	0.1574	−3.5	13.49
0.1500	0.2211	0.1634	−3.5	12.33
0.1550	0.2364	0.1696	−3.5	11.30
0.1600	0.2520	0.1758	−3.5	10.39
0.1650	0.2677	0.1821	−3.5	9.57
0.1700	0.2838	0.1885	−3.5	8.83
0.1750	0.3001	0.1950	−3.5	8.16
0.1800	0.3167	0.2017	−3.5	7.55
0.1850	0.3336	0.2084	−3.5	6.99
0.1900	0.3508	0.2153	−3.5	6.48
0.1950	0.3683	0.2223	−3.5	6.00
0.2000	0.3862	0.2295	−3.5	5.56
0.2050	0.4045	0.2368	−3.5	5.15
0.2100	0.4232	0.2443	−3.5	4.77
0.2150	0.4423	0.2519	−3.5	4.41
0.2200	0.4619	0.2598	−3.5	4.08
0.2250	0.4820	0.2678	−3.5	3.76
0.2300	0.5026	0.2760	−3.5	3.46
0.2350	0.5238	0.2845	−3.5	3.18
0.2400	0.5457	0.2933	−3.5	2.91
0.2450	0.5682	0.3023	−3.5	2.66
0.2500	0.5915	0.3116	−3.5	2.42
0.2550	0.6157	0.3213	−3.5	2.18
0.2600	0.6408	0.3313	−3.5	1.96
0.2601	0.6414	0.3316	−3.5	1.96

μ	$n = b/b_w = 4$		$n = b/b_w = 5$		$n = b/b_w = 6$	
	ξ	ω	ξ	ω	ξ	ω
0.1386	0.1866	0.1498	0.1864	0.1498	0.1862	0.1498
0.1450	0.2249	0.1575	0.2442	0.1576	0.2442	0.1576
0.1500	0.2559	0.1637	0.2919	0.1639	0.2919	0.1639
0.1550	0.2878	0.1701	0.3421	0.1706	0.3421	0.1706
0.1600	0.3209	0.1767	0.3953	0.1777	0.3953	0.1777
0.1650	0.3551	0.1835	0.4521	0.1853	0.4521	0.1853
0.1700	0.3908	0.1907	0.5131	0.1934	0.5131	0.1934

Table U.7 T-Section; $f_{ck} \leq 50\,\text{N/mm}^2$; $w = h_f/d = 0.20$

$n = b/b_w = 2$				
μ	ξ	ω	$1000\ \varepsilon_c$	$1000\ \varepsilon_s$
0.1800	0.2500	0.2000	−3.5	10.50
0.1850	0.2657	0.2063	−3.5	9.67
0.1900	0.2818	0.2127	−3.5	8.92
0.1950	0.2980	0.2192	−3.5	8.24
0.2000	0.3146	0.2258	−3.5	7.63
0.2050	0.3314	0.2326	−3.5	7.06
0.2100	0.3486	0.2394	−3.5	6.54
0.2150	0.3661	0.2464	−3.5	6.06
0.2200	0.3840	0.2536	−3.5	5.62
0.2250	0.4022	0.2609	−3.5	5.20
0.2300	0.4208	0.2683	−3.5	4.82
0.2350	0.4399	0.2760	−3.5	4.46
0.2400	0.4594	0.2838	−3.5	4.12
0.2450	0.4794	0.2918	−3.5	3.80
0.2500	0.5000	0.3000	−3.5	3.50
0.2550	0.5211	0.3085	−3.5	3.22
0.2600	0.5429	0.3172	−3.5	2.95
0.2650	0.5653	0.3261	−3.5	2.69
0.2700	0.5886	0.3354	−3.5	2.45
0.2750	0.6126	0.3450	−3.5	2.21
0.2800	0.6376	0.3551	−3.5	1.99
0.2807	0.6414	0.3566	−3.50	1.96

μ	$n = b/b_w = 4$		$n = b/b_w = 5$		$n = b/b_w = 6$	
	ξ	ω	ξ	ω	ξ	ω
0.1800	0.2500	0.2000	0.2500	0.2000	0.2500	0.2000
0.1850	0.2818	0.2064	0.2899	0.2064	0.2980	0.2064
0.1900	0.3146	0.2129	0.3314	0.2130	0.3486	0.2131
0.1950	0.3486	0.2197	0.3750	0.2200	0.4022	0.2203
0.2000	0.3840	0.2268	0.4208	0.2273	0.4594	0.2279
0.2050	0.4208	0.2342	0.4694	0.2351	0.5211	0.2362
0.2100	0.4594	0.2419	0.5211	0.2434	0.5886	0.2451
0.2150	0.5000	0.2500	0.5769	0.2523	0.6414	0.2522
0.2200	0.5429	0.2586	0.6376	0.2620	–	–
0.2250	0.5886	0.2677	0.6414	0.2626	–	–
0.2300	0.6376	0.2775	–	–	–	–

Table U.8 T-Section; $f_{ck} \leq 50\,\mathrm{N/mm^2}$; $w = h_f/d = 0.25$

$n = b/b_w = 2$				
μ	ξ	ω	$1000\,\varepsilon_c$	$1000\,\varepsilon_s$
0.2188	0.3126	0.2500	–3.5	7.70
0.2250	0.3336	0.2584	–3.5	6.99
0.2300	0.3508	0.2653	–3.5	6.48
0.2350	0.3683	0.2723	–3.5	6.00
0.2400	0.3862	0.2795	–3.5	5.56
0.2450	0.4045	0.2868	–3.5	5.15
0.2500	0.4232	0.2943	–3.5	4.77
0.2550	0.4423	0.3019	–3.5	4.41
0.2600	0.4619	0.3098	–3.5	4.08
0.2650	0.4820	0.3178	–3.5	3.76
0.2700	0.5026	0.3260	–3.5	3.46
0.2750	0.5238	0.3345	–3.5	3.18
0.2800	0.5457	0.3433	–3.5	2.91
0.2850	0.5682	0.3523	–3.5	2.66
0.2900	0.5915	0.3616	–3.5	2.42
0.2950	0.6157	0.3713	–3.5	2.18
0.3000	0.6408	0.3813	–3.5	1.96

μ	$n = b/b_w = 4$		$n = b/b_w = 5$		$n = b/b_w = 6$	
	ξ	ω	ξ	ω	ξ	ω
0.2188	0.3126	0.2500	0.3127	0.2500	0.3127	0.2500
0.2238	0.3466	0.2568	0.3553	0.2569	0.3641	0.2569
0.2288	0.3819	0.2639	0.4001	0.2640	0.4187	0.2642
0.2338	0.4187	0.2712	0.4474	0.2716	0.4772	0.2720
0.2388	0.4571	0.2789	0.4976	0.2796	0.5404	0.2804
0.2438	0.4976	0.2870	0.5515	0.2882	0.6099	0.2897
0.2488	0.5403	0.2956	0.6098	0.2976	0.6415	0.2939
0.2538	0.5858	0.3047	0.6414	0.3026	-	-
0.2588	0.6347	0.3144	-	-	-	-
0.2594	0.6414	0.3158	-	-	-	-

Table U.9 T-Section; $f_{ck} \leq 50\,\text{N/mm}^2$; $w = h_f/d = 0.30$

$n = b/b_w = 2$				
μ	ξ	ω	$1000\ \varepsilon_c$	$1000\ \varepsilon_s$
0.2500	0.3573	0.2929	-3.5	6.30
0.2550	0.3750	0.3000	-3.5	5.83
0.2600	0.3930	0.3072	-3.5	5.40
0.2650	0.4115	0.3146	-3.5	5.01
0.2700	0.4303	0.3221	-3.5	4.63
0.2750	0.4496	0.3298	-3.5	4.28
0.2800	0.4694	0.3378	-3.5	3.96
0.2850	0.4897	0.3459	-3.5	3.65
0.2900	0.5105	0.3542	-3.5	3.36
0.2950	0.5319	0.3628	-3.5	3.08
0.3000	0.5540	0.3716	-3.5	2.82
0.3050	0.5769	0.3807	-3.5	2.57
0.3100	0.6005	0.3902	-3.5	2.33
0.3150	0.6250	0.4000	-3.5	2.10
0.3182	0.6414	0.4066	-3.5	1.96

μ	$n = b/b_w = 4$		$n = b/b_w = 5$		$n = b/b_w = 6$	
	ξ	ω	ξ	ω	ξ	ω
0.2500	0.3400	0.2930	0.3314	0.2930	0.3230	0.2931
0.2550	0.3750	0.3000	0.3750	0.3000	0.3750	0.3000
0.2600	0.4115	0.3073	0.4208	0.3073	0.4303	0.3074
0.2650	0.4496	0.3149	0.4694	0.3151	0.4897	0.3153
0.2700	0.4897	0.3229	0.5211	0.3234	0.5540	0.3239
0.2750	0.5319	0.3314	0.5769	0.3323	0.6250	0.3333
0.2800	0.5769	0.3404	0.6376	0.3420	0.6414	0.3355
0.2850	0.6250	0.3500	0.6415	0.3426	-	-
0.2900	0.6414	0.3533	-	-	-	-

U.23–U.27—Interaction Diagrams ν − μ of Rectangular Cross-Section with Reinforcement in the Four Corners Under Axial Force and Biaxial Bending Moments

Interaction diagrams $\nu - \mu$ for axial force and biaxial bending moments; rectangular cross-section with reinforcement in the four corners.

They are calculated referring to the generalized parabola—rectangle law of concrete [3.1.7 (1)] and the elastic-perfectly plastic diagram of steel [3.2.7].

The diagrams have 8 sectors, each characterized by a dimensionless axial force value $\nu = 0.0$; $\nu = 0.2$; …; $\nu = 1.4$; in each sector the bending moments relative to the two main directions are combined.

The diagrams are:

U23—Rectangular Section; $f_{ck} \leq 50$ N/mm^2

U24—Rectangular Section; $f_{ck} = 60$ N/mm^2

U25—Rectangular Section; $f_{ck} = 70$ N/mm^2

U26—Rectangular Section; $f_{ck} = 80$ N/mm^2

U27—Rectangular Section; $f_{ck} = 90$ N/mm^2

Serviceability Limit State

Tables for Rectangular Cross-Sections: ρ–ξ–i, D′/d = 0.10, α_e = 15 and α_e = 9 (α_e Notional Modular Ratio)

Tables representing the geometric characteristics in dimensionless form, for rectangular sections in bending in state II (cracked). The main parameter, ρ, is the geometric ratio of the tensile reinforcement. The tables are developed up to $\rho = 0.04$. The tables are given as a function of ρ', the geometric ratio of the compressed reinforcement.

$$\rho = \frac{A_S}{bd}, \quad \rho' = \frac{A'_S}{bd}$$

$d'/d = 0.1$. Each table provides, as a function of $\alpha_e = 15$ (concrete strength $f_{ck} \leq 50$) and $\alpha_e = 9$ (concrete strength $f_{ck} > 50$ N/mm^2), the following correlations.

$$\rho - \xi - i, \text{ with } \xi = \frac{x}{d} \text{ e } i = \frac{I}{bd^3}$$

The notional modular ratio, α_e, is assumed according to the criteria of Sect. 11.2.

To facilitate the calculation of the stresses σ_c and σ_s, the values $\frac{\xi}{i}$ and $\frac{1-\xi}{i}$ are also reported.

The tables are (Tables E.1, E.2, E.3 and E.4):

E.1—$\rho' = 0$

E.2—$\rho' = 0.1\ \rho$

E.3—$\rho' = 0.2\ \rho$

E.4—$\rho' = 0.3\ \rho$

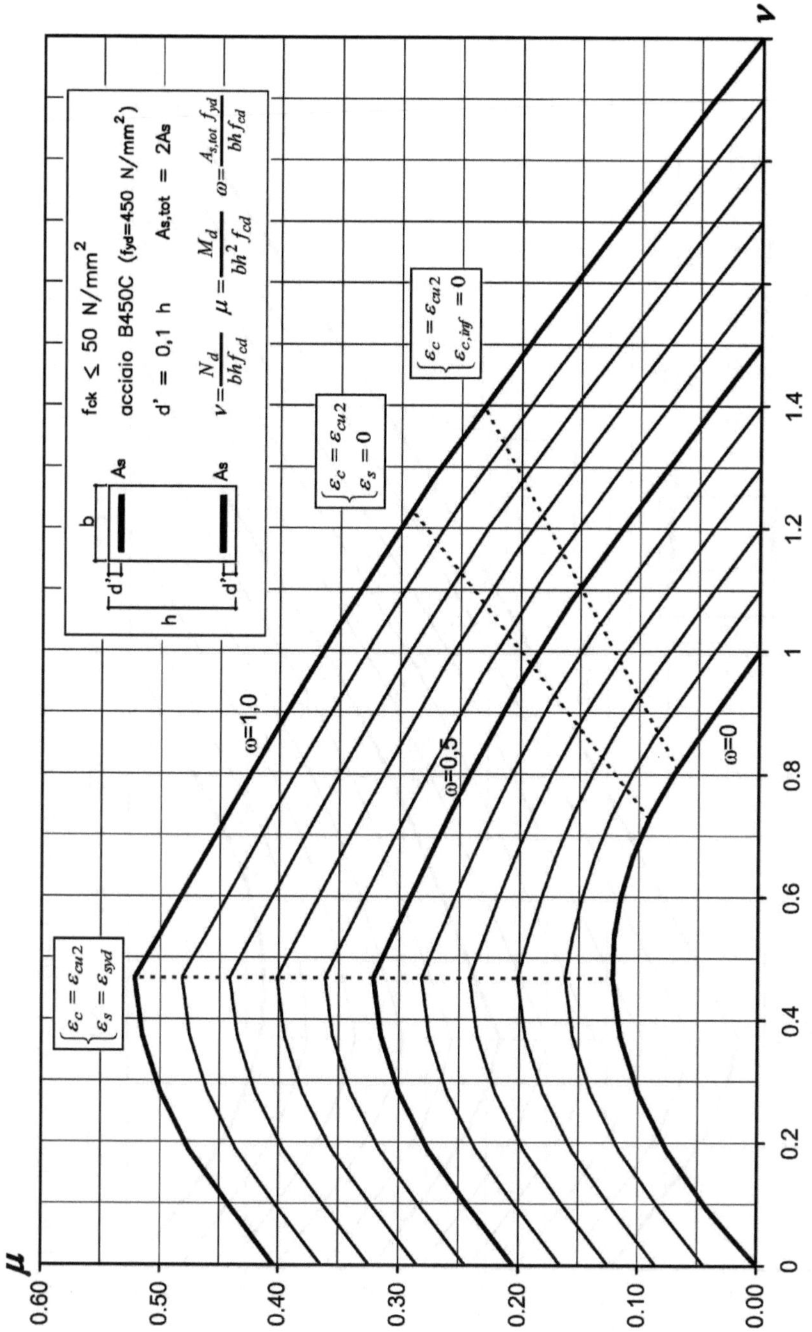

Diagram U.10 Rectangular section; $f_{ck} \leq 50\,\text{N/mm}^2$; steel B450C;$d'/d = 0.1$

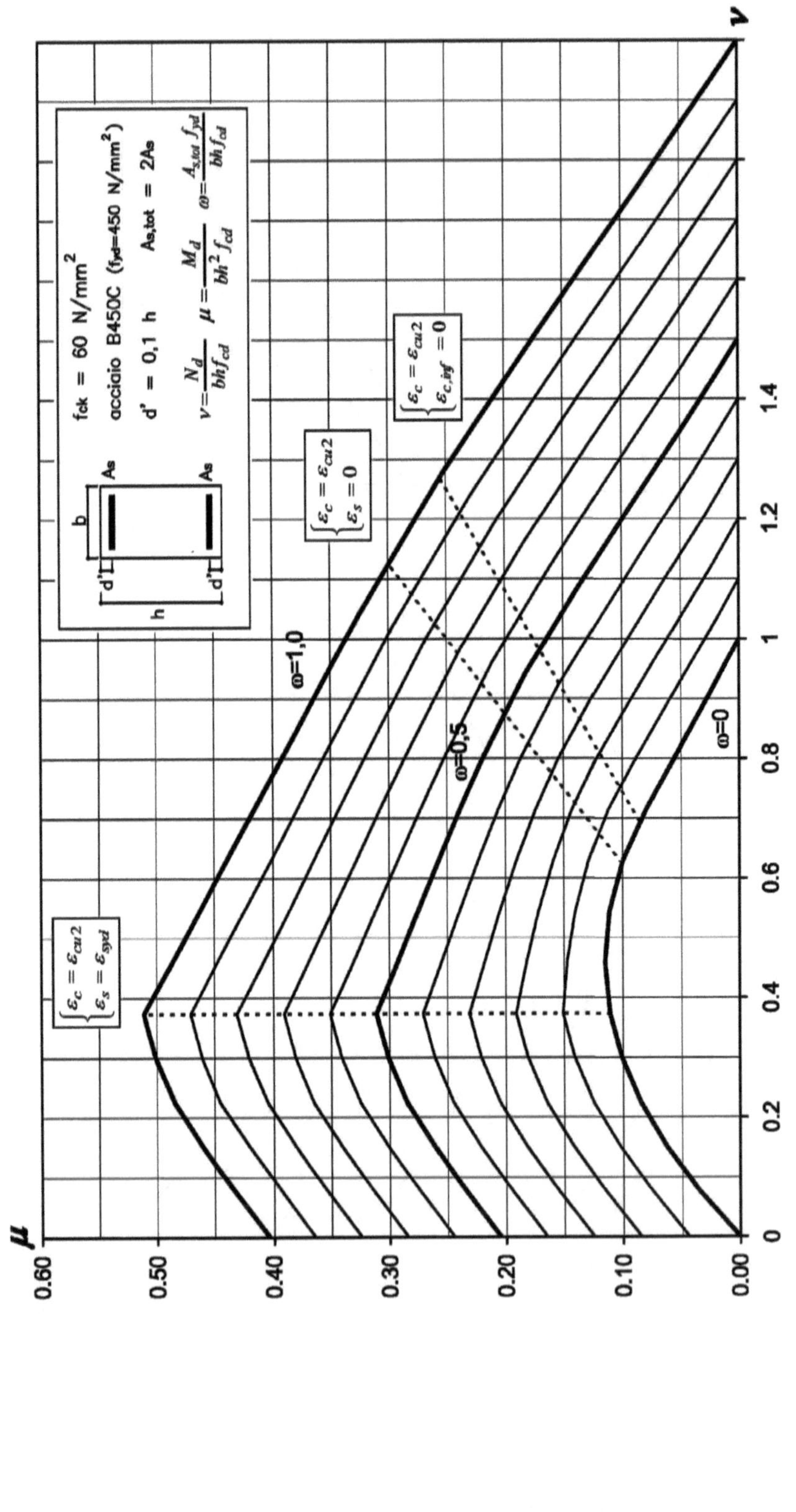

Diagram U.11 Rectangular section; $f_{ck} = 60\,N/mm^2$; steel B450C; $d'/d = 0.1$

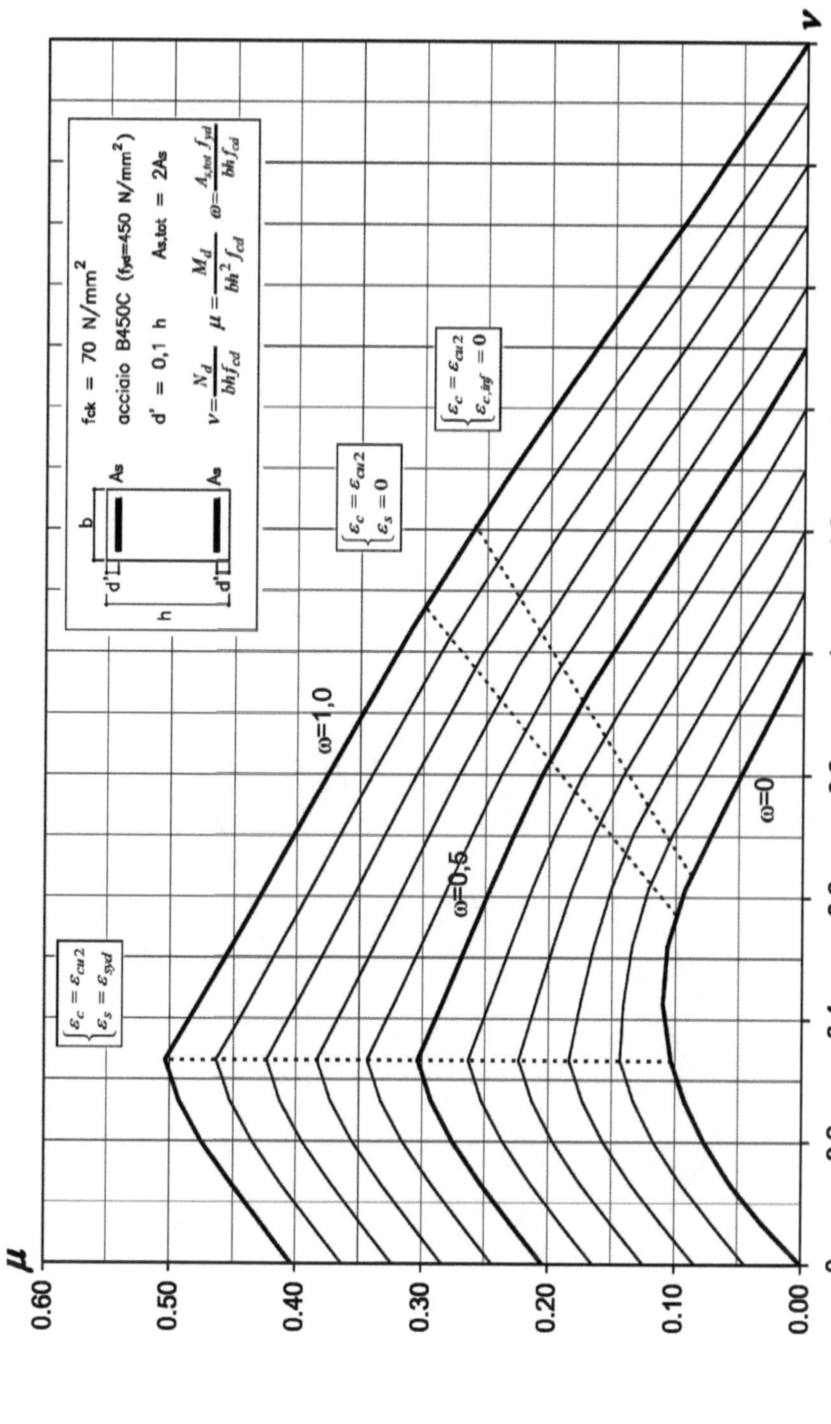

Diagram U.12 Rectangular section; $f_{ck} = 70\,\text{N/mm}^2$; steel B450C; $d'/d = 0.1$

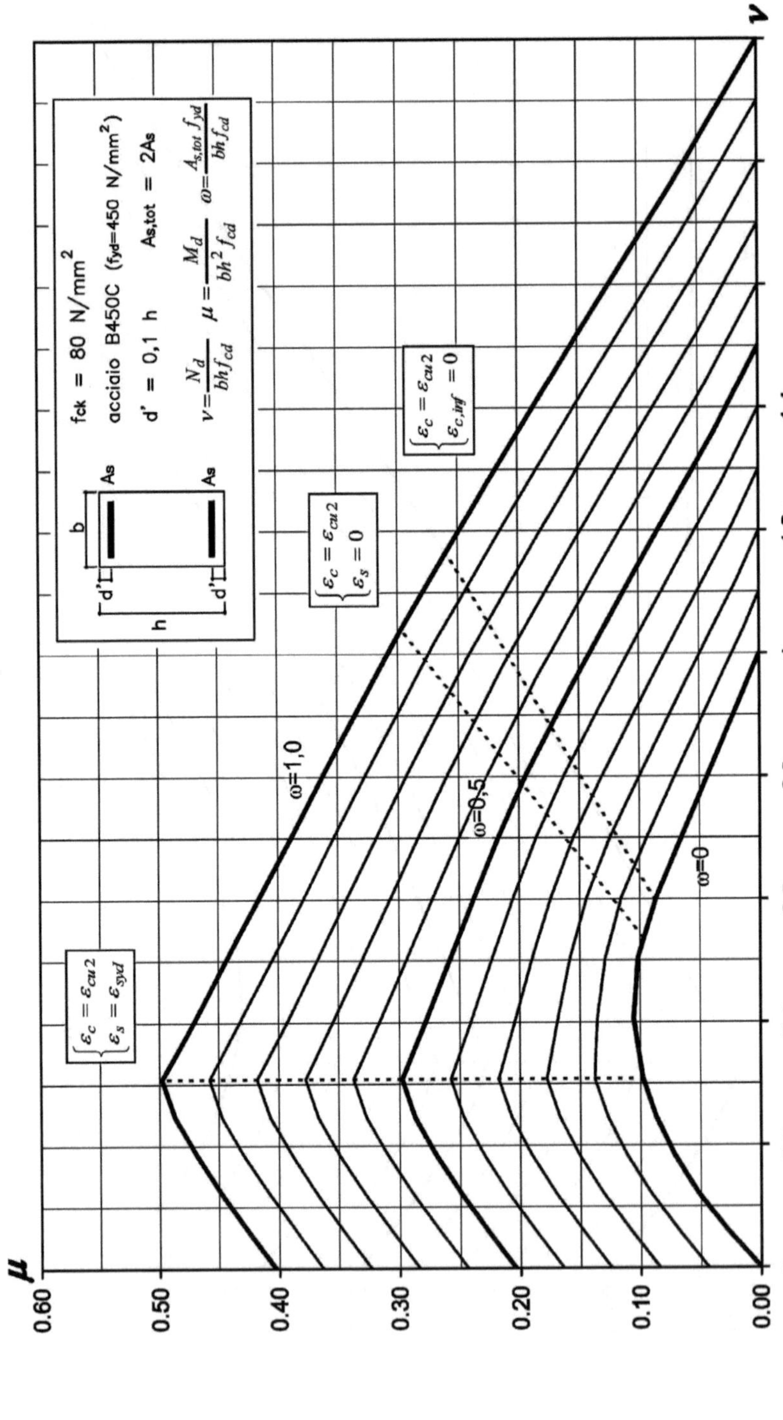

Diagram U.13 Rectangular section; $f_{ck} = 80\,\mathrm{N/mm^2}$; steel B450C; $d'/d = 0.1$

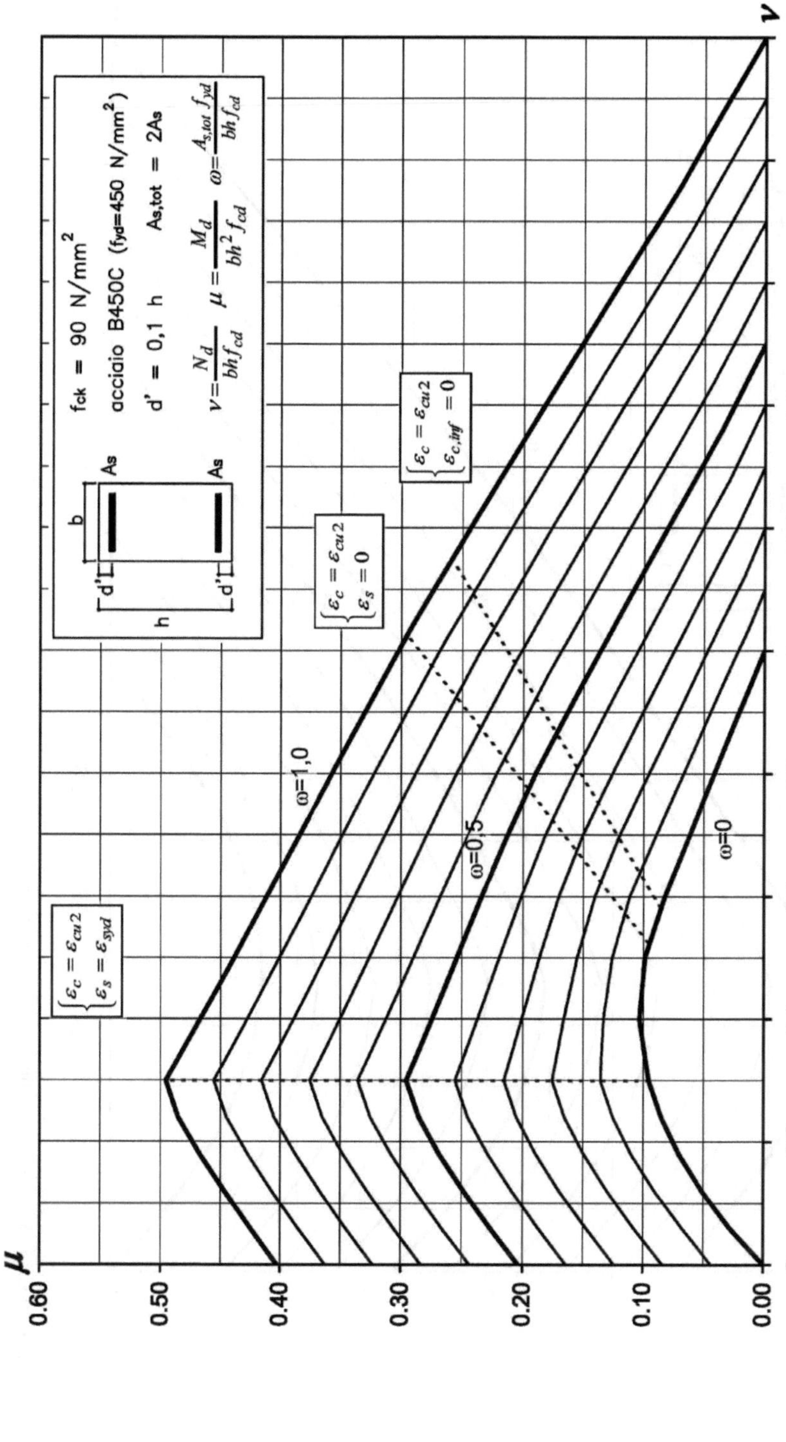

Diagram U.14 Rectangular section; $f_{ck} = 90\,\text{N/mm}^2$; steel B450C; $d'/d = 0.1$

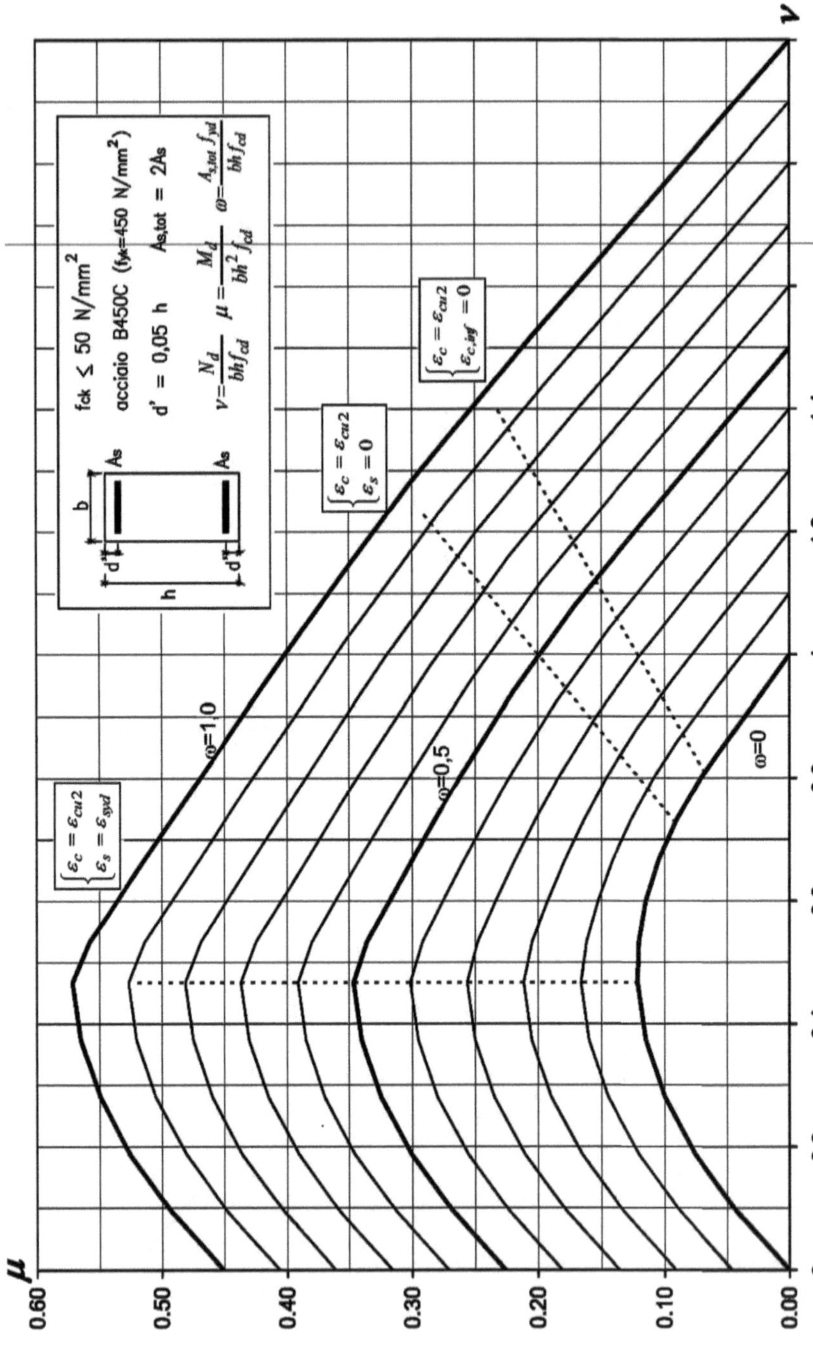

Diagram U.15 Rectangular section; $f_{ck} \leq 50\,\text{N/mm}^2$; steel B450C; $d'/d = 0.05$

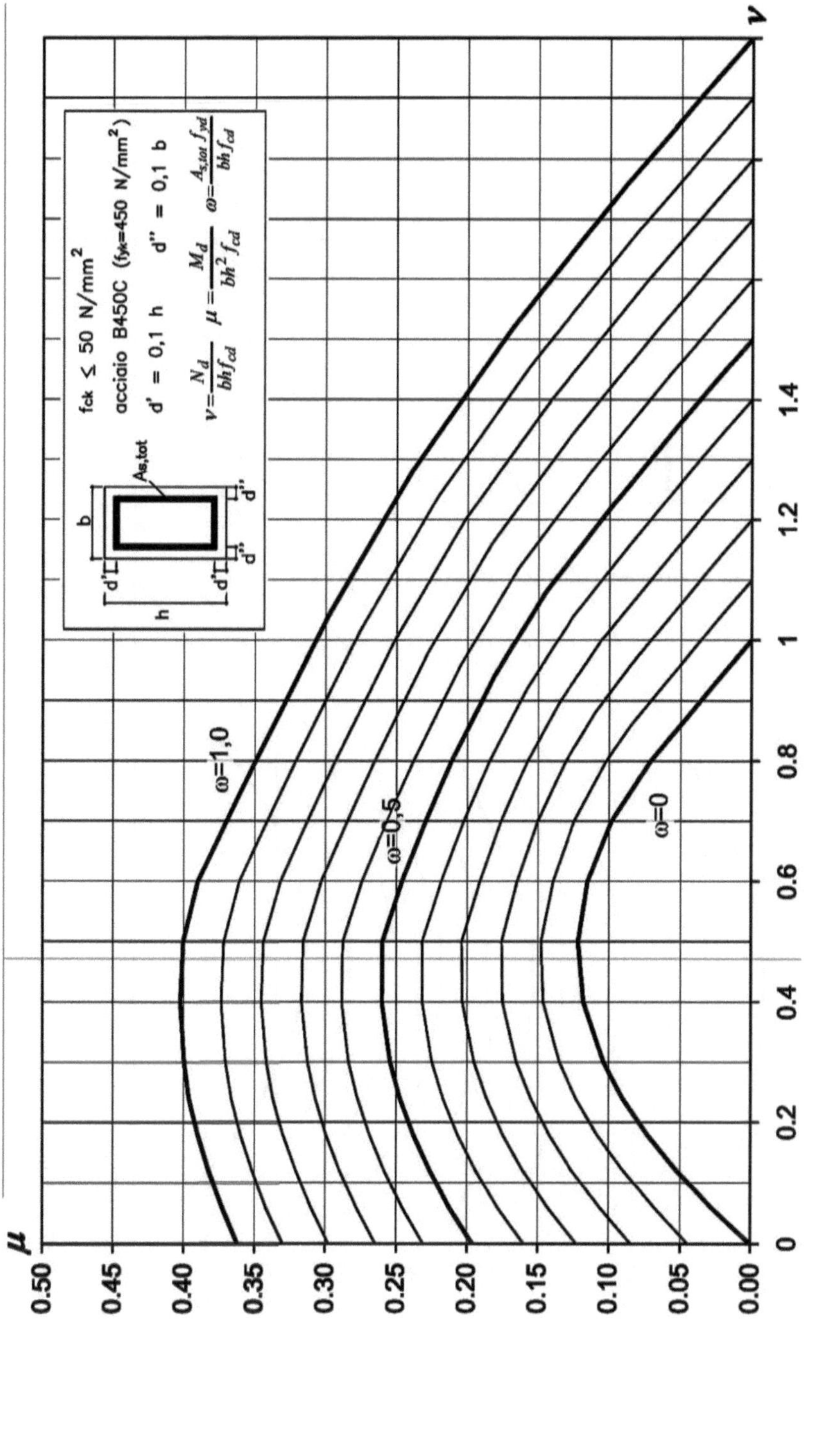

Diagram U.16 Rectangular section; reinforcement distributed on 4 sides; $f_{ck} \leq 50\,\text{N/mm}^2$; steel B450; $d'/d = 0.1$

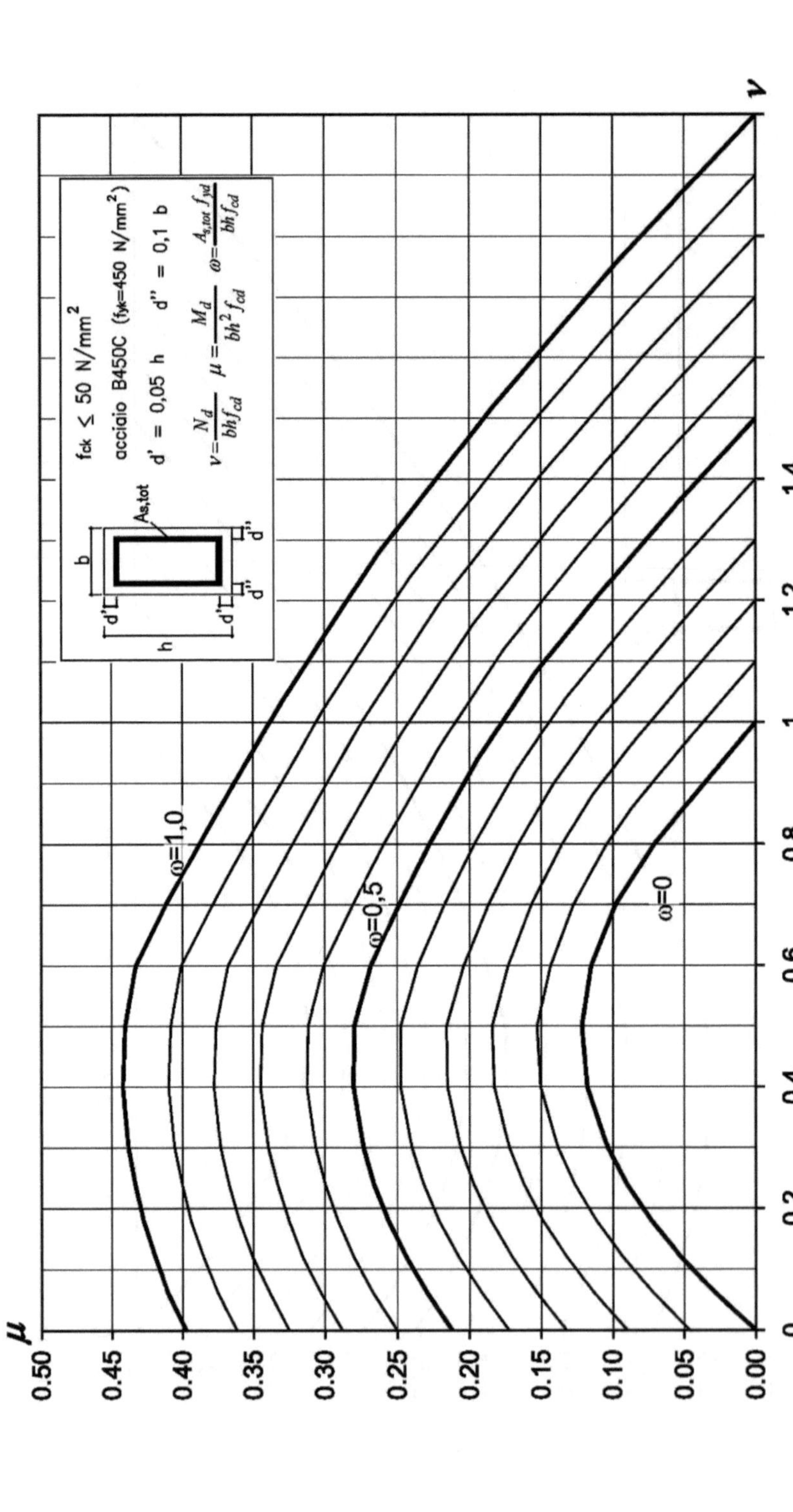

Diagram U.17 Rectangular section; reinforcement distributed on 4 sides; $f_{ck} \leq 50\,\mathrm{N/mm^2}$; steel B450C; $d'/d = 0.05$

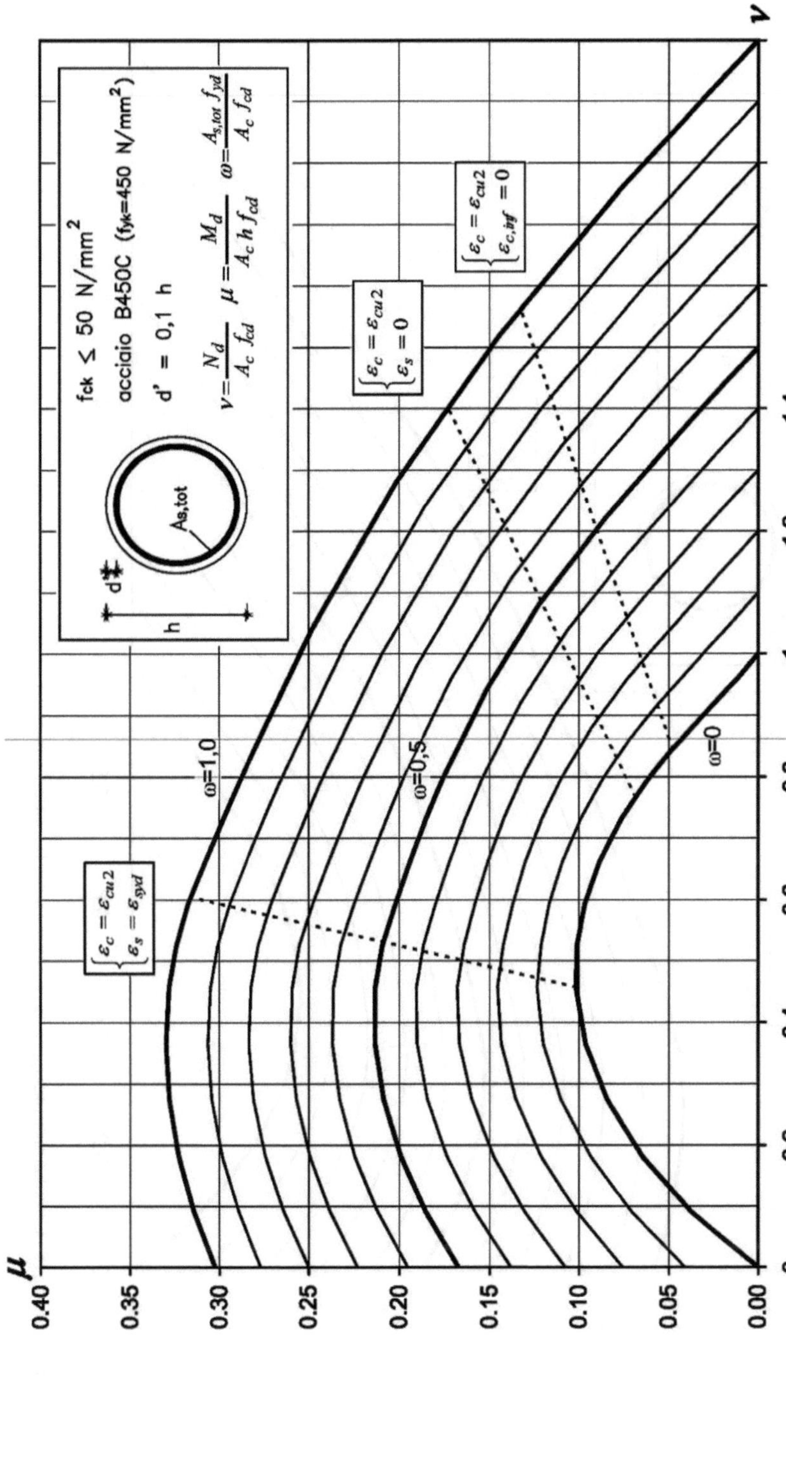

Diagram U.18 Circular section; $f_{ck} \leq 50\,\mathrm{N/mm^2}$; steel B450C; $d'/h = 0.1$

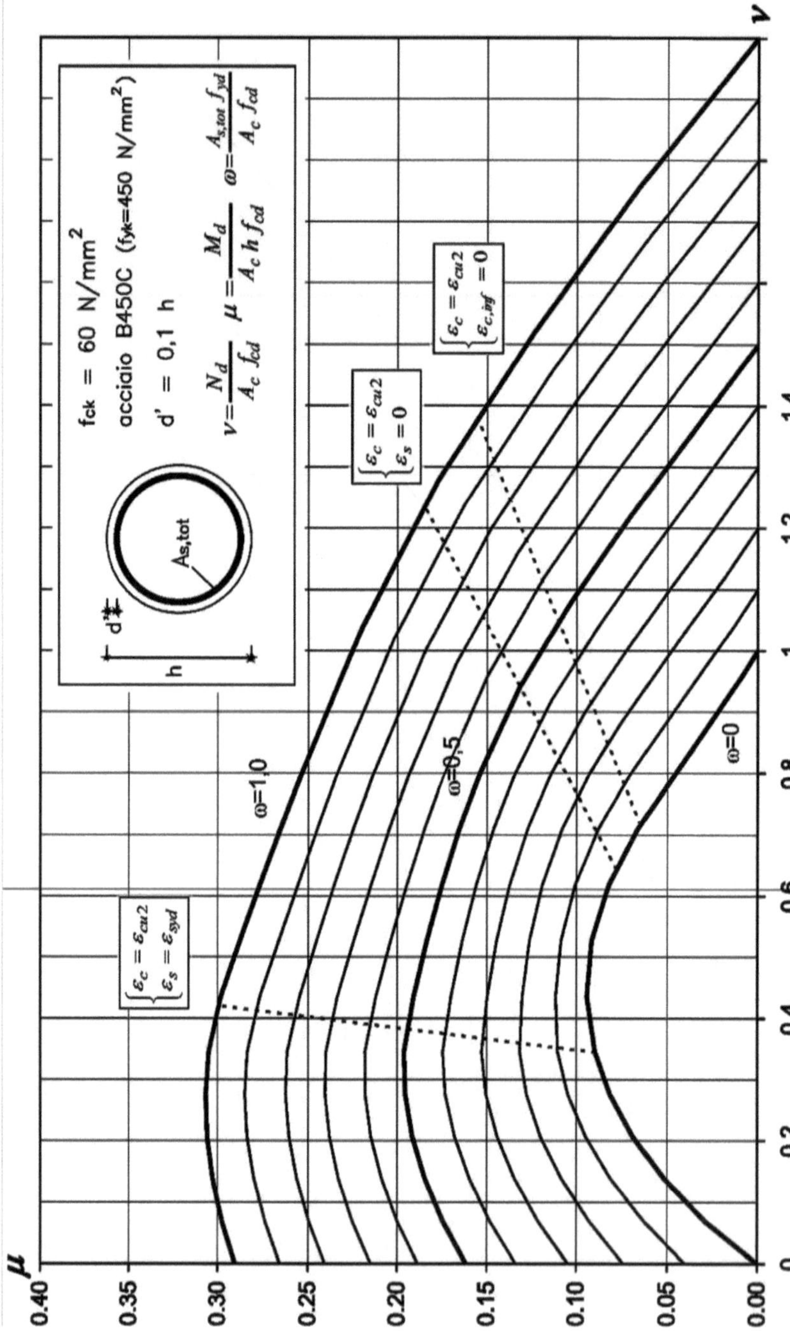

Diagram U.19 Circular section; $f_{ck} = 60\,\text{N/mm}^2$; steel B450C; $d'/h = 0.1$

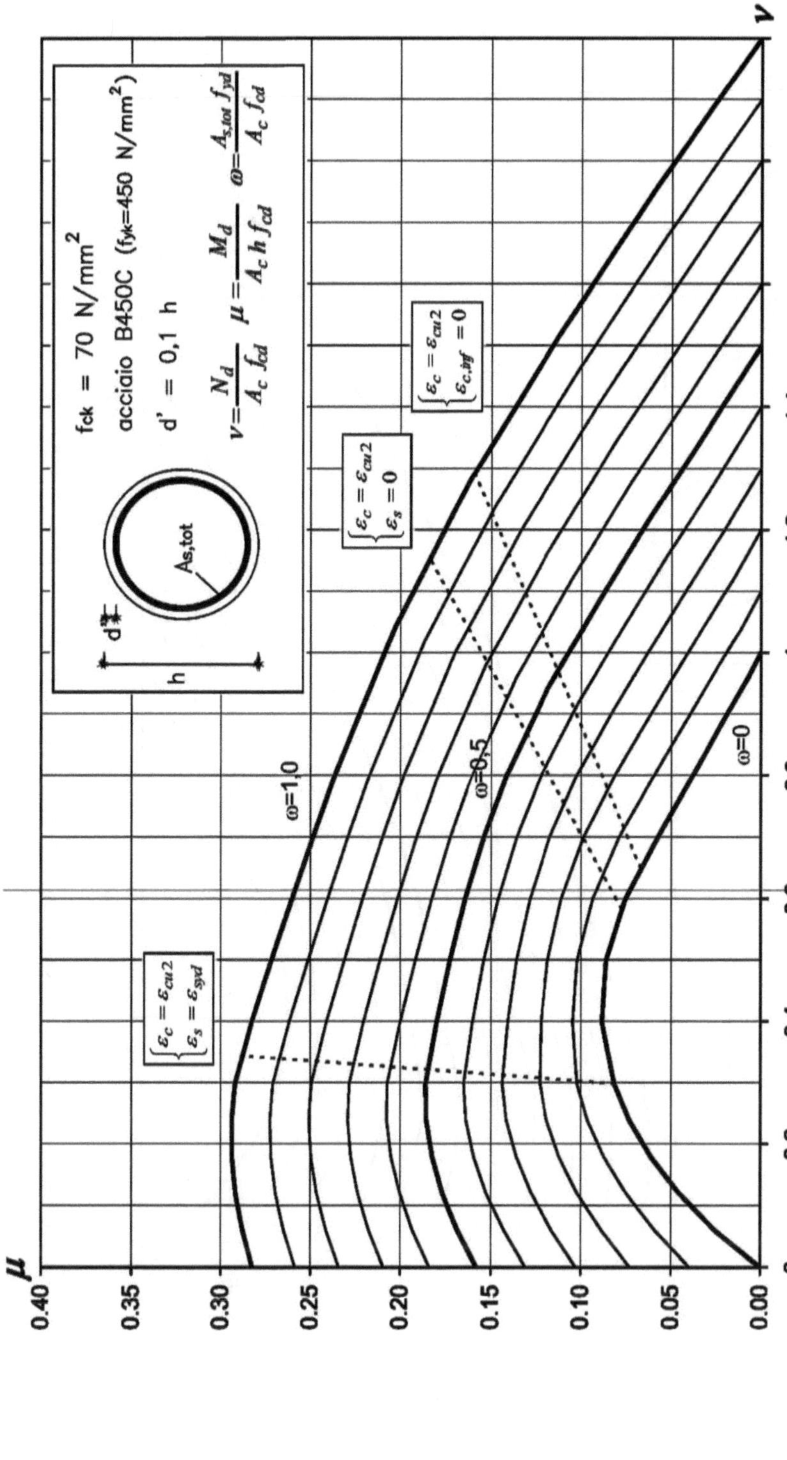

Diagram U.20 Circular section; $f_{ck} = 70\,\mathrm{N/mm^2}$; steel B450C; $d'/h = 0.1$

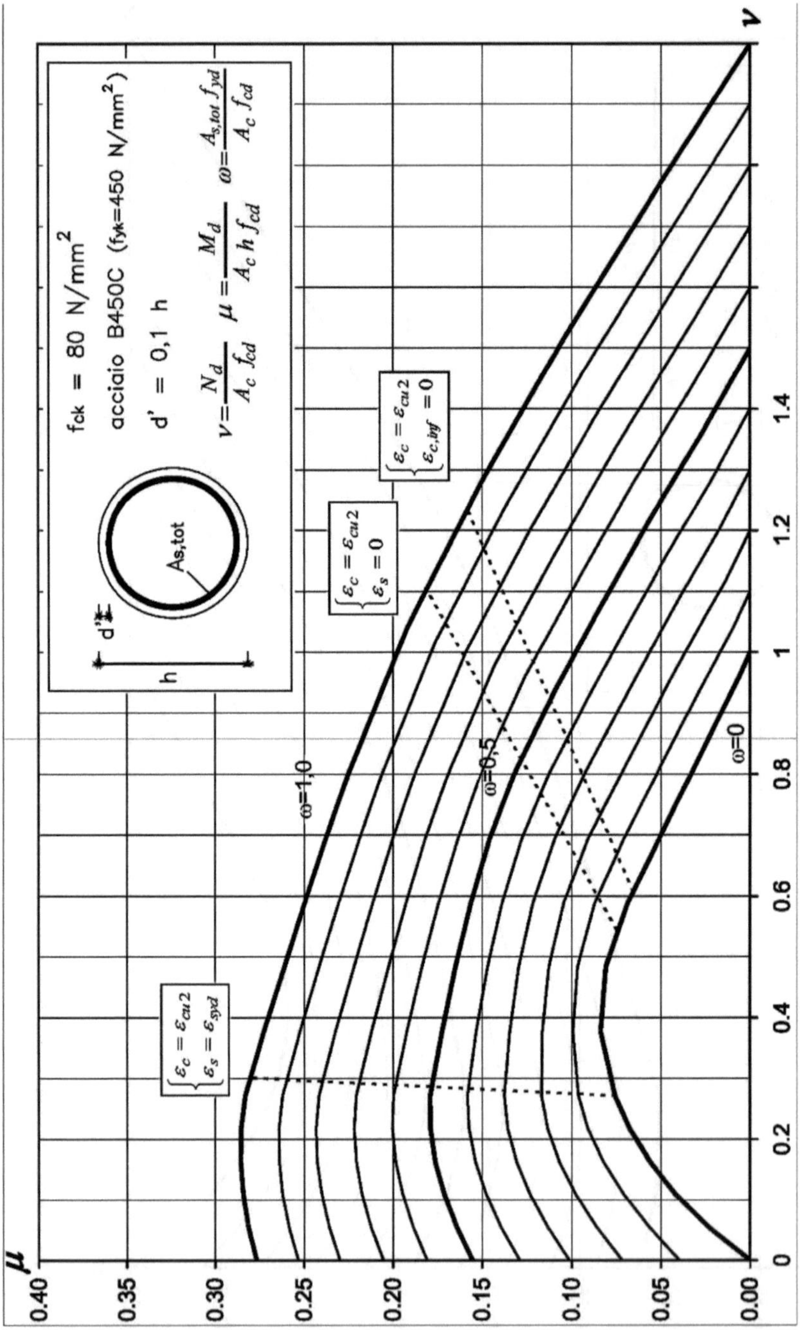

Diagram U.21 Circular section; $f_{ck} = 80\,\mathrm{N/mm^2}$; steel B450C; $d'/h = 0.1$

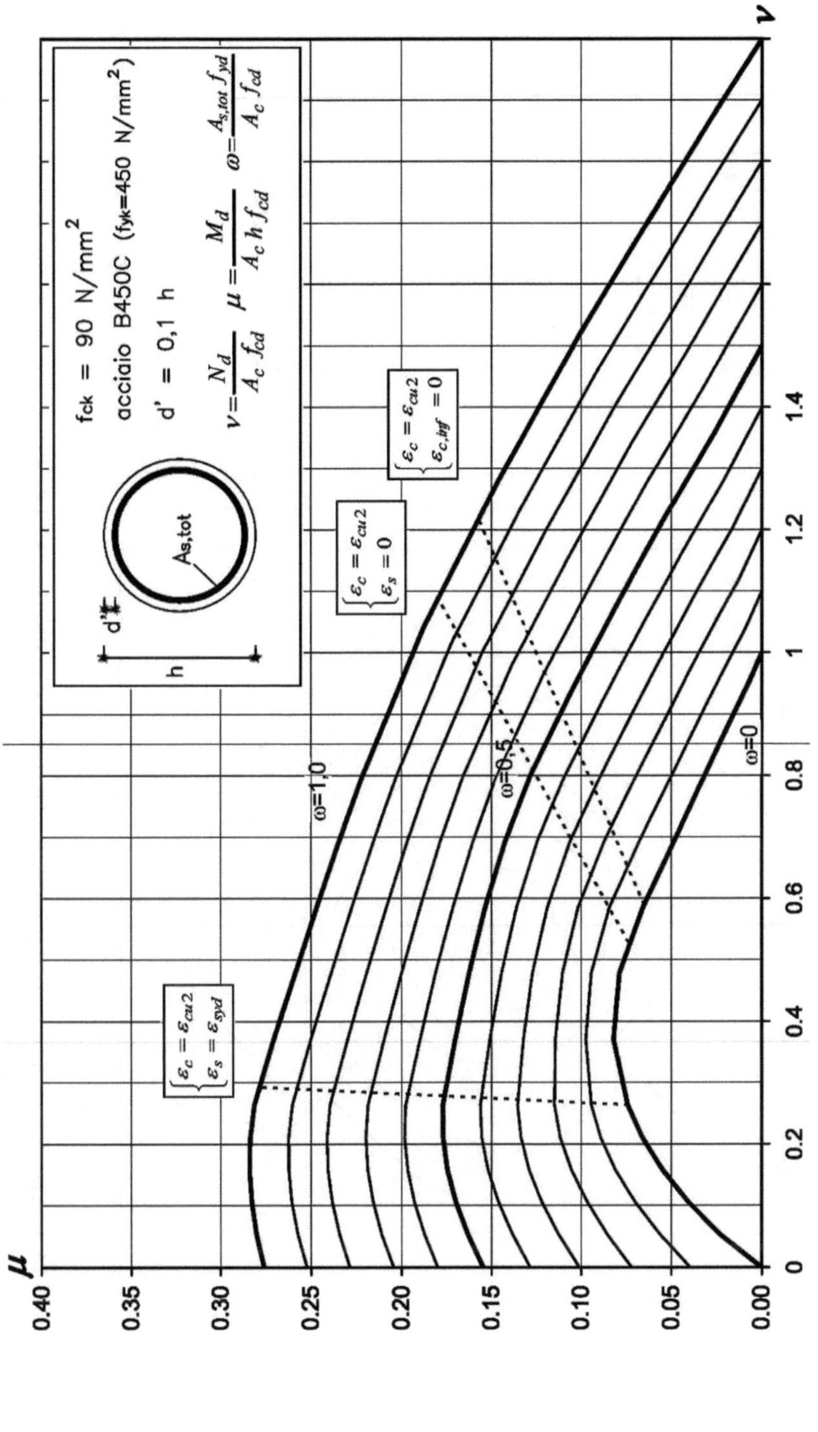

Diagram U.22 Circular section; $f_{ck} = 90 \text{N/mm}^2$; steel B450C; $d'/h = 0.1$

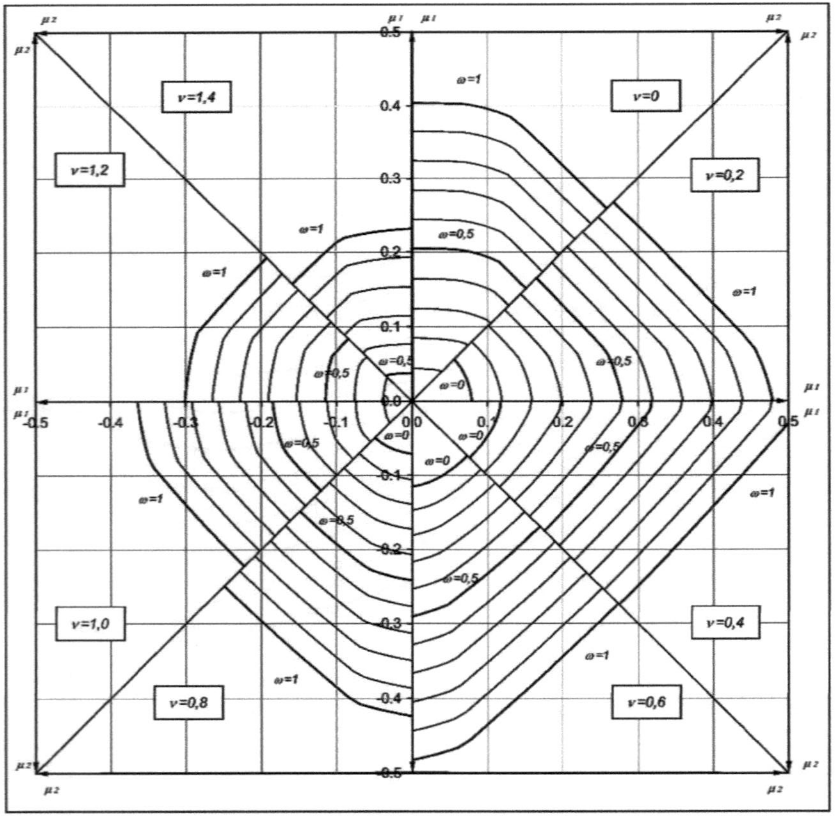

Interaction diagram
biaxial bending

concrete $f_{ck} \leq 50$ N/mm^2
steel B450C $f_{yk} = 450$ N/mm^2

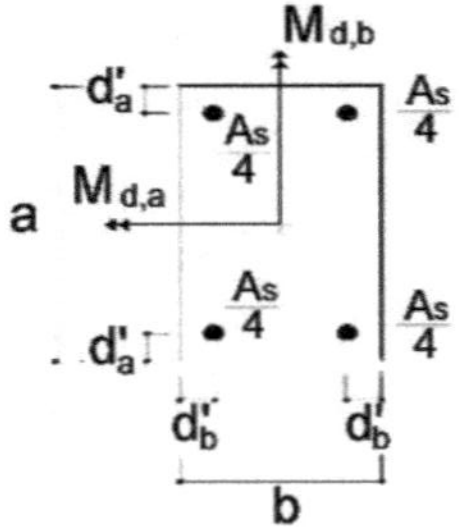

$d'_a = 0,1\ a$

$d'_b = 0,1\ b$

$A_c = a \times b$

$$\nu = \frac{N_d}{A_c f_{cd}} \qquad \omega = \frac{A_s f_{yd}}{A_c f_{cd}}$$

$$\mu_a = \frac{M_{d,a}}{A_c\, a\, f_{cd}} \qquad \mu_b = \frac{M_{d,b}}{A_c\, b\, f_{cd}}$$

se $\mu_a > \mu_b$: $\mu_1 = \mu_a$ $\mu_2 = \mu_b$

se $\mu_b > \mu_a$: $\mu_1 = \mu_b$ $\mu_2 = \mu_a$

Diagram U.23 Rectangular section; reinforcements set in the 4 corners; $f_{ck} \leq 50\,\text{N/mm}^2$; steel B450C

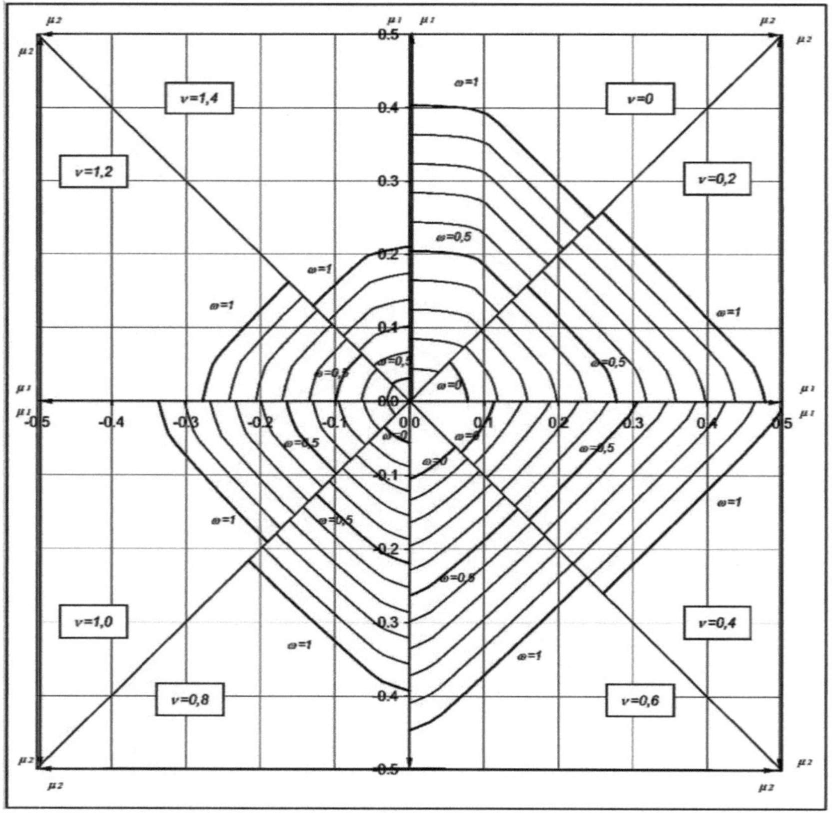

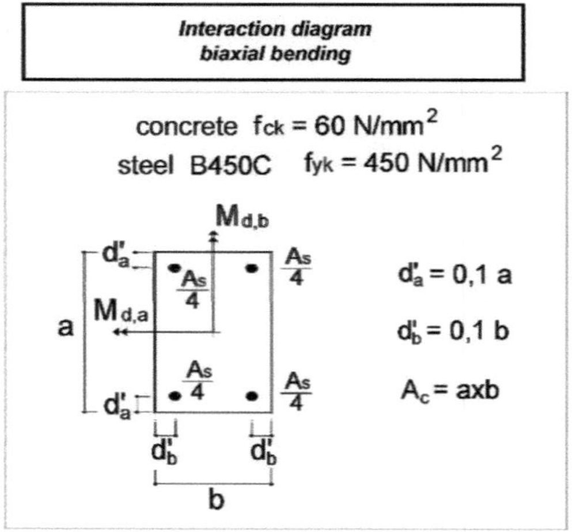

$$\nu = \frac{N_d}{A_c f_{cd}} \qquad \omega = \frac{A_s f_{yd}}{A_c f_{cd}}$$

$$\mu_a = \frac{M_{d,a}}{A_c a f_{cd}} \qquad \mu_b = \frac{M_{d,b}}{A_c b f_{cd}}$$

$$se \quad \mu_a > \mu_b : \quad \mu_1 = \mu_a \quad \mu_2 = \mu_b$$

$$se \quad \mu_b > \mu_a : \quad \mu_1 = \mu_b \quad \mu_2 = \mu_a$$

Diagram U.24 Rectangular section; reinforcements set in the 4 corners; $f_{ck} = 60\,\text{N/mm}^2$; steel B450C

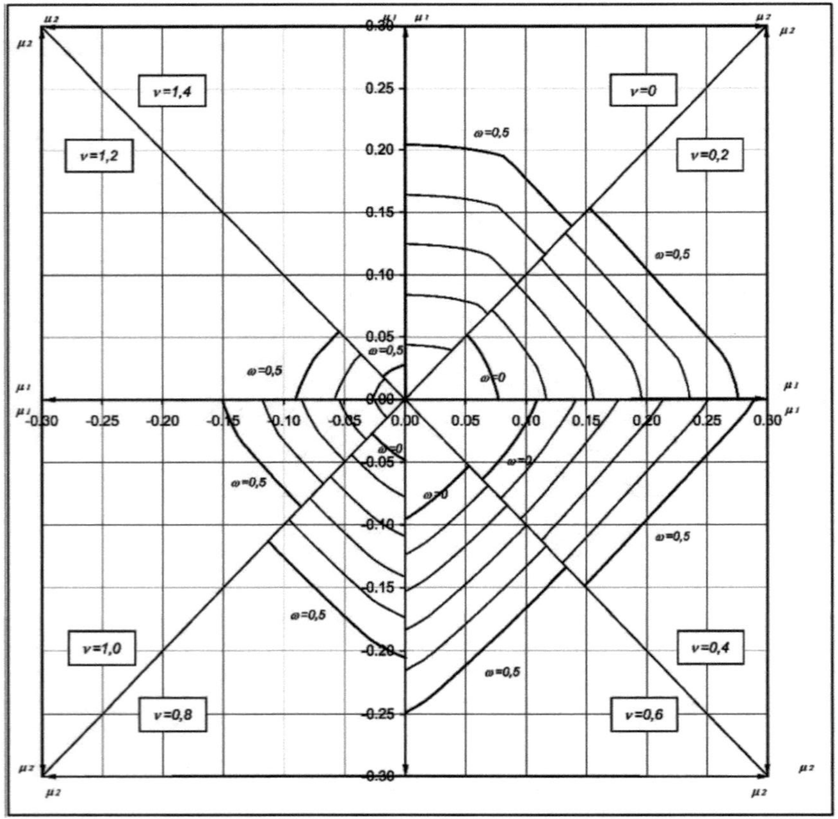

Interaction diagram
biaxial bending

concrete f_{ck} = 70 N/mm^2
steel B450C f_{yk} = 450 N/mm^2

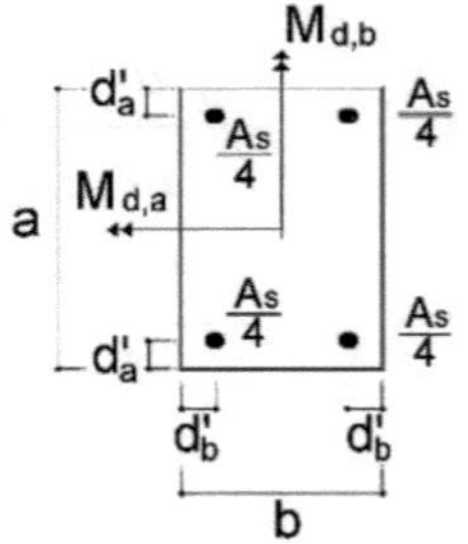

$d'_a = 0,1\ a$

$d'_b = 0,1\ b$

$A_c = a \times b$

$$\nu = \frac{N_d}{A_c f_{cd}} \qquad \omega = \frac{A_s f_{yd}}{A_c f_{cd}}$$

$$\mu_a = \frac{M_{d,a}}{A_c a\, f_{cd}} \qquad \mu_b = \frac{M_{d,b}}{A_c b\, f_{cd}}$$

$$se \quad \mu_a > \mu_b : \quad \mu_1 = \mu_a \quad \mu_2 = \mu_b$$

$$se \quad \mu_b > \mu_a : \quad \mu_1 = \mu_b \quad \mu_2 = \mu_a$$

Diagram U.25 Rectangular section; reinforcements set in the 4 corners; $f_{ck} = 70\,\mathrm{N/mm^2}$; steel B450C

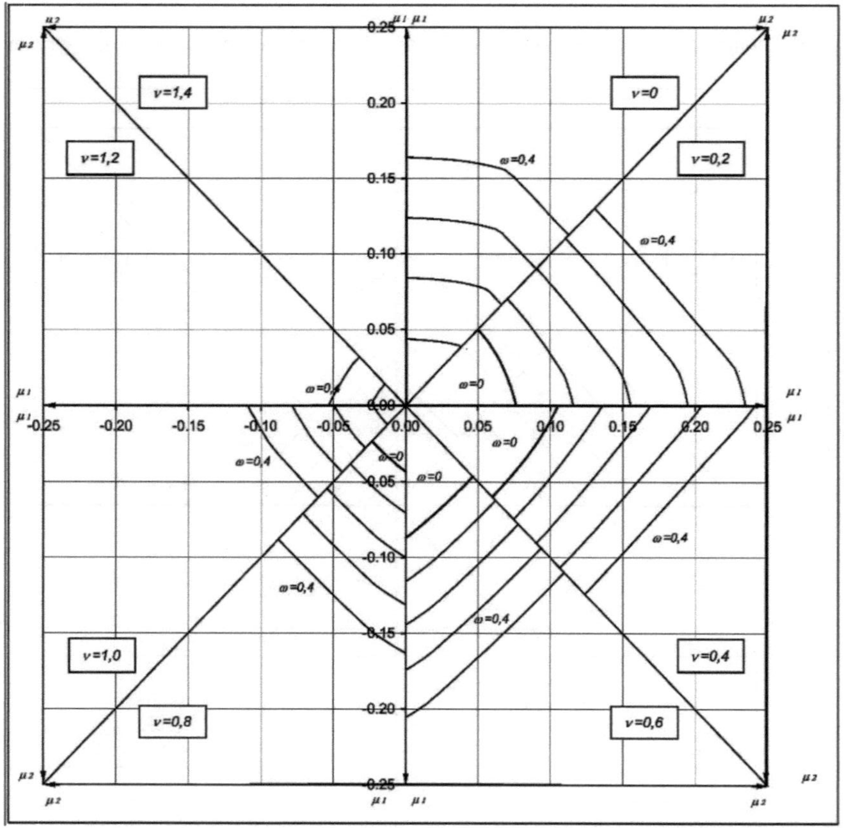

Interaction diagram
biaxial bending

concrete f_{ck} = 80 N/mm^2
steel B450C f_{yk} = 450 N/mm^2

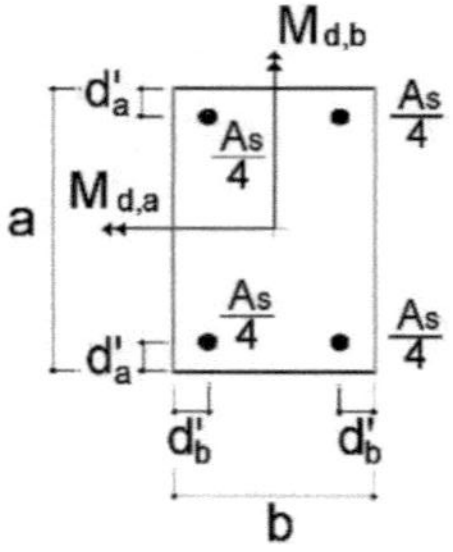

$d'_a = 0,1\ a$

$d'_b = 0,1\ b$

$A_c = a \times b$

$$\nu = \frac{N_d}{A_c f_{cd}} \qquad \omega = \frac{A_s f_{yd}}{A_c f_{cd}}$$

$$\mu_a = \frac{M_{d,a}}{A_c a f_{cd}} \qquad \mu_b = \frac{M_{d,b}}{A_c b f_{cd}}$$

$$se \quad \mu_a > \mu_b : \quad \mu_1 = \mu_a \quad \mu_2 = \mu_b$$

$$se \quad \mu_b > \mu_a : \quad \mu_1 = \mu_b \quad \mu_2 = \mu_a$$

Diagram U.26 Rectangular section; reinforcements set in the 4 corners; $f_{ck} = 80\,\text{N/mm}^2$; steel B450C

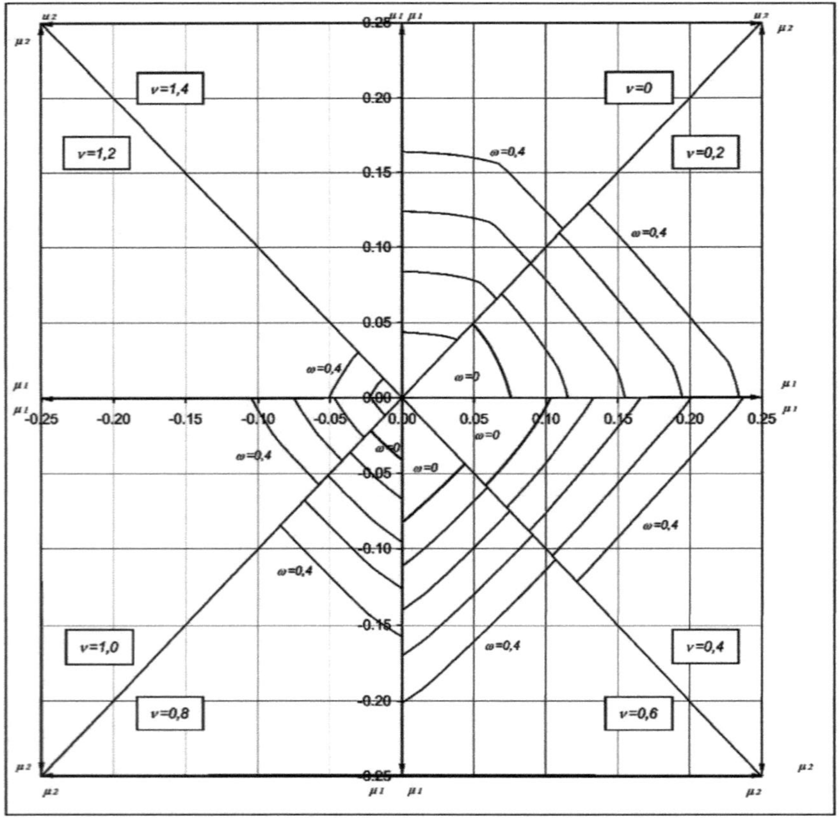

Interaction diagram
biaxial bending

concrete f_{ck} = 90 N/mm^2

steel B450C f_{yk} = 450 N/mm^2

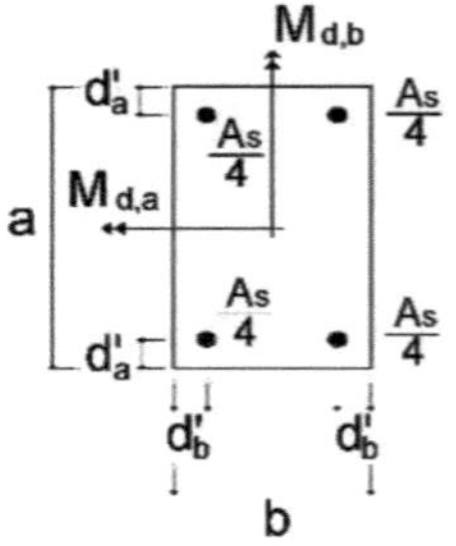

$d'_a = 0{,}1\ a$

$d'_b = 0{,}1\ b$

$A_c = a \times b$

$$\nu = \frac{N_d}{A_c f_{cd}} \qquad \omega = \frac{A_s f_{yd}}{A_c f_{cd}}$$

$$\mu_a = \frac{M_{d,a}}{A_c a\, f_{cd}} \qquad \mu_b = \frac{M_{d,b}}{A_c b\, f_{cd}}$$

$$se \quad \mu_a > \mu_b : \quad \mu_1 = \mu_a \quad \mu_2 = \mu_b$$

$$se \quad \mu_b > \mu_a : \quad \mu_1 = \mu_b \quad \mu_2 = \mu_a$$

Diagram U.27 Rectangular section; reinforcements set in the 4 corners; $f_{ck} = 90\,\mathrm{N/mm^2}$; steel B450C

Table E.1 Rectangular section; $\rho' = 0$; $d'/d = 0.1$

$\alpha_e = 15$ ($f_{ck} \leq 50$ N/mm^2)				ρ	$\alpha_e = 9$ ($f_{ck} \leq 50$ N/mm^2)			
ξ	i	ξ/i	$(1-\xi)/i$		ξ	i	ξ/i	$(1-\xi)/i$
0.1908	0.0170	11.193	47.463	**0.0015**	0.1514	0.0109	13.915	78.010
0.2168	0.0218	9.945	35.930	**0.0020**	0.1726	0.0140	12.296	58.947
0.2584	0.0305	8.471	24.316	**0.0030**	0.2069	0.0199	10.381	39.781
0.2916	0.0384	7.598	18.461	**0.0040**	0.2347	0.0254	9.244	30.136
0.3195	0.0456	7.006	14.923	**0.0050**	0.2584	0.0305	8.471	24.316
0.3437	0.0523	6.572	12.549	**0.0060**	0.2790	0.0353	7.902	20.418
0.3651	0.0585	6.237	10.844	**0.0070**	0.2975	0.0399	7.462	17.620
0.3844	0.0644	5.968	9.558	**0.0080**	0.3142	0.0442	7.109	15.514
0.4019	0.0699	5.747	8.553	**0.0090**	0.3296	0.0483	6.818	13.869
0.4179	0.0752	5.560	7.746	**0.0100**	0.3437	0.0523	6.572	12.549
0.4327	0.0801	5.401	7.082	**0.0110**	0.3569	0.0561	6.361	11.465
0.4464	0.0848	5.263	6.527	**0.0120**	0.3691	0.0597	6.178	10.558
0.4592	0.0893	5.142	6.055	**0.0130**	0.3807	0.0633	6.017	9.789
0.4712	0.0936	5.035	5.649	**0.0140**	0.3916	0.0667	5.874	9.128
0.4825	0.0977	4.939	5.296	**0.0150**	0.4019	0.0699	5.747	8.553
0.4932	0.1016	4.853	4.986	**0.0160**	0.4116	0.0731	5.631	8.049
0.5033	0.1054	4.775	4.712	**0.0170**	0.4209	0.0762	5.527	7.603
0.5129	0.1090	4.704	4.467	**0.0180**	0.4298	0.0791	5.431	7.205
0.5220	0.1125	4.639	4.248	**0.0190**	0.4383	0.0820	5.344	6.849
0.5307	0.1159	4.579	4.050	**0.0200**	0.4464	0.0848	5.263	6.527
0.5389	0.1191	4.524	3.870	**0.0210**	0.4542	0.0875	5.189	6.235
0.5469	0.1223	4.472	3.706	**0.0220**	0.4617	0.0902	5.120	5.969
0.5545	0.1253	4.425	3.556	**0.0230**	0.4689	0.0928	5.055	5.726
0.5617	0.1282	4.381	3.418	**0.0240**	0.4758	0.0953	4.995	5.502
0.5687	0.1311	4.339	3.290	**0.0250**	0.4825	0.0977	4.939	5.296
0.5755	0.1338	4.300	3.173	**0.0260**	0.4890	0.1001	4.886	5.106
0.5819	0.1365	4.264	3.063	**0.0270**	0.4953	0.1024	4.837	4.929
0.5882	0.1391	4.230	2.962	**0.0280**	0.5013	0.1047	4.790	4.764
0.5942	0.1416	4.197	2.867	**0.0290**	0.5072	0.1069	4.746	4.611
0.6000	0.1440	4.167	2.778	**0.0300**	0.5129	0.1090	4.704	4.467
0.6056	0.1464	4.138	2.694	**0.0310**	0.5184	0.1111	4.664	4.333
0.6111	0.1487	4.110	2.616	**0.0320**	0.5238	0.1132	4.626	4.207
0.6163	0.1509	4.084	2.543	**0.0330**	0.5290	0.1152	4.590	4.088
0.6214	0.1531	4.059	2.473	**0.0340**	0.5340	0.1172	4.556	3.976
0.6264	0.1552	4.036	2.407	**0.0350**	0.5389	0.1191	4.524	3.870

(continued)

Table E.1 (continued)

$\alpha_e = 15$ ($f_{ck} \le 50$ N/mm^2)				ρ	$\alpha_e = 9$ ($f_{ck} \le 50$ N/mm^2)			
ξ	i	ξ/i	$(1-\xi)/i$		ξ	i	ξ/i	$(1-\xi)/i$
0.6312	0.1573	4.013	2.345	**0.0360**	0.5437	0.1210	4.492	3.770
0.6358	0.1593	3.992	2.286	**0.0370**	0.5484	0.1229	4.463	3.675
0.6403	0.1613	3.971	2.230	**0.0380**	0.5530	0.1247	4.434	3.585
0.6447	0.1632	3.951	2.177	**0.0390**	0.5574	0.1265	4.407	3.499
0.6490	0.1650	3.932	2.127	**0.0400**	0.5617	0.1282	4.381	3.418

Table E.2 Rectangular section; $\rho' = 0.1\,\rho$; $d'/d = 0.1$

$\alpha_e = 15$ ($f_{ck} \le 50$ N/mm^2)				ρ	$\alpha_e = 9$ ($f_{ck} \le 50$ N/mm^2)			
ξ	i	ξ/i	$(1-\xi)/i$		ξ	i	ξ/i	$(1-\xi)/i$
0.1899	0.0171	11.125	47.468	**0.0015**	0.1510	0.0109	13.872	78.023
0.2154	0.0218	9.862	35.928	**0.0020**	0.1719	0.0140	12.239	58.956
0.2560	0.0306	8.364	24.304	**0.0030**	0.2057	0.0200	10.303	39.782
0.2883	0.0386	7.472	18.441	**0.0040**	0.2330	0.0255	9.151	30.129
0.3154	0.0460	6.862	14.897	**0.0050**	0.2560	0.0306	8.364	24.304
0.3387	0.0528	6.412	12.519	**0.0060**	0.2762	0.0355	7.784	20.400
0.3593	0.0593	6.062	10.810	**0.0070**	0.2941	0.0401	7.333	17.599
0.3777	0.0654	5.779	9.521	**0.0080**	0.3103	0.0445	6.969	15.489
0.3944	0.0711	5.545	8.514	**0.0090**	0.3251	0.0488	6.668	13.842
0.4097	0.0766	5.346	7.704	**0.0100**	0.3387	0.0528	6.412	12.519
0.4237	0.0819	5.174	7.038	**0.0110**	0.3514	0.0567	6.193	11.432
0.4367	0.0869	5.024	6.481	**0.0120**	0.3631	0.0605	6.001	10.524
0.4488	0.0918	4.891	6.008	**0.0130**	0.3742	0.0642	5.832	9.753
0.4601	0.0964	4.772	5.601	**0.0140**	0.3846	0.0677	5.681	9.090
0.4707	0.1009	4.665	5.247	**0.0150**	0.3944	0.0711	5.545	8.514
0.4806	0.1052	4.568	4.936	**0.0160**	0.4037	0.0745	5.422	8.008
0.4901	0.1094	4.479	4.660	**0.0170**	0.4125	0.0777	5.310	7.561
0.4990	0.1135	4.397	4.415	**0.0180**	0.4210	0.0809	5.207	7.162
0.5075	0.1174	4.321	4.194	**0.0190**	0.4290	0.0839	5.112	6.804
0.5155	0.1213	4.251	3.995	**0.0200**	0.4367	0.0869	5.024	6.481
0.5232	0.1250	4.186	3.815	**0.0210**	0.4440	0.0898	4.942	6.189
0.5305	0.1286	4.125	3.650	**0.0220**	0.4511	0.0927	4.866	5.922
0.5375	0.1322	4.067	3.499	**0.0230**	0.4579	0.0955	4.795	5.678
0.5442	0.1356	4.013	3.361	**0.0240**	0.4644	0.0982	4.728	5.453
0.5506	0.1390	3.962	3.233	**0.0250**	0.4707	0.1009	4.665	5.247
0.5568	0.1423	3.913	3.115	**0.0260**	0.4767	0.1035	4.606	5.055

(continued)

Table E.2 (continued)

$\alpha_e = 15$ ($f_{ck} \leq 50$ N/mm^2)				ρ	$\alpha_e = 9$ ($f_{ck} \leq 50$ N/mm^2)			
ξ	i	ξ/i	$(1-\xi)/i$		ξ	i	ξ/i	$(1-\xi)/i$
0.5628	0.1455	3.867	3.005	**0.0270**	0.4826	0.1061	4.549	4.878
0.5685	0.1487	3.824	2.903	**0.0280**	0.4882	0.1086	4.496	4.713
0.5740	0.1518	3.782	2.807	**0.0290**	0.4937	0.1111	4.445	4.559
0.5793	0.1548	3.742	2.718	**0.0300**	0.4990	0.1135	4.397	4.415
0.5844	0.1578	3.704	2.635	**0.0310**	0.5041	0.1159	4.351	4.280
0.5893	0.1607	3.668	2.556	**0.0320**	0.5091	0.1182	4.307	4.153
0.5941	0.1635	3.633	2.482	**0.0330**	0.5139	0.1205	4.265	4.033
0.5987	0.1663	3.599	2.412	**0.0340**	0.5186	0.1228	4.224	3.921
0.6032	0.1691	3.567	2.346	**0.0350**	0.5232	0.1250	4.186	3.815
0.6075	0.1718	3.536	2.284	**0.0360**	0.5276	0.1272	4.149	3.714
0.6117	0.1745	3.505	2.225	**0.0370**	0.5319	0.1293	4.113	3.619
0.6158	0.1771	3.476	2.169	**0.0380**	0.5361	0.1315	4.078	3.528
0.6197	0.1797	3.448	2.116	**0.0390**	0.5402	0.1336	4.045	3.443
0.6236	0.1823	3.421	2.065	**0.0400**	0.5442	0.1356	4.013	3.361

Table E.3 Rectangular section; $\rho' = 0.2\,\rho$; $d'/d = 0.1$

$\alpha_e = 15$ ($f_{ck} \leq 50$ N/mm^2)				ρ	$\alpha_e = 9$ ($f_{ck} \leq 50$ N/mm^2)			
ξ	i	ξ/i	$(1-\xi)/i$		ξ	i	ξ/i	$(1-\xi)/i$
0.1889	0.0171	11.059	47.473	**0.0015**	0.1505	0.0109	13.830	78.036
0.2140	0.0219	9.781	35.926	**0.0020**	0.1712	0.0141	12.183	58.965
0.2538	0.0307	8.261	24.292	**0.0030**	0.2045	0.0200	10.228	39.782
0.2852	0.0388	7.350	18.423	**0.0040**	0.2312	0.0255	9.060	30.123
0.3114	0.0463	6.725	14.874	**0.0050**	0.2538	0.0307	8.261	24.292
0.3339	0.0533	6.261	12.491	**0.0060**	0.2734	0.0356	7.669	20.385
0.3537	0.0600	5.898	10.779	**0.0070**	0.2908	0.0403	7.208	17.580
0.3713	0.0663	5.604	9.488	**0.0080**	0.3065	0.0448	6.835	15.467
0.3872	0.0723	5.358	8.479	**0.0090**	0.3208	0.0492	6.525	13.817
0.4017	0.0780	5.148	7.667	**0.0100**	0.3339	0.0533	6.261	12.491
0.4150	0.0836	4.966	7.000	**0.0110**	0.3460	0.0573	6.034	11.403
0.4273	0.0889	4.807	6.442	**0.0120**	0.3573	0.0612	5.835	10.493
0.4387	0.0941	4.665	5.968	**0.0130**	0.3679	0.0650	5.658	9.720
0.4494	0.0990	4.537	5.560	**0.0140**	0.3779	0.0687	5.500	9.056
0.4593	0.1039	4.421	5.205	**0.0150**	0.3872	0.0723	5.358	8.479
0.4687	0.1086	4.316	4.893	**0.0160**	0.3961	0.0758	5.228	7.972
0.4775	0.1132	4.219	4.617	**0.0170**	0.4045	0.0792	5.110	7.524

(continued)

Table E.3 (continued)

$\alpha_e = 15$ ($f_{ck} \leq 50$ N/mm^2)				ρ	$\alpha_e = 9$ ($f_{ck} \leq 50$ N/mm^2)			
ξ	i	ξ/i	$(1-\xi)/i$		ξ	i	ξ/i	$(1-\xi)/i$
0.4858	0.1176	4.129	4.371	**0.0180**	0.4125	0.0825	5.001	7.124
0.4937	0.1220	4.046	4.150	**0.0190**	0.4201	0.0857	4.900	6.765
0.5012	0.1263	3.969	3.951	**0.0200**	0.4273	0.0889	4.807	6.442
0.5083	0.1304	3.897	3.770	**0.0210**	0.4343	0.0920	4.720	6.149
0.5151	0.1345	3.829	3.605	**0.0220**	0.4409	0.0951	4.638	5.881
0.5215	0.1385	3.765	3.454	**0.0230**	0.4473	0.0981	4.561	5.637
0.5277	0.1425	3.704	3.315	**0.0240**	0.4534	0.1010	4.489	5.412
0.5336	0.1463	3.647	3.187	**0.0250**	0.4593	0.1039	4.421	5.205
0.5393	0.1501	3.593	3.069	**0.0260**	0.4650	0.1067	4.357	5.013
0.5447	0.1538	3.541	2.959	**0.0270**	0.4705	0.1095	4.296	4.835
0.5500	0.1575	3.491	2.857	**0.0280**	0.4758	0.1123	4.238	4.670
0.5550	0.1611	3.444	2.762	**0.0290**	0.4809	0.1150	4.182	4.515
0.5598	0.1647	3.399	2.673	**0.0300**	0.4858	0.1176	4.129	4.371
0.5645	0.1682	3.356	2.589	**0.0310**	0.4906	0.1203	4.079	4.236
0.5690	0.1717	3.314	2.510	**0.0320**	0.4952	0.1229	4.031	4.109
0.5733	0.1751	3.274	2.437	**0.0330**	0.4997	0.1254	3.984	3.989
0.5775	0.1785	3.235	2.367	**0.0340**	0.5040	0.1279	3.940	3.876
0.5816	0.1818	3.198	2.301	**0.0350**	0.5083	0.1304	3.897	3.770
0.5855	0.1851	3.162	2.239	**0.0360**	0.5124	0.1329	3.855	3.669
0.5893	0.1884	3.128	2.180	**0.0370**	0.5164	0.1353	3.816	3.574
0.5930	0.1916	3.094	2.124	**0.0380**	0.5202	0.1377	3.777	3.483
0.5965	0.1948	3.062	2.071	**0.0390**	0.5240	0.1401	3.740	3.397
0.6000	0.1980	3.030	2.020	**0.0400**	0.5277	0.1425	3.704	3.315

Table E.4 Rectangular section; $\rho' = 0.3\,\rho$; $d'/d = 0.1$

$\alpha_e = 15$ ($f_{ck} \leq 50$ N/mm^2)				ρ	$\alpha_e = 9$ ($f_{ck} \leq 50$ N/mm^2)			
ξ	i	ξ/i	$(1-\xi)/i$		ξ	i	ξ/i	$(1-\xi)/i$
0.1880	0.0171	10.994	47.478	**0.0015**	0.1501	0.0109	13.788	78.049
0.2126	0.0219	9.702	35.925	**0.0020**	0.1706	0.0141	12.128	58.974
0.2515	0.0308	8.160	24.282	**0.0030**	0.2033	0.0200	10.154	39.783
0.2821	0.0390	7.233	18.406	**0.0040**	0.2295	0.0256	8.972	30.118
0.3075	0.0466	6.594	14.852	**0.0050**	0.2515	0.0308	8.160	24.282
0.3292	0.0538	6.118	12.467	**0.0060**	0.2706	0.0358	7.558	20.370
0.3482	0.0606	5.744	10.752	**0.0070**	0.2875	0.0406	7.088	17.562

(continued)

Table E.4 (continued)

$\alpha_e = 15$ ($f_{ck} \leq 50$ N/mm^2)				ρ	$\alpha_e = 9$ ($f_{ck} \leq 50$ N/mm^2)			
ξ	i	ξ/i	$(1-\xi)/i$		ξ	i	ξ/i	$(1-\xi)/i$
0.3651	0.0671	5.439	9.459	**0.0080**	0.3027	0.0451	6.706	15.446
0.3803	0.0734	5.184	8.448	**0.0090**	0.3165	0.0495	6.389	13.794
0.3941	0.0794	4.966	7.635	**0.0100**	0.3292	0.0538	6.118	12.467
0.4067	0.0852	4.776	6.967	**0.0110**	0.3409	0.0579	5.884	11.377
0.4183	0.0908	4.609	6.408	**0.0120**	0.3517	0.0619	5.678	10.465
0.4291	0.0962	4.459	5.933	**0.0130**	0.3619	0.0658	5.496	9.691
0.4391	0.1015	4.325	5.524	**0.0140**	0.3714	0.0696	5.332	9.026
0.4485	0.1067	4.203	5.169	**0.0150**	0.3803	0.0734	5.184	8.448
0.4572	0.1118	4.092	4.857	**0.0160**	0.3887	0.0770	5.049	7.940
0.4655	0.1167	3.989	4.580	**0.0170**	0.3967	0.0805	4.926	7.491
0.4733	0.1215	3.894	4.334	**0.0180**	0.4043	0.0840	4.812	7.091
0.4806	0.1263	3.806	4.113	**0.0190**	0.4115	0.0874	4.707	6.732
0.4876	0.1309	3.724	3.914	**0.0200**	0.4183	0.0908	4.609	6.408
0.4942	0.1355	3.647	3.733	**0.0210**	0.4249	0.0941	4.517	6.114
0.5004	0.1400	3.574	3.568	**0.0220**	0.4312	0.0973	4.431	5.846
0.5064	0.1444	3.506	3.417	**0.0230**	0.4372	0.1005	4.351	5.601
0.5121	0.1488	3.442	3.279	**0.0240**	0.4429	0.1036	4.275	5.376
0.5176	0.1531	3.380	3.151	**0.0250**	0.4485	0.1067	4.203	5.169
0.5228	0.1574	3.322	3.033	**0.0260**	0.4538	0.1097	4.135	4.977
0.5278	0.1616	3.267	2.923	**0.0270**	0.4589	0.1128	4.070	4.799
0.5326	0.1657	3.214	2.821	**0.0280**	0.4639	0.1157	4.009	4.633
0.5372	0.1698	3.164	2.726	**0.0290**	0.4687	0.1186	3.950	4.479
0.5416	0.1738	3.115	2.637	**0.0300**	0.4733	0.1215	3.894	4.334
0.5459	0.1778	3.069	2.554	**0.0310**	0.4777	0.1244	3.841	4.199
0.5500	0.1818	3.025	2.475	**0.0320**	0.4820	0.1272	3.789	4.072
0.5539	0.1857	2.982	2.402	**0.0330**	0.4862	0.1300	3.740	3.952
0.5577	0.1896	2.941	2.332	**0.0340**	0.4902	0.1328	3.692	3.839
0.5614	0.1935	2.901	2.267	**0.0350**	0.4942	0.1355	3.647	3.733
0.5650	0.1973	2.863	2.205	**0.0360**	0.4980	0.1382	3.603	3.632
0.5684	0.2011	2.826	2.146	**0.0370**	0.5016	0.1409	3.560	3.537
0.5717	0.2049	2.790	2.090	**0.0380**	0.5052	0.1436	3.519	3.446
0.5750	0.2086	2.756	2.037	**0.0390**	0.5087	0.1462	3.480	3.361
0.5781	0.2123	2.722	1.987	**0.0400**	0.5121	0.1488	3.442	3.279

Tables for T Sections: $\rho-\xi-i$

Tables representing the geometric characteristics in dimensionless form, for T sections in bending in state II (cracked).

Only the tensile reinforcement is considered and $\alpha_e = 15\left(f_{ck} \leq 50\,\text{N/mm}^2\right)$.

The tables start from ρ values associated to a neutral axis greater than the thickness of the flange and are developed up to the limit of tensile reinforcement $A_S = 0.04\,A_c$ [9.2.1.1(3)].

The Tables are:

E5—$\alpha_\text{e} = 15$; $h_f\,/d = 0.15$
E6—$\alpha_\text{e} = 15$; $h_f\,/d = 0.20$
E7—$\alpha_\text{e} = 15$; $h_f\,/d = 0.25$
with:
$\rho = A_s/(bd)$
$\xi = x/d$
$i = I/\left(bd^3\right)$
b flange width
b_w web width
h_f flange thickness
$n = b/b_w$ (Table E.5, E.6 and E.7).

Interaction Diagrams $\nu-\mu$ of Rectangular Cross-Sections with Double Symmetrical Reinforcement

Interaction diagrams $\nu-\mu$ as a function of ρ, geometric reinforcement ratio, for rectangular sections with symmetrical reinforcement. The diagrams represent the achievement of the stress limits $\sigma_c = 0.6\,f_{ck}$ and/or $\sigma_s = 0.8\,f_{yk}$. The dimensionless parameters are expressed as a function of f_{ck}. The diagrams are drawn for $f_{ck} = 20$, 25, 30 and 40 N/mm ^{2} since, for higher concrete strengths, the ULS checks prevail over the SLS and make superfluous to perform the SLS checks in the service load combinations.

The diagrams are:
E8—$d'/h = 0.10$; $f_{ck} =$ 20 N/mm^2
E9—$d'/h = 0.10$; $f_{ck} =$ 25 N/mm^2
E10—$d'/h = 0.10$; $f_{ck} =$ 30 N/mm^2
E11—$d'/h = 0.10$; $f_{ck} =$ 40 N/mm^2
The diagrams considers the maximum value $\rho = A_{S,\text{tot}}/(bh) = 0.04$, as defined for the columns [9.5.2(3)] (Diagrams E.8, E.9, E.10 and E.11).

Table E.5 T-section; $h_f/d = 0.15$; $\alpha_e = 15\ (f_{ck} \leq 50\,\text{N/mm}^2)$

$n = b/b_w = 2$				ρ	$n = b/b_w = 4$			
ξ	i	ξ/i	$(1-\xi)/i$		ξ	i	ξ/i	$(1-\xi)/i$
0.1514	0.0109	13.910	77.980	**0.0009**	0.1514	0.0109	13.915	78.010
0.1590	0.0120	13.299	70.361	**0.0010**	0.1590	0.0119	13.309	70.380
0.1930	0.0170	11.325	47.360	**0.0015**	0.1942	0.0170	11.406	47.316
0.2220	0.0217	10.207	35.776	**0.0020**	0.2252	0.0217	10.375	35.687
0.2474	0.0261	9.464	28.795	**0.0025**	0.2529	0.0261	9.710	28.678
0.2700	0.0303	8.923	24.126	**0.0030**	0.2780	0.0301	9.236	23.992
0.2905	0.0341	8.507	20.782	**0.0035**	0.3008	0.0339	8.878	20.637
0.3091	0.0378	8.174	18.268	**0.0040**	0.3217	0.0374	8.594	18.117
0.3263	0.0413	7.900	16.309	**0.0045**	0.3411	0.0408	8.362	16.154
0.3423	0.0446	7.669	14.737	**0.0050**	0.3591	0.0440	8.168	14.581
0.3571	0.0478	7.471	13.449	**0.0055**	0.3758	0.0470	8.003	13.293
0.3710	0.0508	7.298	12.374	**0.0060**	0.3915	0.0498	7.861	12.218
0.3841	0.0537	7.147	11.462	**0.0065**	0.4062	0.0525	7.736	11.307
0.3964	0.0565	7.012	10.678	**0.0070**	0.4201	0.0551	7.625	10.525
0.4080	0.0592	6.891	9.998	**0.0075**	0.4332	0.0576	7.526	9.846
0.4191	0.0618	6.783	9.402	**0.0080**	0.4456	0.0599	7.438	9.252
0.4296	0.0643	6.684	8.875	**0.0085**	0.4574	0.0622	7.357	8.727
0.4396	0.0667	6.594	8.406	**0.0090**	0.4686	0.0643	7.284	8.259
0.4492	0.0690	6.511	7.985	**0.0095**	0.4793	0.0664	7.217	7.841
0.4583	0.0712	6.435	7.606	**0.0100**	0.4895	0.0684	7.156	7.463
0.4671	0.0734	6.364	7.262	**0.0105**	0.4992	0.0703	7.099	7.121

(continued)

Table E.5 (continued)

$n = b/b_w = 2$				ρ	$n = b/b_w = 4$			
ξ	i	ξ/i	$(1-\xi)/i$		ξ	i	ξ/i	$(1-\xi)/i$
0.4755	0.0755	6.299	6.949	**0.0110**	0.5085	0.0722	7.046	6.810
0.4835	0.0775	6.238	6.663	**0.0115**	0.5174	0.0739	6.997	6.526
0.4913	0.0795	6.181	6.400	**0.0120**	0.5260	0.0757	6.952	6.265
0.4988	0.0814	6.128	6.158	**0.0125**	0.5342	0.0773	6.909	6.024
0.5060	0.0833	6.078	5.934	**0.0130**	0.5421	0.0789	6.869	5.802
0.5130	0.0851	6.030	5.726	**0.0135**	0.5497	0.0805	6.832	5.596
0.5197	0.0868	5.986	5.533	**0.0140**	0.5571	0.0820	6.797	5.404
0.5262	0.0885	5.944	5.353	**0.0145**	0.5641	0.0834	6.763	5.226
0.5325	0.0902	5.904	5.184	**0.0150**	–	–	–	–
0.5386	0.0918	5.867	5.026	**0.0155**	–	–	–	–
0.5445	0.0934	5.831	4.878	**0.0160**	–	–	–	–
0.5502	0.0949	5.797	4.739	**0.0165**	–	–	–	–
0.5558	0.0964	5.764	4.608	**0.0170**	–	–	–	–
0.5612	0.0979	5.734	4.484	0.0175	–	–	–	–
0.5664	0.0993	5.704	4.366	0.0180	–	–	–	–
0.5715	0.1007	5.676	4.255	0.0185	–	–	–	–
0.5765	0.1021	5.649	4.150	0.0190	–	–	–	–
0.5813	0.1034	5.623	4.050	0.0195	–	–	–	–
0.5906	0.1060	5.574	3.864	**0.0205**	–	–	–	–
0.5951	0.1072	5.551	3.777	**0.0210**	–	–	–	–
0.5995	0.1084	5.529	3.694	**0.0215**	–	–	–	–
0.6037	0.1096	5.508	3.615	**0.0220**	–	–	–	–
0.6079	0.1108	5.488	3.540	**0.0225**	–	–	–	–
0.6119	0.1119	5.468	3.468	**0.0230**	–	–	–	–
$n = b/b_w = 5$				ρ	$n = b/b_w = 6$			
ξ	i	ξ/i	$(1-\xi)/i$		ξ	i	ξ/i	$(1-\xi)/i$
0.1514	0.0109	13.915	78.010	**0.0009**	0.1514	0.0109	13.925	78.064
0.1590	0.0119	13.310	70.380	**0.0010**	0.1590	0.0119	13.320	70.426
0.1945	0.0170	11.423	47.305	**0.0015**	0.1947	0.0170	11.441	47.324
0.2260	0.0217	10.412	35.665	**0.0020**	0.2265	0.0217	10.442	35.669
0.2542	0.0260	9.765	28.649	**0.0025**	0.2551	0.0260	9.809	28.643
0.2798	0.0301	9.309	23.959	**0.0030**	0.2811	0.0300	9.364	23.946
0.3032	0.0338	8.965	20.601	**0.0035**	0.3049	0.0338	9.031	20.584
0.3247	0.0374	8.694	18.079	**0.0040**	0.3269	0.0373	8.769	18.059
0.3446	0.0407	8.474	16.114	**0.0045**	0.3472	0.0406	8.558	16.092

(continued)

Table E.5 (continued)

$n = b/b_w = 5$				ρ	$n = b/b_w = 6$			
ξ	i	ξ/i	$(1-\xi)/i$		ξ	i	ξ/i	$(1-\xi)/i$
0.3631	0.0438	8.291	14.540	**0.0050**	0.3660	0.0437	8.382	14.516
0.3804	0.0468	8.135	13.251	**0.0055**	0.3837	0.0466	8.233	13.226
0.3965	0.0496	8.001	12.175	**0.0060**	0.4001	0.0494	8.105	12.149
0.4117	0.0522	7.883	11.264	**0.0065**	0.4156	0.0520	7.993	11.238
0.4260	0.0548	7.780	10.482	**0.0070**	0.4302	0.0545	7.894	10.455
0.4395	0.0572	7.687	9.804	**0.0075**	0.4440	0.0569	7.807	9.776
0.4523	0.0595	7.605	9.210	**0.0080**	0.4570	0.0591	7.728	9.182
0.4644	0.0617	7.530	8.685	**0.0085**	0.4694	0.0613	7.657	8.657
0.4759	0.0638	7.462	8.218	**0.0090**	0.4811	0.0634	7.593	8.190
0.4868	0.0658	7.399	7.799	**0.0095**	0.4923	0.0653	7.534	7.771
0.4973	0.0677	7.342	7.422	**0.0100**	0.5029	0.0672	7.481	7.394
0.5073	0.0696	7.290	7.081	**0.0105**	0.5131	0.0690	7.431	7.053
0.5168	0.0714	7.241	6.770	**0.0110**	0.5228	0.0708	7.385	6.742
0.5259	0.0731	7.196	6.486	**0.0115**	0.5321	0.0725	7.343	6.458
0.5347	0.0747	7.154	6.225	**0.0120**	–	–	–	–
0.5431	0.0763	7.114	5.985	**0.0125**	–	–	–	–

Table E.6 T-section; $h_f/d = 0.20$; $\alpha_e = 15$ $\left(f_{ck} \leq 50\text{N/mm}^2\right)$

$n = b/b_w = 2$				ρ	$n = b/b_w = 4$			
ξ	i	ξ/i	$(1-\xi)/i$		ξ	i	ξ/i	$(1-\xi)/i$
0.2171	0.0218	9.9585	35.9174	**0.0020**	0.2172	0.0218	9.9658	35.9109
0.2404	0.0263	9.1551	28.9296	**0.0025**	0.2412	0.0263	9.1884	28.9044
0.2615	0.0305	8.5823	24.2432	**0.0030**	0.2633	0.0304	8.6475	24.1998
0.2807	0.0344	8.1481	20.8813	**0.0035**	0.2837	0.0344	8.2456	20.8227
0.2984	0.0382	7.8045	18.3513	**0.0040**	0.3026	0.0381	7.9330	18.2807
0.3148	0.0418	7.5240	16.3778	**0.0045**	0.3203	0.0417	7.6814	16.2979
0.3301	0.0453	7.2896	14.7951	**0.0050**	0.3369	0.0451	7.4737	14.7079
0.3444	0.0486	7.0898	13.4972	**0.0055**	0.3525	0.0483	7.2986	13.4045
0.3578	0.0517	6.9171	12.4134	**0.0060**	0.3673	0.0514	7.1485	12.3163
0.3705	0.0548	6.7659	11.4945	**0.0065**	0.3812	0.0543	7.0182	11.3942
0.3825	0.0577	6.6321	10.7055	**0.0070**	0.3944	0.0571	6.9037	10.6026
0.3939	0.0605	6.5127	10.0204	**0.0075**	0.4069	0.0598	6.8022	9.9157
0.4047	0.0632	6.4052	9.4200	**0.0080**	0.4188	0.0624	6.7114	9.3139
0.4151	0.0658	6.3079	8.8892	**0.0085**	0.4302	0.0649	6.6296	8.7822
0.4249	0.0683	6.2193	8.4167	**0.0090**	0.4410	0.0673	6.5554	8.3091

(continued)

Table E.6 (continued)

$n = b/b_w = 2$				ρ	$n = b/b_w = 4$			
ξ	i	ξ/i	$(1-\xi)/i$		ξ	i	ξ/i	$(1-\xi)/i$
0.4344	0.0708	6.1382	7.9933	**0.0095**	0.4514	0.0696	6.4879	7.8853
0.4434	0.0731	6.0635	7.6116	**0.0100**	0.4613	0.0718	6.4260	7.5034
0.4521	0.0754	5.9946	7.2658	**0.0105**	0.4709	0.0739	6.3690	7.1576
0.4604	0.0776	5.9308	6.9509	**0.0110**	0.4800	0.0760	6.3165	6.8428
0.4684	0.0798	5.8713	6.6629	**0.0115**	0.4888	0.0780	6.2677	6.5552
0.4761	0.0819	5.8159	6.3986	**0.0120**	0.4973	0.0799	6.2224	6.2912
0.4836	0.0839	5.7641	6.1551	**0.0125**	0.5054	0.0818	6.1801	6.0481
0.4908	0.0859	5.7154	5.9300	**0.0130**	0.5133	0.0836	6.1406	5.8235
0.4977	0.0878	5.6697	5.7213	**0.0135**	0.5208	0.0853	6.1035	5.6153
0.5045	0.0897	5.6266	5.5272	**0.0140**	0.5281	0.0870	6.0686	5.4218
0.5110	0.0915	5.5859	5.3463	**0.0145**	0.5352	0.0887	6.0358	5.2415
0.5173	0.0932	5.5474	5.1772	**0.0150**	0.5421	0.0903	6.0049	5.0730
0.5234	0.0950	5.5109	5.0188	**0.0155**	0.5487	0.0918	5.9756	4.9153
0.5293	0.0967	5.4763	4.8701	**0.0160**	0.5551	0.0933	5.9479	4.7673
0.5351	0.0983	5.4434	4.7302	**0.0165**	–	–	–	–
0.5406	0.0999	5.4121	4.5984	**0.0170**	–	–	–	–
0.5461	0.1015	5.3822	4.4740	**0.0175**	–	–	–	–
0.5514	0.1030	5.3537	4.3563	**0.0180**	–	–	–	–
0.5565	0.1045	5.3264	4.2449	**0.0185**	–	–	–	–
0.5615	0.1059	5.3003	4.1392	**0.0190**	–	–	–	–
0.5664	0.1074	5.2754	4.0388	**0.0195**	–	–	–	–
0.5711	0.1088	5.2514	3.9433	**0.0200**	–	–	–	–
0.5758	0.1101	5.2284	3.8524	**0.0205**	–	–	–	–
0.5803	0.1115	5.2063	3.7657	**0.0210**	–	–	–	–
0.5847	0.1128	5.1851	3.6829	**0.0215**	–	–	–	–
0.5890	0.1140	5.1646	3.6038	**0.0220**	–	–	–	–
0.5932	0.1153	5.1449	3.5282	**0.0225**	–	–	–	–
0.5973	0.1165	5.1259	3.4557	**0.0230**	–	–	–	–
0.6013	0.1177	5.1076	3.3863	**0.0235**	–	–	–	–
0.6053	0.1189	5.0899	3.3196	0.0240	–	–	–	–
$n = b/b_w = 5$				ρ	$n = b/b_w = 6$			
ξ	i	ξ/i	$(1-\xi)/i$		ξ	i	ξ/i	$(1-\xi)/i$
0.2173	0.0218	9.9673	35.9096	0.0020	0.2173	0.0218	9.9683	35.9087
0.2414	0.0263	9.1954	28.8991	0.0025	0.2415	0.0262	9.2002	28.8956
0.2637	0.0304	8.6615	24.1904	0.0030	0.2639	0.0304	8.6710	24.1840
0.2843	0.0344	8.2669	20.8098	0.0035	0.2848	0.0344	8.2814	20.8009

(continued)

Table E.6 (continued)

$n = b/b_w = 5$				ρ	$n = b/b_w = 6$			
ξ	i	ξ/i	$(1-\xi)/i$		ξ	i	ξ/i	$(1-\xi)/i$
0.3036	0.0381	7.9614	18.2648	0.0040	0.3042	0.0381	7.9809	18.2539
0.3216	0.0417	7.7167	16.2796	0.0045	0.3224	0.0417	7.7409	16.2670
0.3385	0.0450	7.5153	14.6877	0.0050	0.3396	0.0450	7.5441	14.6736
0.3544	0.0482	7.3462	13.3827	0.0055	0.3557	0.0482	7.3792	13.3675
0.3694	0.0513	7.2017	12.2933	0.0060	0.3709	0.0512	7.2387	12.2771
0.3836	0.0542	7.0766	11.3701	0.0065	0.3853	0.0541	7.1174	11.3532
0.3971	0.0570	6.9670	10.5777	0.0070	0.3990	0.0569	7.0112	10.5601
0.4099	0.0597	6.8700	9.8901	0.0075	0.4120	0.0596	6.9175	9.8720
0.4221	0.0622	6.7835	9.2878	0.0080	0.4244	0.0621	6.8340	9.2692
0.4337	0.0647	6.7057	8.7557	0.0085	0.4362	0.0645	6.7591	8.7368
0.4448	0.0670	6.6353	8.2822	0.0090	0.4475	0.0669	6.6914	8.2630
0.4554	0.0693	6.5712	7.8581	0.0095	0.4582	0.0691	6.6299	7.8387
0.4656	0.0715	6.5127	7.4761	0.0100	0.4685	0.0713	6.5738	7.4565
0.4753	0.0736	6.4589	7.1301	0.0105	0.4784	0.0734	6.5223	7.1104
0.4847	0.0756	6.4093	6.8153	0.0110	0.4879	0.0754	6.4749	6.7954
0.4936	0.0776	6.3634	6.5275	0.0115	0.4970	0.0773	6.4310	6.5076
0.5023	0.0795	6.3207	6.2635	0.0120	0.5058	0.0792	6.3903	6.2436
0.5106	0.0813	6.2810	6.0205	0.0125	0.5142	0.0810	6.3525	6.0005
0.5186	0.0831	6.2439	5.7959	0.0130	0.5224	0.0827	6.3171	5.7759
0.5263	0.0848	6.2091	5.5877	0.0135	–	–	–	–
0.5338	0.0864	6.1765	5.3943	0.0140	–	–	–	–

Table E.7 T-section; $h_f/d = 0.25$; $\alpha_e = 15$ $\left(f_{ck} \leq 50\,\mathrm{N/mm^2}\right)$

$n = b/b_w = 2$				ρ	$n = b/b_w = 4$			
ξ	i	ξ/i	$(1-\xi)/i$		ξ	i	ξ/i	$(1-\xi)/i$
0.2584	0.0305	8.4727	24.3145	**0.0030**	0.2584	0.0305	8.4737	24.3135
0.2763	0.0345	8.0027	20.9623	**0.0035**	0.2766	0.0345	8.0111	20.9551
0.2929	0.0384	7.6344	18.4329	**0.0040**	0.2936	0.0384	7.6544	18.4175
0.3083	0.0420	7.3362	16.4562	**0.0045**	0.3096	0.0420	7.3697	16.4324
0.3228	0.0455	7.0888	14.8687	**0.0050**	0.3248	0.0455	7.1364	14.8372
0.3365	0.0489	6.8793	13.5655	**0.0055**	0.3391	0.0489	6.9412	13.5272
0.3494	0.0521	6.6991	12.4764	**0.0060**	0.3527	0.0521	6.7751	12.4321
0.3616	0.0553	6.5421	11.5525	**0.0065**	0.3657	0.0551	6.6316	11.5030
0.3731	0.0583	6.4038	10.7587	**0.0070**	0.3780	0.0581	6.5063	10.7048
0.3841	0.0612	6.2807	10.0693	**0.0075**	0.3898	0.0609	6.3958	10.0116

(continued)

Table E.7 (continued)

$n = b/b_w = 2$				ρ	$n = b/b_w = 4$			
ξ	i	ξ/i	$(1-\xi)/i$		ξ	i	ξ/i	$(1-\xi)/i$
0.3946	0.0640	6.1704	9.4648	**0.0080**	0.4011	0.0637	6.2974	9.4039
0.4047	0.0667	6.0708	8.9305	**0.0085**	0.4119	0.0663	6.2091	8.8668
0.4143	0.0693	5.9803	8.4547	**0.0090**	0.4222	0.0689	6.1294	8.3887
0.4235	0.0718	5.8977	8.0283	**0.0095**	0.4321	0.0713	6.0570	7.9602
0.4323	0.0743	5.8218	7.6439	**0.0100**	0.4416	0.0737	5.9910	7.5741
0.4408	0.0766	5.7519	7.2956	**0.0105**	0.4508	0.0760	5.9304	7.2243
0.4490	0.0790	5.6872	6.9784	**0.0110**	0.4597	0.0782	5.8746	6.9059
0.4569	0.0812	5.6272	6.6884	**0.0115**	0.4682	0.0804	5.8230	6.6149
0.4645	0.0834	5.5712	6.4222	**0.0120**	0.4764	0.0825	5.7752	6.3478
0.4719	0.0855	5.5190	6.1770	**0.0125**	0.4843	0.0845	5.7306	6.1019
0.4790	0.0876	5.4700	5.9503	**0.0130**	0.4920	0.0865	5.6891	5.8746
0.4858	0.0896	5.4240	5.7401	**0.0135**	0.4994	0.0884	5.6502	5.6639
0.4925	0.0915	5.3807	5.5447	**0.0140**	0.5066	0.0902	5.6137	5.4681
0.4989	0.0934	5.3399	5.3625	**0.0145**	0.5135	0.0920	5.5794	5.2857
0.5052	0.0953	5.3013	5.1923	**0.0150**	0.5203	0.0938	5.5471	5.1152
0.5113	0.0971	5.2648	5.0328	**0.0155**	0.5268	0.0955	5.5166	4.9556
0.5172	0.0989	5.2302	4.8831	**0.0160**	0.5331	0.0971	5.4877	4.8058
0.5229	0.1006	5.1972	4.7424	**0.0165**	0.5393	0.0988	5.4604	4.6650
0.5284	0.1023	5.1659	4.6097	**0.0170**	0.5453	0.1003	5.4345	4.5324
0.5339	0.1039	5.1361	4.4845	**0.0175**	0.5511	0.1019	5.4098	4.4072
0.5391	0.1056	5.1076	4.3661	**0.0180**	–	–	–	–
0.5443	0.1071	5.0805	4.2540	**0.0185**	–	–	–	–
0.5493	0.1087	5.0545	4.1477	**0.0190**	–	–	–	–
0.5541	0.1102	5.0296	4.0467	**0.0195**	–	–	–	–
0.5589	0.1117	5.0057	3.9506	**0.0200**	–	–	–	–
0.5635	0.1131	4.9828	3.8592	**0.0205**	–	–	–	–
0.5681	0.1145	4.9608	3.7720	**0.0210**	–	–	–	–
0.5725	0.1159	4.9397	3.6888	**0.0215**	–	–	–	–
0.5768	0.1173	4.9194	3.6092	**0.0220**	–	–	–	–
0.5810	0.1186	4.8998	3.5332	**0.0225**	–	–	–	–
0.5852	0.1199	4.8810	3.4603	**0.0230**	–	–	–	–
0.5892	0.1212	4.8628	3.3905	**0.0235**	–	–	–	–
0.5931	0.1224	4.8452	3.3236	**0.0240**	–	–	–	–
0.5970	0.1236	4.8283	3.2593	**0.0245**	–	–	–	–
0.6008	0.1249	4.8119	3.1975	**0.0250**	–	–	–	–

(continued)

Table E.7 (continued)

$n = b/b_w = 5$				ρ	$n = b/b_w = 6$			
ξ	i	ξ/i	$(1-\xi)/i$		ξ	i	ξ/i	$(1-\xi)/i$
0.2585	0.0305	8.4739	24.3134	**0.0030**	0.2585	0.0305	8.4740	24.3132
0.2766	0.0345	8.0129	20.9536	**0.0035**	0.2767	0.0345	8.0140	20.9527
0.2937	0.0384	7.6586	18.4142	**0.0040**	0.2938	0.0384	7.6614	18.4121
0.3099	0.0420	7.3768	16.4274	**0.0045**	0.3101	0.0420	7.3815	16.4240
0.3252	0.0455	7.1466	14.8304	**0.0050**	0.3255	0.0455	7.1535	14.8259
0.3397	0.0488	6.9545	13.5189	**0.0055**	0.3401	0.0488	6.9636	13.5132
0.3535	0.0520	6.7915	12.4224	**0.0060**	0.3540	0.0520	6.8027	12.4157
0.3666	0.0551	6.6512	11.4921	**0.0065**	0.3672	0.0551	6.6645	11.4846
0.3791	0.0581	6.5288	10.6928	**0.0070**	0.3798	0.0580	6.5443	10.6845
0.3911	0.0609	6.4212	9.9986	**0.0075**	0.3919	0.0609	6.4386	9.9897
0.4025	0.0636	6.3255	9.3901	**0.0080**	0.4035	0.0636	6.3449	9.3806
0.4135	0.0663	6.2399	8.8523	**0.0085**	0.4146	0.0662	6.2612	8.8422
0.4240	0.0688	6.1628	8.3735	**0.0090**	0.4252	0.0687	6.1858	8.3630
0.4340	0.0712	6.0929	7.9445	**0.0095**	0.4354	0.0712	6.1176	7.9336
0.4437	0.0736	6.0291	7.5579	**0.0100**	0.4452	0.0735	6.0555	7.5466
0.4531	0.0759	5.9708	7.2077	**0.0105**	0.4546	0.0758	5.9987	7.1961
0.4621	0.0781	5.9171	6.8890	**0.0110**	0.4637	0.0780	5.9466	6.8771
0.4707	0.0802	5.8676	6.5976	**0.0115**	0.4725	0.0801	5.8985	6.5855
0.4791	0.0823	5.8217	6.3303	**0.0120**	0.4809	0.0822	5.8540	6.3180
0.4871	0.0843	5.7791	6.0841	**0.0125**	0.4891	0.0841	5.8127	6.0716
0.4949	0.0862	5.7393	5.8566	**0.0130**	0.4970	0.0861	5.7743	5.8439
0.5025	0.0881	5.7022	5.6457	**0.0135**	0.5046	0.0879	5.7383	5.6330
0.5098	0.0900	5.6674	5.4498	**0.0140**	0.5120	0.0898	5.7047	5.4369
0.5169	0.0917	5.6347	5.2672	**0.0145**	0.5192	0.0915	5.6732	5.2542
0.5237	0.0935	5.6039	5.0966	**0.0150**	0.5261	0.0932	5.6435	5.0835
0.5303	0.0951	5.5749	4.9369	**0.0155**	–	–	–	–
0.5368	0.0968	5.5475	4.7871	**0.0160**	–	–	–	–

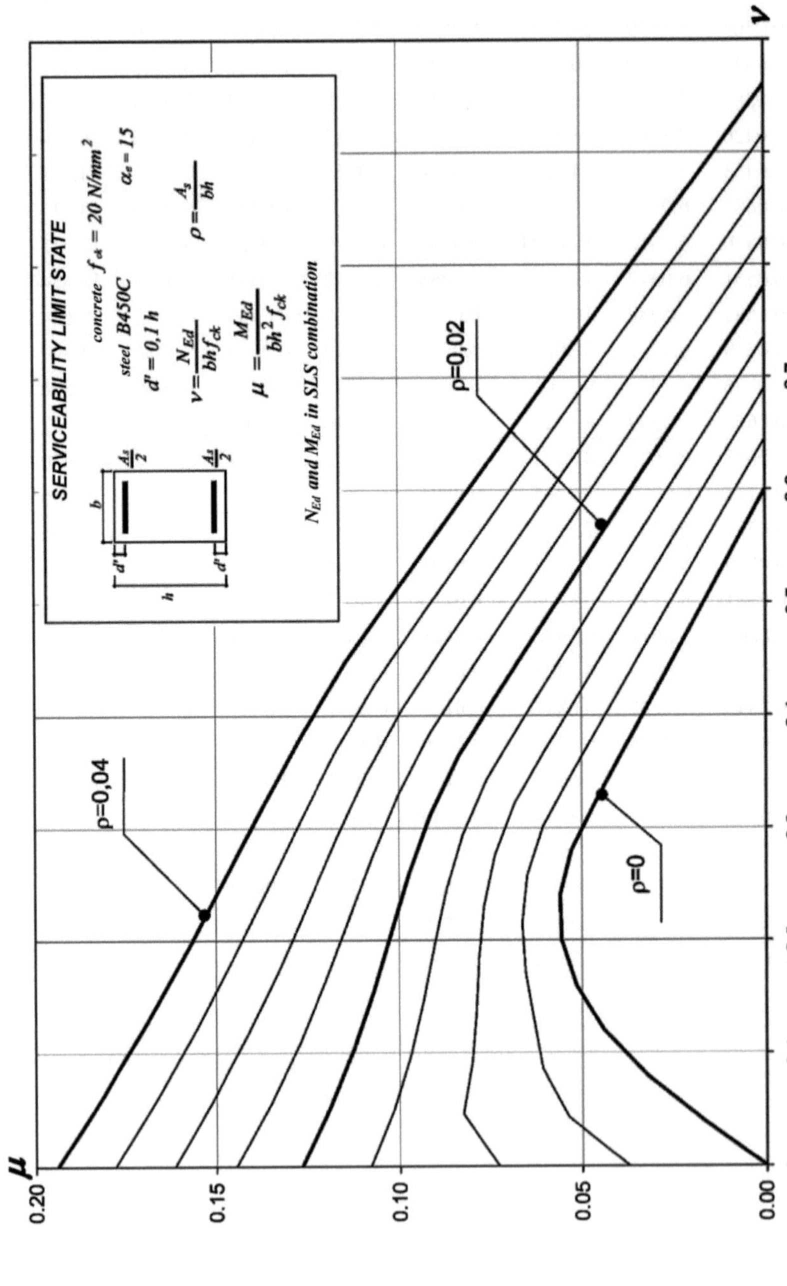

Diagram E.8 Rectangular section; $f_{ck} = 20\,\text{N/mm}^2$; $d'/h = 0.10$

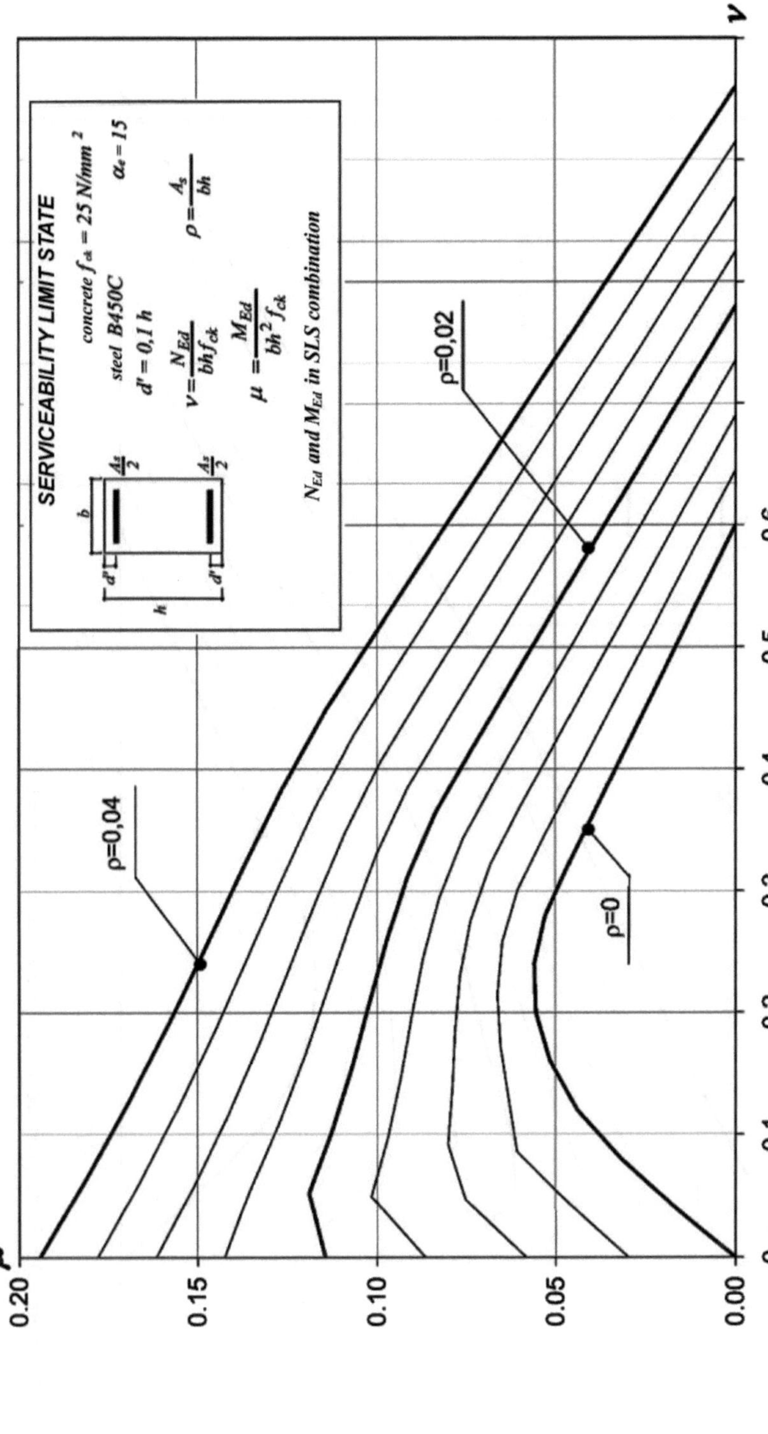

Diagram E.9 Rectangular section; $f_{ck} = 25\,\text{N/mm}^2$; $d'/h = 0.10$

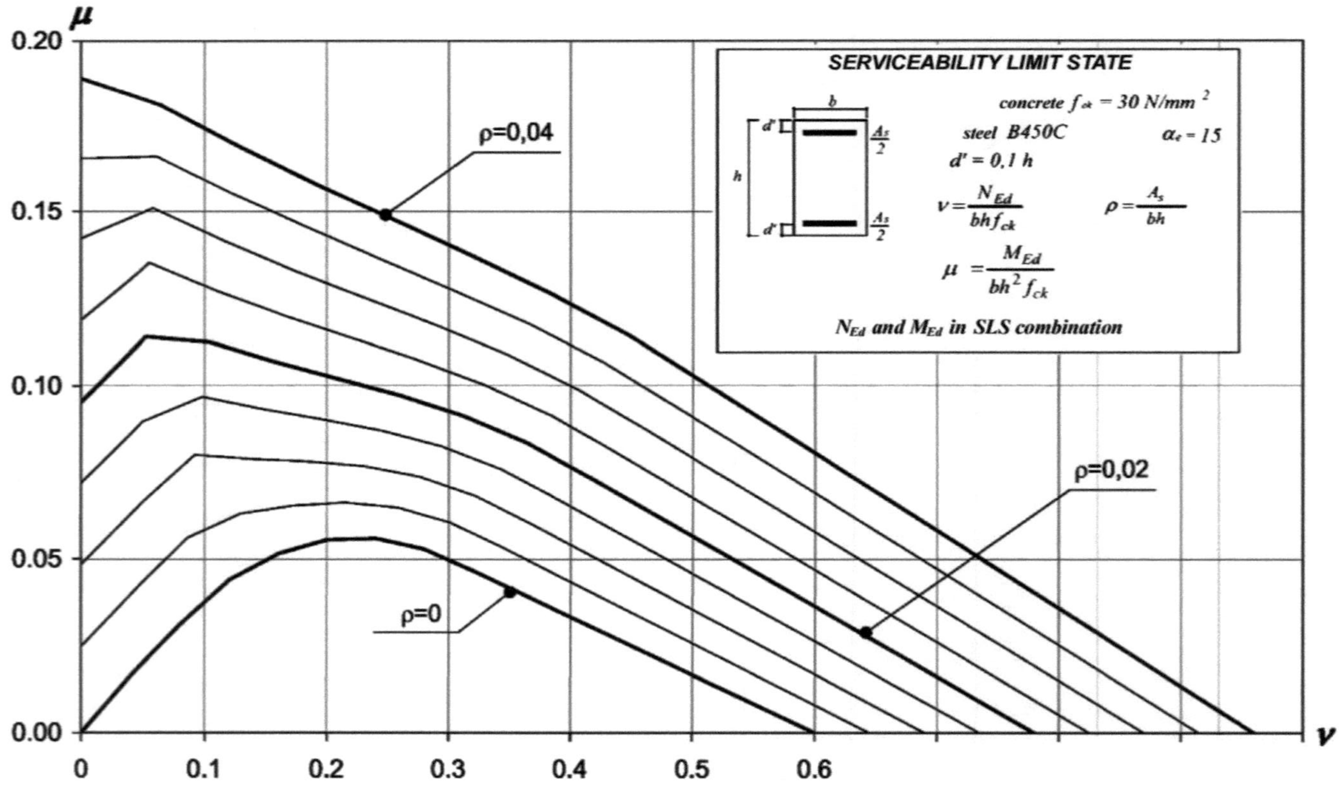

Diagram E.10 Rectangular section; $f_{ck} = 30\,\text{N/mm}^2$; $d'/h = 0.10$

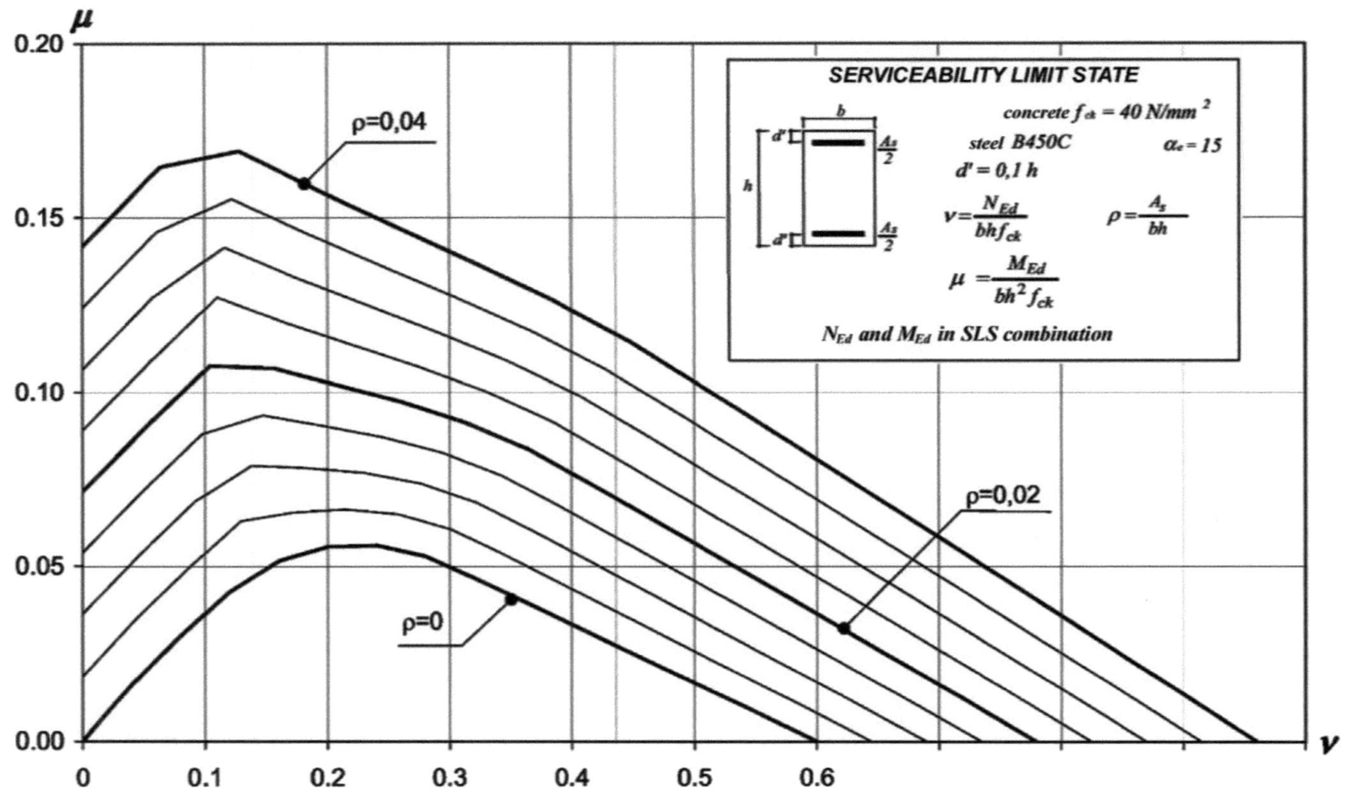

Diagram E.11 Rectangular section; $f_{ck} = 40\,\mathrm{N/mm^2}$; $d'/h = 0.10$

MIX
Papier aus verantwortungsvollen Quellen
Paper from responsible sources
FSC® C105338

If you have any concerns about our products,
you can contact us on
ProductSafety@springernature.com

In case Publisher is established outside the EU,
the EU authorized representative is:
Springer Nature Customer Service Center GmbH
Europaplatz 3, 69115 Heidelberg, Germany

Printed by Libri Plureos GmbH
in Hamburg, Germany